3.5 SOLUTION BY DETERMINANTS

$$\begin{vmatrix} a & b \\ c & d \end{vmatrix} = ad - bc$$

$$\begin{vmatrix} a_1 & b_1 & c_1 \\ a_2 & b_2 & c_2 \\ a_3 & b_3 & c_3 \end{vmatrix} = a_1 \begin{vmatrix} b_2 & c_2 \\ b_3 & c_3 \end{vmatrix} - b_1 \begin{vmatrix} a_2 & c_2 \\ a_3 & c_3 \end{vmatrix}$$
$$+ c_1 \begin{vmatrix} a_2 & b_2 \\ a_3 & b_3 \end{vmatrix}$$

The solution of $\begin{cases} ax + by = e \\ cx + dy = f \end{cases}$ is $x = \dfrac{D_x}{D}$ and $y = \dfrac{D_y}{D}$,

where $D = \begin{vmatrix} a & b \\ c & d \end{vmatrix}$, $D_x = \begin{vmatrix} e & b \\ f & d \end{vmatrix}$, $D_y = \begin{vmatrix} a & e \\ c & f \end{vmatrix}$.

4.1 LINEAR INEQUALITIES

Trichotomy property: $a < b$, $a = b$, or $a > b$

Transitive property: If $a < b$ and $b < c$, then $a < c$.

If a and b are real numbers and $a < b$, then

$a + c < b + c$ $\qquad a - c < b - c$

$ac < bc$ $(c > 0)$ $\qquad ac > bc$ $(c < 0)$

$\frac{a}{c} < \frac{b}{c}$ $(c > 0)$ $\qquad \frac{a}{c} > \frac{b}{c}$ $(c < 0)$

$c < x < d$ is equivalent to $c < x$ and $x < d$.

4.2 EQUATIONS AND INEQUALITIES WITH ABSOLUTE VALUES

Absolute value: $\begin{cases} \text{If } x \geq 0, \text{ then } |x| = x. \\ \text{If } x < 0, \text{ then } |x| = -x. \end{cases}$

If $k > 0$, $|x| = k$ is equivalent to $x = k$ or $x = -k$.

$|a| = |b|$ is equivalent to $a = b$ or $a = -b$.

If $k > 0$, then

$|x| < k$ is equivalent to $-k < x < k$.

$|x| > k$ is equivalent to $x < -k$ or $x > k$.

5.3 MULTIPLYING POLYNOMIALS

$a(b + c + d + \cdots) = ab + ac + ad + \cdots$

$(x + y)^2 = x^2 + 2xy + y^2$

$(x - y)^2 = x^2 - 2xy + y^2$

$(x + y)(x - y) = x^2 - y^2$

5.4–5.7 FACTORING POLYNOMIALS

$ax + bx = x(a + b)$

$(a + b)x + (a + b)y = (a + b)(x + y)$

$ax + ay + cx + cy = a(x + y) + c(x + y)$
$$= (x + y)(a + c)$$

$x^2 - y^2 = (x + y)(x - y)$

$x^3 + y^3 = (x + y)(x^2 - xy + y^2)$

$x^3 - y^3 = (x - y)(x^2 + xy + y^2)$

$x^2 + 2xy + y^2 = (x + y)(x + y)$

$x^2 - 2xy + y^2 = (x - y)(x - y)$

Test for factorability: $ax^2 + bx + c$ will factor with integer coefficients if $b^2 - 4ac$ is a perfect square.

5.8 SOLVING EQUATIONS BY FACTORING

Zero-factor theorem: If $xy = 0$, then $x = 0$ or $y = 0$.

6.1 RATIONAL FUNCTIONS AND SIMPLIFYING RATIONAL EXPRESSIONS

Division by 0 is undefined.

$\frac{ak}{bk} = \frac{a}{b}$ $(b \neq 0, k \neq 0)$

6.2 MULTIPLYING AND DIVIDING RATIONAL EXPRESSIONS

$\frac{a}{b} \cdot \frac{c}{d} = \frac{ac}{bd}$ $(b \neq 0, d \neq 0)$

$\frac{a}{b} \div \frac{c}{d} = \frac{a}{b} \cdot \frac{d}{c}$ $(b \neq 0, d \neq 0, c \neq 0)$

6.3 ADDING AND SUBTRACTING RATIONAL EXPRESSIONS

$\frac{a}{b} + \frac{c}{b} = \frac{a + c}{b}$ $(b \neq 0)$ and $\frac{a}{b} - \frac{c}{b} = \frac{a - c}{b}$ $(b \neq 0)$

6.6 PROPORTION AND VARIATION

Direct variation: $y = kx$

Inverse variation: $y = \frac{k}{x}$

Joint variation: $y = kxz$

Combined variation: $y = \frac{kx}{z}$

6.8 SYNTHETIC DIVISION (OPTIONAL)

Remainder theorem: If $P(x)$ is divided by $x - r$, the remainder is $P(r)$.

Factor theorem: If $P(x)$ is divided by $x - r$, $P(r) = 0$, if and only if $x - r$ is a factor of $P(x)$.

Concepts
of Intermediate
Algebra

**AN EARLY FUNCTIONS
APPROACH**

Books in the Gustafson/Frisk Series

Concepts of Intermediate Algebra

AN EARLY FUNCTIONS APPROACH

R. David Gustafson

Rock Valley College

Brooks/Cole Publishing Company

I(T)P™ An International Thomson Publishing Company

Pacific Grove • Albany • Bonn • Boston • Cincinnati • Detroit • London • Madrid • Melbourne
Mexico City • New York • Paris • San Francisco • Singapore • Tokyo • Toronto • Washington

A ROBERT W. PIRTLE BOOK

Sponsoring Editor: *Robert W. Pirtle*
Marketing Team: *Patrick Farrant & Jean Thompson*
Editorial Assistant: *Linda Row*
Production Editor: *Ellen Brownstein*
Production Service: *Hoyt Publishing Services*
Manuscript Editor: *David Hoyt*
Permissions Editor: *Carline Haga*
Interior Design: *E. Kelly Shoemaker*

Cover Design: *Vernon T. Boes*
Cover Art: *Laura Militzer Bryant*
Art Coordinator: *David Hoyt*
Interior Illustration: *Lori Heckelman*
Photo Editor: *Kathleen Olson*
Typesetting: *The Clarinda Company*
Cover Printing: *Color Dot Graphics, Inc.*
Printing and Binding: *Quebecor-Hawkins*

COPYRIGHT © 1996 by Brooks/Cole Publishing Company
A division of International Thomson Publishing Inc.
I(T)P The ITP logo is a trademark under license.

For more information, contact:

BROOKS/COLE PUBLISHING COMPANY
511 Forest Lodge Road
Pacific Grove, CA 93950
USA

International Thomson Publishing Europe
Berkshire House 168-173
High Holborn
London WC1V 7AA
England

Thomas Nelson Australia
102 Dodds Street
South Melbourne, 3205
Victoria, Australia

Nelson Canada
1120 Birchmount Road
Scarborough, Ontario
Canada M1K 5G4

International Thomson Editores
Campos Eliseos 385, Piso 7
Col. Polanco
11560 México D. F. México

International Thomson Publishing GmbH
Königswinterer Strasse 418
53227 Bonn
Germany

International Thomson Publishing Asia
221 Henderson Road
#05-10 Henderson Building
Singapore 0315

International Thomson Publishing Japan
Hirakawacho Kyowa Building, 3F
2-2-1 Hirakawacho
Chiyoda-ku, Tokyo 102
Japan

Printed in the United States of America

10 9 8 7 6 5 4 3 2 1

Library of Congress Cataloging-in-Publication Data

Gustafson, R. David (Roy David), [date]
    Concepts of intermediate algebra / R. David Gustafson.
        p.   cm.
    Includes index.
    ISBN 0-534-33859-3
    1. Algebra. I. Title.
QA154.2.G8734   1995
512.9—dc20   95-24684
                                                     CIP

To
  Daniel
  Tyler
  Spencer
and the new generation of students.

Concepts of Intermediate Algebra: An Early Functions Approach presents all of the topics usually associated with a second course in algebra. It is designed to prepare students for additional work in college algebra, finite mathematics, statistics, or precalculus. The text will also prepare students for employment in the 21st century.

In keeping with the spirit of the NCTM Standards and the recommendations of the American Mathematical Association of Two-Year Colleges, this text emphasizes conceptual understanding, an early treatment of functions, problem solving, geometry, and use of technology. It also includes many algebraic applications from statistics. Skill topics such as factoring, algebraic fractions, and radicals have been deemphasized, but the coverage of these topics remains comprehensive and thorough. It is much easier for instructors to leave out material than it is for them to provide missing material.

It has been my intention to write a text that

- is relevant and easy to understand
- stresses the concept of function throughout
- uses real-life applications to motivate students to solve problems
- develops critical thinking skills in all students
- reduces the content overlap between beginning algebra and intermediate algebra courses
- develops the necessary skills for students to continue their study of mathematics

■ ORGANIZATION

To accomplish these goals, it was necessary to reorder the sequence of topics traditionally found in intermediate algebra texts. In *Concepts of Intermediate Algebra*, the first four chapters cover linear topics. Chapter 1 includes the properties of real numbers, exponents, linear equations, and problem solving. Chapter 2 covers graphing, slope, writing equations of lines, and functions, with strong emphasis on reading information from graphs. Technology is introduced in Chapter 2 and then integrated throughout the rest of the text; graphing devices are utilized for performing and exploring mathematics. Chapters 3 and 4 include systems of linear equations, inequalities, absolute value, systems of inequalities, and linear programming, with considerable emphasis on problem solving.

Chapter 5 covers polynomial functions, arithmetic of polynomials (no division), and factoring. Chapter 6 includes rational functions, arithmetic of rational expressions, dividing polynomials, and synthetic division. Chapter 7 introduces the square root and cube root functions, radicals, rational exponents, and radical equations. All three chapters strongly emphasize problem solving and technology.

Chapters 8 and 9 cover quadratic functions and equations, complex numbers, quadratic inequalities, algebra of functions, exponential functions, and logarithmic functions. These chapters also emphasize problem solving and technology.

Chapter 10 includes miscellaneous topics from which an instructor can pick and choose.

■ FEATURES

The following pedagogical features have proven to be effective in the classroom.

Four-Color Design   The text has a functional four-color design. For easy reference, definitions are boxed in purple, theorems are boxed in yellow, and strategies are boxed in tan. In addition, color is used to highlight terms and expressions that you would point to in a classroom discussion.

Examples and Exercises   The text contains more than 500 worked examples, with extensive explanatory notes to make them easy to follow. There are also more than 5000 carefully graded exercises; answers to the odd-numbered exercises are provided as an appendix in the Student Edition. Answers to all of the Chapter Review Exercises, Chapter Tests, and Cumulative Review Exercises also appear in the answer appendix.

Real-Life Application Problems   The text contains a wealth of real-life application problems. Each one has a title to identify its subject matter. An extensive index of applications is included.

Geometry   Geometric topics are integrated throughout the text. Topics covered include angles, parallel and perpendicular lines, triangles, similarity, parallelograms, and areas and volumes.

Mathematics in the Workplace   Each chapter opens with a short discussion of a career that uses mathematics. A sample application problem is given, along with a page reference showing where the problem appears in the chapter.

Student Warnings   **WARNING!**   symbols appear throughout the text to warn students of common errors.

Oral Exercises   Each exercise set is preceded by a short set of oral exercises, which help instructors check student understanding before dismissing the class.

Writing and Discussion   Each exercise set is followed by Writing Exercises to reinforce student understanding of definitions and basic concepts, and Something to Think About exercises to provide opportunities for classroom discussion.

Review   Constant review is a particular strength of the text. There are Review Exercises at the end of each exercise set and at the end of each chapter, and Cumulative Review Exercises appear after every two chapters. Key formulas and ideas are displayed inside the front and back covers of the text for easy reference.

Technology   Technology is integrated throughout the text. All graphing-calculator art has been generated with a Texas Instruments TI-82 graphing calculator. Keystrokes for graphing calculators are not given in the text, but they are available for most popular calculators in a separate supplement.

After mathematical ideas are taught, technology is used for reinforcement and further discovery. Technology supplements the course; it is not the basis of the course.

Statistics   Accent on Statistics sections introduce students to some basic concepts of statistics through real-life examples. Some topics discussed are the mean, median, and mode; frequency distributions; regression; sample sizes; and variance and standard deviation.

Group Learning   Each chapter concludes with a Problems and Projects section that can be used for group learning or extended individual assignments. The application problems here are similar to those found in the text, but they usually have an extra twist. These problems should take between 10 and 20 minutes to complete.

The projects are designed to make connections between several topics and help students see the big picture. The projects may take an hour or more to complete.

Videotapes   Videotapes teach many examples in the text. These examples are marked with the symbol ⬤⬤ in the text.

Computer-Aided Instruction   BCX tutorial software provides students with interactive examples and drill. A reporting system is included.

■ EXAMPLES OF THE FEATURES IN THE TEXT

---

18   CHAPTER 1 BASIC CONCEPTS

■ ■ ■ ■ ■ ■ ■ ■ ■ **Measures of Central Tendency**

ACCENT ON
STATISTICS

There are three types of averages that are commonly used in statistics: the **mean**, the **median**, and the **mode**.

> **Mean**
> The **mean** of several values is the sum of those values divided by the number of values.
> $$\text{Mean} = \frac{\text{sum of the values}}{\text{number of values}}$$

EXAMPLE 6   **Football** Figure 1-14 shows the gains and losses made by a running back on seven plays. Find the mean number of yards per carry.

Solution   To find the mean number of yards per carry, we add the numbers and divide by 7.

$$\frac{-8 + (+2) + (-6) + (+6) + (+4) + (-7) + (-5)}{7} = \frac{-14}{7} = -2$$

The running back averaged -2 yards (or lost 2 yards) per carry.

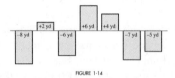

FIGURE 1-14   ■

> **Median**
> The **median** of several values is the middle value. To find the median,
> 1. Arrange the values in increasing order.
> 2. If there is an odd number of values, choose the middle value.
> 3. If there is an even number of v___
>   values.
>
> **Mode**
> The **mode** of several values is the va___

EXAMPLE 7   Ten workers in a small business have r___

$2500, $1750, $2415, $3240, $279___

Find **a.** the median and **b.** the m___

Solution   **a.** To find the median, we first arrang___

$1750, $2415, $2415, $2415, $___

◀ Accent on Statistics sections are included throughout the text.

Geometry problems are abundant.
▼

1.6 USING EQUATIONS TO SOLVE PROBLEMS   **63**

ILLUSTRATION 7

23. Enclosing a swimming pool A woman wants to enclose the swimming pool shown in Illustration 8 and have a walkway of uniform width all the way around. How wide will the walkway be if the woman uses 180 feet of fencing?

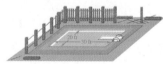

ILLUSTRATION 8

24. Framing a picture An artist wants to frame the picture shown in Illustration 9 with a frame 2 inches wide. How wide will the framed picture be if the artist uses 70 inches of framing material?

ILLUSTRATION 9

25. Supplementary angles If one of two supplementary angles is 35° larger than the other, find the measure of the smaller angle.

27. Complementary angles If one of two complementary angles is 22° greater than the other, find the measure of the larger angle.

28. Complementary angles in a right triangle Explain why the acute angles in a right triangle are complementary. In Illustration 11, find the measure of angle A.

ILLUSTRATION 11

29. Supplementary angles and parallel lines In Illustration 12, lines r and s are cut by a third line l to form angles 1 and 2. When lines r and s are parallel, angles 1 and 2 are supplementary. If the measure of angle 1 is $(x + 50)°$ and the measure of angle 2 is $(2x - 20)°$ and lines r and s are parallel, find x.

30. In Illustration 12, find the measure of angle 3. (*Hint:* See Exercise 29.)

ILLUSTRATION 12

31. Vertical angles When two lines intersect as in Illustration 13, four angles are formed. Angles that are side-by-side, such as angles 1 and 2, are called *adjacent angles*. Angles that are nonadjacent, such as angles 1 and 3 or angles 2 and 4, are called *vertical angles*. From geometry, we know that if two lines intersect, vertical angles have the same measure. If

Application problems are in two ▶
columns and have a title.

Application art is representational. ▶

138 CHAPTER 2 GRAPHS, EQUATIONS OF LINES, AND FUNCTIONS

FIGURE 2-48 ∎

■ SOLVING EQUATIONS WITH A GRAPHING DEVICE

EXAMPLE 6   Solve the equation $2(x - 3) + 3 = 7$.

Solution   To use a graphing device to solve the equation $2(x - 3) + 3 = 7$, we graph both the left-hand side and the right-hand side of the equation in the same window, as shown in Figure 2-49(a). We then trace to find the coordinates of the point where the two graphs intersect, as shown in Figure 2-49(b). We can then zoom and trace again to get Figure 2-49(c). Solving the equation algebraically will show that the coordinates of the intersection point are indeed $(5, 7)$, and $x = 5$.

(a)   (b)   (c)

FIGURE 2-49 ∎

Orals   **1.** Describe a parabola.
   2. Describe the graph of $f(x) = |x| + $
   3. Describe the graph of $xy = 9$.
   4. Tell why the choice of a viewing w

EXERCISE 2.6

In Exercises 1–8, graph each equation by plotting points.

1. $f(x) = x^2 - 3$ **2.** $f(x) = -x^2 + 2$ **3.** $f(x) = (x - $

◀ Technology is included throughout.

Each chapter includes a Problems ▶
and Projects section for extended
study and group learning.

246   CHAPTER 4 INEQUALITIES

■ ■ ■ ■ ■ ■ ■ ■ **PROBLEMS AND PROJECTS**

1. Your employer provides two medical benefit plans. Plan A allows an unlimited number of doctor visits per year for $18.50 per month per person. In Plan B, you pay $6.50 per month per person plus $15 for each doctor visit after three visits in a given year. Which plan is best for you, and why?

2. At the beginning of the semester, your English teacher gave you a formula for figuring your final average: The final average is 25% of the final exam grade, plus 10% of the homework average, plus 65% of the exam average, which includes the term paper (counting as two exam grades.) Before finals, your homework average is 88, your term paper grade is 78, and your exam grades are 82, 85, 73, and 69. If you want to receive a grade of B ($80 \leq B \leq 89$), what is the lowest grade you can make on the final exam?

3. The cost of making a new medicine is 6.5¢ per pill. As an experimental drug manufacturer, you can only sell the pills for 6¢ each. However, after distributing 5000 pills, you will receive from the FDA a subsidy of 10% of the amount you receive from the sales of all pills over 5000. Find the range of distribution that produces a profit.

4. A heating company makes and installs duct work. To make and install duct work above ground costs $6 per linear foot; underground, it costs $12 per linear foot. It takes 1.6 hours to make and install duct work either above or below ground. The company has $2400 available for expenses and 360 hours of time. The company bookkeeper has figured that there is a profit of $3.40 per linear foot above ground and $2.80 per linear foot underground. If a job requires duct work both above and below ground, find the conditions for maximum profit. What would the maximum profit be?

PROJECT 1   A farmer is building a machine shed onto his barn, as shown in Illustration 1. The shed is to be 12 feet wide, and of course h_2 must be no more than 20 feet. In order for all of the shed to be useful for storing machinery, h_1 must be at least 6 feet. For the roof to shed rain and melting snow adequately, the slope of the roof must be at least $\frac{1}{2}$, but to be easily shingled, it must have a slope that is no greater than 1.

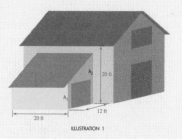

ILLUSTRATION 1

$$r_1 r_2 = rr_2 + rr_1 \qquad \text{Simplify each fraction.}$$
$$r_1 r_2 = r(r_2 + r_1) \qquad \text{Factor out } r \text{ on the right-hand side.}$$
$$r = \frac{r_1 r_2}{r_2 + r_1} \qquad \text{Divide both sides by } r_2 + r_1 \text{ and use the symmetric property of equality.} \blacksquare$$

■ **PROBLEM SOLVING**

Many applications lead to rational equations.

EXAMPLE 6 **Drywalling a house** A contractor knows that his best crew can drywall a house in 4 days and that his second crew can drywall the same house in 5 days. One day must be allowed for the plaster coat to dry. If the contractor uses both crews, can the house be ready for painting in 4 days?

① *Analyze the Problem* Because 1 day is necessary for drying, the drywallers must complete their work in 3 days. Since the first crew can drywall the house in 4 days, it can do $\frac{1}{4}$ of the job in 1 day. Since the second crew can drywall the house in 5 days, it can do $\frac{1}{5}$ of the job in 1 day. If it takes x days for both crews to finish the house, together they can do $\frac{1}{x}$ of the job in 1 day. The amount of work the first crew can do in 1 day plus the amount of work the second crew can do in 1 day equals the amount of work both crews can do in 1 day working together.

② *Form an Equation* If x represents the number of days it takes for both crews to drywall the house, we can form the equation

What crew 1 can do in 1 day		what crew 2 can do in 1 day		what they can do together in 1 day.

③ *Solve the Equation*
$$\frac{1}{4} \quad + \quad \frac{1}{5} \quad = \quad \frac{1}{x}$$

$$20x\left(\frac{1}{4} + \frac{1}{5}\right) = 20x\left(\frac{1}{x}\right) \qquad \text{Multiply both sides by } 20x.$$
$$5x + 4x = 20 \qquad \text{Remove parentheses and simplify.}$$
$$9x = 20 \qquad \text{Combine like terms.}$$
$$x = \frac{20}{9} \qquad \text{Divide both sides by 9.}$$

④ *State the Conclusion* Since it will take only $2\frac{2}{9}$ days for both crews to drywall the house, and it takes 1 day for drying, it will be ready for painting in $3\frac{2}{9}$ days, which is less than 4 days. ■

⑤ *Check the Result* Check the solution.

EXAMPLE 7 **Uniform motion problem** A man drove 200 miles to a convention. Because of road construction, his average speed on [...] than his average speed going to the conv[...] how fast did he drive in each direction[...]

◄ The text uses a five-step approach to problem solving.

◄ Word equations are used before math equations.

---

c. $4^7 xy^4$ is a monomial of degree 5. Because the sum of the exponents on the variables is 5.

d. 3 is a monomial of degree 0. $3 = 3x^0$, and if $x \neq 0$, the degree of $3x^0$ is 0. ■

WARNING! Since $a \neq 0$ in the previous definition, 0 has no defined degree.

Degree of a Polynomial
The **degree of a polynomial** is the same as the degree of the term in the polynomial with largest degree.

EXAMPLE 2 **a.** $3x^5 + 4x^2 + 7$ is a trinomial of degree 5. Because the largest degree of the three monomials is 5.

b. $7x^2 y^8 - 3xy$ is a binomial of degree 10.

c. $3x + 2y - xy$ is a trinomial of degree 2.

d. $18x^2 y^3 - 12x^7 y^2 + 3x^9 y^3 - 3$ is a polynomial of degree 12. ■

If the exponents on the variable in a polynomial in one variable decrease as we move from left to right, we say that they are written in descending order. If the exponents increase as we move from left to right, we say that they are written in ascending order.

EXAMPLE 3 Write the exponents of $7x^2 - 5x^4 + 3x + 2x^3 - 1$ in **a.** descending order and **b.** ascending order.

Solution **a.** $-5x^4 + 2x^3 + 7x^2 + 3x - 1$ **b.** $-1 + 3x + 7x^2 + 2x^3 - 5x^4$ ■

In the following polynomial, the exponents on x are in descending order, and the exponents on y are in ascending order.

$$7x^4 - 2x^3 y + 4x^2 y^2 - 8xy^3 + 12y^4$$

Warning symbols appear throughout. ▶

Definitions are clearly boxed. ▶

Video logos show which examples are included in the videotapes. ▶

■ SUPPLEMENTS FOR THE INSTRUCTOR

Annotated Instructor's Edition

The Annotated Instructor's Edition has the answer to every exercise printed next to that exercise.

Test Manual
Jerry Frang

The *Test Manual* contains four ready-to-use forms of every chapter test. Two of the tests are free-response and two are multiple-choice.

Complete Solutions Manual
Michael G. Weldon

The *Complete Solutions Manual* contains worked-out solutions for all text exercises.

Transparencies

A set of four-color transparencies of key graphics from the text, covering all topics in the course, is available to assist the instructor in the classroom.

Video Tutorial Series

A set of book-specific videotapes is available without charge for adoptions of 100 books or more. The videotapes feature David Gustafson, Diane Koenig, and Robert Clark, who take students step-by-step through the key concepts from the text and explain first-hand the reasoning behind their approach. Also included are the solutions of all examples marked in the text with the symbol [◉◉].

Computerized Testing Software

Available with the text are two extensive electronic question banks, one free-response and one multiple-choice. Each bank is available for either DOS and Windows or Mac platforms. The testing program gives you all of the features of a state-of-the-art word processor and more, including the ability to see all technical symbols, fonts, and formatting on screen just the way they will appear when printed. The question banks can be edited.

EXPTEST™ (DOS and Windows) runs on IBM and compatible computers. Chariot Microtest III (Mac) runs on Macintosh computers.

Printed Test Bank

The *Printed Test Bank* contains both extensive question banks, multiple-choice and free-response, that are available on the Computerized Testing Software.

Technical Support

Toll-free technical support is available for all Brooks/Cole software products: (800)241-2661 or Email: support@brookscole.com

■ SUPPLEMENTS FOR THE STUDENT

Student Solutions Manual
Michael G. Welden

The *Student Solutions Manual* gives complete solutions for all odd-numbered exercises in the text. It is available for sale at your college bookstore.

Study Guide with Practice Tests
George Grisham

The *Study Guide with Practice Tests* provides more explanation, worked examples, practice problems, and practice tests. An answer section is also included. It is available for sale at your college bookstore.

Explorations in Beginning and Intermediate Algebra Using the TI-82
Deborah J. Cochener
and *Bonnie M. Hodge*

Explorations in Beginning and Intermediate Algebra Using the TI-82 takes a four-fold approach to each unit: graphs, equations, patterns, and verbalization/writing. Guided exploration can be done individually or in group sessions. Also included are Applications, Troubleshooting sections, and Do's and Don'ts sections. A correlation chart for Gustafson's *Concepts of Intermediate Algebra* is available. This useful supplement is available for sale at your college bookstore.

Spanish Study Guide
Alberto Beron

Written in Spanish, this *Study Guide* reviews key ideas, provides completely worked examples, and offers additional exercises. It is available for sale at your college bookstore.

Brooks/Cole Exerciser 2.0
Laurel Technical Services

BCX is book-specific tutorial software, available for DOS, DOS/Windows, and Mac platforms, that instructs and drills students on problems similar to those found in the text. There is a set of BCX questions for every section in the text. BCX provides hints to students and, if they cannot answer a question correctly on two tries, BCX displays the complete solution. BCX also monitors student progress and includes a printed or on-line reporting system. BCX is complimentary to professors upon adoption.

TEMATH 2.1
Robert Kowalczyk
and *Adam Hausknecht*

This Macintosh software allows students to graph functions, find roots and asymptotes, and much more. It is available for sale at your college bookstore.

Technical Support

Toll-free technical support is available for any Brooks/Cole software product: (800)214-2661 or Email: support@brookscole.com

Congratulations. You now own a state-of-the-art textbook that has been written especially for you. To use the book properly, read it carefully, do the exercises, and check your progress with the Review Exercises and the Chapter Tests. Be sure to read and use the hints on studying algebra listed below.

A *Student Solutions Manual,* available for sale at your college bookstore, contains solutions to the odd-numbered exercises. A *Study Guide* that contains additional explanations, worked examples, and practice problems is also available for sale.

When you finish this course, consider keeping your text. It is the single reference source that will keep at your fingertips the information that you have learned. You may need this reference material in future mathematics, science, or business courses.

■ HINTS ON STUDYING ALGEBRA

The phrase "Practice makes perfect" is not quite true. It is *perfect* practice that makes perfect. For this reason, it is important that you learn how to study algebra to get the most out of this course.

Although we all learn differently, there are some hints on how to study algebra that most students find useful. Here is a list of some things you should consider.

Plan a Strategy for Success   To get where you want to be, you need a goal and a plan. Set a goal of passing this course with a grade of A or B. To meet this goal, you must have a plan, which should include several points:

- getting ready for class,
- attending class,
- doing homework,
- arranging for special help when you need it, and
- having a strategy for taking tests.

Getting Ready for Class   To get the most out of every class period, you will need to prepare. One of the best things that you can do is to read the material in the text before your instructor discusses it. You may not understand all of what you read, but you will be better able to understand it when your instructor presents the material in class.

Be sure to do your work every day. If you get behind and attend class without understanding prior material, you will be lost, and your classroom time will be wasted. Even worse, you will become frustrated and discouraged. Promise yourself that you will always prepare for class, and then keep your promise.

Attending Class   The classroom experience is your opportunity to learn from your instructor. Make the most of it by attending every class. Sit near the front of the room where you can see and hear well and where you won't be distracted. It is your responsibility to follow the instructor's discussion, even though that might be hard work.

Pay attention and jot down the important things that your instructor says. However, do not spend so much time taking notes that you fail to concentrate on what your instructor is explaining. It is much better to listen and understand than merely to copy solutions to problems.

Don't be afraid to ask questions. If something is unclear to you, it is probably unclear to other students as well. They will appreciate your willingness to ask. Besides, asking questions will make you an active participant in class. This will help you pay attention and keep you alert and involved.

Doing Homework   Everyone knows that it requires practice to excel at tennis, master a musical instrument, or learn a foreign language. It also requires practice to learn algebra. Since *practice* in algebra is the homework, homework is your opportunity to experiment with ideas and practice skills.

It is very important to pick a definite time to study and do homework. Make a schedule and stick to it. Try to study in a place that is comfortable and quiet. If you can, do some homework shortly after class, or at least before you forget what was discussed in class. This quick follow-up will help you remember the concepts and skills your instructor taught that day.

Study Sessions   Each study session should include three parts:

1. Begin every study session with a review period. Look over previous chapters and see if you can do a few problems from previous sections. Keeping old skills alive will greatly reduce the amount of time that you will need to cram for tests.

2. After reviewing, read the assigned material. Resist the temptation to dive into the problems without reading and understanding the examples. Instead, work the examples with pencil and paper. Only after you completely understand the underlying principles behind them should you try to work the problems.

Once you begin to work the problems, check your answers with the printed answers in the back of the text. If one of your answers differs from the printed answer, see if you can reconcile the two. Sometimes correct answers have more than one form. If you still believe that your answer is wrong, compare your work to the example in the text that most closely resembles the problem and try to find your mistake. If you cannot find an error, consult the *Student Solutions Manual*. If nothing works, mark the problem and ask about it during the next class meeting.

3. After completing the written assignment, read the next section. That preview will be helpful when you hear the material discussed during the next class period.

You probably know that the rule of thumb for doing homework is two hours of homework for every hour spent in class. If mathematics is hard for you, plan on spending even more time on homework.

To make homework more enjoyable, study with one or more of your friends. The interaction will clarify ideas and help you remember them. If you must study alone, try talking to yourself. A good study technique is to explain the material to yourself.

Arranging for Special Help   Take advantage of any special help that is available from your instructor. Often, your instructor can clear up difficulties in a very short time.

Find out if your college has a free tutoring program. Peer tutors can be of great help. Be sure to use the videotapes and BCX tutorial software.

Taking Tests   Students often get nervous before taking a test, because they are afraid that they won't do well. The most common reason for this fear is that students are not confident that they know the material.

To build confidence in your ability to work tests, rework many of the problems in the exercise sets, work the Review Exercises at the end of each chapter, and work the Chapter Tests. Check all answers with the answers printed at the back of the text.

Then guess what the instructor will ask, make up your own tests, and work them. Once you know your instructor, you will be surprised at how good you can get at picking test questions. With this preparation, you will have some idea of what will be on the test. This will build confidence in your ability to do well.

When taking a test, work slowly and deliberately. Scan the test and work the easy problems first. Tackle the harder problems last.

I wish you well.

■■■■■■■■■ ACKNOWLEDGMENTS

I am grateful to the following people who have reviewed the book at various stages of its development:

Randall Allbritton
Daytona Beach Community College

Dona Boccio
Queensborough Community College

Edgar M. Chandler
Scottsdale Community College

Floyd Downs
Arizona State University-Tempe

Jim Edmondson
Santa Barbara Community College

Mary Lou Hammond
Spokane Community College

Pam Hunt
Paris Community College

Martha Ann Larkin
Southern Utah University

Eric Leung
Harrisburg Area Community College-Lebanon

Christine Panoff
University of Michigan-Flint

Charles Peveto
North Harris Community College

Shirley Thompson
North Lake College

Lenore Vest
Lower Columbia College

I wish to thank Diane Koenig and Robert Hessel, who read the entire manuscript and worked every problem. I am also grateful to George Mader, who wrote the projects, and Wayne Miller, Mary Anne Petruska, and Sharon Louvier, who wrote the problems in the Problems and Projects sections. I also thank George Grisham, Michael Welden, Rob Clark, Jerry Frang, and Laurel Technical Services, who prepared the ancillary materials.

I give special thanks to Robert Pirtle, Ellen Brownstein, David Hoyt, Kelly Shoemaker, Lori Heckelman, Kathleen Olson, Elizabeth Rammel, and Linda Row for their assistance in the production process.

R. David Gustafson

7 Rational Exponents and Radicals   *398*

8 Quadratic Functions, Inequalities, and Algebra of Functions   *458*

9 Exponential and Logarithmic Functions   *529*

Basic Concepts

MATHEMATICS IN THE
WORKPLACE

Computer Systems Analyst

Computer systems analysts help businesses and scientific research organizations develop computer systems to process and interpret data. Using techniques such as cost accounting, sampling, and mathematical model building, they analyze information and often present the results to management in charts and diagrams. The demand for systems analysts is growing.

SAMPLE APPLICATION ■ The process of sorting records into sequential order is a common task in electronic data processing. One sorting technique, called a *selection sort*, requires C comparisons to sort N records, where C and N are related by the formula

$$C = \frac{N(N - 1)}{2}$$

How many comparisons are necessary to sort 10,000 records?
See Exercise 84 in Exercise 1.2.

What is algebra? Mathematics is the study of relationships between quantities and magnitudes, as well as the rules that govern these relationships. The language of mathematics is *algebra*. The word *algebra* comes from the title of a book written by the Arabian mathematician Al-Khowarazmi around A.D. 800. Its title, *Ihm al-jabr wa' l muqabalah*, means restoration and reduction, a process then used to solve equations.

In the branch of mathematics called *arithmetic*, we worked with expressions that involved numbers:

$$15 + 32, \qquad 37 - 15, \qquad 12 \times 57, \qquad 23\overline{)367.32}, \qquad \text{and} \qquad \frac{6}{11}$$

In algebra, we will work with expressions and equations like

$$x + 45, \qquad \frac{8x}{3} - \frac{y}{2}, \qquad \frac{1}{2}bh, \qquad \frac{a}{b} = \frac{c}{d}, \qquad \text{and} \qquad 3x + 2y = 12$$

which involve both letters and numbers, where the letters stand for unknown quantities and are called **variables**. The use of variables in algebra enables us to write general statements about the rules of arithmetic. For example, we know that the same sum results regardless of the order in which we add two numbers. In arithmetic, we can only give specific examples to illustrate this fact: $2 + 5 = 5 + 2 = 7$ or $4 + 8 = 8 + 4 = 12$. In algebra, we can use variables to concisely express the general rule:

> If a and b represent any two numbers, then $a + b = b + a$.

It is the use of variables that distinguishes algebra from arithmetic.

1.1 Sets and the Real Number System
■ SET NOTATION ■ SUBSETS OF SETS ■ UNION AND INTERSECTION OF SETS ■ SETS OF NUMBERS
■ ORDERING OF THE REAL NUMBERS ■ INTERVALS ■ ABSOLUTE VALUE OF A NUMBER

■ SET NOTATION

A **set** is a collection of objects. To denote a set using the **roster method**, we enclose a list of its **elements** with braces.

EXAMPLE 1

a. $\{1, 2, 3, 4\}$ denotes the set with elements 1, 2, 3, and 4.

b. $\{a, b, c\}$ denotes the set with elements a, b, and c.   ■

To indicate that 3 is an element of the set $\{1, 2, 3, 4\}$, we write

$3 \in \{1, 2, 3, 4\}$ Read $\in$ as "is an element of."

To show that 5 is not an element of the set, we write $5 \notin \{1, 2, 3, 4\}$.
The expression

$A = \{a, e, i, o, u\}$

means that A is the set containing the vowels a, e, i, o, and u. This set can also be expressed in **set-builder notation** as

$B = \{x \mid x \text{ is a vowel in the English alphabet.}\}$ Read as "B is the set of all letters x such that x represents a vowel in the English alphabet."

Since sets A and B have the same elements, we say that they are equal.

> **Equality of Sets**
> Two sets are **equal** when they have exactly the same elements. If sets A and B are equal, we write $A = B$.

■ SUBSETS OF SETS

If each element in set B is also an element in set A, we say that set B is a **subset** of A, and we write $B \subseteq A$. If the interior of circle B, shown in Figure 1-1, represents the elements in set B, and the interior of circle A represents the elements in set A, the figure shows that every element in set B is also an element in set A.

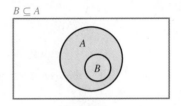

FIGURE 1-1

EXAMPLE 2 **a.** $\{1, 3, 5\} \subseteq \{1, 2, 3, 4, 5\}$ Because each element in the first set is also an element of the second set.

b. $\{a, b, c\} \subseteq \{a, b, c, d, e\}$ Because each element in the first set is also an element of the second set.

c. $\{2, 3, 5\} \subseteq \{2, 3, 5\}$ Because each element in the first set is also an element of the second set.   ∎

Part **c** of Example 2 illustrates that any set is a subset of itself.

A set with no elements is called the **empty set**, and is denoted as $\emptyset$. Thus, $\emptyset = \{\quad\}$. The empty set is considered to be a subset of every set.

 WARNING! The empty set $\emptyset$ is a set with no elements. It does not represent the number 0. $\{0\}$ is not empty. It has one element, the number 0.

■ UNION AND INTERSECTION OF SETS

If we join the elements in set A with the elements in set B, we obtain the **union** of A and B, denoted as $A \cup B$. The elements of $A \cup B$ are either elements in A, or elements in B, or both, as shown in Figure 1-2(a).

The set of elements that are common to set A and set B is called the **intersection** of A and B, denoted as $A \cap B$. The elements of $A \cap B$ are elements that are in both A and B, as shown in Figure 1-2(b).

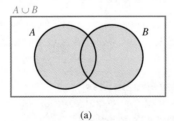
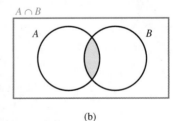

(a)   (b)

FIGURE 1-2

Since $\emptyset$ is the empty set, the following statements are true.

$A \cup \emptyset = A$ The elements in set A joined with no elements is set A.

$A \cap \emptyset = \emptyset$ Since $\emptyset$ has no elements, sets A and $\emptyset$ have no elements in common.

EXAMPLE 3   If $A = \{a, e, i, o, u\}$ and $B = \{a, b, c, d, e\}$, find **a.** $A \cup B$, **b.** $A \cap B$, and **c.** $(A \cap \emptyset) \cup B$.

Solution   **a.** $A \cup B = \{a, b, c, d, e, i, o, u\}$ **b.** $A \cap B = \{a, e\}$

   c. $(A \cap \emptyset) \cup B = \emptyset \cup B$ Do the work in parentheses first.: $A \cap \emptyset = \emptyset$.

$= B$

$= \{a, b, c, d, e\}$ ∎

■ SETS OF NUMBERS

In algebra, we will work with many different sets of numbers. For example,

- To express the number of bedrooms in a house, we use counting numbers: 1, 2, 3, 4, 5, and so on.
- To express interest rates, we use decimals: $5\% = 0.05$ or $7.5\% = 0.075$.
- To express temperatures below zero, we use negative numbers: $5°$ below $0 = -5°$.

In the following definitions, each group of three dots, called an **ellipsis**, shows that the numbers continue forever.

Natural Numbers
The **natural numbers** are the numbers that we use for counting:
$$\{1, 2, 3, 4, 5, 6, 7, 8, 9, \ldots\}$$

Whole Numbers
The **whole numbers** are the natural numbers together with 0:
$$\{0, 1, 2, 3, 4, 5, 6, 7, 8, 9, \ldots\}$$

Integers
The **integers** are the natural numbers, 0, and the negatives of the natural numbers:
$$\{\ldots, -4, -3, -2, -1, 0, 1, 2, 3, 4, \ldots\}$$

In Figure 1-3, we graph each of these sets from -6 to 6 on a number line.

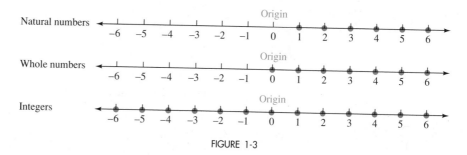

FIGURE 1-3

In Figure 1-3, to each number x there corresponds a point on the number line called its **graph**. Furthermore, to each point there corresponds a number x called its **coordinate**. Numbers to the left of 0 are **negative numbers**, and numbers to the right of 0 are **positive numbers**. Positive numbers are sometimes written with a + sign. For example, $5 = +5$ and $10 = +10$.

 WARNING!   0 is neither positive nor negative.

Integers that are divisible by 2 are called **even integers**. Integers that are not divisible by 2 are called **odd integers**.

EXAMPLE 4

a. The even integers between -9 and 9 are $\{-8, -6, -4, -2, 0, 2, 4, 6, 8\}$.

b. The odd integers between -8 and 8 are $\{-7, -5, -3, -1, 1, 3, 5, 7\}$.   ∎

Two important subsets of the natural numbers are the **prime** and **composite numbers**.

Prime Numbers
The **prime numbers** are the natural numbers greater than 1 that are divisible only by themselves and 1.

Composite Numbers
The **composite numbers** are the natural numbers greater than 1 that are not prime numbers.

Figure 1-4 shows the graphs of the primes and composites that are less than 15.

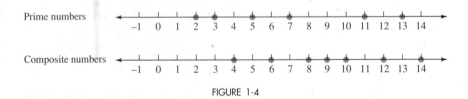

FIGURE 1-4

 WARNING!   1 is the only natural number that is neither prime nor composite.

To find the coordinates of more points on the number line, we need the **rational numbers**.

Rational Numbers
The **rational numbers** are the numbers that can be written in the form $\frac{a}{b}$ ($b \neq 0$), where a and b are integers.

Each of the following numbers is an example of a rational number.

$$\frac{8}{5}, \quad \frac{2}{3}, \quad -\frac{44}{23}, \quad \frac{-315}{476}, \quad 0 = \frac{0}{7}, \quad \text{and} \quad 17 = \frac{17}{1}$$

Because each number has an integer numerator and a nonzero integer denominator.

WARNING!   Note that $\frac{0}{5} = 0$, because $5 \cdot 0 = 0$. However, the fraction $\frac{5}{0}$ is undefined, because there is no number that when multiplied by 0 gives 5.

The fraction $\frac{0}{0}$ is indeterminate because all numbers when multiplied by 0 give 0. Remember that the denominator of a fraction cannot be 0.

EXAMPLE 5

a. The integer -7 is a rational number, because it can be written in the form $\frac{-7}{1}$. Note that all integers are rational numbers.

b. 0.125 is a rational number, because it can be written in the form $\frac{1}{8}$.

c. -0.666. . . is a rational number, because it can be written in the form $-\frac{2}{3}$.

Every rational number can be written as a decimal that either terminates or repeats in a block of digits.

EXAMPLE 6

Change each fraction to decimal form and tell whether the decimal terminates or repeats: **a.** $\frac{3}{4}$ and   **b.** $\frac{421}{990}$.

Solution   **a.** To change $\frac{3}{4}$ to a decimal, we divide 3 by 4 to obtain 0.75. Because the decimal 0.75 ends, it is a terminating decimal.

b. To change $\frac{421}{990}$ to a decimal, we divide 421 by 990 to obtain 0.4252525. . . . The quotient 0.4252525 . . . is a repeating decimal, because the block of digits 25 repeats forever. This decimal can be written as $0.4\overline{25}$, where the overbar indicates the repeating block of digits.

Rational numbers provide coordinates for many points on the number line that lie between the integers (see Figure 1-5). Note that all integers are also rational numbers.

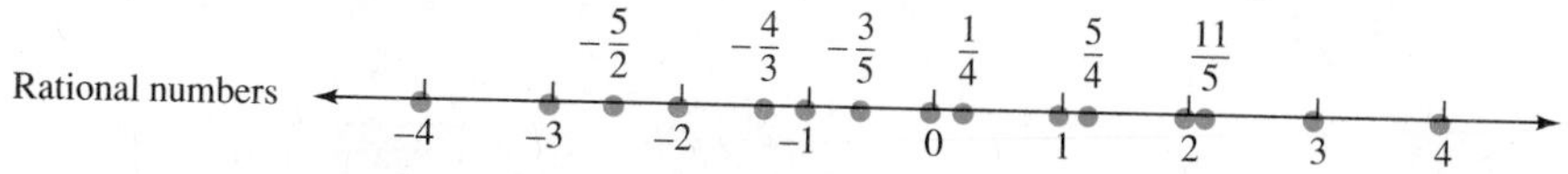

Rational numbers $-\frac{5}{2}$ $-\frac{4}{3}$ $-\frac{3}{5}$ $\frac{1}{4}$ $\frac{5}{4}$ $\frac{11}{5}$
-4 -3 -2 -1 0 1 2 3 4

FIGURE 1-5

There are points on the number line whose coordinates are not rational numbers. The coordinates of these points are called **irrational numbers**.

Irrational Numbers

The **irrational numbers** are the numbers whose decimal forms are nonterminating, nonrepeating decimals.

Some examples of irrational numbers are

0.31 331 3331 . . . $\sqrt{3} = 1.7320508076$. . . $\pi = 3.141592653$ . . .

The union of the set of rational numbers and the set of irrational numbers is the set of **real numbers**, denoted by $\Re$.

> **Real Numbers**
> The **real numbers** are the numbers that can be expressed as either a terminating, a repeating, or a nonrepeating decimal.

The number line in Figure 1-6 shows several points on the number line and their real-number coordinates. The points whose coordinates are real numbers fill up the number line.

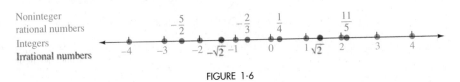

FIGURE 1-6

Figure 1-7 shows how several of the previous sets of numbers are related.

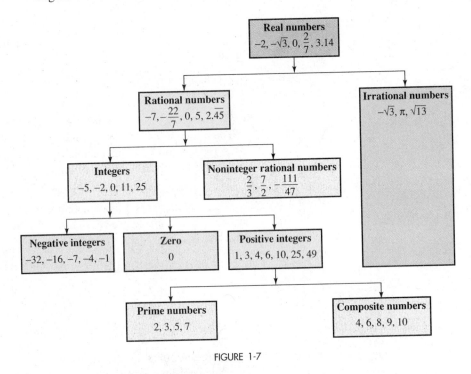

FIGURE 1-7

■ ORDERING OF THE REAL NUMBERS

As we move from left to right on the number line, the coordinates of the points get larger. On a number line,

- If a point is to the right of another, its coordinate is the greater.
- If a point is to the left of another, its coordinate is the smaller.

EXAMPLE 7

a. $4 > -2$ (4 is greater than -2), because 4 lies to the right of -2.

b. $-5 < -1$ (-5 is less than -1), because -5 lies to the left of -1.

c. $-7 > -10$ (-7 is greater than -10), because -7 lies to the right of -10. ■

Two other common inequality symbols are $\leq$ (is less than or equal to) and $\geq$ (is greater than or equal to).

EXAMPLE 8

a. $15 \leq 15$, because $15 = 15$.   **b.** $20 \geq -19$, because $20 > -19$. ■

Any inequality can be written so that the inequality symbol points in the opposite direction. For example, $-3 \leq 9$ can be written as $9 \geq -3$.

For any two real numbers a and b, the **trichotomy property** points out that either

$$a < b \quad \text{or} \quad a = b \quad \text{or} \quad a > b$$

For any real numbers a, b, and c, the **transitive property** points out that

If $a < b$ and $b < c$, then $a < c$.

The transitive property also applies for the $>$, $\leq$, and $\geq$ symbols.

EXAMPLE 9

If x is a real number, then

a. $x < 5$ or $x = 5$ or $x > 5$, by the trichotomy property.

b. If $x > 12$ and $12 > 5$, then $x > 5$, by the transitive property. ■

■ INTERVALS

The graphs of sets of real numbers are portions of the number line called **intervals**.

EXAMPLE 10

Graph the real numbers x such that   **a.** $x > 3$ and   **b.** $x \leq -3$.

Solution   **a.** The graph of $x > 3$ is an arrow including all points with coordinates greater than 3 (see Figure 1-8(a)). The parenthesis at 3 indicates that the point with coordinate 3 is not included. Because the interval has no endpoints, it is called an **open interval**. This interval is denoted as $(3, \infty)$, where the symbol ∞ is read as "infinity." The parentheses show that no endpoints are included.

b. The graph of $x \leq -3$ is an arrow including the point at -3 and all points with coordinates less than -3 (see Figure 1-8(b)). The bracket at -3 indicates that the point at -3 is included. Because the interval has only one endpoint, it is called a **half-open** interval. It is denoted as $(-\infty, -3]$, where the bracket shows that the endpoint with coordinate -3 is included.

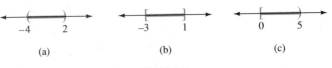

FIGURE 1-8

EXAMPLE 11   Graph the real numbers x such that   **a.** $-4 < x < 2$,   **b.** $-3 \leq x \leq 1$,   and
c. $0 \leq x < 5$.

Solution   **a.** Figure 1-9(a) shows the graph of all real numbers between -4 and 2. The graph
is the open interval denoted as $(-4, 2)$.

b. Figure 1-9(b) shows the graph of all real numbers from -3 to 1. Since the
graph includes both endpoints, it represents a **closed interval**. This closed
interval is denoted as $[-3, 1]$. The brackets indicate that both endpoints are
included.

c. Figure 1-9(c) shows the graph of all real numbers from 0 to 5, not including 5.
This half-open interval is denoted as $[0, 5)$.

FIGURE 1-9

EXAMPLE 12   If $A = (-2, 4)$ and $B = [1, 5)$, find the graphs of   **a.** $A \cup B$ and   **b.** $A \cap B$.

Solution   **a.** The union of intervals A and B is the set of all real numbers that are elements
of either set A or set B or both. Numbers between -2 and 4 are in set A, and
numbers between 1 and 5 (including 1) are in set B. All numbers between -2
and 5 are in at least one of these sets. Figure 1-10(a) shows that

$$A \cup B = (-2, 4) \cup [1, 5) = (-2, 5)$$

b. The intersection of intervals A and B is the set of all real numbers that are elements of both set A and set B. The numbers that are in both of these sets are the
numbers between 1 and 4 (including 1). Figure 1-10(b) shows that

$$A \cap B = (-2, 4) \cap [1, 5) = [1, 4)$$

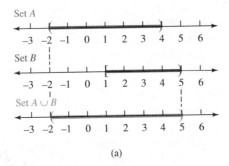

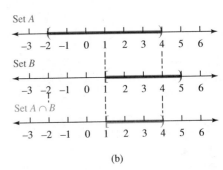

FIGURE 1-10

■ ABSOLUTE VALUE OF A NUMBER

The **absolute value** of any real number a, denoted as $|a|$, is the distance on a number line between 0 and the point with coordinate a. For example, the points shown in Figure 1-11 with coordinates of 3 and -3 both lie 3 units from 0. Thus, $|3| = |-3| = 3$.

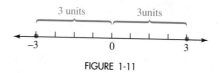

FIGURE 1-11

In general, for any real number a, $|a| = |-a|$.

The absolute value of a number can be defined more formally.

Absolute Value

For any real number x, $\begin{cases} \text{If } x \geq 0, \text{ then } |x| = x. \\ \text{If } x < 0, \text{ then } |x| = -x. \end{cases}$

If x is positive or 0, then x is its own absolute value. However, if x is negative, then $-x$ (which is a positive number) is the absolute value of x. Thus, $|x| \geq 0$ for all real numbers x.

EXAMPLE 13

a. $|3| = 3$ **b.** $|-4| = 4$

c. $|0| = 0$ **d.** $-|-8| = -(8) = -8$ Note that $|-8| = 8$. ■

Orals *If $A = \{1, 2, 4\}$ and $B = \{3, 4, 5\}$, find*

1. $A \cup B$ **2.** $A \cap B$

3. Is $A \subseteq B$? **4.** Is $B \cap \emptyset = \emptyset$?

5. List the first four prime numbers.

6. Define a rational number.

7. Find $|-6|$. **8.** Find $|10|$.

EXERCISE 1.1

In Exercises 1–8, $A = \{1, 3, 5, 9\}$, $B = \{3, 5, 9, 11\}$, and $C = \{x \mid x \text{ is an odd integer}\}$. Insert either a $\in$ or a $\subseteq$ symbol to make a true statement.

1. A ☐ C **2.** 3 ☐ A **3.** 9 ☐ B **4.** $\{3, 5\}$ ☐ C

5. $\emptyset$ ☐ B **6.** 9 ☐ C **7.** A ☐ A **8.** $\{3\}$ ☐ B

In Exercises 9–16, $A = \{1, 3, 5\}$, $B = \{2, 4, 6, 8, 10\}$, and $C = \{1, 3, 5, 7, 9\}$. Find each set.

9. $A \cup B$ **10.** $A \cap C$ **11.** $B \cap C$ **12.** $A \cup C$

13. $(C \cap \emptyset) \cap B$ **14.** $B \cup (\emptyset \cap C)$ **15.** $(A \cup B) \cap C$ **16.** $(A \cap B) \cup C$

In Exercises 17–20, list the elements in each set, if possible.

17. $\{x \mid x$ is a prime number less than 10.$\}$

18. $\{x \mid x$ is an integer between 3 and 7.$\}$

19. $\{x \mid x$ is a number that is both rational and irrational.$\}$

20. $\{x \mid x$ is a composite number between 13 and 19.$\}$

In Exercises 21–32, list the elements in the set $\left\{ -3, 0, \dfrac{2}{3}, 1, \sqrt{3}, 2, 9 \right\}$ *that satisfy the given condition.*

21. natural number

22. whole number

23. integer

24. rational number

25. irrational number

26. real number

27. even natural number

28. odd integer

29. prime number

30. composite number

31. odd composite number

32. even prime number

In Exercises 33–36, graph each set on the number line.

33. The set of prime numbers less than 8.

34. The set of integers between -9 and 0.

35. The set of odd integers between 10 and 20.

36. The set of composite numbers less than 10.

In Exercises 37–40, change each fraction to a decimal and classify the result as a terminating or a repeating decimal.

37. $\dfrac{7}{8}$ **38.** $\dfrac{7}{3}$ **39.** $-\dfrac{11}{15}$ **40.** $-\dfrac{19}{16}$

In Exercises 41–48, insert either a $<$ or a $>$ symbol to make a true statement.

41. $5 \quad 9$ **42.** $9 \quad 0$ **43.** $-5 \quad -10$ **44.** $-3 \quad 10$

45. $-7 \quad 7$ **46.** $0 \quad -5$ **47.** $6 \quad -6$ **48.** $-6 \quad -2$

In Exercises 49–56, write each statement with the inequality symbol pointing in the opposite direction.

49. $19 > 12$ **50.** $-3 \geq -5$ **51.** $-6 \leq -5$ **52.** $-10 < 13$

53. $5 \geq -3$ **54.** $0 \leq 12$ **55.** $-10 < 0$ **56.** $-4 > -8$

In Exercises 57–68, graph each interval on the number line.

57. $x > 3$ **58.** $x < 0$ **59.** $-3 < x < 2$ **60.** $-5 \leq x < 2$

61. $0 < x \leq 5$ **62.** $-4 \leq x \leq -2$ **63.** $(-2, \infty)$ **64.** $(-\infty, 4]$

65. $[-6, 9]$ **66.** $(-1, 3)$ **67.** $(-2, 4]$ **68.** $[-5, 2]$

In Exercises 69–76, A, B, C, and D are intervals with $A = [-4, 4]$, $B = (0, 6)$, $C = [2, 8)$, and $D = (-3, 4]$. Graph each set.

69. $A \cap C$

70. $B \cup C$

71. $A \cup C$

72. $A \cap B$

73. $A \cap B \cap C$

74. $A \cup B \cup C$

75. $A \cup B \cup D$

76. $B \cap C \cap D$

In Exercises 77–84, write each expression without using absolute value symbols. Simplify the result when possible.

77. $|20|$

78. $|-20|$

79. $-|-6|$

80. $-|8|$

81. $|-5| + |-2|$

82. $|12| + |-4|$

83. $|-5| \cdot |4|$

84. $|-6| \cdot |-3|$

85. Find x if $|x| = 3$.

86. Find x if $|x| = 7$.

87. What numbers x are equal to their own absolute values?

88. What numbers x when added to their own absolute values give a sum of 0?

89. How many integers have an absolute value that is less than 50?

90. How many odd integers have an absolute value between 20 and 40?

Writing Exercises

Write a paragraph using your own words.

1. Distinguish between "is an element of" and "is a subset of."

2. Explain why every integer is a rational number, but not every rational number is an integer.

3. Explain why the union of the set of primes and the set of composites is not the set of natural numbers.

4. Is the absolute value of a number always positive? Explain.

Something to Think About

1. List every subset of $\{1, 2\}$. Don't forget to include $\emptyset$ and the set itself. How many subsets are there?

2. List every subset of $\{1, 2, 3\}$. How many are there? Can you guess how many subsets $\{1, 2, 3, 4\}$ has?

3. If $(1, 5) \cap (a, b) = (a, b)$, what do you know about a and b?

4. If $(1, 5) \cup (a, b) = (a, b)$, what do you know about a and b?

Review Exercises

To simplify a fraction, factor the numerator and the denominator and divide out common factors. For example, $\dfrac{12}{18} = \dfrac{6 \cdot 2}{6 \cdot 3} = \dfrac{\cancel{6} \cdot 2}{\cancel{6} \cdot 3} = \dfrac{2}{3}$. Simplify each fraction.

1. $\dfrac{6}{8}$

2. $\dfrac{15}{20}$

3. $\dfrac{32}{40}$

4. $\dfrac{56}{72}$

To multiply fractions, multiply the numerators and multiply the denominators. To divide fractions, invert the divisor and multiply. Always simplify the result if possible.

5. $\dfrac{1}{4} \cdot \dfrac{3}{5}$

6. $\dfrac{3}{5} \cdot \dfrac{20}{27}$

7. $\dfrac{2}{3} \div \dfrac{3}{7}$

8. $\dfrac{3}{5} \div \dfrac{9}{15}$

To add (or subtract) fractions, write each fraction with a common denominator and then add (or subtract) the numerators and keep the same denominator. Always simplify the result if possible.

9. $\dfrac{5}{9} + \dfrac{4}{9}$ **10.** $\dfrac{16}{7} - \dfrac{2}{7}$ **11.** $\dfrac{2}{3} + \dfrac{4}{5}$ **12.** $\dfrac{7}{9} - \dfrac{2}{5}$

1.2 Arithmetic and Properties of Real Numbers

■ ADDING REAL NUMBERS ■ SUBTRACTING REAL NUMBERS ■ MULTIPLYING REAL NUMBERS
■ DIVIDING REAL NUMBERS ■ ORDER OF OPERATIONS ■ ACCENT ON STATISTICS ■ EVALUATING
ALGEBRAIC EXPRESSIONS ■ PROPERTIES OF REAL NUMBERS

In this section, we will show how to add, subtract, multiply, and divide real numbers. We will then discuss several properties of real numbers.

■ ADDING REAL NUMBERS

When two numbers are added, we call the result their **sum**. To find the sum of $+2$ and $+3$, we can represent the numbers with arrows, as shown in Figure 1-12(a). Since the endpoint of the second arrow is at $+5$, we have $+2 + (+3) = +5$.

To add -2 and -3, we can draw arrows as shown in Figure 1-12(b). Since the endpoint of the second arrow is at -5, we have $(-2) + (-3) = -5$.

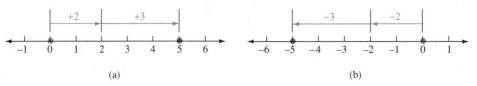

(a)   (b)

FIGURE 1-12

To add -6 and $+2$, we can draw arrows as shown in Figure 1-13(a). Since the endpoint of the second arrow is at -4, we have $(-6) + (+2) = -4$.

To add $+7$ and -4, we can draw arrows as shown in Figure 1-13(b). Since the endpoint of the final arrow is at $+3$, we have $(+7) + (-4) = +3$.

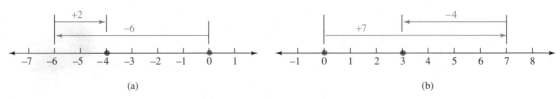

(a)   (b)

FIGURE 1-13

These examples suggest the following rules.

Adding Real Numbers
With like signs: Add the absolute values of the numbers and keep the common sign.
With unlike signs: Subtract the absolute values of the numbers (the smaller from the larger) and keep the sign of the number with the larger absolute value.

EXAMPLE 1 (b, d)

a. $+4 + (+6) = +10$

Add the absolute values and use the common sign: $4 + 6 = +10$.

b. $-5 + (-3) = -8$

Add the absolute values and use the common sign: $-(5 + 3) = -8$.

c. $+9 + (-5) = +4$

Subtract the absolute values and use a $+$ sign: $+(9 - 5) = +4$.

d. $-12 + (+5) = -7$

Subtract the absolute values and use a $-$ sign: $-(12 - 5) = -7$.

■ SUBTRACTING REAL NUMBERS

When two numbers are subtracted, we call the result their **difference**. To find a difference, we can always change the subtraction into an equivalent addition. For example, the subtraction $7 - 4$ is equivalent to the addition $7 + (-4)$, because they have the same answer:

$$7 - 4 = 3 \quad \text{and} \quad 7 + (-4) = 3$$

Thus, to subtract two numbers, we change the sign of the number being subtracted and add.

Subtracting Real Numbers
If a and b are real numbers, then $a - b = a + (-b)$.

EXAMPLE 2 (a, c)

a. $12 - 4 = 12 + (-4)$
  $= 8$

Change the sign of 4 and add.

b. $-13 - 5 = -13 + (-5)$
  $= -18$

Change the sign of 5 and add.

c. $-14 - (-6) = -14 + (+6)$
  $= -8$

Change the sign of -6 and add.

■ MULTIPLYING REAL NUMBERS

When two numbers are multiplied, we call the result their **product**. We can find the product of 5 and 4 by using 4 in an addition five times:

$$5(4) = 4 + 4 + 4 + 4 + 4 = 20$$

We can find the product of 5 and -4 by using -4 in an addition five times:

$$5(-4) = (-4) + (-4) + (-4) + (-4) + (-4) = -20$$

Since multiplication by a negative number can be defined as repeated subtraction, we can find the product of -5 and 4 by using 4 in a subtraction five times:

$$-5(4) = -4 - 4 - 4 - 4 - 4$$
$$= -4 + (-4) + (-4) + (-4) + (-4)$$
$$= -20$$

Change the sign of each 4 and add.

We can find the product of -5 and -4 by using -4 in a subtraction five times:

$$-5(-4) = -(-4) - (-4) - (-4) - (-4) - (-4)$$
$$= 4 + 4 + 4 + 4 + 4 \qquad \text{Change the sign of each } -4 \text{ and add.}$$
$$= 20$$

The products $5(4)$ and $-5(-4)$ both equal $+20$, and the products $5(-4)$ and $-5(4)$ both equal -20. These results suggest the first two of the following rules.

Multiplying Real Numbers
With like signs: Multiply their absolute values. The product is positive.
With unlike signs: Multiply their absolute values. The product is negative.
Multiplication by 0: If x is any real number, then $x \cdot 0 = 0 \cdot x = 0$.

 (a, b, c)   **EXAMPLE 3**

a. $4(-7) = -28$ Multiply the absolute values: $4 \cdot 7 = 28$. Since the signs are unlike, the product is negative.

b. $-5(-6) = +30$ Multiply the absolute values: $5 \cdot 6 = 30$. Since the signs are alike, the product is positive.

c. $-7(6) = -42$ Multiply the absolute values: $7 \cdot 6 = 42$. Since the signs are unlike, the product is negative.

d. $8(6) = +48$ Multiply the absolute values: $8 \cdot 6 = 48$. Since the signs are alike, the product is positive. ∎

■ DIVIDING REAL NUMBERS

When two numbers are divided, we call the result their **quotient**. In the division $\frac{x}{y} = q$ $(y \neq 0)$, the quotient q is a number such that $y \cdot q = x$. We can use this relationship to find rules for dividing real numbers. We consider four divisions:

$$\frac{+10}{+2} = +5 \quad \text{because } +2(+5) = +10 \qquad \frac{-10}{-2} = +5 \quad \text{because } -2(+5) = -10$$

$$\frac{-10}{+2} = -5 \quad \text{because } +2(-5) = -10 \qquad \frac{+10}{-2} = -5 \quad \text{because } -2(-5) = +10$$

These results suggest the first two rules for dividing real numbers.

Dividing Real Numbers
With like signs: Divide their absolute values. The quotient is positive.
With unlike signs: Divide their absolute values. The quotient is negative.
Division by 0: Division by 0 is undefined.

WARNING! If $x \neq 0$, then $\frac{0}{x} = 0$. However, $\frac{x}{0}$ is undefined for any nonzero value of x, and $\frac{0}{0}$ is indeterminate.

OO (b, c, d)   **EXAMPLE 4**

a. $\dfrac{36}{18} = +2$ Divide the absolute values: $\frac{36}{18} = 2$. Since the signs are alike, the quotient is positive.

b. $\dfrac{-44}{11} = -4$ Divide the absolute values: $\frac{44}{11} = 4$. Since the signs are unlike, the quotient is negative.

c. $\dfrac{27}{-9} = -3$ Divide the absolute values: $\frac{27}{9} = 3$. Since the signs are unlike, the quotient is negative.

d. $\dfrac{-64}{-8} = +8$ Divide the absolute values: $\frac{64}{8} = 8$. Since the signs are alike, the quotient is positive. ∎

■ ORDER OF OPERATIONS

To guarantee that calculations will have only one answer, we will do the operations of addition, subtraction, multiplication, and division in the following order.

> **Order of Operations**
> Use the following steps to do all calculations within each pair of grouping symbols (parentheses, brackets, etc.), working from the innermost pair to the outermost pair.
> **1.** Do all multiplications and/or divisions, in order from left to right.
> **2.** Do all additions and/or subtractions, in order from left to right.
> When all grouping symbols have been removed, repeat Steps 1 and 2 to finish the calculation.
>   In a fraction, simplify the numerator and denominator separately. Then simplify the fraction, if possible.

OO (c, d, e)   **EXAMPLE 5**

a. $4 + 2 \cdot 3 = 4 + 6$ Do the multiplication first.
$\qquad\qquad = 10$ Then do the addition.

b. $2(3 + 4) = 2 \cdot 7$ Because of the parentheses, do the addition first.
$\qquad\qquad = 14$

Then do the multiplication.

c. $5(3 - 6) \div 3 + 1 = 5(-3) \div 3 + 1$ Do the subtraction within parentheses.

$\qquad\qquad\qquad = -15 \div 3 + 1$ Then do the multiplication: $5(-3) = -15$.

$\qquad\qquad\qquad = -5 + 1$ Then do the division: $-15 \div 3 = -5$.

$\qquad\qquad\qquad = -4$ Finally, do the addition.

d. $5[3 - 2(6 \div 3 + 1)] = 5[3 - 2(2 + 1)]$ Do the division within parentheses: $6 \div 3 = 2$.

$\qquad\qquad\qquad = 5[3 - 2(3)]$ Do the addition: $2 + 1 = 3$.

$\qquad\qquad\qquad = 5(3 - 6)$ Do the multiplication: $2(3) = 6$.

$\qquad\qquad\qquad = 5(-3)$ Do the subtraction: $3 - 6 = -3$.

$\qquad\qquad\qquad = -15$ Do the multiplication.

e. $\dfrac{4 + 8(3 - 4)}{6 - 2(2)} = \dfrac{-4}{2}$ Simplify the numerator and denominator separately.

$\qquad\qquad\qquad = -2$ ∎

■ ■ ■ ■ ■ ■ ■ ■ ■   **Measures of Central Tendency**

ACCENT ON
STATISTICS

There are three types of averages that are commonly used in statistics: the **mean**, the **median**, and the **mode**.

> **Mean**
> The **mean** of several values is the sum of those values divided by the number of values.
>
> $$\text{Mean} = \frac{\text{sum of the values}}{\text{number of values}}$$

EXAMPLE 6

Football   Figure 1-14 shows the gains and losses made by a running back on seven plays. Find the mean number of yards per carry.

Solution   To find the mean number of yards per carry, we add the numbers and divide by 7.

$$\frac{-8 + (+2) + (-6) + (+6) + (+4) + (-7) + (-5)}{7} = \frac{-14}{7} = -2$$

The running back averaged -2 yards (or lost 2 yards) per carry.

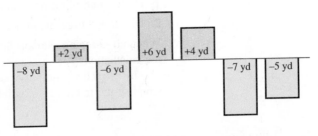

FIGURE 1-14   ■

> **Median**
> The **median** of several values is the middle value. To find the median,
> 1. Arrange the values in increasing order.
> 2. If there is an odd number of values, choose the middle value.
> 3. If there is an even number of values, find the mean of the middle two values.
>
> **Mode**
> The **mode** of several values is the value that occurs most often.

EXAMPLE 7

Ten workers in a small business have monthly salaries of

$2500, $1750, $2415, $3240, $2790, $3240, $2650, $2415, $2415, $2650

Find   **a.** the median   and   **b.** the mode   of the distribution.

Solution   **a.** To find the median, we first arrange the scores in increasing order:

$1750, $2415, $2415, $2415, $2500, $2650 , $2650, $2790, $3240, $3240

Because there is an even number of scores, the median will be the mean of the middle two scores, $2500 and $2650. Thus, the median is

$$\text{Median} = \frac{\$2500 + \$2650}{2} = \$2575$$

b. Since the salary $2415 occurs most often, it is the mode. ∎

If two different numbers in a distribution tie for occurring most often, the distribution is called *bimodal*.

Although the mean is probably the most common measure of average, the median and the mode are frequently used. For example, workers' salaries are often compared to the median (average) salary. To say that the modal (average) shoe size is 9 means that a shoe size of 9 occurs more often than any other size.

■ EVALUATING ALGEBRAIC EXPRESSIONS

Variables and numbers can be combined with the operations of arithmetic to produce **algebraic expressions**. To evaluate algebraic expressions, we substitute numbers for the variables and carry out the arithmetic.

EXAMPLE 8   If $a = 2$, $b = -3$, and $c = -5$, evaluate **a.** $a + bc$ and   **b.** $\dfrac{ab + 3c}{b(c - a)}$.

Solution   We substitute 2 for a, -3 for b, and -5 for c and simplify.

a. $a + bc = 2 + (-3)(-5)$

$= 2 + (15)$

$= 17$

b. $\dfrac{ab + 3c}{b(c - a)} = \dfrac{2(-3) + 3(-5)}{-3(-5 - 2)}$

$= \dfrac{-6 + (-15)}{-3(-7)}$

$= \dfrac{-21}{21}$

$= -1$ ∎

Table 1-1 (on page 20) shows the formulas for the perimeters of several geometric figures.

EXAMPLE 9   Find the perimeter of the rectangle shown in Figure 1-15.

Solution   We substitute 2.75 for l and 1.25 for w into the formula $P = 2l + 2w$ and simplify.

$P = 2l + 2w$

$P = 2(2.75) + 2(1.25)$

$= 5.5 + 2.5$

$= 8.00$

2.75 m

1.25 m

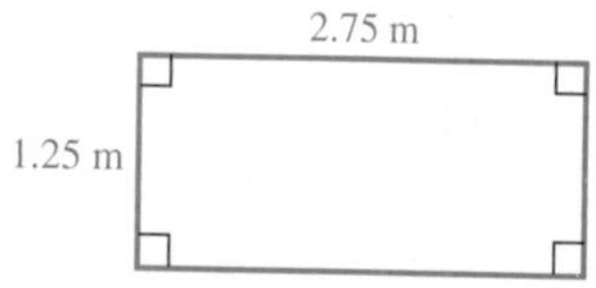

FIGURE 1-15

The perimeter is 8 meters. ∎

Figure	Name	Perimeter/ Circumference
	Square	$P = 4s$
	Rectangle	$P = 2l + 2w$
	Triangle	$P = a + b + c$
	Trapezoid	$P = a + b + c + d$
	Circle	$C = \pi D$ (π is approximately 3.1416)

TABLE 1-1

■ PROPERTIES OF REAL NUMBERS

There are several properties of equality that we will use throughout the book.

Properties of Equality

If a, b, and c are real numbers, then

Reflexive property $a = a$

Symmetric property   If $a = b$, then $b = a$.

Transitive property   If $a = b$ and $b = c$, then $a = c$.

Substitution property   If $a = b$, then b can be substituted for a in any algebraic expression to obtain an equivalent expression.

EXAMPLE 10

a. By the reflexive property: $5 = 5$.

b. By the symmetric property: If $4 = 3 + 1$, then $3 + 1 = 4$.

c. By the transitive property: If $10 = 5 + 5$ and $5 + 5 = 2 \cdot 5$, then $10 = 2 \cdot 5$.

d. By the substitution property: If $x + \frac{1}{2} = x \cdot y$ and $x = \frac{5}{2}$, then $\frac{5}{2} + \frac{1}{2} = \frac{5}{2} \cdot y$. ■

Because of the closure properties, the sum, difference, product, and quotient of any two real numbers is another real number (provided there are no divisions by 0).

The Closure Properties

If a and b are real numbers, then

$a + b$ is a real number. $a - b$ is a real number.

ab is a real number. $\frac{a}{b}$ is a real number $(b \neq 0)$

The associative properties enable us to group the numbers in a sum or a product any way that we wish and still get the same result.

The Associative Properties

If a, b, and c are real numbers, then

$(a + b) + c = a + (b + c)$ The associative property of addition.

$(ab)c = a(bc)$ The associative property of multiplication.

EXAMPLE 11

a. $(2 + 3) + 4 = 5 + 4$ **b.** $2 + (3 + 4) = 2 + 7$
$\qquad\qquad\quad = 9$ $\qquad\qquad\qquad = 9$

c. $(2 \cdot 3) \cdot 4 = 6 \cdot 4$ **d.** $2 \cdot (3 \cdot 4) = 2 \cdot 12$
$\qquad\qquad\quad = 24$ $\qquad\qquad\qquad = 24$

 WARNING!   Subtraction and division are not associative, because different groupings give different results. For example,

$(8 - 4) - 2 = 4 - 2 = 2$ but $8 - (4 - 2) = 8 - 2 = 6$

$(8 \div 4) \div 2 = 2 \div 2 = 1$ but $8 \div (4 \div 2) = 8 \div 2 = 4$

The commutative properties enable us to add or multiply two numbers in either order and obtain the same result.

The Commutative Properties

If a and b are real numbers, then

$a + b = b + a$ The commutative property of addition.

$ab = ba$ The commutative property of multiplication.

EXAMPLE 12

a. $2 + 3 = 5$ and $3 + 2 = 5$ **b.** $7 \cdot 9 = 63$ and $9 \cdot 7 = 63$

 WARNING!   Subtraction and division are not commutative, because doing these operations in different orders will give different results. For example,

$8 - 4 = 4$ but $4 - 8 = -4$

$8 \div 4 = 2$ but $4 \div 8 = \frac{1}{2}$

The distributive property enables us to evaluate many expressions involving a multiplication and an addition. We can add first and then multiply, or multiply first and then add.

The Distributive Property
If a, b, and c are real numbers, then

$$a(b + c) = ab + ac$$

EXAMPLE 13

$$2(3 + 7) = 2 \cdot 10 \quad \text{and} \quad 2(3 + 7) = 2 \cdot 3 + 2 \cdot 7$$
$$= 20 \qquad\qquad\qquad\qquad = 6 + 14$$
$$= 20 \qquad \blacksquare$$

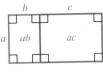

FIGURE 1-16

We can interpret the distributive property geometrically. Since the area of the largest rectangle in Figure 1-16 is the product of its width a and its length $b + c$, its area is $a(b + c)$. The areas of the two smaller rectangles are ab and ac. Since the area of the largest rectangle is equal to the sum of the areas of the smaller rectangles, we have $a(b + c) = ab + ac$.

A more general form of the distributive property is the **extended distributive property**.

$$a(b + c + d + e + \cdots) = ab + ac + ad + ae + \cdots$$

EXAMPLE 14

Use the distributive property to write each expression without parentheses: **a.** $2(x + 3)$ and   **b.** $2(x + y - 7)$.

Solution   **a.** $2(x + 3) = 2x + 2 \cdot 3$ **b.** $2(x + y - 7) = 2x + 2y - 2 \cdot 7$
$$= 2x + 6 \qquad\qquad\qquad\qquad\qquad = 2x + 2y - 14 \qquad \blacksquare$$

The real numbers 0 and 1 have important special properties.

Properties of 0 and 1
Additive identity: The sum of 0 and any number is the number itself.

$$0 + a = a + 0 = a$$

Multiplication property of 0: The product of any number and 0 is 0.

$$a \cdot 0 = 0 \cdot a = 0$$

Multiplicative identity: The product of 1 and any number is the number itself.

$$1 \cdot a = a \cdot 1 = a$$

EXAMPLE 15

a. $7 + 0 = 7$ **b.** $7(0) = 0$ **c.** $1(5) = 5$ **d.** $(-7)1 = -7$ $\blacksquare$

If the sum of two numbers is 0, the numbers are called **additive inverses, negatives**, or **opposites** of each other. For example, 6 and -6 are negatives, because $6 + (-6) = 0$.

Additive Inverse Property

For every real number a, there is one real number $-a$ such that

$$a + (-a) = -a + a = 0$$

The symbol $-(-6)$ means "the negative of negative 6." Because the sum of two numbers that are negatives is 0, we have

$$-6 + [-(-6)] = 0 \quad \text{and} \quad -6 + 6 = 0$$

Because -6 has only one additive inverse, it follows that $-(-6) = 6$. In general, we have the following rule.

The Double Negative Rule

If a represents any real number, then $-(-a) = a$.

If the product of two numbers is 1, the numbers are called **multiplicative inverses** or **reciprocals** of each other.

Multiplicative Inverse Property

For every nonzero real number a, there is one real number $\frac{1}{a}$ such that

$$a \cdot \frac{1}{a} = \frac{1}{a} \cdot a = 1 \quad (a \neq 0)$$

EXAMPLE 16

a. Since $5\left(\frac{1}{5}\right) = 1$, the numbers 5 and $\frac{1}{5}$ are reciprocals.

b. Since $\frac{3}{2}\left(\frac{2}{3}\right) = 1$, the numbers $\frac{3}{2}$ and $\frac{2}{3}$ are reciprocals.

c. Since $-0.25(-4) = 1$, the numbers -0.25 and -4 are reciprocals.

d. The reciprocal of 0 does not exist, because $\frac{1}{0}$ does not exist. ∎

Orals *Do each operation.*

1. $+2 + (-4)$ 　　　　　　　　　　　**2.** $-5 - 2$

3. $7(-4)$ 　　　　　　　　　　　　　**4.** $(-7)(-4)$

5. $3 + (-3)(2)$ 　　　　　　　　　　**6.** $\dfrac{-4 + (-2)}{-3 + 5}$

EXERCISE 1.2

In Exercises 1–28, do the operations.

1. $-3 + (-5)$ 　　**2.** $2 + (+8)$ 　　**3.** $-7 + 2$ 　　**4.** $3 + (-5)$

5. $-3 - 4$ 　　**6.** $-11 - (-17)$ 　　**7.** $-33 - (-33)$ 　　**8.** $14 - (-13)$

9. $-2(6)$ 　　**10.** $3(-5)$ 　　**11.** $-3(-7)$ 　　**12.** $-2(-5)$

13. $\dfrac{-8}{4}$

14. $\dfrac{25}{-5}$

15. $\dfrac{-16}{-4}$

16. $\dfrac{-5}{-25}$

17. $\dfrac{1}{2} + \left(-\dfrac{1}{3}\right)$

18. $-\dfrac{3}{4} + \left(-\dfrac{1}{5}\right)$

19. $\dfrac{1}{2} - \left(-\dfrac{3}{5}\right)$

20. $\dfrac{1}{26} - \dfrac{11}{13}$

21. $\dfrac{1}{3} - \dfrac{1}{2}$

22. $\dfrac{7}{8} - \left(-\dfrac{3}{4}\right)$

23. $\left(-\dfrac{3}{5}\right)\left(\dfrac{10}{7}\right)$

24. $\left(-\dfrac{6}{7}\right)\left(-\dfrac{5}{12}\right)$

25. $\dfrac{3}{4} \div \left(-\dfrac{3}{8}\right)$

26. $-\dfrac{3}{5} \div \dfrac{7}{10}$

27. $-\dfrac{16}{5} \div \left(-\dfrac{10}{3}\right)$

28. $-\dfrac{5}{24} \div \dfrac{10}{3}$

In Exercises 29–36, use signed numbers to solve each problem.

29. Earning money One day Scott earned $22.25 tutoring mathematics and $39.75 tutoring physics. How much did he earn?

30. Losing weight During an illness, Wendy lost 13.5 pounds. She then dieted and lost another 11.5 pounds. How much did she lose?

31. Changing temperatures The temperature rose 17° in 1 hour and then dropped 13° in the next hour. Find the overall change in temperature.

32. Displaying the flag Before the American flag is displayed at half-mast, it should first be raised to the top of the flagpole. How far has the flag in Illustration 1 traveled if it is now at half-mast?

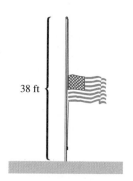

38 ft

ILLUSTRATION 1

33. Changing temperatures If the temperature has been dropping 4° each hour, how much warmer was it 3 hours ago?

34. Playing slot machines In Las Vegas, Harry lost $30 per hour playing the slot machines. How much did he lose after gambling for 15 hours?

35. Filling a pool The flow of water from a pipe is filling a pool at the rate of 23 gallons per minute. How much less water was in the pool 5 hours ago?

36. Draining a pool If a drain is emptying a pool at the rate of 12 gallons per minute, how much more water was in the pool 2 hours ago?

In Exercises 37–56, do the operations.

37. $3 + 4 \cdot 5$

38. $5 \cdot 3 - 6 \cdot 4$

39. $3 - 2 - 1$

40. $5 - 3 - 1$

41. $3 - (2 - 1)$

42. $5 - (3 - 1)$

43. $2 - 3 \cdot 5$

44. $6 + 4 \cdot 7$

45. $8 \div 4 \div 2$

46. $100 \div 10 \div 5$

47. $8 \div (4 \div 2)$

48. $100 \div (10 \div 5)$

49. $2 + 6 \div 3 - 5$

50. $6 - 8 \div 4 - 2$

51. $(2 + 6) \div (3 - 5)$

52. $(6 - 8) \div (4 - 2)$

53. $\dfrac{3(8 + 4)}{2 \cdot 3 - 9}$

54. $\dfrac{5(4 - 1)}{3 \cdot 2 + 5 \cdot 3}$

55. $\dfrac{100(2 - 4)}{1000 \div 10 \div 10}$

56. $\dfrac{8(3) - 4(6)}{5(3) + 3(-7)}$

In Exercises 57–60, use a calculator to help solve the following problems.

57. Military science An army retreated 2300 meters. After regrouping, it moved forward 1750 meters. The next day, it gained another 1875 meters. Find the army's net gain (or loss).

58. Grooming horses John earned $8 an hour for grooming horses. After working for 8 hours, he had $94. How much did he have before he started work?

59. Managing a checkbook Sally started with $437.37 in a checking account. One month, she had deposits of $125.18, $137.26, and $145.56. That same month, she had withdrawals of $117.11, $183.49, and $122.89. Find her ending balance.

60. Stock averages Illustration 2 shows the daily advances and declines of the Dow Jones average for one week. Find the total gain or loss for the week.

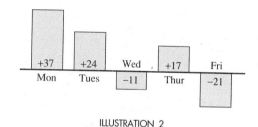

ILLUSTRATION 2

In Exercises 61–63, use the following distribution of numbers: 7, 5, 9, 10, 8, 6, 6, 7, 9, 12, 9.

61. Find the mean. **62.** Find the median. **63.** Find the mode.

In Exercises 64–66, use the following distribution of numbers: 8, 12, 23, 12, 10, 16, 26, 12, 14, 8, 16, 23.

64. Find the median. **65.** Find the mode. **66.** Find the mean.

In Exercises 67–72, use a calculator.

67. Selling clothes If a clerk in a dress shop had the sales shown in Illustration 3 for one week, find the mean of daily sales.

Monday	$1525
Tuesday	$ 785
Wednesday	$1628
Thursday	$1214
Friday	$ 917
Saturday	$1197

ILLUSTRATION 3

68. Size of viruses Illustration 4 gives the approximate lengths (in centimicrons) of the viruses that cause five common diseases. Find the mean length of the viruses.

Polio	2.5
Influenza	105.1
Pharyngitis	74.9
Chicken pox	137.4
Yellow fever	52.6

ILLUSTRATION 4

69. Calculating grades A student has scores of 75, 82, 87, 80, and 76 on five exams. Find his average (mean) score.

70. Averaging weights The offensive line of a football team usually has two guards, two tackles, and a center. If the guards weigh 298 and 287 pounds, the tackles weigh 310 and 302 pounds, and the center weighs 303 pounds, find the average (mean) weight of the offensive line.

71. Analyzing ads The businessman who ran the ad that is shown in Illustration 5 earns $100,000 and employs four students who earn $10,000 each. Is the ad honest?

HIRING

Hard-working, intelligent students
Good pay: average wage of
$28,000

ILLUSTRATION 5

72. Averaging grades A student has grades of 78%, 85%, 88%, and 96%. There is one test left, and the student needs to average 90% to earn an A. Does he have a chance?

73. Give three applications where the median would be the most appropriate average to use.

74. Give three applications where the mode would be the most appropriate average to use.

In Exercises 75–82, a = 3, b = −2, c = −1, and d = 2. Evaluate each expression.

75. $ab + cd$

76. $ad + bc$

77. $a(b + c)$

78. $d(b + a)$

79. $\dfrac{ad + c}{cd + b}$

80. $\dfrac{ab + d}{bd + a}$

81. $\dfrac{ac - bd}{cd - ad}$

82. $\dfrac{bc - ad}{bd + ac}$

 In Exercises 83–84, use this information. Sorting records is a common task in data processing. A selection sort requires C comparisons to sort N records, where C and N are related by the formula

$$C = \frac{N(N - 1)}{2}$$

83. How many comparisons are needed to sort 200 records?

84. How many comparisons are needed to sort 10,000 records?

In Exercises 85–88, use a calculator to help solve each problem.

85. Perimeter of a square   Find the perimeter of the square shown in Illustration 6.

86. Circumference of a circle   Find the circumference of the circle shown in Illustration 7.

87. Perimeter of a triangle   Find the perimeter of a triangle with sides that are 23.5, 37.2, and 39.7 feet long.

88. Perimeter of a trapezoid   Find the perimeter of a trapezoid with sides that are 43.27, 47.37, 50.21, and 52.93 centimeters long.

7.5 cm

ILLUSTRATION 6

25 m

ILLUSTRATION 7

In Exercises 89–94, tell which property of equality justifies each statement.

89. If $a = b + c$, then $b + c = a$.

90. If $x = y + z$ and $z = 3$, then $x = y + 3$.

91. $a + b + c = a + b + c$

92. If $a = 37$ and $37 = b$, then $a = b$.

93. If $x = y + z$ and $y + z = 10$, then $x = 10$.

94. $(a + b) + c = (a + b) + c$.

In Exercises 95–108, tell which property of the real numbers justifies each statement.

95. $3(4)$ is a real number.

96. $5 + 5$ is a real number.

97. $3 + 7 = 7 + 3$

98. $2(9 \cdot 13) = (2 \cdot 9)13$

99. $3(2 + 5) = 3 \cdot 2 + 3 \cdot 5$

100. $1 \cdot 3 = 3 \cdot 1$

101. $81 + 0 = 81$

102. $3(9 + 2) = 3 \cdot 9 + 3 \cdot 2$

103. $5 \cdot \dfrac{1}{5} = 1$

104. $3 + (9 + 0) = (9 + 0) + 3$

105. $a + (7 + 8) = (a + 7) + 8$

106. $1 \cdot 3 = 3$

107. $(2 \cdot 3) \cdot 4 = 4 \cdot (2 \cdot 3)$

108. $8 + (-8) = 0$

 In Exercises 109–112, use a calculator to verify each statement. Identify the property of real numbers that is being illustrated.

109. $(37.9 + 25.2) + 14.3 = 37.9 + (25.2 + 14.3)$

110. $7.1(3.9 + 8.8) = 7.1 \cdot 3.9 + 7.1 \cdot 8.8$

111. $2.73(4.534 + 57.12) = 2.73 \cdot 4.534 + 2.73 \cdot 57.12$

112. $(6.789 + 345.1) + 27.347 = (345.1 + 6.789) + 27.347$

In Exercises 113–118, use a calculator to verify each statement. To find the negative of a number, enter the number and press the +/− key. To find a reciprocal of a number, enter the number and press the 1/x key.

113. $3568.45 + (-3568.45) = 0$

114. $-0.00573 + (0.00573) = 0$

115. $-(-32.432) = 32.432$

116. The reciprocal of 0.03125 is 32.

117. $13.9 \cdot \dfrac{1}{13.9} = 1$

118. $\dfrac{3.7}{2.9} = \dfrac{1}{\dfrac{2.9}{3.7}}$

Writing Exercises *Write a paragraph using your own words.*

1. The symmetric property of equality is often confused with the commutative properties. Why do you think this is so?

2. Explain how to add numbers with opposite signs.

Something to Think About

1. Pick five numbers and find their mean. Add 7 to each of the numbers to get five new numbers, and find their mean. What do you discover? Is this property always true?

2. Take the original five numbers in Exercise 1 and multiply each one by 7 to get five new numbers, and find their mean. What do you discover? Is this property always true?

Review Exercises *Graph each interval on a number line.*

1. $x > 4$

2. $x < -5$

3. $(2, 10]$

4. $[-4, 4]$

5. A man bought 32 gallons of gasoline at $1.29 per gallon and 3 quarts of oil at $1.35 per quart. The sales tax was included in the price of the gasoline, but 5% sales tax was added to the cost of the oil. Find the total cost.

6. On an adjusted income of $57,760, a woman must pay taxes according to the schedule shown in Table 1. Compute the tax bill.

Over	but not over		of the amount over
$0	$ 22,750	· · · · · 15%	$0
22,750	55,100	$3,412.50 + 28%	22,750
55,100	115,000	12,470.50 + 31%	55,100

TABLE 1

1.3 Exponents

■ PROPERTIES OF EXPONENTS ■ THE ZERO EXPONENT ■ NEGATIVE EXPONENTS ■ CALCULATORS
■ ORDER OF OPERATIONS ■ EVALUATING FORMULAS

Exponents indicate repeated multiplication. For example,

$$y^2 = y \cdot y \qquad \text{Read } y^2 \text{ as "y to the second power" or "y squared."}$$
$$z^3 = z \cdot z \cdot z \qquad \text{Read } z^3 \text{ as "z to the third power" or "z cubed."}$$
$$x^4 = x \cdot x \cdot x \cdot x \qquad \text{Read } x^4 \text{ as "x to the fourth power."}$$

These examples suggest the following definition.

> **Natural-Number Exponents**
> If n is a natural number, then
> $$x^n = \overbrace{x \cdot x \cdot x \cdot \cdots \cdot x}^{n \text{ factors of } x}$$

The exponential expression x^n is called a **power of x**, and we read it as "x to the nth power." In this expression, x is called the **base** and n is called the **exponent**.

$$\text{Base} \longrightarrow x^n \longleftarrow \text{Exponent}$$

A natural-number exponent tells how many times the base of an exponential expression is to be used as a factor in a product.

 (c, d, e)

EXAMPLE 1

a. $2^5 = 2 \cdot 2 \cdot 2 \cdot 2 \cdot 2$
    $= 32$

b. $(-2)^5 = (-2)(-2)(-2)(-2)(-2)$
    $= -32$

c. $-4^4 = -(4^4)$
    $= -(4 \cdot 4 \cdot 4 \cdot 4)$
    $= -256$

d. $(-4)^4 = (-4)(-4)(-4)(-4)$
    $= 256$

e. $\left(\frac{1}{2}a\right)^3 = \left(\frac{1}{2}a\right)\left(\frac{1}{2}a\right)\left(\frac{1}{2}a\right)$
    $= \frac{1}{8}a^3$

f. $\left(-\frac{1}{5}b\right)^2 = \left(-\frac{1}{5}b\right)\left(-\frac{1}{5}b\right)$
    $= \frac{1}{25}b^2$ ■

WARNING! Note the difference between $-x^n$ and $(-x)^n$.

$$-x^n = -\overbrace{(x \cdot x \cdot x \cdot \cdots \cdot x)}^{n \text{ factors of } x} \qquad \text{and} \qquad (-x)^n = \overbrace{(-x)(-x)(-x) \cdot \cdots \cdot (-x)}^{n \text{ factors of } -x}$$

Also, note the difference between ax^n and $(ax)^n$.

$$ax^n = a \cdot \overbrace{x \cdot x \cdot x \cdot \cdots \cdot x}^{n \text{ factors of } x} \qquad \text{and} \qquad (ax)^n = \overbrace{(ax)(ax)(ax) \cdot \cdots \cdot (ax)}^{n \text{ factors of } ax}$$

■ PROPERTIES OF EXPONENTS

Since x^5 means that x is to be used as a factor five times, and x^3 means that x is to be used as a factor three times, $x^5 \cdot x^3$ means that x will be used as a factor eight times.

$$x^5x^3 = \overbrace{x \cdot x \cdot x \cdot x \cdot x}^{5 \text{ factors of } x} \cdot \overbrace{x \cdot x \cdot x}^{3 \text{ factors of } x} = \overbrace{x \cdot x \cdot x \cdot x \cdot x \cdot x \cdot x \cdot x}^{8 \text{ factors of } x}$$

In general,

$$x^m x^n = \overbrace{x \cdot x \cdot x \cdots x}^{m \text{ factors of } x} \cdot \overbrace{x \cdot x \cdots x}^{n \text{ factors of } x} = \overbrace{x \cdot x \cdot x \cdot x \cdots x}^{m + n \text{ factors of } x}$$

Thus, *to multiply exponential expressions with the same base, we keep the same base and add the exponents.*

> **Product Rule of Exponents**
> If m and n are natural numbers, then
> $$x^m x^n = x^{m+n}$$

 WARNING! The product rule of exponents applies only to exponential expressions with the same base. The expression $x^5 y^3$, for example, cannot be simplified, because the bases of the exponential expressions are different.

 (c, d)   **EXAMPLE 2**

a. $x^{11}x^5 = x^{11+5}$
$\qquad\;\; = x^{16}$

b. $a^5 a^4 a^3 = (a^5 a^4)a^3$
$\qquad\qquad = a^9 a^3$
$\qquad\qquad = a^{12}$

c. $a^2 b^3 a^3 b^2 = a^2 a^3 b^3 b^2$
$\qquad\qquad = a^5 b^5$

d. $-8x^4\left(\dfrac{1}{4}x^3\right) = -8\left(\dfrac{1}{4}\right)x^4 x^3$
$\qquad\qquad\qquad\; = -2x^7$ ■

To find another property, we simplify $(x^4)^3$, which means x^4 cubed or $x^4 \cdot x^4 \cdot x^4$.

$$(x^4)^3 = x^4 \cdot x^4 \cdot x^4 = \overbrace{x \cdot x \cdot x \cdot x}^{x^4} \overbrace{x \cdot x \cdot x \cdot x}^{x^4} \overbrace{x \cdot x \cdot x \cdot x}^{x^4} = x^{12}$$

In general, we have

$$(x^m)^n = \overbrace{x^m \cdot x^m \cdot x^m \cdots x^m}^{n \text{ factors of } x^m} = \overbrace{x \cdot x \cdot x \cdot x \cdot x \cdots x}^{mn \text{ factors of } x} = x^{mn}$$

Thus, *to raise an exponential expression to a power, we keep the same base and multiply the exponents.*

To find a third property, we square $3x$ and get

$$(3x)^2 = 3x \cdot 3x = 3 \cdot 3 \cdot x \cdot x = 3^2 x^2 = 9x^2$$

In general, we have

$$(xy)^n = \overbrace{(xy)(xy)(xy) \cdot \cdots \cdot (xy)}^{n \text{ factors of } xy} = \overbrace{xxx \cdots \cdot x}^{n \text{ factors of } x} \cdot \overbrace{yyy \cdots \cdot y}^{n \text{ factors of } y} = x^n y^n$$

To find a fourth property, we cube $\frac{x}{3}$ and get

$$\left(\frac{x}{3}\right)^3 = \frac{x}{3} \cdot \frac{x}{3} \cdot \frac{x}{3} = \frac{x \cdot x \cdot x}{3 \cdot 3 \cdot 3} = \frac{x^3}{3^3} = \frac{x^3}{27}$$

In general, we have

$$\left(\frac{x}{y}\right)^n = \overbrace{\left(\frac{x}{y}\right)\left(\frac{x}{y}\right)\left(\frac{x}{y}\right) \cdot \cdots \cdot \left(\frac{x}{y}\right)}^{n \text{ factors of } \frac{x}{y}} \quad (y \neq 0)$$

$$= \frac{\overbrace{x \, x \, x \cdot \cdots \cdot x}^{n \text{ factors of } x}}{\underbrace{y \, y \, y \cdot \cdots \cdot y}_{n \text{ factors of } y}}$$

Multiply the numerators and multiply the denominators.

$$= \frac{x^n}{y^n}$$

The previous three results are called the power rules of exponents.

> **Power Rules of Exponents**
> If m and n are natural numbers, then
>
> $$(x^m)^n = x^{mn} \qquad (xy)^n = x^n y^n \qquad \left(\frac{x}{y}\right)^n = \frac{x^n}{y^n} \quad (y \neq 0)$$

EXAMPLE 3

a. $(3^2)^3 = 3^{2 \cdot 3}$
$= 3^6$
$= 729$

b. $(x^{11})^5 = x^{11 \cdot 5}$
$= x^{55}$

c. $(x^2 x^3)^6 = (x^5)^6$
$= x^{30}$

d. $(x^2)^4 (x^3)^2 = x^8 x^6$
$= x^{14}$

 (b, d)

EXAMPLE 4

a. $(x^2 y)^3 = (x^2)^3 y^3$
$= x^6 y^3$

b. $(x^3 y^4)^4 = (x^3)^4 (y^4)^4$
$= x^{12} y^{16}$

c. $\left(\dfrac{x}{y^2}\right)^4 = \dfrac{x^4}{(y^2)^4}$ $\quad(y \neq 0)$ $\qquad$ **d.** $\left(\dfrac{x^3}{y^4}\right)^2 = \dfrac{(x^3)^2}{(y^4)^2}$ $\quad(y \neq 0)$

$\qquad\qquad\qquad = \dfrac{x^4}{y^8}$ $\qquad\qquad\qquad\qquad\qquad = \dfrac{x^6}{y^8}$ $\qquad\qquad\qquad\qquad\qquad\qquad$ ■

■ THE ZERO EXPONENT

Since the rules for exponents hold for exponents of 0, we have

$$x^0 x^n = x^{0+n} = x^n = 1x^n$$

Because $x^0 x^n = 1x^n$, it follows that $x^0 = 1$ $\quad(x \neq 0)$.

> **Zero Exponent**
> If $x \neq 0$, then $x^0 = 1$.

In general, *any nonzero base raised to the* 0th *power is* 1.

EXAMPLE 5 If all bases are nonzero, then

a. $5^0 = 1$ $\quad$ **b.** $(-7)^0 = 1$ $\quad$ **c.** $(3ax^3)^0 = 1$ $\quad$ **d.** $\left(\tfrac{1}{2}x^5 y^7 z^9\right)^0 = 1$ $\qquad$ ■

 WARNING! 0^0 is undefined.

■ NEGATIVE EXPONENTS

Since the rules for exponents are true for negative integer exponents, we have

$$x^{-n} x^n = x^{-n+n} = x^0 = 1 \quad(x \neq 0)$$

Because $x^{-n} \cdot x^n = 1$ and $\dfrac{1}{x^n} \cdot x^n = 1$, we define x^{-n} to be the reciprocal of x^n.

> **Negative Exponents**
> If n is an integer and $x \neq 0$, then
> $$x^{-n} = \dfrac{1}{x^n} \qquad \text{and} \qquad \dfrac{1}{x^{-n}} = x^n$$

 WARNING! By the definition of negative exponents, the base cannot be 0. Thus, an expression such as 0^{-5} is undefined.

(b, c, d)　　EXAMPLE 6　　**a.** $5^{-2} = \dfrac{1}{5^2} = \dfrac{1}{25}$　　　　　　　　**b.** $10^{-3} = \dfrac{1}{10^3} = \dfrac{1}{1000}$

c. $(2x)^{-3} = \dfrac{1}{(2x)^3} = \dfrac{1}{8x^3}$ $(x \neq 0)$　　　**d.** $3x^{-1} = 3 \cdot \dfrac{1}{x} = \dfrac{3}{x}$ $(x \neq 0)$ ∎

EXAMPLE 7　　**a.** $x^{-5}x^3 = x^{-5+3}$　　　　　　**b.** $(x^{-3})^{-2} = x^{(-3)(-2)}$

$= x^{-2}$　　　　　　　　　　$= x^6$

$= \dfrac{1}{x^2}$　　∎

To develop a rule for dividing exponential expressions, we proceed as follows:

$$\frac{x^m}{x^n} = x^m\left(\frac{1}{x^n}\right) = x^m x^{-n} = x^{m+(-n)} = x^{m-n}$$

Thus, *to divide two exponential expressions with the same nonzero base, we keep the same base and subtract the exponent in the denominator from the exponent in the numerator.*

> **Quotient Rule**
> If m and n are integers, then
> $$\frac{x^m}{x^n} = x^{m-n} \quad (x \neq 0)$$

(f)　　EXAMPLE 8　　**a.** $\dfrac{a^5}{a^3} = a^{5-3}$　　**b.** $\dfrac{x^{-5}}{x^{11}} = x^{-5-11}$　　**c.** $\dfrac{x^4 x^3}{x^{-5}} = \dfrac{x^7}{x^{-5}}$

$= a^2$　　　　　　$= x^{-16}$　　　　　　$= x^{7-(-5)}$

$= \dfrac{1}{x^{16}}$　　　　　　$= x^{12}$

d. $\dfrac{(x^2)^3}{(x^3)^2} = \dfrac{x^6}{x^6}$　　**e.** $\dfrac{x^2 y^3}{xy^4} = x^{2-1}y^{3-4}$　　**f.** $\left(\dfrac{a^{-2}b^3}{a^2 a^3 b^4}\right)^3 = \left(\dfrac{a^{-2}b^3}{a^5 b^4}\right)^3$

$= x^{6-6}$　　　　　　$= xy^{-1}$　　　　　　$= (a^{-2-5}b^{3-4})^3$

$= x^0$　　　　　　$= \dfrac{x}{y}$　　　　　　$= (a^{-7}b^{-1})^3$

$= 1$　　　　　　　　　　　　$= \left(\dfrac{1}{a^7 b}\right)^3$

$= \dfrac{1}{a^{21}b^3}$ ∎

Because of the next theorem, a fraction raised to a negative power can be inverted and then raised to a positive power.

Theorem

If n is an integer, then

$$\left(\frac{x}{y}\right)^{-n} = \left(\frac{y}{x}\right)^{n} \quad (x \neq 0, y \neq 0)$$

(d)

EXAMPLE 9

a. $\left(\dfrac{2}{3}\right)^{-4} = \left(\dfrac{3}{2}\right)^{4}$

$\qquad = \dfrac{81}{16}$

b. $\left(\dfrac{y^2}{x^3}\right)^{-3} = \left(\dfrac{x^3}{y^2}\right)^{3}$

$\qquad = \dfrac{x^9}{y^6}$

c. $\left(\dfrac{2x^2}{3y^{-3}}\right)^{-4} = \left(\dfrac{3y^{-3}}{2x^2}\right)^{4}$

$\qquad = \dfrac{81y^{-12}}{16x^8}$

$\qquad = \dfrac{81}{16x^8} \cdot y^{-12}$

$\qquad = \dfrac{81}{16x^8} \cdot \dfrac{1}{y^{12}}$

$\qquad = \dfrac{81}{16x^8 y^{12}}$

d. $\left(\dfrac{a^{-2}b^3}{a^2 a^3 b^4}\right)^{-3} = \left(\dfrac{a^2 a^3 b^4}{a^{-2}b^3}\right)^{3}$

$\qquad = \left(\dfrac{a^5 b^4}{a^{-2}b^3}\right)^{3}$

$\qquad = (a^{5-(-2)}b^{4-3})^3$

$\qquad = (a^7 b)^3$

$\qquad = a^{21}b^3$

We summarize the rules of exponents as follows.

Properties of Exponents

If there are no divisions by 0, then for all integers m and n,

$$x^m x^n = x^{m+n} \qquad (x^m)^n = x^{mn} \qquad (xy)^n = x^n y^n \qquad \left(\frac{x}{y}\right)^n = \frac{x^n}{y^n}$$

$$x^0 = 1 \ (x \neq 0) \qquad x^{-n} = \frac{1}{x^n} \qquad \frac{x^m}{x^n} = x^{m-n} \qquad \left(\frac{x}{y}\right)^{-n} = \left(\frac{y}{x}\right)^n$$

The same rules apply to exponents that are variables.

(c, d)

EXAMPLE 10

Simplify each expression. Assume that $a \neq 0$ and $x \neq 0$.

a. $\dfrac{a^n a}{a^2} = a^{n+1-2}$

$\qquad = a^{n-1}$

b. $\dfrac{x^3 x^2}{x^n} = x^{3+2-n}$

$\qquad = x^{5-n}$

c. $\left(\dfrac{x^n}{x^2}\right)^2 = \dfrac{x^{2n}}{x^4}$

$\qquad = x^{2n-4}$

d. $\dfrac{a^n a^{-3}}{a^{-1}} = a^{n+(-3)-(-1)}$

$\qquad = a^{n-3+1}$

$\qquad = a^{n-2}$

■ CALCULATORS

To find powers of numbers with a scientific calculator, we use the y^x key. For example, to find 5.37^4, we enter these numbers and press these keys:

5.37 $\boxed{y^x}$ 4 $\boxed{=}$ Some calculators have an $\boxed{x^y}$ key.

The display will read 831.5668016. To use a graphing calculator, we enter these numbers and press these keys:

5.37 $\boxed{\wedge}$ 4 $\boxed{\text{ENTER}}$ Some calculators have an $\boxed{x^y}$ or a $\boxed{y^x}$ key.

If neither of these methods works, consult your owner's manual.

■ ORDER OF OPERATIONS

When simplifying expressions containing exponents, we find powers before performing additions and multiplications.

EXAMPLE 11   If $x = 2$, $y = -3$, find the value of $3x + 2y^3$.

Solution
$$
\begin{aligned}
3x + 2y^3 &= 3(2) + 2(-3)^3 & &\text{Substitute 2 for } x \text{ and } -3 \text{ for } y. \\
&= 3(2) + 2(-27) & &\text{First find the power: } (-3)^3 = -27. \\
&= 6 - 54 & &\text{Then do the multiplications.} \\
&= -48 & &\text{Then do the subtraction.}
\end{aligned}
$$

■ EVALUATING FORMULAS

Table 1-2 (on page 35) shows the formulas used to compute areas and volumes of many geometric figures.

EXAMPLE 12   Find the volume of the sphere shown in Figure 1-17.

Solution   The formula for the volume of a sphere is $V = \frac{4}{3}\pi r^3$. Since a radius is half as long as a diameter, the radius of the sphere is half of 20 centimeters, or 10 centimeters.

$$V = \frac{4}{3}\pi r^3$$

$$V = \frac{4}{3}\pi(10)^3 \qquad \text{Substitute 10 for } r.$$

$$\approx 4188.790205 \qquad \text{Press these keys on a calculator: } 10 \;\boxed{y^x}\; 3 \;\boxed{=}\; \boxed{\times}\; \pi \;\boxed{\times}\; 4 \;\boxed{\div}\; 3 \;\boxed{=}.$$

To two decimal places, the volume is 4188.79 cm³.

20 cm

FIGURE 1-17

Figure	Name	Area	Figure	Name	Volume
	Square	$A = s^2$		Cube	$V = s^3$
	Rectangle	$A = lw$		Rectangular solid	$V = lwh$
	Circle	$A = \pi r^2$		Sphere	$V = \dfrac{4}{3}\pi r^3$
	Triangle	$A = \dfrac{1}{2}bh$		Cylinder	$V = Bh*$
	Trapezoid	$A = \dfrac{1}{2}h(b_1 + b_2)$		Cone	$V = \dfrac{1}{3}Bh*$
				Pyramid	$V = \dfrac{1}{3}Bh*$

*B represents the area of the base.

TABLE 1-2

Orals   *Simplify each expression.*

1. 4^2 **2.** 3^3 **3.** x^2x^3 **4.** y^3y^4

5. 17^0 **6.** $(x^2)^3$ **7.** $(a^2b)^3$ **8.** $\left(\dfrac{b}{a^2}\right)^2$

9. 5^{-2} **10.** $(x^{-2})^{-1}$ **11.** $\dfrac{x^5}{x^2}$ **12.** $\dfrac{x^2}{x^5}$

EXERCISE 1.3

In Exercises 1–8, identify the base and the exponent.

1. 5^3

2. -7^2

3. $-x^5$

4. $(-t)^4$

5. $2b^6$

6. $(3xy)^5$

7. $(-mn^2)^3$

8. $(-p^2q)^2$

In Exercises 9–90, simplify each expression. Assume that no denominators are zero.

9. 3^2

10. 3^4

11. -3^2

12. -3^4

13. $(-3)^2$

14. $(-3)^3$

15. $(-2x)^5$

16. $(-3a)^3$

17. 5^{-2}

18. 5^{-4}

19. -5^{-2}

20. -5^{-4}

21. $(-5)^{-2}$

22. $(-5)^{-4}$

23. 8^0

24. 9^0

25. -8^0

26. -9^0

27. $(-8)^0$

28. $(-9)^0$

29. $-(2x)^7$

30. $(-2a)^4$

31. $(-2x)^6$

32. $(-3y)^5$

33. x^2x^3

34. y^3y^4

35. k^0k^7

36. x^8x^{11}

37. $x^2x^3x^5$

38. $y^3y^7y^2$

39. p^9pp^0

40. z^7z^0z

41. aba^3b^4

42. $x^2y^3x^3y^2$

43. $(-x)^2y^4x^3$

44. $-x^2y^7y^3x^{-2}$

45. $(x^4)^7$

46. $(y^7)^5$

47. $(b^{-8})^9$

48. $(z^{12})^2$

49. $(x^3y^2)^4$

50. $(x^2y^5)^2$

51. $(r^{-3}s)^3$

52. $(m^5n^2)^{-3}$

53. $(a^2a^3)^4$

54. $(bb^2b^3)^4$

55. $(-d^2)^3(d^{-3})^3$

56. $(c^3)^2(c^4)^{-2}$

57. $(3x^3y^4)^3$

58. $\left(\dfrac{1}{2}a^2b^5\right)^4$

59. $\left(-\dfrac{1}{3}mn^2\right)^6$

60. $(-3p^2q^3)^5$

61. $(x^{-2}yx^3y^4)^2$

62. $(-a^2b^{-4}a^3b^2)^3$

63. $(a^2b)^{-2}(ab^{-3})^4$

64. $(p^2q^{-2})^3(q^2)^{-2}$

65. $\left(\dfrac{a^3}{b^2}\right)^5$

66. $\left(\dfrac{a^2}{b^3}\right)^4$

67. $\left(\dfrac{a^{-3}}{b^{-2}}\right)^{-2}$

68. $\left(\dfrac{k^{-3}}{k^{-4}}\right)^{-1}$

69. $\dfrac{a^8}{a^3}$

70. $\dfrac{c^7}{c^2}$

71. $\dfrac{c^{12}c^5}{c^{10}}$

72. $\dfrac{a^{33}}{a^2a^3}$

73. $\dfrac{m^9m^{-2}}{(m^2)^3}$

74. $\dfrac{a^{10}a^{-3}}{a^5a^{-2}}$

75. $\dfrac{1}{a^{-4}}$

76. $\dfrac{3}{b^{-5}}$

77. $\dfrac{3m^5m^{-7}}{m^2m^{-5}}$

78. $\dfrac{(2a^{-2})^3}{a^3a^{-4}}$

79. $\left(\dfrac{4a^{-2}b}{3ab^{-3}}\right)^3$

80. $\left(\dfrac{2ab^{-3}}{3a^{-2}b^2}\right)^2$

81. $\left(\dfrac{3a^{-2}b^2}{17a^2b^3}\right)^0$

82. $\dfrac{a^0+b^0}{2(a+b)^0}$

83. $\left(\dfrac{-2a^4b}{a^{-3}b^2}\right)^{-3}$

84. $\left(\dfrac{-3x^4y^2}{-9x^5y^{-2}}\right)^{-2}$

85. $\left(\dfrac{2a^3b^2}{3a^{-3}b^2}\right)^{-3}$

86. $\left(\dfrac{3x^5y^2}{6x^5y^{-2}}\right)^{-4}$

87. $\dfrac{(3x^2)^{-2}}{x^3x^{-4}x^0}$

88. $\dfrac{y^{-3}y^{-4}y^0}{(2y^{-2})^3}$

89. $\dfrac{-3x^{-2}y^2}{(-2x^{-3})^0}$

90. $\dfrac{-4x^{-2}x^2(y^0)^2}{(-4x^2y^{-4})^0}$

In Exercises 91–98, simplify each expression.

91. $\dfrac{a^n a^3}{a^4}$

92. $\dfrac{b^9 b^7}{b^n}$

93. $\left(\dfrac{b^n}{b^3}\right)^3$

94. $\left(\dfrac{a^2}{a^n}\right)^4$

95. $\dfrac{a^{-n}a^2}{a^3}$

96. $\dfrac{a^n a^{-2}}{a^4}$

97. $\dfrac{a^{-n}a^{-2}}{a^{-4}}$

98. $\dfrac{a^n}{a^{-3}a^5}$

 In Exercises 99–106, use a calculator to find each value.

99. 3.4^4

100. 1.23^6

101. 0.0537^4

102. 0.2345^4

103. -6.25^3

104. -4.17^4

105. $(-25.1)^5$

106. $(-0.35)^4$

 In Exercises 107–114, use a calculator to verify that each statement is true.

107. $(3.68)^0 = 1$

108. $(2.1)^4(2.1)^3 = (2.1)^7$

109. $(7.2)^2(2.7)^2 = [(7.2)(2.7)]^2$

110. $(3.7)^2 + (4.8)^2 \neq (3.7 + 4.8)^2$

111. $(3.2)^2(3.2)^{-2} = 1$

112. $[(5.9)^3]^2 = (5.9)^6$

113. $(7.23)^{-3} = \dfrac{1}{(7.23)^3}$

114. $\left(\dfrac{5.4}{2.7}\right)^{-4} = \left(\dfrac{2.7}{5.4}\right)^4$

In Exercises 115–122, evaluate each expression when $x = -2$ and $y = 3$.

115. $x^2 y^3$

116. $x^3 y^2$

117. $\dfrac{x^{-3}}{y^3}$

118. $\dfrac{x^2}{y^{-3}}$

119. $(xy^2)^{-2}$

120. $-y^3 x^{-2}$

121. $(-yx^{-1})^3$

122. $(-y)^3 x^{-2}$

In Exercises 123–130, find the area of each figure. Round all answers to the nearest unit.

123.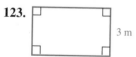

124.

125.

126.

127.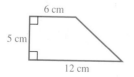

128.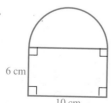

129.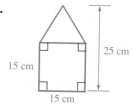

130.

In Exercises 131–138, find the volume of each figure. Round all answers to the nearest unit.

131.

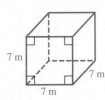

7 m

7 m

7 m

132.

40 cm

133.

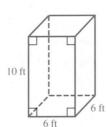

10 ft

6 ft

6 ft

134.

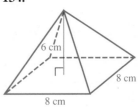

6 cm

8 cm

8 cm

135.

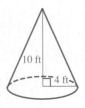

10 ft

4 ft

136.

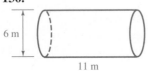

6 m

11 m

137.

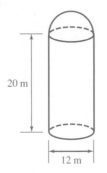

20 m

12 m

138.

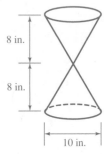

8 in.

8 in.

10 in.

139. Construct an example using numbers to show that
$x^m + x^n \neq x^{m+n}$.

140. Construct an example using numbers to show that
$x^m + y^m \neq (x + y)^m$.

Writing Exercises   *Write a paragraph using your own words.*

1. Explain why a positive number raised to a negative power is positive.

2. Explain the rules that determine the order in which operations are performed.

3. In the definition of x^{-1}, x cannot be 0. Why not?

4. Explain why $(xyz)^2 = x^2y^2z^2$.

Something to Think About   **1.** Simplify: $2^{-1} + 3^{-1} - 4^{-1}$

2. Simplify: $(3^{-1} + 4^{-1})^{-2}$

Review Exercises   *If $a = 4$, $b = -2$, and $c = 5$, find each value.*

1. $a + b + c$

2. $a - 2b - c$

3. $\dfrac{ab + 2c}{a + b}$

4. $\dfrac{ac - bc}{6ab + b}$

1.4 Scientific Notation

■ USING SCIENTIFIC NOTATION TO SIMPLIFY COMPUTATIONS ■ CALCULATORS ■ SIGNIFICANT DIGITS ■ PROBLEM SOLVING

Very large and very small numbers occur often in science. For example, the speed of light is approximately 29,980,000,000 centimeters per second, and the mass of a hydrogen atom is approximately 0.0000000000000000000001673 gram. With exponents, we can write these numbers more compactly by using **scientific notation**.

> **Scientific Notation**
> A number is written in **scientific notation** when it is written in the form $N \times 10^n$, where $1 \le |N| < 10$ and n is an integer.

EXAMPLE 1   Change   **a.** 29,980,000,000,   **b.** 0.0000000000000000000001673,   and **c.** −0.0013   to scientific notation.

Solution   **a.** The number 2.998 is between 1 and 10. To get 29,980,000,000, the decimal point in 2.998 must be moved ten places to the right. This can be done by multiplying 2.998 by 10^{10}.

$$29{,}980{,}000{,}000 = 2.998 \times 10^{10}$$

b. The number 1.673 is between 1 and 10. To get 0.0000000000000000000001673, the decimal point in 1.673 must be moved twenty-four places to the left. This can be done by multiplying 1.673 by 10^{-24}.

$$0.0000000000000000000001673 = 1.673 \times 10^{-24}$$

c. The absolute value of −1.3 is between 1 and 10. To get −0.0013, we move the decimal point in −1.3 three places to the left by multiplying by 10^{-3}.

$$-0.0013 = -1.3 \times 10^{-3}$$

EXAMPLE 2   Change   **a.** 3.7×10^5 and   **b.** 1.1×10^{-3} to standard notation.

Solution   **a.** Since multiplication by 10^5 moves the decimal point 5 places to the right,

$$3.7 \times 10^5 = 370{,}000$$

b. Since multiplication by 10^{-3} moves the decimal point 3 places to the left,

$$1.1 \times 10^{-3} = 0.0011$$

Each of the following numbers is written in both scientific and standard notation. In each case, the exponent gives the number of places that the decimal point moves, and the sign of the exponent indicates the direction that it moves:

$5.32 \times 10^4 = 5\,3\,2\,0\,0$ $6.45 \times 10^7 = 6\,4\,5\,0\,0\,0\,0\,0$

       4 places to the right   7 places to the right

$2.37 \times 10^{-4} = 0.\,0\,0\,0\,2\,3\,7$ $9.234 \times 10^{-2} = 0.\,0\,9\,2\,3\,4$

       4 places to the left   2 places to the left

$4.89 \times 10^0 = 4.89$

       No movement of the decimal point

EXAMPLE 3   Change   **a.** 47.2×10^{-3} and   **b.** 0.063×10^{-2} to scientific notation.

Solution Since the first factors are not between 1 and 10, neither number is in scientific notation. However, we can change them to scientific notation as follows:

a. $47.2 \times 10^{-3} = (4.72 \times 10^1) \times 10^{-3}$ Write 47.2 in scientific notation.

$$= 4.72 \times (10^1 \times 10^{-3})$$

$$= 4.72 \times 10^{-2}$$

b. $0.063 \times 10^{-2} = (6.3 \times 10^{-2}) \times 10^{-2}$ Write 0.063 in scientific notation.

$$= 6.3 \times (10^{-2} \times 10^{-2})$$

$$= 6.3 \times 10^{-4}$$

■ USING SCIENTIFIC NOTATION TO SIMPLIFY COMPUTATIONS

Scientific notation is useful when multiplying and dividing very large or very small numbers.

EXAMPLE 4   Use scientific notation to simplify $\dfrac{(0.00000064)(24{,}000{,}000{,}000)}{(400{,}000{,}000)(0.0000000012)}$.

Solution After changing each number into scientific notation, we can do the arithmetic on the numbers and the exponential expressions separately.

$$\frac{(0.00000064)(24{,}000{,}000{,}000)}{(400{,}000{,}000)(0.0000000012)} = \frac{(6.4 \times 10^{-7})(2.4 \times 10^{10})}{(4 \times 10^{8})(1.2 \times 10^{-9})}$$

$$= \frac{(6.4)(2.4)}{(4)(1.2)} \cdot \frac{10^{-7}10^{10}}{10^{8}10^{-9}}$$

$$= 3.2 \times 10^4$$

In standard notation, the result is 32,000.

■ CALCULATORS

Scientific and graphing calculators often give answers in scientific notation. For example, if we use a calculator to find 301.2^8, the display will read

| 6.77391496 ^{19} | On a scientific calculator. |
| 301.2 $\wedge$ 8 6.773914961E19 | On a graphing calculator. |

In either case, the answer is given in scientific notation and is to be interpreted as

$$6.77391496 \times 10^{19}$$

Numbers can also be entered into a calculator in scientific notation. For example, to enter 24,000,000,000 (which is 2.4×10^{10} in scientific notation), we enter these numbers and press these keys:

2.4 **EXP** 10   On a scientific calculator.

2.4 **EE** 10   On a graphing calculator.

To use a calculator to simplify

$$\frac{(24{,}000{,}000{,}000)(0.00000006495)}{0.00000004824}$$

we must enter each number in scientific notation, because each number has too many digits to be entered directly. In scientific notation, the three numbers are

$$2.4 \times 10^{10} \qquad 6.495 \times 10^{-8} \qquad 4.824 \times 10^{-8}$$

To use a calculator to simplify the fraction, we enter these numbers and press these keys:

2.4 EXP 10 × 6.495 EXP 8 +/− ÷ 4.824 EXP 8 +/− =

The display will read 3.231343284 ^{10} . In standard notation, the answer is 32,313,432,840.

The steps are similar on a graphing calculator.

■ SIGNIFICANT DIGITS

If we measure the length of a rectangle and report the length to be 45 centimeters, we have rounded to the nearest centimeter. If we measure more carefully and find the length to be 45.2 centimeters, we have rounded to the nearest tenth of a centimeter. We say that the second measurement is more accurate than the first because 45.2 has three significant digits, but 45 has only two significant digits.

It is not always easy to know how many significant digits a number has. For example, 270 might be accurate to two or three significant digits. If 270 is rounded to the nearest ten, the number has two significant digits. If 270 is rounded to the nearest unit, it has three significant digits. This ambiguity does not exist when a number is written in scientific notation.

Finding Significant Digits
If a number M is written in scientific notation as $N \times 10^n$, where $1 \leq |N| < 10$ and n is an integer, the number of significant digits in M is the same as the number of digits in N.

In a problem where measurements are multiplied or divided, the final result should be rounded so that the answer has the same number of significant digits as the least accurate measurement.

■ PROBLEM SOLVING

EXAMPLE 5 The earth is approximately 93,000,000 miles from the sun, and Jupiter is approximately 484,000,000 miles from the sun. Assuming the alignment shown in Figure 1-18, how long would it take a spaceship traveling at 7500 miles per hour to fly from the earth to Jupiter?

Solution When the planets are aligned as shown in the figure, the distance between the earth and Jupiter is (484,000,000 − 93,000,000) miles or 391,000,000 miles. To find the length of time in hours for the trip, we divide the distance by the rate.

$$\frac{391,000,000 \text{ mi}}{7500 \frac{\text{mi}}{\text{hr}}} = \frac{3.91 \times 10^8}{7.5 \times 10^3} \frac{\text{mi}}{\frac{\text{mi}}{\text{hr}}}$$

There are three significant digits in the numerator and two significant digits in the denominator.

$$\approx 0.5213333 \times 10^5 \text{ mi} \cdot \frac{\text{hr}}{\text{mi}}$$

$$\approx 52,133.33 \text{ hr}$$

Since there are 24×365 hours in a year, we can change this result from hours to years by dividing 52,133.33 by (24×365).

$$\frac{52{,}133.33 \text{ hr}}{(24 \times 365) \frac{\text{hr}}{\text{yr}}} \approx 5.9512934 \text{ hr} \cdot \frac{\text{yr}}{\text{hr}} \approx 5.9512934 \text{ yr}$$

After rounding to two significant digits, we see that the trip will take about 6.0 years.

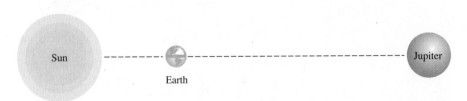

FIGURE 1-18 ■

Orals *Give each number in scientific notation.*

1. 352   **2.** 5130

3. 0.002   **4.** 0.00025

Give each number in standard notation.

5. 3.5×10^2 **6.** 4.3×10^3

7. 2.7×10^{-1} **8.** 8.5×10^{-2}

EXERCISE 1.4

In Exercises 1–20, write each numeral in scientific notation.

1. 3900   **2.** 1700   **3.** 0.0078   **4.** 0.068

5. $-45{,}000$ **6.** $-547{,}000$ **7.** -0.00021 **8.** -0.00078

9. 17,600,000   **10.** 89,800,000   **11.** 0.0000096   **12.** 0.000046

13. 323×10^5 **14.** 689×10^9 **15.** 6200×10^{-7} **16.** 765×10^{-5}

17. 0.0527×10^5 **18.** 0.0298×10^3 **19.** 0.0317×10^{-2} **20.** 0.0012×10^{-3}

In Exercises 21–32, write each numeral in standard notation.

21. 2.7×10^2 **22.** 7.2×10^3 **23.** 3.23×10^{-3} **24.** 6.48×10^{-2}

25. 7.96×10^5 **26.** 9.67×10^6 **27.** 3.7×10^{-4} **28.** 4.12×10^{-5}

29. 5.23×10^0 **30.** 8.67×10^0 **31.** 23.65×10^6 **32.** 75.6×10^{-5}

In Exercises 33–36, write each numeral in scientific notation and do the operations. Give all answers in scientific notation.

33. $\dfrac{(4000)(30{,}000)}{0.0006}$ **34.** $\dfrac{(0.0006)(0.00007)}{21{,}000}$

35. $\dfrac{(640{,}000)(2{,}700{,}000)}{120{,}000}$ **36.** $\dfrac{(0.0000013)(0.00009)}{0.00039}$

In Exercises 37–42, write each numeral in scientific notation and do the operations. Give all answers in standard notation.

37. $\dfrac{(0.006)(0.008)}{0.0012}$

38. $\dfrac{(600)(80,000)}{120,000}$

39. $\dfrac{(220,000)(0.000009)}{0.00033}$

40. $\dfrac{(0.00024)(96,000,000)}{640,000,000}$

41. $\dfrac{(320,000)^2(0.0009)}{12,000^2}$

42. $\dfrac{(0.000012)^2(49,000)^2}{0.021}$

In Exercises 43–56, use scientific notation to find each answer. Round all answers to the proper number of significant digits.

43. Speed of sound The speed of sound in air is 3.31×10^4 centimeters per second. Find the speed of sound in centimeters per hour.

44. Volume of a tank Find the volume of the tank shown in Illustration 1.

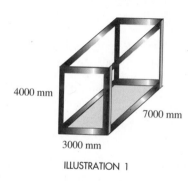

4000 mm

7000 mm

3000 mm

ILLUSTRATION 1

45. Mass of protons If the mass of one proton is 0.0000000000000000000000000167248 gram, find the mass of 1 million protons.

46. Speed of light The speed of light in a vacuum is about 30,000,000,000 centimeters per second. Find the speed of light in miles per hour. (*Hint:* 160,000 cm ≈ 1 mile. Read ≈ as "approximately equal to.")

47. Distance to the moon The moon is about 235,000 miles from the earth. Find this distance in inches.

48. Distance to the sun The sun is about 149,700,000 kilometers from the earth. Find this distance in miles. (*Hint:* 1 km ≈ 0.6214 mile.)

49. Solar flares Solar flares often produce immense loops of glowing gas ejected from the sun's surface. The flare in Illustration 2 extends about 95,000 kilometers into space. Express this distance in miles. (*Hint:* 1 km ≈ 0.6214 mi.)

Frank Rossotto/The Stock Market

ILLUSTRATION 2

50. Distance to the moon The moon is about 378,196 kilometers from the earth. Express this distance in inches. (*Hint:* 1 km ≈ 0.6214 mile.)

51. Angstroms per inch One **angstrom** is 0.0000001 millimeter, and one inch is 25.4 millimeters. Find the number of angstroms in one inch.

52. Range of a comet One **astronomical unit** (AU) is the distance from the earth to the sun—about 9.3×10^7 miles. Halley's Comet ranges from 0.6 to 18 AU from the sun. Express this range in miles.

53. Flight to Pluto The planet Pluto is approximately 3,574,000,000 miles from the earth. If a spaceship can travel 18,000 miles per hour, how long will it take to reach Pluto?

54. Light year Light travels about 300,000,000 meters per second. A **light year** is the distance that light can travel in one year. How many meters are in one light year?

55. Distance to Alpha Centauri Light travels about 186,000 miles per second. A **parsec** is 3.26 light years. The star Alpha Centauri is 1.3 parsecs from the earth. Express this distance in miles.

56. Life of a comet The mass of the comet shown in Illustration 3 is about 10^{16} grams. When the comet is close to the sun, matter evaporates at the rate of 10^7 grams per second. Calculate the life of the comet if it appears every 50 years and spends ten days close to the sun.

Frank Rossotto/The Stock Market

ILLUSTRATION 3

In Exercises 57–62, use a scientific calculator to evaluate each expression. Round each answer to the appropriate number of significant digits.

57. $23{,}437^3$

58. 0.00034^4

59. $(63{,}480)(893{,}322)$

60. $(0.0000413)(0.0000049)^2$

61. $\dfrac{(69.4)^8(73.1)^2}{(0.0043)^3}$

62. $\dfrac{(0.0031)^4(0.0012)^5}{(0.0456)^{-7}}$

Writing Exercises *Write a paragraph using your own words.*

1. Explain how to change a number from standard notation to scientific notation.

2. Explain how to change a number from scientific notation to standard notation.

Something to Think About

1. Find the highest power of 2 that can be evaluated with a scientific calculator.

2. Find the highest power of 7 that can be evaluated with a scientific calculator.

Review Exercises *Write each fraction as a terminating or repeating decimal.*

1. $\dfrac{3}{4}$

2. $\dfrac{4}{5}$

3. $\dfrac{13}{9}$

4. $\dfrac{14}{11}$

5. A man raises 3 to the second power, 4 to the third power, and 2 to the fourth power and finds their sum. What number does he obtain?

6. If $a = -2$, $b = -3$, and $c = 4$, find the value of $\dfrac{5ab - 4ac - 2}{3bc + abc}$.

1.5 Solving Equations

■ PROPERTIES OF EQUALITY ■ SOLVING LINEAR EQUATIONS ■ COMBINING LIKE TERMS
■ IDENTITIES AND IMPOSSIBLE EQUATIONS ■ REPEATING DECIMALS ■ FORMULAS

An **equation** is a statement indicating that two quantities are equal. The equation $2 + 4 = 6$ is true, and the equation $2 + 4 = 7$ is false. If an equation has a variable (say, x), it can be either true or false, depending on the value of x. For example, if $x = 1$, the equation $7x - 3 = 4$ is true.

$$7(1) - 3 = 4 \qquad \text{Substitute 1 for } x.$$
$$7 - 3 = 4$$
$$4 = 4$$

However, the equation is false for all other values of x. Since 1 makes the equation true, we say that it *satisfies* the equation.

The set of numbers that satisfies an equation is called its **solution set**. The elements of the solution set are called **solutions** or **roots** of the equation. Finding the solution set of an equation is called *solving the equation.*

EXAMPLE 1   Determine whether 3 is a solution of the equation $2x + 4 = 10$.

Solution   We substitute 3 for x and see whether it satisfies the equation.

$$2x + 4 = 10$$
$$2(3) + 4 \overset{?}{=} 10 \qquad \text{Substitute 3 for } x.$$
$$6 + 4 \overset{?}{=} 10 \qquad \text{First do the multiplication on the left-hand side.}$$
$$10 = 10 \qquad \text{Then do the addition.}$$

Since $10 = 10$, the number 3 satisfies the equation. It is a solution.   ■

■ PROPERTIES OF EQUALITY

To solve an equation, we replace the equation with simpler ones, all having the same solution set. Such equations are called **equivalent equations**.

> **Equivalent Equations**
> Equations with the same solution set are called **equivalent equations**.

We continue to replace each resulting equation with an equivalent one until we have isolated the variable on one side of an equation. To isolate the variable, we can use the following properties.

> **Addition and Subtraction Properties of Equality**
> If a, b, and c are real numbers and $a = b$, then
> $$a + c = b + c \qquad \text{and} \qquad a - c = b - c$$

In words, *If any quantity is added to (or subtracted from) both sides of an equation, a new equation is formed that is equivalent to the original equation.*

> **Multiplication and Division Properties of Equality**
> If a, b, and c are real numbers, $a = b$, and $c \neq 0$, then
> $$ac = bc \qquad \text{and} \qquad \frac{a}{c} = \frac{b}{c}$$

In words, *If both sides of an equation are multiplied (or divided) by the same non-zero quantity, a new equation is formed that is equivalent to the original equation.*

■ SOLVING LINEAR EQUATIONS

The easiest equations to solve are **linear equations**.

> **Linear Equation**
> A **linear equation in one variable** (say, x) is any equation that can be written in the form
> $$ax + c = 0 \qquad (a \text{ and } c \text{ are real numbers and } a \neq 0)$$

EXAMPLE 2   Solve $2x + 8 = 0$.

Solution   To solve the equation, we will isolate x on the left-hand side of the equation.

$$2x + 8 = 0$$

$$2x + 8 - 8 = 0 - 8 \qquad \text{To eliminate 8 from the left-hand side, subtract 8 from both sides.}$$

$$2x = -8 \qquad \text{Simplify.}$$

$$\frac{2x}{2} = \frac{-8}{2} \qquad \text{To eliminate 2 from the left-hand side, divide both sides by 2.}$$

$$x = -4 \qquad \text{Simplify.}$$

Check: We substitute -4 for x and verify that it satisfies the original equation.

$$2x + 8 = 0$$

$$2(-4) + 8 \stackrel{?}{=} 0 \qquad \text{Substitute } -4 \text{ for } x.$$

$$-8 + 8 \stackrel{?}{=} 0$$

$$0 = 0$$

Since -4 satisfies the original equation, it is the solution. The solution set is $\{-4\}$. ■

EXAMPLE 3   Solve the equation $3(x - 2) = 20$.

Solution   We isolate x on the left-hand side.

$$3(x - 2) = 20$$

$$3x - 6 = 20 \qquad \text{Use the distributive property to remove parentheses.}$$

$$3x - 6 + 6 = 20 + 6 \qquad \text{To eliminate } -6 \text{ from the left-hand side, add 6 to both sides.}$$

$$3x = 26 \qquad \text{Simplify.}$$

$$\frac{3x}{3} = \frac{26}{3} \qquad \text{To eliminate the 3 from the left-hand side, divide both sides by 3.}$$

$$x = \frac{26}{3} \qquad \text{Simplify.}$$

Check: $3(x - 2) = 20$

$$3\left(\frac{26}{3} - 2\right) \stackrel{?}{=} 20$$ Substitute $\frac{26}{3}$ for x.

$$3\left(\frac{26}{3} - \frac{6}{3}\right) \stackrel{?}{=} 20$$ Get a common denominator: $2 = \frac{6}{3}$.

$$3\left(\frac{20}{3}\right) \stackrel{?}{=} 20$$ Combine the fractions: $\frac{26}{3} - \frac{6}{3} = \frac{20}{3}$.

$$20 = 20$$ Simplify.

Since $\dfrac{26}{3}$ satisfies the equation, it is the solution. The solution set is $\left\{\dfrac{26}{3}\right\}$. ∎

■ COMBINING LIKE TERMS

To solve more complicated equations, we will need to combine like terms. An **algebraic term** is either a number or the product of numbers (called **constants**) and variables. Some examples of terms are $3x$, $-7y$, y^2, and 8. The **numerical coefficients** of the terms are 3, -7, 1, and 8 (8 can be written as $8x^0$), respectively.

In algebraic expressions, terms are separated by $+$ and $-$ signs. For example, the expression $3x^2 + 2x - 4$ has three terms, and the expression $3x + 7y$ has two terms.

Terms with the same variables with the same exponents are called **like terms** or **similar terms**:

$5x$ and $6x$ are like terms $27x^2y^3$ and $-326x^2y^3$ are like terms

$4x$ and $-17y$ are unlike terms $15x^2y$ and $6xy^2$ are unlike terms

By using the distributive law, we can combine like terms. For example,

$$5x + 6x = (5 + 6)x = 11x \qquad \text{and} \qquad 32y - 16y = (32 - 16)y = 16y$$

To combine like terms, *we add or subtract their numerical coefficients and keep the same variables with the same exponents.*

EXAMPLE 4   Solve the equation $3(2x - 1) = 2x + 9$.

Solution

$$3(2x - 1) = 2x + 9$$

$$6x - 3 = 2x + 9$$ Use the distributive property to remove parentheses.

$$6x - 3 + 3 = 2x + 9 + 3$$ To eliminate -3 from the left-hand side, add 3 to both sides.

$$6x = 2x + 12$$ Combine like terms.

$$6x - 2x = 2x + 12 - 2x$$ To eliminate $2x$ from the right-hand side, subtract $2x$ from both sides.

$$4x = 12$$ Combine like terms.

$$x = 3$$ To eliminate 4 from the left-hand side, divide both sides by 4.

Check: $3(2x - 1) = 2x + 9$

$$3(2 \cdot 3 - 1) \stackrel{?}{=} 2 \cdot 3 + 9 \qquad \text{Substitute 3 for } x.$$

$$3(5) \stackrel{?}{=} 6 + 9$$

$$15 = 15$$

Since 3 satisfies the equation, it is the solution. The solution set is {3}.   ∎

To solve linear equations, we can follow these steps.

Steps for Solving Linear Equations

1. If an equation contains fractions, multiply both sides of the equation by a number that will eliminate the denominators.

2. Use the distributive property to remove all sets of parentheses and combine like terms.

3. Use the addition and subtraction properties to get all variables on one side of the equation and all numbers on the other side. Combine like terms, if necessary.

4. Use the multiplication and division properties to make the coefficient of the variable equal to 1.

5. Check the result by replacing the variable with the possible solution and verifying that the number satisfies the equation.

EXAMPLE 5   Solve the equation $\dfrac{5}{3}(x - 3) = \dfrac{3}{2}(x - 2) + 2$.

Solution   *Step 1:* Since 6 is the smallest number that can be divided by both 2 and 3, we multiply both sides of the equation by 6 to eliminate the fractions:

$$\frac{5}{3}(x - 3) = \frac{3}{2}(x - 2) + 2$$

$$6\left[\frac{5}{3}(x - 3)\right] = 6\left[\frac{3}{2}(x - 2) + 2\right] \qquad \text{To eliminate the fractions, multiply both sides by 6.}$$

$$10(x - 3) = 9(x - 2) + 6 \cdot 2 \qquad \text{Use the distributive property on the right-hand side.}$$

Step 2: We use the distributive property to remove parentheses and then combine like terms.

$$10x - 30 = 9x - 18 + 12$$

$$10x - 30 = 9x - 6$$

Step 3: We use the addition and subtraction properties by adding 30 to both sides and subtracting $9x$ from both sides.

$$10x - 30 - 9x + 30 = 9x - 6 - 9x + 30$$

$$x = 24 \qquad \text{Combine like terms.}$$

Since the coefficient of x in the above equation is 1, Step 4 is unnecessary.

Step 5: We check by substituting 24 for x in the original equation and simplifying:

$$\frac{5}{3}(x - 3) = \frac{3}{2}(x - 2) + 2$$

$$\frac{5}{3}(24 - 3) \stackrel{?}{=} \frac{3}{2}(24 - 2) + 2$$

$$\frac{5}{3}(21) \stackrel{?}{=} \frac{3}{2}(22) + 2$$

$$5(7) \stackrel{?}{=} 3(11) + 2$$

$$35 = 35$$

Since 24 satisfies the equation, it is the solution. The solution set is $\{24\}$. ∎

EXAMPLE 6   Solve the equation $\dfrac{x + 2}{5} - 4x = \dfrac{8}{5} - \dfrac{x + 9}{2}$.

Solution

$$\frac{x + 2}{5} - 4x = \frac{8}{5} - \frac{x + 9}{2}$$

$$10\left(\frac{x + 2}{5} - 4x\right) = 10\left(\frac{8}{5} - \frac{x + 9}{2}\right)$$ To eliminate the fractions, multiply both sides by 10.

$$2(x + 2) - 40x = 2(8) - 5(x + 9)$$ Remove the red parentheses.

$$2x + 4 - 40x = 16 - 5x - 45$$ Remove parentheses.

$$-38x + 4 = -5x - 29$$ Combine like terms.

$$-33x = -33$$ Add $5x$ and -4 to both sides.

$$\frac{-33x}{-33} = \frac{-33}{-33}$$ Divide both sides by -33.

$$x = 1$$ Simplify.

Check: $\dfrac{x + 2}{5} - 4x = \dfrac{8}{5} - \dfrac{x + 9}{2}$

$$\frac{1 + 2}{5} - 4(1) \stackrel{?}{=} \frac{8}{5} - \frac{1 + 9}{2}$$ Substitute 1 for x.

$$\frac{3}{5} - 4 \stackrel{?}{=} \frac{8}{5} - 5$$

$$\frac{3}{5} - \frac{20}{5} \stackrel{?}{=} \frac{8}{5} - \frac{25}{5}$$

$$-\frac{17}{5} = -\frac{17}{5}$$

Since 1 satisfies the equation, it is the solution. The solution set is $\{1\}$. ∎

■ IDENTITIES AND IMPOSSIBLE EQUATIONS

The equations discussed so far have been **conditional equations.** For these equations, some numbers x satisfy the equation and others do not. An **identity** is an equation that is satisifed by every number x for which both sides of the equation are defined.

EXAMPLE 7 Solve the equation $2(x - 1) + 4 = 4(1 + x) - (2x + 2)$.

Solution

$$2(x - 1) + 4 = 4(1 + x) - (2x + 2)$$

$$2x - 2 + 4 = 4 + 4x - 2x - 2 \qquad \text{Use the distributive property to remove parentheses.}$$

$$2x + 2 = 2x + 2 \qquad \text{Combine like terms.}$$

The result, $2x + 2 = 2x + 2$, will be a true equation for every value of x. Since every number x satisfies the equation, it is an identity. ∎

An **impossible equation** or a **contradiction** is an equation that has no solution.

EXAMPLE 8 Solve the equation $\dfrac{x - 1}{3} + 4x = \dfrac{3}{2} + \dfrac{13x - 2}{3}$.

Solution

$$\frac{x - 1}{3} + 4x = \frac{3}{2} + \frac{13x - 2}{3}$$

$$6\left(\frac{x - 1}{3} + 4x\right) = 6\left(\frac{3}{2} + \frac{13x - 2}{3}\right) \qquad \text{To eliminate the fractions, multiply both sides by 6.}$$

$$2(x - 1) + 6(4x) = 3(3) + 2(13x - 2) \qquad \text{Use the distributive property to remove the red parentheses.}$$

$$2x - 2 + 24x = 9 + 26x - 4 \qquad \text{Remove parentheses.}$$

$$26x - 2 = 26x + 5 \qquad \text{Combine like terms.}$$

$$-2 = 5 \qquad \text{Subtract } 26x \text{ from both sides.}$$

Since $-2 = 5$ is false, no number x can satisfy the equation. Its solution set is $\emptyset$. ∎

■ **REPEATING DECIMALS**

By using equations, we can change repeating decimals to fractional form. For example, to write $0.2\,\overline{54}$ as a fraction, we note that the decimal has a repeating block of two digits and then form an equation by setting x equal to the decimal.

1. $x = 0.2\,54\,54\,54 \ldots$

We then form another equation by multiplying both sides of Equation 1 by 10^2, which is 100.

2. $100x = 25.4\,54\,54\,54 \ldots$

We can subtract each side of Equation 1 from the corresponding side of Equation 2 to obtain

$$100x = 25.4\,54\,54\,54 \ldots$$
$$\underline{x = 0.2\,54\,54\,54 \ldots}$$
$$99x = 25.2$$

Finally, we solve $99x = 25.2$ for x and simplify the fraction.

$$x = \frac{25.2}{99} = \frac{25.2 \cdot 10}{99 \cdot 10} = \frac{252}{990} = \frac{18 \cdot 14}{18 \cdot 55} = \frac{14}{55}$$

We can use a calculator to verify that the decimal representation of $\frac{14}{55}$ is $0.2\,\overline{54}$.

The key step in the solution was multiplying both sides of Equation 1 by 10^2. If there had been n digits in the repeating block of the decimal, we would have multiplied both sides of Equation 1 by 10^n.

■ **FORMULAS**

Suppose we want to find the heights of several triangles whose areas and bases are known. It would be tedious to substitute values of A and b into the formula $A = \frac{1}{2}bh$ and then repeatedly solve the formula for h. It is easier to solve for h first and then substitute values for A and b and compute h directly.

To solve a formula for a variable means to isolate that variable on one side of the equation and isolate all other quantities on the other side.

EXAMPLE 9   Solve the formula $A = \frac{1}{2}bh$ for h.

Solution

$$A = \frac{1}{2}bh$$

$$2A = bh \qquad \text{To eliminate the fraction, multiply both sides by 2.}$$

$$\frac{2A}{b} = h \qquad \text{To isolate } h, \text{ divide both sides by } b.$$

$$h = \frac{2A}{b} \qquad \text{Use the symmetric property of equality.} \qquad ■$$

EXAMPLE 10   For simple interest, the formula $A = p + prt$ gives the amount of money in an account at the end of a specific time. A represents the amount, p the principal, r the rate of interest, and t the time. We can solve the formula for t as follows:

Solution

$$A = p + prt$$

$$A - p = prt \qquad \text{To isolate the term involving } t, \text{ subtract } p \text{ from both sides.}$$

$$\frac{A - p}{pr} = t \qquad \text{To isolate } t, \text{ divide both sides by } pr.$$

$$t = \frac{A - p}{pr} \qquad \text{Use the symmetric property of equality.} \qquad ■$$

EXAMPLE 11   The formula $F = \frac{9}{5}C + 32$ converts degrees Celsius to degrees Fahrenheit. Solve the formula for C.

Solution

$$F = \frac{9}{5}C + 32$$

$$F - 32 = \frac{9}{5}C \qquad \text{To isolate the term involving } C, \text{ subtract } 32 \text{ from both sides.}$$

$$\frac{5}{9}(F - 32) = \frac{5}{9}\left(\frac{9}{5}C\right) \qquad \text{To isolate } C, \text{ multiply both sides by } \frac{5}{9}.$$

$$\frac{5}{9}(F - 32) = C \qquad \frac{5}{9} \cdot \frac{9}{5} = 1.$$

$$C = \frac{5}{9}(F - 32) \qquad \text{Use the symmetric property of equality.}$$

To convert degrees Fahrenheit to degrees Celsius, we can use the formula
$C = \frac{5}{9}(F - 32)$. ■

Orals *Combine like terms.*

1. $5x + 4x$

2. $7s^2 - 5s^2$

Tell whether each number is a solution of $2x + 5 = 13$.

3. 3   **4.** 4   **5.** 5   **6.** 6

Solve each equation.

7. $3x - 2 = 7$

8. $\frac{1}{2}x - 1 = 5$

9. $\dfrac{x - 2}{3} = 1$

10. $\dfrac{x + 3}{2} = 3$

EXERCISE 1.5

In Exercises 1–4, tell whether 5 is a solution of each equation.

1. $3x + 2 = 17$ **2.** $7x - 2 = 33$ **3.** $\frac{3}{5}x - 5 = -2$ **4.** $\frac{2}{5}x + 12 = 8$

In Exercises 5–20, solve each equation.

5. $x + 6 = 8$ **6.** $y - 7 = 3$ **7.** $a - 5 = 20$ **8.** $b + 4 = 18$

9. $2u = 6$ **10.** $3v = 12$ **11.** $\frac{x}{4} = 7$ **12.** $\frac{x}{6} = 8$

13. $3x + 1 = 3$ **14.** $8x - 2 = 13$ **15.** $2x + 1 = 13$ **16.** $2x - 4 = 16$

17. $3(x - 4) = -36$ **18.** $4(x + 6) = 84$ **19.** $3(r - 4) = -4$ **20.** $4(s - 5) = -3$

In Exercises 21–28, tell whether the terms are like terms. If they are, combine them.

21. $2x, 6x$ **22.** $-3x, 5y$ **23.** $-5xy, -7yz$ **24.** $-3t^2, 12t^2$

25. $3x^2, -5x^2$ **26.** $5y^2, 7xy$ **27.** $xy, 3xt$ **28.** $-4x, -5x$

In Exercises 29–58, solve each equation.

29. $3a - 22 = -2a - 7$ **30.** $a + 18 = 6a - 3$

31. $2(2x + 1) = 15 + 3x$ **32.** $-2(x + 5) = 30 - x$

33. $3(y - 4) - 6 = y$ **34.** $2x + (2x - 3) = 5$

35. $5(5 - a) = 37 - 2a$ **36.** $4a + 17 = 7(a + 2)$

37. $4(y + 1) = -2(4 - y)$ **38.** $5(r + 4) = -2(r - 3)$

39. $2(a - 5) - (3a + 1) = 0$ **40.** $8(3a - 5) - 4(2a + 3) = 12$

41. $3(y - 5) + 10 = 2(y + 4)$ **42.** $2(5x + 2) = 3(3x - 2)$

43. $9(x + 2) = -6(4 - x) + 18$ **44.** $3(x + 2) - 2 = -(5 + x) + x$

45. $\frac{1}{2}x - 4 = -1 + 2x$ **46.** $2x + 3 = \frac{2}{3}x - 1$

47. $\dfrac{x}{2} - \dfrac{x}{3} = 4$ **48.** $\dfrac{x}{2} + \dfrac{x}{3} = 10$

49. $\dfrac{x}{6} + 1 = \dfrac{x}{3}$

50. $\dfrac{3}{2}(y + 4) = \dfrac{20 - y}{2}$

51. $5 - \dfrac{x + 2}{3} = 7 - x$

52. $3x - \dfrac{2(x + 3)}{3} = 16 - \dfrac{x + 2}{2}$

53. $\dfrac{4x - 2}{2} = \dfrac{3x + 6}{3}$

54. $\dfrac{t + 4}{2} = \dfrac{2t - 3}{3}$

55. $\dfrac{a + 1}{3} + \dfrac{a - 1}{5} = \dfrac{2}{15}$

56. $\dfrac{2z + 3}{3} + \dfrac{3z - 4}{6} = \dfrac{z - 2}{2}$

57. $\dfrac{5a}{2} - 12 = \dfrac{a}{3} + 1$

58. $\dfrac{5a}{6} - \dfrac{5}{2} = -\dfrac{1}{2} - \dfrac{a}{6}$

In Exercises 59–66, solve each equation. If the equation is an identity or an impossible equation, so indicate.

59. $4(2 - 3t) + 6t = -6t + 8$

60. $2x - 6 = -2x + 4(x - 2)$

61. $\dfrac{a + 1}{4} + \dfrac{2a - 3}{4} = \dfrac{a}{2} - 2$

62. $\dfrac{y - 8}{5} + 2 = \dfrac{2}{5} - \dfrac{y}{3}$

63. $3(x - 4) + 6 = -2(x + 4) + 5x$

64. $2(x - 3) = \dfrac{3}{2}(x - 4) + \dfrac{x}{2}$

65. $y(y + 2) + 1 = y^2 + 2y + 1$

66. $x(x - 3) = x^2 - 2x + 1 - (5 + x)$

In Exercises 67–70, write each repeating decimal number as a fraction. Simplify the answer when possible.

67. $0.\overline{3}$ **68.** $0.\overline{29}$ **69.** $-0.34\,\overline{89}$ **70.** $-2.3\,\overline{47}$

In Exercises 71–88, solve each formula for the indicated variable.

71. $A = lw$ for w

72. $p = 4s$ for s

73. $V = \dfrac{1}{3}Bh$ for B

74. $A = \dfrac{1}{2}bh$ for b

75. $I = prt$ for t

76. $I = prt$ for r

77. $p = 2l + 2w$ for w

78. $p = 2l + 2w$ for l

79. $A = \dfrac{1}{2}h(B + b)$ for B

80. $A = \dfrac{1}{2}h(B + b)$ for b

81. $y = mx + b$ for x

82. $y = mx + b$ for m

83. $l = a + (n - 1)d$ for n

84. $l = a + (n - 1)d$ for d

85. $S = \dfrac{a - lr}{1 - r}$ for l

86. $C = \dfrac{5}{9}(F - 32)$ for F

87. $S = \dfrac{n(a + l)}{2}$ for l

88. $S = \dfrac{n(a + l)}{2}$ for n

89. Force of gravity   The masses of the two objects in Illustration 1 are m and M. The force of gravitation, F, between the masses is

$$F = \dfrac{GmM}{d^2}$$

where G is a constant and d is the distance between them. Solve for m.

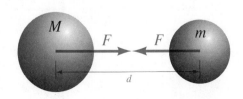

ILLUSTRATION 1

90. Thermodynamics   In thermodynamics, the Gibbs free-energy function is given by the formula $G = U - TS + pV$. Solve for S.

91. Converting temperatures   Solve the formula $F = \frac{9}{5}C + 32$ for C, and find the Celsius temperatures that correspond to Fahrenheit temperatures of $32°$, $70°$, and $212°$.

92. Doubling money   A man intends to invest \$1000 at simple interest. Solve the formula $A = p + prt$ for t and find how long it will take to double his money at the rates of 5%, 7%, and 10%.

93. Cost of electricity   The cost of electricity in a certain city is given by the formula $C = 0.07n + 6.50$, where C is the cost and n is the number of kilowatt hours used. Solve for n and find the number of kilowatt hours used for costs of \$49.97, \$76.50, and \$125.

94. Cost of water   A monthly water bill in a certain city is calculated by using the formula $n = \frac{5000C - 17,500}{6}$ where n is the number of gallons used and C is the monthly cost. Solve for C and compute the bill for quantities used of 500, 1200, and 2500 gallons.

95. Ohm's law   The formula $E = IR$, called **Ohm's law**, is used in electronics. Solve for R and then calculate the resistance R if the voltage E is 56 volts and the current I is 7 amperes. (Resistance has units of ohms.)

96. Earning interest   An amount P, invested at a simple interest rate r, will grow to an amount A in t years according to the formula $A = P(1 + rt)$. Solve for P. Suppose a man invested some money at 5.5%. If after 5 years, he had \$6693.75 on deposit, what amount did he originally invest?

97. Angles of a polygon   A regular polygon has n equal sides and n equal angles. The measure a of an interior angle is given by $a = 180°\left(1 - \frac{2}{n}\right)$. Solve for n. Find the number of sides of the regular polygon in Illustration 2, if an interior angle is $135°$.

ILLUSTRATION 2

98. Power loss   Illustration 3 is the schematic diagram of a resistor connected to a voltage source of 60 volts. As a result, the resistor dissipates power in the form of heat. The power P lost when a voltage V is placed across a resistance R is given by the formula

$$P = \frac{E^2}{R}$$

Solve for R. If P is 4.8 watts and E is 60 volts, find R.

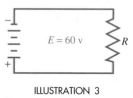

ILLUSTRATION 3

Writing Exercises   *Write a paragraph using your own words.*

1. Explain the difference between a conditional equation, an identity, and an impossible equation.

2. Explain how you would solve an equation.

Something to Think About

1. Is 0.9 equal to 1?

2. Is $0.\overline{9}$ equal to 1?

Review Exercises   *Simplify each expression.*

1. $(-4)^3$

2. -3^3

3. $\left(\dfrac{x + y}{x - y}\right)^0$

4. $(x^2 x^3)^4$

5. $\left(\dfrac{x^2 x^5}{x^3}\right)^2$

6. $\left(\dfrac{x^4 y^3}{x^5 y}\right)^3$

7. $(2x)^{-3}$

8. $\left(\dfrac{x^2}{y^5}\right)^{-4}$

1.6   Using Equations to Solve Problems

■ RECREATION PROBLEMS ■ BUSINESS PROBLEMS ■ GEOMETRIC PROBLEMS ■ LEVER PROBLEMS

In this section, we will solve application problems. When we translate the words of a problem into mathematics, we are creating a *mathematical model* of the problem. To create these models, we can use the following chart to translate certain words into mathematical operations.

Addition (+)	Subtraction (−)	Multiplication (·)	Division (÷)
added to	subtracted from	multiplied by	divided by
plus	difference	product	quotient
the sum of	less than	times	ratio
more than	less	of	half
increased by	decreased by	twice	

We can then change English phrases into algebraic expressions.

English phrase	Algebraic expression
2 added to some number	$x + 2$
The difference between two numbers	$x - y$
5 times some number	$5x$
The product of 925 and some number	$925x$
5% of some number	$0.05x$
The sum of twice a number and 10	$2x + 10$
The quotient (or ratio) of two numbers	$\frac{x}{y}$

Once we know how to change phrases into algebraic expressions, we can solve many problems. The following list of steps provides an excellent strategy for solving problems.

Strategy for Problem Solving

1. *Analyze the problem* by reading it carefully to understand the given facts. What information is given? What vocabulary is given? What are you asked to find? Often a diagram will help you visualize the facts of the problem.

2. *Form an equation* by picking a variable to represent the quantity to be found. Then express all other quantities mentioned as expressions involving the variable. Finally, write an equation expressing a quantity in two different ways.

3. *Solve the equation.*

4. *State the conclusion.*

5. *Check the result* in the words of the problem.

■ RECREATION PROBLEMS

EXAMPLE 1   **Cutting a rope**   A mountain climber wants to cut a rope 213 feet long into three pieces. If each piece is to be two feet longer than the previous one, where should he make the cuts?

Analyze the Problem   If x represents the length of the shortest piece, the climber wants the lengths of the three pieces to be

$$x, \quad x + 2, \quad \text{and} \quad x + 4$$

feet long. He knows that the sum of these three lengths can be expressed in two ways: as $x + (x + 2) + (x + 4)$, and as 213.

Form an Equation   Let x represent the length of the first piece of rope.
Then $x + 2$ represents the length of the second piece,
and $x + 4$ represents the length of the third piece.

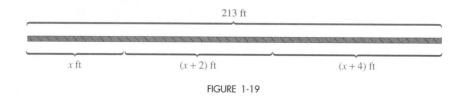

213 ft

x ft $(x + 2)$ ft $(x + 4)$ ft

FIGURE 1-19

From Figure 1-19, we see that the sum of the individual pieces must equal the total length of the rope.

The length of the first piece	+	the length of the second piece	+	the length of the third piece	=	the total length of the rope.

Solve the Equation

$$x \quad + \quad x + 2 \quad + \quad x + 4 \quad = \quad 213$$

$3x + 6 = 213$	Combine like terms.
$3x = 207$	Subtract 6 from both sides.
$x = 69$	Divide both sides by 3.
$x + 2 = 71$	
$x + 4 = 73$	

State the Conclusion   He should make cuts 69 feet from one end and 73 feet from the other end, to get lengths of 69 feet, 71 feet, and 73 feet.

Check the Result   Each length is two feet longer than the previous length, and the sum of the lengths is 213 feet.   ■

■ BUSINESS PROBLEMS

When the regular price of merchandise is reduced, the amount of reduction is called **markdown** (or discount).

$$\boxed{\text{Sale price}} \quad = \quad \boxed{\text{regular price}} \quad - \quad \boxed{\text{markdown}}$$

Usually, the markdown is expressed as a percent of the regular price.

$$\boxed{\text{Markdown}} \quad = \quad \boxed{\text{percent of markdown}} \quad \cdot \quad \boxed{\text{regular price}}$$

EXAMPLE 2

Finding the percent of markdown A home theater system is on sale for $777. If the list price was $925, find the percent of markdown.

Analyze the Problem In this case, $777 is the sale price, $925 is the regular price, and the markdown is the product of $925 and the percent of markdown.

Form an Equation We can let r represent the percent of markdown, expressed as a decimal. We then substitute $777 for the sale price and $925 for the regular price in the formula

$$\boxed{\text{Sale price}} \quad = \quad \boxed{\text{regular price}} \quad - \quad \boxed{\text{markdown}}$$

Solve the Equation
$$777 \qquad = \qquad 925 \qquad - \qquad r \cdot 925$$

$$777 = 925 - 925r$$
$$-148 = -925r \qquad \text{Subtract 925 from both sides.}$$
$$0.16 = r \qquad \text{Divide both sides by } -925.$$

State the Conclusion The percent of markdown is 16%.

Check the Result Since the markdown is 16% of $925, or $148, the sale price is $925 − $148, or $777.

EXAMPLE 3

Portfolio analysis A college foundation owns stock in IBC (selling at $54 per share), GS (selling at $65 per share), and ATB (selling at $105 per share). The foundation owns equal shares of GS and IBC, but five times as many shares of ATB.

If this portfolio is worth $450,800, how many shares of each type does the foundation own?

Analyze the Problem The value of the IBC stock plus the value of the GS stock plus the value of the ATB stock must equal $450,800.

- If x represents the number of shares of IBC, then $54x$ is the value of that stock.
- Since the foundation has equal numbers of shares of GS and of IBC, x also represents the number of shares of GS. The value of this stock is $65x$.
- Since the foundation owns five times as many shares of ATB, it owns $5x$ shares of ATB. The value of this stock is $105(5x)$.

We set the sum of these values equal to $450,800.

Form an Equation   We let x represent the number of shares of IBC.
Then x also represents the number of shares of GS,
and $5x$ represents the number of shares of ATB.

The value of IBC stock	+	the value of GS stock	+	the value of ATB stock	=	the total value of the stock.

Solve the Equation   54x +   65x +   105($5x$)   =   450,800

$$54x + 65x + 525x = 450{,}800 \qquad 105(5x) = 525x.$$
$$644x = 450{,}800 \qquad \text{Combine like terms.}$$
$$x = 700 \qquad \text{Divide both sides by 644.}$$

State the Conclusion   The foundation owns 700 shares of IBC, 700 shares of GS, and 5(700), or 3500, shares of ABT.

Check the Result   The value of 700 shares of IBC at $54 per share is $37,800.
The value of 700 shares of GS at $65 per share is $45,500.
The value of 3500 shares of ATB at $105 per share is $367,500.

The sum of these values is $450,800.   ■

■ GEOMETRIC PROBLEMS

Figure 1-20 illustrates several geometric figures. A **right angle** is an angle whose measure is 90°. A **straight angle** is an angle whose measure is 180°. An **acute angle** is an angle whose measure is greater than 0° but less than 90°.

If the sum of two angles equals 90°, the angles are called **complementary**, and each angle is called the **complement** of the other. If the sum of two angles

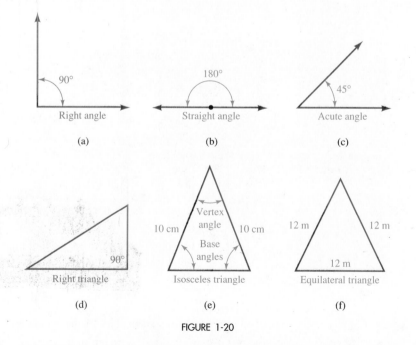

FIGURE 1-20

equals 180°, the angles are called **supplementary**, and each angle is called the **supplement** of the other.

A **right triangle** is a triangle with one right angle. An **isosceles triangle** is a triangle with two sides of equal measure that meet to form the **vertex angle**. The angles opposite the equal sides, called the **base angles**, are also equal. An **equilateral triangle** is a triangle with three equal sides and three equal angles.

EXAMPLE 4   **Angles in a triangle**   If the vertex angle of the isosceles triangle shown in Figure 1-20(e) measures 64°, find the measure of each base angle.

Analyze the Problem   We are given that the vertex angle measures 64°. If we let $x°$ represent the measure of one base angle, then the measure of the other base angle is also $x°$. Thus, the sum of the angles in the triangle is $x° + x° + 64°$. Because the sum of the measures of the angles of any triangle is 180°, we know that $x° + x° + 64°$ is equal to 180°.

Form an Equation   We can form the equation

The measure of one base angle	+	the measure of the other base angle	+	the measure of the vertex angle	=	180°.

Solve the Equation

$$x \quad + \quad x \quad + \quad 64 \quad = \quad 180$$

$$
\begin{aligned}
x + x + 64 &= 180 \\
2x + 64 &= 180 \qquad \text{Combine like terms.} \\
2x &= 116 \qquad \text{Subtract 64 from both sides.} \\
x &= 58 \qquad \text{Divide both sides by 2.}
\end{aligned}
$$

State the Conclusion   The measure of each base angle is 58°.

Check the Result   The sum of the measures of each base angle and the vertex angle is 180°:

$$58° + 58° + 64° = 180°$$ ∎

EXAMPLE 5   **Designing a kennel**   A man has 28 meters of fencing to make a rectangular kennel. If he wants the kennel to be 6 feet longer than it is wide, find its dimensions.

Analyze the Problem   The perimeter, P, of a rectangle is the distance around it. If w is chosen to represent the width of the kennel, then $w + 6$ represents its length. (See Figure 1-21.) The perimeter can be expressed either as $2w + 2(w + 6)$ or as 28.

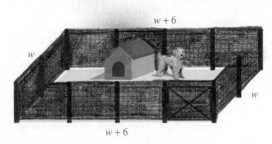

FIGURE 1-21

Form an Equation   We let w represent the width of the kennel. Then $w + 6$ represents its length.

Two widths	+	two lengths	=	the perimeter.

Solve the Equation $2 \cdot w$ $+$ $2 \cdot (w + 6)$ $=$ 28

$$2w + 2w + 12 = 28 \qquad \text{Use the distributive property to remove parentheses.}$$
$$4w + 12 = 28 \qquad \text{Combine like terms.}$$
$$4w = 16 \qquad \text{Subtract 12 from both sides.}$$
$$w = 4 \qquad \text{Divide both sides by 4.}$$
$$w + 6 = 10$$

State the Conclusion   The dimensions of the kennel are 4 meters by 10 meters.

Check the Result   If a kennel has a width of 4 meters and a length of 10 meters, its length is 6 meters longer than its width, and the perimeter is 2(4) meters + 2(10) meters = 28 meters. ∎

■ LEVER PROBLEMS

EXAMPLE 6   **Engineering**   Design engineers must position two hydraulic cylinders as in Figure 1-22 to balance a 9500-pound force at point A. The first cylinder at the end of the lever exerts a 3500-pound force. Where should the design engineers position the second cylinder, which is capable of exerting a 5500-pound force?

Analyze the Problem   From physics, the lever will be in balance when the force of the first cylinder multiplied by its distance from the pivot (also called the **fulcrum**), added to the second cylinder's force multiplied by its distance from the fulcrum, is equal to the product of the 9500-pound force and its distance from the fulcrum.

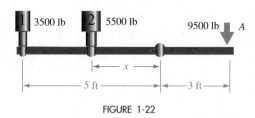

FIGURE 1-22

Form an Equation   We let x represent the distance from the larger cylinder to the fulcrum.

Force of cylinder 1, times its distance	+	force of cylinder 2, times its distance	=	force to be balanced, times its distance.

Solve the Equation $3500 \cdot 5$ $+$ $5500x$ $=$ $9500 \cdot 3$

$$17{,}500 + 5500x = 28{,}500$$
$$5500x = 11{,}000 \qquad \text{Subtract 17,500 from both sides.}$$
$$x = 2 \qquad \text{Divide both sides by 5500.}$$

State the Conclusion   The design must specify that the second cylinder be positioned 2 feet from the fulcrum.

Check the Result $3500 \cdot 5 + 5500 \cdot 2 = 17{,}500 + 11{,}000 = 28{,}500$
$9500 \cdot 3 = 28{,}500$ ∎

Orals   *Find each value.*

1. 20% of 500

2. $33\frac{1}{3}\%$ of 600

If a stock costs $54, find the cost of

3. 5 shares

4. *x* shares

Find the area of the rectangle with the given dimensions.

5. 6 meters long, 4 meters wide

6. *l* meters long, $(l - 5)$ meters wide

EXERCISE 1.6

In Exercises 1–44, solve each problem.

1. Cutting a rope   A 60-foot rope is cut into four pieces with each successive piece being twice as long as the previous one. Find the length of the longest piece.

2. Cutting a cable   A 186-foot cable is to be cut into four pieces. Find the length of each piece if each successive piece is 3 feet longer than the previous one.

3. Cutting a board   The carpenter in Illustration 1 saws a board into two pieces. He wants one piece to be 1 foot longer than twice the length of the shorter piece. Find the length of each piece.

ILLUSTRATION 1

4. Cutting a beam   A 30-foot steel beam is to be cut into two pieces. The longer piece is to be 2 feet more than 3 times as long as the shorter piece. Find the length of each piece.

5. Buying a TV and a VCR   See Illustration 2. If the TV costs $55 more than the VCR, how much does the TV cost?

BUY *BOTH* FOR
$655

ILLUSTRATION 2

6. Buying golf clubs   The cost of a set of golf clubs is $590. If the irons cost $40 more than the woods, find the cost of the irons.

7. Buying a washer and dryer   Find the percent of markdown of the sale in Ilustration 3.

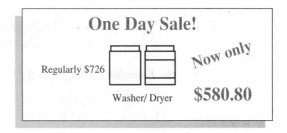

One Day Sale!

Regularly $726

Washer/ Dryer

Now only

$580.80

ILLUSTRATION 3

8. **Buying furniture**   A bedroom set regularly sells for $983. If it is on sale for $737.25, what is the percent of markdown?

9. **Buying a calculus book**   A bookstore buys a used calculus book for $12 and sells it for $40. Find the percent of markup.

10. **Selling stuffed animals**   The owner of a gift shop buys stuffed animals for $18 and sells them for $30. Find the percent of markup.

11. **Value of an IRA**   In an Individual Retirement Account (IRA) valued at $53,900, a student has 500 shares of stock, some in Big Bank Corporation and some in Safe Savings and Loan. If Big Bank sells for $115 per share and Safe Savings sells for $97 per share, how many shares of each does the student own?

12. **Assets of a pension fund**   A pension fund owns 12,000 shares in mutual stock funds and mutual bond funds. Currently, the stock funds sell for $12 per share, and the bond funds sell for $15 per share. How many shares of each does the fund own if the value of the securities is $165,000?

13. **Selling calculators**   Last month, a bookstore ran the ad shown in Illustration 4 and sold 85 calculators, generating $3875 in sales. How many of each type of calculator did the bookstore sell?

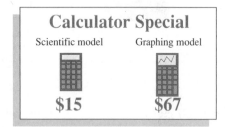

Calculator Special

Scientific model   Graphing model

$15 $67

ILLUSTRATION 4

14. **Selling grass seed**   A seed company sells two grades of grass seed. A 100-pound bag of a mixture of rye and Kentucky bluegrass sells for $245, and a 100-pound bag of bluegrass sells for $347. How many bags are sold in a week when the receipts for 19 bags are $5369?

15. **Buying roses**   A man with $21.25 stops after work to buy some roses for his wife's birthday. If each rose costs $1.25 and there is a delivery charge of $5, how many roses can he buy?

16. **Renting a truck**   To move to Wisconsin, a man can rent a truck for $29.95 per day plus 19¢ per mile. If he keeps the truck for one day, how many miles can he drive for a cost of $77.45?

17. **Car rental**   While waiting for his car to be repaired, a man rents a car for $12 per day plus 10 cents per mile. If he keeps the car for 2 days, how many miles can he drive for a total cost of $30? How many miles can he drive for a total cost of $36?

18. **Computing salaries**   A student earns $17 per day for delivering overnight packages. She is paid $5 per day plus 60¢ for each package delivered. How many more deliveries must she make each day to increase her daily earnings to $23?

19. **Finding dimensions**   The rectangular garden shown in Illustration 5 is twice as long as it is wide. Find its dimensions.

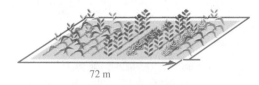

72 m

ILLUSTRATION 5

20. **Finding dimensions**   The width of a rectangular swimming pool is one-third its length. If its perimeter is 96 meters, find the dimensions of the pool.

21. **Fencing a pasture**   A farmer has 624 feet of fencing to enclose the pasture shown in Illustration 6. Because a river runs along one side, fencing will be needed on only three sides. Find the dimensions of the pasture if its length is double its width.

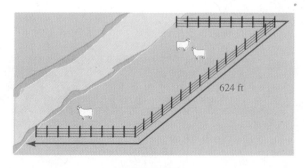

624 ft

ILLUSTRATION 6

22. **Fencing a pen**   A man has 150 feet of fencing to build the pen shown in Illustration 7. If one end is a square, find the outside dimensions.

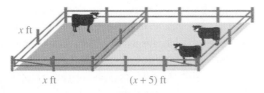

ILLUSTRATION 7

23. Enclosing a swimming pool A woman wants to enclose the swimming pool shown in Illustration 8 and have a walkway of uniform width all the way around. How wide will the walkway be if the woman uses 180 feet of fencing?

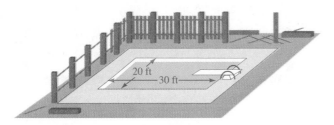

ILLUSTRATION 8

24. Framing a picture An artist wants to frame the picture shown in Illustration 9 with a frame 2 inches wide. How wide will the framed picture be if the artist uses 70 inches of framing material?

ILLUSTRATION 9

25. Supplementary angles If one of two supplementary angles is 35° larger than the other, find the measure of the smaller angle.

26. Supplementary angles Refer to Illustration 10 and find x.

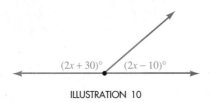

ILLUSTRATION 10

27. Complementary angles If one of two complementary angles is 22° greater than the other, find the measure of the larger angle.

28. Complementary angles in a right triangle Explain why the acute angles in a right triangle are complementary. In Illustration 11, find the measure of angle A.

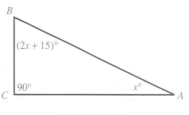

ILLUSTRATION 11

29. Supplementary angles and parallel lines In Illustration 12, lines r and s are cut by a third line l to form angles 1 and 2. When lines r and s are parallel, angles 1 and 2 are supplementary. If the measure of angle 1 is $(x + 50)°$ and the measure of angle 2 is $(2x - 20)°$ and lines r and s are parallel, find x.

30. In Illustration 12, find the measure of angle 3. (*Hint:* See Exercise 29.)

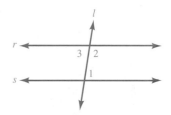

ILLUSTRATION 12

31. Vertical angles When two lines intersect as in Illustration 13, four angles are formed. Angles that are side-by-side, such as angles 1 and 2, are called *adjacent angles*. Angles that are nonadjacent, such as angles 1 and 3 or angles 2 and 4, are called *vertical angles*. From geometry, we know that if two lines intersect, vertical angles have the same measure. If the measure of angle 1 is $(3x + 10)°$ and the measure of angle 3 is $(5x - 10)°$, find x.

ILLUSTRATION 13

32. If the measure of angle 2 in Illustration 13 is $(6x + 20)°$ and the measure of angle 4 is $(8x - 20)°$, find the measure of angle 1. (See Exercise 31.)

33. Angles of an equilateral triangle   Find the measure of each angle in an equilateral triangle.

34. Angles of a quadrilateral   The sum of the angles of any four-sided figure (called a *quadrilateral*) is 360°. The quadrilateral shown in Illustration 14 has two equal base angles. Find x.

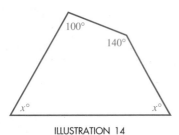

ILLUSTRATION 14

35. Height of a triangle   If the height of a triangle with a base of 8 inches is tripled, its area is increased by 96 square inches. Find the height of the triangle.

36. Engineering design   The width, w, of the flange in the engineering drawing in Illustration 15 has not yet been determined. Find w so that the area of the rectangular portion is exactly one-half of the total area.

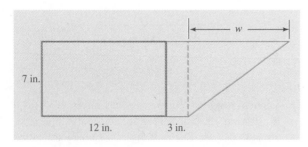

ILLUSTRATION 15

37. Balancing a seesaw   A seesaw is 20 feet long, and the fulcrum is in the center. If an 80-pound boy sits at one end, how far will the boy's 160-pound father have to sit from the fulcrum to balance the seesaw?

38. Establishing equilibrium   Two forces—110 pounds and 88 pounds—are applied to opposite ends of an 18-foot lever. How far from the greater force must the fulcrum be placed so that the lever is balanced?

39. Moving a stone   A woman uses a 10-foot bar to lift a 210-pound stone. If she places another rock 3 feet from the stone to act as the fulcrum, how much force must she exert to move the stone?

40. Lifting a car   A 350-pound football player brags that he can lift a 2500-pound car. If he uses a 12-foot bar with the fulcrum placed 3 feet from the car, will he be able to lift the car?

41. Balancing a lever   Forces are applied to a lever as indicated in Illustration 16. Find x, the distance of the smallest force from the fulcrum.

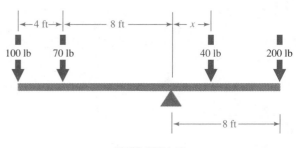

ILLUSTRATION 16

42. Balancing a seesaw   Jim and Bob sit at opposite ends of an 18-foot seesaw, with the fulcrum at its center. Jim weighs 160 pounds, and Bob weighs 200 pounds. Kim sits 4 feet in front of Jim, and the seesaw balances. How much does Kim weigh?

43. Temperature scales   The Celsius and Fahrenheit temperature scales are related by the equation $C = \frac{5}{9}(F - 32)$. At what temperature will a Fahrenheit and a Celsius thermometer give the same reading?

44. Installing solar heating   One solar panel in Illustration 17 is to be 3 feet wider than the other, but to be equally efficient, they must have the same area. Find the width of each.

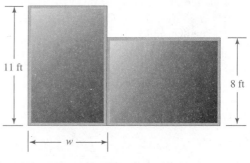

ILLUSTRATION 17

Writing Exercises   *Write a paragraph using your own words.*

1. Explain the steps for solving an applied problem.

2. Explain how to check the solution of an applied problem.

3. Give an argument to show that if two lines intersect, adjacent angles are supplementary.

4. Give an argument to show that if two lines intersect, the pairs of vertical angles have the same measure.

Something to Think About

1. Find the distance x required to balance the lever in the illustration.

2. Interpret the answer to Question 1.

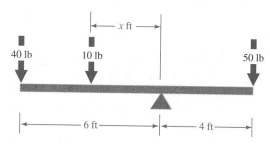

3. Give an argument to show that if two lines intersect, adjacent angles are supplementary.

4. Give an argument to show that if two lines intersect, each pair of vertical angles has the same measure.

Review Exercises   *Simplify each expression.*

1. $\left(\dfrac{3x^{-3}}{4x^2}\right)^{-4}$

2. $\left(\dfrac{r^{-3}s^2}{r^2r^3s^{-4}}\right)^{-5}$

3. $\dfrac{a^m a^3}{a^2}$

4. $\left(\dfrac{b^n}{b^3}\right)^3$

1.7 More Applications of Equations

■ INVESTMENT PROBLEMS ■ UNIFORM MOTION PROBLEMS ■ MIXTURE PROBLEMS

■ INVESTMENT PROBLEMS

EXAMPLE 1   **Investing money** A professor has $15,000 to invest for one year, some at 8% and the rest at 7%. If she wants to earn $1110 from these investments, how much should she invest at each rate?

Analyze the Problem   We will add the interest from the 8% investment to the interest from the 7% investment and set the sum equal to the total interest earned.

Simple interest is computed by the formula $i = prt$, where i is the interest earned, p is the principal, r is the annual interest rate, and t is the length of time the principal is invested. Thus, if $\$x$ is invested at 8% for one year, the interest earned is $\$0.08x$. If the remaining $\$(15{,}000 - x)$ is invested at 7%, the amount earned on that investment is $\$0.07(15{,}000 - x)$. The sum of these amounts should equal $\$1110$.

Form an Equation We can let x represent the number of dollars invested at 8%.
Then $15,000 - x$ represents the number of dollars invested at 7%.

The interest earned at 8%	+	the interest earned at 7%	=	the total interest.

Solve the Equation
$$0.08x \quad + \quad 0.07(15,000 - x) \quad = \quad 1110$$

$8x + 7(15,000 - x) = 111,000$	To eliminate the decimals, multiply both sides by 100.
$8x + 105,000 - 7x = 111,000$	Use the distributive property to remove parentheses.
$x + 105,000 = 111,000$	Combine like terms.
$x = 6000$	Subtract 105,000 from both sides.
$15,000 - x = 9000$	

State the Conclusion She should invest $6000 at 8% and $9000 at 7%.

Check the Result The interest on $6000 is 0.08($6000) = $480.
The interest earned on $9000 is 0.07($9000) = $630.
The total interest is $1110.

■ UNIFORM MOTION PROBLEMS

EXAMPLE 2 **Travel time** A car leaves Rockford, traveling toward Wausau at the rate of 55 miles per hour. At the same time, another car leaves Wausau, traveling toward Rockford at the rate of 50 miles per hour. How long will it take them to meet if the cities are 157.5 miles apart?

Analyze the Problem In this case, the cars are traveling toward each other, as shown in Figure 1-23(a).

Uniform motion problems are based on the formula $d = rt$, where d is distance, r is rate, and t is time. We can organize the given information as in the chart shown in Figure 1-23(b).

We know that one car is traveling at 55 miles per hour and that the other is going 50 miles per hour. We also know that they travel for the same amount of time—say, t hours. Thus, the distance that the faster car travels is $55t$ miles, and the distance that the slower car travels is $50t$ miles. The sum of these distances equals 157.5 miles, the distance between the cities.

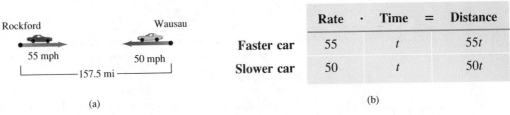

	Rate ·	**Time** =	**Distance**
Faster car	55	t	$55t$
Slower car	50	t	$50t$

(a)   (b)

FIGURE 1-23

Form an Equation   We can let t represent the time that each car travels.
Then $55t$ represents the distance traveled by the faster car,
and $50t$ represents the distance traveled by the slower car.

The distance the faster car goes	+	the distance the slower car goes	=	the distance between cities.

Solve the Equation $55t$ $+$ $50t$ $=$ 157.5

$105t = 157.5$ Combine like terms.

$t = 1.5$ Divide both sides by 105.

State the Conclusion   The two cars will meet in $1\frac{1}{2}$ hours.

Check the Result   The faster car travels $1.5(55) = 82.5$ miles.
The slower car travels $1.5(50) = 75$ miles.
The total distance traveled is 157.5 miles.   ∎

■ MIXTURE PROBLEMS

EXAMPLE 3

Mixing nuts   The owner of a candy store notices that 20 pounds of gourmet cashews are getting stale. They did not sell because of their high price of $12 per pound. The store owner decides to mix peanuts with the cashews and lower the price per pound. If peanuts sell for $3 per pound, how many pounds of peanuts must be mixed with the cashews to make a mixture that could be sold for $6 per pound?

Analyze the Problem   This problem is based on the formula $V = pn$, where V represents value, p represents the price per pound, and n represents the number of pounds.

We can let x represent the number of pounds of peanuts to be used and enter the known information in the chart shown in Figure 1-24. The value of the cashews plus the value of the peanuts will be equal to the value of the mixture.

	Price $\cdot$	Number of pounds	=	Value
Cashews	12	20		240
Peanuts	3	x		$3x$
Mixture	6	$20 + x$		$6(20 + x)$

FIGURE 1-24

Form an Equation   We can let x represent the number of pounds of peanuts to be used. Then $20 + x$ represents the number of pounds in the mixture.

The value of the cashews	+	the value of the peanuts	=	the value of the mixture.

Solve the Equation

$$240 \quad + \quad 3x \quad = \quad 6(20 + x)$$

$240 + 3x = 120 + 6x$ Use the distributive property to remove parentheses.

$120 = 3x$ Subtract $3x$ and 120 from both sides.

$40 = x$ Divide both sides by 3.

State the Conclusion   The store owner should mix 40 pounds of peanuts with the 20 pounds of cashews.

Check the Result   The cashews are valued at $\$12(20) = \$240$.
The peanuts are valued at $\$3(40) = \$120$.
The mixture is valued at $\$6(60) = \$360$.

The value of the cashews plus the value of the peanuts equals the value of the mixture. ■

EXAMPLE 4   **Milk production**   A container is partially filled with 12 liters of whole milk containing 4% butterfat. How much 1% milk must be added to get a mixture that is 2% butterfat?

Analyze the Problem   If the first container shown in Figure 1-25 contains 12 liters of 4% milk, it contains $0.04(12)$ liters of butterfat. To this container, we will add the contents of the second container, which holds $0.01l$ liters of butterfat.

    The sum of these two amounts of butterfat $(0.04(12) + 0.01l)$ will be the amount of butterfat in the third container, which is $0.02(12 + l)$ liters of butterfat.

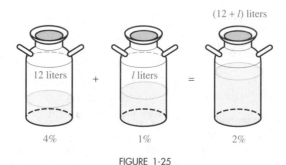

FIGURE 1-25

Form an Equation   We can let l represent the number of liters of 1% milk to be added. Then

The amount of butterfat in 12 liters of 4% milk	+	the amount of butterfat in l liters of 1% milk	=	the amount of butterfat in $(12 + l)$ liters of 2% mixture.

Solve the Equation

$$0.04(12) \quad + \quad 0.01l \quad = \quad 0.02(12 + l)$$

$4(12) + 1l = 2(12 + l)$ Multiply both sides by 100.

$48 + l = 24 + 2l$ Use the distributive property to remove parentheses.

$24 = l$ Subtract 24 and l from both sides.

State the Conclusion   Thus, 24 liters of 1% milk should be added to get a mixture that is 2% butterfat.

Check the Result   12 liters of 4% milk contains 0.48 liters of butterfat.
24 liters of 1% milk contains 0.24 liters of butterfat.
This gives a total of 36 liters of a mixture that contains 0.72 liters of butterfat, which is a 2% solution. ∎

Orals   *Assume all investments are for one year.*

1. How much interest will $1500 earn if invested at 6%?

2. How much interest will $x earn if invested at 5%?

3. If $x of $30,000 is invested at 5%, how much is left to be invested at 6%?

4. If Brazil nuts are worth $x per pound, how much will 20 pounds be worth?

5. If whole milk is 4% butterfat, how much butterfat is in 2 gallons?

EXERCISE 1.7

In Exercises 1–30, solve each problem.

1. Investing money   Lured by the advertisement in Illustration 1, a woman invested $12,000, some in a money market account and the rest in a 5-year CD. How much was invested in each account if the income from both investments is $1060 per year?

> **First Republic Savings and Loan**
>
> | Account | Rate |
> |---------|------|
> | NOW | 5.5% |
> | Savings | 7.5% |
> | Money market | 8.0% |
> | Checking | 4.0% |
> | 5-year CD | 9.0% |

ILLUSTRATION 1

2. Investing money   A man invested $14,000, some at 7% and some at 10% annual interest. The annual income from these investments was $1280. How much did he invest at each rate?

3. Supplemental income   A teacher wants to earn $1500 per year in supplemental income from a cash gift of $16,000. She puts $6000 in a credit union that pays 7% annual interest. What rate must she earn on the remainder to achieve her goal?

4. Inheriting money   Paul split an inheritance between two investments, one paying 7% annual interest and the other 10%. He invested twice as much in the 10% investment as he did in the 7% investment. If his combined annual income from the two investments was $4050, how much did he inherit?

5. Investing money   Maria has some money to invest. If she could invest $3000 more, she could qualify for an 11% investment. Otherwise, she could invest the money at 7.5% annual interest. If the 11% investment would yield twice as much annual income as the 7.5% investment, how much does she have on hand to invest?

6. Supplemental income   A bus driver wants to earn $3500 per year in supplemental income from an inheritance of $40,000. If the driver invests $10,000 in a mutual fund paying 8%, what rate must he earn on the remainder to achieve his goal?

7. Concert receipts   For a jazz concert, student tickets were $2 each and adult tickets were $4 each. If 200 tickets were sold and the total receipts were $750, how many student tickets were sold?

8. School play   At a school play, 140 tickets were sold, with total receipts of $290. If adult tickets cost $2.50 each and student tickets cost $1.50 each, how many adult tickets were sold?

9. **Computing time** One car leaves Chicago headed for Cleveland, a distance of 343 miles. At the same time, a second car leaves Cleveland headed toward Chicago. If the first car averages 50 miles per hour and the second car averages 48 miles per hour, how long will it take the cars to meet?

10. **Cycling** A cyclist leaves Las Vegas riding at the rate of 18 miles per hour. One hour later, a car leaves Las Vegas going 45 miles per hour in the same direction. How long will it take the car to overtake the cyclist?

11. **Computing distance** At 2 P.M., two cars leave Eagle River, WI, one headed north and one headed south. If the car headed north averages 50 miles per hour and the car headed south averages 60 miles per hour, when will the cars be 165 miles apart?

12. **Running a marathon race** Two marathon runners leave the starting gate, one running 12 miles per hour and the other 10 miles per hour. If they maintain the pace, how long will it take for them to be one-quarter of a mile apart?

13. **Riding a jet ski** A jet ski can go 12 miles per hour in still water. If a rider goes upstream for 3 hours against a current of 4 miles per hour, how long will it take the rider to return? (*Hint:* Upstream speed is $(12 - 4)$ mph; how far can the rider go in 3 hours?)

14. **Taking a walk** Sarah walked north at the rate of 3 miles per hour and returned at the rate of 4 miles per hour. How many miles did she walk if the round trip took 3.5 hours?

15. **Computing travel time** Grant traveled a distance of 400 miles in 8 hours. Part of the time, his rate of speed was 45 miles per hour; the rest of the time, his rate of speed was 55 miles per hour. How long did Grant travel at each rate?

16. **Riding a motorboat** The motorboat in Illustration 2 can go 18 miles per hour in still water. If it can make a trip downstream in 4 hours and it takes 5 hours to return, find the speed of the current.

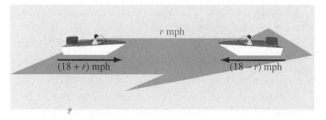

ILLUSTRATION 2

17. **Mixing candies** The owner of a candy store wants to make a 30-pound mixture of two candies to sell for $1 per pound. If one candy sells for 95¢ per pound and the other for $1.10 per pound, how many pounds of each should be used?

18. **Computing selling price** A mixture of candy is made to sell for 89¢ per pound. If 32 pounds of a cheaper candy, selling for 80¢ per pound, are used along with 12 pounds of a more expensive candy, find the price per pound of the better candy.

19. **Diluting solutions** In Illustration 3, how much water should be added to 20 ounces of a 15% solution of alcohol to dilute it to a 10% solution?

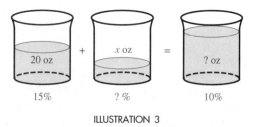

ILLUSTRATION 3

20. **Increasing concentration** How much water must be boiled away to increase the concentration of 300 gallons of a salt solution from 2% to 3%?

21. **Making whole milk** Cream is approximately 22% butterfat. How many gallons of cream must be mixed with milk testing at 2% butterfat to get 20 gallons of milk containing 4% butterfat?

22. **Mixing solutions** How much acid must be added to 60 grams of a solution that is 65% acid to obtain a new solution that is 75% acid?

23. **Raising grades** A student had a score of 70% on a test that contained 30 questions. To improve his score, the instructor agreed to let him work 15 additional questions. How many must he get right to raise his grade to 80%?

24. **Raising grades** On a second exam, the student in Exercise 23 earned a score of 60% on a 20-question test. This time, the instructor allowed him to work 20 extra problems to improve his score. How many must he get right to raise his grade to 70%?

25. **Computing grades** Before the final, Maria had earned a total of 375 points on four tests. To receive an A in the course, she must have 90% of a possible total of 450 points. Find the lowest number of points that she can earn on the final and still receive an A.

26. Computing grades A student has earned a total of 435 points on five algebra tests. To receive a B in the course, he must have 80% of a possible total of 600 points. Find the lowest number of points that the student can make on the final and still receive a B.

27. Managing a bookstore A bookstore sells a calculus book for $65. If the bookstore makes a profit of 30% on each book, what does the bookstore pay the publisher for each book? (*Hint:* The retail price = the wholesale price + the markup.)

28. Managing a bookstore A bookstore sells a textbook for $39.20. If the bookstore makes a profit of 40% on each sale, what does the bookstore pay the publisher for each book? (*Hint:* The retail price = the wholesale price + the markup.)

29. Making furniture A woodworker wants to put two partitions crosswise in a drawer that is 28 inches

deep. (See Illustration 4.) He wants to place the partitions so that the spaces created increase by 3 inches from front to back. If the thickness of each partition is $\frac{1}{2}$ inch, how far from the front end should he place the first partition?

30. Building shelves A carpenter wants to put four shelves on an 8-foot wall so that the five spaces created decrease by 6 inches as we move up the wall. (See Illustration 5.) If the thickness of each shelf is $\frac{3}{4}$ inch, how far will the bottom shelf be from the floor?

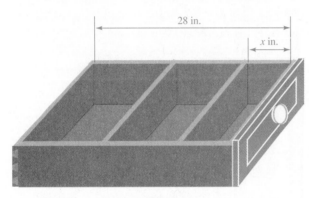

ILLUSTRATION 4

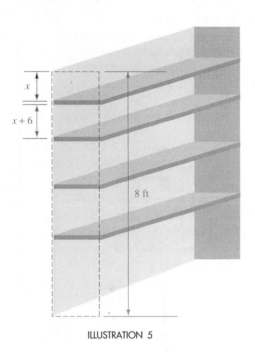

ILLUSTRATION 5

Writing Exercises *Write a paragraph using your own words.*

1. What do you find most difficult in solving application problems, and why?

2. Which type of application problem do you find easiest, and why?

Something to Think About **1.** Discuss the difficulties in solving this problem:
A man drives 100 miles at 30 miles per hour. How fast should he drive on the return trip to average 60 mph for the entire trip?

2. What difficulties do you encounter when solving this problem:
Adult tickets cost $4; student tickets cost $2. Sales of 71 tickets brought in $245. How many of each were sold?

Review Exercises *Solve each equation.*

1. $9x - 3 = 6x$

2. $7a + 2 = 12 - 4(a - 3)$

3. $\dfrac{8(y - 5)}{3} = 2(y - 4)$

4. $\dfrac{t - 1}{3} = \dfrac{t + 2}{6} + 2$

■ ■ ■ ■ ■ ■ ■ ■ ■   **PROBLEMS AND PROJECTS**

1. You are the production manager at Mt. Desert Little Theater. Opening night for the play, "The Great Texas Tuna," was a sellout. The next morning, you counted the receipts for the opening night. For the 110-seat theater, the play grossed $512 from senior-citizen tickets costing $3 each and all other tickets costing $5 each. The production must pay a 7% royalty on the sales of the non-senior-citizen tickets. How much do you owe in royalties for the opening night:

2. A chemist needed the molecular weight (in g/mole) of a solid. From the lab manuals, the chemist was able to ascertain that a third and a half of the molecular weight together was 90 g/mole. Find the molecular weight of the solid.

3. Most of the time, the starting weight or volume of a drug is printed on the bottle, but in the experimental unit of GTHM, someone failed to note the starting volume of Serum V. It was known that the first dosage was $\frac{1}{7}$ of the original amount, and the second was $\frac{1}{3}$ of what was left. After the two doses were administered, there was 24cc of the serum left. This experimental drug is billed to the NSF at $35 per cc. How much did each dose cost?

4. Archimedes stated that if you were given a long enough lever, you could move the world. You have contracted to build a mountain cabin. A large rock on the construction site must be removed. To move the rock, a 15-foot lever with the fulcrum located 3 feet from the rock and a combined weight of 350 pounds is needed. The truck you are going to use to carry the rock away needs to be able to carry at least how much weight?

5. A landscape architect comes into your sheet metal shop to have an outdoor flower box constructed. The architect has the following information concerning his client's desires. The rectangular box has to be 12 feet long and 5 feet wide and be able to hold 8 cubic yards of organic material. How deep should you make the box?

6. Harvester's Candies and Nuts wanted to make spring baskets. After making the baskets, they had 27 pounds of leftover pecan halves, which sell for $6.30 per pound. There were also 25 pounds of leftover cashews, which sell for $8 per pound. To get rid of the leftovers, Harvester's wants to sell mixed bags of pecans and cashews for $7 per one-pound bag. How many ounces of pecans and cashews will there be in each bag? How many bags can be made?

PROJECT 1   You are a member of the Parchdale City Planning Council. The biggest debate in years is taking place over whether or not the town should build an emergency reservoir for use in especially dry periods. The proposed site has room enough for a conical reservoir of diameter 141 feet and depth 85.3 feet.

As long as Parchdale's main water supply is working, the reservoir will be kept full. When a water emergency occurs, however, the reservoir will lose water to evaporation as well as supplying the town with water. The company that has designed the reservoir has given the town the following information.

If D equals the number of consecutive days the reservoir of volume V is used as Parchdale's water supply, then the total amount of water lost to evaporation after D days will be

$$.1V \cdot \left(\frac{D - .7}{D} \right)^2$$

So the volume of water left in the reservoir after it has been used for D days is

$$\text{Volume} = V - (\text{usage per day}) \cdot D - (\text{evaporation})$$

Parchdale uses about 57,500 cubic feet of water per day under emergency conditions. The majority of the council members feel that building the reservoir is a good idea if it will supply the town with water for a week. Otherwise, they will vote against building the reservoir.

a. Calculate the volume of water the reservoir will hold. Remember to use significant digits correctly. Express the answer in scientific notation.

b. Show that the reservoir will supply Parchdale with water for a full week.

c. How much water per day could Parchdale use if the proposed reservoir was to supply the city with water for ten days?

PROJECT 2

Theresa manages the local outlet of Hook n' Slice, a nationwide sporting goods retailer. She recently hired you to help solve some problems. Be sure to provide explanations of how you arrive at your solutions.

a. The company has recently directed that all shipments of golf clubs be sold at a retail price of three times what they cost the company. A shipment of golf clubs arrived yesterday, and they have not yet been labeled. This morning, the main office of Hook n' Slice called to say that the clubs should be sold at 35% off their retail prices. Rather than label each club with its retail price, calculate the sale price, and then relabel the club, Theresa wonders how to go directly from the cost of the club to the sale price. That is, if C is the cost to the company of a certain club, what is the sale price for that club?

 1. Develop a formula and answer this question for her.

 2. Now try out your formula on a club that cost the company $26. Compare your answer with what you find by following the company procedure of first computing the retail price and then reducing that by 35%.

b. All golf bags have been sale-priced at 20% off for the past few weeks. Theresa has all of them in a large rack, marked with a sign that reads "20% off retail price." The company has now directed that this sale price be reduced by 35%. Rather than relabel every bag with its new sale price, Theresa would like to simply change the sign to say "_____% off retail price."

 1. Determine what number Theresa should put in the blank.

 2. Check your answer by computing the *final* sale price for a golf bag with a $100 retail price in two ways:

  • by using the percentage you told Theresa should go on the sign, and
  • by doing the two separate price reductions.

 3. Would the sign read the same if the retail price were first reduced by 35%, and then the sale price were reduced by 20%? Explain.

■ Chapter Summary

Key Words

absolute value (1.1)	half-open interval (1.1)	prime numbers (1.1)
additive inverses (1.2)	identity (1.5)	product (1.2)
algebraic expressions (1.2)	impossible equation (1.5)	quotient (1.2)
algebraic terms (1.5)	integers (1.1)	rational numbers (1.1)
base (1.3)	intersection of sets (1.1)	real numbers (1.1)
base angles (1.6)	intervals (1.1)	reciprocal (1.2)
closed interval (1.1)	irrational numbers (1.1)	root of an equation (1.5)
composite numbers (1.1)	isosceles triangle (1.6)	roster method (1.1)
conditional equations (1.5)	like terms (1.5)	scientific notation (1.4)
constants (1.5)	linear equations (1.5)	set (1.1)
contradiction (1.5)	markdown (1.6)	set-builder notation (1.1)
coordinate (1.1)	mean (1.2)	similar terms (1.5)
difference (1.2)	median (1.2)	solution of an equation (1.5)
element of a set (1.1)	mode (1.2)	solution set (1.5)
ellipsis (1.1)	multiplicative inverses (1.2)	subset (1.1)
empty set (1.1)	natural numbers (1.1)	sum (1.2)
equation (1.5)	negative numbers (1.1)	transitive property (1.1)
equivalent equations (1.5)	negatives (1.2)	trichotomy property (1.1)
even integers (1.1)	numerical coefficients (1.5)	union of sets (1.1)
exponent (1.3)	odd integers (1.1)	variables (1.1)
extended distributive	open interval (1.1)	vertex angle (1.6)
property (1.2)	opposites (1.2)	whole numbers (1.1)
fulcrum (1.6)	positive numbers (1.1)	
graph (1.1)	power of x (1.3)	

Key Ideas

(1.1) Two sets are equal if they have the same elements.

Natural numbers:	$1, 2, 3, 4, 5, \ldots$
Whole numbers:	$0, 1, 2, 3, 4, 5, \ldots$
Integers:	$\ldots, -4, -3, -2, -1, 0, 1,$ $2, 3, 4, \ldots$
Even integers:	$\ldots, -6, -4, -2, 0, 2, 4, 6, \ldots$
Odd integers:	$\ldots, -5, -3, -1, 1, 3, 5, \ldots$
Prime numbers:	$2, 3, 5, 7, 11, 13, \ldots$
Composite numbers:	$4, 6, 8, 9, 10, 12, \ldots$
Rational numbers:	numbers that can be written as $\frac{a}{b}$ $(b \neq 0)$, where a and b are integers

Irrational numbers: numbers that can be written as nonterminating, nonrepeating decimals

Real numbers: numbers that can be written as decimals

Properties of inequality:
$a < b$, $a = b$, or $a > b$ (trichotomy property)
If $a < b$ and $b < c$, then $a < c$.   (transitive property)

Absolute value: $\begin{cases} \text{If } x \geq 0, \text{ then } |x| = x. \\ \text{If } x < 0, \text{ then } |x| = -x. \end{cases}$

(1.2) Adding and subtracting real numbers:
With like signs: Add their absolute values and keep the same sign.
With unlike signs: Subtract their absolute values and keep the sign of the number with the greater absolute value.
$x - y$ is equivalent to $x + (-y)$.

Multiplying and dividing real numbers:
With like signs: Multiply (or divide) their absolute values. The sign is positive.
With unlike signs: Multiply (or divide) their absolute values. The sign is negative.
If a is any nonzero number, $\frac{a}{0}$ is undefined; $\frac{0}{0}$ is indeterminate.

Properties of equality:
Reflexive property: $a = a$
Symmetric property: If $a = b$, then $b = a$.
Transitive property: If $a = b$ and $b = c$, then $a = c$.
Substitution property: If $a = b$, then b can be substituted for a in any expression to get an equivalent expression.

Order of operations:

1. Unless parentheses indicate otherwise, do multiplications and/or divisions first, in order from left to right.

2. Then do the additions and/or subtractions, from left to right.

3. In a fraction, simplify the numerator and denominator separately. Then simplify the fraction, if possible.

Closure properties:
$a + b$ is a real number.
$a - b$ is a real number.
ab is a real number.
$\frac{a}{b}$ $(b \neq 0)$ is a real number.

Associative properties:
$(a + b) + c = a + (b + c)$
$(ab)c = a(bc)$

Commutative properties:
$a + b = b + a$
$ab = ba$

Distributive property:
$a(b + c) = ab + ac$

0 is the **additive identity:**
$a + 0 = 0 + a = a$

1 is the **multiplicative identity:**
$a \cdot 1 = 1 \cdot a = a$

$-a$ is the **negative** (or **additive inverse**) of a:
$a + (-a) = 0$

Double negative rule:
$-(-a) = a$

If $a \neq 0$, then $\frac{1}{a}$ is the **reciprocal** (or **multiplicative inverse**) of a:
$a\left(\frac{1}{a}\right) = 1$

(1.3) Properties of exponents: If there are no divisions by 0, then

$$\overbrace{x^n = x \cdot x \cdot x \cdot \cdots \cdot x}^{n \text{ factors of } x}$$

$$x^m x^n = x^{m+n} \qquad (x^m)^n = x^{mn} \qquad (xy)^n = x^n y^n \qquad \left(\frac{x}{y}\right)^n = \frac{x^n}{y^n}$$

$$x^0 = 1 \qquad x^{-n} = \frac{1}{x^n} \qquad \frac{x^m}{x^n} = x^{m-n} \qquad \left(\frac{x}{y}\right)^{-n} = \left(\frac{y}{x}\right)^n$$

(1.4) Scientific notation: $N \times 10^n$, where $1 \leq |N| < 10$ and n is an integer.

(1.5) If a and b are real numbers and $a = b$, then
$$a + c = b + c, \qquad a - c = b - c,$$
$$ac = bc, \qquad \frac{a}{c} = \frac{b}{c} \quad (c \neq 0)$$

(1.6–1.7) To solve a word problem, express a quantity in two ways to form an equation. Then solve the equation.

Chapter 1 Review Exercises

In Exercises 1–8, $A = \{1, 2, 4, 6, 8, 9\}$, $B = \{1, 2, 4, 9\}$, and $C = \{3, 5, 7, 9\}$. Tell whether each statement is true. If a statement is false, rewrite it to make it true.

1. $4 \in A$

2. $B \subseteq C$

3. $B \in A$

4. $\{3, 7\} \subseteq C$

5. $\emptyset \subseteq B$

6. $\emptyset \in C$

7. $\{2, 8\} \subseteq A$

8. $\{4, 8\} \in A$

In Exercises 9–12, assume that A = {1, 3, 5} and B = {4, 5, 6}. Find each set.

9. $A \cup B$ **10.** $A \cap B$ **11.** $A \cup (B \cap \emptyset)$ **12.** $(A \cup \emptyset) \cap B$

In Exercises 13–24, list the elements in the set $\left\{-4, -\frac{2}{3}, 0, 1, 2, \pi, 4\right\}$ *that satisfy the given condition.*

13. whole number   **14.** natural number   **15.** rational number   **16.** integer

17. irrational number   **18.** real number   **19.** negative number   **20.** positive number

21. prime number   **22.** composite number   **23.** even integer   **24.** odd integer

25. Graph the set of prime numbers between 20 and 30.   **26.** Graph the set of composite numbers between 5 and 13.

In Exercises 27–34, graph each interval on the number line.

27. $x \geq -4$ **28.** $-2 < x \leq 6$ **29.** $(-2, 3)$ **30.** $[2, 6]$

31. $(2, \infty)$ **32.** $(-\infty, -1)$ **33.** $(-2, 4] \cup (0, 6)$ **34.** $(-2, 4] \cap (0, 6)$

In Exercises 35–38, write each expression without using absolute value symbols.

35. $|0|$ **36.** $|-1|$ **37.** $|8|$ **38.** $-|8|$

In Exercises 39–54, do the operations and simplify when possible.

39. $3 + (-5)$ **40.** $5 - 3$ **41.** $-2 + 5$ **42.** $-3 - 5$

43. $-8 - (-3)$ **44.** $7 - (-9)$ **45.** $4(-3)$ **46.** $-3(8)$

47. $-4(3 - 6)$ **48.** $3[8 - (-1)]$ **49.** $\dfrac{-8}{2}$ **50.** $\dfrac{8}{-4}$

51. $\dfrac{-16}{-4}$ **52.** $\dfrac{-25}{-5}$ **53.** $\dfrac{3 - 8}{10 - 5}$ **54.** $\dfrac{-32 - 8}{6 - 16}$

In Exercises 55–58, consider the numbers 12, 13, 14, 14, 15, 15, 15, 17, 19, 20.

55. ▦ Find the mean.   **56.** Find the median.

57. Find the mode.   **58.** Could the mean, median, and mode of a group of numbers be the same?

In Exercises 59–62, a = 5, b = -2, c = -3, and d = 2. Simplify each expression.

59. $\dfrac{3a - 2b}{cd}$ **60.** $\dfrac{3b + 2d}{ac}$ **61.** $\dfrac{ab + cd}{c(b - d)}$ **62.** $\dfrac{ac - bd}{a(d + c)}$

In Exercises 63–74, tell which property of equality or property of real numbers justifies each statement.

63. $3(4 + 2) = 3 \cdot 4 + 3 \cdot 2$ **64.** If $3 = 2 + 1$, then $2 + 1 = 3$.

65. $3 + (x + 7) = (x + 7) + 3$ **66.** $3 + (x + 7) = (3 + x) + 7$

67. $3 + 0 = 3$

68. $3 + (-3) = 0$

69. $xy = xy$

70. $5(3) = 3(5)$

71. $3(xy) = (3x)y$

72. $3x \cdot 1 = 3x$

73. $a\left(\dfrac{1}{a}\right) = 1 \ (a \neq 0)$

74. If $x = 7$ and $7 = y$, then $x = y$.

In Exercises 75–102, simplify each expression. Write all answers without using negative exponents.

75. 3^6

76. -2^6

77. $(-4)^3$

78. -5^{-4}

79. $(3x^4)(-2x^2)$

80. $(-x^5)(3x^3)$

81. $x^{-4}x^3$

82. $x^{-10}x^{12}$

83. $(3x^2)^3$

84. $(4x^4)^4$

85. $(-2x^2)^5$

86. $-(-3x^3)^5$

87. $(x^2)^{-5}$

88. $(x^{-4})^{-5}$

89. $(3x^{-3})^{-2}$

90. $(2x^{-4})^4$

91. $\dfrac{x^6}{x^4}$

92. $\dfrac{x^{12}}{x^7}$

93. $\dfrac{a^7}{a^{12}}$

94. $\dfrac{a^4}{a^7}$

95. $\dfrac{y^{-3}}{y^4}$

96. $\dfrac{y^5}{y^{-4}}$

97. $\dfrac{x^{-5}}{x^{-4}}$

98. $\dfrac{x^{-6}}{x^{-9}}$

99. $(3x^2y^3)^2$

100. $(-3a^3b^2)^{-4}$

101. $\left(\dfrac{3x^2}{4y^3}\right)^{-3}$

102. $\left(\dfrac{4y^{-2}}{5y^{-3}}\right)^3$

In Exercises 103–104, write each numeral in scientific notation.

103. $19{,}300{,}000{,}000$

104. 0.0000000273

In Exercises 105–106, write each numeral in standard notation.

105. 7.2×10^7

106. 8.3×10^{-9}

In Exercises 107–114, solve each equation.

107. $5x + 12 = 37$

108. $-3x - 7 = 20$

109. $4(y - 1) = 28$

110. $3(x + 7) = 42$

111. $13(x - 9) - 2 = 7x - 5$

112. $\dfrac{8(x - 5)}{3} = 2(x - 4)$

113. $\dfrac{3y}{4} - 13 = -\dfrac{y}{3}$

114. $\dfrac{2y}{5} + 5 = \dfrac{14y}{10}$

In Exercises 115–118, solve for the quantity indicated.

115. $V = \dfrac{4}{3}\pi r^3$ for r^3

116. $V = \dfrac{1}{3}\pi r^2 h$ for h

117. $v = \dfrac{1}{6}ab(x + y)$ for x

118. $V = \pi h^2\left(r - \dfrac{h}{3}\right)$ for r.

In Exercises 119–124, solve each problem.

119. Carpentry   A carpenter wants to cut a 20-foot rafter so that one piece is 3 times as long as the other. Where should he cut the board?

120. Geometry   A rectangle is 4 meters longer than it is wide. If the perimeter of the rectangle is 28 meters, find its area.

121. Balancing a seesaw   Sue weighs 48 pounds, and her father weighs 180 pounds. If Sue sits on one end of a 20-foot long seesaw with the fulcrum in the middle, how far from the fulcrum should her father sit to balance the seesaw?

122. Investment problem Sally has $25,000 to invest. She invests some money at 10% interest and the rest at 9%. If her total annual income from these two investments is $2430, how much does she invest at each rate?

123. Mixing solutions How much water must be added to 20 liters of a 12% alcohol solution to dilute it to an 8% solution?

124. Motion problem A car and a motorcycle both leave from the same point and travel in the same direction. (See Illustration 1.) The car travels at an average rate of 55 miles per hour, and the motorcycle at an average rate of 40 miles per hour. How long will it take before the vehicles are 5 miles apart?

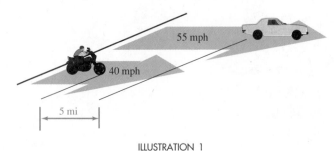

ILLUSTRATION 1

■ Chapter 1 Test

In Problems 1–4, A = {1, 2, 3, 4}, B = {3, 4, 5, 6}, and C = {5, 6, 7, 8}. Insert either a ∈ or a ⊆ symbol to make a true statement.

1. 3   *A*   **2.** ∅   *B*   **3.** *A* ∩ *B*   *B*   **4.** {7, 8}   *C*

In Problems 5–6, A = {1, 2, 3, 4}, B = {3, 4, 5, 6}, and C = {5, 6, 7, 8}. Find each set.

5. *A* ∪ *B*

6. (*A* ∪ ∅) ∩ *C*

In Problems 7–8, let A = $\left\{-2, 0, 1, \frac{6}{5}, 2, \sqrt{7}, 5\right\}$.

all kinds

7. What numbers in *A* are natural numbers?

8. What numbers in *A* are irrational numbers?

In Problems 9–10, graph each set on the number line.

9. The set of odd integers from −4 to 6.

10. The set of prime numbers less than 12.

In Problems 11–12, graph each set on the number line.

11. [−2, 4)

12. [−1, 5] ∩ [−2, 3)

In Problems 13–14, write each expression without using absolute value symbols.

13. −|8|

14. |−5|

In Problems 15–18, do the operations.

15. 7 + (−5)   **16.** −5(−4)   **17.** $\dfrac{12}{-3}$ **18.** $-4 - \dfrac{-15}{3}$

In Problems 19–20, consider the numbers −2, 0, 2, −2, 3, −1, −1, 1, 1, 2.

19. Find the mean.

20. Find the median.

In Problems 21–24, a = 2, b = −3, and c = 4. Evaluate each expression.

21. *ab*   **22.** *a* + *bc*   **23.** *ab* − *bc*   **24.** $\dfrac{-3b + a}{ac - b}$

In Problems 25–26, tell which property of real numbers justifies each statement.

25. $3 + 5 = 5 + 3$

26. $a(b + c) = ab + ac$

In Problems 27–30, simplify each expression. Write all answers without using negative exponents. Assume that no denominators are zero.

27. $x^3 x^5$

28. $(x^2 y^3)^3$

29. $(m^{-4})^2$

30. $\left(\dfrac{m^2 n^3}{m^4 n^{-2}} \right)^{-2}$

In Problems 31–32, write each numeral in scientific notation.

31. $4{,}700{,}000$

32. 0.00000023

In Problems 33–34, write each numeral in standard notation.

33. 6.53×10^5

34. 24.5×10^{-3}

In Problems 35–36, solve each equation.

35. $9(x + 4) + 4 = 4(x - 5)$

36. $\dfrac{y - 1}{5} + 2 = \dfrac{2y - 3}{3}$

37. Solve $P = L + \dfrac{s}{f}i$ for i.

38. A rectangle has a perimeter of 26 centimeters and is 5 centimeters longer than it is wide. Find its area.

39. Bob invests part of $10,000 at 9% annual interest and the rest at 8%. If his annual income from these investments is $860, how much does he invest at 8%?

40. How many liters of water are needed to dilute 20 liters of a 5% salt solution to a 1% solution?

Graphs,
Equations of Lines,
and Functions

2

Economist

Economists study the way a society uses resources such as land, labor, raw materials, and machinery to provide goods and services. Some economists are theoreticians who use mathematical models to explain the causes of recession and inflation. Most economists, however, are concerned with practical applications of economic policy in a particular area.

SAMPLE APPLICATION ■ An electronics firm manufactures tape recorders, receiving $120 for each recorder it makes. If x represents the number of recorders produced, the income received is determined by the *revenue function*

$$R(x) = 120x$$

The manufacturer has fixed costs of $12,000 per month and variable costs of $57.50 for each tape recorder manufactured. Thus, the *cost function* is

$$C(x) = \text{variable costs} + \text{fixed costs}$$
$$= 57.50x + 12,000$$

How many recorders must the company sell for revenue to equal cost? See Exercise 69 in Exercise 2.5.

2.1 Tables and Graphs

■ READING DATA FROM TABLES ■ READING BAR GRAPHS ■ READING PIE GRAPHS ■ READING LINE GRAPHS ■ ACCENT ON STATISTICS

Business and statistical information is often presented in the form of tables or graphs. For example, the *table*, *bar graph*, and *pie graph* shown in Figure 2-1 show how a worker spent $3000 in monthly income during the month of March. In the pie graph, the size of each region represents the percent of $3000 that was spent in each category.

It is easy to see from any part of Figure 2-1 that the largest amount of money was spent for housing, and that the least amount of money was spent on entertainment and on savings. However, this information is most easily seen in the graphs.

Category	Money spent
Housing	$900
Food	$750
Clothing	$450
Transportation	$600
Entertainment	$150
Savings	$150

(a)

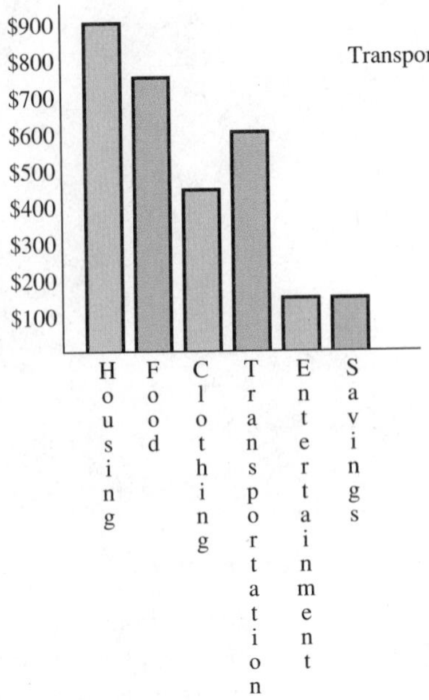

(b)

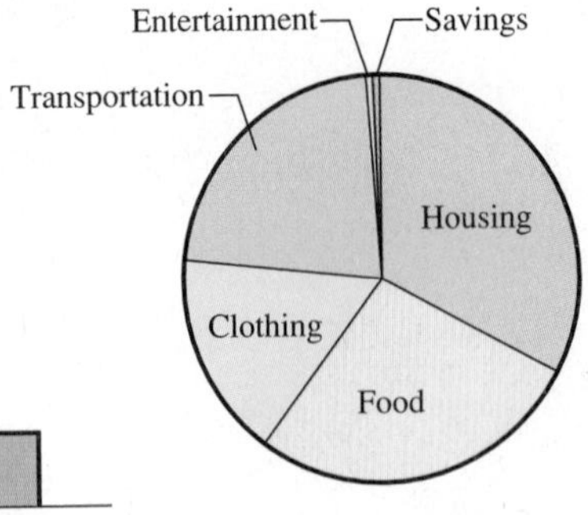

(c)

Distribution of monthly income for March

FIGURE 2-1

■ READING DATA FROM TABLES

EXAMPLE 1 Find the cost of sending a $10\frac{1}{2}$-pound package by priority mail to a person living in zone 6.

Solution Postal rates for priority mail are shown in Figure 2-2.

U.S. Priority Mail Weight, up to but not exceeding___pound(s)	Local 1, 2, and 3	Zones				
		4	5	6	7	8
1	3.00	3.00	3.00	3.00	3.00	3.00
2	3.00	3.00	3.00	3.00	3.00	3.00
3	4.00	4.00	4.00	4.00	4.00	4.00
4	5.00	5.00	5.00	5.00	5.00	5.00
5	6.00	6.00	6.00	6.00	6.00	6.00
6	6.35	6.90	7.10	7.20	7.80	8.00
7	6.65	7.80	8.10	8.40	9.20	9.80
8	6.95	8.70	9.05	9.50	10.40	11.60
9	7.40	9.35	10.00	10.60	11.30	13.00
10	7.85	10.00	10.75	11.40	12.15	14.05
11	8.25	10.65	11.45	12.20	13.00	15.10
12	8.70	11.30	12.20	13.00	13.90	16.15

FIGURE 2-2

To find the cost of sending the $10\frac{1}{2}$-pound package, we find the row in the table for a package that does not exceed 11 pounds and the column for zone 6. Since the intersection of this row and column occurs at 12.20, it will cost $12.20 to send the package. ■

Many magazines and newspapers, such as *USA TODAY*, use graphs to present statistical information.

■ READING BAR GRAPHS

EXAMPLE 2   Figure 2-3 shows the applications programs that are most frequently used in a household with a computer. Refer to the graph and answer the following questions.

a. Which computer application is most frequently used?

b. What fraction of U.S. households now have a computer?

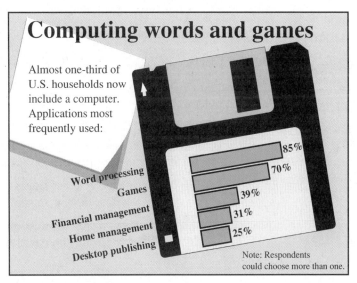

FIGURE 2-3
Copyright, *USA Today.* Reprinted with permission.

Solution   **a.** Since the longest bar follows Word processing, it is the most frequently used application.

b. From a note on the graph, almost one-third of U.S. households now have a computer. ■

EXAMPLE 3   Figure 2-4 shows the frequency of arrests for different types of traffic violations in two consecutive years. Refer to the graph and answer the following questions.

a. In 1995, how many arrests were made for failure to yield?

b. In what categories did arrests decrease?

c. In what categories did arrests increase?

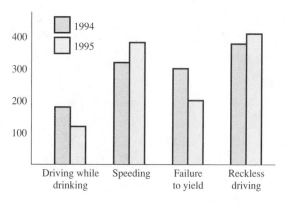

Arrests for moving violations

FIGURE 2-4

d. In 1995, which violation occurred most often?

e. In 1994, which violation occurred least often?

Solution **a.** The third pair of bars shows the arrests for failure to yield. From the key, we see that the blue bar in each pair shows the arrests made in 1995. The height of the blue bar, measured against the vertical scale to the left of the graph, is about 200. Approximately 200 arrests were made in 1995 for failure to yield.

b. We look for the categories where the blue bar is shorter than the red bar. This occurs for the categories of Driving while drinking and Failure to yield.

c. We look for the categories where the blue bar is longer than the red bar. This occurs for the categories of Speeding and Reckless driving.

d. The blue bars represent arrests made in 1995. Since the tallest blue bar occurs above Reckless driving, this violation occurred most often.

e. The red bars represent arrests made in 1994. Since the shortest red bar occurs above Driving while drinking, this violation occurred least often. ■

■ READING PIE GRAPHS

EXAMPLE 4 The pie graph in Figure 2-5 shows the various fuels used for generating electricity. Refer to the graph and answer the following questions.

a. What percent of electricity is produced by nuclear power?

b. What percent of electricity is produced by either coal or gas?

c. What percent of electricity is produced by sources other than coal and gas?

Solution **a.** From the red portion of the graph, we see that 21.2% of the electricity is produced by nuclear power.

b. From the blue portion of the graph, we see that 56.9% of the electricity is produced by coal. From the green portion, we see that 9.0% is produced by gas. The amount produced by coal or gas is the sum of these two amounts:

$$56.9\% + 9.0\% = 65.9\%$$

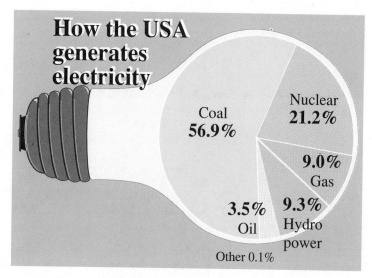

FIGURE 2-5
Copyright, *USA Today.* Reprinted with permission.

c. To find the amount produced by sources other than coal and gas, we can subtract 65.9% from 100% to get 34.1%. We can obtain the same result by adding
the percents from all of the sources except for coal and gas:

$$21.2\% + 9.3\% + 0.1\% + 3.5\% = 34.1\%$$ ■

■ READING LINE GRAPHS

Line graphs are often used to describe how quantities change with time. From such
a graph, we can determine when a quantity is increasing and when it is decreasing.

EXAMPLE 5
The line graph in Figure 2-6 shows how automobile production has varied in the
years since 1900. Refer to the graph and answer the following questions:

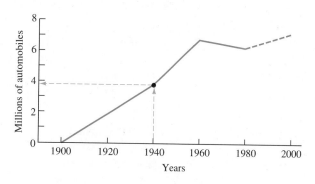

Automobile production

FIGURE 2-6

a. How many cars were made in 1940?

b. How many cars were made in 1950?

c. Over which 20-year span did car production increase most rapidly?

d. When did production decrease?

e. Why is a dashed line used for the portion of the graph between 1980 and 2000?

Solution a. To find the number of cars made in 1940, we find the height of the graph at the point directly above the year 1940. We follow the dashed line from the year 1940 straight up to the graph and then over to the vertical scale. There, we read that approximately 3.7 million cars were made in 1940.

b. To find the number of cars made in 1950, we find the year 1950 on the horizontal line at the bottom of the graph. Although the year 1950 is not shown, it lies halfway between 1940 and 1960. From that point, we move up to the graph and then to the left, where we read that a little more than 5 million cars were made.

c. Between 1940 and 1960 the rise of the graph is the greatest. During those years, the production of cars increased most rapidly.

d. The graph falls between the years 1960 and 1980, indicating that car production decreased between 1960 and 1980.

e. Because the year 2000 is still in the future, the production levels indicated by the graph are only a projection, and the dashed line shows that the numbers are not certain. The projected manufacturing level in the year 2000 is 7.2 million cars, but that number may turn out to be wrong. ■

EXAMPLE 6   The line graph in Figure 2-7 shows the annual earnings of three automobile manufacturers: General Motors, Ford, and Chrysler. From the graph, answer the following questions:

a. How much money did General Motors make in 1982?

b. Which company lost the most money in 1990?

c. In what years did the earnings of Ford Motor Corporation surpass those of General Motors?

d. Which company's earnings increased the most during the years 1984 to 1988?

Solution a. From the key, we see that the earnings of General Motors are represented by the red line. To find the company's earnings in 1982, we move vertically upward from the year 1982 until we reach the red line. We then move to the left until we reach the number 1 on the vertical scale. Because that scale indicates *billions* of dollars, we see that General Motors earned $1 billion in 1982.

b. To find which company lost the most money in 1990, we move upward from the year 1990 until we reach the lowest of the three graphs. Since that graph is a red line, which represents the earnings of General Motors, General Motors lost the most money, about $2 billion.

c. From the key, Ford's earnings are represented by the blue line, and General Motors' earnings are represented by the red line. We then refer to the graph to find which years the blue line is above the red line. This happened in 1986, 1987, and 1988, and again in 1990 and 1991. In those years, Ford's earnings were more than the earnings of General Motors.

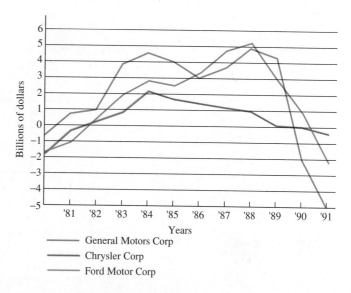

Automobile manufacturers' corporate earnings

FIGURE 2-7

d. We find the years 1984 and 1988 on the horizontal axis and see which of the three graphs has risen the most during that time. Chrysler's earnings (the green line) steadily decreased during that time, and General Motors' earnings increased slightly. The earnings of Ford Motor Corporation increased the most, from approximately $3 billion in 1984 to more than $5 billion in 1988. ■

ACCENT ON STATISTICS

Frequency Distributions

In statistics a *distribution* is any set of values, and the *frequency* of each value is the number of times it occurs. For example, suppose we wish to learn about the course loads taken by students at Rock Valley College during the spring semester. To estimate the course loads, we survey 20 students and ask how many courses they are taking, with the following results:

2 courses, 5 courses, 3 courses, 4 courses, 4 courses,
1 course, 4 courses, 3 courses, 2 courses, 5 courses,
3 courses, 4 courses, 3 courses, 1 course, 2 courses,
2 courses, 4 courses, 4 courses, 5 courses, 3 courses

To make the information easier to read at a glance, we can organize the information in the frequency table shown in Figure 2-8(a) and use the information in the table to construct the bar graph shown in Figure 2-8(b). Notice how much easier it is to read the information from the table or graph than from the above list.

In statistics, a *discrete variable* is a variable that has only whole numbers for its values. If x represents the number of courses taken by a student during the spring semester, x can only be a whole number such as 1, 2, 3, and so on. Thus, x is a discrete variable. A bar graph is the most convenient graph to depict a distribution of values of a discrete variable.

Course Load	Frequency
One course	2
Two courses	4
Three courses	5
Four courses	6
Five courses	3

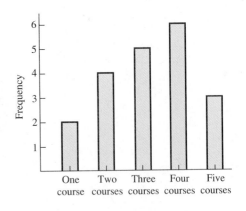

(a)   (b)

FIGURE 2-8

EXERCISE 2.1

In Exercises 1–2, refer to the table in Figure 2-2 on page 82.

1. Find the cost of sending a package weighing $5\frac{1}{2}$ pounds to zone 4.

2. Find the cost of sending a package weighing 10.4 pounds to zone 8.

1994 Tax Rate Schedules

Schedule X—Use if your filing status is **Single**

If the amount on Form 1040, line 37, is: Over—	*But not over—*	Enter on Form 1040, line 38	*of the amount over—*
$ 0	$ 22,750	... 15%	$ 0
22,750	55,100	$3,412.50 + 28%	22,750
55,100	115,000	12,470.50 + 31%	55,100
115,000	250,000	31,039.50 + 36%	115,000
250,000	...	79,639.50 + 39.6%	250,000

Schedule Y-1—Use if your filing status is **Married filing jointly** or **Qualifying widow(er)**

If the amount on Form 1040, line 37, is: Over—	*But not over—*	Enter on Form 1040, line 38	*of the amount over—*
$ 0	$ 38,000	... 15%	$ 0
38,000	91,850	$5,700.00 + 28%	38,000
91,850	140,000	20,778.00 + 31%	91,850
140,000	250,000	35,704.50 + 36%	140,000
250,000	...	75,304.50 + 39.6%	250,000

ILLUSTRATION 1

In Exercises 3–6, refer to the tax tables in Illustration 1.

3. A student with a taxable income of $27,600 is married and files jointly. Compute his tax.

4. A woman with a taxable income of $79,250 is single. Compute her tax.

5. A man with a taxable income of $53,000 is single. If he gets married, he will gain more deductions, which will reduce his taxable income by $2000. If he gets married and files a joint return, how much will he save in taxes?

6. Carlos has a taxable income of $53,000 and his wife has a taxable income of $75,000. If they file jointly, what is their tax?

In Exercises 7–10, refer to Illustration 2.

7. What is the most dangerous occupation?

8. How many truck drivers are injured each year?

9. What percent of injuries in the total work force are suffered by nursing aides and orderlies?

10. What percent of injuries in the total work force are suffered by construction laborers?

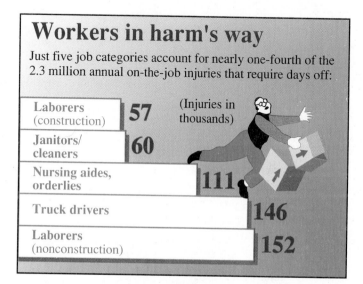

ILLUSTRATION 2
Copyright, *USA Today.* Reprinted with permission.

In Exercises 11–14, refer to Illustration 3.

11. The world production of lead in 1970 was about equal to the production of zinc in another year. What year was that?

12. In what year was the production of zinc about two-thirds the production of lead?

13. In what year was the production of zinc over twice that of lead?

14. By how many metric tons did the production of zinc decrease between 1970 and 1990?

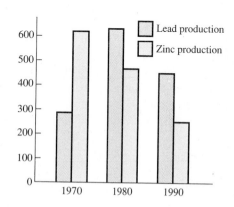

World lead and zinc production, in thousands of metric tons

ILLUSTRATION 3

In Exercises 15–18, refer to Illustration 4.

15. What percent of passengers pick a flight because of cost?
16. What percent of passengers did not respond to the survey?
17. Assuming 250,000 flights, how many would have been chosen primarily because they were considered to be safe?

18. Assuming 350,000 flights, how many would have been chosen primarily because they would earn the passenger frequent flier miles?

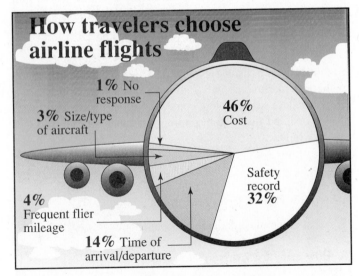

ILLUSTRATION 4
Copyright, *USA Today*. Reprinted with permission.

In Exercises 19–22, refer to Illustration 5.

19. What percent of credit card purchases were made with Visa or MasterCard?
20. What percent of credit card purchases were made with a Discover or an American Express card?
21. What percent of the market do the competitors of Visa have?
22. What percent of credit card purchases were made with a name-brand card?

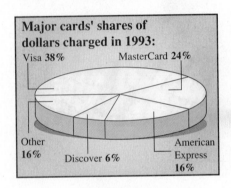

ILLUSTRATION 5
Copyright, *USA Today*. Reprinted with permission.

In Exercises 23–28, refer to Illustration 6.

23. Which runner ran faster at the start of the race?
24. Which runner stopped to rest first?
25. Which runner dropped the baton and had to go back and get it?
26. At what times is runner 1 stopped and runner 2 running?
27. Describe what is happening at time D.

28. Which runner won the race?

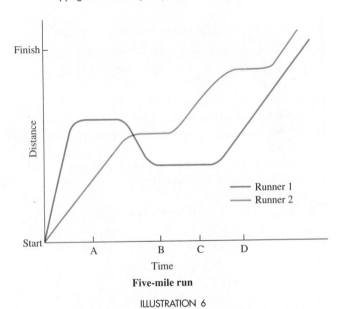

ILLUSTRATION 6

In Exercises 29–30, find the unknown fractional part of each pie graph.

29.

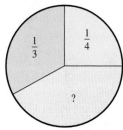

30.

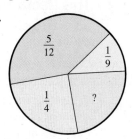

In Exercises 31–32, the marital status of ten faculty members is as follows: single, married, married, widowed, married, divorced, married, divorced, single, divorced.

31. Complete the following frequency table.

Marital status	Frequency
Single	
Married	
Divorced	
Widowed	

32. Use the information in Exercise 31 to construct a bar graph.

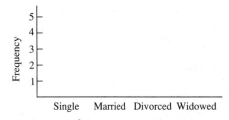

Writing Exercises *Write a paragraph using your own words.*

1. Tell which kind of graph you find easiest to read.

2. Define a discrete variable.

Something to Think About

1. If two people are single, will they always save on taxes if they get married and file jointly? See the tables in Exercises 3–6.

2. If a single man has a taxable income of $80,000, what percent of his income goes to taxes?

Review Exercises *Do the operations.*

1. $4 + 2 \cdot 3^2$

2. $2^2 + 3 \cdot 4$

3. $\dfrac{4(3 + 2) - 5}{12 - 2(3)}$

4. $\dfrac{(4^2 - 3) + 7}{5(2 + 3) - 1}$

5. Write the prime numbers between 10 and 30.

6. Write the first ten composite numbers.

2.2   The Rectangular Coordinate System

■ THE RECTANGULAR COORDINATE SYSTEM ■ GRAPHING LINEAR EQUATIONS ■ GRAPHING LINES PARALLEL TO THE *X*- AND *Y*-AXES ■ APPLICATIONS ■ THE MIDPOINT FORMULA ■ GRAPHING DEVICES

$y = -\frac{1}{2}x + 4$

x	y
-4	6
-2	5
0	4
2	3
4	2

FIGURE 2-9

In algebra, we also present information in tables and graphs. For example, several solutions of the equation $y = -\frac{1}{2}x + 4$ are listed in the table shown in Figure 2-9. Each pair of *x*- and *y*-values in the table is a solution of the equation, because each pair satisfies the equation. To show that the pair $x = 2$ and $y = 3$ satisfies the equation, we substitute 2 for *x* and 3 for *y* into the equation and simplify.

$$y = -\frac{1}{2}x + 4$$

$$3 = -\frac{1}{2}(2) + 4 \qquad \text{Substitute 2 for } x \text{ and 3 for } y.$$

$$3 = -1 + 4$$

$$3 = 3$$

Since the left-hand side equals the right-hand side, the equation is satisfied.

Before we can present the information in the table shown in Figure 2-9 in graphical form, we need to discuss the rectangular coordinate system.

■ THE RECTANGULAR COORDINATE SYSTEM

The **rectangular coordinate system** consists of two perpendicular number lines that divide the plane into four quadrants that are numbered as in Figure 2-10. The horizontal number line is called the **x-axis**, and the vertical number line is called the **y-axis**. The *x*- and *y*-axes intersect at the **origin**, which is the 0 point on each axis. The positive direction on the *x*-axis is to the right, and the positive direction on the *y*-axis is upward.

To plot (or graph) the point associated with the pair $x = 2$ and $y = 3$, denoted by the pair (2, 3), we start at the origin and count 2 units to the right and then 3 units up, as in Figure 2-11. Point *P* (which lies in the first quadrant) is the graph of the pair (2, 3). The pair (2, 3) gives the **coordinates** of point *P*.

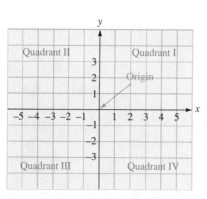

FIGURE 2-10

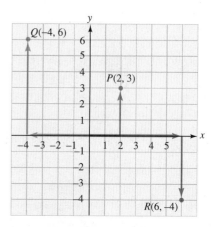

FIGURE 2-11

To plot point Q with coordinates $(-4, 6)$, we start at the origin and count 4 units to the left and then 6 units up. Point Q lies in the second quadrant. Point R with coordinates $(6, -4)$ lies in the fourth quadrant.

 WARNING!   The pairs $(-4, 6)$ and $(6, -4)$ represent different points. The first one is in the second quadrant, and the second one is in the fourth quadrant. Since order is important when graphing pairs of real numbers, such pairs are called **ordered pairs**.

The first coordinate in the ordered pair $(-4, 6)$ is the **x-coordinate**, and the second coordinate is the **y-coordinate**.

■ GRAPHING LINEAR EQUATIONS

The **graph of the equation** $y = -\frac{1}{2}x + 4$ is the graph of all points (x, y) on the rectangular coordinate system whose coordinates satisfy the equation.

EXAMPLE 1   Graph the equation $y = -\dfrac{1}{2}x + 4$.

Solution   To graph the equation, we plot the five ordered pairs listed in the table shown in Figure 2-12. These five points appear to lie on the line shown in the figure. In fact, if we were to plot hundreds of pairs that satisfied the equation, it would become obvious that the resulting points would lie on the line.

When we say the graph of an equation is a line, we imply two things:

1. Every point with coordinates that satisfy the equation will lie on the line.

2. Every point on the line will have coordinates that will satisfy the equation.

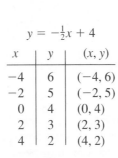

$$y = -\tfrac{1}{2}x + 4$$

x	y	(x, y)
-4	6	$(-4, 6)$
-2	5	$(-2, 5)$
0	4	$(0, 4)$
2	3	$(2, 3)$
4	2	$(4, 2)$

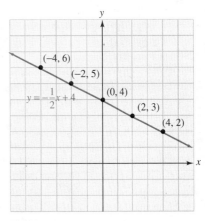

FIGURE 2-12   ■

When the graph of an equation is a line, we call the equation a **linear equation**. Linear equations are often written in the form $Ax + By = C$, where A, B, and C stand for specific numbers (called **constants**) and x and y are variables.

EXAMPLE 2   Graph the equation $3x + 2y = 12$.

Solution   We can pick values for either x or y, substitute them into the equation, and solve for the other variable. For example, if $x = 2$, then we have

$$3x + 2y = 12$$
$$3(2) + 2y = 12 \qquad \text{Substitute 2 for } x.$$
$$6 + 2y = 12 \qquad \text{Simplify.}$$
$$2y = 6 \qquad \text{Subtract 6 from both sides.}$$
$$y = 3 \qquad \text{Divide both sides by 2.}$$

The ordered pair $(2, 3)$ satisfies the equation. If $y = 6$, we have

$$3x + 2y = 12$$
$$3x + 2(6) = 12 \qquad \text{Substitute 6 for } y.$$
$$3x + 12 = 12 \qquad \text{Simplify.}$$
$$3x = 0 \qquad \text{Subtract 12 from both sides.}$$
$$x = 0 \qquad \text{Divide both sides by 3.}$$

A second ordered pair that satisfies the equation is $(0, 6)$.

These pairs and others that satisfy the equation are shown in Figure 2-13. After we plot each pair, we see that they lie on a line. The graph of the equation is the line shown in the figure.

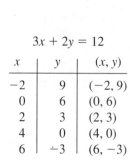

$$3x + 2y = 12$$

x	y	(x, y)
-2	9	$(-2, 9)$
0	6	$(0, 6)$
2	3	$(2, 3)$
4	0	$(4, 0)$
6	-3	$(6, -3)$

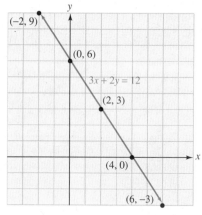

FIGURE 2-13

In Example 2, the graph crossed the y-axis at the point with coordinates $(0, 6)$, called the **y-intercept**, and crossed the x-axis at the point with coordinates $(4, 0)$, called the **x-intercept**. In general, we have the following definitions.

Intercepts of a Line
The **y-intercept** of a line is the point $(0, b)$ where the line intersects the y-axis. To find b, substitute 0 for x in the equation of the line and solve for y.

The **x-intercept** of a line is the point $(a, 0)$ where the line intersects the x-axis. To find a, substitute 0 for y in the equation of the line and solve for x.

EXAMPLE 3   Use the *x*- and *y*-intercepts to graph the equation $2x + 5y = 10$.

Solution   To find the *y*-intercept, we substitute 0 for *x* and solve for *y*:

$$2x + 5y = 10$$
$$2(0) + 5y = 10 \qquad \text{Substitute 0 for } x.$$
$$5y = 10 \qquad \text{Simplify.}$$
$$y = 2 \qquad \text{Divide both sides by 5.}$$

The *y*-intercept is the point $(0, 2)$. To find the *x*-intercept, we substitute 0 for *y* and solve for *x*:

$$2x + 5y = 10$$
$$2x + 5(0) = 10 \qquad \text{Substitute 0 for } y.$$
$$2x = 10 \qquad \text{Simplify.}$$
$$x = 5 \qquad \text{Divide both sides by 2.}$$

The *x*-intercept is the point $(5, 0)$.

Although two points are enough to draw the line, it is a good idea to find and plot a third point as a check. To find the coordinates of a third point, we can substitute any convenient number (such as -5) for *x* and solve for *y*:

$$2x + 5y = 10$$
$$2(-5) + 5y = 10 \qquad \text{Substitute } -5 \text{ for } x.$$
$$-10 + 5y = 10 \qquad \text{Simplify.}$$
$$5y = 20 \qquad \text{Add 10 to both sides.}$$
$$y = 4 \qquad \text{Divide both sides by 5.}$$

The line will also pass through the point $(-5, 4)$.

A table of ordered pairs and the graph of $2x + 5y = 10$ are shown in Figure 2-14.

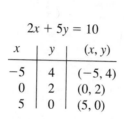

$2x + 5y = 10$

x	y	(x, y)
−5	4	(−5, 4)
0	2	(0, 2)
5	0	(5, 0)

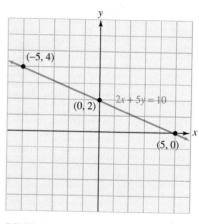

FIGURE 2-14

EXAMPLE 4   Graph the equation $y = 3x + 4$.

Solution   **Find the y-intercept:**   **Find the x-intercept:**

If $x = 0$, then   If $y = 0$, then

$y = 3x + 4$	
$y = 3(0) + 4$	Substitute 0 for x.
$y = 4$	Simplify.

The y-intercept is the point $(0, 4)$.

$y = 3x + 4$	
$0 = 3x + 4$	Substitute 0 for y.
$-4 = 3x$	Subtract 4 from both sides.
$-\dfrac{4}{3} = x$	Divide both sides by 3.

The x-intercept is the point $\left(-\frac{4}{3}, 0\right)$.

To find the coordinates of a third point, we let $x = 1$ and find that the corresponding value of $y = 7$. The point $(1, 7)$ also lies on the graph, as shown in Figure 2-15.

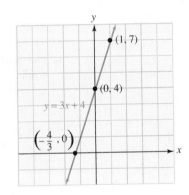

$y = 3x + 4$

x	y	(x, y)
$-\frac{4}{3}$	0	$\left(-\frac{4}{3}, 0\right)$
0	4	$(0, 4)$
1	7	$(1, 7)$

FIGURE 2-15

■ GRAPHING LINES PARALLEL TO THE X- AND Y-AXES

EXAMPLE 5   Graph the equations   **a.** $y = 3$ and   **b.** $x = -2$.

Solution   **a.** Since the equation $y = 3$ does not contain x, the numbers chosen for x have no effect on y. The value of y is always 3.

After plotting the pairs (x, y) shown in Figure 2-16(a), we see that the graph is a horizontal line, parallel to the x-axis, with a y-intercept of $(0, 3)$. The line has no x-intercept.

b. Since the equation $x = -2$ does not contain y, the value of y can be any number.

After plotting the pairs (x, y) shown in Figure 2-16(b), we see that the graph is a vertical line, parallel to the y-axis, with an x-intercept of $(-2, 0)$. The line has no y-intercept.

$y = 3$

x	y	(x, y)
-3	3	$(-3, 3)$
0	3	$(0, 3)$
2	3	$(2, 3)$
4	3	$(4, 3)$

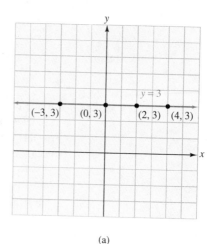

(a)

$x = -2$

x	y	(x, y)
-2	-2	$(-2, -2)$
-2	0	$(-2, 0)$
-2	2	$(-2, 2)$
-2	3	$(-2, 3)$

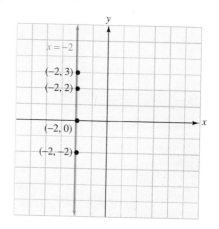

(b)

FIGURE 2-16

The results of Example 5 suggest the following facts.

Equations of Lines Parallel to the Coordinate Axes

If a and b are real numbers, then

The graph of the equation $x = a$ is a vertical line with x-intercept at $(a, 0)$. If $a = 0$, the line is the y-axis.

The graph of the equation $y = b$ is a horizontal line with y-intercept at $(0, b)$. If $b = 0$, the line is the x-axis.

■ APPLICATIONS

EXAMPLE 6   The following table gives the number of miles (y) that a bus can be driven on x gallons of gas. Plot the ordered pairs and estimate how far a bus can be driven on 9 gallons.

x	2	3	4	5	6
y	12	18	24	30	36

Solution   Since the distances driven are rather large numbers, we plot the points on a coordinate system where the unit distance on the x-axis is larger than the unit distance on the y-axis. After plotting each ordered pair as in Figure 2-17, we see that the points lie on a line.

To estimate how far the bus can go on 9 gallons, we find 9 on the x-axis, move up the graph, and then move to the left to locate a y-value of 54. The bus can be driven approximately 54 miles on 9 gallons of gas.

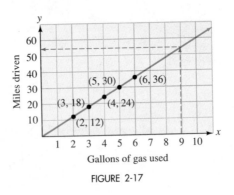

FIGURE 2-17   ■

EXAMPLE 7   **Depreciation**   A computer purchased for \$4750 is expected to depreciate according to the formula $y = -950x + 4750$, where y is the value of the computer after x years. When will the computer be worthless?

Solution   The computer will be worthless when the value (y) of the computer is 0. To find x when $y = 0$, we substitute 0 for y and solve for x.

$$y = -950x + 4750$$
$$0 = -950x + 4750$$
$$-4750 = -950x \qquad \text{Subtract 4750 from both sides.}$$
$$5 = x \qquad \text{Divide both sides by } -950.$$

The computer will be worthless in 5 years.   ■

■ THE MIDPOINT FORMULA

To distinguish betwen the coordinates of two points on a line, we often use subscript notation. Point $P(x_1, y_1)$ is read as "point P with coordinates of x sub 1 and y sub 1." Point $Q(x_2, y_2)$ is read as "point Q with coordinates of x sub 2 and y sub 2."

If point M in Figure 2-18 lies midway between points $P(x_1, y_1)$ and $Q(x_2, y_2)$, point M is called the **midpoint** of segment PQ. To find the coordinates of M, we find the mean of the x-coordinates and the mean of the y-coordinates of P and Q.

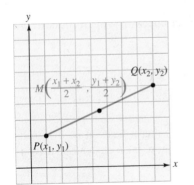

FIGURE 2-18

Midpoint Formula

The **midpoint** of the line segment with endpoints at $P(x_1, y_1)$ and $Q(x_2, y_2)$ is the point M with coordinates of

$$\left(\frac{x_1 + x_2}{2}, \frac{y_1 + y_2}{2} \right)$$

EXAMPLE 8

Find the midpoint of the line segment joining $P(-2, 3)$ and $Q(3, -5)$.

Solution

To find the midpoint, we find the mean of the x-coordinates and the mean of the y-coordinates:

$$\frac{x_1 + x_2}{2} = \frac{-2 + 3}{2} \quad \text{and} \quad \frac{y_1 + y_2}{2} = \frac{3 + (-5)}{2}$$

$$= \frac{1}{2} \qquad\qquad\qquad = -1$$

The midpoint of segment PQ is the point $M\left(\frac{1}{2}, -1\right)$. ∎

■ **GRAPHING DEVICES**

So far, we have graphed linear equations by finding ordered pairs, plotting points, and drawing a line through those points. Graphing is much easier if we use a graphing calculator or computer software.

Graphing devices have a window to display graphs (see Figure 2-19). To see the proper picture of a graph, we must decide on the minimum and maximum values for the x- and y-coordinates. A window with standard settings of

$$\text{Xmin} = -10 \qquad \text{Xmax} = 10 \qquad \text{Ymin} = -10 \qquad \text{Ymax} = 10$$

will produce a graph where the value of x is in the interval $[-10, 10]$, and the value of y is in the interval $[-10, 10]$.

FIGURE 2-19

Photo courtesy of Texas Instruments.

EXAMPLE 9

Use a graphing device to graph $3x + 2y = 12$.

Solution

Before we can enter the equation into a graphing device, we must solve the equation for y.

$$3x + 2y = 12$$

$$2y = -3x + 12 \qquad \text{Subtract } 3x \text{ from both sides.}$$

$$y = -\frac{3}{2}x + 6 \qquad \text{Divide both sides by 2.}$$

To graph the equation, we enter the right-hand side of the equation after a symbol like $Y_1 =$ or $f(x) =$. After entering the right-hand side, the display should look like

$$Y_1 = -(3/2)X + 6 \qquad \text{or} \qquad f(x) = -(3/2)X + 6$$

We then execute a GRAPH command to get the graph shown in Figure 2-20(a). To show more detail, we can redraw the graph in a different window. A window with settings of $[-1, 5]$ for x and $[-2, 7]$ for y will give the graph shown in Figure 2-20(b).

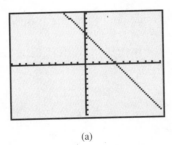   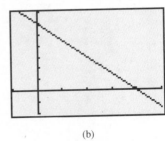

(a)   (b)

FIGURE 2-20

We can use a trace command to find the coordinates of any point on a graph. After activating the trace feature, a flashing cursor will appear on the graph. The coordinates of the cursor will also appear on the screen.

To find the x-intercept of the graph of $y = -\frac{5}{2}x - \frac{7}{2}$, we graph the equation using window settings of $[-10, 10]$ for x and $[-10, 10]$ for y and trace to get Figure 2-21(a). We can then move the cursor along the line toward the x-intercept until we arrive at a point with coordinates of $x = -1.489362$ and $y = .22340426$, as in Figure 2-21(b).

To get better results, we can zoom in to get a magnified picture, trace again, and move the cursor to the point $(-1.382979, -.0425532)$, as in Figure 2-21(c). Since the y-coordinate is nearly 0, this point is nearly the x-intercept. We can achieve better results with repeated zooms.

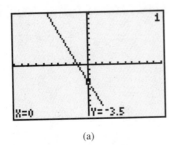   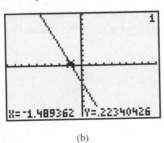

(a)   (b)   (c)

FIGURE 2-21

Orals   *Find the x- and y-intercepts of each line.*

1. $x + y = 3$ **2.** $3x + y = 6$

3. $x + 4y = 8$ **4.** $3x - 4y = 12$

Tell whether the graphs of the equations are horizontal or vertical.

5. $x = -6$ **6.** $y = 8$

Find the midpoint of a line segment with endpoints at

7. $(2, 4), (6, 8)$ **8.** $(-4, 6), (4, -8)$

EXERCISE 2.2

In Exercises 1–8, plot each point on the rectangular coordinate system shown in Illustration 1.

1. $A(4, 3)$

2. $B(-2, 1)$

3. $C(3, -2)$

4. $D(-2, -3)$

5. $E(0, 5)$

6. $F(-4, 0)$

7. $G(-3, 0)$

8. $H(0, 3)$

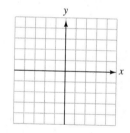

ILLUSTRATION 1

In Exercises 9–16, give the coordinates of each point shown in Illustration 2.

9. A

10. B

11. C

12. D

13. E

14. F

15. G

16. H

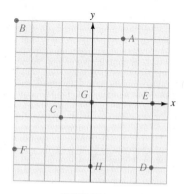

ILLUSTRATION 2

In Exercises 17–20, complete the table of solutions for each equation.

17. $y = -x + 4$

x	y
-1	
0	
2	

18. $y = x - 2$

x	y
-2	
0	
4	

19. $y = 2x - 3$

x	y
-1	
0	
3	

20. $y = -\dfrac{1}{2}x + \dfrac{5}{2}$

x	y
-3	
-1	
3	

In Exercises 21–24, graph each equation. See Exercises 17–20.

21. $y = -x + 4$

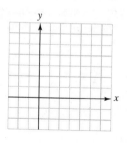

22. $y = x - 2$

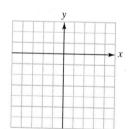

23. $y = 2x - 3$

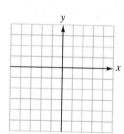

24. $y = -\dfrac{1}{2}x + \dfrac{5}{2}$

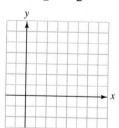

In Exercises 25–36, graph each equation.

25. $3x + 4y = 12$

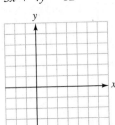

26. $4x - 3y = 12$

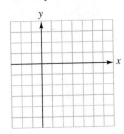

27. $y = -3x + 2$

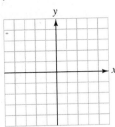

28. $y = 2x - 3$

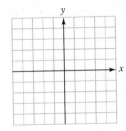

29. $3y = 6x - 9$

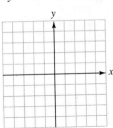

30. $2x = 4y - 10$

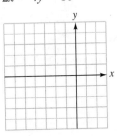

31. $3x + 4y - 8 = 0$

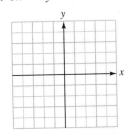

32. $-2y - 3x + 9 = 0$

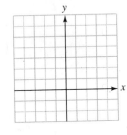

33. $x = 3$

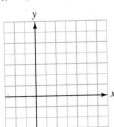

34. $y = -4$

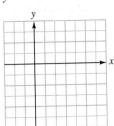

35. $-3y + 2 = 5$

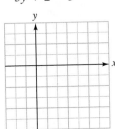

36. $-2x + 3 = 11$

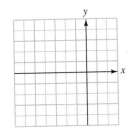

37. Hourly wages The following table gives the amount y (in dollars) that a student can earn for working x hours. Plot the ordered pairs and estimate how much the student will earn for working 8 hours.

x	2	4	5	6
y	12	24	30	36

38. Distance traveled The following table shows the distance y (in miles) that a biker can travel in x hours. Plot the ordered pairs and estimate how far the biker can go in 8 hours.

x	2	4	5	6
y	30	60	75	90

39. Value of a car The following table shows the value y (in dollars) of a car that is x years old. Plot the ordered pairs and estimate the value of the car when it is 4 years old.

x	0	1	3
y	15,000	12,000	6000

40. Earning interest The following table shows the amount y (in dollars) in a bank account drawing simple interest left on deposit for x years. Plot the ordered pairs and estimate the value of the account in 6 years.

x	0	1	4
y	1000	1050	1200

41. House appreciation A house purchased for $125,000 is expected to appreciate according to the formula $y = 7500x + 125,000$, where y is the value of the house after x years. Find the value of the house 5 years later.

42. Car depreciation A car purchased for $17,000 is expected to depreciate according to the formula $y = -1360x + 17,000$. When will the car be worthless?

43. Demand equation The number of television sets that consumers buy depends on price. The higher the price, the fewer TVs people will buy. The equation that relates price to the number of TVs sold at that price is called a **demand equation**. If the demand equation for a 13-inch TV is $p = -\frac{1}{10}q + 170$, where p is the price and q is the number of TVs sold at that price, how many TVs will be sold at a price of $150?

44. Supply equation The number of television sets that manufacturers produce depends on price. The higher the price, the more TVs manufacturers will produce. The equation that relates price to the number of TVs produced at that price is called a **supply equation**. If the supply equation for a 13-inch TV is $p = \frac{1}{10}q + 130$, where p is the price and q is the number of TVs produced for sale at that price, how many TVs will be produced if the price is $150?

45. Meshing gears The rotational speed, V, of a large gear (with N teeth) is related to the speed, v, of the smaller gear (with n teeth) by the equation $V = \frac{nv}{N}$. If the larger gear in Illustration 3 is making 60 revolutions per minute, how fast is the smaller gear spinning?

ILLUSTRATION 3

46. Crime prevention The number, n, of incidents of family violence requiring police response appears to be related to d, the money spent on crisis intervention, by the equation $n = 430 - 0.005d$. What expenditure would reduce the number of incidents to 350?

In Exercises 47–56, find the midpoint of segment PQ.

47. $P(0, 0)$, $Q(6, 8)$

48. $P(10, 12)$, $Q(0, 0)$

49. $P(6, 8)$, $Q(12, 16)$

50. $P(10, 4)$, $Q(2, -2)$

51. $P(2, 4)$, $Q(5, 8)$

52. $P(5, 9)$, $Q(8, 13)$

53. $P(-2, -8)$, $Q(3, 4)$

54. $P(-5, -2)$, $Q(7, 3)$

55. $Q(-3, 5)$, $P(-5, -5)$

56. $Q(2, -3)$, $P(4, -8)$

57. Finding the endpoint of a segment If $M(-2, 3)$ is the midpoint of segment PQ and the coordinates of P are $(-8, 5)$, find the coordinates of Q.

58. Finding the endpoint of a segment If $M(6, -5)$ is the midpoint of segment PQ and the coordinates of Q are $(-5, -8)$, find the coordinates of P.

In Exercises 59–62, use a graphing device to graph each equation, and then find the x-coordinate of the x-intercept to the nearest hundredth.

59. $y = 3.7x - 4.5$

60. $y = \frac{3}{5}x + \frac{5}{4}$

61. $1.5x - 3y = 7$

62. $0.3x + y = 7.5$

Writing Exercises *Write a paragraph using your own words.*

1. Explain how to graph a line using the intercept method.

2. Explain how to determine the quadrant in which the point $P(a, b)$ lies.

Something to Think About

1. If the line $y = ax + b$ passes through only quadrants I and II, what can be known about the constants a and b?

2. What are the coordinates of three points that divide the line segment joining $P(a, b)$ and $Q(c, d)$ into four equal parts?

Review Exercises   *Graph each interval on the number line.*

1. $[-3, 2) \cup (-2, 3]$

2. $(-1, 4) \cap [-2, 2]$

3. $[-3, -2) \cap (2, 3]$

4. $[-4, -3) \cup (2, 3]$

2.3 Slope of a Nonvertical Line

■ INTERPRETATION OF SLOPE ■ HORIZONTAL AND VERTICAL LINES ■ SLOPES OF PARALLEL LINES
■ SLOPES OF PERPENDICULAR LINES

An extended service offered by a computer online company costs $2 per month plus $3 for each hour of connect time. The table shown in Figure 2-22(a) gives the cost (y) for certain numbers of hours (x) of connect time. If we construct a graph from this data, we get the line shown in Figure 2-22(b).

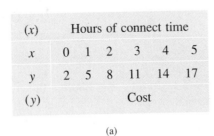

(x)	Hours of connect time					
x	0	1	2	3	4	5
y	2	5	8	11	14	17
(y)	Cost					

(a)

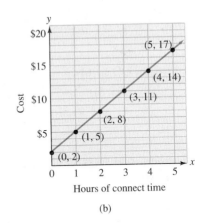

(b)

FIGURE 2-22

From the graph, we can see that if x changes from 0 to 1, y changes from 2 to 5. As x changes from 1 to 2, y changes from 5 to 8, and so on. The ratio of the change in y divided by the change in x is the constant 3.

$$\frac{\text{change in } y}{\text{change in } x} = \frac{5 - 2}{1 - 0} = \frac{8 - 5}{2 - 1} = \frac{11 - 8}{3 - 2} = \frac{14 - 11}{4 - 3} = \frac{17 - 14}{5 - 4} = \frac{3}{1} = 3$$

The ratio of the change in y divided by the change in x between any two points on any line is always a constant. This constant rate of change is called the **slope** of the line.

Slope of a Nonvertical Line
The **slope of the nonvertical line** passing through points $P(x_1, y_1)$ and $Q(x_2, y_2)$ is

$$m = \frac{\text{change in } y}{\text{change in } x} = \frac{y_2 - y_1}{x_2 - x_1} \quad (x_2 \neq x_1)$$

EXAMPLE 1   Find the slope of the line shown in Figure 2-23.

Solution   We can let $P(x_1, y_1) = P(-2, 4)$ and $Q(x_2, y_2) = Q(3, -4)$. Then

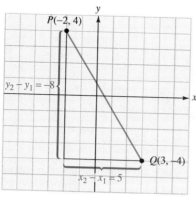

$$
\begin{aligned}
m &= \frac{\text{change in } y}{\text{change in } x} \\[4pt]
&= \frac{y_2 - y_1}{x_2 - x_1} \\[4pt]
&= \frac{-4 - 4}{3 - (-2)} \quad \text{Substitute } -4 \text{ for } y_2, \, 4 \text{ for } y_1, \, 3 \text{ for } x_2, \text{ and } -2 \text{ for } x_1. \\[4pt]
&= \frac{-8}{5} \\[4pt]
&= -\frac{8}{5}
\end{aligned}
$$

FIGURE 2-23

We found the slope of the line to be $-\frac{8}{5}$. We would obtain the same result if we had let $P(x_1, y_1) = P(3, -4)$ and $Q(x_2, y_2) = Q(-2, 4)$. ∎

 WARNING!   When calculating slope, always subtract the y values and the x values in the same order.

$$m = \frac{y_2 - y_1}{x_2 - x_1} \quad \text{or} \quad m = \frac{y_1 - y_2}{x_1 - x_2}$$

Note that

$$m \neq \frac{y_2 - y_1}{x_1 - x_2} \quad \text{and} \quad m \neq \frac{y_1 - y_2}{x_2 - x_1}$$

The change in y (often denoted as Δy) is the **rise** of the line between points P and Q. The change in x (often denoted as Δx) is the **run**. Using this terminology, we can define slope to be the ratio of the rise to the run:

$$m = \frac{\Delta y}{\Delta x} = \frac{\text{rise}}{\text{run}} \quad (\Delta x \neq 0)$$

EXAMPLE 2   Find the slope of the line determined by $3x - 4y = 12$.

Solution   We first find the coordinates of two points on the line.

• If $x = 0$, then $y = -3$, and the point $(0, -3)$ is on the line.
• If $y = 0$, then $x = 4$, and the point $(4, 0)$ is on the line.

We then refer to Figure 2-24 and find the slope of the line between $P(0, -3)$ and $Q(4, 0)$.

$$m = \frac{\text{change in } y}{\text{change in } x}$$

$$= \frac{y_2 - y_1}{x_2 - x_1}$$

$$= \frac{0 - (-3)}{4 - 0} \qquad \text{Substitute 0 for } y_2, \\ -3 \text{ for } y_1, 4 \text{ for} \\ x_2, \text{ and 0 for } x_1.$$

$$= \frac{3}{4}$$

The slope of the line is $\dfrac{3}{4}$.

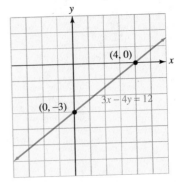

FIGURE 2-24   ∎

∎ INTERPRETATION OF SLOPE

Many application problems involve equations of lines and their slopes.

EXAMPLE 3

Cost of carpet If carpet costs \$25 per square yard, the total cost (c) of n square yards is the price per square yard times the number of square yards purchased:

c	$=$	cost per square yard	$\cdot$	the number of square yards
c	$=$	25	$\cdot$	n

Graph the equation $c = 25n$ and interpret the slope of the line.

Solution   We can graph the equation on a coordinate system with a vertical c-axis and a horizontal n-axis. Figure 2-25 shows a table of ordered pairs and the graph.

$c = 25n$

x	y	(x, y)
10	250	(10, 250)
20	500	(20, 500)
30	750	(30, 750)
40	1000	(40, 1000)
50	1250	(50, 1250)

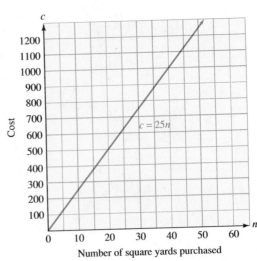

FIGURE 2-25

If we pick the points (30, 750) and (50, 1250) to find the slope, we have

$$m = \frac{\Delta c}{\Delta n}$$

$$= \frac{c_2 - c_1}{n_2 - n_1}$$

$$= \frac{1250 - 750}{50 - 30} \qquad \text{Substitute 1250 for } c_2 \text{, 750 for } c_1 \text{, 50 for } n_2 \text{, and 30 for } n_1.$$

$$= \frac{500}{20}$$

$$= 25$$

The slope of 25 (in dollars/square yard) is the ratio of the change in the cost to the change in the number of square yards purchased. It is the cost per square yard of the carpet. ∎

EXAMPLE 4   **Rate of descent**   It takes a skier 25 minutes to complete the course shown in Figure 2-26. Find his average rate of descent in feet per minute.

Solution   To find the average rate of descent, we must find the ratio of the change in altitude to the change in time. To find this ratio, we calculate the slope of the line passing through the points (0, 12000) and (25, 8500).

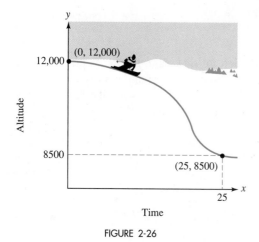

$$\text{average rate of descent} = \frac{12{,}000 - 8500}{0 - 25}$$

$$= \frac{3500}{-25}$$

$$= -140$$

The average rate of descent is -140 ft/min.

FIGURE 2-26

■ HORIZONTAL AND VERTICAL LINES

If $P(x_1, y_1)$ and $Q(x_2, y_2)$ are points on the horizontal line shown in Figure 2-27(a), then $y_1 = y_2$, and the numerator of the fraction

$$\frac{y_2 - y_1}{x_2 - x_1} \qquad \text{On a horizontal line, } x_2 \neq x_1.$$

is 0. Thus, the value of the fraction is 0, and the slope of the horizontal line is 0.

If $P(x_1, y_1)$ and $Q(x_2, y_2)$ are two points on the vertical line shown in Figure 2-27(b), then $x_1 = x_2$, and the denominator of the fraction

$$\frac{y_2 - y_1}{x_2 - x_1} \qquad \text{On a vertical line, } y_2 \neq y_1.$$

is 0. Since the denominator of a fraction cannot be 0, a vertical line has no defined slope.

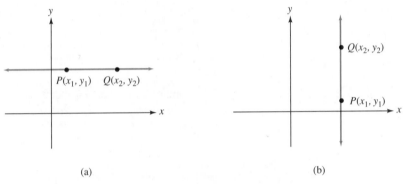

FIGURE 2-27

Horizontal and Vertical Lines

All horizontal lines (lines with equations of the form $y = b$) have a slope of 0.

Vertical lines (lines with equations of the form $x = a$) have no defined slope.

If a line rises as we follow it from left to right, as in Figure 2-28(a), its slope is positive. If a line drops as we follow it from left to right, as in Figure 2-28(b), its slope is negative. If a line is horizontal, as in Figure 2-28(c), its slope is 0. If a line is vertical, as in Figure 2-28(d), it has no defined slope.

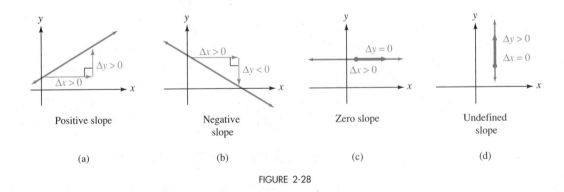

FIGURE 2-28

■ SLOPES OF PARALLEL LINES

To see a relationship between parallel lines and their slopes, we refer to the parallel lines l_1 and l_2 shown in Figure 2-29, with slopes of m_1 and m_2, respectively. Because right triangles ABC and DEF are similar, it follows that

$$m_1 = \frac{\Delta y \text{ of } l_1}{\Delta x \text{ of } l_1}$$

$$= \frac{\Delta y \text{ of } l_2}{\Delta x \text{ of } l_2}$$

$$= m_2$$

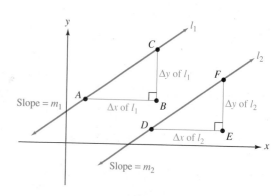

FIGURE 2-29

This shows that if two nonvertical lines are parallel, they have the same slope. It is also true that when two lines have the same slope, they are parallel.

Slopes of Parallel Lines
Nonvertical parallel lines have the same slope, and lines having the same slope are parallel.

Since vertical lines are parallel, lines with no defined slope are parallel.

EXAMPLE 5 The lines in Figure 2-30 are parallel. Find y.

Solution Since the lines are parallel, they have equal slopes. To find y, we find the slope of each line, set them equal, and solve the resulting equation.

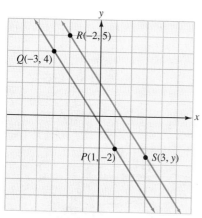

FIGURE 2-30

Slope of PQ	**Slope of RS**	
$\dfrac{-2 - 4}{1 - (-3)}$	$=$	$\dfrac{y - 5}{3 - (-2)}$

$$\frac{-6}{4} = \frac{y - 5}{5}$$

$-30 = 4(y - 5)$ Multiply both sides by 20.

$-30 = 4y - 20$ Use the distributive property.

$-10 = 4y$ Add 20 to both sides.

$-\dfrac{5}{2} = y$ Divide both sides by 4 and simplify.

Thus, $y = -\dfrac{5}{2}$.

■ SLOPES OF PERPENDICULAR LINES

Two real numbers a and b are called **negative reciprocals** if $ab = -1$. For example,

$$-\frac{4}{3} \quad \text{and} \quad \frac{3}{4}$$

are negative reciprocals, because $-\frac{4}{3}\left(\frac{3}{4}\right) = -1$.

The following statements relate perpendicular lines and their slopes.

Slopes of Perpendicular Lines

If two nonvertical lines are perpendicular, their slopes are negative reciprocals.

If the slopes of two lines are negative reciprocals, the lines are perpendicular.

Because a horizontal line is perpendicular to a vertical line, a line with a slope of 0 is perpendicular to a line with no defined slope.

EXAMPLE 6   Are the lines shown in Figure 2-31 perpendicular?

Solution   We find the slopes of the lines and see whether they are negative reciprocals.

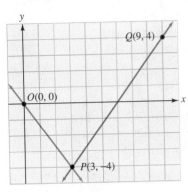

FIGURE 2-31

$$\text{Slope of } OP = \frac{\Delta y}{\Delta x} \qquad \text{and} \qquad \text{Slope of } PQ = \frac{\Delta y}{\Delta x}$$

$$= \frac{y_2 - y_1}{x_2 - x_1} \qquad\qquad\qquad = \frac{y_2 - y_1}{x_2 - x_1}$$

$$= \frac{-4 - 0}{3 - 0} \qquad\qquad\qquad = \frac{4 - (-4)}{9 - 3}$$

$$= -\frac{4}{3} \qquad\qquad\qquad\qquad = \frac{8}{6}$$

$$\qquad\qquad\qquad\qquad\qquad\qquad = \frac{4}{3}$$

Since their slopes are not negative reciprocals, the lines are not perpendicular. ∎

Orals   *Find the slope of the line passing through*

1. $(0, 0), (1, 3)$ **2.** $(0, 0), (3, 6)$

3. Are lines with slopes of -2 and $\frac{8}{-4}$ parallel?

4. Find the negative reciprocal of -0.2.

5. Are lines with slopes of -2 and $\frac{1}{2}$ perpendicular?

EXERCISE 2.3

In Exercises 1–12, find the slope of the line that passes through the given points, if possible.

1. $(0, 0), (3, 9)$ **2.** $(9, 6), (0, 0)$ **3.** $(-1, 8), (6, 1)$ **4.** $(-5, -8), (3, 8)$

5. $(3, -1), (-6, 2)$ **6.** $(0, -8), (-5, 0)$ **7.** $(7, 5), (-9, 5)$ **8.** $(2, -8), (3, -8)$

9. $(-7, -5), (-7, -2)$ **10.** $(3, -5), (3, 14)$

11. $(a, b) (b, a)$ **12.** $(a, b), (-b, -a)$

In Exercises 13–20, find the slope of the line determined by each equation.

13. $3x + 2y = 12$

14. $2x - y = 6$

15. $3x = 4y - 2$

16. $x = y$

17. $y = \dfrac{x - 4}{2}$

18. $x = \dfrac{3 - y}{4}$

19. $4y = 3(y + 2)$

20. $x + y = \dfrac{2 - 3y}{3}$

21. Rate of growth When a college started an aviation program, the administration predicted enrollments using a straight-line method. If the enrollment during the first year was 12, and the enrollment during the fifth year was 26, find the rate of growth per year (the slope of the line). (See Illustration 1.)

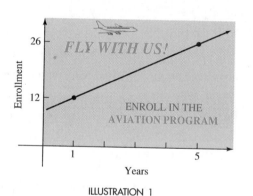

ILLUSTRATION 1

22. Rate of growth A small business predicts sales according to a straight-line method. If sales were $50,000 in the first year and $110,000 in the third year, find the rate of growth in sales per year (the slope of the line).

23. Rate of decrease The price of computer technology has been dropping steadily for the past ten years. If a desktop PC cost $6700 ten years ago, and the same computing power cost $2200 three years ago, find the rate of decrease per year. (Assume a straight-line model).

24. Hospital costs Illustration 2 shows the changing mean daily cost for a hospital room. Find the rate of change per year of the portion of the room cost that is absorbed by the hospital.

	Cost passed on to patient	Total cost to the hospital
1980	$130	$245
1985	214	459
1990	295	670

ILLUSTRATION 2

In Exercises 25–30, tell whether the slope of the line in each graph is positive, negative, 0, or undefined.

25.

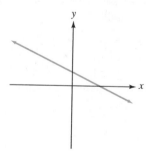

26.

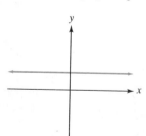

27.

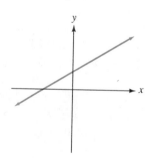

28.

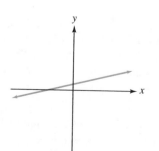

29.

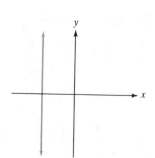

30.

In Exercises 31–36, tell whether the lines with the given slopes are parallel, perpendicular, or neither.

31. $m_1 = 3, m_2 = -\dfrac{1}{3}$

32. $m_1 = \dfrac{1}{4}, m_2 = 4$

33. $m_1 = 4, m_2 = 0.25$

34. $m_1 = -5, m_2 = \dfrac{1}{-0.2}$

35. $m_1 = \dfrac{a}{b}, m_2 = \left(\dfrac{b}{a}\right)^{-1}$

36. $m_1 = \dfrac{c}{d}, m_2 = \dfrac{d}{c}$

In Exercises 37–42, tell whether the line PQ is parallel or perpendicular to a line with a slope of -2.

37. $P(3, 4), Q(4, 2)$

38. $P(6, 4), Q(8, 5)$

39. $P(-2, 1), Q(6, 5)$

40. $P(3, 4), Q(-3, -5)$

41. $P(5, 4), Q(6, 6)$

42. $P(-2, 3), Q(4, -9)$

In Exercises 43–48, find the slopes of lines PQ and PR and tell whether the points P, Q, and R lie on the same line. (Hint: Two lines with the same slope and a point in common must be the same line.)

43. $P(-2, 4), Q(4, 8), R(8, 12)$

44. $P(6, 10), Q(0, 6), R(3, 8)$

45. $P(-4, 10), Q(-6, 0), R(-1, 5)$

46. $P(-10, -13), Q(-8, -10), R(-12, -16)$

47. $P(-2, 4), Q(0, 8), R(2, 12)$

48. $P(8, -4), Q(0, -12), R(8, -20)$

49. Find the equation of the x-axis and its slope.

50. Find the equation of the y-axis and its slope, if any.

51. Geometry Show that points with coordinates of $(-3, 4)$, $(4, 1)$, and $(-1, -1)$ are vertices of a right triangle.

52. Geometry Show that a triangle with vertices at $(0, 0)$, $(12, 0)$, and $(13, 12)$ is not a right triangle.

53. Geometry A square has vertices at points $(a, 0)$, $(0, a)$, $(-a, 0)$, and $(0, -a)$, where $a \neq 0$. Show that its adjacent sides are perpendicular.

54. Geometry If $b \neq 0$, show that the points $(2b, 4)$, (b, b), and $(4, 0)$ are vertices of a right triangle.

55. Geometry Show that the points $(0, 0)$, $(0, a)$, (b, c), and $(b, a + c)$ are the vertices of a parallelogram. (*Hint:* Opposite sides of a parallelogram are parallel.)

56. Geometry If $b \neq 0$, show that the points $(0, 0)$, $(0, b)$, $(8, b + 2)$, and $(12, 3)$ are the vertices of a trapezoid. (*Hint:* A **trapezoid** is a four-sided figure with exactly two sides parallel.)

57. The points $(3, a)$, $(5, 7)$, and $(7, 10)$ lie on a line. Find a.

58. The line passing through points $A(1, 3)$ and $B(-2, 7)$ is perpendicular to the line passing through points $C(4, b)$ and $D(8, -1)$. Find b.

Writing Exercises   *Write a paragraph using your own words.*

1. Explain why a vertical line has no defined slope.

2. Explain how to determine from their slopes whether two lines are parallel, perpendicular, or neither.

Something to Think About   **1.** Find the slope of the line $Ax + By = C$. Follow the procedure of Example 2.

2. Follow Example 2 to find the slope of the line $y = mx + b$.

Review Exercises *Simplify each expression. Write all answers without negative exponents.*

1. $(x^3y^2)^3$

2. $\left(\dfrac{x^5}{x^3}\right)^3$

3. $(x^{-3}y^2)^{-4}$

4. $\left(\dfrac{x^{-6}}{y^3}\right)^{-4}$

5. $\left(\dfrac{3x^2y^3}{8}\right)^0$

6. $\left(\dfrac{x^3x^{-7}y^{-6}}{x^4y^{-3}y^{-2}}\right)^{-2}$

2.4 Writing Equations of Lines

■ POINT–SLOPE FORM OF THE EQUATION OF A LINE ■ SLOPE–INTERCEPT FORM OF THE EQUATION OF A LINE ■ USING SLOPE AS AN AID IN GRAPHING ■ GENERAL FORM OF THE EQUATION OF A LINE ■ STRAIGHT-LINE DEPRECIATION ■ ACCENT ON STATISTICS

■ POINT–SLOPE FORM OF THE EQUATION OF A LINE

Suppose that line l shown in Figure 2-32 has a slope of m and passes through the point $P(x_1, y_1)$. If $Q(x, y)$ is a second point on line l, we have

$$m = \frac{y - y_1}{x - x_1}$$

If we multiply both sides by $x - x_1$, we have

1. $y - y_1 = m(x - x_1)$

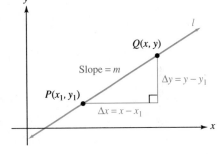

FIGURE 2-32

Because Equation 1 displays the coordinates of the point (x_1, y_1) on the line and the slope m of the line, it is called the **point–slope form** of the equation of a line.

Point–Slope Form
The equation of the line passing through $P(x_1, y_1)$ and with slope m is
$$y - y_1 = m(x - x_1)$$

EXAMPLE 1   Write the equation of a line with a slope of $-\frac{2}{3}$ and passing through $P(-4, 5)$.

Solution   We substitute $-\frac{2}{3}$ for m, -4 for x_1, and 5 for y_1 in the point–slope form and simplify.

$$y - y_1 = m(x - x_1)$$

$$y - 5 = -\frac{2}{3}[x - (-4)] \qquad \text{Substitute } -\tfrac{2}{3} \text{ for } m, -4 \text{ for } x_1, \text{ and } 5 \text{ for } y_1.$$

$$y - 5 = -\frac{2}{3}(x + 4) \qquad -(-4) = 4.$$

$$y - 5 = -\frac{2}{3}x - \frac{8}{3} \qquad \text{Use the distributive property to remove parentheses.}$$

$$y = -\frac{2}{3}x + \frac{7}{3} \qquad \text{Add 5 to both sides and simplify.}$$

The equation of the line is $y = -\frac{2}{3}x + \frac{7}{3}$. ■

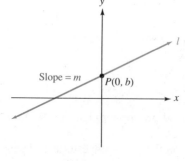

EXAMPLE 2   Write the equation of the line passing through $P(-5, 4)$ and $Q(8, -6)$.

Solution   First we find the slope of the line:

$$m = \frac{y_2 - y_1}{x_2 - x_1}$$

$$= \frac{-6 - 4}{8 - (-5)} \qquad \text{Substitute } -6 \text{ for } y_2, 4 \text{ for } y_1, 8 \text{ for } x_2, \text{ and } -5 \text{ for } x_1.$$

$$= -\frac{10}{13}$$

Because the line passes through both P and Q, we can choose either point and substitute its coordinates into the point–slope form. If we choose $P(-5, 4)$, we substitute -5 for x_1, 4 for y_1, and $-\frac{10}{13}$ for m and proceed as follows.

$$y - y_1 = m(x - x_1)$$

$$y - 4 = -\frac{10}{13}[x - (-5)] \qquad \text{Substitute } -\tfrac{10}{13} \text{ for } m, -5 \text{ for } x_1, \text{ and } 4 \text{ for } y_1.$$

$$y - 4 = -\frac{10}{13}(x + 5) \qquad -(-5) = 5.$$

$$y - 4 = -\frac{10}{13}x - \frac{50}{13} \qquad \text{Use the distributive property to remove parentheses.}$$

$$y = -\frac{10}{13}x + \frac{2}{13} \qquad \text{Add 4 to both sides and simplify.}$$

The equation of the line is $y = -\frac{10}{13}x + \frac{2}{13}$. ■

■ SLOPE–INTERCEPT FORM OF THE EQUATION OF A LINE

Since the y-intercept of the line shown in Figure 2-33 is the point $(0, b)$, we can write the equation of the line by substituting 0 for x_1 and b for y_1 in the point–slope form and simplifying.

$$y - y_1 = m(x - x_1)$$

$$y - b = m(x - 0)$$

$$y - b = mx$$

2. $\qquad y = mx + b \qquad$ Add b to both sides.

FIGURE 2-33

$y = mx + b$ linear function

$f(x) = mx + b$

$f(x) = y$

Because Equation 2 displays the slope m and the y-coordinate b of the y-intercept, it is called the **slope–intercept form** of the equation of a line.

> ### Slope–Intercept Form
> The equation of the line with slope m and y-intercept $(0, b)$ is
> $$y = mx + b$$

EXAMPLE 3   Use the slope–intercept form to write the equation of the line with slope 4 that passes through the point $P(5, 9)$.

Solution   Since we are given that $m = 4$ and that the pair $(5, 9)$ satisfies the equation, we can substitute 5 for x, 9 for y, and 4 for m in the equation $y = mx + b$ and solve for b.

$$y = mx + b$$
$$9 = 4(5) + b \qquad \text{Substitute 9 for } y, 4 \text{ for } m, \text{ and } 5 \text{ for } x.$$
$$9 = 20 + b \qquad \text{Simplify.}$$
$$-11 = b \qquad \text{Subtract 20 from both sides.}$$

Because $m = 4$ and $b = -11$, the equation of the line is $y = 4x - 11$. ■

■ USING SLOPE AS AN AID IN GRAPHING

It is easy to graph a linear equation when it is written in slope–intercept form. For example, to graph $y = \frac{4}{3}x - 2$, we note that $b = -2$ and that the y-intercept is $P(0, b) = P(0, -2)$. (See Figure 2-34.)

Because the slope of the line is $\frac{\Delta y}{\Delta x} = \frac{4}{3}$, we can locate another point Q on the line by starting at point P and counting 3 units to the right and 4 units up. The change in x from point P to point Q is $\Delta x = 3$, and the corresponding change in y is $\Delta y = 4$. The line joining points P and Q is the graph of the equation.

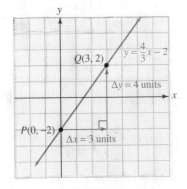

FIGURE 2-34

EXAMPLE 4   Find the slope and the y-intercept of the line with the equation $2(x - 3) = -3(y + 5)$. Then graph the line.

Solution   We write the equation in the form $y = mx + b$ to find the slope m and the y-intercept $(0, b)$.

$$2(x - 3) = -3(y + 5)$$
$$2x - 6 = -3y - 15 \qquad \text{Use the distributive property to remove parentheses.}$$
$$2x + 3y - 6 = -15 \qquad \text{Add } 3y \text{ to both sides.}$$
$$3y - 6 = -2x - 15 \qquad \text{Subtract } 2x \text{ from both sides.}$$
$$3y = -2x - 9 \qquad \text{Add 6 to both sides.}$$
$$y = -\frac{2}{3}x - 3 \qquad \text{Divide both sides by 3.}$$

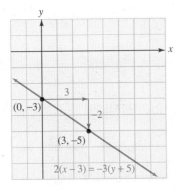

FIGURE 2-35

The slope of the line is $-\frac{2}{3}$, and the y-intercept is $(0, -3)$. To draw the graph, we plot the y-intercept $(0, -3)$ and then find a second point on the line by moving 3 units to the right and 2 units down to locate the point $(3, -5)$. We draw a line through the two points to obtain the graph shown in Figure 2-35. ■

EXAMPLE 5   Show that the lines represented by $4x + 8y = 10$ and $2x = 12 - 4y$ are parallel.

Solution   We solve each equation for y to see that the lines are distinct and that their slopes are equal.

$$
\begin{array}{c|c}
4x + 8y = 10 & 2x = 12 - 4y \\
8y = -4x + 10 & 4y = -2x + 12 \\
y = -\dfrac{1}{2}x + \dfrac{5}{4} & y = -\dfrac{1}{2}x + 3
\end{array}
$$

Since the slopes of the two lines are each $-\frac{1}{2}$, but the lines have different y-intercepts, the lines are parallel. ∎

EXAMPLE 6   Show that the lines represented by $4x + 8y = 10$ and $4x - 2y = 21$ are perpendicular.

Solution   We solve each equation for y to see whether the slopes of their straight-line graphs are negative reciprocals.

$$
\begin{array}{c|c}
4x + 8y = 10 & 4x - 2y = 21 \\
8y = -4x + 10 & -2y = -4x + 21 \\
y = -\dfrac{1}{2}x + \dfrac{5}{4} & y = 2x - \dfrac{21}{2}
\end{array}
$$

Since the slopes of the lines are $-\frac{1}{2}$ and 2 (which are negative reciprocals), the lines are perpendicular. ∎

EXAMPLE 7   Write the equation of the line passing through $P(-2, 5)$ and parallel to the line $y = 8x - 3$.

Solution   Since the slope of the line given by $y = 8x - 3$ is the coefficient of x, the slope is 8. Since the desired equation is to have a graph that is parallel to the graph of $y = 8x - 3$, its slope must also be 8.

We substitute -2 for x_1, 5 for y_1, and 8 for m in the point–slope form and simplify.

$$
\begin{array}{ll}
y - y_1 = m(x - x_1) & \\
y - 5 = 8[x - (-2)] & \text{Substitute 5 for } y_1, \text{ 8 for } m, \text{ and } -2 \text{ for } x_1. \\
y - 5 = 8(x + 2) & -(-2) = 2. \\
y - 5 = 8x + 16 & \text{Use the distributive property to remove parentheses.} \\
y = 8x + 21 & \text{Add 5 to both sides.}
\end{array}
$$

The equation of the desired line is $y = 8x + 21$. ∎

EXAMPLE 8   Write the equation of a line passing through $P(-2, 5)$ and perpendicular to the line $y = 8x - 3$.

Solution   The slope of the given line is 8. Thus, the slope of the desired line must be $-\frac{1}{8}$, which is the negative reciprocal of 8.

We substitute -2 for x_1, 5 for y_1, and $-\frac{1}{8}$ for m into the point–slope form and simplify.

$$y - y_1 = m(x - x_1)$$

$$y - 5 = -\frac{1}{8}[x - (-2)] \qquad \text{Substitute 5 for } y_1, -\tfrac{1}{8} \text{ for } m, \text{ and } -2 \text{ for } x_1.$$

$$y - 5 = -\frac{1}{8}(x + 2) \qquad -(-2) = 2.$$

$$8y - 40 = -1(x + 2) \qquad \text{Multiply both sides by 8.}$$

$$8y - 40 = -x - 2 \qquad \text{Use the distributive property to remove parentheses.}$$

$$x + 8y - 40 = -2 \qquad \text{Add } x \text{ to both sides.}$$

$$x + 8y = 38 \qquad \text{Add 40 to both sides.}$$

The equation of the line is $x + 8y = 38$. ■

■ GENERAL FORM OF THE EQUATION OF A LINE

Any linear equation that is written in the form $Ax + By = C$, where A, B, and C are constants, is said to be written in **general form**.

WARNING! When writing equations in general form, it is customary to clear the equation of fractions and make A positive.

It is also customary to make A, B, and C as small as possible. For example, the equation $6x + 12y = 24$ can be changed to $x + 2y = 4$ by dividing both sides by 6.

> **Finding the Slope and y-Intercept from the General Form**
> If A, B, and C are real numbers and $B \neq 0$, then the graph of the equation
> $$Ax + By = C$$
> is a nonvertical line with slope of $-\dfrac{A}{B}$ and a y-intercept of $\left(0, \dfrac{C}{B}\right)$.

You will be asked to justify the previous results in the exercises. You will also be asked to show that if $B = 0$, the equation $Ax + By = C$ represents a vertical line with x-intercept of $\left(\frac{C}{A}, 0\right)$.

EXAMPLE 9 Show that the lines represented by $4x + 3y = 7$ and $3x - 4y = 12$ are perpendicular.

Solution To show that the lines are perpendicular, we will show that their slopes are negative reciprocals. The first equation, $4x + 3y = 7$, is written in general form, with $A = 4$, $B = 3$, and $C = 7$. Because the equation is written in general form, the slope of the line is

$$m_1 = -\frac{A}{B} = -\frac{4}{3}$$

The second equation, $3x - 4y = 12$, is also written in general form, with $A = 3$, $B = -4$, and $C = 12$. The slope of this line is

$$m_2 = -\frac{A}{B} = -\frac{3}{-4} = \frac{3}{4}$$

Since the slopes of the two lines are negative reciprocals, they are perpendicular. ■

We summarize the various forms for the equation of a line as follows.

General form of a linear equation	$Ax + By = C$ A and B cannot both be 0.
Slope–intercept form of a linear equation	$y = mx + b$ The slope is m, and the y-intercept is $(0, b)$.
Point–slope form of a linear equation	$y - y_1 = m(x - x_1)$ The slope is m, and the line passes through (x_1, y_1).
A horizontal line	$y = b$ The slope is 0, and the y-intercept is $(0, b)$.
A vertical line	$x = a$ There is no defined slope, and the x-intercept is $(a, 0)$.

■ STRAIGHT-LINE DEPRECIATION

For tax purposes, many businesses use *straight-line depreciation* to find the declining value of aging equipment.

EXAMPLE 10

Value of a lathe A machine shop buys a lathe for $1970 and expects it to last for ten years. The lathe can then be sold as scrap for an estimated *salvage value* of $270. If y represents the value of the lathe after x years of use, and y and x are related by the equation of a line,

a. Find the equation of the line.

b. Find the value of the lathe after $2\frac{1}{2}$ years.

c. Find the economic meaning of the y-intercept of the line.

d. Find the economic meaning of the slope of the line.

Solution **a.** To find the equation of the line, we first calculate its slope and then use the point–slope form to find its equation.

When the lathe is new, its age x is 0, and its value y is $1970. When the lathe is 10 years old, $x = 10$, and its value is $y = $270. Since the line passes through the points $(0, 1970)$ and $(10, 270)$, as shown in Figure 2-36, the slope of the line is

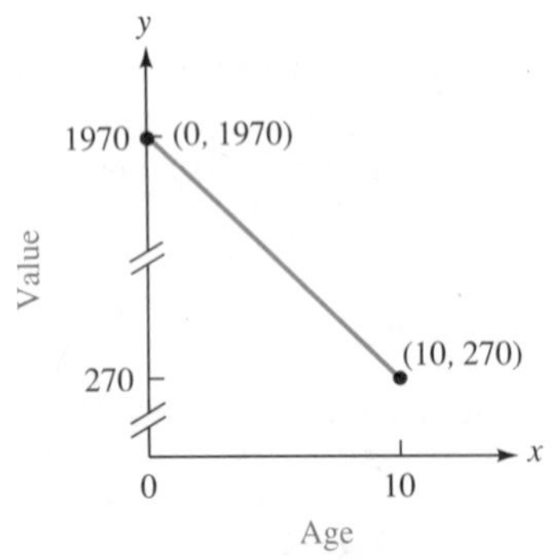

$$
\begin{aligned}
m &= \frac{y_2 - y_1}{x_2 - x_1} \\
&= \frac{270 - 1970}{10 - 0} \\
&= \frac{-1700}{10} \\
&= -170
\end{aligned}
$$

FIGURE 2-36

To find the equation of the line, we substitute -170 for m, 0 for x_1, and 1970 for y_1 in the point–slope form and simplify.

$$y - y_1 = m(x - x_1)$$
$$y - 1970 = -170(x - 0)$$

3. $\qquad y = -170x + 1970$

The current value y of the lathe is related to its age x by the equation $y = -170x + 1970$.

b. To find the value of the lathe after $2\frac{1}{2}$ years, we substitute 2.5 for x in Equation 3 and solve for y.

$$y = -170x + 1970$$
$$y = -170(2.5) + 1970$$
$$ = -425 + 1970$$
$$ = 1545$$

In $2\frac{1}{2}$ years, the lathe will be worth $1545.

c. The y-intercept of the graph is $(0, b)$, where b is the value of y when $x = 0$.

$$y = -170x + 1970$$
$$y = -170(0) + 1970 \qquad \text{Substitute 0 for } x.$$
$$y = 1970$$

Thus, b is the value of a 0-year-old lathe, which is the lathe's original cost.

d. Each year, the value of the lathe decreases by $170, because the slope of the line is -170. The slope of the depreciation line is the *annual depreciation rate*. ◼

■ ■ ■ ■ ■ ■ ■ ■ ■ ■ **Curve Fitting**

ACCENT ON
STATISTICS

In statistics, the process of using one variable to predict another is called *regression*. For example, if we know a man's height, we can make a good prediction about his weight, because taller men usually weigh more than shorter men.

Figure 2-37 shows the results of sampling ten men at random and finding their heights and weights. The graph of the ordered pairs (h, w) is called a *scattergram*.

To write a *prediction equation* (sometimes called a *regression equation*), we must find the equation of the line that comes closer to all of the points in the scattergram than any other possible line. In statistics, there are exact methods for finding this equation. However, here we only approximate this equation.

To write an approximation of the regression equation, we place a straightedge on the scattergram shown in Figure 2-37 and draw the line joining two points that seems to best fit all of the points. In the figure, line PQ is drawn, where point P has coordinates of $(66, 140)$ and point Q has coordinates of $(75, 210)$.

Man	Height (*h*) in inches	Weight (*w*) in pounds
1	66	140
2	68	150
3	68	165
4	70	180
5	70	165
6	71	175
7	72	200
8	74	190
9	75	210
10	75	215

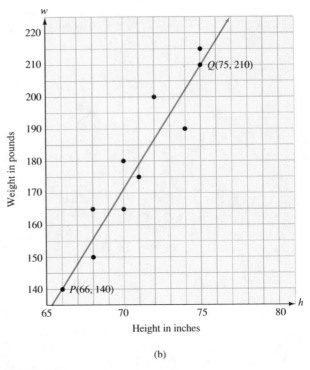

(a)   (b)

FIGURE 2-37

Our approximation of the regression equation will be the equation of the line passing through points P and Q. To find the equation of this line, we first find its slope.

$$m = \frac{y_2 - y_1}{x_2 - x_1}$$

$$= \frac{210 - 140}{75 - 66}$$

$$= \frac{70}{9}$$

We can then use point–slope form to find the equation of the line.

$$y - y_1 = m(x - x_1)$$

$$y - 140 = \frac{70}{9}(x - 66) \qquad \text{Choose } (66, 140) \text{ for } (x_1, y_1).$$

$$y = \frac{70}{9}x - \frac{4620}{9} + 140 \qquad \begin{array}{l}\text{Remove parentheses and add 140 to}\\ \text{both sides.}\end{array}$$

1. $$y = \frac{70}{9}x - \frac{3360}{9} \qquad \text{Add } -\frac{4620}{9} \text{ and 140.}$$

Our approximation of the regression equation is $y = \frac{70}{9}x - \frac{3360}{9}$.

To predict the weight of a man who is 73 inches tall, for example, we substitute 73 for x in Equation 1 and simplify.

$$y = \frac{70}{9}x - \frac{3360}{9}$$

$$y = \frac{70}{9}(73) - \frac{3360}{9}$$

$$y = 194.444. . .$$

We would predict that a 73-inch-tall man chosen at random will weigh about 194 pounds.

■ ■ ■ ■ ■ ■ ■ ■ ■ ■

Orals *Write the point–slope form of the equation of a line with m = 2, passing through the given point.*

1. $(2, 3)$ **2.** $(-3, 8)$

Write the equation of a line with m = −3 and y-intercept of

3. $(0, 5)$ **4.** $(0, -7)$

Tell whether the lines are parallel or perpendicular.

5. $y = 3x - 4, \ y = 3x + 5$

6. $y = -3x + 7, \ y = \frac{1}{3}x - 1$

EXERCISE 2.4

In Exercises 1–4, use point–slope form to write the equation of the line with the given properties. Write each equation in general form.

1. $m = 5$, passing through $P(0, 7)$ **2.** $m = -8$, passing through $P(0, -2)$

3. $m = -3$, passing through $P(2, 0)$ **4.** $m = 4$, passing through $P(-5, 0)$

In Exercises 5–6, use point–slope form to write the equation of each line. Write the equation in general form.

5.

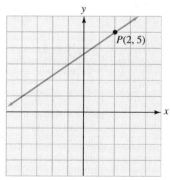

6.

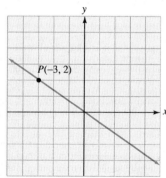

In Exercises 7–10, use point–slope form to write the equation of the line passing through the two given points. Write each equation in slope–intercept form.

7. $P(0, 0), Q(4, 4)$ **8.** $P(-5, -5), Q(0, 0)$

9. $P(3, 4), Q(0, -3)$ **10.** $P(4, 0), Q(6, -8)$

In Exercises 11–12, use point–slope form to write the equation of each line. Write each answer in slope–intercept form.

11.

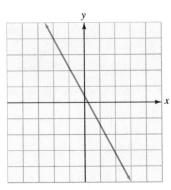

12.

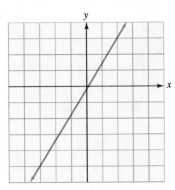

In Exercises 13–20, use the slope–intercept form to write the equation of the line with the given properties. Write each equation in slope–intercept form.

13. $m = 3$, $b = 17$

14. $m = -2$, $b = 11$

15. $m = -7$, passing through $P(7, 5)$

16. $m = 3$, passing through $P(-2, -5)$

17. $m = 0$, passing through $P(2, -4)$

18. $m = -7$, passing through the origin

19. passing through $P(6, 8)$ and $Q(2, 10)$

20. passing through $P(-4, 5)$ and $Q(2, -6)$

In Exercises 21–26, write each equation in slope–intercept form to find the slope and the y-intercept. Then use the slope and y-intercept to draw the line.

21. $y + 1 = x$

22. $x + y = 2$

23. $x = \dfrac{3}{2}y - 3$

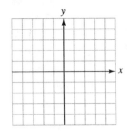

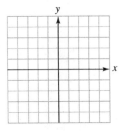

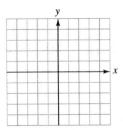

24. $x = -\dfrac{4}{5}y + 2$

25. $3(y - 4) = -2(x - 3)$

26. $-4(2x + 3) = 3(3y + 8)$

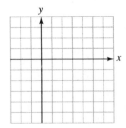

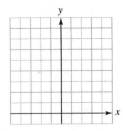

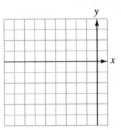

In Exercises 27–32, find the slope and the y-intercept of the line determined by the given equation.

27. $3x - 2y = 8$

28. $-2x + 4y = 12$

29. $-2(x + 3y) = 5$

30. $5(2x - 3y) = 4$

31. $x = \dfrac{2y - 4}{7}$

32. $3x + 4 = -\dfrac{2(y - 3)}{5}$

In Exercises 33–38, investigate the properties of slope and the y-intercept by experimenting with the following problems.

33. Graph $y = mx + 2$ for several positive values of m. What do you notice?

34. Graph $y = mx + 2$ for several negative values of m. What do you notice?

35. Graph $y = 2x + b$ for several increasing positive values of b. What do you notice?

36. Graph $y = 2x + b$ for several decreasing negative values of b. What do you notice?

37. How does the graph of $y = \frac{1}{2}x + 5$ compare to the graph of $y = \frac{1}{2}x - 5$?

38. How does the graph of $y = \frac{1}{2}x - 5$ compare to the graph of $y = \frac{1}{2}x$?

In Exercises 39–50, tell whether the graphs of each pair of equations are parallel, perpendicular, or neither.

39. $y = 3x + 4$, $y = 3x - 7$

40. $y = 4x - 13$, $y = \frac{1}{4}x + 13$

41. $x + y = 2$, $y = x + 5$

42. $x = y + 2$, $y = x + 3$

43. $y = 3x + 7$, $2y = 6x - 9$

44. $2x + 3y = 9$, $3x - 2y = 5$

45. $x = 3y + 4$, $y = -3x + 7$

46. $3x + 6y = 1$, $y = \frac{1}{2}x$

47. $y = 3$, $x = 4$

48. $y = -3$, $y = -7$

49. $x = \frac{y - 2}{3}$, $3(y - 3) + x = 0$

50. $2y = 8$, $3(2 + y) = 3(x + 2)$

In Exercises 51–56, write the equation of the line that passes through the given point and is parallel to the given line. Write the answer in slope–intercept form.

51. $P(0, 0)$, $y = 4x - 7$

52. $P(0, 0)$, $x = -3y - 12$

53. $P(2, 5)$, $4x - y = 7$

54. $P(-6, 3)$, $y + 3x = -12$

55. $P(4, -2)$, $x = \frac{5}{4}y - 2$

56. $P(1, -5)$, $x = -\frac{3}{4}y + 5$

In Exercises 57–62, write the equation of the line that passes through the given point and is perpendicular to the given line. Write the answer in slope–intercept form.

57. $P(0, 0)$, $y = 4x - 7$

58. $P(0, 0)$, $x = -3y - 12$

59. $P(2, 5)$, $4x - y = 7$

60. $P(-6, 3)$, $y + 3x = -12$

61. $P(4, -2)$, $x = \frac{5}{4}y - 2$

62. $P(1, -5)$, $x = -\frac{3}{4}y + 5$

In Exercises 63–66, use the method of Example 9 to find if the graphs determined by each pair of equations are parallel, perpendicular, or neither.

63. $4x + 5y = 20$, $5x - 4y = 20$

64. $9x - 12y = 17$, $3x - 4y = 17$

65. $2x + 3y = 12$, $6x + 9y = 32$

66. $5x + 6y = 30$, $6x + 5y = 24$

67. Find the equation of the line perpendicular to the line $y = 3$ and passing through the midpoint of the segment joining $(2, 4)$ and $(-6, 10)$.

68. Find the equation of the line parallel to the line $y = -8$ and passing through the midpoint of the segment joining $(-4, 2)$ and $(-2, 8)$.

69. Find the equation of the line parallel to the line $x = 3$ and passing through the midpoint of the segment joining $(2, -4)$ and $(8, 12)$.

70. Find the equation of the line perpendicular to the line $x = 3$ and passing through the midpoint of the segment joining $(-2, 2)$ and $(4, -8)$.

71. Solve $Ax + By = C$ for y and thereby show that the slope of its graph is $-\frac{A}{B}$ and its y-intercept is $\left(0, \frac{C}{B}\right)$.

72. Show that the x-intercept of the graph of $Ax + By = C$ is $\left(\frac{C}{A}, 0\right)$.

In Exercises 73–83, assume straight-line depreciation or straight-line appreciation.

73. Finding a depreciation equation   A taxicab was purchased for \$24,300. Its salvage value at the end of its 7-year useful life is expected to be \$1900. Find the depreciation equation.

74. Finding a depreciation equation   A small business purchases the computer system shown in Illustration 1. It will be depreciated over a 4-year period, when its salvage value will be \$300. Find the depreciation equation.

$3900

ILLUSTRATION 1

75. Finding an appreciation equation   An apartment building was purchased for \$475,000, excluding the cost of land. The owners expect the building to double in value in 10 years. Find the appreciation equation.

76. Finding an appreciation equation   A house purchased for \$112,000 is expected to double in value in 12 years. Find its appreciation equation.

77. Finding a depreciation equation   Find the depreciation equation for the TV in the want ad in Illustration 2.

For Sale: 3-year-old 54-inch TV, \$1,900 new. Asking \$1,190. Call 875-5555. Ask for Mike.

ILLUSTRATION 2

78. Depreciating a word processor   A word processor cost \$555 when new and is expected to be worth \$80 after 5 years. What will it be worth after 3 years?

79. Finding salvage value   A copier cost \$1050 when new and will be depreciated at the rate of \$120 per year. If the useful life of the copier is 8 years, find its salvage value.

80. Finding annual rate of depreciation   A truck that cost \$27,600 when new will have no salvage value after 12 years. Find its annual rate of depreciation.

81. Finding the value of antiques   An antique table is expected to appreciate \$40 each year. If the table will be worth \$450 in 2 years, what will it be worth in 13 years?

82. Finding the value of antiques   An antique clock is expected to be worth \$350 after 2 years and \$530 after 5 years. What will the clock be worth after 7 years?

83. Finding the purchase price of real estate   A cottage that was purchased 3 years ago is now appraised at \$47,700. If the property has been appreciating \$3500 per year, find its original purchase price.

84. Charges for computer repair   A computer repair company charges a fixed amount, plus an hourly rate, for a service call. Use the information in Illustration 3 to find the hourly rate.

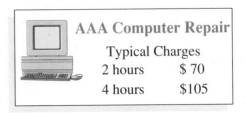

AAA Computer Repair

Typical Charges

2 hours	\$ 70
4 hours	\$105

ILLUSTRATION 3

85. Charges for automobile repair An auto repair shop charges an hourly rate, plus the cost of parts. If the cost of labor for a $1\frac{1}{2}$-hour radiator repair is $69, find the cost of labor for a 5-hour transmission overhaul.

86. Finding printer charges A printer charges a fixed setup cost, plus $1 for every 100 copies. If 700 copies cost $52, how much will it cost to print 1000 copies?

87. Predicting fires A local fire department recognizes that city growth and the number of reported fires are related by a linear equation. City records show that 300 fires were reported in a year when the local population was 57,000 persons, and 325 fires were reported in a year when the population was 59,000 persons. How many fires can be expected when the population reaches 100,000 persons?

88. Estimating the cost of rain gutter A neighbor says that an installer of rain gutter charges $60, plus a dollar amount per foot. If the neighbor paid $435 for the installation of 250 feet of gutter, how much will it cost you to have 300 feet installed?

Writing Exercises   *Write a paragraph using your own words.*

1. Explain how to find the equation of a line passing through two given points.

2. In straight-line depreciation, explain why the slope of the line is called the *rate of depreciation.*

Something to Think About

1. The graph of $y = ax + b$ passes through quadrants I, II, and IV. What can be known about the constants a and b?

2. The graph of $Ax + By = C$ passes through the quadrants I and IV only. What is known about the constants A, B, and C?

Review Exercises   *Solve each equation.*

1. $3(x + 2) + x = 5x$

2. $12b + 6(3 - b) = b + 3$

3. $\dfrac{5(2 - x)}{3} - 1 = x + 5$

4. $\dfrac{r - 1}{3} = \dfrac{r + 2}{6} + 2$

5. Mixing alloys In 60 ounces of alloy for watch cases, there are 20 ounces of gold. How much copper must be added to the alloy so that a watch case weighing 4 ounces, made from the new alloy, will contain exactly 1 ounce of gold?

6. Mixing coffee To make a mixture of 80 pounds of coffee worth $272, a grocer mixes coffee worth $3.25 a pound with coffee worth $3.85 a pound. How many pounds of the cheaper coffee should the grocer use?

2.5 Introduction to Functions

■ FUNCTION NOTATION ■ FINDING DOMAINS AND RANGES ■ THE VERTICAL LINE TEST ■ LINEAR FUNCTIONS

In this section, we will discuss one of the most important concepts in mathematics: the concept of **function.**

> **Function**
> A **function** is a correspondence between the elements of one set (called the **domain**) and the elements of another set (called the **range**), where exactly one element in the range corresponds to each element in the domain.

Here are some examples of correspondences that are functions.

To every house, there corresponds exactly one address.

To every telephone, there corresponds exactly one telephone number.

To every value of x in the equation $y = -\frac{1}{2}x + 4$, there corresponds exactly one value of y.

In the first example, the domain is the set of houses, and the range is the set of addresses. In the second example, the domain is the set of telephones, and the range is the set of telephone numbers. In the third example, the domain is the set of possible values of x (the real numbers), and the range is the set of possible values of y (also the real numbers). We say that the equation $y = -\frac{1}{2}x + 2$ defines y to be a function of x.

EXAMPLE 1   In each correspondence, give the domain and range and then tell whether the correspondence is a function.

a. To every country, there corresponds a flag.

b. To every face, there correspond two eyes.

c. To every real number, there corresponds its square.

Solution   **a.** The domain is the set of countries, and the range is a set of flags. Since every country has a single flag, the correspondence is a function.

b. The domain is the set of faces. The range is the set of eyes. Since every face has more than one eye, the correspondence is not a function.

c. The domain is the set of real numbers. Since the square of any real number is nonnegative, the range is the set of nonnegative real numbers. Because every real number has a single square, the correspondence is a function.   ■

EXAMPLE 2   Do the equations   **a.** $y = 4x - 3$ and   **b.** $y^2 = x$ define y to be a function of x?

Solution   **a.** For a function to exist, every possible value of x in the domain (the set of real numbers) must determine a single value y in the range (the set of real numbers). To find y in the equation $y = 4x - 3$, we multiply x by 4 and then subtract 3. Since this arithmetic always gives a single result, each choice of x determines a single value y. Thus, the equation $y = 4x - 3$ does define y to be a function of x.

b. For a function to exist, each number x in the domain (the set of nonnegative real numbers) must determine a single value y in the range (the set of real numbers). If we let $x = 16$, for example, y could be 4 or -4, because $4^2 = 16$ and $(-4)^2 = 16$. Since more than one value of y is determined when $x = 16$, the equation does not represent a function.   ■

The concept of function is illustrated graphically in Figure 2-38. To each value in the domain, there corresponds exactly one value in the range.

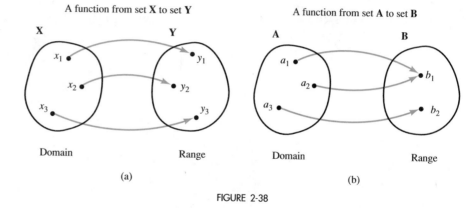

FIGURE 2-38

Every correspondence between the elements of one set and the elements of another set is called a **relation**. If a correspondence is not a function, it is still a relation.

> ### Relation
> A **relation** is a correspondence between the elements of one set (called the **domain**) and the elements of another set (called the **range**), where one or more elements in the range correspond to each element in the domain.

 WARNING! A function is always a relation, but a relation is not necessarily a function.

■ FUNCTION NOTATION

There is a special notation that we will use to denote functions.

> ### Function Notation
> The notation $y = f(x)$ denotes that y is a function of x.

The notation $y = f(x)$ is read as "y equals f of x." Note that y and $f(x)$ are two notations for the same quantity. Thus, the equations $y = 4x + 3$ and $f(x) = 4x + 3$ are equivalent.

 WARNING! The notation $f(x)$ does not mean "f times x."

The notation $y = f(x)$ provides a way of denoting the value of y (called the **dependent variable**) that corresponds to some number x (called the **independent variable**). For example, if $y = f(x)$, the value of y that is determined by $x = 3$ is denoted by $f(3)$.

(b, d)

EXAMPLE 3

Let $f(x) = 4x + 3$. Find **a.** $f(3)$, **b.** $f(-1)$, **c.** $f(0)$, and **d.** $f(r)$.

Solution

a. We replace x with 3:

$$f(x) = 4x + 3$$
$$f(3) = 4(3) + 3$$
$$= 12 + 3$$
$$= 15$$

b. We replace x with -1:

$$f(x) = 4x + 3$$
$$f(-1) = 4(-1) + 3$$
$$= -4 + 3$$
$$= -1$$

c. We replace x with 0:

$$f(x) = 4x + 3$$
$$f(0) = 4(0) + 3$$
$$= 3$$

d. We replace x with r:

$$f(x) = 4x + 3$$
$$f(r) = 4r + 3$$

To see why function notation is helpful, we consider the following sentences:

1. In the equation $y = 4x + 3$, find the value of y when x is 3.

2. In the equation $f(x) = 4x + 3$, find $f(3)$.

Statement 2, which uses $f(x)$ notation, is much more concise.

We can think of a function as a machine that takes some input x and turns it into some output $f(x)$, as shown in Figure 2-39(a). The machine shown in Figure 2-39(b) turns the input number 2 into the output value -3 and turns the input number 6 into the output value -11. The set of numbers that we can put into the machine is the domain of the function, and the set of numbers that comes out is the range.

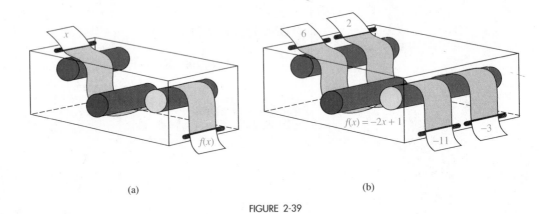

(a)

(b)

FIGURE 2-39

The letter f used in the notation $y = f(x)$ represents the word *function*. However, other letters can be used to represent functions. The notations $y = g(x)$ and $y = h(x)$ also denote functions involving the independent variable x.

In Example 4, the equation $y = g(x) = x^2 - 2x$ determines a function, because every possible value of x gives a single value of $g(x)$.

(a, c)

EXAMPLE 4

Let $g(x) = x^2 - 2x$. Find **a.** $g\left(\frac{2}{5}\right)$, **b.** $g(s)$, **c.** $g(s^2)$, and **d.** $g(-t)$.

Solution **a.** We replace x with $\frac{2}{5}$:

$$g(x) = x^2 - 2x$$

$$g\left(\frac{2}{5}\right) = \left(\frac{2}{5}\right)^2 - 2\left(\frac{2}{5}\right)$$

$$= \frac{4}{25} - \frac{4}{5}$$

$$= -\frac{16}{25}$$

b. We replace x with s:

$$g(x) = x^2 - 2x$$

$$g(s) = s^2 - 2s$$

c. We replace x with s^2:

$$g(x) = x^2 - 2x$$

$$g(s^2) = (s^2)^2 - 2s^2$$

$$= s^4 - 2s^2$$

d. We replace x with $-t$:

$$g(x) = x^2 - 2x$$

$$g(-t) = (-t)^2 - 2(-t)$$

$$= t^2 + 2t$$ ■

⊙⊙ (b)

EXAMPLE 5 Let $f(x) = 4x - 1$. Find **a.** $f(3) + f(2)$ and **b.** $f(a) - f(b)$.

Solution **a.** We find $f(3)$ and $f(2)$ separately.

$$\begin{aligned} f(x) &= 4x - 1 \\ f(3) &= 4(3) - 1 \\ &= 12 - 1 \\ &= 11 \end{aligned} \qquad \begin{aligned} f(x) &= 4x - 1 \\ f(2) &= 4(2) - 1 \\ &= 8 - 1 \\ &= 7 \end{aligned}$$

We then add the results to obtain $f(3) + f(2) = 11 + 7 = 18$.

b. We find $f(a)$ and $f(b)$ separately.

$$\begin{aligned} f(x) &= 4x - 1 \\ f(a) &= 4a - 1 \end{aligned} \qquad \begin{aligned} f(x) &= 4x - 1 \\ f(b) &= 4b - 1 \end{aligned}$$

We then subtract the results to obtain

$$f(a) - f(b) = (4a - 1) - (4b - 1) = 4a - 1 - 4b + 1 = 4a - 4b$$ ■

■ FINDING DOMAINS AND RANGES

EXAMPLE 6 Find the domain and range of the function defined by **a.** the ordered pairs $(-2, 4)$, $(0, 6)$, $(2, 8)$ and **b.** the equation $y = \frac{1}{x-2}$.

Solution **a.** The ordered pairs set up a correspondence between x and y where a single value of y corresponds to each x.

$x \longrightarrow y$
$-2 \longrightarrow 4$
$0 \longrightarrow 6$
$2 \longrightarrow 8$

The domain is the set of numbers x: $\{-2, 0, 2\}$. The range is the set of values y: $\{4, 6, 8\}$.

b. Since any number except 2 can be substituted for x in the equation $y = \frac{1}{x-2}$, the domain is $\{x \mid x \text{ is any real number except 2.}\}$. In interval notation, the domain is $(-\infty, 2) \cup (2, \infty)$.

Since a fraction with a numerator of 1 cannot be 0, the range is $\{y \mid y \text{ is any real number except 0.}\}$. In interval notation, the range is $(-\infty, 0) \cup (0, \infty)$. ■

The *graph of a function* is the graph of the ordered pairs $(x, f(x))$ that define the function. For the graph in Figure 2-40, the domain is shown on the x-axis, and the range is shown on the y-axis. For any x in the domain, there corresponds a value $y = f(x)$ in the range.

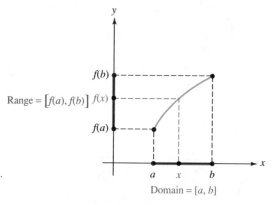

FIGURE 2-40

EXAMPLE 7 Find the domain and range of the function defined by $y = -2x + 1$.

Solution We graph the equation as in Figure 2-41. Since every real number x on the x-axis determines a corresponding value of y, the domain is the interval $(-\infty, \infty)$ shown on the x-axis. Since the values of y can be any real number on the y-axis, the range is the interval $(-\infty, \infty)$ shown on the y-axis.

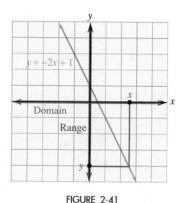

FIGURE 2-41 ■

■ THE VERTICAL LINE TEST

The **vertical line test** can be used to determine whether the graph of an equation represents a function. If any vertical line intersects a graph more than once, the graph cannot represent a function, because to one number x there would correspond more than one value of y.

The graph in Figure 2-42(a) represents a function, because every vertical line that intersects the graph does so exactly once. The graph in Figure 2-42(b) does not represent a function, because some vertical lines intersect the graph more than once.

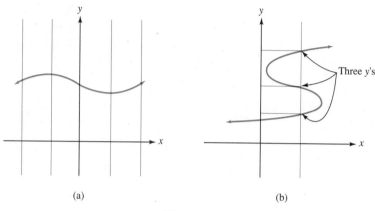

(a)   (b)

FIGURE 2-42

■ LINEAR FUNCTIONS

In Section 2.2, we graphed equations whose graphs were lines. These equations define basic functions, called **linear functions**.

Linear Function
A **linear function** is a function defined by an equation that can be written in the form
$$f(x) = mx + b \qquad \text{or} \qquad y = mx + b$$
where m is the slope of its line graph and $(0, b)$ is the y-intercept.

EXAMPLE 8

Solve the equation $3x + 2y = 10$ for y to show that it defines a linear function. Then graph it to find its domain and range.

Solution   We solve the equation for y as follows:

$$3x + 2y = 10$$
$$2y = -3x + 10 \qquad \text{Subtract } 3x \text{ from both sides.}$$
$$y = -\frac{3}{2}x + 5 \qquad \text{Divide both sides by 2.}$$

Because the given equation can be written in the form $y = mx + b$, it defines a linear function. The slope of its line graph is $-\frac{3}{2}$, and the y-intercept is $(0, 5)$. The graph appears in Figure 2-43. From the graph, we can see that both the domain and the range are the interval $(-\infty, \infty)$.

5, 7, 9, 11

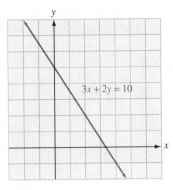

FIGURE 2-43

Orals   *Tell whether each equation or inequality determines y to be a function of x.*

1. $y = 2x + 1$ **2.** $y \geq 2x$ **3.** $y^2 = x$

If $f(x) = 2x + 1$, find

4. $f(0)$ **5.** $f(1)$ **6.** $f(-2)$

EXERCISE 2.5

In Exercises 1–4, give the domain and range of each relation and tell whether the relation is a function.

1. *Batter*   *Most Home Runs*
Roger Maris ⟶ 61
Babe Ruth ⟶ 60
Jimmy Foxx ⟶ 58
Hank Greenberg ⟶ 58
Hack Wilson ⟶ 56

2. *Golfer*   *Tournament Champion*
Hale Irwin ⟶ United States Open
Jack Nicklaus ⟶ The Masters
Greg Norman ⟶ The British Open

3. Green ⟶ 1
Red ⟶ 5
Yellow ⟶ 10
Blue ⟶ 15

4. Pie ⟶ 2nd place
Cake ⟶ 1st place
Bread ⟶ 3rd place
Brownies ⟶ 4th place

In Exercises 5–12, tell whether each equation determines y to be a function of x.

5. $y = 2x + 3$ **6.** $y = 4x - 1$ **7.** $y = 2x^2$ **8.** $y^2 = x + 1$
9. $y = 3 + 7x^2$ **10.** $y^2 = 3 - 2x$ **11.** $x = |y|$ **12.** $y = |x|$

In Exercises 13–20, find $f(3)$ and $f(-1)$.

13. $f(x) = 3x$ **14.** $f(x) = -4x$ **15.** $f(x) = 2x - 3$ **16.** $f(x) = 3x - 5$
17. $f(x) = 7 + 5x$ **18.** $f(x) = 3 + 3x$ **19.** $f(x) = 9 - 2x$ **20.** $f(x) = 12 + 3x$

In Exercises 21–28, find $f(2)$ and $f(3)$.

21. $f(x) = x^2$ **22.** $f(x) = x^2 - 2$ **23.** $f(x) = x^3 - 1$ **24.** $f(x) = x^3$

25. $f(x) = (x + 1)^2$ **26.** $f(x) = (x - 3)^2$ **27.** $f(x) = 2x^2 - x$ **28.** $f(x) = 5x^2 + 2x$

In Exercises 29–36, find $f(2)$ and $f(-2)$.

29. $f(x) = |x| + 2$ **30.** $f(x) = |x| - 5$ **31.** $f(x) = x^2 - 2$ **32.** $f(x) = x^2 + 3$

33. $f(x) = \dfrac{1}{x + 3}$ **34.** $f(x) = \dfrac{3}{x - 4}$ **35.** $f(x) = \dfrac{x}{x - 3}$ **36.** $f(x) = \dfrac{x}{x^2 + 2}$

In Exercises 37–40, find $g(w)$ and $g(w + 1)$.

37. $g(x) = 2x$ **38.** $g(x) = -3x$ **39.** $g(x) = 3x - 5$ **40.** $g(x) = 2x - 7$

In Exercises 41–48, $f(x) = 2x + 1$. Find each value.

41. $f(3) + f(2)$ **42.** $f(1) - f(-1)$ **43.** $f(b) - f(a)$ **44.** $f(b) + f(a)$

45. $f(b) - 1$ **46.** $f(b) - f(1)$ **47.** $f(0) + f\left(-\frac{1}{2}\right)$ **48.** $f(a) + f(2a)$

In Exercises 49–52, find the domain and range of each function.

49. $\{(-2, 3), (4, 5), (6, 7)\}$ **50.** $\{(0, 2), (1, 2), (3, 4)\}$ **51.** $f(x) = \dfrac{1}{x - 4}$ **52.** $f(x) = \dfrac{5}{x + 1}$

In Exercises 53–56, each graph represents a relation. Give the domain and range of each relation and use the vertical line test to tell whether it is a function.

53. 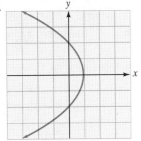 **54.** 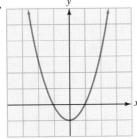 **55.** 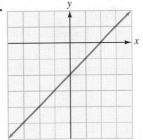 **56.**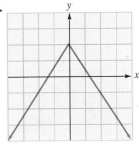

In Exercises 57–60, draw the graph of each linear function. Give the domain and range.

57. $f(x) = 2x - 1$ **58.** $f(x) = -x + 2$ **59.** $2x - 3y = 6$ **60.** $3x + 2y = -6$

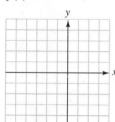   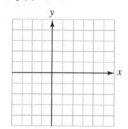   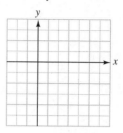   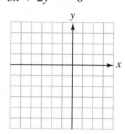

In Exercises 61–64, tell whether each equation defines a linear function.

61. $y = 3x^2 + 2$ **62.** $y = \dfrac{x - 3}{2}$ **63.** $x = 3y - 4$ **64.** $x = \dfrac{8}{y}$

65. Ballistics A bullet shot straight upward is s feet above the ground after t seconds, where $s = f(t) = -16t^2 + 256t$. Find the height of the bullet 3 seconds after it is shot.

66. Artillery fire A mortar shell is s feet above the ground after t seconds, where $s = f(t) = -16t^2 + 512t + 64$. Find the height of the shell 20 seconds after it is fired.

67. Conversion from degrees Celsius to degrees Fahrenheit The temperature in degrees Fahrenheit that is equivalent to a temperature in degrees Celsius is given by the function $F(C) = \frac{9}{5}C + 32$. Find the Fahrenheit temperature that is equivalent to 25° C.

68. Conversion from degrees Fahrenheit to degrees Celsius The temperature in degrees Celsius that is equivalent to a temperature in degrees Fahrenheit is given by the function $C(F) = \frac{5}{9}F - \frac{160}{9}$. Find the Celsius temperature that is equivalent to 14° F.

69. Selling tape recorders An electronics firm manufactures tape recorders, receiving $120 for each recorder it makes. If x represents the number of recorders produced, the income received is determined by the *revenue function* $R(x) = 120x$. The manufacturer has fixed costs of $12,000 per month and variable costs of $57.50 for each recorder manufactured. Thus, the *cost function* is $C(x) = 57.50x + 12,000$. How many recorders must the company sell for revenue to equal cost?

70. Selling tires A tire company manufactures premium tires, receiving $130 for each tire it makes. If the manufacturer has fixed costs of $15,512.50 per month and variable costs of $93.50 for each tire manufactured, how many tires must the company sell for revenue to equal cost? (*Hint:* See Exercise 69.)

In Exercises 71–76, follow the directions.

71. Graph $y = 3(x + k) + 1$ for several positive values of k. What do you notice?

72. Graph $y = 3(x + k) + 1$ for several negative values of k. What do you notice?

73. Graph $y = 3(kx) + 1$ for several values of k, where $k > 1$. What do you notice?

74. Graph $y = 3(kx) + 1$ for several values of k, where $0 < k < 1$. What do you notice?

75. Graph $y = 3(kx) + 1$ for several values of k, where $k < -1$. What do you notice?

76. Graph $y = 3(kx) + 1$ for several values of k, where $-1 < k < 0$. What do you notice?

Writing Exercises   *Write a paragraph using your own words.*

1. Explain why a relation is not always a function.

2. Explain the concepts of range and domain.

Something to Think About   *Let $f(x) = 2x + 1$ and $g(x) = x^2$. Assume that $f(x) \neq 0$ and $g(x) \neq 0$.*

1. Is $f(x) + g(x) = g(x) + f(x)$?   **2.** Is $f(x) - g(x) = g(x) - f(x)$?

Review Exercises   *Solve each equation.*

1. $\dfrac{y + 2}{2} = 4(y + 2)$ **2.** $\dfrac{3z - 1}{6} - \dfrac{3z + 4}{3} = \dfrac{z + 3}{2}$

3. $\dfrac{2}{x - 3} - 1 = -\dfrac{1}{3}$ **4.** $\dfrac{5}{x} + \dfrac{6}{x(x + 2)} = \dfrac{-3}{x + 2}$

2.6 Graphs of Other Functions

■ SOLVING EQUATIONS WITH A GRAPHING DEVICE

In this section, we will graph many functions whose graphs are not straight lines.

EXAMPLE 1   Graph the function $y = f(x) = x^2 - 2$.

Solution   We make a table of ordered pairs $(x, f(x))$ by picking values for x and finding the corresponding values of $f(x)$. For example, if $x = -3$, then $f(x) = 7$:

$$f(x) = x^2 - 2$$
$$f(-3) = (-3)^2 - 2 \qquad \text{Substitute } -3 \text{ for } x.$$
$$= 9 - 2$$
$$= 7$$

After plotting the points listed in the table shown in Figure 2-44(a) and joining them with a smooth curve, we get the **parabola** that is shown in Figure 2-44(b).

If we graph the equation with a graphing device with settings of $[-5, 5]$ for x and $[-5, 5]$ for y, we obtain Figure 2-44(c). In interval notation, the domain of the function $f(x) = x^2 - 2$ is $(-\infty, \infty)$, and the range is $[-2, \infty)$.

$y = f(x) = x^2 - 2$

x	$f(x)$	$(x, f(x))$
-3	7	$(-3, 7)$
-2	2	$(-2, 2)$
-1	-1	$(-1, -1)$
0	-2	$(0, -2)$
1	-1	$(1, -1)$
2	2	$(2, 2)$
3	7	$(3, 7)$

(a)

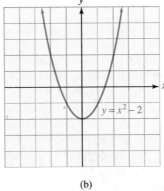

(b)

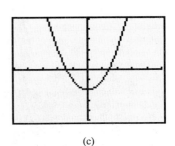

(c)

FIGURE 2-44

EXAMPLE 2 Graph the function $y = f(x) = -|x| + 3$.

Solution Again we make a table of ordered pairs $(x, f(x))$ by picking values for x and finding the corresponding values of $f(x)$. For example, if $x = -2$, then $f(x) = 1$:

$$f(x) = -|x| + 3$$
$$f(-2) = -|-2| + 3$$
$$= -(2) + 3 \qquad |-2| = 2.$$
$$= 1$$

After plotting the points in the table shown in Figure 2-45(a), we obtain the graph shown in Figure 2-45(b). If we graph the equation with a graphing device with settings of $[-5, 5]$ for x and $[-5, 5]$ for y, we obtain Figure 2-45(c). In interval notation, the domain of the function $f(x) = -|x| + 3$ is $(-\infty, \infty)$, and the range is $(-\infty, 3]$.

$y = f(x) = -|x| + 3$

x	$f(x)$	$(x, f(x))$
-3	0	$(-3, 0)$
-2	1	$(-2, 1)$
-1	2	$(-1, 2)$
0	3	$(0, 3)$
1	2	$(1, 2)$
2	1	$(2, 1)$
3	0	$(3, 0)$

(a)

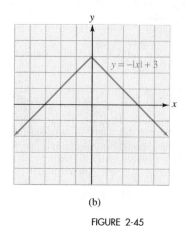

(b)

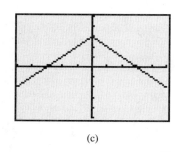

(c)

FIGURE 2-45

EXAMPLE 3 Use a graphing device to graph the function defined by the equation $xy = 4$.

Solution Before we can enter the equation $xy = 4$ into a graphing device, we must solve the equation for y:

$$xy = 4$$
$$y = \frac{4}{x} \qquad \text{Divide both sides by } x.$$

To graph the equation $y = \frac{4}{x}$ with settings of $[10, 20]$ for x and $[10, 20]$ for y, we enter the values and enter the right-hand side of the equation. When we press GRAPH, we get the blank window shown in Figure 2-46(a). Thus, we need to graph the equation again using a different window.

To graph the equation using $[0, 5]$ for x and $[0, 5]$ for y, we enter the values. There is no need to re-enter the equation. When we press GRAPH, we get the graph shown in Figure 2-46(b). Here, only part of the graph appears. To see the whole graph, we need a larger window.

To graph the equation using $[-6, 6]$ for x and $[-6, 6]$ for y, we enter the values and press GRAPH. This time, we see both branches of the curve, called a **hyperbola**, shown in Figure 2-46(c). The domain of the function $f(x) = \frac{4}{x}$ is the interval $(-\infty, 0) \cup (0, \infty)$, and the range is $(-\infty, 0) \cup (0, \infty)$.

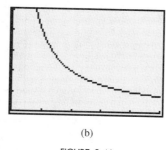
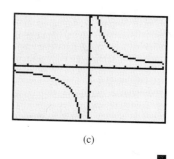

(a)   (b)   (c)

FIGURE 2-46

The choice of x- and y-values makes a big difference in how a graph looks in the viewing window. One of the challenges of using graphing devices is finding the right values to give an appropriate viewing window.

 WARNING! When using a graphing device, be sure the viewing window does not show a misleading graph.

EXAMPLE 4 Graph the function $y = f(x) = -2x^2 + 15$.

Solution To graph $y = f(x) = -2x^2 + 15$ using $[1, 5]$ for x and $[-5, 5]$ for y, we enter the values, enter the equation, and press GRAPH to get the graph shown in Figure 2-47(a). Although the graph appears to be a line, it is not. We are seeing a part of a parabola that only appears to be straight. If we pick a different window, using $[-5, 5]$ for x and $[-5, 20]$ for y, we will see a more complete picture of the parabola, as shown in Figure 2-47(b). In interval notation, the domain of the function $f(x) = -2x^2 + 15$ is $(-\infty, \infty)$, and the range is $(-\infty, 15]$.

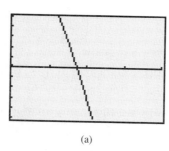
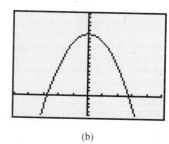

(a)   (b)

FIGURE 2-47

EXAMPLE 5 Graph the function $y = f(x) = |x - 2|$.

Solution To graph the function $y = f(x) = |x - 2|$, we must pick an appropriate viewing window. Since absolute values are always nonnegative, the minimum value of y in this equation will be 0, and this occurs when $x = 2$. To place the point $(2, 0)$ near the center of the viewing window, we pick values of $[-3, 7]$ for x and $[-2, 8]$ for y.

We enter the right-hand side of the equation and press GRAPH, producing the graph shown in Figure 2-48. In interval notation, the domain of the function $f(x) = |x - 2|$ is $(-\infty, \infty)$, and the range is $[0, \infty)$.

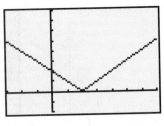

FIGURE 2-48

■ SOLVING EQUATIONS WITH A GRAPHING DEVICE

EXAMPLE 6   Solve the equation $2(x - 3) + 3 = 7$.

Solution   To use a graphing device to solve the equation $2(x - 3) + 3 = 7$, we graph both the left-hand side and the right-hand side of the equation in the same window, as shown in Figure 2-49(a). We then trace to find the coordinates of the point where the two graphs intersect, as shown in Figure 2-49(b). We can then zoom and trace again to get Figure 2-49(c). Solving the equation algebraically will show that the coordinates of the intersection point are indeed $(5, 7)$, and $x = 5$.

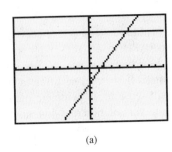
(a)

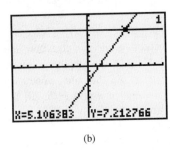
(b)

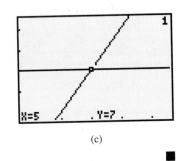
(c)

FIGURE 2-49

Orals   **1.** Describe a parabola.
        2. Describe the graph of $f(x) = |x| + 3$.
        3. Describe the graph of $xy = 9$.
        4. Tell why the choice of a viewing window is important.

EXERCISE 2.6

In Exercises 1–8, graph each equation by plotting points.

1. $f(x) = x^2 - 3$

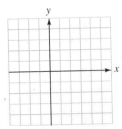

2. $f(x) = -x^2 + 2$

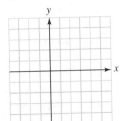

3. $f(x) = (x - 1)^2$

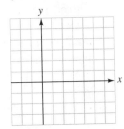

4. $f(x) = -(x - 1)^2$

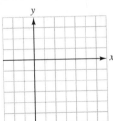

5. $f(x) = -|x| + 2$

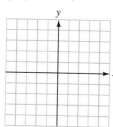

6. $f(x) = |x + 3|$

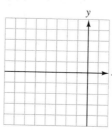

7. $f(x) = |x - 1| - 2$

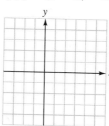

8. $f(x) = -|x + 2| + 4$

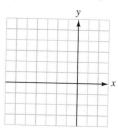

In Exercises 9–16, use a graphing device to graph each function. Use values of $[-5, 5]$ for x and $[-5, 5]$ for y. Compare the graphs with your handmade graphs in Exercises 1–8.

9. $f(x) = x^2 - 3$

10. $f(x) = -x^2 + 2$

11. $f(x) = (x - 1)^2$

12. $f(x) = -(x - 1)^2$

13. $f(x) = -|x| + 2$

14. $f(x) = |x + 3|$

15. $f(x) = |x - 1| - 2$

16. $f(x) = -|x + 2| + 4$

In Exercises 17–24, graph each function using values of $[-4, 4]$ for x and $[-4, 4]$ for y. The graph is not what it appears to be. Pick a better viewing window and find the true graph.

17. $f(x) = x^2 + 8x$

18. $f(x) = x^2 - 8x$

19. $f(x) = |x + 5|$

20. $f(x) = |x - 5|$

21. $f(x) = x^2 - 9x + 14$

22. $f(x) = x^2 + 9x + 14$

23. $f(x) = x^3 + 8$

24. $f(x) = x^3 - 12$

In Exercises 25–32, solve each equation for y and use a graphing device to graph it.

25. $2x - 5y = 10$

26. $3x + 4y = 12$

27. $x^2 + y = 6$

28. $x^2 - y = -4$

29. $|x| - 2y = 4$

30. $|x| + \frac{1}{2}y = 2$

31. $xy = 12$

32. $xy = -6$

In Exercises 33–38, use a graphing device to solve each equation.

33. $3x + 6 = 0$

34. $7x - 21 = 0$

35. $4(x - 1) = 3x$

36. $4(x - 3) - x = x - 6$

37. $11x + 6(3 - x) = 3$

38. $2(x + 2) = 2(1 - x) + 10$

In Exercises 39–46, graph each parabola and find the coordinates of its highest or lowest point. This point is called the vertex of the parabola.

39. $y = x^2 + 4x$

40. $y = x^2 - 2x$

41. $y = -x^2 + 6x$

42. $y = -x^2 + 4x$

43. $y = x^2 + 4x - 5$

44. $y = x^2 - 5x - 2$

45. $y = -x^2 - 4x + 2$

46. $y = -x^2 - 4x + 1$

Writing Exercises   *Write a paragraph using your own words.*

1. Explain how to graph an equation by plotting points.

2. Explain why the correct choice of x- and y-values is important when using a graphing device.

Something to Think About

1. Use a graphing device with settings of $[-10, 10]$ *for x and* $[-10, 10]$ *for y to graph*

  a. $y = x^2$
  b. $y = x^2 + 1$
  c. $y = x^2 + 2$
  What do you notice?

2. Use a graphing device with settings of $[-10, 10]$ *for x and* $[-10, 10]$ *for y to graph*

  a. $y = -|x|$
  b. $y = -|x| - 1$
  c. $y = -|x| - 2$
  What do you notice?

Review Exercises

1. List the prime numbers between 40 and 50.

2. State the associative property of addition.

3. State the commutative property of multiplication.

4. What is the additive identity element?

5. What is the multiplicative identity element?

6. Find the multiplicative inverse of $\frac{5}{3}$.

■ ■ ■ ■ ■ ■ ■ ■ ■ ■ PROBLEMS AND PROJECTS

1. A fax machine is purchased for $795. Six years later, it has a salvage value of $105. Find the depreciation equation. How much will the fax be worth in 4 years? Find the annual rate of depreciation.

2. The revenue function of a certain product is given by $R = 125 + 45x - 0.01x^2$, where x represents the number of units the company produces. Use a graphing device to find the maximum revenue. What production level will give the maximum revenue? Give each answer to the nearest ten.

3. Big Buck Outdoor Company has a revenue equation given by $R = x + 15$. Its cost equation is given by $C = 40 - 2x$. Use a graphing device to find the company's equilibrium point. Give the answer to the nearest hundredth.

4. A bolt fired from a crossbow is s feet above the ground t seconds after it is released. Its height is given by $s(t) = -16t^2 + 375t + 81$. Use a graphing device to find the height of the bolt 6 seconds and 14 seconds after it was released. Find the maximum height of the bolt. How long it takes for the bolt to hit the ground?

PROJECT 1   The Board of Administrators of Boondocks County has hired your consulting firm to plan a highway. A new highway is to be built in the outback section of the county, an area with a rugged terrain where road building is hard and expensive. The Board has hired you because they want to get everything right the first time.

The two main roads in the outback section are Highway N, running in a straight line north and south, and Highway E, running in a straight line east and west. These two highways meet at an intersection that the locals call Four Corners. The only other county road in the area is Slant Road, which runs in a straight line from northwest to southeast, cutting across Highway N north of Four Corners and Highway E east of Four Corners.

The county clerk is unable to find an official map of the area, but there is an old sketch made by the original road designer. It shows that if a rectangular coordinate system is set up using Highways N and E as the axes and Four Corners as the origin, then the equation of the line representing Slant Road is

$$2x + 3y = 12 \qquad \text{(where the unit length is 1 mile)}$$

Given this information, the county wants you to do the following:

a. Update the current information by giving the coordinates of the intersections of Slant Road with Highway N and Highway E.

b. Plan a new highway, Country Drive, that will begin 1 mile north of Four Corners and run in a straight line in a generally northeasterly direction, intersecting Slant Road at right angles. The county wants to know the equation of the line representing Country Drive. You should also state the domain on which this equation is valid as a representation of Country Drive.

PROJECT 2   You are representing your branch of the large Buy-from-Us corporation at the company's regional meeting, and you are looking forward to presenting your revenue and cost reports to the other branch representatives. But now disaster strikes! The graphs you had planned to present, containing cost and revenue information for this year and last year, are unlabeled! You cannot immediately recognize which graphs represent costs, which represent revenues, and which represent which year. Without these graphs, your presentation will not be effective.

The only other information you have with you is in the notes you made for your talk. From these you are able to glean the following financial data about your branch.

1. All cost and revenue figures on the graphs are rounded to the nearest $50,000.

2. Costs for the fourth quarter of last year were $400,000.

3. Revenue was not above $400,000 for any quarter last year.

4. Last year, your branch lost money during the first quarter.

5. This year, your branch made money during three of the four quarters.

6. Profit during the second quarter of this year was $150,000.

And, of course, you know that Profit = revenue − cost.

With this information, you must match each of the graphs on the next page with one of the following titles:

Costs, This Year   Costs, Last Year

Revenues, This Year   Revenues, Last Year

You should be sure to have sound reasons for your choices—reasons that guarantee that no other arrangement of the titles will fit the data. The *last* thing you want to do is present incorrect information to the company bigwigs!

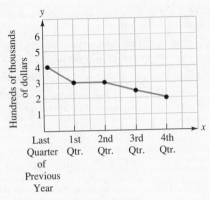

ILLUSTRATION 1

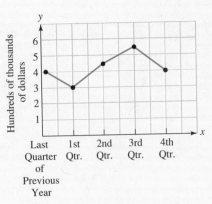

ILLUSTRATION 2

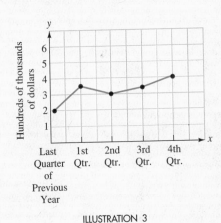

ILLUSTRATION 3

ILLUSTRATION 4

■ Chapter Summary

KEY WORDS

constants (2.2)

coordinates (2.2)

dependent variable (2.5)

domain (2.5)

function (2.5)

general form of a line (2.4)

graph of an equation (2.2)

hyperbola (2.6)

independent variable (2.5)

linear equation (2.2)

linear function (2.5)

midpoint (2.2)

negative reciprocal (2.3)

ordered pairs (2.2)

origin (2.2)

parabola (2.6)

point–slope form of a line (2.4)

range (2.5)

rectangular coordinate
  system (2.2)

relation (2.5)

rise (2.3)

run (2.3)

slope–intercept form of a
  line (2.4)

slope of a nonvertical
  line (2.3)

vertical line test (2.5)

x-axis (2.2)

x-coordinate (2.2)

x-intercept (2.2)

y-axis (2.2)

y-coordinate (2.2)

y-intercept (2.2)

Key Ideas

(2.1) Information can be conveniently presented in tables and graphs.

(2.2) The graph of $x = a$ is a vertical line with x-intercept at $(a, 0)$.

The graph of $y = b$ is a horizontal line with y-intercept at $(0, b)$.

Midpoint formula: If the endpoints of a line segment are $P(x_1, y_1)$ and $Q(x_2, y_2)$, the midpoint of segment PQ is

$$M\left(\frac{x_1 + x_2}{2}, \frac{y_1 + y_2}{2}\right)$$

(2.3) **The slope of a nonvertical line**:

$$m = \frac{\Delta y}{\Delta x} = \frac{y_2 - y_1}{x_2 - x_1} \qquad (x_2 \neq x_1)$$

Horizontal lines have a slope of 0.

Vertical lines have no defined slope.

Nonvertical parallel lines have the same slope, and lines with the same slope are parallel.

If two nonvertical lines are perpendicular, their slopes are negative reciprocals.

If the slopes of two lines are negative reciprocals, the lines are perpendicular.

(2.4) **Equations of a line**:
  Point–slope form: $y - y_1 = m(x - x_1)$
  Slope–intercept form: $y = mx + b$
  General form: $Ax + By = C$

(2.5) The **vertical line test** can be used to determine whether a graph represents a function.

(2.6) Graphs of nonlinear equations are not lines.

■ Chapter 2 Review Exercises

In Review Exercises 1–4, refer to Illustration 1.

1. What is the most popular size of rental car?

2. What percent of the public rents luxury cars?

3. What percent of the public rents either a full-size or a luxury car?

4. Why are midsize cars most popular?

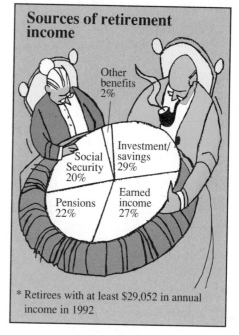

Sources of retirement income

* Retirees with at least $29,052 in annual income in 1992

ILLUSTRATION 2

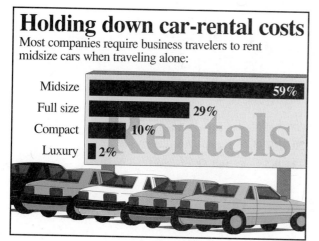

Holding down car-rental costs

Most companies require business travelers to rent midsize cars when traveling alone:

Midsize	59%
Full size	29%
Compact	10%
Luxury	2%

ILLUSTRATION 1

In Review Exercises 5–8, refer to Illustration 2.

5. What is the greatest source of income for retirees?

6. What percent of retiree income comes from social security and pensions?

7. What percent of retirees have earned income?

8. Can we conclude that only 20% of retirees get social security?

In Review Exercises 9–10, assume that the numbers of hours worked each day by students enrolled in a mathematics class are as follows: 5, 4, 5, 6, 4, 3, 0, 3, 4, 5, 3, 2, 3, 2, 0, 3, 3, 4, 5, 3, 3, 4, 3, 5, 4, 3, 4, 3, 4, 3.

9. Construct a frequency table.

Hours	Frequency

10. Construct a bar graph for the frequency distribution.

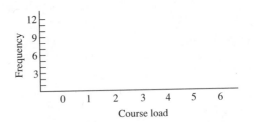

In Review Exercises 11–18, graph each equation.

11. $x + y = 4$

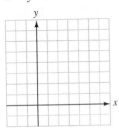

12. $2x - y = 8$

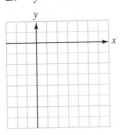

13. $y = 3x + 4$

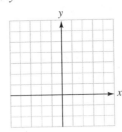

14. $x = 4 - 2y$

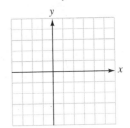

15. $y = 4$

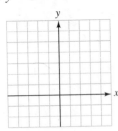

16. $x = -2$

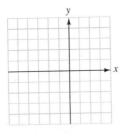

17. $2(x + 3) = x + 2$

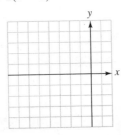

18. $3y = 2(y - 1)$

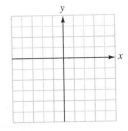

In Review Exercises 19–22, find the slope of the line passing through points P and Q.

19. $P(2, 5)$ and $Q(5, 8)$

20. $P(-3, -2)$ and $Q(6, 12)$

21. $P(-3, 4)$ and $Q(-5, -6)$

22. $P(5, -4)$ and $Q(-6, -9)$

In Review Exercises 23–26, find the slope of the graph of each equation, if one exists.

23. $2x - 3y = 18$

24. $2x + y = 8$

25. $-2(x - 3) = 10$

26. $3y + 1 = 7$

In Review Exercises 27–30, tell whether the lines with the given slopes are parallel, perpendicular, or neither.

27. $m_1 = 4$, $m_2 = -\dfrac{1}{4}$

28. $m_1 = 0.5$, $m_2 = \dfrac{1}{2}$

29. $m_1 = 0.5$, $m_2 = -\dfrac{1}{2}$

30. $m_1 = 5$, $m_2 = -0.2$

In Review Exercises 31–34, write the equation of the line with the given properties. Write each answer in general form.

31. Slope of 3; passing through $P(-8, 5)$.

32. Passing through $(-2, 4)$ and $(6, -9)$.

33. Passing through $(-3, -5)$; parallel to the graph of $3x - 2y = 7$.

34. Passing through $(-3, -5)$; perpendicular to the graph of $3x - 2y = 7$.

In Review Exercises 35–38, tell whether each equation determines y to be a function of x.

35. $y = 6x - 4$

36. $y = 4 - x$

37. $y^2 = x$

38. $|y| = x^2$

In Review Exercises 39–42, assume that $f(x) = 3x + 2$ and $g(x) = x^2 - 4$. Find each value.

39. $f(-3)$

40. $g(8)$

41. $g(-2)$

42. $f(5)$

In Review Exercises 43–48, find the domain and range of each function or relation.

43. $f(x) = 4x - 1$

44. $f(x) = 3x - 10$

45. $f(x) = x^2 + 1$

46. $f(x) = \dfrac{4}{2 - x}$

47. $f(x) = \dfrac{x}{x - 3}$

48. $y^2 = 4x$

In Review Exercises 49–52, use the vertical line test to determine whether each graph represents a function.

49.

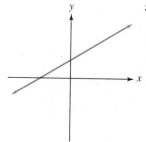

50.

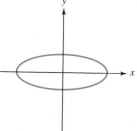

51.

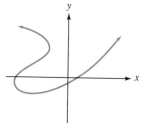

52.

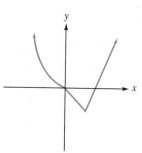

In Review Exercises 53–56, tell which equations define linear functions.

53. $y = 3x + 2$

54. $y = \dfrac{x + 5}{4}$

55. $4x - 3y = 12$

56. $y = x^2 - 25$

In Review Exercises 57–60, graph each function.

57. $f(x) = x^2 - 3$

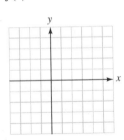

58. $f(x) = |x| - 4$

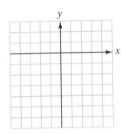

59. $f(x) = \dfrac{10}{x}$

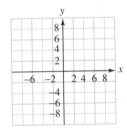

60. $f(x) = x^2 - 6x$

In Review Exercises 61–64, use a graphing device to graph each function. Compare the results to Exercises 57–60.

61. $f(x) = x^2 - 3$

62. $f(x) = |x| - 4$

63. $f(x) = \dfrac{10}{x}$

64. $f(x) = x^2 - 6x$

Chapter 2 Test

In Problems 1–2, refer to Illustration 1.

1. Which investment has provided the greatest return?

2. Which investment has provided the lowest return?

3. Graph the equation $2x - 5y = 10$.

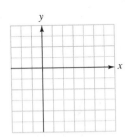

4. Find the x- and y-intercepts of the graph of $y = \dfrac{x - 3}{5}$.

In Problems 5–8, find the slope of each line, if possible.

5. The line through $P(-2, 4)$ and $Q(6, 8)$

6. The graph of $2x - 3y = 8$

7. The graph of $x = 12$

8. The graph of $y = 12$

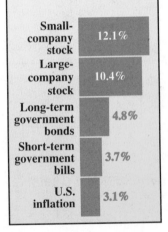

Stocks offer best odds for growth

Stocks traditionally have been the best way to stay ahead of inflation, which saps the purchasing power of retirees. How the average annual return of different investments has changed compared with inflation from 1926-1991:

Small-company stock 12.1%
Large-company stock 10.4%
Long-term government bonds 4.8%
Short-term government bills 3.7%
U.S. inflation 3.1%

ILLUSTRATION 1

9. Write the equation of the line with slope of $\frac{2}{3}$ that passes through $P(4, -5)$. Give the answer in slope–intercept form.

10. Write the equation of the line that passes through $P(-2, 6)$ and $Q(-4, -10)$. Give the answer in general form.

11. Find the slope and the y-intercept of the graph of $-2(x - 3) = 3(2y + 5)$.

12. Determine whether the graphs of $4x - y = 12$ and $y = \frac{1}{4}x + 3$ are parallel, perpendicular, or neither.

13. Determine whether the graphs of $y = -\frac{2}{3}x + 4$ and $2y = 3x - 3$ are parallel, perpendicular, or neither.

14. Write the equation of the line that passes through the origin and is parallel to the graph of $y = \frac{3}{2}x - 7$.

15. Write the equation of the line that passes through $P(-3, 6)$ and is perpendicular to the graph of $y = -\frac{2}{3}x - 7$.

16. Does $|y| = x$ define y to be a function of x?

17. Find the domain and range of the function $f(x) = |x|$.

18. Find the domain and range of the function $f(x) = x^3$.

In Problems 19–22, $f(x) = 3x + 1$ and $g(x) = x^2 - 2$. Find each value.

19. $f(3)$

20. $g(0)$

21. $f(a)$

22. $g(-x)$

In Problems 23–24, tell whether each graph represents a function.

23.

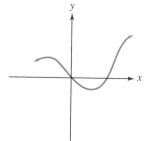

24.

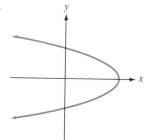

25. Graph $f(x) = -x^2 + 3$.

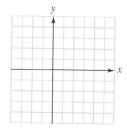

26. Graph $f(x) = -|x + 2|$.

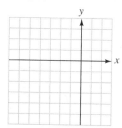

■ Cumulative Review Exercises (Chapters 1–2)

In Exercises 1–10, tell which numbers in the set $\left\{-2, 0, 1, 2, \frac{13}{12}, 6, 7, \sqrt{5}, \pi\right\}$ are in each category.

1. natural numbers

2. whole numbers

3. rational numbers

4. irrational numbers

5. negative numbers

6. real numbers

7. prime numbers

8. composite numbers

9. even numbers

10. odd numbers

In Exercises 11–12, graph each interval on the number line.

11. $-2 < x \leq 5$

12. $[-5, 0) \cap [-3, 6]$

In Exercises 13–14, simplify each expression.

13. $-|5| + |-3|$

14. $\dfrac{|-5| + |-3|}{-|4|}$

In Exercises 15–18, do the operations.

15. $2 + 4 \cdot 5$

16. $\dfrac{8 - 4}{2 - 4}$

17. $20 \div (-10 \div 2)$

18. $\dfrac{6 + 3(6 + 4)}{2(3 - 9)}$

In Exercises 19–20, $x = 2$ and $y = -3$. Evaluate each expression.

19. $-x - 2y$

20. $\dfrac{x^2 - y^2}{2x + y}$

In Exercises 21–24, tell which property of real numbers justifies each statement.

21. If $3 = x$ and $x = y$, then $3 = y$.

22. $3(x + y) = 3x + 3y$

23. $(a + b) + c = c + (a + b)$

24. $(ab)c = a(bc)$

In Exercises 25–28, simplify each expression. Assume that all variables are positive numbers, and write all answers without negative exponents.

25. $(x^2y^3)^4$

26. $\dfrac{c^4 c^8}{(c^5)^2}$

27. $\left(-\dfrac{a^3 b^{-2}}{ab}\right)^{-1}$

28. $\left(\dfrac{-3a^3 b^{-2}}{6a^{-2} b^3}\right)^0$

29. Change 0.00000497 to scientific notation.

30. Change 9.32×10^8 to standard notation.

In Exercises 31–34, solve each equation.

31. $2x - 5 = 11$

32. $\dfrac{2x - 6}{3} = x + 7$

33. $4(y - 3) + 4 = -3(y + 5)$

34. $2x - \dfrac{3(x - 2)}{2} = 7 - \dfrac{x - 3}{3}$

35. Change 0.875 to a common fraction.

36. Change $0.\overline{45}$ to a common fraction.

In Exercises 37–38, solve each formula.

37. $S = \dfrac{n(a + l)}{2}$ for a

38. $A = \dfrac{1}{2}h(b_1 + b_2)$ for h

39. The sum of three consecutive even integers is 90. Find the integers.

40. A rectangle is three times as long as it is wide. If its perimeter is 112 centimeters, find its dimensions.

41. Graph $2x - 3y = 6$ and tell whether it defines a function.

42. Find the slope of a line passing through $P(-2, 5)$ and $Q(8, -9)$.

43. Write the equation of the line passing through $P(-2, 5)$ and $Q(8, -9)$.

44. Write the equation of the line passing through $P(-2, 3)$ and parallel to the graph of $3x + y = 8$.

In Exercises 45–48, $f(x) = 3x^2 + 2$ and $g(x) = 2x - 1$. Evaluate each expression.

45. $f(-1)$

46. $g(0)$

47. $g(t)$

48. $f(-r)$

In Exercises 49–50, graph each equation and tell whether it is a function. If it is a function, give the domain and range.

49. $y = -x^2 + 1$

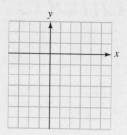

50. $y = \left| \dfrac{1}{2}x - 3 \right|$

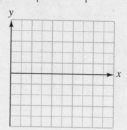

Systems of Equations

3

Electrical/Electronic Engineer

Electrical engineers design, develop, test, and supervise the manufacture of electronic equipment. Electrical engineers who work with electronic equipment are often called *electronic engineers.*

SAMPLE APPLICATION ■ In a radio, an inductor and a capacitor are used in a resonant circuit to select a desired radio station at a frequency f and reject all others. The inductance L and the capacitance C determine the inductive reactance X_L and the capacitive reactance X_C of that circuit, where

$$X_L = 2\pi f L \qquad \text{and} \qquad X_C = \frac{1}{2\pi f C}$$

The radio station selected will be at the frequency f where $X_L = X_C$. Write a formula for f in terms of L and C.
See Exercise 71 in Exercise 3.2.

We have considered linear equations with the variables x and y. We found that each equation had infinitely many solutions (x, y), and that we could graph each equation on the rectangular coordinate system. In this chapter, we will discuss many **systems of linear equations** involving two or three equations.

3.1   Solution by Graphing

■ CONSISTENT SYSTEMS ■ INCONSISTENT SYSTEMS ■ DEPENDENT EQUATIONS
■ GRAPHING DEVICES

In the pair of equations

$$\begin{cases} x + 2y = 4 \\ 2x - y = 3 \end{cases}$$ The pair of equations is called a system of equations.

there are infinitely many ordered pairs (x, y) that satisfy the first equation and infinitely many ordered pairs (x, y) that satisfy the second equation. However, there is only one ordered pair (x, y) that satisfies both equations. The process of finding this ordered pair is called *solving the system*.

We follow these steps to solve a system of two equations in two variables by graphing.

> ### The Graphing Method
>
> **1.** On a single set of coordinate axes, graph each equation.
>
> **2.** Find the coordinates of the point (or points) where the graphs intersect. These coordinates give the solution of the system.
>
> **3.** If the graphs have no point in common, the system has no solution.
>
> **4.** If the graphs of the equations coincide, the system has infinitely many solutions.
>
> **5.** Check the solution in both of the original equations.

■ CONSISTENT SYSTEMS

When a system of equations has a solution (as in Example 1), the system is called a **consistent system.**

EXAMPLE 1   Solve the system $\begin{cases} x + 2y = 4 \\ 2x - y = 3 \end{cases}$.

Solution   We graph both equations on a single set of coordinate axes, as shown in Figure 3-1.

Although infinitely many ordered pairs (x, y) satisfy $x + 2y = 4$, and infinitely many ordered pairs (x, y) satisfy $2x - y = 3$, only the coordinates of the point where the graphs intersect satisfy both equations. Since the intersection point has coordinates of $(2, 1)$, the solution is the ordered pair $(2, 1)$ or $x = 2$ and $y = 1$.

To check the solution, we substitute 2 for x and 1 for y in each equation and verify that the ordered pair $(2, 1)$ satisfies each equation.

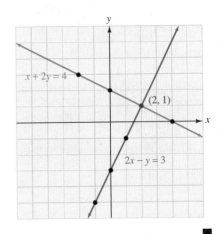

$x + 2y = 4$				$2x - y = 3$		
x	y	(x, y)		x	y	(x, y)
4	0	$(4, 0)$		1	-1	$(1, -1)$
0	2	$(0, 2)$		0	-3	$(0, -3)$
-2	3	$(-2, 3)$		-1	-5	$(-1, -5)$

FIGURE 3-1

■ INCONSISTENT SYSTEMS

When a system has no solution (as is the case in Example 2), it is called an **inconsistent system**.

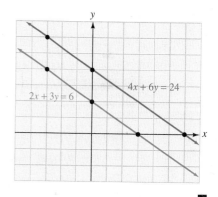

EXAMPLE 2   Solve the system $\begin{cases} 2x + 3y = 6 \\ 4x + 6y = 24 \end{cases}$.

Solution   We graph both equations on the same set of coordinate axes, as shown in Figure 3-2. In this example, the graphs are parallel, because the slopes of the two lines are equal and their y-intercepts are different. We can see that the slope of each line is $-\frac{2}{3}$ by writing each equation in slope–intercept form.

Since the graphs are parallel lines, the lines do not intersect, and the system does not have a solution. It is an inconsistent system.

$2x + 3y = 6$		
x	y	(x, y)
3	0	$(3, 0)$
0	2	$(0, 2)$
-3	4	$(-3, 4)$

$4x + 6y = 24$		
x	y	(x, y)
6	0	$(6, 0)$
0	4	$(0, 4)$
-3	6	$(-3, 6)$

FIGURE 3-2

■ DEPENDENT EQUATIONS

When the equations of a system have different graphs (as in Examples 1 and 2), the equations are called **independent equations**. Two equations with the same graph (as in Example 3) are called **dependent equations**.

EXAMPLE 3   Solve the system $\begin{cases} 2y - x = 4 \\ 2x + 8 = 4y \end{cases}$.

Solution   We graph each equation on the same set of coordinate axes, as shown in Figure 3-3. Since the graphs coincide, the system has infinitely many solutions. Any ordered pair (x, y) that satisfies one equation satisfies the other also.

From the tables of ordered pairs shown in Figure 3-3, we see that $(-4, 0)$ and $(0, 2)$ satisfy both equations. We can find infinitely many more solutions by finding additional ordered pairs (x, y) that satisfy either equation.

Because the two equations have the same graph, they are dependent equations.

$2y - x = 4$				$2x + 8 = 4y$		
x	y	(x, y)		x	y	(x, y)
-4	0	$(-4, 0)$		-4	0	$(-4, 0)$
0	2	$(0, 2)$		0	2	$(0, 2)$

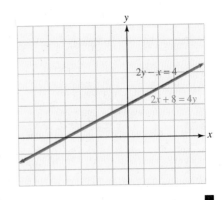

FIGURE 3-3

■

We summarize the possibilities that can occur when two equations, each with two variables, are graphed.

If the lines are different and intersect, the equations are independent and the system is consistent. **One solution exists.**

If the lines are different and parallel, the equations are independent and the system is inconsistent. **No solution exists.**

If the lines coincide, the equations are dependent and the system is consistent. **Infinitely many solutions exist.**

Two systems are called **equivalent** if they have the same solution set. In Example 4, we solve a more difficult system by changing it into a simpler equivalent system.

EXAMPLE 4   Solve the system $\begin{cases} \frac{3}{2}x - y = \frac{5}{2} \\ x + \frac{1}{2}y = 4 \end{cases}$.

Solution   We multiply both sides of $\frac{3}{2}x - y = \frac{5}{2}$ by 2 to eliminate the fractions and obtain the equation $3x - 2y = 5$. We multiply both sides of $x + \frac{1}{2}y = 4$ by 2 to eliminate the fractions and obtain the equation $2x + y = 8$.

The new system

$$\begin{cases} 3x - 2y = 5 \\ 2x + y = 8 \end{cases}$$

is equivalent to the original system and is easier to solve, because it has no fractions. If we graph each equation of the new system, as in Figure 3-4, we see that the two lines intersect at the point $(3, 2)$. Verify that $x = 3$ and $y = 2$ satisfy each equation in the original system.

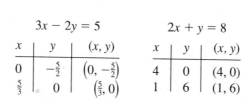

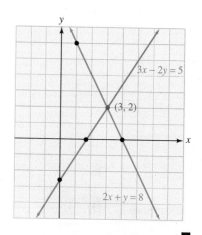

$3x - 2y = 5$

x	y	(x, y)
0	$-\frac{5}{2}$	$\left(0, -\frac{5}{2}\right)$
$\frac{5}{3}$	0	$\left(\frac{5}{3}, 0\right)$

$2x + y = 8$

x	y	(x, y)
4	0	$(4, 0)$
1	6	$(1, 6)$

FIGURE 3-4   ■

■ GRAPHING DEVICES

The graphing method has limitations. First, the method is limited to equations with two variables. Systems with three or more variables cannot be solved graphically. Second, it is often difficult to find exact solutions graphically. However, the trace and zoom capabilities of most graphing devices enable us to get very good approximations of such solutions.

EXAMPLE 5   Use a graphing device to solve the system $\begin{cases} 3x + 2y = 12 \\ 2x - 3y = 12 \end{cases}$.

Solution   To enter the equations, we must first solve them for y to get the following equivalent system:

$$\begin{cases} y = -\frac{3}{2}x + 6 \\ y = \frac{2}{3}x - 4 \end{cases}$$

If we use settings of $[-10, 10]$ for x and $[-10, 10]$ for y, the graphs of the equations will look like those in Figure 3-5(a). If we zoom in on the intersection point of the two lines and then trace, we will get the approximate solution shown in Figure 3-5(b). To get better results, we can do additional zooms. Verify that the exact solution is $x = \frac{60}{13}$ and $y = -\frac{12}{13}$.

On some calculators, we can use the INTERSECT feature to get an approximate solution.

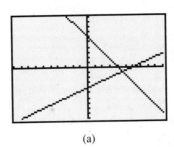
(a)

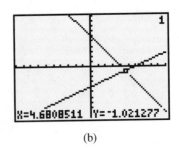
(b)

FIGURE 3-5

Orals *Tell whether the following systems will have one solution, no solution, or infinitely many solutions.*

1. $\begin{cases} y = 2x \\ y = 2x + 5 \end{cases}$

2. $\begin{cases} y = 2x \\ y = x + x \end{cases}$

3. $\begin{cases} y = 2x \\ y = -2x \end{cases}$

4. $\begin{cases} y = 2x + 1 \\ 2x = y \end{cases}$

EXERCISE 3.1

In Exercises 1–4, tell whether the ordered pair is a solution of the system of equations.

1. $(1, 2)$; $\begin{cases} y = 2x \\ y = \dfrac{1}{2}x + \dfrac{3}{2} \end{cases}$

2. $(-1, 2)$; $\begin{cases} y = 3x + 5 \\ y = x + 4 \end{cases}$

3. $(2, -3)$; $\begin{cases} y = \dfrac{1}{2}x - 2 \\ 3x + 2y = 0 \end{cases}$

4. $(-4, 3)$; $\begin{cases} 4x - y = -19 \\ 3x + 2y = -6 \end{cases}$

In Exercises 5–24, solve each system by graphing, if possible.

5. $\begin{cases} x + y = 6 \\ x - y = 2 \end{cases}$

6. $\begin{cases} x - y = 4 \\ 2x + y = 5 \end{cases}$

7. $\begin{cases} 2x + y = 1 \\ x - 2y = -7 \end{cases}$

8. $\begin{cases} 3x - y = -3 \\ 2x + y = -7 \end{cases}$

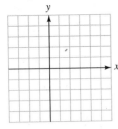
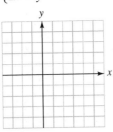
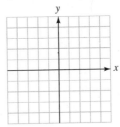
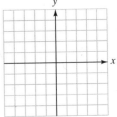

9. $\begin{cases} x = 13 - 4y \\ 3x = 4 + 2y \end{cases}$

10. $\begin{cases} 3x = 7 - 2y \\ 2x = 2 + 4y \end{cases}$

11. $\begin{cases} x = 3 - 2y \\ 2x + 4y = 6 \end{cases}$

12. $\begin{cases} 3x = 5 - 2y \\ 3x + 2y = 7 \end{cases}$

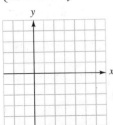
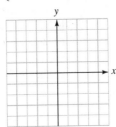
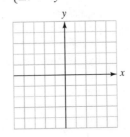
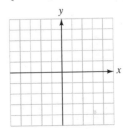

13. $\begin{cases} x = 2 \\ y = \dfrac{4 - x}{2} \end{cases}$

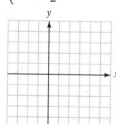

14. $\begin{cases} y = -2 \\ x = \dfrac{4 + 3y}{2} \end{cases}$

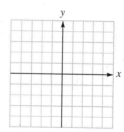

15. $\begin{cases} y = 3 \\ x = 2 \end{cases}$

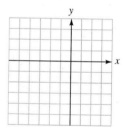

16. $\begin{cases} 2x + 3y = -15 \\ 2x + y = -9 \end{cases}$

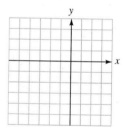

17. $\begin{cases} x = \dfrac{11 - 2y}{3} \\ y = \dfrac{11 - 6x}{4} \end{cases}$

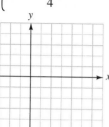

18. $\begin{cases} x = \dfrac{1 - 3y}{4} \\ y = \dfrac{12 + 3x}{2} \end{cases}$

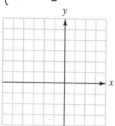

19. $\begin{cases} \dfrac{5}{2}x + y = \dfrac{1}{2} \\ 2x - \dfrac{3}{2}y = 5 \end{cases}$

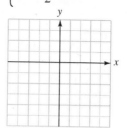

20. $\begin{cases} \dfrac{5}{2}x + 3y = 6 \\ y = \dfrac{24 - 10x}{12} \end{cases}$

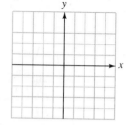

21. $\begin{cases} x = \dfrac{5y - 4}{2} \\ x - \dfrac{5}{3}y + \dfrac{1}{3} = 0 \end{cases}$

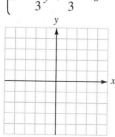

22. $\begin{cases} 2x = 5y - 11 \\ 3x = 2y \end{cases}$

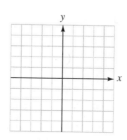

23. $\begin{cases} x = -\dfrac{3}{2}y \\ x = \dfrac{3}{2}y - 2 \end{cases}$

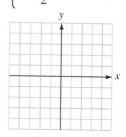

24. $\begin{cases} x = \dfrac{3y - 1}{4} \\ y = \dfrac{4 - 8x}{3} \end{cases}$

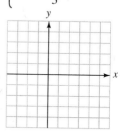

In Exercises 25–28, use a graphing device to solve each system. Give answers to the nearest hundredth.

25. $\begin{cases} y = 3.2x - 1.5 \\ y = -2.7x - 3.7 \end{cases}$

26. $\begin{cases} y = -0.45x + 5 \\ y = 5.55x - 13.7 \end{cases}$

27. $\begin{cases} 1.7x + 2.3y = 3.2 \\ y = 0.25x + 8.95 \end{cases}$

28. $\begin{cases} 2.75x = 12.9y - 3.79 \\ 7.1x - y = 35.76 \end{cases}$

Writing Exercises   *Write a paragraph using your own words.*

1. Explain how to solve a system of two equations in two variables.

2. Can a system of two equations in two variables have exactly two solutions? Why or why not?

Something to Think About

1. Form an independent system of equations with the solution of $(-5, 2)$.

2. Form a dependent system of equations with a solution of $(-5, 2)$.

Review Exercises   *Write each number in scientific notation.*

1. 93,000,000   **2.** 0.0000000236   **3.** 345×10^2 **4.** 752×10^{-5}

3.2 Solution by Elimination

■ THE SUBSTITUTION METHOD ■ THE ADDITION METHOD ■ AN INCONSISTENT SYSTEM
■ A SYSTEM WITH INFINITELY MANY SOLUTIONS ■ PROBLEM SOLVING ■ BREAK-EVEN
ANALYSIS ■ PARALLELOGRAMS

The graphing method provides a way to visualize the process of solving systems of equations. However, it cannot be used to solve systems of higher order, such as three equations that each have three variables. In this section, we will discuss algebraic methods that will enable us to solve such systems.

■ THE SUBSTITUTION METHOD

To solve a system of two equations (each with two variables) by substitution, we use the following steps.

The Substitution Method

1. If necessary, solve one equation for one of its variables, preferably a variable with a coefficient of 1.

2. Substitute the resulting expression for the variable obtained in Step 1 into the other equation and solve that equation.

3. Find the value of the other variable by substituting the value of the variable found in Step 2 into any equation containing both variables.

4. State the solution.

5. Check the solution in both of the original equations.

EXAMPLE 1   Solve the system $\begin{cases} 4x + y = 13 \\ -2x + 3y = -17 \end{cases}$.

Solution   *Step 1:* We solve the first equation for y, because y has a coefficient of 1 and no fractions are introduced.

$$4x + y = 13$$

1. $y = -4x + 13$ Subtract $4x$ from both sides.

Step 2: We then substitute $-4x + 13$ for y in the second equation of the system and solve for x.

$$-2x + 3y = -17$$
$$-2x + 3(-4x + 13) = -17 \qquad \text{Substitute } -4x + 13 \text{ for } y.$$
$$-2x - 12x + 39 = -17 \qquad \text{Use the distributive property to remove parentheses.}$$
$$-14x = -56 \qquad \text{Combine like terms and subtract 39 from both sides.}$$
$$x = 4 \qquad \text{Divide both sides by } -14.$$

Step 3: To find y, we substitute 4 for x in Equation 1 and simplify:

$$y = -4x + 13$$
$$= -4(4) + 13 \qquad \text{Substitute 4 for } x.$$
$$= -3$$

Step 4: The solution is $x = 4$ and $y = -3$, or just $(4, -3)$. If we graphed the equations of the system, they would intersect at the point $(4, -3)$.

Step 5: To verify that this solution satisfies each equation, we substitute $x = 4$ and $y = -3$ into each equation in the system and simplify.

$$4x + y = 13 \qquad\qquad -2x + 3y = -17$$
$$4(4) + (-3) = 13 \qquad\qquad -2(4) + 3(-3) = -17$$
$$16 - 3 = 13 \qquad\qquad -8 - 9 = -17$$
$$13 = 13 \qquad\qquad -17 = -17$$

Since the ordered pair $(4, -3)$ satisfies each equation of the system, the solution $(4, -3)$ checks. ∎

EXAMPLE 2 Solve the system $\begin{cases} \frac{4}{3}x + \frac{1}{2}y = -\frac{2}{3} \\ \frac{1}{2}x + \frac{2}{3}y = \frac{5}{3} \end{cases}$.

Solution First, we find an equivalent system without fractions by multiplying both sides of each equation by 6, obtaining the system

2. $\begin{cases} 8x + 3y = -4 \\ 3x + 4y = 10 \end{cases}$
3.

Because no variable in either equation has a coefficient of 1, it is impossible to avoid fractions when solving for a variable. We begin by solving Equation 3 for x.

$$3x + 4y = 10$$
$$3x = -4y + 10 \qquad \text{Subtract } 4y \text{ from both sides.}$$
4. $$x = -\frac{4}{3}y + \frac{10}{3} \qquad \text{Divide both sides by 3.}$$

We then substitute $-\frac{4}{3}y + \frac{10}{3}$ for x in Equation 2 and solve for y.

$$8x + 3y = -4$$

$$8\left(-\frac{4}{3}y + \frac{10}{3}\right) + 3y = -4 \qquad \text{Substitute } -\frac{4}{3}y + \frac{10}{3} \text{ for } x.$$

$$-\frac{32}{3}y + \frac{80}{3} + 3y = -4 \qquad \text{Use the distributive property to remove parentheses.}$$

$$-32y + 80 + 9y = -12 \qquad \text{Multiply both sides by 3.}$$

$$-23y = -92 \qquad \text{Combine terms and subtract 80 from both sides.}$$

$$y = 4 \qquad \text{Divide both sides by } -23.$$

We can find x by substituting 4 for y in Equation 4 and simplifying:

$$x = -\frac{4}{3}y + \frac{10}{3}$$

$$x = -\frac{4}{3}(4) + \frac{10}{3} \qquad \text{Substitute 4 for } y.$$

$$= -\frac{6}{3}$$

$$= -2$$

The solution is the ordered pair $(-2, 4)$. Verify that this solution satisfies each equation in the original system.   ∎

■ THE ADDITION METHOD

In the addition method, we combine the equations of the system in a way that will eliminate the terms involving one of the variables.

The Addition Method

1. Write both equations of the system in general form.
2. Multiply the terms of one or both of the equations by constants chosen to make the coefficients of x (or y) differ only in sign.
3. Add the equations and solve the equation that results, if possible.
4. Substitute the value obtained in Step 3 into either of the original equations and solve for the remaining variable.
5. State the solution obtained in Steps 3 and 4.
6. Check the solution in both of the original equations.

EXAMPLE 3   Use the addition method to solve the system $\begin{cases} 4x + y = 13 \\ -2x + 3y = -17 \end{cases}.$

Solution   *Step 1:* This is the system discussed in Example 1. Since both equations are already written in general form, Step 1 is unnecessary.

Step 2: To solve the system by addition, we multiply the second equation by 2 to make the coefficients of x differ only in sign.

5. $\begin{cases} 4x + y = 13 \\ -4x + 6y = -34 \end{cases}$

Step 3: When these equations are added, the terms involving x drop out, and we get

$$7y = -21$$
$$y = -3 \qquad \text{Divide both sides by 7.}$$

Step 4: To find x, we substitute -3 for y in either of the original equations and solve for x. If we use Equation 5, we have

$$4x + y = 13$$
$$4x + (-3) = 13 \qquad \text{Substitute } -3 \text{ for } y.$$
$$4x = 16 \qquad \text{Add 3 to both sides.}$$
$$x = 4 \qquad \text{Divide both sides by 4.}$$

Step 5: The solution is $x = 4$ and $y = -3$, or just $(4, -3)$.

Step 6: The check was completed in Example 1. ∎

EXAMPLE 4 Use the addition method to solve the system $\begin{cases} \frac{4}{3}x + \frac{1}{2}y = -\frac{2}{3} \\ \frac{1}{2}x + \frac{2}{3}y = \frac{5}{3} \end{cases}$.

Solution This system was discussed in Example 2. To solve it by addition, we find an equivalent system with no fractions by multiplying both sides of each equation by 6 to obtain

6. $\begin{cases} 8x + 3y = -4 \\ \textbf{7.} \quad 3x + 4y = 10 \end{cases}$

To make the y-terms drop out when we add the equations, we multiply both sides of Equation 6 by 4 and both sides of Equation 7 by -3 to get

$$\begin{cases} 32x + 12y = -16 \\ -9x - 12y = -30 \end{cases}$$

When these equations are added, the y-terms drop out, and we get

$$23x = -46$$
$$x = -2 \qquad \text{Divide both sides by 23.}$$

To find y, we substitute -2 for x in either Equation 6 or Equation 7. If we substitute -2 for x in Equation 7, we get

$$3x + 4y = 10$$
$$3(-2) + 4y = 10 \qquad \text{Substitute } -2 \text{ for } x.$$
$$-6 + 4y = 10 \qquad \text{Simplify.}$$
$$4y = 16 \qquad \text{Add 6 to both sides.}$$
$$y = 4 \qquad \text{Divide both sides by 4.}$$

The solution is the ordered pair $(-2, 4)$. ∎

■ AN INCONSISTENT SYSTEM

EXAMPLE 5   Solve the system $\begin{cases} y = 2x + 4 \\ 8x - 4y = 7 \end{cases}$.

Solution   Because the first equation is already solved for y, we use the substitution method.

$$8x - 4y = 7$$
$$8x - 4(2x + 4) = 7 \qquad \text{Substitute } 2x + 4 \text{ for } y.$$

We then solve for x:

$$8x - 8x - 16 = 7 \qquad \text{Use the distributive property to remove parentheses.}$$
$$-16 = 7 \qquad \text{Combine like terms.}$$

But

$$-16 \neq 7$$

This shows that the equations in the system are independent and that the system is inconsistent. Since the system has no solution, the graphs of the equations in the system would be parallel. ■

■ A SYSTEM WITH INFINITELY MANY SOLUTIONS

EXAMPLE 6   Solve the system $\begin{cases} 4x + 6y = 12 \\ -2x - 3y = -6 \end{cases}$.

Solution   Since the equations are written in general form, we use the addition method. To make the x-terms drop out when we add the equations, we multiply both sides of the second equation by 2 to get

$$\begin{cases} 4x + 6y = 12 \\ -4x - 6y = -12 \end{cases}$$

After adding the left-hand sides and the right-hand sides, we get

$$0x + 0y = 0$$
$$0 = 0$$

Here, both the x-terms and the y-terms drop out. The true statement $0 = 0$ shows that the equations in this system are dependent and that the system is consistent.

Note that the equations of the system are equivalent, because when the second equation is multiplied by -2, it becomes the first equation. The line graphs of these equations will coincide. Since any ordered pair that satisfies one of the equations also satisfies the other, there are infinitely many solutions.

To find some of these solutions, we can solve either equation in the original system for y. For example, we will solve $4x + 6y = 12$ for y.

$$4x + 6y = 12$$
$$6y = -4x + 12 \qquad \text{Subtract } 4x \text{ from both sides.}$$
$$y = -\frac{2}{3}x + 2 \qquad \text{Divide both sides by 6 and simplify.}$$

Since we have found the values of y in terms of x, every solution of the system has the form $(x, y) = \left(x, -\frac{2}{3}x + 2\right)$, where x can be any real number. For example,

If $x = 0$, then $y = 2$, and a solution is $(0, 2)$.

If $x = 3$, then $y = 0$, and a solution is $(3, 0)$.

If $x = 6$, then $y = -2$, and a solution is $(6, -2)$. ∎

■ PROBLEM SOLVING

To solve problems using two variables, we follow the same problem-solving strategy discussed in Chapter 1, except that we form two equations instead of one.

EXAMPLE 7	**Retail sales** A store advertised two types of cordless telephones, one selling for $67 and the other for $100. If the receipts from the sale of 36 phones totaled $2940, how many of each type were sold?
Analyze the Problem	We can let x represent the number of phones sold for $67 and let y represent the number of phones sold for $100. Then the receipts for the sale of the lower-priced phones are $67x$, and the receipts for the sale of the higher-priced phones are $100y$.
Form Two Equations	The information of the problem gives the following two equations:

The number of lower-priced phones	+	the number of higher-priced phones	=	the total number of phones.
x	+	y	=	36

The receipts from the lower-priced phones	+	the receipts from the higher-priced phones	=	the total receipts.
$67x$	+	$100y$	=	2940

Solve the System   To find out how many of each type of phone were sold, we must solve the following system:

8. **9.** $\begin{cases} x + y = 36 \\ 67x + 100y = 2940 \end{cases}$

We multiply both sides of Equation 8 by -100, add the resulting equation to Equation 9, and solve for x:

$$\begin{array}{r} -100x - 100y = -3600 \\ \underline{67x + 100y = 2940} \\ -33x = -660 \\ x = 20 \end{array}$$ Divide both sides by -33.

To find y, we substitute 20 for x in Equation 8 and solve for y:

$x + y = 36$

$20 + y = 36$ Substitute 20 for x.

$y = 16$ Subtract 20 from both sides.

State the Conclusion   The store sold 20 of the lower-priced phones and 16 of the higher-priced phones.

Check the Result   If 20 of one type were sold and 16 of the other type were sold, a total of 36 phones were sold.

Since the value of the lower-priced phones is 20($67) = $1340 and the value of the higher-priced phones is 16($100) = $1600, the total receipts are $2940. ■

EXAMPLE 8   **Mixing solutions**   How many ounces of a 5% saline solution and how many ounces of a 20% saline solution must be mixed together to obtain 50 ounces of a 15% saline solution?

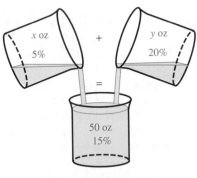

FIGURE 3-6

Analyze the Problem   We can let x represent the number of ounces of the 5% solution and let y represent the number of ounces of the 20% solution that are to be mixed. Then the amount of salt in the 5% solution is $0.05x$, and the amount of salt in the 20% solution is $0.20y$. (See Figure 3-6.)

Form Two Equations   The information of the problem gives the following two equations:

The number of ounces of 5% solution	+	the number of ounces of 20% solution	=	the total number of ounces in the mixture.
x	+	y	=	50

The salt in the 5% solution	+	the salt in the 20% solution	=	the salt in the mixture.
$0.05x$	+	$0.20y$	=	$0.15(50)$

Solve the System   To find out how many ounces of each are needed, we solve the following system:

10. $\begin{cases} x + y = 50 \\ 0.05x + 0.20y = 7.5 \end{cases}$ $0.15(50) = 7.5.$
11.

To solve this system by substitution, we can solve Equation 10 for y

$x + y = 50$

12. $y = 50 - x$ Subtract x from both sides.

and then substitute $50 - x$ for y in Equation 11.

$$0.05x + 0.20y = 7.5$$

$0.05x + 0.20(50 - x) = 7.5$ Substitute $50 - x$ for y.

$5x + 20(50 - x) = 750$ Multiply both sides by 100.

$5x + 1000 - 20x = 750$ Use the distributive property to remove parentheses.

$-15x = -250$ Combine like terms and subtract 1000 from both sides.

$x = \dfrac{-250}{-15}$ Divide both sides by -15.

$x = \dfrac{50}{3}$ Simplify $\frac{-250}{-15}$.

To find y, we can substitute $\frac{50}{3}$ for x in Equation 12:

$$y = 50 - x$$

$= 50 - \dfrac{50}{3}$ Substitute $\frac{50}{3}$ for x.

$= \dfrac{100}{3}$

State the Conclusion   To obtain 50 ounces of a 15% solution, we must mix $\frac{50}{3}$ $\left(\text{or } 16\frac{2}{3}\right)$ ounces of the 5% solution with $\frac{100}{3}$ $\left(\text{or } 33\frac{1}{3}\right)$ ounces of the 20% solution.

Check the Result   We note that $16\frac{2}{3}$ ounces of solution plus $33\frac{1}{3}$ ounces of solution equals the required 50 ounces of solution. We also note that 5% of $16\frac{2}{3} \approx 0.83$, and 20% of $33\frac{1}{3} \approx 6.67$, giving a total of 7.5, which is 15% of 50.   ■

■ BREAK-EVEN ANALYSIS

Running a machine involves both *setup costs* and *unit costs*. Setup costs include the cost of preparing a machine to do a certain job. Unit costs depend on the number of items to be manufactured, including the costs of raw materials and labor.

     Suppose that a certain machine has a setup cost of $600 and a unit cost of $3. If x items are manufactured using this machine, the cost will be

     Cost $= 600 + 3x$ Cost = setup cost + unit costs.

Furthermore, suppose that a larger and more efficient machine has a setup cost of $800 and a unit cost of $2. The cost of manufacturing x items using this machine is

     Cost using larger machine $= 800 + 2x$

The *break-even point* is the number of units x that need to be manufactured to make the cost the same using either machine. It can be found by setting the two costs equal to each other and solving for x.

     $600 + 3x = 800 + 2x$

           $x = 200$ Subtract 600 and $2x$ from both sides.

The break-even point is 200 units, because the cost using either machine is $1200:

Cost on small machine $= 600 + 3x$
$$= 600 + 3(200)$$
$$= 600 + 600$$
$$= 1200$$

Cost on larger machine $= 800 + 2x$
$$= 800 + 2(200)$$
$$= 800 + 400$$
$$= 1200$$

EXAMPLE 9   One machine has a setup cost of $400 and a unit cost of $1.50, and another machine has a setup cost of $500 and a unit cost of $1.25. Find the break-even point.

Analyze the Problem   The cost C_1 of manufacturing x units on machine 1 is $1.50x + $400 (the number of units manufactured times $1.50, plus the setup cost of $400). The cost C_2 of manufacturing the same number of units on machine 2 is $1.25x + $500 (the number of units manufactured times $1.25, plus the setup cost of $500). The break-even point occurs when the costs are equal ($C_1 = C_2$).

Form Two Equations   If x represents the number of items to be manufactured, the cost C_1 using machine 1 is

The cost using machine 1	$=$	the cost of manufacturing x units	$+$	the setup cost.
C_1	$=$	$1.5x$	$+$	400

The cost C_2 using machine 2 is

The cost using machine 2	$=$	the cost of manufacturing x units	$+$	the setup cost.
C_2	$=$	$1.25x$	$+$	500

Solve the System   To find the break-even point, we must solve the system $\begin{cases} C_1 = 1.5x + 400 \\ C_2 = 1.25x + 500 \end{cases}$.

Since the break-even point occurs when $C_1 = C_2$, we can substitute $1.5x + 400$ for C_2 to get

$$1.5x + 400 = 1.25x + 500$$
$$1.5x = 1.25x + 100 \qquad \text{Subtract 400 from both sides.}$$
$$0.25x = 100 \qquad \text{Subtract } 1.25x \text{ from both sides.}$$
$$x = 400 \qquad \text{Divide both sides by 0.25.}$$

State the Conclusion   The break-even point is 400 units.

Check the Result   The cost using machine 1 is $400 + 1.5(400) = 400 + 600 = 1000$.
The cost using machine 2 is $500 + 1.25(400) = 500 + 500 = 1000$.
Since the costs are equal, the break-even point is 400. ■

■ PARALLELOGRAMS

A **parallelogram** is a four-sided figure with its opposite sides parallel. (See Figure 3-7(a).) Here are some important facts about parallelograms.

1. Opposite sides of a parallelogram have the same length.

2. Opposite angles of a parallelogram have the same measure.

3. Consecutive angles of a parallelogram are supplementary.

4. A diagonal of a parallelogram (see Figure 3-7(b)) divides the parallelogram into two *congruent triangles*—triangles with the same shape and same area.

5. In Figure 3-7(b), angles 1 and 2, and angles 3 and 4, are called pairs of *alternate interior angles*. When a diagonal intersects two parallel sides of a parallelogram, all pairs of alternate interior angles have the same measure.

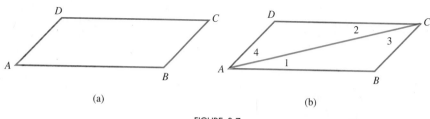

(a)   (b)

FIGURE 3-7

EXAMPLE 10   Refer to the parallelogram shown in Figure 3-8 and find the values of x and y.

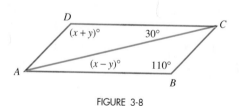

FIGURE 3-8

Solution   Since diagonal AC intersects two parallel sides, the alternate interior angles that are formed have the same measure. Thus, $(x - y)° = 30°$. Since opposite angles of a parallelogram have the same measure, we known that $(x + y)° = 110°$. We can form the following system of equations and solve it by addition.

1. $\begin{cases} x - y = 30 \\ x + y = 110 \end{cases}$
2.

$$2x = 140 \qquad \text{Add Equations 1 and 2.}$$
$$x = 70 \qquad \text{Divide both sides by 2.}$$

We can substitute 70 for x in Equation 2 and solve for y.

$$x + y = 110$$
$$70 + y = 110 \qquad \text{Substitute 70 for } x.$$
$$y = 40 \qquad \text{Subtract 70 from both sides.}$$

Thus, $x = 70$ and $y = 40$.

Orals   *Solve each system for x.*

1. $\begin{cases} y = 2x \\ x + y = 6 \end{cases}$

2. $\begin{cases} y = -x \\ 2x + y = 4 \end{cases}$

3. $\begin{cases} x - y = 6 \\ x + y = 2 \end{cases}$

4. $\begin{cases} x + y = 4 \\ 2x - y = 5 \end{cases}$

EXERCISE 3.2

In Exercises 1–12, solve each system by substitution, if possible.

1. $\begin{cases} y = x \\ x + y = 4 \end{cases}$

2. $\begin{cases} y = x + 2 \\ x + 2y = 16 \end{cases}$

3. $\begin{cases} x - y = 2 \\ 2x + y = 13 \end{cases}$

4. $\begin{cases} x - y = -4 \\ 3x - 2y = -5 \end{cases}$

5. $\begin{cases} x + 2y = 6 \\ 3x - y = -10 \end{cases}$

6. $\begin{cases} 2x - y = -21 \\ 4x + 5y = 7 \end{cases}$

7. $\begin{cases} 3x = 2y - 4 \\ 6x - 4y = -4 \end{cases}$

8. $\begin{cases} 8x = 4y + 10 \\ 4x - 2y = 5 \end{cases}$

9. $\begin{cases} 3x - 4y = 9 \\ x + 2y = 8 \end{cases}$

10. $\begin{cases} 3x - 2y = -10 \\ 6x + 5y = 25 \end{cases}$

11. $\begin{cases} 2x + 2y = -1 \\ 3x + 4y = 0 \end{cases}$

12. $\begin{cases} 5x + 3y = -7 \\ 3x - 3y = 7 \end{cases}$

In Exercises 13–24, solve each system by addition, if possible.

13. $\begin{cases} x - y = 3 \\ x + y = 7 \end{cases}$

14. $\begin{cases} x + y = 1 \\ x - y = 7 \end{cases}$

15. $\begin{cases} 2x + y = -10 \\ 2x - y = -6 \end{cases}$

16. $\begin{cases} x + 2y = -9 \\ x - 2y = -1 \end{cases}$

17. $\begin{cases} 2x + 3y = 8 \\ 3x - 2y = -1 \end{cases}$

18. $\begin{cases} 5x - 2y = 19 \\ 3x + 4y = 1 \end{cases}$

19. $\begin{cases} 4x + 9y = 8 \\ 2x - 6y = -3 \end{cases}$

20. $\begin{cases} 4x + 6y = 5 \\ 8x - 9y = 3 \end{cases}$

21. $\begin{cases} 8x - 4y = 16 \\ 2x - 4 = y \end{cases}$

22. $\begin{cases} 2y - 3x = -13 \\ 3x - 17 = 4y \end{cases}$

23. $\begin{cases} x = \dfrac{3}{2}y + 5 \\ 2x - 3y = 8 \end{cases}$

24. $\begin{cases} x = \dfrac{2}{3}y \\ y = 4x + 5 \end{cases}$

In Exercises 25–32, solve each system by any method.

25. $\begin{cases} \dfrac{x}{2} + \dfrac{y}{2} = 6 \\ \dfrac{x}{2} - \dfrac{y}{2} = -2 \end{cases}$

26. $\begin{cases} \dfrac{x}{2} - \dfrac{y}{3} = -4 \\ \dfrac{x}{2} + \dfrac{y}{9} = 0 \end{cases}$

27. $\begin{cases} \dfrac{3}{4}x + \dfrac{2}{3}y = 7 \\ \dfrac{3}{5}x - \dfrac{1}{2}y = 18 \end{cases}$

28. $\begin{cases} \dfrac{2}{3}x - \dfrac{1}{4}y = -8 \\ \dfrac{1}{2}x - \dfrac{3}{8}y = -9 \end{cases}$

29. $\begin{cases} \dfrac{3x}{2} - \dfrac{2y}{3} = 0 \\ \dfrac{3x}{4} + \dfrac{4y}{3} = \dfrac{5}{2} \end{cases}$

30. $\begin{cases} \dfrac{3x}{5} + \dfrac{5y}{3} = 2 \\ \dfrac{6x}{5} - \dfrac{5y}{3} = 1 \end{cases}$

31. $\begin{cases} \dfrac{2}{5}x - \dfrac{1}{6}y = \dfrac{7}{10} \\ \dfrac{3}{4}x - \dfrac{2}{3}y = \dfrac{19}{8} \end{cases}$

32. $\begin{cases} \dfrac{5}{6}x + \dfrac{2}{3}y = \dfrac{7}{6} \\ \dfrac{10}{7}x - \dfrac{4}{9}y = \dfrac{17}{21} \end{cases}$

In Exercises 33–36, solve each system for x and y. Solve for $\frac{1}{x}$ and $\frac{1}{y}$ first.

33. $\begin{cases} \dfrac{1}{x} + \dfrac{1}{y} = \dfrac{5}{6} \\ \dfrac{1}{x} - \dfrac{1}{y} = \dfrac{1}{6} \end{cases}$

34. $\begin{cases} \dfrac{1}{x} + \dfrac{1}{y} = \dfrac{9}{20} \\ \dfrac{1}{x} - \dfrac{1}{y} = \dfrac{1}{20} \end{cases}$

35. $\begin{cases} \dfrac{1}{x} + \dfrac{2}{y} = -1 \\ \dfrac{2}{x} - \dfrac{1}{y} = -7 \end{cases}$

36. $\begin{cases} \dfrac{3}{x} - \dfrac{2}{y} = -30 \\ \dfrac{2}{x} - \dfrac{3}{y} = -30 \end{cases}$

In Exercises 37–56, use two variables and two equations to solve each problem.

37. Merchandising   A pair of shoes and a sweater cost $98. If the sweater costs $16 more than the shoes, how much does the sweater cost?

38. Merchandising   A sporting goods salesperson sells 2 fishing reels and 5 rods for $270. The next day, the salesperson sells 4 reels and 2 rods for $220. How much does each cost?

39. Electronics   Two resistors in the voltage divider circuit in Illustration 1 have a total resistance of 1375 ohms. To provide the required voltage, R_1 must be 125 ohms greater than R_2. Find both resistances.

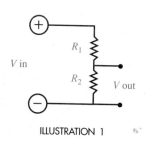

ILLUSTRATION 1

40. Stowing baggage   A small aircraft can carry 950 pounds of baggage, distributed between two storage compartments. On one flight, the plane is fully loaded, with 150 pounds more baggage in one compartment than the other. How much is stowed in each compartment?

41. Geometry problem   The rectangular field in Illustration 2 is surrounded by 72 meters of fencing. If the field is partitioned as shown, a total of 88 meters of fencing is required. Find the dimensions of the field.

42. Geometry   In a right triangle, one acute angle is 15° greater than two times the other acute angle. Find the difference between the angles.

43. Investment income   Part of $8000 was invested at 10% interest and the rest at 12%. If the annual income from these investments was $900, how much was invested at each rate?

ILLUSTRATION 2

44. Investment income   Part of $12,000 was invested at 6% interest and the rest at 7.5%. If the annual income from these investments was $810, how much was invested at each rate?

45. Mixing a solution   How many ounces of the two alcohol solutions in Illustration 3 must be mixed to obtain 100 ounces of a 12.2% solution?

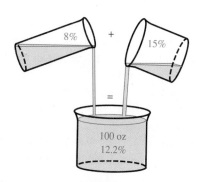

ILLUSTRATION 3

46. Mixing candy   How many pounds of each kind of candy shown in Illustration 4 on page 170 must be mixed to obtain 60 pounds of candy that is worth $3 per pound?

47. Travel   A car travels 50 miles in the same time that a plane travels 180 miles. The speed of the plane is 143 miles per hour faster than the speed of the car. Find the speed of the car.

ILLUSTRATION 4

Mix	Protein	Carbohydrates
Mix *A*	12%	9%
Mix *B*	15%	5%

ILLUSTRATION 6

48. Travel   A car and a truck leave Rockford at the same time, heading in opposite directions. When they are 350 miles apart, the car has gone 70 miles farther than the truck. How far has the car traveled?

49.   **Making bicycles**   A bicycle manufacturer builds racing bikes and mountain bikes, with the per-unit manufacturing costs shown in Illustration 5. The company has budgeted $15,900 for labor and $13,075 for materials. How many bicycles of each type can be built?

Model	Cost of materials	Cost of labor
racing	$55	$60
mountain	$70	$90

ILLUSTRATION 5

50. Feeding cattle   A farmer keeps some animals on a strict diet. Each animal is to receive 15 grams of protein and 7.5 grams of carbohydrates. The farmer uses two food mixes with the nutrients shown in Illustration 6. How many grams of each mix should be used to provide the correct nutrients for each animal?

51. Milling brass plates   Two machines can mill a brass plate. One machine has a setup cost of $300 and a cost per plate of $2. The other machine has a setup cost of $500 and a cost per plate of $1. Find the break-even point.

52. Printing books   A printer has two presses. One has a setup cost of $210 and can print the pages of a certain book for $5.98. The other press has a setup cost of $350 and can print the pages of the same book for $5.95. Find the break-even point.

53. Managing a computer store   The manager of a computer store knows that fixed costs are $8925 per month and that the unit cost is $850 for every computer sold. If he can sell all the computers he can get for $1275 each, how many computers must he sell each month to break even?

54. Managing a beauty shop   A beauty shop specializing in permanents has fixed costs of $2101.20 per month. The owner estimates that the cost of each permanent is $23.60. This covers labor, chemicals, and electricity. If her shop can give as many permanents as she wants at a price of $44 each, how many must be given each month to break even?

55. Running a small business   A person invests $18,375 to set up a small business to produce a piece of computer software that will sell for $29.95. If each piece can be produced for $5.45, how many pieces must be sold to break even?

56. Running a record company   Three people invest $35,000 each to start a record company that will produce reissues of classic jazz. Each release will be a set of three CDs, retailing for $15 per disc. If each set can be produced for $18.95, how many sets must be sold for the investors to make a profit?

In Exercises 57–60, a paint manufacturer can choose between two processes for manufacturing house paint, with monthly costs as shown in the following table. Assume the paint sells for $18 per gallon.

57. Find the break-even point for process *A*.

58. Find the break-even point for process *B*.

59. If expected sales are 6000 gallons per month, which process should the company use?

60. If expected sales are 7000 gallons per month, which process should the company use?

Process	Fixed costs	Unit cost (per gallon)
A	$32,500	$13
B	$80,600	$ 5

In Exercises 61–66, a manufacturer of automobile water pumps is considering retooling for one of two manufacturing processes, with monthly fixed costs and unit costs as indicated in the following table. Each water pump can be sold for $50.

61. Find the break-even point for process *A*.

62. Find the break-even point for process *B*.

Process	Fixed costs	Unit cost
A	$12,390	$29
B	$20,460	$17

63. If expected sales are 550 per month, which process should be used?

64. If expected sales are 600 per month, which process should be used?

66. At what monthly sales level is process *B* better?

65. If expected sales are 650 per month, which process should be used?

67. Geometry If two angles are supplementary, their sum is 180°. If the difference between two supplementary angles is 110°, find the measure of each angle.

68. Geometry If two angles are complementary, their sum is 90°. If one of two complementary angles is 16° greater than the other, find the measure of each angle.

In Exercises 69–70, Illustrations 7 and 8 are parallelograms.

69. Find *x* and *y* in Illustration 7.

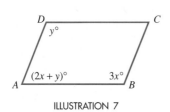

ILLUSTRATION 7

70. Find *x* and *y* in Illustration 8.

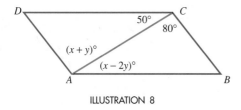

ILLUSTRATION 8

71. Selecting radio frequencies In a radio, an inductor and a capacitor are used in a resonant circuit to select a desired radio station at a frequency *f* and reject all others. The inductance *L* and the capacitance *C* determine the inductive reactance X_L and the capacitive reactance X_C of that circuit, where

$$X_L = 2\pi f L \qquad \text{and} \qquad X_C = \frac{1}{2\pi f C}$$

The radio station selected will be at the frequency *f* where $X_L = X_C$. Write a formula for f^2 in terms of *L* and *C*.

72. Choosing salary plans A sales clerk can choose from two salary options: 1. a straight 7% commission; or 2. $150 per month + 2% commission. How much would the clerk have to sell for each plan to produce the same monthly paycheck?

Writing Exercises   *Write a paragraph using your own words.*

1. Tell which method you would use to solve the following system. Why?

$$\begin{cases} y = 3x + 1 \\ 3x + 2y = 12 \end{cases}$$

2. Tell which method you would use to solve the following system. Why?

$$\begin{cases} 2x + 4y = 9 \\ 3x - 5y = 20 \end{cases}$$

Something to Think About
1. Under what conditions will a system of two equations in two variables be inconsistent?

2. Under what conditions will the equations of a system of two equations in two variables be dependent?

Review Exercises *Simplify each expression. Write all answers without using negative exponents.*

1. $(a^2a^3)^2(a^4a^2)^2$ **2.** $\left(\dfrac{a^2b^3c^4d}{ab^2c^3d^4}\right)^{-3}$ **3.** $\left(\dfrac{-3x^3y^4}{x^{-5}y^3}\right)^{-4}$ **4.** $\dfrac{3t^0 - 4t^0 + 5}{5t^0 + 2t^0}$

3.3 Solution of Three Equations in Three Variables

■ A CONSISTENT SYSTEM ■ AN INCONSISTENT SYSTEM ■ SYSTEMS WITH DEPENDENT EQUATIONS
■ PROBLEM SOLVING

We now extend the definition of a linear equation to include equations of the form $ax + by + cz = d$. The solution of a system of three linear equations with three variables is an ordered triple of numbers. For example, the solution of the system

$$\begin{cases} 2x + 3y + 4z = 20 \\ 3x + 4y + 2z = 17 \\ 3x + 2y + 3z = 16 \end{cases}$$

is the triple (1, 2, 3), since each equation is satisfied when $x = 1$, $y = 2$, and $z = 3$.

$$2x + 3y + 4z = 20 \qquad\qquad 3x + 4y + 2z = 17 \qquad\qquad 3x + 2y + 3z = 16$$
$$2(1) + 3(2) + 4(3) = 20 \qquad 3(1) + 4(2) + 2(3) = 17 \qquad 3(1) + 2(2) + 3(3) = 16$$
$$2 + 6 + 12 = 20 \qquad\qquad 3 + 8 + 6 = 17 \qquad\qquad 3 + 4 + 9 = 16$$
$$20 = 20 \qquad\qquad\qquad 17 = 17 \qquad\qquad\qquad 16 = 16$$

The graph of an equation of the form $ax + by + cz = d$ is a flat surface called a *plane*. A system of three linear equations in three variables is consistent or inconsistent, depending on how the three planes corresponding to the three equations intersect. Figure 3-9 illustrates some of the possibilities.

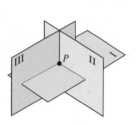

The three planes intersect at a single point P: One solution

(a)

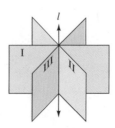

The three planes have a line l in common: An infinite number of solutions

(b)

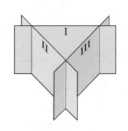

The three planes have no point in common: No solutions

(c)

FIGURE 3-9

To solve a system of three linear equations in three variables, we follow these steps.

> **Solving Three Equations in Three Variables**
> 1. Pick any two equations and eliminate a variable.
> 2. Pick a different pair of equations and eliminate the same variable.
> 3. Solve the resulting pair of two equations in two variables.
> 4. To find the value of the third variable, substitute the values of the two variables found in Step 3 into any equation containing all three variables and solve the equation.
> 5. Check the solution in all three of the original equations.

■ A CONSISTENT SYSTEM

EXAMPLE 1   Solve the system $\begin{cases} 2x + y + 4z = 12 \\ x + 2y + 2z = 9 \\ 3x - 3y - 2z = 1 \end{cases}$.

Solution   We are given the system

1. $\begin{cases} 2x + y + 4z = 12 \end{cases}$
2. $\begin{cases} x + 2y + 2z = 9 \end{cases}$
3. $\begin{cases} 3x - 3y - 2z = 1 \end{cases}$

If we pick Equations 2 and 3 and add them, the variable z is eliminated:

2. $x + 2y + 2z = 9$
3. $\underline{3x - 3y - 2z = 1}$
4. $4x - y \qquad\; = 10$

We now pick a different pair of equations (Equations 1 and 3) and eliminate z again. If each side of Equation 3 is multiplied by 2 and the resulting equation is added to Equation 1, z is again eliminated:

1. $2x + y + 4z = 12$
   $\underline{6x - 6y - 4z = 2}$
5. $8x - 5y \qquad\; = 14$

Equations 4 and 5 form a system of two equations in two variables:

4. $\begin{cases} 4x - y = 10 \end{cases}$
5. $\begin{cases} 8x - 5y = 14 \end{cases}$

To solve this system, we multiply Equation 4 by -5 and add the resulting equation to Equation 5 to eliminate y:

   $-20x + 5y = -50$
5. $\underline{\quad 8x - 5y = \quad 14}$
   $-12x \qquad\quad = -36$

   $x = 3$ Divide both sides by -12.

To find y, we substitute 3 for x in any equation containing x and y (such as Equation 5) and solve for y:

5. $8x - 5y = 14$

$8(3) - 5y = 14$ Substitute 3 for x.

$24 - 5y = 14$ Simplify.

$-5y = -10$ Subtract 24 from both sides.

$y = 2$ Divide both sides by -5.

To find z, we substitute 3 for x and 2 for y in an equation containing x, y, and z (such as Equation 1) and solve for z:

1. $2x + y + 4z = 12$

$2(3) + 2 + 4z = 12$ Substitute 3 for x and 2 for y.

$8 + 4z = 12$ Simplify.

$4z = 4$ Subtract 8 from both sides.

$z = 1$ Divide both sides by 4.

The solution of the system is $(x, y, z) = (3, 2, 1)$. Verify that these values satisfy each equation in the original system.   ■

■ AN INCONSISTENT SYSTEM

EXAMPLE 2   Solve the system $\begin{cases} 2x + y - 3z = -3 \\ 3x - 2y + 4z = 2 \\ 4x + 2y - 6z = -7 \end{cases}$.

Solution   We are given the system of equations

1. $\begin{cases} 2x + y - 3z = -3 \\ 3x - 2y + 4z = 2 \\ 4x + 2y - 6z = -7 \end{cases}$
2.
3.

We can multiply Equation 1 by 2 and add the resulting equation to Equation 2 to eliminate y:

$4x + 2y - 6z = -6$

2. $\underline{3x - 2y + 4z = 2}$

4. $7x - 2z = -4$

We now add Equations 2 and 3 to eliminate y again:

2. $3x - 2y + 4z = 2$

3. $\underline{4x + 2y - 6z = -7}$

5. $7x - 2z = -5$

Equations 4 and 5 form the system

4. $\begin{cases} 7x - 2z = -4 \\ 7x - 2z = -5 \end{cases}$
5.

Since $7x - 2z$ cannot equal both -4 and -5, this system is inconsistent, and it has no solution.   ■

■ SYSTEMS WITH DEPENDENT EQUATIONS

When the equations in a system of two equations in two variables are dependent, the system always has infinitely many solutions. This is not always true for systems of three equations in three variables. In fact, a system can have dependent equations and still be inconsistent. Figure 3-10 illustrates the different possibilities.

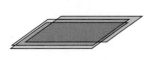
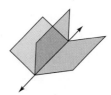
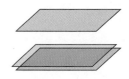

When three planes coincide, the equations are dependent, and there is an infinite number of solutions.

When three planes intersect in a common line, the equations are dependent, and there is an infinite number of solutions.

When two planes coincide and are parallel to a third plane, the system is inconsistent, and there are no solutions.

(a)   (b)   (c)

FIGURE 3-10

EXAMPLE 3 Solve the system $\begin{cases} 3x - 2y + z = -1 \\ 2x + y - z = 5 \\ 5x - y = 4 \end{cases}$.

Solution We can add the first two equations to get

$$3x - 2y + z = -1$$
$$\underline{2x + y - z = 5}$$
1. $5x - y = 4$

Since Equation 1 is the same as the third equation of the system, the equations of the system are dependent. From a graphical perspective, the equations represent three planes that intersect in a common line, as shown in Figure 3-10(b). Thus, there will be an infinite number of solutions.

To write the general solution of this system, we can solve Equation 1 for y to get

$$5x - y = 4$$
$$-y = -5x + 4 \qquad \text{Subtract } 5x \text{ from both sides.}$$
$$y = 5x - 4 \qquad \text{Multiply both sides by } -1.$$

We can substitute $5x - 4$ for y in the first equation of the system and solve for z to get

$$3x - 2y + z = -1$$
$$3x - 2(5x - 4) + z = -1 \qquad \text{Substitute } 5x - 4 \text{ for } y.$$
$$3x - 10x + 8 + z = -1 \qquad \text{Use the distributive property to remove parentheses.}$$
$$-7x + 8 + z = -1 \qquad \text{Combine like terms.}$$
$$z = 7x - 9 \qquad \text{Add } 7x \text{ and } -8 \text{ to both sides.}$$

Since we have found the values of y and z in terms of x, every solution of the system has the form $(x, 5x - 4, 7x - 9)$, where x can be any real number. For example,

If $x = 1$, then a solution is $(1, 1, -2)$. $5(1) - 4 = 1$, and $7(1) - 9 = -2$.

If $x = 2$, then a solution is $(2, 6, 5)$. $5(2) - 4 = 6$, and $7(2) - 9 = 5$.

If $x = 3$, then a solution is $(3, 11, 12)$. $5(3) - 4 = 11$, and $7(3) - 9 = 12$.

∎

■ PROBLEM SOLVING

EXAMPLE 4   **Manufacturing hammers**   A company manufactures three types of hammers—good, better, and best. The cost of manufacturing each type of hammer is $4, $6, and $7, respectively, and the hammers sell for $6, $9, and $12. Each day, the cost of manufacturing 100 hammers is $520, and the daily revenue from their sales is $810. How many of each type are manufactured?

Analyze the Problem   If we let x represent the number of good hammers, y represent the number of better hammers, and z represent the number of best hammers, we know that

The total number of hammers is $x + y + z$.

The cost of manufacturing good hammers is $4x$ ($4 times x hammers).

The cost of manufacturing better hammers is $6y$ ($6 times y hammers).

The cost of manufacturing best hammers is $7z$ ($7 times z hammers).

The revenue received by selling good hammers is $6x$ ($6 times x hammers).

The revenue received by selling better hammers is $9y$ ($9 times y hammers).

The revenue received by selling best hammers is $12z$ ($12 times z hammers).

Form Three Equations   Since x represents the number of good hammers manufactured, y represents the number of better hammers manufactured, and z represents the number of best hammers manufactured, we have

The number of good hammers	+	the number of better hammers	+	the number of best hammers	=	the total number of hammers.
x	+	y	+	z	=	100

The cost of good hammers	+	the cost of better hammers	+	the cost of best hammers	=	the total cost.
$4x$	+	$6y$	+	$7z$	=	520

The revenue from good hammers	+	the revenue from better hammers	+	the revenue from best hammers	=	the total revenue.
$6x$	+	$9y$	+	$12z$	=	810

Solve the System   We must now solve the system

$$\begin{cases} \textbf{1.} & x + y + z = 100 \\ \textbf{2.} & 4x + 6y + 7z = 520 \\ \textbf{3.} & 6x + 9y + 12z = 810 \end{cases}$$

If we multiply Equation 1 by -7 and add the result to Equation 2, we get

$$\begin{aligned} -7x - 7y - 7z &= -700 \\ \underline{4x + 6y + 7z} &= \underline{520} \\ \textbf{4.} \quad -3x - y & = -180 \end{aligned}$$

If we multiply Equation 1 by -12 and add the result to Equation 3, we get

$$\begin{aligned} -12x - 12y - 12z &= -1200 \\ \underline{6x + 9y + 12z} &= \underline{810} \\ \textbf{5.} \quad -6x - 3y & = -390 \end{aligned}$$

If we multiply Equation 4 by -3 and add it to Equation 5, we get

$$\begin{aligned} 9x + 3y &= 540 \\ \textbf{5.} \quad \underline{-6x - 3y} &= \underline{-390} \\ 3x &= 150 \\ x &= 50 \qquad \text{Divide both sides by 3.} \end{aligned}$$

To find y, we can substitute 50 for x in Equation 4:

$$\begin{aligned} -3x - y &= -180 \\ -3(\mathbf{50}) - y &= -180 \qquad \text{Substitute 50 for } x. \\ -150 - y &= -180 \qquad \text{Multiply.} \\ -y &= -30 \qquad \text{Add 150 to both sides.} \\ y &= 30 \qquad \text{Divide both sides by } -1. \end{aligned}$$

To find z, we can substitute 50 for x and 30 for y in Equation 1:

$$\begin{aligned} \textbf{1.} \quad x + y + z &= 100 \\ \mathbf{50} + \mathbf{30} + z &= 100 \\ z &= 20 \qquad \text{Subtract 80 from both sides.} \end{aligned}$$

State the Conclusion   The company manufactures 50 good hammers, 30 better hammers, and 20 best hammers each day.

Check the Result   Check the solution in each equation in the original system.   ∎

EXAMPLE 5   **Curve fitting**   The equation of the parabola shown in Figure 3-11 is of the form $y = ax^2 + bx + c$. Find the equation of the parabola.

Solution   Since the parabola passes through the points shown in the figure, each pair of coordinates satisfies the equation $y = ax^2 + bx + c$. If we substitute the x- and

y-values of each point into the equation and simplify, we obtain the following system of equations:

1. $a - b + c = 5$
2. $a + b + c = 1$
3. $4a + 2b + c = 2$

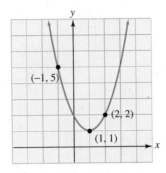

FIGURE 3-11

If we add Equations 1 and 2, we obtain $2a + 2c = 6$. If we multiply Equation 1 by 2 and add the result to Equation 3, we get $6a + 3c = 12$. We can then divide both sides of $2a + 2c = 6$ by 2 and divide both sides of $6a + 3c = 12$ by 3 to get the system

4. $\begin{cases} a + c = 3 \\ 2a + c = 4 \end{cases}$
5.

If we multiply Equation 4 by -1 and add the result to Equation 5, we get $a = 1$. To find c, we can substitute 1 for a in Equation 4 and find that $c = 2$. To find b, we can substitute 1 for a and 2 for c in Equation 2 and find that $b = -2$.

After we substitute these values of a, b, and c into the equation $y = ax^2 + bx + c$, we have the equation of the parabola.

$$y = ax^2 + bx + c$$
$$y = 1x^2 - 2x + 2$$
$$y = x^2 - 2x + 2$$

■

Orals   *Is the triple a solution of the system?*

1. $(1, 1, 1)$, $\begin{cases} 2x + y - 3z = 0 \\ 3x - 2y + 4z = 5 \\ 4x + 2y - 6z = 0 \end{cases}$ **2.** $(2, 0, 1)$, $\begin{cases} 3x + 2y - z = 5 \\ 2x - 3y + 2z = 4 \\ 4x - 2y + 3z = 10 \end{cases}$

EXERCISE 3.3

In Exercises 1–2, tell whether the given triple is a solution of the system.

1. $(2, 1, 1)$, $\begin{cases} x - y + z = 2 \\ 2x + y - z = 4 \\ 2x - 3y + z = 2 \end{cases}$ **2.** $(-3, 2, -1)$, $\begin{cases} 2x + 2y + 3z = -1 \\ 3x + y - z = -6 \\ x + y + 2z = 1 \end{cases}$

In Exercises 3–14, solve each system.

3. $\begin{cases} x + y + z = 4 \\ 2x + y - z = 1 \\ 2x - 3y + z = 1 \end{cases}$ **4.** $\begin{cases} x + y + z = 4 \\ x - y + z = 2 \\ x - y - z = 0 \end{cases}$ **5.** $\begin{cases} 2x + 2y + 3z = 10 \\ 3x + y - z = 0 \\ x + y + 2z = 6 \end{cases}$ **6.** $\begin{cases} x - y + z = 4 \\ x + 2y - z = -1 \\ x + y - 3z = -2 \end{cases}$

7. $\begin{cases} a + b + 2c = 7 \\ a + 2b + c = 8 \\ 2a + b + c = 9 \end{cases}$ **8.** $\begin{cases} 2a + 3b + c = 2 \\ 4a + 6b + 2c = 5 \\ a - 2b + c = 3 \end{cases}$ **9.** $\begin{cases} 2x + y - z = 1 \\ x + 2y + 2z = 2 \\ 4x + 5y + 3z = 3 \end{cases}$ **10.** $\begin{cases} 4x + 3z = 4 \\ 2y - 6z = -1 \\ 8x + 4y + 3z = 9 \end{cases}$

11. $\begin{cases} 2x + 3y + 4z = 6 \\ 2x - 3y - 4z = -4 \\ 4x + 6y + 8z = 12 \end{cases}$ **12.** $\begin{cases} x - 3y + 4z = 2 \\ 2x + y + 2z = 3 \\ 4x - 5y + 10z = 7 \end{cases}$

13. $\begin{cases} x + \dfrac{1}{3}y + z = 13 \\ \dfrac{1}{2}x - y + \dfrac{1}{3}z = -2 \\ x + \dfrac{1}{2}y - \dfrac{1}{3}z = 2 \end{cases}$ **14.** $\begin{cases} x - \dfrac{1}{5}y - z = 9 \\ \dfrac{1}{4}x + \dfrac{1}{5}y - \dfrac{1}{2}z = 5 \\ 2x + y + \dfrac{1}{6}z = 12 \end{cases}$

In Exercises 15–26, solve each problem.

15. Integer problem   The sum of three integers is 18. The third integer is four times the second, and the second integer is 6 more than the first. Find the integers.

16. Integer problem   The sum of three integers is 48. If the first integer is doubled, the sum is 60. If the second integer is doubled, the sum is 63. Find the integers.

17. Geometry problem   The sum of the angles in any triangle is 180°. In triangle ABC, angle A is 100° less than the sum of angles B and C, and angle C is 40° less than twice angle B. Find the measure of each angle.

18. Geometry problem   The sum of the angles of any four-sided polygon is 360°. In the quadrilateral shown in Illustration 1, angle A = angle B, angle C is 20° greater than angle A, and angle D = 40°. Find the angles.

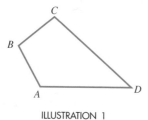

ILLUSTRATION 1

19. Nutritional planning   For each of three foods, 1 unit has the nutrients shown in the table. How many units of each must be used to provide exactly 11 grams of fat, 6 grams of carbohydrate, and 10 grams of protein?

Food	Fat	Carbohydrates	Protein
A	1	1	2
B	2	1	1
C	2	1	2

20. Nutritional planning   For each of three foods, 1 unit has the nutrients shown in the table. How many units of each must be used to provide exactly 14 grams of fat, 9 grams of carbohydrate, and 9 grams of protein?

Food	Fat	Carbohydrates	Protein
A	2	1	2
B	3	2	1
C	1	1	2

21. Making statues   An artist makes three types of ceramic statues at a monthly cost of $650 for 180 statues. The manufacturing costs for the three types are $5, $4, and $3. If the statues sell for $20, $12, and $9, respectively, how many of each type should be made to produce $2100 in monthly revenue?

22. Manufacturing footballs   A factory manufactures three types of footballs at a monthly cost of $2425 for 1125 footballs. The manufacturing costs for the three types of footballs are $4, $3, and $2. These footballs sell for $16, $12, and $10, respectively. How many of each type are manufactured if the monthly profit is $9275? (*Hint:* Profit = income − cost.)

23. Concert tickets Tickets for a concert cost $5, $3, and $2. Twice as many $5 tickets were sold as $2 tickets. The receipts for 750 tickets were $2625. How many of each price ticket were sold?

24. Mixing nuts The owner of a candy store wants to mix some peanuts worth $3 per pound, some cashews worth $9 per pound, and some Brazil nuts worth $9 per pound to get 50 pounds of a mixture that will sell for $6 per pound. She used 15 fewer pounds of cashews than peanuts. How many pounds of each did she use?

25. Chain saw sculpting A northwoods sculptor carves three types of statues with a chain saw. The time required for carving, sanding, and painting a totem pole, a bear, and a deer are shown in the table. How many of each should be produced to use all available labor hours?

	Totem pole	Bear	Deer	Time available
Carving	2 hours	2 hours	1 hour	14 hours
Sanding	1 hour	2 hours	2 hours	15 hours
Painting	3 hours	2 hours	2 hours	21 hours

26. Making clothing A clothing manufacturer makes coats, shirts, and slacks. The time required for cutting, sewing, and packaging each item is shown in the table. How many of each should be made to use all available labor hours?

	Coats	Shirts	Slacks	Time available
Cutting	20 min	15 min	10 min	115 hr
Sewing	60 min	30 min	24 min	280 hr
Packaging	5 min	12 min	6 min	65 hr

27. Curve fitting Find the equation of the parabola shown in Illustration 2.

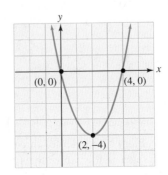

ILLUSTRATION 2

28. Curve fitting Find the equation of the parabola shown in Illustration 3.

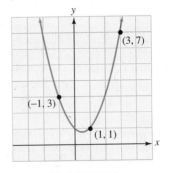

ILLUSTRATION 3

In Exercises 29–30, the equation of a circle is of the form $x^2 + y^2 + cx + dy + e = 0$.

29. Curve fitting Find the equation of the circle shown in Illustration 4.

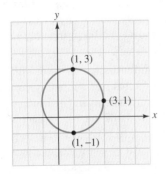

ILLUSTRATION 4

30. Curve fitting Find the equation of the circle shown in Illustration 5.

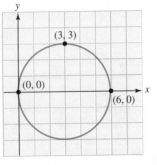

ILLUSTRATION 5

Writing Exercises   *Write a paragraph using your own words.*

1. What makes a system of three equations in three variables inconsistent?

2. What makes the equations of a system of three equations in three variables dependent?

Something to Think About

1. Solve the system

$$\begin{cases} x + y + z + w = 3 \\ x - y - z - w = -1 \\ x + y - z - w = 1 \\ x + y - z + w = 3 \end{cases}$$

2. Solve the system

$$\begin{cases} 2x + y + z + w = 3 \\ x - 2y - z + w = -3 \\ x - y - 2z - w = -3 \\ x + y - z + 2w = 4 \end{cases}$$

Review Exercises   *Consider the line passing through $P(-2, -4)$ and $Q(3, 5)$.*

1. Find the slope of line PQ.

2. Write the equation of line PQ in general form.

Let $f(x) = 2x^2 + 1$. Find each value.

3. $f(0)$ **4.** $f(-2)$ **5.** $f(s)$ **6.** $f(2t)$

3.4 Solution by Matrices

■ GAUSSIAN ELIMINATION ■ SYSTEMS WITH MORE EQUATIONS THAN VARIABLES ■ SYSTEMS WITH MORE VARIABLES THAN EQUATIONS

Another method of solving systems of equations involves a rectangular array of numbers called a **matrix**.

> **Matrix**
> A **matrix** is any rectangular array of numbers.

Some examples of matrices are

$$A = \begin{bmatrix} 1 & 2 & 3 \\ 4 & 5 & 6 \end{bmatrix} \qquad B = \begin{bmatrix} 1 & 2 \\ 3 & 4 \\ 5 & 6 \end{bmatrix} \qquad C = \begin{bmatrix} 2 & 4 & 6 \\ 8 & 10 & 12 \\ 14 & 16 & 18 \end{bmatrix}$$

The numbers in each matrix are called its **elements**. Because matrix A has two rows and three columns, it is called a 2×3 matrix (read "2 by 3" matrix). Matrix B is a 3×2 matrix, because it has three rows and two columns. Matrix C is a 3×3 matrix (three rows and three columns).

Any matrix with the same number of rows and columns is called a **square matrix**. Matrix C is an example of a square matrix.

To show one way we can use matrices to solve systems of linear equations, we consider the following system

$$\begin{cases} x - 2y - z = 6 \\ 2x + 2y - z = 1 \\ -x - y + 2z = 1 \end{cases}$$

which can be represented by the following matrix, called an **augmented matrix**:

$$\begin{bmatrix} 1 & -2 & -1 & \vdots & 6 \\ 2 & 2 & -1 & \vdots & 1 \\ -1 & -1 & 2 & \vdots & 1 \end{bmatrix}$$

The 3×3 matrix to the left of the dashed line, called the **coefficient matrix**, is determined by the coefficients of x, y, and z in the equations of the system. The 3×1 matrix to the right of the dashed line is determined by the constants in the equations. Each row of the augmented matrix represents exactly one equation of the system:

$$\begin{bmatrix} 1 & -2 & -1 & \vdots & 6 \\ 2 & 2 & -1 & \vdots & 1 \\ -1 & -1 & 2 & \vdots & 1 \end{bmatrix} \leftrightarrow \begin{cases} x - 2y - z = 6 \\ 2x + 2y - z = 1 \\ -x - y + 2z = 1 \end{cases}$$

■ GAUSSIAN ELIMINATION

To solve a system by **Gaussian elimination**, we transform the augmented matrix into the following matrix that has all 0s below its main diagonal, which is formed by elements a, e, and h.

$$\begin{bmatrix} a & b & c & \vdots & d \\ 0 & e & f & \vdots & g \\ 0 & 0 & h & \vdots & i \end{bmatrix} \qquad (a, b, c, \ldots, i \text{ are real numbers.})$$

We can often write a matrix in this form, called **triangular form**, by using three operations called **elementary row operations**.

Elementary Row Operations
1. Any two rows of a matrix can be interchanged.

2. Any row of a matrix can be multiplied by a nonzero constant.

3. Any row of a matrix can be changed by adding a nonzero constant multiple of another row to it.

- A type 1 row operation corresponds to interchanging two equations of the system.

- A type 2 row operation corresponds to multiplying both sides of an equation by a nonzero constant.

- A type 3 row operation corresponds to adding a nonzero multiple of one equation to another.

None of these operations will change the solution set of the given system of equations.

After we have written the matrix in triangular form, we can solve the corresponding system of equations by a back substitution process, as shown in Example 1.

EXAMPLE 1   Solve the system $\begin{cases} x - 2y - z = 6 \\ 2x + 2y - z = 1. \\ -x - y + 2z = 1 \end{cases}$

Solution   We can represent the system with the following augmented matrix:

$$\begin{bmatrix} 1 & -2 & -1 & \vdots & 6 \\ 2 & 2 & -1 & \vdots & 1 \\ -1 & -1 & 2 & \vdots & 1 \end{bmatrix}$$

To get 0's under the 1 in the first column, we use a type 3 row operation twice:

Multiply row 1 by -2 and add to row 2.   Multiply row 1 by 1 and add to row 3.

$$\begin{bmatrix} 1 & -2 & -1 & \vdots & 6 \\ 2 & 2 & -1 & \vdots & 1 \\ -1 & -1 & 2 & \vdots & 1 \end{bmatrix} \approx \begin{bmatrix} 1 & -2 & -1 & \vdots & 6 \\ 0 & 6 & 1 & \vdots & -11 \\ -1 & -1 & 2 & \vdots & 1 \end{bmatrix} \approx \begin{bmatrix} 1 & -2 & -1 & \vdots & 6 \\ 0 & 6 & 1 & \vdots & -11 \\ 0 & -3 & 1 & \vdots & 7 \end{bmatrix}$$

The symbol $\approx$ is read as "is row equivalent to." Each of the above matrices represents a system of equations, and they are all equivalent.

To get a 0 under the 6 in the second column of the last matrix, we use another type 3 row operation:

Multiply row 2 by $\frac{1}{2}$ and add to row 3.

$$\begin{bmatrix} 1 & -2 & -1 & \vdots & 6 \\ 0 & 6 & 1 & \vdots & -11 \\ 0 & -3 & 1 & \vdots & 7 \end{bmatrix} \approx \begin{bmatrix} 1 & -2 & -1 & \vdots & 6 \\ 0 & 6 & 1 & \vdots & -11 \\ 0 & 0 & \frac{3}{2} & \vdots & \frac{3}{2} \end{bmatrix}$$

We can use a type 2 row operation to simplify further.

Multiply row 3 by $\frac{2}{3}$, the reciprocal of $\frac{3}{2}$.

$$\begin{bmatrix} 1 & -2 & -1 & \vdots & 6 \\ 0 & 6 & 1 & \vdots & -11 \\ 0 & 0 & \frac{3}{2} & \vdots & \frac{3}{2} \end{bmatrix} \approx \begin{bmatrix} 1 & -2 & -1 & \vdots & 6 \\ 0 & 6 & 1 & \vdots & -11 \\ 0 & 0 & 1 & \vdots & 1 \end{bmatrix}$$

The final matrix represents the system of equations

1. $\begin{cases} x - 2y - z = 6 \\ 0x + 6y + z = -11 \\ 0x + 0y + z = 1 \end{cases}$
2.
3.

From Equation 3, we can read that $z = 1$. To find y, we substitute 1 for z in Equation 2 and solve for y:

2. $6y + z = -11$

$6y + \mathbf{1} = -11$ Substitute 1 for z.

$6y = -12$ Subtract 1 from both sides.

$y = -2$ Divide both sides by 6.

Thus, $y = -2$. To find x, we substitute 1 for z and -2 for y in Equation 1 and solve for x:

1.
$$x - 2y - z = 6$$
$$x - 2(-2) - 1 = 6 \qquad \text{Substitute 1 for } z \text{ and } -2 \text{ for } y.$$
$$x + 3 = 6 \qquad \text{Simplify.}$$
$$x = 3 \qquad \text{Subtract 3 from both sides.}$$

Thus, $x = 3$. The solution to the given system is $(3, -2, 1)$. Verify that this triple satisfies each equation of the original system. ■

■ SYSTEMS WITH MORE EQUATIONS THAN VARIABLES

We can use matrices to solve systems with more equations than variables.

EXAMPLE 2 Solve the system $\begin{cases} x + y = -1 \\ 2x - y = 7 \\ -x + 2y = -8 \end{cases}$.

Solution This system can be represented by a 3×3 augmented matrix:

$$\begin{bmatrix} 1 & 1 & \vdots & -1 \\ 2 & -1 & \vdots & 7 \\ -1 & 2 & \vdots & -8 \end{bmatrix}$$

To get 0's under the 1 in the first column, we do a type 3 row operation twice:

$$\begin{array}{cc} \text{Multiply row 1 by } -2 & \text{Multiply row 1 by 1} \\ \text{and add to row 2.} & \text{and add to row 3.} \end{array}$$

$$\begin{bmatrix} 1 & 1 & \vdots & -1 \\ 2 & -1 & \vdots & 7 \\ -1 & 2 & \vdots & -8 \end{bmatrix} \approx \begin{bmatrix} 1 & 1 & \vdots & -1 \\ 0 & -3 & \vdots & 9 \\ -1 & 2 & \vdots & -8 \end{bmatrix} \approx \begin{bmatrix} 1 & 1 & \vdots & -1 \\ 0 & -3 & \vdots & 9 \\ 0 & 3 & \vdots & -9 \end{bmatrix}$$

We can do other row operations to get a 0 under the -3 in the second column, and then 1 in the second row, second column.

$$\begin{array}{cc} \text{Multiply row 2 by 1} & \text{Multiply row 2} \\ \text{and add to row 3.} & \text{by } -\frac{1}{3}. \end{array}$$

$$\begin{bmatrix} 1 & 1 & \vdots & -1 \\ 0 & -3 & \vdots & 9 \\ 0 & 3 & \vdots & -9 \end{bmatrix} \approx \begin{bmatrix} 1 & 1 & \vdots & -1 \\ 0 & -3 & \vdots & 9 \\ 0 & 0 & \vdots & 0 \end{bmatrix} \approx \begin{bmatrix} 1 & 1 & \vdots & -1 \\ 0 & 1 & \vdots & -3 \\ 0 & 0 & \vdots & 0 \end{bmatrix}$$

The final matrix represents the system

$$\begin{cases} x + y = -1 \\ 0x + y = -3 \\ 0x + 0y = 0 \end{cases}$$

The third equation can be discarded, because $0x + 0y = 0$ for all x and y. From the second equation, we can read that $y = -3$. To find x, we substitute -3 for y in the first equation and solve for x:

$$x + y = -1$$
$$x - 3 = -1 \qquad \text{Substitute } -3 \text{ for } y.$$
$$x = 2 \qquad \text{Add 3 to both sides.}$$

The solution is $(2, -3)$. Verify that this solution satisfies all three equations of the original system. ■

If the last row of the final matrix in Example 2 had been of the form $0x + 0y = k$, where $k \neq 0$, the system would have no solution. No values of x and y could make the expression $0x + 0y$ equal to a nonzero constant k.

■ SYSTEMS WITH MORE VARIABLES THAN EQUATIONS

We can also solve many systems that have more variables than equations.

EXAMPLE 3   Solve the system $\begin{cases} x + y - 2z = -1 \\ 2x - y + z = -3 \end{cases}$.

Solution   We start by doing a type 3 row operation to get a 0 under the 1 in the first column.

Multiply row 1 by -2
and add to row 2.

$$\begin{bmatrix} 1 & 1 & -2 & \vdots & -1 \\ 2 & -1 & 1 & \vdots & -3 \end{bmatrix} \approx \begin{bmatrix} 1 & 1 & -2 & \vdots & -1 \\ 0 & -3 & 5 & \vdots & -1 \end{bmatrix}$$

We then do a type 2 row operation:

Multiply row 2 by $-\frac{1}{3}$.

$$\begin{bmatrix} 1 & 1 & -2 & \vdots & -1 \\ 0 & -3 & 5 & \vdots & -1 \end{bmatrix} \approx \begin{bmatrix} 1 & 1 & -2 & \vdots & -1 \\ 0 & 1 & -\frac{5}{3} & \vdots & \frac{1}{3} \end{bmatrix}$$

The final matrix represents the system

$$\begin{cases} x + y - 2z = -1 \\ \quad\quad y - \dfrac{5}{3}z = \dfrac{1}{3} \end{cases}$$

We add $\frac{5}{3}z$ to both sides of the second equation to obtain

$$y = \frac{1}{3} + \frac{5}{3}z$$

We have not found a specific value for y. However, we have found y in terms of z. To find a value of x in terms of z, we substitute $\frac{1}{3} + \frac{5}{3}z$ for y in the first equation and simplify to get

$$x + y - 2z = -1$$
$$x + \frac{1}{3} + \frac{5}{3}z - 2z = -1 \qquad \text{Substitute } \tfrac{1}{3} + \tfrac{5}{3}z \text{ for } y.$$
$$x + \frac{1}{3} - \frac{1}{3}z = -1 \qquad \text{Combine like terms.}$$
$$x - \frac{1}{3}z = -\frac{4}{3} \qquad \text{Subtract } \tfrac{1}{3} \text{ from both sides.}$$
$$x = -\frac{4}{3} + \frac{1}{3}z \qquad \text{Add } \tfrac{1}{3}z \text{ to both sides.}$$

A solution to this system must have the form

$$\left(-\frac{4}{3} + \frac{1}{3}z, \quad \frac{1}{3} + \frac{5}{3}z, \quad z\right)$$

for all values of z. This system has infinitely many solutions, a different one for each value of z. For example,

- If $z = 0$, the corresponding solution is $\left(-\frac{4}{3}, \frac{1}{3}, 0\right)$.
- If $z = 1$, the corresponding solution is $(-1, 2, 1)$.

Verify that both of these solutions satisfy each equation of the given system. ■

Orals   *Consider the system* $\begin{cases} 3x + 2y = 8 \\ 4x - 3y = 6 \end{cases}$

1. Find the coefficient matrix.

2. Find the augmented matrix.

Tell whether each matrix is in triangular form.

3. $\begin{bmatrix} 4 & 1 & 5 \\ 0 & 2 & 7 \\ 0 & 0 & 4 \end{bmatrix}$
   4. $\begin{bmatrix} 8 & 5 & 2 \\ 0 & 4 & 5 \\ 0 & 7 & 0 \end{bmatrix}$

EXERCISE 3.4

In Exercises 1–4, use a row operation to find the missing number in the second matrix.

1. $\begin{bmatrix} 2 & 1 & 1 \\ 5 & 4 & 1 \end{bmatrix}$
$\begin{bmatrix} 2 & 1 & 1 \\ 3 & 3 & \end{bmatrix}$

2. $\begin{bmatrix} -1 & 3 & 2 \\ 1 & -2 & 3 \end{bmatrix}$
$\begin{bmatrix} -1 & 3 & 2 \\ & 1 & 5 \end{bmatrix}$

3. $\begin{bmatrix} 3 & -2 & 1 \\ -1 & 2 & 4 \end{bmatrix}$
$\begin{bmatrix} 3 & -2 & 1 \\ -2 & 4 & \end{bmatrix}$

4. $\begin{bmatrix} 2 & 1 & -3 \\ 2 & 6 & 1 \end{bmatrix}$
$\begin{bmatrix} 6 & 3 & \\ 2 & 6 & 1 \end{bmatrix}$

In Exercises 5–16, use matrices to solve each system of equations. Each system has one solution.

5. $\begin{cases} x + y = 2 \\ x - y = 0 \end{cases}$

6. $\begin{cases} x + y = 3 \\ x - y = -1 \end{cases}$

7. $\begin{cases} x + 2y = -4 \\ 2x + y = 1 \end{cases}$

8. $\begin{cases} 2x - 3y = 16 \\ -4x + y = -22 \end{cases}$

9. $\begin{cases} 3x + 4y = -12 \\ 9x - 2y = 6 \end{cases}$

10. $\begin{cases} 5x - 4y = 10 \\ x - 7y = 2 \end{cases}$

11. $\begin{cases} x + y + z = 6 \\ x + 2y + z = 8 \\ x + y + 2z = 9 \end{cases}$

12. $\begin{cases} x - y + z = 2 \\ x + 2y - z = 6 \\ 2x - y - z = 3 \end{cases}$

13. $\begin{cases} 2x + y + 3z = 3 \\ -2x - y + z = 5 \\ 4x - 2y + 2z = 2 \end{cases}$

14. $\begin{cases} 3x + 2y + z = 8 \\ 6x - y + 2z = 16 \\ -9x + y - z = -20 \end{cases}$

15. $\begin{cases} 3x - 2y + 4z = 4 \\ x + y + z = 3 \\ 6x - 2y - 3z = 10 \end{cases}$

16. $\begin{cases} 2x + 3y - z = -8 \\ x - y - z = -2 \\ -4x + 3y + z = 6 \end{cases}$

In Exercises 17–24, use matrices to solve each system of equations. If a system has no solution, so indicate.

17. $\begin{cases} x + y = 3 \\ 3x - y = 1 \\ 2x + y = 4 \end{cases}$
18. $\begin{cases} x - y = -5 \\ 2x + 3y = 5 \\ x + y = 1 \end{cases}$
19. $\begin{cases} 2x - y = 4 \\ x + 3y = 2 \\ -x - 4y = -2 \end{cases}$
20. $\begin{cases} 3x - 2y = 5 \\ x + 2y = 7 \\ -3x - y = -11 \end{cases}$

21. $\begin{cases} 2x + y = 7 \\ x - y = 2 \\ -x + 3y = -2 \end{cases}$
22. $\begin{cases} 3x - y = 2 \\ -6x + 3y = 0 \\ -x + 2y = -4 \end{cases}$
23. $\begin{cases} x + 3y = 7 \\ x + y = 3 \\ 3x + y = 5 \end{cases}$
24. $\begin{cases} x + y = 3 \\ x - 2y = -3 \\ x - y = 1 \end{cases}$

In Exercises 25–28, use matrices to solve each system of equations.

25. $\begin{cases} x + 2y + 3z = -2 \\ -x - y - 2z = 4 \end{cases}$

26. $\begin{cases} 2x - 4y + 3z = 6 \\ -4x + 6y + 4z = -6 \end{cases}$

27. $\begin{cases} x - y = 1 \\ y + z = 1 \\ x + z = 2 \end{cases}$

28. $\begin{cases} x + z = 1 \\ x + y = 2 \\ 2x + y + z = 3 \end{cases}$

In Exercises 29–32, remember these facts from geometry:
Two angles whose measures add to 90° are complementary.
Two angles whose measures add to 180° are supplementary.
The sum of the angles of a triangle is 180°.

29. Geometry One angle is 46° larger than its complement. Find the angles.

30. Geometry One angle is 28° larger than its supplement. Find the angles.

31. Geometry In Illustration 1, angle B is 25° more than angle A, and angle C is 5° less than twice angle A. Find each angle in the triangle.

32. Geometry In Illustration 2, angle A is 10° less than angle B, and angle B is 10° less than angle C. Find each angle in the triangle.

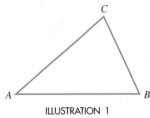

ILLUSTRATION 1

ILLUSTRATION 2

In Exercises 33–34, remember that the equation of a parabola is of the form $y = ax^2 + bx + c$.

33. Curve fitting Find the equation of the parabola passing through the points $(0, 1)$, $(1, 2)$, and $(-1, 4)$.

34. Curve fitting Find the equation of the parabola passing through the points $(0, 1)$, $(1, 1)$, and $(-1, -1)$.

Writing Exercises *Write a paragraph using your own words.*

1. Explain how to check the solution of a system of equations.

2. Explain how to perform a type 3 row operation.

Something to Think About

1. If the system represented by

$$\begin{bmatrix} 1 & 1 & 0 & | & 1 \\ 0 & 0 & 1 & | & 2 \\ 0 & 0 & 0 & | & k \end{bmatrix}$$ has no

solution, what do you know about k?

2. Is it possible for a system with fewer equations than variables to have no solution? Illustrate.

3.5 Solution by Determinants

■ DETERMINANTS ■ CRAMER'S RULE ■ GRAPHING DEVICES

Closely related to the concept of a matrix is the **determinant**.

■ DETERMINANTS

A determinant is a number that is associated with a square matrix. For any square matrix A, the symbol $|A|$ represents the determinant of matrix A.

> **Value of a 2 × 2 Determinant**
>
> If a, b, c, and d are numbers, the **determinant** of the matrix $\begin{bmatrix} a & b \\ c & d \end{bmatrix}$ is
>
> $$\begin{vmatrix} a & b \\ c & d \end{vmatrix} = ad - bc$$

The determinant of a 2×2 matrix is the number that is equal to the product of the numbers on the major diagonal

$$\begin{vmatrix} a & b \\ c & d \end{vmatrix}$$

minus the product of the numbers on the other diagonal

$$\begin{vmatrix} a & b \\ c & d \end{vmatrix}$$

(b)   **EXAMPLE 1**   Evaluate the determinants   **a.** $\begin{vmatrix} 3 & 2 \\ 6 & 9 \end{vmatrix}$ and **b.** $\begin{vmatrix} -5 & \frac{1}{2} \\ -1 & 0 \end{vmatrix}$.

Solution   **a.** $\begin{vmatrix} 3 & 2 \\ 6 & 9 \end{vmatrix} = 3(9) - 2(6)$ **b.** $\begin{vmatrix} -5 & \frac{1}{2} \\ -1 & 0 \end{vmatrix} = -5(0) - \frac{1}{2}(-1)$

$= 27 - 12$ $= 0 + \frac{1}{2}$

$= 15$ $= \frac{1}{2}$

A 3×3 determinant is evaluated by expanding by **minors**.

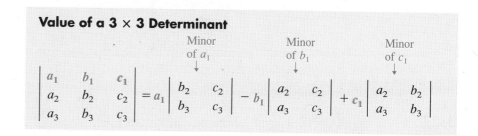

Value of a 3 × 3 Determinant

$$\begin{vmatrix} a_1 & b_1 & c_1 \\ a_2 & b_2 & c_2 \\ a_3 & b_3 & c_3 \end{vmatrix} = a_1 \begin{vmatrix} b_2 & c_2 \\ b_3 & c_3 \end{vmatrix} - b_1 \begin{vmatrix} a_2 & c_2 \\ a_3 & c_3 \end{vmatrix} + c_1 \begin{vmatrix} a_2 & b_2 \\ a_3 & b_3 \end{vmatrix}$$

with "Minor of a_1", "Minor of b_1", "Minor of c_1" labels above.

To find the minor of a_1, we find the determinant formed by crossing out the elements of the matrix that are in the same row and column as a_1:

$$\begin{vmatrix} a_1 & b_1 & c_1 \\ a_2 & b_2 & c_2 \\ a_3 & b_3 & c_3 \end{vmatrix} \qquad \text{The minor of } a_1 \text{ is } \begin{vmatrix} b_2 & c_2 \\ b_3 & c_3 \end{vmatrix}.$$

To find the minor of b_1, we cross out the elements of the matrix that are in the same row and column as b_1:

$$\begin{vmatrix} a_1 & b_1 & c_1 \\ a_2 & b_2 & c_2 \\ a_3 & b_3 & c_3 \end{vmatrix} \qquad \text{The minor of } b_1 \text{ is } \begin{vmatrix} a_2 & c_2 \\ a_3 & c_3 \end{vmatrix}.$$

To find the minor of c_1, we cross out the elements of the matrix that are in the same row and column as c_1:

$$\begin{vmatrix} a_1 & b_1 & c_1 \\ a_2 & b_2 & c_2 \\ a_3 & b_3 & c_3 \end{vmatrix} \qquad \text{The minor of } c_1 \text{ is } \begin{vmatrix} a_2 & b_2 \\ a_3 & b_3 \end{vmatrix}.$$

EXAMPLE 2   Evaluate the determinant $\begin{vmatrix} 1 & 3 & -2 \\ 2 & 1 & 3 \\ 1 & 2 & 3 \end{vmatrix}$.

Solution   We expand the determinant along the first row.

$$\begin{vmatrix} 1 & 3 & -2 \\ 2 & 1 & 3 \\ 1 & 2 & 3 \end{vmatrix} = 1 \begin{vmatrix} 1 & 3 \\ 2 & 3 \end{vmatrix} - 3 \begin{vmatrix} 2 & 3 \\ 1 & 3 \end{vmatrix} + (-2) \begin{vmatrix} 2 & 1 \\ 1 & 2 \end{vmatrix}$$

with "Minor of 1", "Minor of 3", "Minor of -2" labels above.

$$= 1(3 - 6) - 3(6 - 3) - 2(4 - 1)$$
$$= -3 - 9 - 6$$
$$= -18$$

We can evaluate a 3×3 determinant by expanding it along any row or column. To determine the signs between the terms of the expansion of a 3×3 determinant, we use the following array of signs:

Array of Signs for a 3 × 3 Determinant

$$
\begin{array}{ccc}
+ & - & + \\
- & + & - \\
+ & - & +
\end{array}
$$

EXAMPLE 3   Evaluate the determinant $\begin{vmatrix} 1 & 3 & -2 \\ 2 & 1 & 3 \\ 1 & 2 & 3 \end{vmatrix}$ by expanding on the middle column.

Solution   This is the determinant of Example 2. To expand it along the middle column, we use the signs of the middle column of the array of signs:

$$
\begin{vmatrix} 1 & 3 & -2 \\ 2 & 1 & 3 \\ 1 & 2 & 3 \end{vmatrix} = -3 \overset{\text{Minor of 3}}{\begin{vmatrix} 2 & 3 \\ 1 & 3 \end{vmatrix}} + 1 \overset{\text{Minor of 1}}{\begin{vmatrix} 1 & -2 \\ 1 & 3 \end{vmatrix}} - 2 \overset{\text{Minor of 2}}{\begin{vmatrix} 1 & -2 \\ 2 & 3 \end{vmatrix}}
$$

$$
\begin{aligned}
&= -3(6 - 3) + 1[3 - (-2)] - 2[3 - (-4)] \\
&= -3(3) + 1(5) - 2(7) \\
&= -9 + 5 - 14 \\
&= -18
\end{aligned}
$$

As expected, we get the same value as in Example 2.   ■

■ CRAMER'S RULE

The method of using determinants to solve systems of equations is called **Cramer's rule**, named after the 18th-century mathematician Gabriel Cramer. To develop Cramer's rule, we consider the system

$$
\begin{cases} ax + by = e \\ cx + dy = f \end{cases}
$$

where x and y are variables and a, b, c, d, e, and f are constants.

If we multiply both sides of the first equation by d and multiply both sides of the second equation by $-b$, we can add the equations and eliminate y:

$$
\begin{aligned}
adx + bdy &= ed \\
-bcx - bdy &= -bf \\
\hline
adx - bcx &= ed - bf
\end{aligned}
$$

To solve for x, we use the distributive property to write $adx - bcx$ as $(ad - bc)x$ on the left-hand side and divide each side by $ad - bc$:

$$
\begin{aligned}
(ad - bc)x &= ed - bf \\
x &= \frac{ed - bf}{ad - bc} \qquad (ad - bc \neq 0)
\end{aligned}
$$

We can find y in a similar manner. After eliminating the variable x, we get

$$y = \frac{af - ec}{ad - bc} \qquad (ad - bc \neq 0)$$

Determinants provide an easy way of remembering these formulas. Note that the denominator for both x and y is

$$\begin{vmatrix} a & b \\ c & d \end{vmatrix} = ad - bc$$

The numerators can be expressed as determinants also:

$$x = \frac{ed - bf}{ad - bc} = \frac{\begin{vmatrix} e & b \\ f & d \end{vmatrix}}{\begin{vmatrix} a & b \\ c & d \end{vmatrix}} \quad \text{and} \quad y = \frac{af - ec}{ad - bc} = \frac{\begin{vmatrix} a & e \\ c & f \end{vmatrix}}{\begin{vmatrix} a & b \\ c & d \end{vmatrix}}$$

If we compare these formulas with the original system

$$\begin{cases} ax + by = e \\ cx + dy = f \end{cases}$$

we note that in the expressions for x and y above, the denominator determinant is formed by using the coefficients a, b, c, and d of the variables in the equations. The numerator determinants are the same as the denominator determinant, except that the column of coefficients of the variable for which we are solving is replaced with the column of constants e and f.

Cramer's Rule for Two Equations in Two Variables

The solution of the system $\begin{cases} ax + by = e \\ cx + dy = f \end{cases}$ is given by

$$x = \frac{D_x}{D} = \frac{\begin{vmatrix} e & b \\ f & d \end{vmatrix}}{\begin{vmatrix} a & b \\ c & d \end{vmatrix}} \quad \text{and} \quad y = \frac{D_y}{D} = \frac{\begin{vmatrix} a & e \\ c & f \end{vmatrix}}{\begin{vmatrix} a & b \\ c & d \end{vmatrix}}$$

If every determinant is 0, the system is consistent but the equations are dependent.

If $D = 0$ and D_x or D_y is nonzero, the system is inconsistent.

EXAMPLE 4   Use Cramer's rule to solve the system $\begin{cases} 4x - 3y = 6 \\ -2x + 5y = 4 \end{cases}$.

Solution   The value of x is the quotient of two determinants. The denominator determinant is made up of the coefficients of x and y:

$$D = \begin{vmatrix} 4 & -3 \\ -2 & 5 \end{vmatrix}$$

To solve for x, we form the numerator determinant from the denominator determinant by replacing its first column (the coefficients of x) with the column of constants (6 and 4).

To solve for y, we form the numerator determinant from the denominator determinant by replacing the second column (the coefficients of y) with the column of constants (6 and 4).

To find the values of x and y, we evaluate each determinant:

$$x = \frac{\begin{vmatrix} 6 & -3 \\ 4 & 5 \end{vmatrix}}{\begin{vmatrix} 4 & -3 \\ -2 & 5 \end{vmatrix}} = \frac{6(5) - (-3)(4)}{4(5) - (-3)(-2)} = \frac{30 + 12}{20 - 6} = \frac{42}{14} = 3$$

$$y = \frac{\begin{vmatrix} 4 & 6 \\ -2 & 4 \end{vmatrix}}{\begin{vmatrix} 4 & -3 \\ -2 & 5 \end{vmatrix}} = \frac{4(4) - 6(-2)}{4(5) - (-3)(-2)} = \frac{16 + 12}{20 - 6} = \frac{28}{14} = 2$$

The solution of this system is (3, 2). Verify that $x = 3$ and $y = 2$ satisfy each equation in the given system. ∎

EXAMPLE 5   Use Cramer's rule to solve the system $\begin{cases} 7x = 8 - 4y \\ 2y = 3 - \frac{7}{2}x \end{cases}$.

Solution   We multiply both sides of the second equation by 2 to eliminate the fraction and write the system in the form

$$\begin{cases} 7x + 4y = 8 \\ 7x + 4y = 6 \end{cases}$$

When we attempt to use Cramer's rule to solve this system for x, we obtain

$$x = \frac{\begin{vmatrix} 8 & 4 \\ 6 & 4 \end{vmatrix}}{\begin{vmatrix} 7 & 4 \\ 7 & 4 \end{vmatrix}} = \frac{8}{0} \qquad \text{which is undefined}$$

Since the denominator determinant is 0 and the numerator determinant is not 0, the system is inconsistent. It has no solutions.

We can see directly from the system that it is inconsistent. For any values of x and y, it is impossible that 7 times x plus 4 times y could be both 8 and 6. ∎

Cramer's Rule for Three Equations in Three Variables

The solution of the system $\begin{cases} ax + by + cz = j \\ dx + ey + fz = k \\ gx + hy + iz = l \end{cases}$ is given by

$$x = \frac{D_x}{D}, \qquad y = \frac{D_y}{D}, \qquad \text{and} \qquad z = \frac{D_z}{D}$$

where

$$D = \begin{vmatrix} a & b & c \\ d & e & f \\ g & h & i \end{vmatrix} \qquad D_x = \begin{vmatrix} j & b & c \\ k & e & f \\ l & h & i \end{vmatrix}$$

$$D_y = \begin{vmatrix} a & j & c \\ d & k & f \\ g & l & i \end{vmatrix} \qquad D_z = \begin{vmatrix} a & b & j \\ d & e & k \\ g & h & l \end{vmatrix}$$

If every determinant is 0, the system is consistent but the equations are dependent.

If $D = 0$ and D_x or D_y or D_z is nonzero, the system is inconsistent.

EXAMPLE 6 Use Cramer's rule to solve the system $\begin{cases} 2x + y + 4z = 12 \\ x + 2y + 2z = 9 \\ 3x - 3y - 2z = 1 \end{cases}$.

Solution The denominator determinant is the determinant formed by the coefficients of the variables. The numerator determinants are formed by replacing the coefficients of the variable being solved for by the column of constants. We form the quotients for x, y, and z and evaluate the determinants:

$$x = \frac{\begin{vmatrix} 12 & 1 & 4 \\ 9 & 2 & 2 \\ 1 & -3 & -2 \end{vmatrix}}{\begin{vmatrix} 2 & 1 & 4 \\ 1 & 2 & 2 \\ 3 & -3 & -2 \end{vmatrix}} = \frac{12\begin{vmatrix} 2 & 2 \\ -3 & -2 \end{vmatrix} - 1\begin{vmatrix} 9 & 2 \\ 1 & -2 \end{vmatrix} + 4\begin{vmatrix} 9 & 2 \\ 1 & -3 \end{vmatrix}}{2\begin{vmatrix} 2 & 2 \\ -3 & -2 \end{vmatrix} - 1\begin{vmatrix} 1 & 2 \\ 3 & -2 \end{vmatrix} + 4\begin{vmatrix} 1 & 2 \\ 3 & -3 \end{vmatrix}} = \frac{12(2) - (-20) + 4(-29)}{2(2) - (-8) + 4(-9)} = \frac{-72}{-24} = 3$$

$$y = \frac{\begin{vmatrix} 2 & 12 & 4 \\ 1 & 9 & 2 \\ 3 & 1 & -2 \end{vmatrix}}{\begin{vmatrix} 2 & 1 & 4 \\ 1 & 2 & 2 \\ 3 & -3 & -2 \end{vmatrix}} = \frac{2\begin{vmatrix} 9 & 2 \\ 1 & -2 \end{vmatrix} - 12\begin{vmatrix} 1 & 2 \\ 3 & -2 \end{vmatrix} + 4\begin{vmatrix} 1 & 9 \\ 3 & 1 \end{vmatrix}}{-24} = \frac{2(-20) - 12(-8) + 4(-26)}{-24} = \frac{-48}{-24} = 2$$

$$z = \frac{\begin{vmatrix} 2 & 1 & 12 \\ 1 & 2 & 9 \\ 3 & -3 & 1 \end{vmatrix}}{\begin{vmatrix} 2 & 1 & 4 \\ 1 & 2 & 2 \\ 3 & -3 & -2 \end{vmatrix}} = \frac{2 \begin{vmatrix} 2 & 9 \\ -3 & 1 \end{vmatrix} - 1 \begin{vmatrix} 1 & 9 \\ 3 & 1 \end{vmatrix} + 12 \begin{vmatrix} 1 & 2 \\ 3 & -3 \end{vmatrix}}{-24} = \frac{2(29) - (-26) + 12(-9)}{-24} = \frac{-24}{-24} = 1$$

The solution of this system is $(3, 2, 1)$. ■

■ GRAPHING DEVICES

Many software packages and graphing devices have matrix and determinant capabilities. To evaluate the determinant of a square matrix with a graphing device, we enter a matrix-edit mode and then enter each element of the matrix, one at a time. Then we select the DET option from a menu. The value of the determinant will appear on the screen.

Orals   *Evaluate each determinant.*

1. $\begin{vmatrix} 2 & 1 \\ 1 & 1 \end{vmatrix}$ **2.** $\begin{vmatrix} 0 & 2 \\ 1 & 1 \end{vmatrix}$ **3.** $\begin{vmatrix} 0 & 1 \\ 0 & 1 \end{vmatrix}$

When using Cramer's rule to solve the system $\begin{cases} x + 2y = 5 \\ 2x - y = 4 \end{cases}$,

4. Write the denominator determinant for x.

5. Write the numerator determinant for x.

6. Write the numerator determinant for y.

EXERCISE 3.5

In Exercises 1–18, evaluate each determinant.

1. $\begin{vmatrix} 2 & 3 \\ -2 & 1 \end{vmatrix}$ **2.** $\begin{vmatrix} 3 & -2 \\ -2 & 4 \end{vmatrix}$ **3.** $\begin{vmatrix} -1 & 2 \\ 3 & -4 \end{vmatrix}$

4. $\begin{vmatrix} -1 & -2 \\ -3 & -4 \end{vmatrix}$ **5.** $\begin{vmatrix} x & y \\ y & x \end{vmatrix}$ **6.** $\begin{vmatrix} x+y & y-x \\ x & y \end{vmatrix}$

7. $\begin{vmatrix} 1 & 0 & 1 \\ 0 & 1 & 0 \\ 1 & 1 & 1 \end{vmatrix}$ **8.** $\begin{vmatrix} 1 & 2 & 0 \\ 0 & 1 & 2 \\ 0 & 0 & 1 \end{vmatrix}$ **9.** $\begin{vmatrix} -1 & 2 & 1 \\ 2 & 1 & -3 \\ 1 & 1 & 1 \end{vmatrix}$

10. $\begin{vmatrix} 1 & 2 & 3 \\ 1 & 2 & 3 \\ 1 & 2 & 3 \end{vmatrix}$ **11.** $\begin{vmatrix} 1 & -2 & 3 \\ -2 & 1 & 1 \\ -3 & -2 & 1 \end{vmatrix}$ **12.** $\begin{vmatrix} 1 & 1 & 2 \\ 2 & 1 & -2 \\ 3 & 1 & 3 \end{vmatrix}$

13. $\begin{vmatrix} 1 & 2 & 3 \\ 4 & 5 & 6 \\ 7 & 8 & 9 \end{vmatrix}$

14. $\begin{vmatrix} 1 & 4 & 7 \\ 2 & 5 & 8 \\ 3 & 6 & 9 \end{vmatrix}$

15. $\begin{vmatrix} a & 2a & -a \\ 2 & -1 & 3 \\ 1 & 2 & -3 \end{vmatrix}$

16. $\begin{vmatrix} 1 & 2b & -3 \\ 2 & -b & 2 \\ 1 & 3b & 1 \end{vmatrix}$

17. $\begin{vmatrix} 1 & a & b \\ 1 & 2a & 2b \\ 1 & 3a & 3b \end{vmatrix}$

18. $\begin{vmatrix} a & b & c \\ 0 & b & c \\ 0 & 0 & c \end{vmatrix}$

In Exercises 19–44, use Cramer's rule to solve each system of equations, if possible.

19. $\begin{cases} x + y = 6 \\ x - y = 2 \end{cases}$

20. $\begin{cases} x - y = 4 \\ 2x + y = 5 \end{cases}$

21. $\begin{cases} 2x + y = 1 \\ x - 2y = -7 \end{cases}$

22. $\begin{cases} 3x - y = -3 \\ 2x + y = -7 \end{cases}$

23. $\begin{cases} 2x + 3y = 0 \\ 4x - 6y = -4 \end{cases}$

24. $\begin{cases} 4x - 3y = -1 \\ 8x + 3y = 4 \end{cases}$

25. $\begin{cases} y = \dfrac{-2x + 1}{3} \\ 3x - 2y = 8 \end{cases}$

26. $\begin{cases} 2x + 3y = -1 \\ x = \dfrac{y - 9}{4} \end{cases}$

27. $\begin{cases} y = \dfrac{11 - 3x}{2} \\ x = \dfrac{11 - 4y}{6} \end{cases}$

28. $\begin{cases} x = \dfrac{12 - 6y}{5} \\ y = \dfrac{24 - 10x}{12} \end{cases}$

29. $\begin{cases} x = \dfrac{5y - 4}{2} \\ y = \dfrac{3x - 1}{5} \end{cases}$

30. $\begin{cases} y = \dfrac{1 - 5x}{2} \\ x = \dfrac{3y + 10}{4} \end{cases}$

31. $\begin{cases} x + y + z = 4 \\ x + y - z = 0 \\ x - y + z = 2 \end{cases}$

32. $\begin{cases} x + y + z = 4 \\ x - y + z = 2 \\ x - y - z = 0 \end{cases}$

33. $\begin{cases} x + y + 2z = 7 \\ x + 2y + z = 8 \\ 2x + y + z = 9 \end{cases}$

34. $\begin{cases} x + 2y + 2z = 10 \\ 2x + y + 2z = 9 \\ 2x + 2y + z = 1 \end{cases}$

35. $\begin{cases} 2x + y - z = 1 \\ x + 2y + 2z = 2 \\ 4x + 5y + 3z = 3 \end{cases}$

36. $\begin{cases} 4x + 3z = 4 \\ 2y - 6z = -1 \\ 8x + 4y + 3z = 9 \end{cases}$

37. $\begin{cases} 2x + y + z = 5 \\ x - 2y + 3z = 10 \\ x + y - 4z = -3 \end{cases}$

38. $\begin{cases} 3x + 2y - z = -8 \\ 2x - y + 7z = 10 \\ 2x + 2y - 3z = -10 \end{cases}$

39. $\begin{cases} 2x + 3y + 4z = 6 \\ 2x - 3y - 4z = -4 \\ 4x + 6y + 8z = 12 \end{cases}$

40. $\begin{cases} x - 3y + 4z - 2 = 0 \\ 2x + y + 2z - 3 = 0 \\ 4x - 5y + 10z - 7 = 0 \end{cases}$

41. $\begin{cases} x + y = 1 \\ \dfrac{1}{2}y + z = \dfrac{5}{2} \\ x - z = -3 \end{cases}$

42. $\begin{cases} 3x + 4y + 14z = 7 \\ -\dfrac{1}{2}x - y + 2z = \dfrac{3}{2} \\ x + \dfrac{3}{2}y + \dfrac{5}{2}z = 1 \end{cases}$

43. $\begin{cases} 2x - y + 4z + 2 = 0 \\ 5x + 8y + 7z = -8 \\ x + 3y + z + 3 = 0 \end{cases}$

44. $\begin{cases} \dfrac{1}{2}x + y + z + \dfrac{3}{2} = 0 \\ x + \dfrac{1}{2}y + z - \dfrac{1}{2} = 0 \\ x + y + \dfrac{1}{2}z + \dfrac{1}{2} = 0 \end{cases}$

In Exercises 45–48, evaluate each determinant and solve the resulting equation.

45. $\begin{vmatrix} x & 1 \\ 3 & 2 \end{vmatrix} = 1$

46. $\begin{vmatrix} x & -x \\ 2 & -3 \end{vmatrix} = -5$

47. $\begin{vmatrix} x & -2 \\ 3 & 1 \end{vmatrix} = \begin{vmatrix} 4 & 2 \\ x & 3 \end{vmatrix}$

48. $\begin{vmatrix} x & 3 \\ x & 2 \end{vmatrix} = \begin{vmatrix} 3 & 2 \\ 1 & 1 \end{vmatrix}$

49. Making investments A student wants to average a 6.6% return by investing $20,000 in the three stocks listed in the table. Because HiTech is considered to be a high-risk investment, he wants to invest three times as much in SaveTel and HiGas as he invests in HiTech. How much should he invest in each stock?

50. Making investments A woman wants to average a $7\frac{1}{3}$% return by investing $30,000 in the three certificates of deposit listed in the table. She wants to invest five times as much in the highest-rate CD as in the 6% CD. How much should she invest in each CD?

Stock	Rate of return
HiTech	10%
SaveTel	5%
HiGas	6%

Type of CD	Rate of return
12 month	6%
24 month	7%
36 month	8%

In Exercises 51–54, use a calculator or a software package with matrix capabilities to evaluate each determinant.

51. $\begin{vmatrix} 2 & -3 & 4 \\ -1 & 2 & 4 \\ 3 & -3 & 1 \end{vmatrix}$

52. $\begin{vmatrix} -3 & 2 & -5 \\ 3 & -2 & 6 \\ 1 & -3 & 4 \end{vmatrix}$

53. $\begin{vmatrix} 2 & 1 & -3 \\ -2 & 2 & 4 \\ 1 & -2 & 2 \end{vmatrix}$

54. $\begin{vmatrix} 4 & 2 & -3 \\ 2 & -5 & 6 \\ 2 & 5 & -2 \end{vmatrix}$

Writing Exercises *Write a paragraph using your own words.*

1. Tell how to find the minor of an element of a determinant.

2. Tell how to find x when solving a system of linear equations by Cramer's rule.

Something to Think About **1.** Show that $\begin{vmatrix} x & y & 1 \\ -2 & 3 & 1 \\ 3 & 5 & 1 \end{vmatrix} = 0$ is the equation of the line passing through $(-2, 3)$ and $(3, 5)$.

2. Show that $\dfrac{1}{2}\begin{vmatrix} 0 & 0 & 1 \\ 3 & 0 & 1 \\ 0 & 4 & 1 \end{vmatrix}$ is the area of the triangle with vertices at $(0, 0)$, $(3, 0)$, and $(0, 4)$.

Determinants with more than 3 rows and 3 columns can be evaluated by expanding them by minors. The sign array for a 4 × 4 determinant is

$$\begin{matrix} + & - & + & - \\ - & + & - & + \\ + & - & + & - \\ - & + & - & + \end{matrix}$$

Evaluate each determinant.

3.
$$\begin{vmatrix} 1 & 0 & 2 & 1 \\ 2 & 1 & 1 & 3 \\ 1 & 1 & 1 & 1 \\ 2 & 1 & 1 & 1 \end{vmatrix}$$

4.
$$\begin{vmatrix} 1 & 2 & -1 & 1 \\ -2 & 1 & 3 & -1 \\ 0 & 1 & 1 & 2 \\ 2 & 0 & 3 & 1 \end{vmatrix}$$

Review Exercises *Solve each equation.*

1. $3(x + 2) - (2 - x) = x - 5$

2. $\dfrac{3}{7}x = 2(x + 11)$

3. $\dfrac{5}{3}(5x + 6) - 10 = 0$

4. $5 - 3(2x - 1) = 2(4 + 3x) - 24$

■ ■ ■ ■ ■ ■ ■ ■ ■ **PROBLEMS AND PROJECTS**

1. Determine graphically if it is possible for the following two functions to represent the revenue and cost functions of a company. If it is, find the break-even point.

$$R(x) = 35x + 72, \quad C = -350 + 70x$$

2. A farmer wants to build three adjacent rectangular pens for chickens, turkeys, and pigs. Each pen is to be the same size, and the outer perimeter of the pens is to be 150 meters. If a total of 200 meters of fencing will be needed, find the dimensions of the turkey pen.

3. How many pints of thinner (selling at $2.50 per pint) and how many pints of gloss white (selling at $4.52 per pint) are needed to make $1\frac{3}{5}$ gallons of a mix (selling at $3.84 per pint) if the mix has twice as much paint as thinner?

4. An accountant has forgotten to bring the computer disk showing her company's monthly profits for last year. However, she does have reports showing units sold and profits for February (1000 units sold, $8500), June (2070 units sold, $12,400), and September (1050 units sold, $8750). Assuming a parabolic shape, what profit curve should the accountant reconstruct to show the monthly profits?

PROJECT 1 The number of units of a product that will be produced, and the number of units that will be sold, depends on the unit price of the product. As the unit price gets higher, the product will be produced in greater quantity, because the producer will make more money on each item. The *supply* of the product will grow, and we say that supply *is a function of* (or *depends on*) the unit price. As the price rises, fewer consumers will buy the product, and the *demand* will decrease. The demand for the product is also a function of the unit price.

In this project, we will assume that both supply and demand are *linear* functions of the unit price. Thus, the graph of supply (the *y*-coordinate) versus price (the *x*-coordinate) is a line with positive slope. The graph of the demand function is a line with negative slope. Because these two lines cannot be parallel, they must intersect. The price at which supply equals demand is called the *market price:* At this price, the same number of units of the product will be sold as are manufactured.

You work for HeckuvaDeal Soda Pop Inc. Your boss has given you the task of analyzing the sales figures for the past year for the soda market in a small city. You have been provided with the following supply and demand functions. (Supply and demand are measured in cases per week; p, the price per case, is measured in dollars.)

The demand for soda is $D(p) = 19{,}000 - 2200p$

The supply of soda is $S(p) = 3000 + 1080p$

Both functions are true for values of p from \$3.50 to \$5.75.

Graph both functions on the same set of coordinate axes, being sure to label each graph and include any other important information. Then write a report for your supervisor that answers the following questions. Include any supporting work.

a. Explain why producers will be able to sell all of the soda they make when the price is \$3.50 per case. How much money will the producers take in from these sales?

b. How much money will producers take in from sales when the price is \$5.75 per case? How much soda will have to be warehoused? (That is, how much extra soda will have been made?)

c. Find the market price for soda (to the nearest cent). How many cases per week will be sold at this price? How much money will the producers take in from sales at the market price?

d. Explain why prices always tend toward the market price. That is, explain why the unit price will rise if the demand is greater than supply, and why the unit price will fall if supply is greater than demand.

PROJECT 2   Goodstuff Produce Company has two large water canals that feed the irrigation ditches on its fruit farm. One of these canals runs directly north and south, and the other runs directly east and west. The canals cross at the center of the farm property (the origin) and divide the farm into four quadrants. The company is interested in digging some new irrigation ditches in a portion of the northeast quadrant. You have been hired to plan the layout of the new system.

Your design is to make use of ditch Z, which is already present. This ditch runs from a point 300 meters north of the origin to a point 400 meters east of the origin. The owners of Goodstuff want two new ditches dug.

• Ditch A is to begin at a point 100 meters north of the origin and follow a line that travels 3 meters north for every 7 meters it travels east until it intersects ditch Z.

• Ditch B is to run from the origin to ditch Z in such a way that it exactly bisects the area in the northeast quadrant that is south of both ditch Z and ditch A.

You are to provide the equations of the lines that the three ditches follow, as well as the exact location of the gates that will be installed where the ditches intersect one another. Be sure to provide explanations and organized work that will clearly display the desired information and assure the owners of Goodstuff that they will get exactly what they want.

■ Chapter Summary

KEY WORDS

augmented matrix (3.4)
coefficient matrix (3.4)
consistent system of
    equations (3.1)
Cramer's rule (3.5)
dependent equations (3.1)
determinant (3.5)

elementary row operations (3.4)
element of a matrix (3.4)
equivalent systems (3.1)
Gaussian elimination (3.4)
inconsistent system of
    equations (3.1)
independent equations (3.1)

linear system of
    equations (3.1)
matrix (3.4)
minor (3.5)
square matrix (3.4)
triangular form of a
    matrix (3.4)

KEY IDEAS

(3.1) If a system of equations has at least one solution, the system is a **consistent system**. Otherwise, it is an **inconsistent system**.

If the graphs of the equations of a system are distinct, the equations are **independent equations**. Otherwise, the equations are **dependent equations**.

(3.2) Two of the methods used to solve systems of two linear equations in two variables are the **substitution method** and the **addition method**.

(3.3) A system of three linear equations in three variables can be solved by using the addition method.

(3.4) A **matrix** is any rectangular array of numbers.

Many systems of linear equations can be solved by using matrices and the method of **Gaussian elimination**.

(3.5) A **determinant of a square matrix** is a number:

$$\begin{vmatrix} a & b \\ c & d \end{vmatrix} = ad - bc$$

$$\begin{vmatrix} a_1 & b_1 & c_1 \\ a_2 & b_2 & c_2 \\ a_3 & b_3 & c_3 \end{vmatrix} = a_1 \begin{vmatrix} b_2 & c_2 \\ b_3 & c_3 \end{vmatrix}$$

$$- b_1 \begin{vmatrix} a_2 & c_2 \\ a_3 & c_3 \end{vmatrix} + c_1 \begin{vmatrix} a_2 & b_2 \\ a_3 & b_3 \end{vmatrix}$$

Many systems of linear equations can be solved by using Cramer's rule.

■ Chapter 3 Review Exercises

In Review Exercises 1–4, solve each system of equations by the graphing method.

1. $\begin{cases} 2x + y = 11 \\ -x + 2y = 7 \end{cases}$

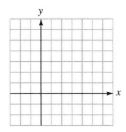

2. $\begin{cases} 3x + 2y = 0 \\ 2x - 3y = -13 \end{cases}$

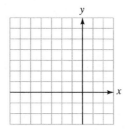

3. $\begin{cases} \dfrac{1}{2}x + \dfrac{1}{3}y = 2 \\ y = 6 - \dfrac{3}{2}x \end{cases}$

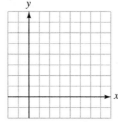

4. $\begin{cases} \dfrac{1}{3}x - \dfrac{1}{2}y = 1 \\ 6x - 9y = 2 \end{cases}$

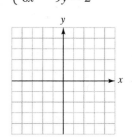

In Review Exercises 5–8, solve each system of equations by substitution.

5. $\begin{cases} y = x + 4 \\ 2x + 3y = 7 \end{cases}$

6. $\begin{cases} y = 2x + 5 \\ 3x - 5y = -4 \end{cases}$

7. $\begin{cases} x + 2y = 11 \\ 2x - y = 2 \end{cases}$

8. $\begin{cases} 2x + 3y = -2 \\ 3x + 5y = -2 \end{cases}$

In Review Exercises 9–14, solve each system of equations by addition.

9. $\begin{cases} x + y = -2 \\ 2x + 3y = -3 \end{cases}$

10. $\begin{cases} 3x + 2y = 1 \\ 2x - 3y = 5 \end{cases}$

11. $\begin{cases} x + \dfrac{1}{2}y = 7 \\ -2x = 3y - 6 \end{cases}$

12. $\begin{cases} y = \dfrac{x - 3}{2} \\ x = \dfrac{2y + 7}{2} \end{cases}$

13. $\begin{cases} x + y + z = 6 \\ x - y - z = -4 \\ -x + y - z = -2 \end{cases}$

14. $\begin{cases} 2x + 3y + z = -5 \\ -x + 2y - z = -6 \\ 3x + y + 2z = 4 \end{cases}$

In Review Exercises 15–18, solve each system of equations by using matrices.

15. $\begin{cases} x + 2y = 4 \\ 2x - y = 3 \end{cases}$

16. $\begin{cases} x + y + z = 6 \\ 2x - y + z = 1 \\ 4x + y - z = 5 \end{cases}$

17. $\begin{cases} x + y = 3 \\ x - 2y = -3 \\ 2x + y = 4 \end{cases}$

18. $\begin{cases} x + 2y + z = 2 \\ 2x + 5y + 4z = 5 \end{cases}$

In Review Exercises 19–22, evaluate each determinant.

19. $\begin{vmatrix} 2 & 3 \\ -4 & 3 \end{vmatrix}$

20. $\begin{vmatrix} -3 & -4 \\ 5 & -6 \end{vmatrix}$

21. $\begin{vmatrix} -1 & 2 & -1 \\ 2 & -1 & 3 \\ 1 & -2 & 2 \end{vmatrix}$

22. $\begin{vmatrix} 3 & -2 & 2 \\ 1 & -2 & -2 \\ 2 & 1 & -1 \end{vmatrix}$

In Review Exercises 23–26, use Cramer's rule to solve each system of equations.

23. $\begin{cases} 3x + 4y = 10 \\ 2x - 3y = 1 \end{cases}$

24. $\begin{cases} 2x - 5y = -17 \\ 3x + 2y = 3 \end{cases}$

25. $\begin{cases} x + 2y + z = 0 \\ 2x + y + z = 3 \\ x + y + 2z = 5 \end{cases}$

26. $\begin{cases} 2x + 3y + z = 2 \\ x + 3y + 2z = 7 \\ x - y - z = -7 \end{cases}$

■ Chapter 3 Test

1. Solve $\begin{cases} 2x + y = 5 \\ y = 2x - 3 \end{cases}$ by graphing.

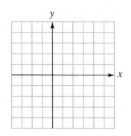

2. Use substitution to solve

$$\begin{cases} 2x - 4y = 14 \\ x = -2y + 7 \end{cases}$$

3. Use addition to solve

$$\begin{cases} 2x + 3y = -5 \\ 3x - 2y = 12 \end{cases}$$

4. Use any method to solve

$$\begin{cases} \dfrac{x}{2} - \dfrac{y}{4} = -4 \\ x + y = -2 \end{cases}$$

In Problems 5–6, consider the system $\begin{cases} 3(x + y) = x - 3 \\ -y = \dfrac{2x + 3}{3} \end{cases}$.

5. Are the equations of the system dependent or independent?

6. Is the system consistent or inconsistent?

In Problems 7–8, use an elementary row operation to find the missing number in the second matrix.

7. $\begin{bmatrix} 1 & 2 & -1 \\ 2 & -2 & 3 \end{bmatrix}, \begin{bmatrix} 1 & 2 & -1 \\ -1 & -8 & \end{bmatrix}$

8. $\begin{bmatrix} -1 & 3 & 6 \\ 3 & -2 & 4 \end{bmatrix}, \begin{bmatrix} -1 & 3 & 6 \\ 5 & & -8 \end{bmatrix}$

In Problems 9–10, consider the system $\begin{cases} x + y + z = 4 \\ x + y - z = 6 \\ 2x - 3y + z = -1 \end{cases}$.

9. Write the augmented matrix that represents the system.

10. Write the coefficient matrix that represents the system.

In Problems 11–12, use matrices to solve each system.

11. $\begin{cases} x + y = 4 \\ 2x - y = 2 \end{cases}$

12. $\begin{cases} x + y = 2 \\ x - y = -4 \\ 2x + y = 1 \end{cases}$

In Problems 13–16, evaluate each determinant.

13. $\begin{vmatrix} 2 & -3 \\ 4 & 5 \end{vmatrix}$

14. $\begin{vmatrix} -3 & -4 \\ -2 & 3 \end{vmatrix}$

15. $\begin{vmatrix} 1 & 2 & 0 \\ 2 & 0 & 3 \\ 1 & -2 & 2 \end{vmatrix}$

16. $\begin{vmatrix} 2 & -1 & 1 \\ 3 & 1 & 0 \\ 0 & 1 & 2 \end{vmatrix}$

In Problems 17–20, consider the system $\begin{cases} x - y = -6 \\ 3x + y = -6 \end{cases}$, *which is to be solved using Cramer's rule.*

17. When solving for x, what is the numerator determinant? (**Don't evaluate it.**)

18. When solving for y, what is the denominator determinant? (**Don't evaluate it.**)

19. Solve the system for x.

20. Solve the system for y.

In Problems 21–22, consider the system $\begin{cases} x + y + z = 4 \\ x + y - z = 6 \\ 2x - 3y + z = -1 \end{cases}$.

21. Solve for x.

22. Solve for z.

Inequalities

4

Statistician

Statisticians devise, carry out, and interpret the numerical results of surveys and experiments. They apply their knowledge to subject areas such as economics, human behavior, natural science, and engineering.

Through the year 2005, employment opportunities will be favorable for people trained in statistics who also have knowledge in a field of application such as manufacturing, engineering, scientific research, or business.

SAMPLE APPLICATION ■ To estimate the mean property tax paid by homeowners living in Rockford, Illinois, a researcher selects a random sample of homeowners and computes the mean tax paid by those homeowners. If the researcher knows that the standard deviation σ of all tax bills is $120, how large must the sample size be before the researcher can be 95% certain that the sample mean will be within $35 of the population mean? See Example 8 in Section 4.1.

4.1 Linear Inequalities

■ PROPERTIES OF INEQUALITIES ■ LINEAR INEQUALITIES ■ COMPOUND INEQUALITIES ■ PROBLEM SOLVING ■ GRAPHING DEVICES ■ ACCENT ON STATISTICS

Inequalities are statements indicating that two quantities are unequal. Inequalities can be recognized by the use of one or more of the following inequality symbols.

> **Inequality Symbols**
> $a < b$ means "a is less than b."
> $a > b$ means "a is greater than b."
> $a \leq b$ means "a is less than or equal to b."
> $a \geq b$ means "a is greater than or equal to b."

To show that 2 is less than 3, we can write $2 < 3$. To show that x is greater than 4 or equal to 4, we can write $x \geq 4$.

By definition, $a < b$ means that "a is less than b," but it also means that $b > a$. If a is to the left of b on the number line, then $a < b$.

In Chapter 1, we saw that many inequalities can be graphed as intervals on a number line. Table 1 shows the possibilities when a and b are real numbers.

Kind of interval	Inequality	Graph	Interval
Open interval	$x > a$		(a, ∞)
	$x < a$		$(-\infty, a)$
	$a < x < b$		(a, b)
Half-open interval	$x \geq a$		$[a, \infty)$
	$x \leq a$		$(-\infty, a]$
	$a \leq x < b$		$[a, b)$
	$a < x \leq b$		$(a, b]$
Closed interval	$a \leq x \leq b$		$[a, b]$

TABLE 1

 WARNING!   Remember that in interval notation, the notation $(-3, 4)$ represents the set of real numbers between -3 and 4, not the coordinates of a point on a graph.

Inequalities like $x + 1 > x$, which are true for all x, are called **absolute inequalities**. Inequalities like $3x + 2 < 8$, which are true for some numbers x but not others, are called **conditional inequalities**.

■ PROPERTIES OF INEQUALITIES

There are several basic properties of inequalities.

The Trichotomy Property
For any real numbers a and b, exactly one of the following statements is true:

$a < b$, $a = b$,   or $a > b$

The trichotomy property indicates that one and only one of three statements is true about any two real numbers. Either

- the first is less than the second,
- the first is equal to the second, or
- the first is greater than the second.

The Transitive Property

If a, b, and c are real numbers with $a < b$ and $b < c$, then $a < c$.

If a, b, and c are real numbers with $a > b$ and $b > c$, then $a > c$.

The first part of the transitive property indicates that

- if a first number is less than a second number and
- the second number is less than a third number, then
- the first number is less than the third.

The second part of the transitive property is similar, with the words "is greater than" substituted for "is less than."

Property 1 of Inequalities

Any real number can be added to (or subtracted from) both sides of an inequality to produce another inequality with the same direction.

Property 1 indicates that any number can be added to both sides of a true inequality to get another true inequality with the same direction (sometimes called the *sense* or *order* of an inequality). For example, if we add 4 to both sides of the inequality $3 < 12$, we get

$$3 + 4 < 12 + 4$$
$$7 < 16$$

and the $<$ symbol is unchanged.

Subtracting 4 from both sides of $3 < 12$ does not change the direction of the inequality either.

$$3 - 4 < 12 - 4$$
$$-1 < 8$$

Property 2 of Inequalities

If both sides of an inequality are multiplied (or divided) by a positive number, another inequality results, with the same direction as the original inequality.

Property 2 indicates that both sides of a true inequality can be multiplied by any positive number to get another true inequality with the same direction.

For example, if both sides of the true inequality $-4 < 6$ are multiplied by $+2$, we get

$$2(-4) < 2(6)$$
$$-8 < 12$$

and the $<$ symbol is unchanged.

Dividing both sides by $+2$ does not change the direction of the inequality either.

$$\frac{-4}{2} < \frac{6}{2}$$
$$-2 < 3$$

Property 3 of Inequalities

If both sides of an inequality are multiplied (or divided) by a negative number, another inequality results, but with the opposite direction of the original inequality.

Property 3 indicates that if both sides of a true inequality are multiplied by a negative number, another true inequality results, but with the opposite direction. For example, if both sides of the true inequality $-4 < 6$ are multiplied by -2, we get

$$-4 < 6$$
$$-2(-4) > -2(6)$$
$$8 > -12$$

and the $<$ symbol changes to an $>$ symbol.

Dividing both sides by -2 also reverses the direction of the inequality.

$$-4 < 6$$
$$\frac{-4}{-2} > \frac{6}{-2}$$
$$2 > -3$$

WARNING! We must remember to change the direction of an inequality symbol every time we multiply or divide by a negative number.

■ LINEAR INEQUALITIES

Linear Inequalities

A **linear inequality** in x is any inequality that can be expressed in one of the following forms, with $a \neq 0$.

$$ax + c < 0 \qquad ax + c > 0 \qquad ax + c \leq 0 \qquad \text{or} \qquad ax + c \geq 0$$

We can solve linear inequalities just as we solve linear equations, but with one exception. If we multiply or divide both sides by a negative number, we must change the direction of the inequality.

EXAMPLE 1 Solve the linear inequality $3(2x - 9) < 9$.

Solution We use the same steps as for solving equations.

$$3(2x - 9) < 9$$

$$6x - 27 < 9 \qquad \text{Use the distributive property to remove parentheses.}$$

$$6x < 36 \qquad \text{Add 27 to both sides.}$$

$$x < 6 \qquad \text{Divide both sides by 6.}$$

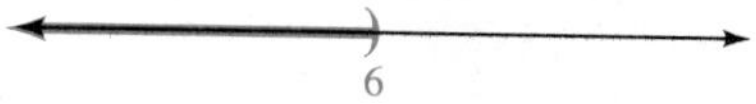

The solution set is the interval $(-\infty, 6)$, whose graph is shown in Figure 4-1. The parenthesis at 6 indicates that 6 is not included in the solution set. ∎

FIGURE 4-1

EXAMPLE 2 Solve the linear inequality $-4(3x + 2) \le 16$.

Solution We use the same steps as for solving equations.

$$-4(3x + 2) \le 16$$

$$-12x - 8 \le 16 \qquad \text{Use the distributive property to remove parentheses.}$$

$$-12x \le 24 \qquad \text{Add 8 to both sides.}$$

$$x \ge -2 \qquad \text{Divide both sides by } -12 \text{ and reverse the } \le \text{ symbol.}$$

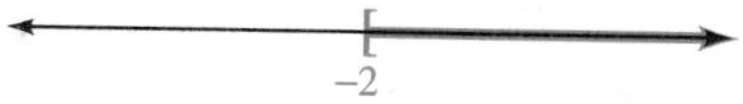

The solution set is the interval $[-2, \infty)$, whose graph is shown in Figure 4-2. The bracket at -2 indicates that -2 is included in the solution set. ∎

FIGURE 4-2

EXAMPLE 3 Solve the linear inequality $\frac{2}{3}(x + 2) > \frac{4}{5}(x - 3)$.

Solution We use the same steps as for solving equations.

$$\frac{2}{3}(x + 2) > \frac{4}{5}(x - 3)$$

$$15 \cdot \frac{2}{3}(x + 2) > 15 \cdot \frac{4}{5}(x - 3) \qquad \text{Multiply both sides by 15, the LCD of 3 and 5.}$$

$$10(x + 2) > 12(x - 3) \qquad \text{Simplify.}$$

$$10x + 20 > 12x - 36 \qquad \text{Use the distributive property to remove parentheses.}$$

$$-2x + 20 > -36 \qquad \text{Subtract } 12x \text{ from both sides.}$$

$$-2x > -56 \qquad \text{Subtract 20 from both sides.}$$

$$x < 28 \qquad \text{Divide both sides by } -2 \text{ and reverse the } > \text{ symbol.}$$

FIGURE 4-3

The solution set is the interval $(-\infty, 28)$, whose graph is shown in Figure 4-3. ∎

■ COMPOUND INEQUALITIES

To say that x is between -3 and 8, we write the inequality

$$-3 < x < 8 \qquad \text{Read as "} -3 \text{ is less than } x \text{ and } x \text{ is less than 8."}$$

This double inequality is called a **compound inequality**, because it is a combination of two inequalities:

$$-3 < x \quad and \quad x < 8$$

The word *and* indicates that both inequalities are true at the same time.

> **Compound Inequalities**
> The compound inequality $c < x < d$ is equivalent to $c < x$ and $x < d$.

WARNING! The inequality $c < x < d$ means $c < x$ and $x < d$. It does not mean $c < x$ or $x < d$.

EXAMPLE 4 Solve the compound inequality $-3 \le 2x + 5 < 7$.

Solution This inequality means that $2x + 5$ is between -3 and 7. We can solve it by isolating x between the inequality symbols.

$$-3 \le 2x + 5 < 7$$
$$-8 \le 2x < 2 \qquad \text{Subtract 5 from all three parts.}$$
$$-4 \le x < 1 \qquad \text{Divide all three parts by 2.}$$

FIGURE 4-4

The solution set is the interval $[-4, 1)$, whose graph is shown in Figure 4-4. ■

EXAMPLE 5 Solve the inequality $x + 3 < 2x - 1 < 4x - 3$.

Solution Here it is impossible to isolate x between the inequality symbols, so we solve each of its linear inequalities separately.

$$x + 3 < 2x - 1 \quad and \quad 2x - 1 < 4x - 3$$
$$4 < x \qquad\qquad\qquad 2 < 2x$$
$$\qquad\qquad\qquad\qquad\qquad 1 < x$$

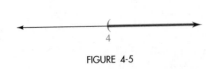

FIGURE 4-5

Only those numbers x where $x > 4$ and $x > 1$ are in the solution set. Since all numbers greater than 4 are also greater than 1, the solutions are the numbers x where $x > 4$. The solution set is the interval $(4, \infty) \cap (1, \infty) = (4, \infty)$, whose graph is shown in Figure 4-5. ■

EXAMPLE 6 Solve the compound inequality $x \le -3$ or $x \ge 8$.

Solution The word *or* in the statement $x \le -3$ or $x \ge 8$ indicates that only one of the inequalities needs to be true to make the statement true. The graph of this inequality is shown in Figure 4-6. The solution set is $(-\infty, -3] \cup [8, \infty)$.

FIGURE 4-6

WARNING! In the statement $x \leq -3$ or $x \geq 8$, it is incorrect to string the equalities together as $8 \leq x \leq -3$, because that would imply that $8 \leq -3$, which is impossible.

■ PROBLEM SOLVING

EXAMPLE 7 Suppose that a long-distance telephone call costs 36¢ for the first three minutes and 11¢ for each additional minute. For how many minutes can a person talk for less than $2?

Analyze the Problem   We can let x represent the total number of minutes that the call can last. Then the cost of the call will be 36¢ for the first three minutes plus 11¢ times the number of additional minutes, where the number of additional minutes is $x - 3$ (the total number of minutes minus 3 minutes). The cost of the call is to be less than $2.

Form an Inequality   With this information, we can form the inequality

The cost of the first three minutes	+	the cost of the additional minutes	<	$2.
.36	+	.11(x − 3)	<	2

Solve the Inequality   We can solve the inequality as follows:

$.36 + .11(x - 3) < 2$	
$36 + 11(x - 3) < 200$	To eliminate the decimal point, multiply both sides by 100.
$36 + 11x - 33 < 200$	Use the distributive property to remove parentheses.
$11x + 3 < 200$	Combine like terms.
$11x < 197$	Subtract 3 from both sides.
$x < 17.\overline{90}$	Divide both sides by 11.

State the Conclusion   Since the phone company does not bill for part of a minute, the longest time the person can talk is 17 minutes.

Check the Result   If the call lasts 17 minutes, the customer will be billed $.36 + $.11(14) = $1.90. If the call lasts 18 minutes, the customer will be billed $.36 + $.11(15) = $2.01.   ■

■ GRAPHING DEVICES

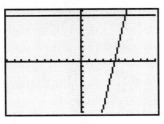

FIGURE 4-7

We can solve linear inequalities with a graphing approach. For example, to solve the inequality $3(2x - 9) < 9$, we can graph $y = 3(2x - 9)$ and $y = 9$ using window settings of $[-10, 10]$ for x and $[-10, 10]$ for y. This produces the graph shown in Figure 4-7. We can then trace to see that the graph of $y = 3(2x - 9)$ is below the graph of $y = 9$ for x values in the interval $(-\infty, 6)$. In this interval, $3(2x - 9) < 9$.

■ ■ ■ ■ ■ ■ ■ ■ ■   **Finding the Size of a Random Sample**

ACCENT ON
STATISTICS

In statistics, researchers often estimate the mean of a population using a random sample taken from the population. To get suitable results, the researcher must decide how large the random sample needs to be.

EXAMPLE 8

A researcher wants to estimate the mean (average) real estate tax paid by homeowners living in Rockford, IL. To do so, he decides to select a *random sample* of homeowners and compute the mean tax paid by the homeowners in that sample. How large must the sample be for the researcher to be 95% certain that his computed sample mean will be within $35 of the true population mean—that is, within $35 of the mean tax paid by all homeowners in the city? Assume that the standard deviation, σ, of all tax bills in the city is known to be $120.

Solution   From elementary statistics, the researcher has the formula

$$\frac{3.84\sigma^2}{N} < E^2$$

where σ^2, called the *population variance*, is the square of the standard deviation, E is the maximum acceptable error, and N is the sample size. The researcher substitutes 120 for σ and 35 for E in the previous formula and solves for N.

$$\frac{3.84(120)^2}{N} < 35^2$$

$$\frac{55{,}296}{N} < 1225 \qquad \text{Simplify.}$$

$$55{,}296 < 1225N \qquad \text{Multiply both sides by } N.$$

$$45.13959184 < N \qquad \text{Divide both sides by 1225.}$$

To be 95% certain that the sample mean will be within $35 of the true population mean, the researcher must sample more than 45.13959184 homeowners. Thus, the sample must contain at least 46 homeowners.   ■

Orals   *Solve each inequality.*

1. $2x < 4$ **2.** $3x + 1 \geq 10$

3. $-3x > 12$ **4.** $-\dfrac{x}{2} \leq 4$

5. $-2 < 2x < 8$ **6.** $3 \leq \dfrac{x}{3} \leq 4$

EXERCISE 4.1

In Exercises 1–14, solve each inequality. Give the result in interval notation and graph the solution set.

1. $x + 4 < 5$ **2.** $x - 5 > 2$

3. $-3x - 1 \leq 5$ **4.** $-2x + 6 \geq 16$

5. $5x - 3 > 7$

6. $7x - 9 < 5$

7. $3(z - 2) \le 2(z + 7)$

8. $5(3 + z) > -3(z + 3)$

9. $-11(2 - b) < 4(2b + 2)$

10. $-9(h - 3) + 2h \le 8(4 - h)$

11. $\dfrac{1}{2}y + 2 \ge \dfrac{1}{3}y - 4$

12. $\dfrac{1}{4}x - \dfrac{1}{3} \le x + 2$

13. $\dfrac{2}{3}x + \dfrac{3}{2}(x - 5) \le x$

14. $\dfrac{5}{9}(x + 3) - \dfrac{4}{3}(x - 3) \ge x - 1$

In Exercises 15–36, solve each compound inequality. Give the result in interval notation and graph the solution set.

15. $-2 < -b + 3 < 5$

16. $2 < -t - 2 < 9$

17. $15 > 2x - 7 > 9$

18. $25 > 3x - 2 > 7$

19. $-6 < -3(x - 4) \le 24$

20. $-4 \le -2(x + 8) < 8$

21. $0 \ge \dfrac{1}{2}x - 4 > 6$

22. $-6 \le \dfrac{1}{3}a + 1 < 0$

23. $0 \le \dfrac{4 - x}{3} \le 2$

24. $-2 \le \dfrac{5 - 3x}{2} \le 2$

25. $x + 3 < 3x - 1 < 2x + 2$

26. $x - 1 \le 2x + 4 \le 3x - 1$

27. $4x \ge -x + 5 \ge 3x - 4$

28. $x + 2 < -\dfrac{1}{3}x < \dfrac{1}{2}x$

29. $5(x + 1) \le 4(x + 3) < 3(x - 1)$

30. $-5(2 + x) < 4x + 1 < 3x$

31. $3x + 2 < 8$ or $2x - 3 > 11$

32. $3x + 4 < -2$ or $3x + 4 > 10$

33. $-4(x + 2) \ge 12$ or $3x + 8 < 11$

34. $5(x - 2) \ge 0$ and $-3x < 9$

35. $x < -3$ and $x > 3$

36. $x < 3$ or $x > -3$

37. If $x > -3$, must it be true that $x^2 > 9$?

38. If $x > 2$, must it be true that $x^2 > 4$?

In Exercises 39–50, solve each problem.

39. Renting a rototiller The cost of renting a rototiller is $15.50 for the first hour and $7.95 for each additional hour. How long can a person have the rototiller if the cost is to be less than $50?

40. Renting a truck The cost of renting a truck is $29.95 for the first hour and $8.95 for each additional hour. How long can a person have the truck if the cost is to be less than $110?

41. Investing money If a woman invests $10,000 at 8% annual interest, how much more must she invest at 9% so that her annual income will exceed $1250?

42. Investing money If a man invests $8900 at 5.5% annual interest, how much more must he invest at 8.75% so that his annual income will be more than $1500?

43. Buying compact discs A student can afford to spend up to $330 on a stereo system and some compact discs. If the stereo costs $175 and the discs are $8.50 each, find the greatest number of discs he can buy.

44. Buying a computer A student who can afford to spend up to $2000 sees the ad shown in Illustration 1. If she buys a computer, find the greatest number of CD-ROMs that she can buy.

Big Sale!!!!

◀ ▌▌▌ $1695.95

All CD-ROMs $19.95

ILLUSTRATION 1

45. Averaging grades A student has scores of 70, 77, and 85 on three exams. What score is needed on a fourth exam to make an average of 80 or better?

46. Averaging grades A student has scores of 70, 79, 85, and 88 on four exams. What score does she need on the fifth exam to keep her average above 80?

47. Planning a work schedule Tom can earn $5 an hour for working at the college library and $9 an hour for construction work. To save time for study, he wants to limit his work to 20 hours a week but wants to earn more than $125. How many hours can he work at the library?

48. Scheduling equipment An excavating company charges $300 an hour for the use of a backhoe and $500 an hour for the use of a bulldozer. (Part of an hour counts as a full hour.) The company employs one operator for 40 hours per week. If the company wants to take in at least $18,500 each week, how many hours per week can it schedule the operator to use a backhoe?

49. Choosing a medical plan A college provides its employees with a choice of the two medical plans shown in Illustration 2. For what size hospital bills is Plan 2 better than Plan 1? (*Hint:* The cost to the employee includes both the deductible payment and the employee's coinsurance payment.)

Plan 1	Plan 2
Employee pays $100 Plan pays 70% of the rest	Employee pays $200 Plan pays 80% of the rest

ILLUSTRATION 2

50. Choosing a medical plan To save costs, the college in Exercise 49 raised the employee deductible, as shown in Illustration 3. For what size hospital bills is Plan 2 better than Plan 1? (*Hint:* The cost to the employee includes both the deductible payment and the employee's coinsurance payment.)

Plan 1	Plan 2
Employee pays $200 Plan pays 70% of the rest	Employee pays $400 Plan pays 80% of the rest

ILLUSTRATION 3

In Exercises 51–54, use a graphing device to solve each inequality.

51. $2x + 3 < 5$

52. $3x - 2 > 4$

53. $5x + 2 \geq -18$

54. $3x - 4 \leq 20$

In Exercises 55–56, solve each problem.

55. Choosing sample size How large would the sample have to be for the researcher in Example 8 to be 95% certain that the true population mean would be within $20 of the sample mean?

56. Choosing sample size How large would the sample have to be for the researcher in Example 8 to be 95% certain that the true population mean would be within $10 of the sample mean?

Writing Exercises *Write a paragraph using your own words.*

1. The techniques for solving linear equations and linear inequalities are similar, yet different. Explain.

2. Explain the concepts of *absolute inequality* and *conditional inequality*.

Something to Think About

1. Which of these relations are transitive? $=, \leq, \not=, \neq$

2. Solve: $\frac{1}{3} > \frac{1}{x}$. The following solution is not correct. Why?

$$\frac{1}{3} > \frac{1}{x}$$

$$3x\left(\frac{1}{3}\right) > \frac{1}{x}(3x) \qquad \text{Multiply both sides by } 3x.$$

$$x > 3 \qquad \text{Simplify.}$$

Review Exercises *Simplify each expression.*

1. $\left(\dfrac{t^3 t^5 t^{-6}}{t^2 t^{-4}}\right)^{-3}$

2. $\left(\dfrac{a^{-2}b^3 a^5 b^{-2}}{a^6 b^{-5}}\right)^{-4}$

3. A man invests $1200 in baking equipment to make pies. Each pie requires $3.40 in ingredients. If the man can sell all the pies he can make for $5.95 each, how many pies will he have to make to earn a profit?

4. A woman invested $15,000, part at 7% annual interest and the rest at 8%. If she earned $2200 in income over a two-year period, how much did she invest at 7%?

4.2 Equations and Inequalities with Absolute Values

■ ABSOLUTE VALUE FUNCTIONS ■ EQUATIONS OF THE FORM $|X| = K$ ■ EQUATIONS WITH TWO ABSOLUTE VALUES ■ INEQUALITIES OF THE FORM $|X| < K$ ■ INEQUALITIES OF THE FORM $|X| > K$ ■ GRAPHING DEVICES

We begin this section by reviewing the definition of the absolute value of x.

Absolute Value
If $x \geq 0$, then $|x| = x$.

If $x < 0$, then $|x| = -x$.

This definition gives a way for associating a nonnegative real number with any real number.

• If $x \geq 0$, then x (which is positive or 0) is its own absolute value.

• If $x < 0$, then $-x$ (which is positive) is the absolute value.

Either way, $|x|$ is positive or 0:

$|x| \geq 0$ for all real numbers x

EXAMPLE 1   Find   **a.** $|9|$,   **b.** $|-5|$,   **c.** $|0|$,   and   **d.** $|2 - \pi|$.

Solution   **a.** Since $9 \geq 0$, 9 is its own absolute value: $|9| = 9$.

b. Since $-5 < 0$, the negative of -5 is the absolute value: $|-5| = -(-5) = 5$.

c. Since $0 \geq 0$, 0 is its own absolute value: $|0| = 0$.

d. Since $\pi \approx 3.14$, it follows that $2 - \pi < 0$. Thus,

$$|2 - \pi| = -(2 - \pi) = \pi - 2 \qquad \blacksquare$$

WARNING!   The placement of a $-$ sign in an expression containing an absolute value symbol is important. For example, $|-19| = 19$, but $-|19| = -19$.

■ **ABSOLUTE VALUE FUNCTIONS**

In Section 2.6, we graphed the equation $y = -|x| + 3$ and obtained the graph shown in Figure 2-45. Since the graph passes the vertical line test, the equation represents a function. It is an example of a broad class of functions called **absolute value functions**.

> **Absolute Value Functions**
> An **absolute value function** is any function defined by an equation of the form $y = |u| + k$, where u is an algebraic expression in x and k is a constant.

EXAMPLE 2   Graph the absolute value function $y = f(x) = \dfrac{1}{2}|x - 1| - 3$.

Solution   We make a table of ordered pairs (x, y), plot the points, and draw the graph, or we use a graphing device. (See Figure 4-8.)

$f(x) = \dfrac{1}{2}|x - 1| - 3$

x	y	(x, y)
-3	-1	$(-3, -1)$
-1	-2	$(-1, -2)$
1	-3	$(1, -3)$
3	-2	$(3, -2)$
5	-1	$(5, -1)$

(a)

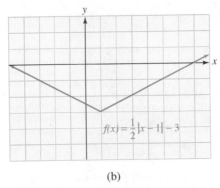

(b)

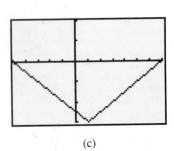

(c)

FIGURE 4-8

$\blacksquare$

■ EQUATIONS OF THE FORM $|X| = K$

In the equation $|x| = 5$, x can be either 5 or -5, because $|5| = 5$ and $|-5| = 5$. In the equation $|x| = 8$, x can be either 8 or -8. In general, the following is true.

> **Absolute Value Equations**
> If $k > 0$, then
> $$|x| = k \quad \text{is equivalent to} \quad x = k \quad \text{or} \quad x = -k$$

FIGURE 4-9

The absolute value of a number represents the distance on the number line that a point is from the origin. The solutions of $|x| = k$ are the coordinates of the two points that lie exactly k units from the origin. (See Figure 4-9.)

The equation $|x - 3| = 7$ indicates that a point on the number line with a coordinate of $x - 3$ is 7 units from the origin. Thus, $x - 3$ can be either 7 or -7.

$$x - 3 = 7 \quad \text{or} \quad x - 3 = -7$$
$$x = 10 \quad | \quad x = -4$$

The solutions of the equation $|x - 3| = 7$ are 10 and -4, as shown in Figure 4-10. If either of these numbers is substituted for x in $|x - 3| = 7$, the equation is satisfied:

$$
\begin{aligned}
|x - 3| &= 7 \\
|10 - 3| &\stackrel{?}{=} 7 \\
|7| &\stackrel{?}{=} 7 \\
7 &= 7
\end{aligned}
\qquad
\begin{aligned}
|x - 3| &= 7 \\
|-4 - 3| &\stackrel{?}{=} 7 \\
|-7| &\stackrel{?}{=} 7 \\
7 &= 7
\end{aligned}
$$

FIGURE 4-10

EXAMPLE 3   Solve the equation $|3x - 2| = 5$.

Solution   We can write $|3x - 2| = 5$ as

$$3x - 2 = 5 \quad \text{or} \quad 3x - 2 = -5$$

and solve each equation for x:

$$
\begin{aligned}
3x - 2 &= 5 \\
3x &= 7 \\
x &= \frac{7}{3}
\end{aligned}
\qquad \text{or} \qquad
\begin{aligned}
3x - 2 &= -5 \\
3x &= -3 \\
x &= -1
\end{aligned}
$$

Verify that both solutions check. ■

EXAMPLE 4   Solve the equation $\left| \dfrac{2}{3}x + 3 \right| + 4 = 10$.

Solution   We first subtract 4 from both sides to isolate the absolute value on the left-hand side.

$$\left| \frac{2}{3}x + 3 \right| + 4 = 10$$

1. $$\left| \frac{2}{3}x + 3 \right| = 6 \qquad \text{Subtract 4 from both sides.}$$

We can now write Equation 1 as

$$\frac{2}{3}x + 3 = 6 \qquad \text{or} \qquad \frac{2}{3}x + 3 = -6$$

and solve each equation for x:

$$\frac{2}{3}x + 3 = 6 \qquad \text{or} \qquad \frac{2}{3}x + 3 = -6$$

$$\frac{2}{3}x = 3 \qquad\qquad\qquad \frac{2}{3}x = -9$$

$$2x = 9 \qquad\qquad\qquad 2x = -27$$

$$x = \frac{9}{2} \qquad\qquad\qquad x = \frac{-27}{2}$$

Verify that both solutions check. ■

WARNING!   Since the absolute value of a quantity cannot be negative, equations like $\left| 7x + \frac{1}{2} \right| = -4$ have no solution. Their solution set is $\emptyset$.

EXAMPLE 5   Solve the equation $\left| \frac{1}{2}x - 5 \right| - 4 = -4$.

Solution   We first add 4 to both sides to isolate the absolute value on the left-hand side.

$$\left| \frac{1}{2}x - 5 \right| - 4 = -4$$

$$\left| \frac{1}{2}x - 5 \right| = 0 \qquad \text{Add 4 to both sides.}$$

Since 0 is the only number whose absolute value is 0, $\frac{1}{2}x - 5$ must be 0, and we have

$$\frac{1}{2}x - 5 = 0$$

$$\frac{1}{2}x = 5 \qquad \text{Add 5 to both sides.}$$

$$x = 10 \qquad \text{Multiply both sides by 2.}$$

Verify that 10 satisfies the original equation. ■

■ EQUATIONS WITH TWO ABSOLUTE VALUES

The equation $|a| = |b|$ is true when $a = b$ or when $a = -b$. For example,

$$|3| = |3| \qquad \text{or} \qquad |3| = |-3|$$

$$3 = 3 \qquad | \qquad 3 = 3$$

In general, the following is true.

> **Equations with Two Absolute Values**
> If a and b represent algebraic expressions, the equation $|a| = |b|$ is equivalent to the pair of equations
>
> $$a = b \qquad \text{or} \qquad a = -b$$

EXAMPLE 6   Solve the equation $|5x + 3| = |3x + 25|$.

Solution   This equation is true when $5x + 3 = 3x + 25$, or when $5x + 3 = -(3x + 25)$. We solve each equation for x.

$$
\begin{array}{rl|l}
5x + 3 = 3x + 25 & \text{or} & 5x + 3 = -(3x + 25) \\
2x = 22 & & 5x + 3 = -3x - 25 \\
x = 11 & & 8x = -28 \\
& & x = -\dfrac{28}{8} \\
& & x = -\dfrac{7}{2}
\end{array}
$$

Verify that both solutions check. ∎

■ INEQUALITIES OF THE FORM $|X| < K$

FIGURE 4-11

The inequality $|x| < 5$ indicates that a point with coordinate x is less than 5 units from the origin. (See Figure 4-11.)

Thus, x is between -5 and 5, and

$$|x| < 5 \qquad \text{is equivalent to} \qquad -5 < x < 5$$

FIGURE 4-12

The solution set of the inequality $|x| < k \ (k > 0)$ includes the coordinates of the points on the number line that are less than k units from the origin. (See Figure 4-12.)

In general, we have the following theorem.

> **Theorem**
> If $k > 0$, then
>
> $$|x| < k \qquad \text{is equivalent to} \qquad -k < x < k$$
> $$|x| \leq k \qquad \text{is equivalent to} \qquad -k \leq x \leq k \quad (k \geq 0)$$

EXAMPLE 7   Solve the inequality $|2x - 3| < 9$.

Solution   We write the inequality as a compound inequality and solve for x.

$$|2x - 3| < 9 \qquad \text{is equivalent to} \qquad -9 < 2x - 3 < 9$$

$$-9 < 2x - 3 < 9$$

$$-6 < 2x < 12 \qquad \text{Add 3 to all three parts.}$$

$$-3 < x < 6 \qquad \text{Divide all parts by 2.}$$

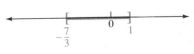

FIGURE 4-13

Any number between -3 and 6, not including either -3 or 6, is in the solution set. This is the interval $(-3, 6)$, whose graph is shown in Figure 4-13. ∎

| EXAMPLE 8 | Solve the inequality $|3x + 2| \leq 5$. |

Solution   We write the expression as a compound inequality and solve for x:

$$|3x + 2| \leq 5 \qquad \text{is equivalent to} \qquad -5 \leq 3x + 2 \leq 5$$

$$-5 \leq 3x + 2 \leq 5$$

$$-7 \leq 3x \leq 3 \qquad \text{Subtract 2 from all three parts.}$$

$$-\frac{7}{3} \leq x \leq 1 \qquad \text{Divide all three parts by 3.}$$

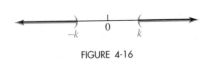

FIGURE 4-14

The solution set is the interval $\left[-\frac{7}{3}, 1\right]$, whose graph is shown in Figure 4-14. ∎

■ INEQUALITIES OF THE FORM $|X| > K$

The inequality $|x| > 5$ indicates that a point with coordinate x is more than 5 units from the origin. (See Figure 4-15.)

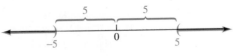

FIGURE 4-15

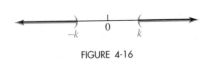

FIGURE 4-16

Thus, $x < -5$ or $x > 5$.

In general, the inequality $|x| > k$ can be interpreted to mean that a point with coordinate x is more than k units from the origin. (See Figure 4-16.)

In general,

$$|x| > k \qquad \text{is equivalent to} \qquad x < -k \quad \text{or} \quad x > k$$

The *or* indicates an either/or situation. It is only necessary for x to satisfy one of the two conditions to be in the solution set.

Theorem
If $k \geq 0$, then
$$|x| > k \qquad \text{is equivalent to} \qquad x < -k \quad \text{or} \quad x > k$$
$$|x| \geq k \qquad \text{is equivalent to} \qquad x \leq -k \quad \text{or} \quad x \geq k$$

EXAMPLE 9 Solve the inequality $|5x - 10| > 20$.

Solution We write the inequality as two separate inequalities and solve each one for x.

$|5x - 10| > 20$ is equivalent to $5x - 10 < -20$ or $5x - 10 > 20$

$5x - 10 < -20$	or	$5x - 10 > 20$	
$5x < -10$		$5x > 30$	Add 10 to both sides.
$x < -2$		$x > 6$	Divide both sides by 5.

Thus, x is either less than -2 or greater than 6:

$$x < -2 \quad \text{or} \quad x > 6$$

The solution set is the interval $(-\infty, -2) \cup (6, \infty)$, whose graph appears in Figure 4-17.

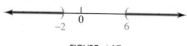

FIGURE 4-17

EXAMPLE 10 Solve the inequality $\left| \dfrac{3 - x}{5} \right| \geq 6$.

Solution We write the inequality as two separate inequalities.

$\left| \dfrac{3 - x}{5} \right| \geq 6$ is equivalent to $\dfrac{3 - x}{5} \leq -6$ or $\dfrac{3 - x}{5} \geq 6$

Then we solve each one for x:

$\dfrac{3 - x}{5} \leq -6$	or	$\dfrac{3 - x}{5} \geq 6$	
$3 - x \leq -30$		$3 - x \geq 30$	Multiply both sides by 5.
$-x \leq -33$		$-x \geq 27$	Subtract 3 from both sides.
$x \geq 33$		$x \leq -27$	Divide both sides by -1 and reverse the direction of the inequality symbol.

The solution set is the interval $(-\infty, -27] \cup [33, \infty)$, whose graph appears in Figure 4-18.

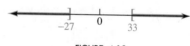

FIGURE 4-18

EXAMPLE 11 Solve the inequality $\left| \dfrac{2}{3}x - 2 \right| - 3 > 6$.

Solution   We begin by adding 3 to both sides to isolate the absolute value on the left-hand side. We then proceed as follows:

$$\left|\frac{2}{3}x - 2\right| - 3 > 6$$

$$\left|\frac{2}{3}x - 2\right| > 9 \qquad \text{Add 3 to both sides to isolate the absolute value.}$$

$$\frac{2}{3}x - 2 < -9 \qquad \text{or} \qquad \frac{2}{3}x - 2 > 9$$

$$\frac{2}{3}x < -7 \qquad\qquad \frac{2}{3}x > 11 \qquad \text{Add 2 to both sides.}$$

$$2x < -21 \qquad\qquad 2x > 33 \qquad \text{Multiply both sides by 3.}$$

$$x < -\frac{21}{2} \qquad\qquad x > \frac{33}{2} \qquad \text{Divide both sides by 2.}$$

The solution set is $\left(-\infty, -\frac{21}{2}\right) \cup \left(\frac{33}{2}, \infty\right)$, whose graph appears in Figure 4-19.

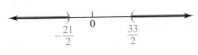

FIGURE 4-19   ■

EXAMPLE 12   Solve the inequality $|3x - 5| \geq -2$.

Solution   Since the absolute value of any number is nonnegative, and since any nonnegative number is larger than -2, the inequality is true for all x. The solution set is the interval $(-\infty, \infty)$, whose graph appears in Figure 4-20.

FIGURE 4-20   ■

■ GRAPHING DEVICES

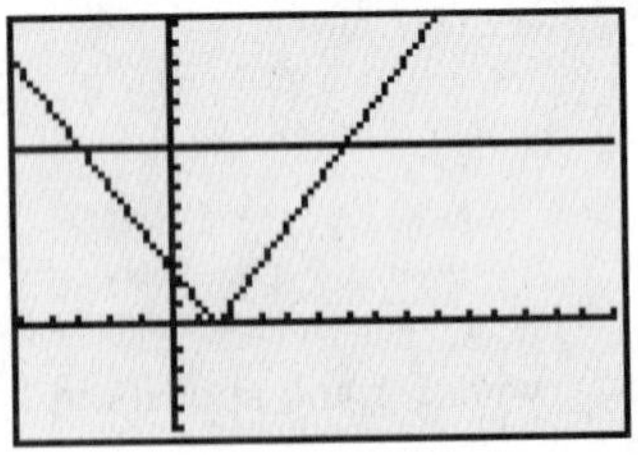

We can also solve absolute value inequalities by a graphing method. For example, to solve the inequality $|2x - 3| < 9$, we graph the equations $y = |2x - 3|$ and $y = 9$ on the same coordinate system. If we choose settings of $[-5, 15]$ for x and $[-5, 15]$ for y, we will get the graph shown in Figure 4-21.

The inequality $|2x - 3| < 9$ will be true for all x-coordinates of points that lie on the graph of $y = |2x - 3|$ and below the graph of $y = 9$. By using the trace feature, we can see that these values of x are in the interval $(-3, 6)$.

FIGURE 4-21

Orals   *Find each value.*

1. $|-5|$ **2.** $-|5|$ **3.** $-|-6|$ **4.** $-|4|$

Solve each equation or inequality.

5. $|x| = 8$ **6.** $|x| = -5$

7. $|x| < 8$ **8.** $|x| > 8$

9. $|x| \geq 4$ **10.** $|x| \leq 7$

EXERCISE 4.2

In Exercises 1–8, find the value of each expression.

1. $|8|$

2. $|-18|$

3. $-|2|$

4. $-|-20|$

5. $-|-30|$

6. $-|25|$

7. $|\pi - 4|$

8. $|2\pi - 4|$

In Exercises 9–12, graph each absolute value function.

9. $f(x) = |x| - 2$

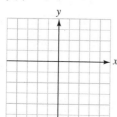

10. $f(x) = -|x| + 1$

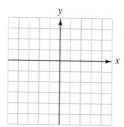

11. $f(x) = -|x + 4|$

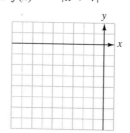

12. $f(x) = |x - 1| + 2$

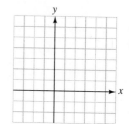

In Exercises 13–34, solve each equation, if possible.

13. $|x| = 4$

14. $|x| = 9$

15. $|x - 3| = 6$

16. $|x + 4| = 8$

17. $|2x - 3| = 5$

18. $|4x - 4| = 20$

19. $|3x + 2| = 16$

20. $|5x - 3| = 22$

21. $\left|\dfrac{7}{2}x + 3\right| = -5$

22. $|2x + 10| = 0$

23. $\left|\dfrac{x}{2} - 1\right| = 3$

24. $\left|\dfrac{4x - 64}{4}\right| = 32$

25. $|3 - 4x| = 5$

26. $|8 - 5x| = 18$

27. $|3x + 24| = 0$

28. $|x - 21| = -8$

29. $\left|\dfrac{3x + 48}{3}\right| = 12$

30. $\left|\dfrac{x}{2} + 2\right| = 4$

31. $|x + 3| + 7 = 10$

32. $|2 - x| + 3 = 5$

33. $\left|\dfrac{3}{5}x - 4\right| - 2 = -2$

34. $\left|\dfrac{3}{4}x + 2\right| + 4 = 4$

In Exercises 35–46, solve each equation, if possible.

35. $|2x + 1| = |3x + 3|$

36. $|5x - 7| = |4x + 1|$

37. $|3x - 1| = |x + 5|$

38. $|3x + 1| = |x - 5|$

39. $|2 - x| = |3x + 2|$

40. $|4x + 3| = |9 - 2x|$

41. $\left|\dfrac{x}{2} + 2\right| = \left|\dfrac{x}{2} - 2\right|$

42. $|7x + 12| = |x - 6|$

43. $\left|x + \dfrac{1}{3}\right| = |x - 3|$

44. $\left|x - \dfrac{1}{4}\right| = |x + 4|$

45. $|3x + 7| = -|8x - 2|$

46. $-|17x + 13| = |3x - 14|$

In Exercises 47–86, solve each inequality. Write the solution set in interval notation and graph it.

47. $|2x| < 8$

48. $|3x| < 27$

49. $|x + 9| \leq 12$

50. $|x - 8| \leq 12$

51. $|3x + 2| \leq -3$

52. $|3x - 2| < 10$

53. $|4x - 1| \leq 7$

54. $|5x - 12| < -5$

55. $|3 - 2x| < 7$

56. $|4 - 3x| \leq 13$

57. $|5x| > 5$

58. $|7x| > 7$

59. $|x - 12| > 24$

60. $|x + 5| \geq 7$

61. $|3x + 2| > 14$

62. $|2x - 5| > 25$

63. $|4x + 3| > -5$

64. $|4x + 3| > 0$

65. $|2 - 3x| \geq 8$

66. $|-1 - 2x| > 5$

67. $-|2x - 3| < -7$

68. $-|3x + 1| < -8$

69. $|8x - 3| > 0$

70. $|7x + 2| > -8$

71. $\left| \dfrac{x - 2}{3} \right| \leq 4$

72. $\left| \dfrac{x - 2}{3} \right| > 4$

73. $|3x + 1| + 2 < 6$

74. $|3x - 2| + 2 \geq 0$

75. $3|2x + 5| \geq 9$

76. $-2|3x - 4| < 16$

77. $|5x - 1| + 4 \leq 0$

78. $-|5x - 1| + 2 < 0$

79. $\left|\dfrac{1}{3}x + 7\right| + 5 > 6$

80. $\left|\dfrac{1}{2}x - 3\right| - 4 < 2$

81. $\left|\dfrac{1}{5}x - 5\right| + 4 > 4$

82. $\left|\dfrac{1}{6}x + 6\right| + 2 < 2$

83. $\left|\dfrac{1}{7}x + 1\right| \leq 0$

84. $|2x + 1| + 2 \leq 2$

85. $\left|\dfrac{x - 5}{10}\right| \leq 0$

86. $\left|\dfrac{3}{5}x - 2\right| + 3 \leq 3$

In Exercises 87–90, write each compound inequality as an inequality using absolute values.

87. $-4 < x < 4$

88. $x < -4$ or $x > 4$

89. $x + 3 < -6$ or $x + 3 > 6$

90. $-5 \leq x - 3 \leq 5$

In Exercises 91–94, solve each problem.

91. Finding temperature ranges The temperatures on a sunny summer day satisfied the inequality $|t - 78°| \leq 8°$, where t is a temperature in degrees Fahrenheit. Express the range of temperatures as a compound inequality.

92. Finding operating temperatures A car CD player has an operating temperature of $|t - 40°| < 80°$, where t is a temperature in degrees Fahrenheit. Express this range of temperatures as a compound inequality.

93. Range of camber angles The specifications for a certain car state that the camber angle c of its wheels should be $0.6° \pm 0.5°$. Express this range with an inequality containing absolute value symbols.

94. Tolerance of a sheet of steel A sheet of steel is to be 0.25 inch thick, with a tolerance of 0.015 inch. Express this specification with an inequality containing absolute value symbols.

Writing Exercises *Write a paragraph using your own words.*

1. Explain how to find the absolute value of a given number.

2. Explain why the equation $|x| + 5 = 0$ has no solution.

3. Explain the use of parentheses and brackets when graphing inequalities.

4. If $k > 0$, explain the differences between the solution sets of $|x| < k$ and $|x| > k$.

Something to Think About

1. For what values of k does $|x| + k = 0$ have exactly two solutions?

2. For what value of k does $|x| + k = 0$ have exactly one solution?

3. Under what conditions is $|x| + |y| > |x + y|$?

4. Under what conditions is $|x| + |y| = |x + y|$?

Review Exercises *Solve each equation or formula.*

1. $3(2a - 1) = 2a$

2. $\dfrac{t}{6} - \dfrac{t}{3} = -1$

3. $\dfrac{5x}{2} - 1 = \dfrac{x}{3} + 12$

4. $4b - \dfrac{b + 9}{2} = \dfrac{b + 2}{5} - \dfrac{8}{5}$

5. $A = p + prt$ for t

6. $P = 2w + 2l$ for l

4.3 Linear Inequalities in Two Variables

■ GRAPHING LINEAR INEQUALITIES ■ GRAPHING COMPOUND INEQUALITIES ■ PROBLEM SOLVING
■ GRAPHING DEVICES

The **graph of a linear inequality** in x and y is the graph of all ordered pairs (x, y) that satisfy the inequality.

> **Linear Inequality**
> A **linear inequality** in x and y is any inequality that can be written in the form
> $$Ax + By < C \quad \text{or} \quad Ax + By > C \quad \text{or} \quad Ax + By \leq C \quad \text{or} \quad Ax + By \geq C$$
> where A, B, and C are real numbers and A and B are not both 0.

Because the inequality $y > 3x + 2$ can be written in the form $-3x + y > 2$, it is an example of a linear inequality in x and y.

■ GRAPHING LINEAR INEQUALITIES

To graph $y > 3x + 2$, we first note that exactly one of the following statements is true:

$$y < 3x + 2, \qquad y = 3x + 2, \qquad \text{or} \qquad y > 3x + 2$$

The graph of the equation $y = 3x + 2$ is the line shown in Figure 4-22(a). The graphs of $y < 3x + 2$ and $y > 3x + 2$ are half-planes, one on each side of that line. The graph of $y = 3x + 2$ is a boundary line separating the two half-planes. It is drawn with a broken line to show that it is not part of the graphs of $y < 3x + 2$ or $y > 3x + 2$.

To find which half-plane is the graph of $y > 3x + 2$, we can substitute the co-ordinates of any point in either half-plane. Since an easy point to use is the origin, we substitute the coordinates of the origin $(0, 0)$ into the inequality and simplify.

$y > 3x + 2$

$0 > 3(0) + 2$ Substitute 0 for x and 0 for y.

$0 \not> 2$

Since the coordinates of the origin do not satisfy the inequality, the origin is not part of the graph of $y > 3x + 2$. Thus, the half-plane on the other side of the broken line is the graph. The graph of $y > 3x + 2$ is shown in Figure 4-22(b).

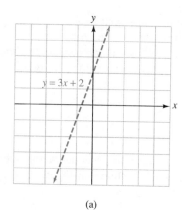
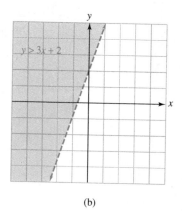

(a)   (b)

FIGURE 4-22

EXAMPLE 1   Graph the inequality $2x - 3y \leq 6$.

Solution   This inequality is the combination of the inequality $2x - 3y < 6$ and the equation $2x - 3y = 6$.

We start by graphing the equation $2x - 3y = 6$ to find the boundary line that separates the graph of $2x - 3y < 6$ from the graph of $2x - 3y > 6$. This time, we draw the solid line shown in Figure 4-23(a), because the graph of $2x - 3y = 6$ is included. To decide which half-plane represents the graph of $2x - 3y < 6$, we check to see whether the coordinates of the origin satisfy the inequality.

$$2x - 3y < 6$$
$$2(0) - 3(0) < 6 \qquad \text{Substitute 0 for } x \text{ and 0 for } y.$$
$$0 < 6$$

Since the coordinates of the origin satisfy the inequality, the origin is in the half-plane that is the graph of $2x - 3y < 6$. The graph is shown in Figure 4-23(b).

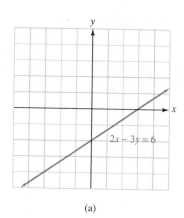
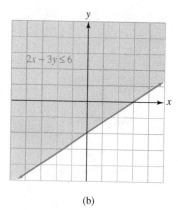

(a)   (b)

FIGURE 4-23   ∎

EXAMPLE 2   Graph the inequality $y < 2x$.

Solution   We graph the equation $y = 2x$, as shown in Figure 4-24(a). Because it is not part of the inequality, we draw the line as a broken line.

To decide which half-plane is the graph of $y < 2x$, we check to see whether the coordinates of some fixed point satisfy the inequality. This time we cannot use the origin as a test point, because the boundary line passes through the origin. However, we can choose a different point—say, $(3, 1)$.

$$y < 2x$$
$$1 < 2(3) \qquad \text{Substitute 1 for } y \text{ and 3 for } x.$$
$$1 < 6$$

Since $1 < 6$ is a true inequality, the point $(3, 1)$ satisfies the inequality and is in the graph, which is shown in Figure 4-24(b).

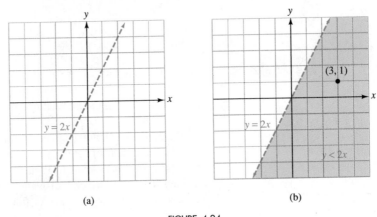

(a)   (b)

FIGURE 4-24   ■

■ GRAPHING COMPOUND INEQUALITIES

EXAMPLE 3   Graph the inequality $2 < x \leq 5$.

Solution   The double inequality $2 < x \leq 5$ is equivalent to the following pair of inequalities:

$$2 < x \qquad \text{and} \qquad x \leq 5$$

The graph of $2 < x \leq 5$ contains all points in the plane that satisfy the inequalities $2 < x$ and $x \leq 5$ simultaneously. These points are in the shaded region of Figure 4-25.

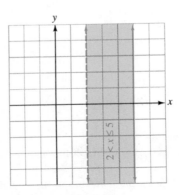

FIGURE 4-25   ■

■ PROBLEM SOLVING

EXAMPLE 4   **Earning money**   Rick has two part-time jobs, one paying $7 per hour and the other paying $5 per hour. He must earn at least $140 per week to pay his expenses while attending college. Write an inequality that shows the various ways he can schedule his time to achieve his goal.

Analyze the Problem   If we let x represent the number of hours he works on the first job, he will earn $7x$ on the first job. If we let y represent the number of hours he works on the second job, he will earn $5y$ on the second job. To achieve his goal, the sum of these two incomes must be at least $140.

Form an Inequality   Since

x represents the number of hours he works on the first job and

y represents the number of hours he works on the second job,

we have

The hourly rate on the first job	·	the hours worked on the first job	+	the hourly rate on the second job	·	the hours worked on the second job	≥	$140
$7	·	x	+	$5	·	y	≥	$140

Solve the Inequality   The graph of the inequality $7x + 5y \geq 140$ is shown in Figure 4-26. Any point in the shaded region indicates a possible way he can schedule his time and earn $140 or more per week. For example, if he works 10 hours on the first job and 15 hours on the second job, he will earn

$$\$7(10) + \$5(15) = \$70 + \$75$$
$$= \$145$$

Since Rick cannot work a negative number of hours, the graph in the figure has no meaning when either x or y is negative, so only the first quadrant is shown. ■

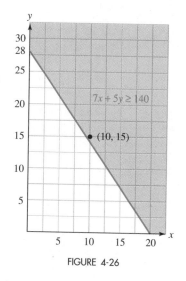

FIGURE 4-26

■ GRAPHING DEVICES

Most graphing devices can graph linear inequalities with a shade feature. Because this feature usually shades only the region between a lower graph and an upper graph, we usually have to enter two equations. Figure 4-27 shows two examples in which the window settings are $[-10, 10]$ for x and $[-10, 10]$ for y.

To graph $y < -3x + 1$, we must enter some equation to act as the lower bound of the shaded region. Since this graph should be outside the viewing window, we pick an equation like $y = -11$ or $y = -12$. We then enter $y = -3x + 1$ as the upper curve and use the shade feature. We will see a graph similar to the one shown in Figure 4-27(a). For details, see your owner's manual.

To graph $y > -3x + 1$, we must enter some equation to act as the upper bound of the shaded region. Since this graph should be outside the viewing window, we pick an equation like $y = 11$ or $y = 12$. We then enter $y = -3x + 1$ as the lower curve and use the shade feature. We will see a graph similar to the one shown in Figure 4-27(b). For details, see your owner's manual.

It is important to note that graphing calculators cannot distinguish between solid and dashed lines to show whether or not the edge of a region is included within the graph.

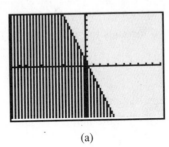   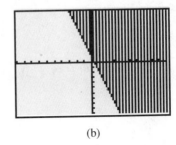

(a)   (b)

FIGURE 4-27

Orals   *Does the given point satisfy* $2x + 3y < 12$?

1. $(0, 0)$ **2.** $(3, 2)$ **3.** $(2, 3)$ **4.** $(-1, 4)$

Does the given point satisfy $3x - 2y \geq 12$?

5. $(0, 0)$ **6.** $(3, 2)$ **7.** $(2, -3)$ **8.** $(5, 1)$

EXERCISE 4.3

In Exercises 1–16, graph each inequality.

1. $y > x + 1$

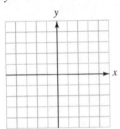

2. $y < 2x - 1$

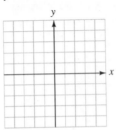

3. $y \geq x$

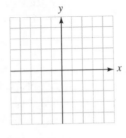

4. $y \leq 2x$

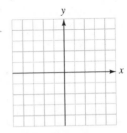

5. $2x + y \leq 6$

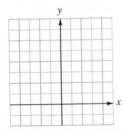

6. $x - 2y \geq 4$

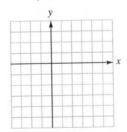

7. $3x \geq -y + 3$

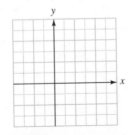

8. $2x \leq -3y - 12$

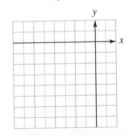

9. $y \geq 1 - \dfrac{3}{2}x$

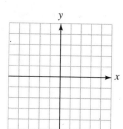

10. $y < \dfrac{1}{3}x - 1$

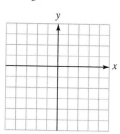

11. $x < 4$

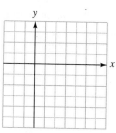

12. $y \geq -2$

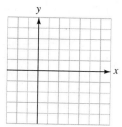

13. $-2 \leq x < 0$

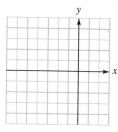

14. $-3 < y \leq -1$

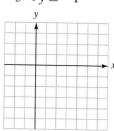

15. $y < -2$ or $y > 3$

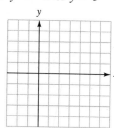

16. $-x \leq 1$ or $x \geq 2$

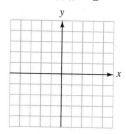

In Exercises 17–26, find the equation of the boundary line or lines. Then give the inequality whose graph is shown.

17.

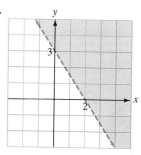

18.

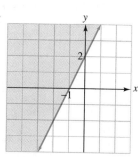

19.

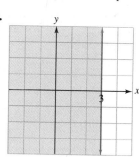

20.

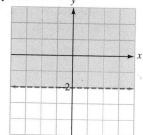

21.

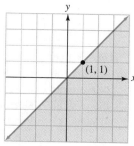

22.

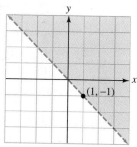

23.

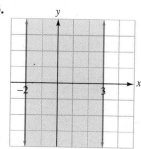

24.

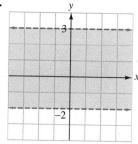

25.

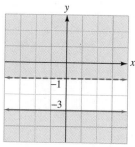

26.

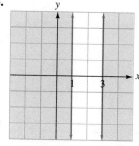

In Exercises 27–32, graph each inequality for nonnegative values of x and y. Then give some ordered pairs that satisfy the inequality.

27. Figuring taxes On average, it takes an accountant 1 hour to complete a simple tax return and 3 hours to complete a complicated return. If the accountant wants to work less than 9 hours per day, find an inequality that shows the possible ways that simple returns (*x*) and complicated returns (*y*) can be completed each day.

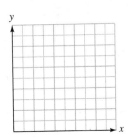

28. Selling trees During a sale, a garden store sold more than $2000 worth of trees. If a 6-foot maple tree costs $100 and a 5-foot pine tree costs $125, find an inequality that shows the possible ways that maple trees (*x*) and pine trees (*y*) were sold.

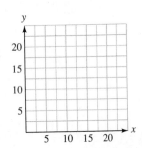

29. Choosing housekeepers One housekeeper charges $6 per hour, and another charges $7 per hour. If Sarah can afford no more than $42 per week to clean her house, find an inequality that shows the possible ways that she can hire the first housekeeper (*x*) and the second housekeeper (*y*).

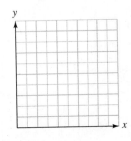

30. Making sporting goods A sporting goods manufacturer allocates at least 1200 units of time per day to make fishing rods and reels. If it takes 10 units of time to make a rod and 15 units of time to make a reel, find an inequality that shows the possible ways to schedule the time to make rods (*x*) and reels (*y*).

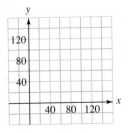

31. Investing in stocks A woman has up to $6000 to invest. If stock in Traffico sells for $50 per share and stock in Cleanco sells for $60 per share, find an inequality that shows the possible ways that she can buy shares of Traffico (*x*) and Cleanco (*y*).

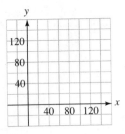

32. Buying concert tickets Tickets to a concert cost $6 for reserved seats and $4 for general admission. If receipts must be at least $10,200 to meet expenses, find an inequality that shows the possible ways that the box office can sell reserved seats (*x*) and general admission tickets (*y*).

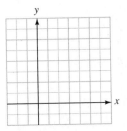

In Exercises 33–36, use a graphing calculator to graph each inequality.

33. $y < 0.27x - 1$ **34.** $y > -3.5x + 2.7$ **35.** $y \geq -2.37x + 1.5$ **36.** $y \leq 3.37x - 1.7$

Writing Exercises *Write a paragraph using your own words.*

1. Explain how to decide where to draw the boundary of the graph of a linear inequality, and whether to draw it as a solid or a broken line.

2. Explain how to decide which side of the boundary of the graph of a linear inequality should be shaded.

Something to Think About **1.** Can an inequality be an identity, one that is satisfied by all (x, y) pairs? Illustrate.

2. Can an inequality have no solutions? Illustrate.

Review Exercises *Solve each system of equations.*

1. $\begin{cases} x + y = 4 \\ x - y = 2 \end{cases}$

2. $\begin{cases} 2x - y = -4 \\ x + 2y = 3 \end{cases}$

3. $\begin{cases} 3x + y = 3 \\ 2x - 3y = 13 \end{cases}$

4. $\begin{cases} 2x - 5y = 8 \\ 5x + 2y = -9 \end{cases}$

4.4 Systems of Inequalities

■ PROBLEM SOLVING

We now consider the graphs of systems of inequalities in the variables x and y. These graphs will usually be the intersection of half-planes.

EXAMPLE 1 Graph the solution set of the system $\begin{cases} x + y \leq 1 \\ 2x - y > 2 \end{cases}$.

Solution On the same set of coordinate axes, we graph each inequality, as shown in Figure 4-28.

The graph of the inequality $x + y \leq 1$ includes the line graph of the equation $x + y = 1$ and all points below it. Since the boundary line is included, we draw it as a solid line.

The graph of the inequality $2x - y > 2$ contains only those points below the graph of the equation $2x - y = 2$. Since the boundary line is not included, we draw it as a broken line.

The area where the half-planes intersect represents the simultaneous solution of the given system of inequalities, because any point in that region has coordinates that will satisfy both inequalities.

$x + y = 1$

x	y	(x, y)
0	1	$(0, 1)$
1	0	$(1, 0)$

$2x - y = 2$

x	y	(x, y)
0	-2	$(0, -2)$
1	0	$(1, 0)$

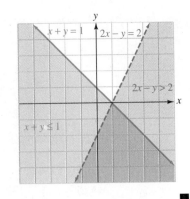

FIGURE 4-28

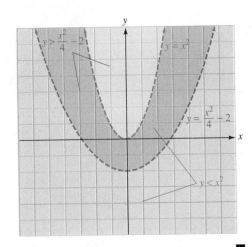

EXAMPLE 2   Graph the solution set of the system $\begin{cases} y < x^2 \\ y > \dfrac{x^2}{4} - 2 \end{cases}$.

Solution   The graph of $y = x^2$ is the parabola shown in Figure 4-29, which opens upward and has its vertex at the origin. The points with coordinates that satisfy the inequality $y < x^2$ are those points that lie below the parabola.

The graph of $y = \dfrac{x^2}{4} - 2$ is a parabola opening upward, with vertex at $(0, -2)$. However, this time the points with coordinates that satisfy the inequality are those points that lie above the parabola. Thus, the graph of the solution set of the system is the area between the parabolas.

$y = x^2$

x	y	(x, y)
0	0	$(0, 0)$
1	1	$(1, 1)$
-1	1	$(-1, 1)$
2	4	$(2, 4)$
-2	4	$(-2, 4)$

$y = \dfrac{x^2}{4} - 2$

x	y	(x, y)
0	-2	$(0, -2)$
2	-1	$(2, -1)$
-2	-1	$(-2, -1)$
4	2	$(4, 2)$
-4	2	$(-4, 2)$

FIGURE 4-29

EXAMPLE 3   Graph the solution set of the system $\begin{cases} x \geq 1 \\ y \geq x \\ 4x + 5y < 20 \end{cases}$.

Solution   The graph of the solution set of the inequality $x \geq 1$ includes those points that lie on the graph of the equation $x = 1$ and to the right, as in Figure 4-30(a).

The graph of the solution set of the inequality $y \geq x$ includes those points that lie on the graph of the equation $y = x$ and above it, as in Figure 4-30(b).

The graph of the solution set of the inequality $4x + 5y < 20$ includes those points that lie below the graph of the equation $4x + 5y = 20$, as in Figure 4-30(c).

If these three graphs are merged onto a single set of coordinate axes, the graph of the original system of inequalities includes those points that lie within the shaded triangle, together with the points on the sides of the triangle drawn as solid lines. (See Figure 4-30(d).)

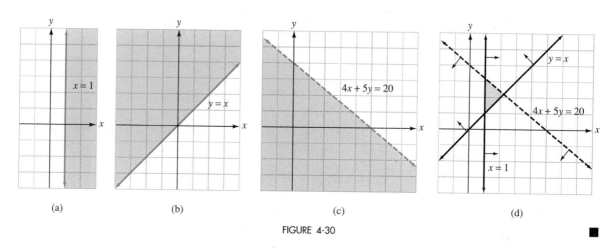

(a)   (b)   (c)   (d)

FIGURE 4-30 ■

■ PROBLEM SOLVING

EXAMPLE 4

Landscaping   A homeowner budgets from \$300 to \$600 for trees and bushes to landscape his yard. After shopping around, he finds that good trees cost \$150 and mature bushes cost \$75. What combinations of trees and bushes can he afford to buy?

Analyze the Problem   If x represents the number of trees purchased, then \$150x will be the cost of the trees. If y represents the number of bushes purchased, then \$75y will be the cost of the bushes. We know that the homeowner wants the sum of these costs to be from \$300 to \$600.

Form Two Inequalities   We let

x represent the number of trees purchased and

y represent the number of bushes purchased.

We can then form the following system of inequalities:

The cost of a tree	·	the number of trees purchased	+	the cost of a bush	·	the number of bushes purchased	≥	\$300.
\$150	·	x	+	\$75	·	y	≥	\$300

The cost of a tree	·	the number of trees purchased	+	the cost of a bush	·	the number of bushes purchased	≤	\$600.
\$150	·	x	+	\$75	·	y	≤	\$600

Solve the System   We graph the system

$$\begin{cases} 150x + 75y \geq 300 \\ 150x + 75y \leq 600 \end{cases}$$

as in Figure 4-31. The coordinates of each point shown in the graph give a possible combination of trees (x) and bushes (y) that can be purchased.

State the Conclusion   These possibilities are

$(0, 4), (0, 5), (0, 6), (0, 7), (0, 8)$

$(1, 2), (1, 3), (1, 4), (1, 5), (1, 6)$

$(2, 0), (2, 1), (2, 2), (2, 3), (2, 4)$

$(3, 0), (3, 1), (3, 2), (4, 0)$

Only these points can be used, because the homeowner cannot buy a portion of a tree.

Check the Result   Check some of the ordered pairs to verify that they satisfy both inequalities.

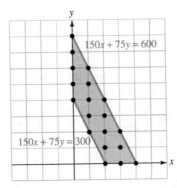

FIGURE 4-31

Orals   *Do the coordinates of* $(1, 1)$ *satisfy both inequalities?*

1. $\begin{cases} y < x + 1 \\ y > x - 1 \end{cases}$

2. $\begin{cases} y > 2x - 3 \\ y < -x + 3 \end{cases}$

EXERCISE 4.4

In Exercises 1–16, graph the solution set of each system of inequalities.

1. $\begin{cases} y < 3x + 2 \\ y < -2x + 3 \end{cases}$

2. $\begin{cases} y \leq x - 2 \\ y \geq 2x + 1 \end{cases}$

3. $\begin{cases} 3x + 2y > 6 \\ x + 3y \leq 2 \end{cases}$

4. $\begin{cases} x + y < 2 \\ x + y \leq 1 \end{cases}$

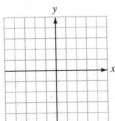
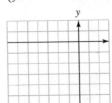
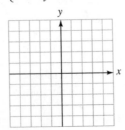
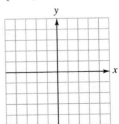

5. $\begin{cases} 3x + y \leq 1 \\ -x + 2y \geq 6 \end{cases}$

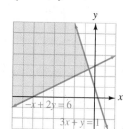

6. $\begin{cases} x + 2y < 3 \\ 2x + 4y < 8 \end{cases}$

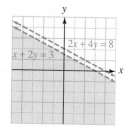

7. $\begin{cases} 2x - y > 4 \\ y < -x^2 + 2 \end{cases}$

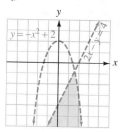

8. $\begin{cases} x \leq y^2 \\ y \geq x \end{cases}$

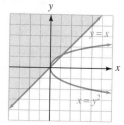

9. $\begin{cases} y > x^2 - 4 \\ y < -x^2 + 4 \end{cases}$

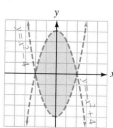

10. $\begin{cases} x \geq y^2 \\ y \geq x^2 \end{cases}$

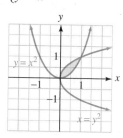

11. $\begin{cases} 2x + 3y \leq 6 \\ 3x + y \leq 1 \\ x \leq 0 \end{cases}$

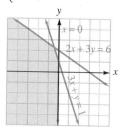

12. $\begin{cases} 2x + y \leq 2 \\ y \geq x \\ x \geq 0 \end{cases}$

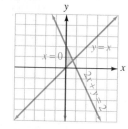

13. $\begin{cases} x - y < 4 \\ y \leq 0 \\ x \geq 0 \end{cases}$

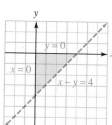

14. $\begin{cases} xy \leq 1 \\ x \geq 0 \\ y \geq 0 \end{cases}$

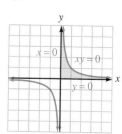

15. $\begin{cases} x \geq 0 \\ y \geq 0 \\ 9x + 3y \leq 18 \\ 3x + 6y \leq 18 \end{cases}$

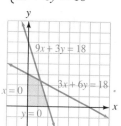

16. $\begin{cases} x + y \geq 1 \\ x - y \leq 1 \\ x - y \geq 0 \\ x \leq 2 \end{cases}$

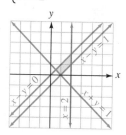

In Exercises 17–20, graph each system of inequalities and give two possible solutions to each problem.

17. Buying compact discs Melodic Music has compact discs on sale for either $10 or $15. If a customer wants to spend at least $30 but no more than $60 on CDs, find a system of inequalities whose graph will show the possible ways a customer can buy $10 CDs (*x*) and $15 CDs (*y*).

1 $10 CD and 2 $15 CDs, 4 $10 CDs, and 1 $15 CD

18. Buying boats Dry Boat Works wholesales aluminum boats for $800 and fiberglass boats for $600. Northland Marina wants to order at least $2400 worth, but no more than $4800 worth, of boats. Find a system of inequalities whose graph will show the possible ways that aluminum boats (*x*) and fiberglass boats (*y*) can be ordered.   4 alum. and 1 fiberglass, 1 alum. and 4 fiberglass

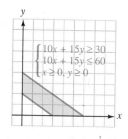

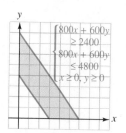

19. **Buying furniture**   A distributor wholesales desk chairs for $150 and side chairs for $100. Best Furniture wants to order no more than $900 worth of chairs, including more side chairs than desk chairs. Find a system of inequalities whose graph will show the possible combinations of desk chairs (x) and side chairs (y) that can be ordered.

20. **Ordering furnace equipment**   J. Bolden Heating Company wants to order no more than $2000 worth of electronic air cleaners and humidifiers from a wholesaler that charges $500 for air cleaners and $200 for humidifiers. If Bolden wants more humidifiers than air cleaners, find a system of inequalities whose graph will show the possible ways that air cleaners (x) and humidifiers (y) can be ordered.

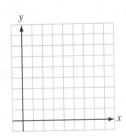

Writing Exercises   *Write a paragraph using your own words.*

1. When graphing a system of linear inequalities, explain how to decide which region to shade.

2. Explain how a system of two linear inequalities might have no solution.

Something to Think About

1. The solution of a system of inequalities in two variables is *bounded* if it is possible to draw a circle around it. Can the solution of two linear inequalities be bounded?

2. The solution of $\begin{cases} y \geq |x| \\ y \leq k \end{cases}$ has an area of 25. Find k.

Review Exercises   *Solve each formula for the given variable.*

1. $A = p + prt$ for r

2. $C = \dfrac{5}{9}(F - 32)$ for F

3. $z = \dfrac{x - \mu}{\sigma}$ for x

4. $P = 2l + 2w$ for w

5. $l = a + (n - 1)d$ for d

6. $z = \dfrac{x - \mu}{\sigma}$ for μ

4.5 Linear Programming

■ APPLICATIONS OF LINEAR PROGRAMMING

Systems of inequalities provide the basis for an area of applied mathematics known as **linear programming**. Linear programming helps answer questions such as "How can a business make as much money as possible?" or "How can I plan a nutritious menu at a school cafeteria at the least cost?" Linear programming was developed during World War II when it became necessary to move huge quantities of men, materials, and supplies as efficiently and economically as possible.

The solutions to such problems depend on certain **constraints**: The business has limited resources; the menu must contain sufficient vitamins and minerals; and so on. Any solution that satisfies the constraints is called a **feasible solution**. In linear programming, the constraints are expressed as a system of linear inequalities, and the quantity that is to be maximized (or minimized) is expressed as a linear function of several variables.

To solve a linear programming problem, we will maximize or minimize a function subject to constraints. For example, suppose that the profit P to be earned by a company is given by the function $P = y + 2x$, subject to the constraints

$$\begin{cases} x + y \geq 1 \\ x - y \leq 1 \\ x - y \geq 0 \\ x \leq 2 \end{cases}$$

We can find the solution of this system and then find the coordinates of each corner of the region R shown in Figure 4-32. We can then rewrite the equation

$$P = y + 2x \qquad \text{in the equivalent form} \qquad y = -2x + P$$

This is the equation for a set of parallel lines, each with a slope of -2 and a y-intercept of P. Many such lines pass through the region R. To decide which of these provides the maximum value of P, we refer to Figure 4-33 and locate the line with the greatest y-intercept. Since line l has the greatest y-intercept and crosses region R at the corner $(2, 2)$, the maximum value of P (subject to the given constraints) is

$$\begin{aligned} P &= y + 2x \\ &= 2 + 2(2) \\ &= 6 \end{aligned}$$

The profit P has a maximum value of 6, subject to the given constraints. This profit occurs when $x = 2$ and $y = 2$.

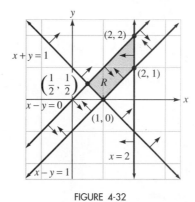

FIGURE 4-32

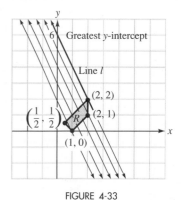

FIGURE 4-33

This example illustrates the following important theorem.

Theorem
If a linear function, subject to the constraints of a system of linear inequalities in two variables, attains a maximum or a minimum value, that value will occur at a corner or along an entire edge of the region R that represents the solution of the system.

In a linear programming problem, the function to be maximized (or minimized) is called the **objective function**. The solution region R for the set of constraints is called the **feasible region**.

EXAMPLE 1   If $P = 2x + 3y$, find the maximum value of P subject to the given constraints:

$$\begin{cases} x \geq 0 \\ y \geq 0 \\ x + y \leq 4 \\ 2x + y \leq 6 \end{cases}$$

Solution   We solve the system of inequalities to find the feasible region R shown in Figure 4-34. The coordinates of its corners are $(0, 0)$, $(3, 0)$, $(0, 4)$, and $(2, 2)$.

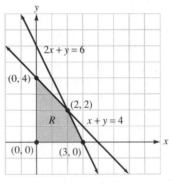

FIGURE 4-34

Because the maximum value of P will occur at a corner of R, we substitute the coordinates of each corner point into the objective function:

$$P = 2x + 3y$$

We then find the one that gives the maximum value of P.

Corner point	$P = 2x + 3y$
$(0, 0)$	$P = 2(0) + 3(0) = 0$
$(3, 0)$	$P = 2(3) + 3(0) = 6$
$(2, 2)$	$P = 2(2) + 3(2) = 10$
$(0, 4)$	$P = 2(0) + 3(4) = 12$

The maximum value $P = 12$ occurs when $x = 0$ and $y = 4$.

EXAMPLE 2   If $P = 3x + 2y$, find the minimum value of P subject to the given constraints:

$$\begin{cases} x + y \geq 1 \\ x - y \leq 1 \\ x - y \geq 0 \\ \quad\;\; x \leq 2 \end{cases}$$

Solution   We refer to the feasible region shown in Figure 4-35, with corners at $\left(\frac{1}{2}, \frac{1}{2}\right)$, $(2, 2)$, $(2, 1)$, and $(1, 0)$. Because the minimum value of P occurs at a corner point of R, we substitute the coordinates of each corner point into the objective function:

$$P = 3x + 2y$$

Then we find which one gives the minimum value of P.

Corner point	$P = 3x + 2y$
$\left(\frac{1}{2}, \frac{1}{2}\right)$	$P = 3\left(\frac{1}{2}\right) + 2\left(\frac{1}{2}\right) = \frac{5}{2}$
$(2, 2)$	$P = 3(2) + 2(2) = 10$
$(2, 1)$	$P = 3(2) + 2(1) = 8$
$(1, 0)$	$P = 3(1) + 2(0) = 3$

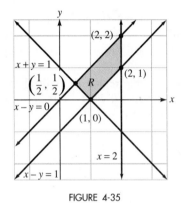

FIGURE 4-35

The minimum value $P = \dfrac{5}{2}$ occurs when $x = \dfrac{1}{2}$ and $y = \dfrac{1}{2}$.

■ APPLICATIONS OF LINEAR PROGRAMMING

EXAMPLE 3   **Maximizing income**   An accountant earns a profit of $60 for preparing an individual tax return and $150 for preparing a commercial tax return. On average, an individual return requires 3 hours of the accountant's time, of which 1 hour is spent working on a computer. Each commercial return requires 4 hours of the accountant's time, of which 2 hours is spent working on a computer. If the accountant's time is limited to 240 hours and the available computer time is limited to 100 hours, how should the accountant's time be divided between individual and commercial returns to maximize his income?

Solution   Suppose that x represents the number of individual returns completed and y represents the number of commercial returns completed. Because each of the x individual returns earns a profit of $60 and each of the y commercial returns earns a profit of $150, the profit function P is given by the equation

$$P = 60x + 150y$$

The following table provides the information about time requirements:

	Individual return	Commercial return	Time available
Accountant's time	3	4	240
Computer time	1	2	100

The profit is subject to the following constraints:

$$\begin{cases} x \geq 0 \\ y \geq 0 \\ 3x + 4y \leq 240 \\ x + 2y \leq 100 \end{cases}$$

The inequalities $x \geq 0$ and $y \geq 0$ indicate that the number of individual and commercial returns cannot be negative.

The inequality $3x + 4y \leq 240$ is a constraint on the accountant's time. Each of the x individual returns takes 3 hours of his time and each of the y commercial returns takes 4 hours. The sum of these two amounts of time must be less than or equal to his available time, which is 240 hours.

The inequality $x + 2y \leq 100$ is a constraint on computer time. Each of the x individual returns takes 1 hour of computer time and each of the y commercial returns takes 2 hours of computer time. The sum of x and $2y$ must be less than or equal to the available time, which is 100 hours.

We graph each of the constraints to find the feasible region R, as in Figure 4-36. The four corners of region R have coordinates of $(0, 0)$, $(80, 0)$, $(40, 30)$, and $(0, 50)$.

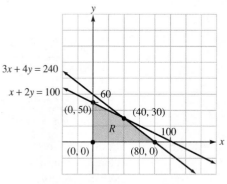

FIGURE 4-36

We substitute the coordinates of each corner point into the profit function

$$P = 60x + 150y$$

to find the maximum profit P.

Corner point	$P = 60x + 150y$
$(0, 0)$	$P = 60(0) + 150(0) = 0$
$(80, 0)$	$P = 60(80) + 150(0) = 4800$
$(40, 30)$	$P = 60(40) + 150(30) = 6900$
$(0, 50)$	$P = 60(0) + 150(50) = 7500$

The accountant will maximize his profit at $7500 if he prepares no individual returns and 50 commercial returns. ■

EXAMPLE 4

Diet problem   Two diet supplements are Vitamix and Megamin. Each Vitamix tablet contains 3 units of calcium, 20 units of Vitamin C, and 40 units of iron and costs $0.50. Each Megamin tablet contains 4 units of calcium, 40 units of Vitamin C, and 30 units of iron and costs $0.60. At least 24 units of calcium, 200 units of Vitamin C, and 120 units of iron are required for the daily needs of a particular patient. How many tablets of each supplement should be taken daily for a minimum cost? Find the daily minimum cost.

Solution   We can let x represent the number of Vitamix tablets to be taken daily and y represent the number of Megamin tablets to be taken daily. We then construct a table displaying the calcium, Vitamin C, iron, and cost information.

	Vitamix	Megamin	Amount required
Calcium	3	4	24
Vitamin C	20	40	200
Iron	40	30	120
Cost	$0.50	$0.60	

Since cost is to be minimized, the cost function is the objective function.

$$C = 0.50x + 0.60y$$

We then write the constraints. Since there are requirements for calcium, Vitamin C, and iron, there is a constraint for each. Note that neither x nor y can be negative.

$$3x + 4y \geq 24 \qquad \text{Calcium}$$
$$20x + 40y \geq 200 \qquad \text{Vitamin C}$$
$$40x + 30y \geq 120 \qquad \text{Iron}$$
$$x \geq 0, \; y \geq 0 \qquad \text{Nonnegative constraints}$$

We graph the inequalities to find the feasible region and the coordinates of its corners, as in Figure 4-37.

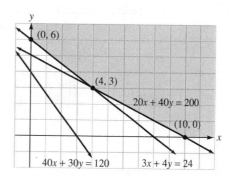

FIGURE 4-37

In this case, the feasible region is not bounded on all sides. The coordinates of the corner points are (0, 6), (4, 3), and (10, 0). To find the minimum cost, we substitute each pair of coordinates into the objective function:

Corner point	$C = 0.50x + 0.60y$
(0, 6)	$C = 0.50(0) + 0.60(6) = 3.60$
(4, 3)	$C = 0.50(4) + 0.60(3) = 3.80$
(10, 0)	$C = 0.50(10) + 0.60(0) = 5.00$

A minimum cost will occur if no Vitamix and 6 Megamin tablets are taken daily. The minimum daily cost is $3.60. ∎

EXAMPLE 5   **Production-schedule problem**   A television program director must schedule comedy skits and musical numbers for prime-time variety shows. Each comedy skit requires 2 hours of rehearsal time, costs $3000, and brings in $20,000 from the show's sponsors. Each musical number requires 1 hour of rehearsal time, costs $6000, and generates $12,000. If 250 hours are available for rehearsal and $600,000 is budgeted for comedy and music, how many segments of each type should be produced to maximize income? Find the maximum income.

Solution   We can let x represent the number of comedy skits and y the number of musical numbers to be scheduled. We then construct a table with information about rehearsal time, production cost, and income generated.

	Comedy	Musical	Available
Rehearsal time (hours)	2	1	250
Cost (in $1000s)	3	6	600
Generated income (in $1000s)	20	12	

Next we write the objective function. Since each of the x comedy skits generates 20 thousand dollars, the income generated by the comedy skits is $20x$ thousand dollars. The musical numbers produce $12y$ thousand dollars. The objective function to be maximized is

$$V = 20x + 12y$$

Then we write the constraints. Since there are limits on rehearsal time and budget, there is a constraint for each. Note that neither x nor y can be negative.

$2x + y \leq 250$ Constraint on rehearsal time

$3x + 6y \leq 600$ Constraint on cost

$x \geq 0, y \geq 0$ Nonnegative constraints

We graph the inequalities to find the feasible region (shown in Figure 4-38) and find the coordinates of each corner point.

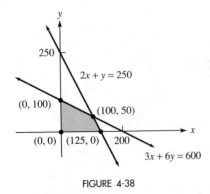

FIGURE 4-38

The coordinates of the corner points of the feasible region are $(0, 0)$, $(0, 100)$, $(100, 50)$, and $(125, 0)$. To find the maximum income, we substitute each pair of coordinates into the objective function:

Corner point	$V = 20x + 12y$
$(0, 0)$	$V = 20(0) + 12(0) = 0$
$(0, 100)$	$V = 20(0) + 12(100) = 1200$
$(100, 50)$	$V = 20(100) + 12(50) = 2600$
$(125, 0)$	$V = 20(125) + 12(0) = 2500$

Maximum income occurs if 100 comedy skits and 50 musical numbers are scheduled. The maximum income will be 2600 thousand dollars, or $2,600,000. ∎

Orals *Evaluate $P = 2x + 5y$ when*

1. $x = 0, y = 5$ 25

2. $x = 2, y = 1$ 9

Find the corner points of the region determined by

3. $\begin{cases} x \geq 0 \\ y \geq 0 \\ x + y \leq 3 \end{cases}$ $(0, 0), (0, 3), (3, 0)$

4. $\begin{cases} x \geq 0 \\ y \geq 0 \\ x + 2y \leq 4 \end{cases}$ $(0, 0),$ •$(0, 2), (4, 0)$

EXERCISE 4.5

In Exercises 1–8, maximize P subject to the following constraints.

1. $P = 2x + 3y$
$\begin{cases} x \geq 0 \\ y \geq 0 \\ x + y \leq 4 \end{cases}$
$P = 12$ at $(0, 4)$

2. $P = 3x + 2y$
$\begin{cases} x \geq 0 \\ y \geq 0 \\ x + y \leq 4 \end{cases}$
$P = 12$ at $(4, 0)$

3. $P = y + \dfrac{1}{2}x$
$\begin{cases} x \geq 0 \\ y \geq 0 \\ 2y - x \leq 1 \\ y - 2x \geq -2 \end{cases}$
$P = \frac{13}{6}$ at $\left(\frac{5}{3}, \frac{4}{3}\right)$

4. $P = 4y - x$
$\begin{cases} x \leq 2 \\ y \geq 0 \\ x + y \geq 1 \\ 2y - x \leq 1 \end{cases}$
$P = 4$ at $\left(2, \frac{3}{2}\right)$

5. $P = 2x + y$
$\begin{cases} y \geq 0 \\ y - x \leq 2 \\ 2x + 3y \leq 6 \\ 3x + y \leq 3 \end{cases}$
$P = \frac{18}{7}$ at $\left(\frac{3}{7}, \frac{12}{7}\right)$

6. $P = x - 2y$
$\begin{cases} x + y \leq 5 \\ y \leq 3 \\ x \leq 2 \\ x \geq 0 \\ y \geq 0 \end{cases}$
$P = 2$ at $(2, 0)$

7. $P = 3x - 2y$
$\begin{cases} x \leq 1 \\ x \geq -1 \\ y - x \leq 1 \\ x - y \leq 1 \end{cases}$
$P = 3$ at $(1, 0)$

8. $P = x - y$
$\begin{cases} 5x + 4y \leq 20 \\ y \leq 5 \\ x \geq 0 \\ y \geq 0 \end{cases}$
$P = 4$ at $(4, 0)$

In Exercises 9–16, minimize P subject to the following constraints.

9. $P = 5x + 12y$
$\begin{cases} x \geq 0 \\ y \geq 0 \\ x + y \leq 4 \end{cases}$
$P = 0$ at $(0, 0)$

10. $P = 3x + 6y$
$\begin{cases} x \geq 0 \\ y \geq 0 \\ x + y \leq 4 \end{cases}$
$P = 0$ at $(0, 0)$

11. $P = 3y + x$
$\begin{cases} x \geq 0 \\ y \geq 0 \\ 2y - x \leq 1 \\ y - 2x \geq -2 \end{cases}$
$P = 0$ at $(0, 0)$

12. $P = 5y + x$
$\begin{cases} x \leq 2 \\ y \geq 0 \\ x + y \geq 1 \\ 2y - x \leq 1 \end{cases}$
$P = 1$ at $(1, 0)$

13. $P = 6x + 2y$
$$\begin{cases} y \geq 0 \\ y - x \leq 2 \\ 2x + 3y \leq 6 \\ 3x + y \leq 3 \end{cases}$$

14. $P = 2y - x$
$$\begin{cases} x \geq 0 \\ y \geq 0 \\ x + y \leq 5 \\ x + 2y \geq 2 \end{cases}$$

15. $P = 2x - 2y$
$$\begin{cases} x \leq 1 \\ x \geq -1 \\ y - x \leq 1 \\ x - y \leq 1 \end{cases}$$

16. $P = y - 2x$
$$\begin{cases} x + 2y \leq 4 \\ 2x + y \leq 4 \\ x + 2y \geq 2 \\ 2x + y \geq 2 \end{cases}$$

In Exercises 17–28, write the objective function and the inequalities that describe the constraints in each problem. Graph the feasible region, showing corner points. Then find the maximum or minimum value of the objective function.

17. Making furniture Two woodworkers, Tom and Carlos, earn a profit of $100 for making a table and $80 for making a chair. On average, Tom must work 3 hours and Carlos 2 hours to make a chair. Tom must work 2 hours and Carlos 6 hours to make a table. If neither wishes to work more than 42 hours per week, how many tables and how many chairs should they make each week to maximize their profit? Find the maximum profit.

18. Making crafts Two artists, Nina and Roberta, make winter yard ornaments. They receive a profit of $80 for each wooden snowman they make and $64 for each wooden Santa Claus. On average, Nina must work 4 hours and Roberta 2 hours to make a snowman. Nina must work 3 hours and Roberta 4 hours to make a Santa Claus. If neither wishes to work more than 20 hours per week, how many of each ornament should they make each week to maximize their profit? Find the maximum profit.

19. Inventories An electronics store manager stocks from 20 to 30 IBM-compatible computers and from 30 to 50 Macintosh computers. There is room in the store to stock up to 60 computers. The manager receives a commission of $50 on the sale of each IBM-compatible computer and $40 on the sale of each Macintosh computer. If the manager can sell all of the computers, how many should she stock to maximize her commissions? Find the maximum commission.

20. Food for a trip Bill packs two foods for a camping trip. One ounce of food A provides 150 calories and 21 units of vitamins. One ounce of food B provides 60 calories and 42 units of vitamins. Every day Bill needs at least 3000 calories and at least 1260 units of vitamins. However, he does not want to carry more than 60 ounces of food for each day of his trip. If food A costs 35¢ per ounce and food B costs 27¢ per ounce, how much of each type of food should he carry to minimize cost? Find the minimum cost.

21. Nutrition The owner of a Fast-Food Burger Bar purchases hamburger from two suppliers. Supplier A's hamburger has 2 units of protein and 20 units of carbohydrate per ounce. Supplier B's hamburger has 8 units of protein and 40 units of carbohydrate per ounce. The mix should have at least 360 units of protein and 2000 units of carbohydrate. If A's hamburger is 10% fat and B's hamburger is 30% fat, how many ounces from each supplier should be used to minimize fat? Find the minimum fat content.

22. Diet problem A diet requires at least 16 units of vitamin C and at least 34 units of vitamin B complex. Two food supplements are available that provide these nutrients in the amounts and costs shown in the table. How much of each should be used to minimize the cost?

Supplement	Vitamin C	Vitamin B	Cost
A	3 units per gram	2 units per gram	3¢ per gram
B	2 units per gram	6 units per gram	4¢ per gram

23. Production problem A company manufactures one type of computer chip that runs at 50 MHz and another that runs at 66 MHz. The company can make a maximum of 50 fast chips per day and a maximum of 100 slow chips per day. It takes 6 hours to make a fast chip and 3 hours to make a slow chip, and the company's employees can provide up to 360 hours of labor per day. If the company makes a profit of $20 on each 66-MHz chip and $27 on each 50-MHz chip, how many of each type should be manufactured to earn the maximum profit?

24. Production problem Manufacturing VCRs and TVs requires the use of the electronics, assembly, and finishing departments of a factory, according to the following schedule:

	Hours for VCR	Hours for TV	Hours available per week
Electronics	3	4	180
Assembly	2	3	120
Finishing	2	1	60

Each VCR has a profit of $40, and each TV has a profit of $32. How many VCRs and TVs should be manufactured weekly to maximize profit? Find the maximum profit.

25. Financial planning A stockbroker has $200,000 to invest in stocks and bonds. She wants to invest at least $100,000 in stocks and at least $50,000 in bonds. If stocks have an annual yield of 9% and bonds have an annual yield of 7%, how much should she invest in each to maximize her income? Find the maximum return.

26. Production problem A small country exports soybeans and flowers. Soybeans require 8 workers per acre; flowers require 12 workers per acre; and 100,000 workers are available. Government contracts require that there be at least 3 times as much soybeans as flowers planted. It costs $250 per acre to plant soybeans and $300 per acre to plant flowers, and there is a budget of $3 million. If the profit from soybeans is $1600 per acre and the profit from flowers is $2000 per acre, how many acres of each crop should be planted to maximize profit? Find the maximum profit.

27. Band trip A high school band trip will require the use of buses and trucks to transport at least 100 students and at least 18 large instruments. Each bus can accommodate 40 students and 3 large instruments and costs $350 to rent. Each truck can accommodate 10 students and 6 large instruments and costs $200 to rent. How many of each type of vehicle should be rented to minimize rental cost? Find the minimum cost.

28. Making ice cream An ice cream store sells two new flavors: Fantasy and Excel. Each barrel of Fantasy requires 4 pounds of nuts and 3 pounds of chocolate and has a profit of $50. Each barrel of Excel requires 4 pounds of nuts and 1 pound of chocolate and has a profit of $40. There are 16 pounds of nuts and 10 pounds of chocolate in stock, and the owner does not want to buy more for this batch. How many barrels of each should be made for a maximum profit? Find the maximum profit.

Writing Exercises   *Write a paragraph using your own words.*

1. What is meant by the constraints of a linear program?

2. What is meant by a feasible solution of a linear program?

Something to Think About

1. Try to construct a linear programming problem. What difficulties do you encounter?

2. Try to construct a linear programming problem that will have a maximum at every point along an edge of the feasibility region.

Review Exercises   *Consider the line passing through P(−2, 4) and Q(5, 7).*

1. Find the slope of line *PQ*.

2. Write the equation of line *PQ* in general form.

3. Write the equation of line *PQ* in slope–intercept form.

4. Write an equation of the line passing through the origin that is parallel to line *PQ*.

■ ■ ■ ■ ■ ■ ■ ■ ■ PROBLEMS AND PROJECTS

1. Your employer provides two medical benefit plans. Plan A allows an unlimited number of doctor visits per year for $18.50 per month per person. In Plan B, you pay $6.50 per month per person plus $15 for each doctor visit after three visits in a given year. Which plan is best for you, and why?

   Plan B is best for 0–12 visits, then Plan A is best.

2. At the beginning of the semester, your English teacher gave you a formula for figuring your final average: The final average is 25% of the final exam grade, plus 10% of the homework average, plus 65% of the exam average, which includes the term paper (counting as two exam grades.) Before finals, your homework average is 88, your term paper grade is 78, and your exam grades are 82, 85, 73, and 69. If you want to receive a grade of B ($80 \le B \le 89$), what is the lowest grade you can make on the final exam?   83.3

3. The cost of making a new medicine is 6.5¢ per pill. As an experimental drug manufacturer, you can only sell the pills for 6¢ each. However, after distributing 5000 pills, you will receive from the FDA a subsidy of 10% of the amount you receive from the sales of all pills over 5000. Find the range of distribution that produces a profit.

   You must sell over 30,000 pills.

4. A heating company makes and installs duct work. To make and install duct work above ground costs $6 per linear foot; underground, it costs $12 per linear foot. It takes 1.6 hours to make and install duct work either above or below ground. The company has $2400 available for expenses and 360 hours of time. The company bookkeeper has figured that there is a profit of $3.40 per linear foot above ground and $2.80 per linear foot underground. If a job requires duct work both above and below ground, find the conditions for maximum profit. What would the maximum profit be?   50 ft above ground, 175 ft below ground; $660

PROJECT 1   A farmer is building a machine shed onto his barn, as shown in Illustration 1. The shed is to be 12 feet wide, and of course h_2 must be no more than 20 feet. In order for all of the shed to be useful for storing machinery, h_1 must be at least 6 feet. For the roof to shed rain and melting snow adequately, the slope of the roof must be at least $\frac{1}{2}$, but to be easily shingled, it must have a slope that is no greater than 1.

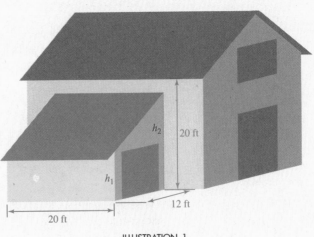

h_2 20 ft

h_1

12 ft

20 ft

ILLUSTRATION 1

a. Represent on a graph all of the possible values for h_1 and h_2, subject to the constraints listed above.

b. The farmer wishes to minimize the construction costs while still making sure that the shed is large enough for his purposes. He does this by setting a lower bound on the volume of the shed (3000 cubic feet) and then minimizing the surface area of the walls that must be built. The volume of the shed can be expressed in a formula that contains h_1 and h_2. Derive this formula and include the volume restriction in your design constraints. Then find the dimensions that will minimize the total area of the two ends of the shed and the outside wall. (The inner wall is already present as a wall of the barn and therefore involves no new cost.)

PROJECT 2   In Chapter 3, we saw that knowing any three points on the graph of a parabolic function is enough to determine the equation of that parabola. It follows that for any three points that could possibly lie on a parabola, there is exactly one parabola that passes through those points, and this parabola will have the equation $y = ax^2 + bx + c$ for appropriate a, b, and c.

a. In order for a set of three points to lie on the graph of a parabolic function, no two of the points can have the same x-coordinate, and not all three can have the same y-coordinate. Explain why we need these restrictions.

b. Suppose that the points $(1,3)$ and $(2,6)$ are on the graph of a parabola. What restrictions would have to be placed on a and b to guarantee that the y-intercept of the parabola has an absolute value of 4 or less? Can $(-1, 8)$ be a third point on such a parabola?

■ Chapter Summary

Key Words

absolute inequalities (4.1)

absolute value function (4.2)

compound inequalities (4.1)

conditional inequalities (4.1)

constraints (4.5)

feasible region (4.5)

feasible solution (4.5)

graph of a linear inequality (4.3)

inequality (4.1)

linear inequality (4.1-4.2)

linear programming (4.5)

objective function (4.5)

Key Ideas

(4.1) Trichotomy property:
$a < b$, $a = b$, or $a > b$

Transitive properties:
If $a < b$ and $b < c$, then $a < c$.
If $a > b$ and $b > c$, then $a > c$.
The relationships $\leq$ and $\geq$ are also transitive.

Properties of inequality:
If a and b are real numbers and $a < b$, then
$a + c < b + c$
$a - c < b - c$

$ac < bc \quad (c > 0)$
$ac > bc \quad (c < 0)$
$\frac{a}{c} < \frac{b}{c} \quad (c > 0)$
$\frac{a}{c} > \frac{b}{c} \quad (c < 0)$

Remember to change the order of the inequality when both sides are multiplied or divided by a negative number.

$c < x < d$ is equivalent to $c < x$ and $x < d$

(4.2) $\begin{cases} \text{If } x \geq 0, \text{ then } |x| = x. \\ \text{If } x < 0, \text{ then } |x| = -x. \end{cases}$

If $k > 0$, $|x| = k$ is equivalent to $x = k$ or $x = -k$

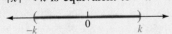

$|a| = |b|$ is equivalent to $a = b$ or $a = -b$

If $k > 0$, then

$\quad |x| < k$ is equivalent to $-k < x < k$

$|x| > k$ is equivalent to $x < -k$ or $x > k$

(4.3) To graph a linear inequality in x and y, graph the boundary line, and then use a test point to decide which side of the boundary should be shaded.

(4.4) Systems of inequalities can be solved by graphing.

(4.5) If a linear function, subject to the constraints of a system of linear inequalities in two variables, attains a maximum or a minimum value, that value will occur at a corner or along an entire edge of the region R that represents the solution of the system.

■ Chapter 4 Review Exercises

In Review Exercises 1–6, solve each inequality. Give each solution set in interval notation and graph it.

1. $5(x - 2) \leq 5$

2. $3x + 4 > 10$

3. $\frac{1}{3}y - 2 \geq \frac{1}{2}y + 2$

4. $\frac{7}{4}(x + 3) < \frac{3}{8}(x - 3)$

5. $3 < 3x + 4 < 10$

6. $4x > 3x + 2 > x - 3$

In Review Exercises 7–10, simplify each absolute value.

7. $|-7|$

8. $|8|$

9. $-|7|$

10. $-|-12|$

In Review Exercises 11–12, graph each function.

11. $f(x) = |x + 1| - 3$

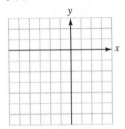

12. $f(x) = -2|x - 2| + 1$

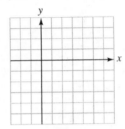

In Review Exercises 13–18, solve and check each equation.

13. $|3x + 1| = 10$

14. $\left| \frac{3}{2}x - 4 \right| = 9$

15. $\left|\dfrac{2-x}{3}\right| = 4$

16. $|3x + 2| = |2x - 3|$

17. $|5x - 4| = |4x - 5|$

18. $\left|\dfrac{3-2x}{2}\right| = \left|\dfrac{3x-2}{3}\right|$

In Review Exercises 19–24, solve each inequality. Give each solution in interval notation and graph it.

19. $|2x + 7| < 3$

20. $|3x - 8| \geq 4$

21. $\left|\dfrac{3}{2}x - 14\right| \geq 0$

22. $\left|\dfrac{2}{3}x + 14\right| < 0$

23. $|5 - 3x| \leq 14$

24. $\left|\dfrac{1-5x}{3}\right| > 7$

In Review Exercises 25–28, graph each inequality.

25. $2x + 3y > 6$

26. $y \leq 4 - x$

27. $-2 < x < 4$

28. $y \leq -2$ or $y > 1$

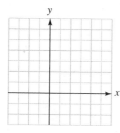
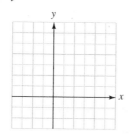
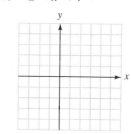
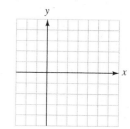

In Review Exercises 29–30, graph the solution set of each system of inequalities.

29. $\begin{cases} y \geq x + 1 \\ 3x + 2y < 6 \end{cases}$

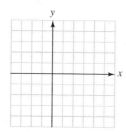

30. $\begin{cases} y \geq x^2 - 4 \\ y < x + 3 \end{cases}$

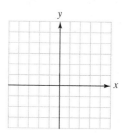

31. Maximize $P = 2x + y$ subject to $\begin{cases} x \geq 0 \\ y \geq 0 \\ x + y \leq 3 \end{cases}$.

32. A company manufactures fertilizers X and Y. Each 50-pound bag requires three ingredients, which are available in the limited quantities shown in the table. The profit on each bag of fertilizer X is \$6, and on each bag of Y, \$5. How many bags of each should be produced to maximize profit?

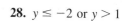

Ingredient	Number of pounds in fertilizer X	Number of pounds in fertilizer Y	Total number of pounds available
Nitrogen	6	10	20,000
Phosphorus	8	6	16,400
Potash	6	4	12,000

■ Chapter 4 Test

In Problems 1–2, graph the solution of each inequality. Also give the solution in interval notation.

1. $-2(2x + 3) \geq 14$

2. $-2 < \dfrac{x - 4}{3} < 4$

In Problems 3–4, write each expression without absolute value symbols.

3. $|5 - 8|$

4. $|4\pi - 4|$

In Problems 5–6, graph each function.

5. $f(x) = \dfrac{1}{2}|x|$

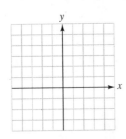

6. $f(x) = -|x + 2|$

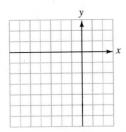

In Problems 7–10, solve each equation.

7. $|2x + 3| = 11$

8. $|4 - 3x| = 19$

9. $|3x + 4| = |x + 12|$

10. $|3 - 2x| = |2x + 3|$

In Problems 11–14, graph the solution of each inequality. Also give the solution in interval notation.

11. $|x + 3| \leq 4$

12. $|2x - 4| > 22$

13. $|4 - 2x| > 2$

14. $|2x - 4| \leq 2$

In Problems 15–16, graph each inequality.

15. $3x + 2y \geq 6$

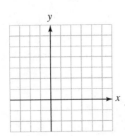

16. $-2 \leq y < 5$

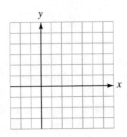

In Problems 17–18, use graphing to solve each system.

17. $\begin{cases} 2x - 3y \geq 6 \\ y \leq -x + 1 \end{cases}$

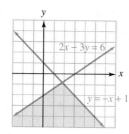

18. $\begin{cases} y \geq x^2 \\ y < x + 3 \end{cases}$

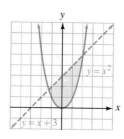

19. Maximize $P = 3x - y$ subject to $\begin{cases} y \geq 1 \\ y \leq 2 \\ y \leq 3x + 1 \\ x \leq 1 \end{cases}$ max is 2 at (1, 1)

◼ Cumulative Review Exercises (Chapters 1–4)

1. Draw a number line and graph the prime numbers from 50 to 60.

2. Find the additive inverse of -5.   5

In Exercises 3–4, let $A = [-5, 5)$ and $B = (-2, 7]$ and graph each interval.

3. $A \cup B$

4. $A \cap B$

In Exercises 5–6, evaluate each expression when $x = 2$ and $y = -4$.

5. $x - xy$ 10

6. $\dfrac{x^2 - y^2}{3x + y}$ -6

In Exercises 7–10, simplify each expression.

7. $(x^2 x^3)^2$ x^{10}

8. $(x^2)^3(x^4)^2$ x^{14}

9. $\left(\dfrac{x^3}{x^5}\right)^{-2}$ x^4

10. $\dfrac{a^2 b^n}{a^n b^2}$ $a^{2-n} b^{n-2}$

11. Write 32,600,000 in scientific notation. 3.26×10^7

12. Write 0.000012 in scientific notation. 1.2×10^{-5}

In Exercises 13–16, solve each equation, if possible.

13. $3x - 6 = 20$ $\frac{26}{3}$

14. $6(x - 1) = 2(x + 3)$ 3

15. $\dfrac{5b}{2} - 10 = \dfrac{b}{3} + 3$ 6

16. $2a - 5 = -2a + 4(a - 2) + 1$ impossible

In Exercises 17–18, tell whether the lines represented by the equations are parallel or perpendicular.

17. $3x + 2y = 12,\quad 2x - 3y = 5$ perpendicular

18. $3x = y + 4,\quad y = 3(x - 4) - 1$ parallel

19. Write the equation of the line passing through $P(-2, 3)$ and perpendicular to the graph of $3x + y = 8$.

20. Solve the formula $A = \dfrac{1}{2}h(b_1 + b_2)$ for h.

In Exercises 21–22, $f(x) = 3x^2 - x$. Find each value.

21. $f(2)$

22. $f(-2)$

23. Use graphing to solve $\begin{cases} 2x + y = 5 \\ x - 2y = 0 \end{cases}$.

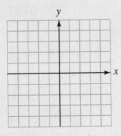

24. Use substitution to solve $\begin{cases} 3x + y = 4 \\ 2x - 3y = -1 \end{cases}$.

25. Use addition to solve $\begin{cases} x + 2y = -2 \\ 2x - y = 6 \end{cases}$.

26. Solve $\begin{cases} \dfrac{x}{10} + \dfrac{y}{5} = \dfrac{1}{2} \\ \dfrac{x}{2} - \dfrac{y}{5} = \dfrac{13}{10} \end{cases}$.

27. Solve $\begin{cases} x + y + z = 1 \\ 2x - y - z = -4 \\ x - 2y + z = 4 \end{cases}$

28. Solve $\begin{cases} x + 2y + 3z = 1 \\ 3x + 2y + z = -1 \\ 2x + 3y + z = -2 \end{cases}$.

29. Evaluate $\begin{vmatrix} 3 & -2 \\ 1 & -1 \end{vmatrix}$.

30. Evaluate $\begin{vmatrix} 2 & 3 & -1 \\ -1 & -1 & 2 \\ 4 & 1 & -1 \end{vmatrix}$.

In Exercises 31–32, use Cramer's rule to solve each system.

31. $\begin{cases} 4x - 3y = -1 \\ 3x + 4y = -7 \end{cases}$

32. $\begin{cases} x - 2y - z = -2 \\ 3x + y - z = 6 \\ 2x - y + z = -1 \end{cases}$

In Exercises 33–34, solve each inequality.

33. $-3(x - 4) \geq x - 32$

34. $-8 < -3x + 1 < 10$

In Exercises 35–36, solve each equation.

35. $|4x - 3| = 9$

36. $|2x - 1| = |3x + 4|$

In Exercises 37–38, solve each inequality.

37. $|3x - 2| \leq 4$

38. $|2x + 3| - 1 > 4$

In Exercises 39–40, use graphing to solve each inequality.

39. $2x - 3y \le 12$

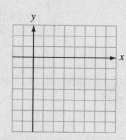

40. $3 > x \ge -2$

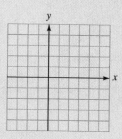

In Exercises 41–42, use graphing to solve each system of inequalities.

41. $\begin{cases} 3x - 2y < 6 \\ y < -x + 2 \end{cases}$

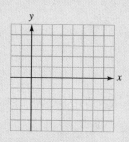

42. $\begin{cases} y < x + 2 \\ 3x + y \le 6 \end{cases}$

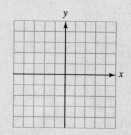

43. To conduct an experiment with mice and rats, a researcher will place the animals into both mazes for the number of minutes shown in Illustration 1. Find the greatest number of animals that can be used in this experiment.

	Time per mouse	Time per rat	Time available
Maze 1	12 min	8 min	240 min
Maze 2	10 min	15 min	300 min

ILLUSTRATION 1

Polynomials and Polynomial Functions

5

Computer Programmer

Computers process vast quantities of information rapidly and accurately when they are given programs to follow. Computer programmers write those programs, which logically list the steps the machine must follow to organize data, solve a problem, or do other tasks. Applications programmers are usually oriented toward business, engineering, or science. System programmers maintain the software that controls a computer system.

The need for computer programmers will remain strong as business, government, schools, and scientific organizations develop new applications and require improvements in the software they already use. In the entertainment field, computer games are booming, giving further opportunities for persons who program those games.

SAMPLE APPLICATION ■ Computers take more time to do multiplications than additions. To make a program run as quickly as possible, a computer programmer wants to write the polynomial $3x^4 + 2x^3 + 5x^2 + 7x + 1$ in a form that requires the fewest number of multiplications. Write the polynomial so that it contains only four multiplications.
See Example 10 in Section 5.4.

MATHEMATICS IN THE WORKPLACE

5.1 Polynomials and Polynomial Functions

■ DEGREE OF A POLYNOMIAL ■ POLYNOMIAL FUNCTIONS ■ GRAPHING POLYNOMIAL FUNCTIONS IN ONE VARIABLE ■ GRAPHING DEVICES

Algebraic terms are expressions that contain constants and/or variables. Some examples are

$$17, \qquad 9x, \qquad 15y^2, \qquad \text{and} \qquad -24x^4y^5$$

The *numerical coefficient* of 17 is 17. The numerical coefficients of the remaining terms are 9, 15, and -24, respectively.

A **polynomial** is the sum of one or more algebraic terms whose variables have whole-number exponents.

Polynomial in One Variable

A **polynomial in one variable** (say, x) is the sum of one or more terms of the form ax^n, where a is a real number and n is a whole number.

The following expressions are polynomials in x.

$$3x^2 + 2x, \qquad \frac{3}{2}x^5 - \frac{7}{3}x^4 - \frac{8}{3}x^3, \qquad \text{and} \qquad 19x^{20} + \sqrt{3}x^{14} + 4.5x^{11} - 17x^2$$

The following expressions are not polynomials.

$$\frac{2x}{x^2 + 1}, \qquad x^{1/2} - 1, \qquad \text{and} \qquad x^{-3} + 2x$$

The first expression is the quotient of two polynomials, and the last two have exponents that are not whole numbers.

Polynomial in Several Variables

A **polynomial in several variables** (say, x, y, and z) is the sum of one or more terms of the form $ax^m y^n z^p$, where a is a real number and m, n, and p are whole numbers.

The following expressions are polynomials in more than one variable.

$$3xy, \qquad 5x^2 y + 2yz^3 - 3xz, \qquad \text{and} \qquad u^2 v^2 w^2 + x^3 y^3 + 1$$

A polynomial with one term is called a **monomial**, a polynomial with two terms is called a **binomial**, and a polynomial with three terms is called a **trinomial**.

Monomials (one term)	Binomials (two terms)	Trinomials (three terms)
$2x^3$	$2x^4 + 5$	$2x^3 + 4x^2 + 3$
$a^2 b$	$-17t^{45} - 3xy$	$3mn^3 - m^2 n^3 + 7n$
$3x^3 y^5 z^2$	$32x^{13} y^5 + 47x^3 yz$	$-12x^5 y^2 + 13x^4 y^3 - 7x^3 y^3$

■ DEGREE OF A POLYNOMIAL

Degree of a Monomial

If $a \neq 0$, the **degree of ax^n** is n. The degree of a monomial containing several variables is the sum of the exponents on those variables.

EXAMPLE 1

a. $3x^4$ is a monomial of degree 4.

b. $-18x^3 y^2 z^{12}$ is a monomial of degree 17. Because the sum of the exponents on the variables is 17.

c. 4^7xy^4 is a monomial of degree 5.   Because the sum of the exponents on the variables is 5.

d. 3 is a monomial of degree 0. $3 = 3x^0$, and if $x \neq 0$, the degree of $3x^0$ is 0. ∎

WARNING!   Since $a \neq 0$ in the previous definition, 0 has no defined degree.

Degree of a Polynomial
The **degree of a polynomial** is the same as the degree of the term in the polynomial with largest degree.

EXAMPLE 2   **a.** $3x^5 + 4x^2 + 7$ is a trinomial of degree 5.   Because the largest degree of the three monomials is 5.

b. $7x^2y^8 - 3xy$ is a binomial of degree 10.

c. $3x + 2y - xy$ is a trinomial of degree 2.

d. $18x^2y^3 - 12x^7y^2 + 3x^9y^3 - 3$ is a polynomial of degree 12. ∎

If the exponents on the variable in a polynomial in one variable decrease as we move from left to right, we say that they are written in descending order. If the exponents increase as we move from left to right, we say that they are written in ascending order.

EXAMPLE 3   Write the exponents of $7x^2 - 5x^4 + 3x + 2x^3 - 1$ in   **a.** descending order   and   **b.** ascending order.

Solution   **a.** $-5x^4 + 2x^3 + 7x^2 + 3x - 1$ **b.** $-1 + 3x + 7x^2 + 2x^3 - 5x^4$ ∎

In the following polynomial, the exponents on x are in descending order, and the exponents on y are in ascending order.

$$7x^4 - 2x^3y + 4x^2y^2 - 8xy^3 + 12y^4$$

■ POLYNOMIAL FUNCTIONS

Polynomials in one variable (say, x) are often denoted by expressions such as

$$P(x)$$ Read $P(x)$ as "P of x."

where the letter within the parentheses represents the variable of the polynomial. Expressions such as

$$P(x) = x^6 + 4x^5 - 3x^2 + x - 2$$

are called **polynomial functions**. To evaluate a polynomial function at specific values of its variable, we use the same process that we used to evaluate functions. For example, to evaluate the polynomial function $P(x)$ at $x = 1$, we substitute 1 for x and simplify.

$$P(x) = x^6 + 4x^5 - 3x^2 + x - 2$$
$$P(1) = (1)^6 + 4(1)^5 - 3(1)^2 + 1 - 2$$
$$= 1 + 4 - 3 + 1 - 2$$
$$= 1$$

Note that to each number x there corresponds a single value $P(x)$.

EXAMPLE 4 (a, b)

Height of a rocket If a toy rocket is launched straight up with an initial velocity of 128 feet per second, its height h (in feet) above the ground after t seconds is given by the polynomial function

$$P(t) = -16t^2 + 128t \qquad \text{The height } h \text{ is the value } P(t).$$

Find the height of the rocket at   **a.** 0 seconds,   **b.** 3 seconds,   and **c.** 7.9 seconds.

Solution   **a.** To find the height at 0 seconds, we substitute 0 for t and simplify.

$$P(t) = -16t^2 + 128t$$
$$P(0) = -16(0)^2 + 128(0)$$
$$= 0$$

At 0 seconds, the rocket is on the ground waiting to be launched.

b. To find the height at 3 seconds, we substitute 3 for t and simplify.

$$P(3) = -16(3)^2 + 128(3)$$
$$= -16(9) + 384$$
$$= -144 + 384$$
$$= 240$$

At 3 seconds, the height of the rocket is 240 feet.

c. To find the height at 7.9 seconds, we substitute 7.9 for t and simplify.

$$P(7.9) = -16(7.9)^2 + 128(7.9)$$
$$= -16(62.41) + 1011.2 \qquad \text{Use a calculator.}$$
$$= -998.56 + 1011.2$$
$$= 12.64$$

At 7.9 seconds, the height of the rocket is only 12.64 feet. It has fallen nearly all the way back to earth. ∎

EXAMPLE 5

For the polynomial function $P(x) = 3x^2 - 2x + 7$, find   **a.** $P(a)$ and   **b.** $P(-2t)$.

Solution   In part **a**, we substitute a for x, and in part **b**, we substitute $-2t$ for x.

a. $P(a) = 3(a)^2 - 2(a) + 7$
$= 3a^2 - 2a + 7$

b. $P(-2t) = 3(-2t)^2 - 2(-2t) + 7$
$= 12t^2 + 4t + 7$ ∎

To evaluate polynomials with more than one variable, we substitute values for the variables in the polynomial and simplify.

EXAMPLE 6   Evaluate $4x^2y - 5xy^3$ at $x = 3$ and $y = -2$.

Solution   We substitute 3 for x and -2 for y and simplify.

$$4x^2y - 5xy^3 = 4(3)^2(-2) - 5(3)(-2)^3$$
$$= 4(9)(-2) - 5(3)(-8)$$
$$= -72 + 120$$
$$= 48$$

■

■ GRAPHING POLYNOMIAL FUNCTIONS IN ONE VARIABLE

In Example 4, we saw that a polynomial function can describe (or model) the flight of a rocket. Since the height (h) of the rocket depends on time (t), we say that the height is a function of time, and we can write $h = f(t)$. We can graph this polynomial function just as we graphed linear functions back in Chapter 2. We plot points and join them with a smooth curve.

EXAMPLE 7   Graph $h = f(t) = -16t^2 + 128t$.

Solution   From Example 4, we have seen that

When $t = 0$, then $h = 0$.
When $t = 3$, then $h = 240$.

These ordered pairs and others that satisfy the equation $h = -16t^2 + 128t$ are given in the table shown in Figure 5-1. Here the ordered pairs are pairs (t, h), where t is the time and h is the height.

We plot these pairs on a horizontal t-axis and a vertical h-axis and join the resulting points to get the parabola shown in the figure. From the figure, we can see that 4 seconds into the flight, the rocket attains a maximum height of 256 feet.

$$h = f(t) = -16t^2 + 128t$$

t	$h = f(t)$	(t, h)
0	0	$(0, 0)$
1	112	$(1, 112)$
2	192	$(2, 192)$
3	240	$(3, 240)$
4	256	$(4, 256)$
5	240	$(5, 240)$
6	192	$(6, 192)$
7	112	$(7, 112)$
8	0	$(8, 0)$

FIGURE 5-1

■

WARNING!   The parabola shown in Figure 5-1 describes the height of the rocket in relation to time. It does not show the path of the rocket. The rocket goes straight up and then comes straight down.

■ GRAPHING DEVICES

We can easily graph polynomial functions with a graphing device. For example, to graph the polynomial function $P(t) = -16t^2 + 128t$, we can use settings of $[0, 8]$ for x and $[0, 260]$ for y to get the parabola shown in Figure 5-2(a).

We can trace to find the height of the rocket at any number of seconds. Figure 5-2(b) shows that the height of the rocket at 1.6170213 seconds is about 165 feet.

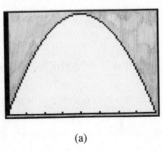

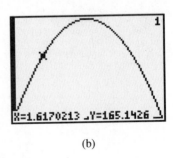

(a)   (b)

FIGURE 5-2

Orals   *Give the degree of each polynomial.*

1. $3x^3$ **2.** $-4x^3y$ **3.** $4x^2y + 2x$ **4.** $3x + 4xyz$

If $P(x) = 2x + 1$, find each value.

5. $P(0)$ **6.** $P(2)$ **7.** $P(-1)$ **8.** $P(-2)$

EXERCISE 5.1

In Exercises 1–8, classify each polynomial as a monomial, a binomial, a trinomial, or none of these.

1. $3x^2$ **2.** $2y^3 + 4y^2$ **3.** $3x^2y - 2x + 3y$ **4.** $a^2 + b^2$

5. $x^2 - y^2$ **6.** $\dfrac{17}{2}x^3 + 3x^2 - x - 4$ **7.** 5 **8.** $8x^3y^5$

In Exercises 9–16, find the degree of each polynomial.

9. $3x^2 + 2$ **10.** $x^{16}y$ **11.** $4x^8 + 3x^2y^4$ **12.** $19x^2y^4 - y^{10}$

13. $4x^2 - 5y^3z^3t^4$ **14.** $7x$ **15.** 121 **16.** $x^2y^3z^4 + z^{12}$

In Exercises 17–20, write each polynomial with the exponents on x in descending order.

17. $3x - 2x^4 + 7 - 5x^2$ **18.** $-x^2 + 3x^5 - 7x + 3x^3$

19. $a^2x - ax^3 + 7a^3x^5 - 5a^3x^2$

20. $4x^2y^7 - 3x^5y^2 + 4x^3y^3 - 2x^4y^6 + 5x^6$

In Exercises 21–24, write each polynomial with the exponents on y in ascending order.

21. $4y^2 - 2y^5 + 7y - 5y^3$ **22.** $y^3 + 3y^2 + 8y^4 - 2$

23. $5x^3y^6 + 2x^4y - 5x^3y^3 + x^5y^7 - 2y^4$

24. $-x^3y^2 + x^2y^3 - 2x^3y + x^7y^6 - 3x^6$

In Exercises 25–28, consider the polynomial $P(x) = 2x^2 + x + 2$. *Find each value.*

25. $P(0)$ **26.** $P(1)$ **27.** $P(-2)$ **28.** $P(-3)$

In Exercises 29–32, the height h, in feet, of a ball shot straight up with an initial velocity of 64 feet per second is given by the polynomial $h = f(t) = -16t^2 + 64t$. *Find the height of the ball after the given number of seconds.*

29. 0 second   **30.** 1 second   **31.** 2 seconds   **32.** 4 seconds

In Exercises 33–36, the number of feet that a car travels before stopping depends on the driver's reaction time and the braking distance. (See Illustration 1.) For one driver, the stopping distance d is given by the polynomial function $d = f(v) = 0.04v^2 + 0.9v$, *where v is the velocity of the car. Find the stopping distances for each of the following speeds.*

33. 30 mph
34. 50 mph
35. 60 mph
36. 70 mph

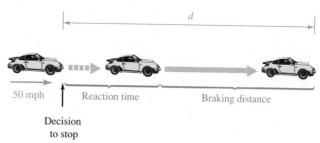

ILLUSTRATION 1

In Exercises 37–40, a rectangular sheet of metal will be used to make a rain gutter by bending up its sides, as shown in Illustration 2. Since a cross section is a rectangle, the cross-sectional area is the product of its length and width, so the capacity, c, of the gutter is a polynomial function of x: $c = f(x) = -2x^2 + 12x$. *Find the capacity for each value of x.*

37. 1 inch
38. 2 inches
39. 3 inches
40. 4 inches

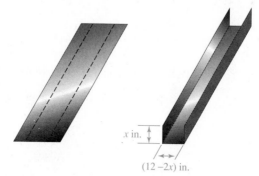

ILLUSTRATION 2

In Exercises 41–48, find each value when $x = 2$ *and* $y = -3$.

41. $x^2 + y^2$ **42.** $x^3 + y^3$ **43.** $x^3 - y^3$ **44.** $x^2 - y^2$

45. $3x^2y + xy^3$ **46.** $8xy - xy^2$ **47.** $-2xy^2 + x^2y$ **48.** $-x^3y - x^2y^2$

In Exercises 49–54, use a calculator to find each value when $x = 3.7$, $y = -2.5$, *and* $z = 8.9$.

49. x^2y **50.** xyz^2 **51.** $\dfrac{x^2}{z^2}$ **52.** $\dfrac{z^3}{y^2}$

53. $\dfrac{x + y + z}{xyz}$ **54.** $\dfrac{x + yz}{xy + z}$

In Exercises 55–62, graph each polynomial function. Check your work with a graphing device.

55. $f(x) = x^2$

56. $f(x) = x^3$

57. $f(x) = -x^2 + 2$

58. $f(x) = \dfrac{1}{2}x^2 - 1$

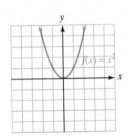
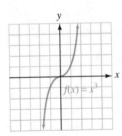
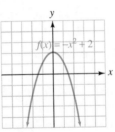
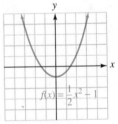

59. $f(x) = -x^3 + x$

60. $f(x) = x^3 - x$

61. $f(x) = x^2 - 2x + 1$

62. $f(x) = x^2 - 2x - 3$

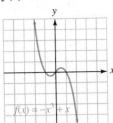
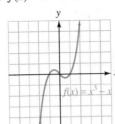
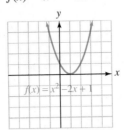
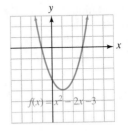

In Exercises 63–66, use a graphing device to graph each polynomial function. Use settings of $[-4, 6]$ for x and $[-5, 5]$ for y.

63. $f(x) = 2.75x^2 - 4.7x + 1.5$

64. $f(x) = -2.5x^2 + 1.7x + 3.2$

65. $f(x) = 0.25x^2 - 0.5x - 2.5$

66. $f(x) = 0.37x^2 - 1.4x + 1.5$

Writing Exercises *Write a paragraph using your own words.*

1. Explain how to find the degree of a polynomial.

2. Explain why
$P(x) = x^6 + 4x^5 - 3x^2 + x - 2$
defines a function.

Something to Think About

1. If $P(x) = x^2 - 5x$, is
$P(2) + P(3) = P(2 + 3)$?   no

2. If $P(x) = x^3 - 3x$, is
$P(2) - P(3) = P(2 - 3)$?   no

3. If $P(x) = x^2 - 2x - 3$, find $P(P(0))$.
12

4. If $P(x) = 2x^2 - x - 5$, find
$P(P(-1))$.   5

Review Exercises *Write each expression with a single exponent.*

1. $a^3 a^2$ a^5

2. $\dfrac{b^3 b^3}{b^4}$ b^2

3. $\dfrac{3(y^3)^{10}}{y^3 y^4}$ $3y^{23}$

4. $\dfrac{4x^{-4}x^5}{2x^{-6}}$ $2x^7$

5. The distance from Mars to the sun is about 114,000,000 miles. Express this number in scientific notation.
1.14×10^8

6. One angstrom is about 0.0000001 millimeter. Express this number in scientific notation.
1×10^{-7}

5.2 Adding and Subtracting Polynomials

■ COMBINING LIKE TERMS ■ ADDING POLYNOMIALS ■ SUBTRACTING POLYNOMIALS ■ ADDING
AND SUBTRACTING MULTIPLES OF POLYNOMIALS

To add and subtract polynomials, we often have to combine like terms.

■ COMBINING LIKE TERMS

We have seen that terms with the same variables that have the same exponents are
called **like** or **similar terms**.

EXAMPLE 1

a. $3x^2$, $5x^2$, and $7x^2$ are like terms, because they have the same variables with the
same exponents.

b. $5x^3y^2$, $17x^3y^2$, and $103x^3y^2$ are like terms, because they have the same variables with the same exponents.

c. $4x^4y^2$, $12xy^5$, and $98x^7y^9$ are not like terms. They have the same variables, but
with different exponents.

d. $3x^4y$ and $5x^4z^2$ are not like terms. They have different variables. ■

We have also seen that the distributive property enables us to combine like
terms.

EXAMPLE 2

a. $3x + 7x = (3 + 7)x$ Use the distributive property.

$\qquad = 10x$ $3 + 7 = 10.$

b. $5x^2y^3 + 22x^2y^3 = (5 + 22)x^2y^3$ Use the distributive property.

$\qquad = 27x^2y^3$ $5 + 22 = 27.$

c. $9xy^4 + 6xy^4 + xy^4 = 9xy^4 + 6xy^4 + 1xy^4$ $xy^4 = 1xy^4.$

$\qquad = (9 + 6 + 1)xy^4$ Use the distributive property.

$\qquad = 16xy^4$ $9 + 6 + 1 = 16.$ ■

The results of Example 2 suggest that to add like terms, *we add their numerical coefficients and keep the same variables with the same exponents.*

 WARNING!   The terms in the following binomials cannot be combined, because
they are not like terms.

$$3x^2 - 5y^2, \qquad -2a^2 + 3a^3, \qquad \text{and} \qquad 5y^2 + 17xy$$

■ ADDING POLYNOMIALS

To add polynomials, we use the distributive property to remove parentheses and
combine like terms, whenever possible.

EXAMPLE 3   Add $3x^2 - 2x + 4$ and $2x^2 + 4x - 3$.

Solution
$$(3x^2 - 2x + 4) + (2x^2 + 4x - 3)$$

$$= 1(3x^2 - 2x + 4) + 1(2x^2 + 4x - 3)$$ Each polynomial has an understood coefficient of 1.

$$= 3x^2 - 2x + 4 + 2x^2 + 4x - 3$$ Use the distributive property to remove parentheses.

$$= 3x^2 + 2x^2 - 2x + 4x + 4 - 3$$ Use the commutative property of addition to get the terms involving x^2 together and the terms involving x together.

$$= 5x^2 + 2x + 1$$ Combine like terms. ∎

EXAMPLE 4   Add $5x^3y^2 + 4x^2y^3$ and $-2x^3y^2 + 5x^2y^3$.

Solution
$$(5x^3y^2 + 4x^2y^3) + (-2x^3y^2 + 5x^2y^3) = 1(5x^3y^2 + 4x^2y^3) + 1(-2x^3y^2 + 5x^2y^3)$$
$$= 5x^3y^2 + 4x^2y^3 - 2x^3y^2 + 5x^2y^3$$
$$= 5x^3y^2 - 2x^3y^2 + 4x^2y^3 + 5x^2y^3$$
$$= 3x^3y^2 + 9x^2y^3$$ ∎

The additions in Examples 3 and 4 can be done by aligning the terms vertically and combining the like terms in each column.

$$
\begin{array}{ll}
3x^2 - 2x + 4 & 5x^3y^2 + 4x^2y^3 \\
\underline{2x^2 + 4x - 3} & \underline{-2x^3y^2 + 5x^2y^3} \\
5x^2 + 2x + 1 & 3x^3y^2 + 9x^2y^3
\end{array}
$$

■ SUBTRACTING POLYNOMIALS

To subtract one monomial from another, we add the negative (or opposite) of the monomial that is to be subtracted.

EXAMPLE 5   **a.** $8x^2 - 3x^2 = 8x^2 + (-3x^2)$ **b.** $3x^2y - 9x^2y = 3x^2y + (-9x^2y)$
$$\qquad\qquad = 5x^2 \qquad\qquad\qquad\qquad\qquad = -6x^2y$$

c. $-5x^5y^3z^2 - 3x^5y^3z^2 = -5x^5y^3z^2 + (-3x^5y^3z^2)$
$$\qquad\qquad\qquad = -8x^5y^3z^2$$ ∎

To subtract polynomials, we use the distributive property to remove parentheses and combine like terms, whenever possible.

EXAMPLE 6   **a.** $(8x^3y + 2x^2y) - (2x^3y - 3x^2y)$

$$= 1(8x^3y + 2x^2y) - 1(2x^3y - 3x^2y)$$ Insert the understood coefficients of 1.

$$= 8x^3y + 2x^2y - 2x^3y + 3x^2y$$ Use the distributive property to remove parentheses.

$$= 6x^3y + 5x^2y$$ Combine like terms.

b. $(3rt^2 + 4r^2t^2) - (8rt^2 - 4r^2t^2 + r^3t^2)$

$= 1(3rt^2 + 4r^2t^2) - 1(8rt^2 - 4r^2t^2 + r^3t^2)$ Insert the understood coefficients of 1.

$= 3rt^2 + 4r^2t^2 - 8rt^2 + 4r^2t^2 - r^3t^2$ Use the distributive property to remove parentheses.

$= -5rt^2 + 8r^2t^2 - r^3t^2$ Combine like terms. ∎

To subtract polynomials in vertical form, we add the negative (or opposite) of the polynomial that is being subtracted.

$$
\begin{array}{r}
8x^3y + 2x^2y \\
- \quad 2x^3y - 3x^2y
\end{array}
\Rightarrow
\begin{array}{r}
8x^3y + 2x^2y \\
+ \quad -2x^3y + 3x^2y \\
\hline
6x^3y + 5x^2y
\end{array}
$$

ADDING AND SUBTRACTING MULTIPLES OF POLYNOMIALS

To add multiples of one polynomial to another, or subtract multiples of one polynomial from another, we use the distributive property to remove parentheses and combine like terms.

EXAMPLE 7   Simplify $3(2x^2 + 4x - 7) - 2(3x^2 - 4x - 5)$.

Solution $3(2x^2 + 4x - 7) - 2(3x^2 - 4x - 5) = 6x^2 + 12x - 21 - 6x^2 + 8x + 10$

$= 20x - 11$ Combine like terms. ∎

Orals   *Combine like terms.*

1. $4x^2 + 5x^2$ **2.** $3y^2 - 5y^2$

Do the operations.

3. $(x^2 + 2x + 1) + (2x^2 - 2x + 1)$
4. $(x^2 + 2x + 1) - (2x^2 - 2x + 1)$
5. $(2x^2 - x - 3) + (x^2 - 3x - 1)$
6. $(2x^2 - x - 3) - (x^2 - 3x - 1)$

EXERCISE 5.2

In Exercises 1–8, tell whether the terms are like or unlike terms. If they are like terms, combine them.

1. $3x, 7x$ **2.** $-8x, 3y$ **3.** $7x, 7y$ **4.** $3mn, 5mn$

5. $3r^2t^3, -8r^2t^3$ **6.** $9u^2v, 10u^2v$ **7.** $9x^2y^3, 3x^2y^2$ **8.** $27x^6y^4z, 8x^6y^4z^2$

In Exercises 9–16, simplify each expression.

9. $8x + 4x$ **10.** $-2y + 16y$ **11.** $5x^3y^2z - 3x^3y^2z$ **12.** $8wxy - 12wxy$

13. $-2x^2y^3 + 3xy^4 - 5x^2y^3$

14. $3ab^4 - 4a^2b^2 - 2ab^4 + 2a^2b^2$

15. $(3x^2y)^2 + 2x^4y^2 - x^4y^2$

16. $(5x^2y^4)^3 - (5x^3y^6)^2$

In Exercises 17–28, do each operation.

17. $(3x^2 + 2x + 1) + (-2x^2 - 7x + 5)$

18. $(-2a^2 - 5a - 7) + (-3a^2 + 7a + 1)$

19. $(-a^2 + 2a + 3) - (4a^2 - 2a - 1)$

20. $(x^2 - 3x + 8) - (3x^2 + x + 3)$

21. $(7y^3 + 4y^2 + y + 3) + (-8y^3 - y + 3)$

22. $(6x^3 + 3x - 2) - (2x^3 + 3x^2 + 5)$

23. $(3x^2 + 4x - 3) + (2x^2 - 3x - 1) - (x^2 + x + 7)$

24. $(-2x^2 + 6x + 5) - (-4x^2 - 7x + 2) - (4x^2 + 10x + 5)$

25. $(3x^3 - 2x + 3) + (4x^3 + 3x^2 - 2) + (-4x^3 - 3x^2 + x + 12)$

26. $(x^4 - 3x^2 + 4) + (-2x^4 - x^3 + 3x^2) + (3x^2 + 2x + 1)$

27. $(3y^2 - 2y + 4) + [(2y^2 - 3y + 2) - (y^2 + 4y + 3)]$

28. $(-t^2 - t - 1) - [(t^2 + 3t - 1) - (-2t^2 + 4)]$

In Exercises 29–32, add the polynomials.

29.
$$\begin{array}{r} 3x^3 - 2x^2 + 4x - 3 \\ -2x^3 + 3x^2 + 3x - 2 \\ \hline 5x^3 - 7x^2 + 7x - 12 \end{array}$$

30.
$$\begin{array}{r} 7a^3 \qquad + 3a + 7 \\ -2a^3 + 4a^2 \qquad - 13 \\ 3a^3 - 3a^2 + 4a + 5 \\ \hline \end{array}$$

31.
$$\begin{array}{r} -2y^4 - 2y^3 + 4y^2 - 3y + 10 \\ -3y^4 + 7y^3 - y^2 + 14y - 3 \\ - 3y^3 - 5y^2 - 5y + 7 \\ -4y^4 + y^3 - 13y^2 + 14y - 2 \\ \hline \end{array}$$

32.
$$\begin{array}{r} 17t^4 + 3t^3 - 2t^2 - 3t + 4 \\ -12t^4 - 2t^3 + 3t^2 - 5t - 17 \\ -2t^4 - 7t^3 + 4t^2 + 12t - 5 \\ 5t^4 + t^3 + 5t^2 - 13t + 12 \\ \hline \end{array}$$

In Exercises 33–36, subtract the bottom polynomial from the top polynomial.

33.
$$\begin{array}{r} 3x^2 - 4x + 17 \\ 2x^2 + 4x - 5 \\ \hline \end{array}$$

34.
$$\begin{array}{r} -2y^2 - 4y + 3 \\ 3y^2 + 10y - 5 \\ \hline \end{array}$$

35.
$$\begin{array}{r} -5y^3 + 4y^2 - 11y + 3 \\ -2y^3 - 14y^2 + 17y - 32 \\ \hline \end{array}$$

36.
$$\begin{array}{r} 17x^4 - 3x^2 - 65x - 12 \\ 23x^4 + 14x^2 + 3x - 23 \\ \hline \end{array}$$

In Exercises 37–50, simplify each expression.

37. $3(x + 2) + 2(x - 5)$

38. $-2(x - 4) + 5(x + 1)$

39. $-6(t - 4) - 5(t - 1)$

40. $4(a + 5) - 3(a - 1)$

41. $2(x^3 + x^2) + 3(2x^3 - x^2)$

42. $3(y^2 + 2y) - 4(y^2 - 4)$

43. $-3(2m - n) + 2(m - 3n)$

44. $5(p - 2q) - 4(2p + q)$

45. $-5(2x^3 + 7x^2 + 4x) - 2(3x^3 - 4x^2 - 4x)$

46. $-3(3a^2 + 4b^3 + 7) + 4(5a^2 - 2b^3 + 3)$

47. $4(3z^2 - 4z + 5) + 6(-2z^2 - 3z + 4) - 2(4z^2 + 3z - 5)$

48. $-3(4x^3 - 2x^2 + 4) - 4(3x^3 + 4x^2 + 3x) + 5(3x - 4)$

49. $5(2a^2 + 4a - 2) - 2(-3a^2 - a + 12) - 2(a^2 + 3a - 5)$

50. $-2(2b^2 - 3b + 3) + 3(3b^2 + 2b - 8) - (3b^2 - b + 4)$

Writing Exercises   *Write a paragraph using your own words.*

1. Explain why the terms x^2y and xy^2 are not like terms.

2. Explain how to recognize like terms and how to add them.

Something to Think About

1. Find the difference when $3x^2 + 4x - 3$ is subtracted from the sum of $-2x^2 - x + 7$ and $5x^2 + 3x - 1$.

2. Find the difference when $8x^3 + 2x^2 - 1$ is subtracted from the sum of $x^2 + x + 2$ and $2x^3 - x + 9$.

3. Find the sum when $2x^2 - 4x + 3$ minus $8x^2 + 5x - 3$ is added to $-2x^2 + 7x - 4$.

4. Find the sum when $7x^3 - 4x$ minus $x^2 + 2$ is added to $5 + 3x$.

Review Exercises   *Solve each inequality. Give the result in interval notation.*

1. $2x + 3 \leq 11$

2. $\dfrac{2}{3}x + 5 > 11$

3. $|x - 4| < 5$

4. $|2x + 1| \geq 7$

5.3 Multiplying Polynomials

■ MULTIPLYING MONOMIALS ■ MULTIPLYING A POLYNOMIAL BY A MONOMIAL ■ MULTIPLYING A POLYNOMIAL BY A POLYNOMIAL ■ THE FOIL METHOD ■ SPECIAL PRODUCTS

In this section, we will discuss how to multiply polynomials.

■ MULTIPLYING MONOMIALS

In Section 1.3, we saw that to multiply one monomial by another, *we multiply the numerical factors and then multiply the variable factors.*

 (a)

EXAMPLE 1   We can use the commutative and associative properties of multiplication to re-arrange the terms and regroup the numbers.

a. $(3x^2)(6x^3) = 3 \cdot x^2 \cdot 6 \cdot x^3$

$\qquad\qquad\quad = (3 \cdot 6)(x^2 \cdot x^3)$

$\qquad\qquad\quad = 18x^5$

b. $(-8x)(2y)(xy) = -8 \cdot x \cdot 2 \cdot y \cdot x \cdot y$

$\qquad\qquad\qquad\quad = (-8 \cdot 2) \cdot x \cdot x \cdot y \cdot y$

$\qquad\qquad\qquad\quad = -16x^2y^2$

c. $(2a^3b)(-7b^2c)(-12ac^4) = 2 \cdot a^3 \cdot b \cdot (-7) \cdot b^2 \cdot c \cdot (-12) \cdot a \cdot c^4$

$\qquad\qquad\qquad\qquad\qquad\quad = 2(-7)(-12) \cdot a^3 \cdot a \cdot b \cdot b^2 \cdot c \cdot c^4$

$\qquad\qquad\qquad\qquad\qquad\quad = 168a^4b^3c^5$

■

■ MULTIPLYING A POLYNOMIAL BY A MONOMIAL

To multiply a polynomial by a monomial, *we use the distributive property and multiply each term of the polynomial by the monomial.*

EXAMPLE 2   **a.** $3x^2(6xy + 3y^2) = 3x^2 \cdot 6xy + 3x^2 \cdot 3y^2$ Use the distributive property.

$$= 18x^3y + 9x^2y^2$$

b. $5x^3y^2(xy^3 - 2x^2y) = 5x^3y^2 \cdot xy^3 - 5x^3y^2 \cdot 2x^2y$
$$= 5x^4y^5 - 10x^5y^3$$

c. $-2ab^2(3bz - 2az + 4z^3) = -2ab^2 \cdot 3bz - (-2ab^2) \cdot 2az + (-2ab^2) \cdot 4z^3$
$$= -6ab^3z + 4a^2b^2z - 8ab^2z^3$$ ■

■ MULTIPLYING A POLYNOMIAL BY A POLYNOMIAL

To multiply a polynomial by another polynomial, we use the distributive property repeatedly.

EXAMPLE 3   **a.** $(3x + 2)(4x + 9) = (3x + 2) \cdot 4x + (3x + 2) \cdot 9$ Use the distributive property.

$$= 12x^2 + 8x + 27x + 18$$ Use the distributive property again.

$$= 12x^2 + 35x + 18$$ Combine like terms.

b. $(2a - b)(3a^2 - 4ab + b^2) = (2a - b)3a^2 - (2a - b)4ab + (2a - b)b^2$
$$= 6a^3 - 3a^2b - 8a^2b + 4ab^2 + 2ab^2 - b^3$$
$$= 6a^3 - 11a^2b + 6ab^2 - b^3$$ ■

The results of Example 3 suggest that to multiply one polynomial by another, *we multiply each term of one polynomial by each term of the other polynomial and combine like terms, when possible.*

In the next example, we organize the work vertically.

EXAMPLE 4   **a.**
$$
\begin{array}{r}
3x + 2 \\
4x + 9 \\
\hline
\end{array}
$$

$4x(3x + 2) \longrightarrow 12x^2 + 8x$

$9(3x + 2) \longrightarrow + 27x + 18$

$ \overline{12x^2 + 35x + 18}$

b.
$$
\begin{array}{r}
3a^2 - 4ab + b^2 \\
2a - b \\
\hline
\end{array}
$$

$2a(3a^2 - 4ab + b^2) \longrightarrow 6a^3 - 8a^2b + 2ab^2$

$-b(3a^2 - 4ab + b^2) \longrightarrow - 3a^2b + 4ab^2 - b^3$

$ \overline{6a^3 - 11a^2b + 6ab^2 - b^3}$ ■

THE FOIL METHOD

When multiplying two binomials, the use of the distributive property requires that each term of one binomial be multiplied by each term of the other binomial. This fact can be emphasized by drawing arrows to show the indicated products. For example, to multiply $3x + 2$ and $x + 4$, we can write

First terms — Last terms

$$
\begin{aligned}
(3x + 2)(x + 4) &= 3x \cdot x + 3x \cdot 4 + 2 \cdot x + 2 \cdot 4 \\
&= 3x^2 + 12x + 2x + 8 \\
&= 3x^2 + 14x + 8 \qquad \text{Combine like terms.}
\end{aligned}
$$

Inner terms
Outer terms

We note that

- the product of the **First** terms is $3x^2$,
- the product of the **Outer** terms is $12x$,
- the product of the **Inner** terms is $2x$, and
- the product of the **Last** terms is 8.

This scheme is often called the **FOIL** method of multiplying two binomials. FOIL is an acronym for **First** terms, **Outer** terms, **Inner** terms, and **Last** terms. Of course, all like terms in the resulting product must be combined.

It is easy to multiply binomials by sight if we use the FOIL method. We find the product of the first terms, then find the products of the outer terms and the inner terms and add them (when possible), and then find the product of the last terms.

EXAMPLE 5   Find the products by sight: **a.** $(2x - 3)(3x + 2)$ and **b.** $(3x + 1)(3x + 4)$, and **c.** $(4x - y)(2x + 3y)$.

Solution **a.** $(2x - 3)(3x + 2) = 6x^2 - 5x - 6$

The middle term of $-5x$ in the result comes from combining the outer and inner products of $+4x$ and $-9x$: $4x + (-9x) = -5x$.

b. $(3x + 1)(3x + 4) = 9x^2 + 15x + 4$

The middle term $+15x$ in the result comes from combining the products $+12x$ and $+3x$: $12x + 3x = 15x$.

c. $(4x - y)(2x + 3y) = 8x^2 + 10xy - 3y^2$

The middle term $+10xy$ in the result comes from combining the products $+12xy$ and $-2xy$: $12xy - 2xy = 10xy$. ∎

SPECIAL PRODUCTS

It is easy to square a binomial by using the FOIL method.

EXAMPLE 6 Find each square: **a.** $(x + y)^2$ and **b.** $(x - y)^2$.

Solution We multiply each term of one binomial by each term of the other binomial, and then combine like terms.

a. $(x + y)^2 = (x + y)(x + y)$

$= x^2 + xy + xy + y^2$ Use the FOIL method.

$= x^2 + 2xy + y^2$ Combine like terms.

We see that the square of the binomial is the square of the first term, plus twice the product of the terms, plus the square of the last term. This product can be illustrated graphically as shown in Figure 5-3.

The area of the large square is the product of its length and width:

$$(x + y)(x + y) = (x + y)^2 .$$

The area of the large square is also the sum of its four pieces:

$$x^2 + xy + xy + y^2 = x^2 + 2xy + y^2$$

So we have $(x + y)^2 = x^2 + 2xy + y^2$.

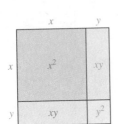

FIGURE 5-3

b. $(x - y)^2 = (x - y)(x - y)$

$= x^2 - xy - xy + y^2$ Use the FOIL method.

$= x^2 - 2xy + y^2$ Combine like terms.

We see that the square of the binomial is the square of the first term, minus twice the product of the terms, plus the square of the last term.

For a geometric interpretation, see Exercise 113.   ◼

EXAMPLE 7 Multiply: $(x + y)(x - y)$.

Solution $(x + y)(x - y) = x^2 - xy + xy - y^2$ Use the FOIL method.

$= x^2 - y^2$ Combine like terms.

From this example, we see that the product of the sum of two quantities and the difference of two quantities is the square of the first quantity minus the square of the second quantity.

For a geometric interpretation, see Exercise 114.   ◼

The products discussed in Examples 6 and 7 are called **special products**. Because they occur so often, it is useful to learn their forms.

Special Product Formulas

$(x + y)^2 = (x + y)(x + y) = x^2 + 2xy + y^2$

$(x - y)^2 = (x - y)(x - y) = x^2 - 2xy + y^2$

$(x + y)(x - y) = x^2 - y^2$

Because $x^2 + 2xy + y^2 = (x + y)^2$ and $x^2 - 2xy + y^2 = (x - y)^2$, the two trinomials are called **perfect square trinomials**.

WARNING! The squares $(x + y)^2$ and $(x - y)^2$ have trinomials for their products. Don't forget to write the middle terms in these products. Remember that

$$(x + y)^2 \neq x^2 + y^2 \qquad \text{and that} \qquad (x - y)^2 \neq x^2 - y^2$$

Also remember that the product $(x + y)(x - y)$ is the binomial $x^2 - y^2$.

At first, the expression $3[x^2 - 2(x + 3)]$ doesn't look like a polynomial. However, if we remove the parentheses and the brackets, it takes on the form of a polynomial.

$$3[x^2 - 2(x + 3)] = 3[x^2 - 2x - 6]$$
$$= 3x^2 - 6x - 18$$

If an expression has one set of grouping symbols that is enclosed within another set, we always eliminate the inner set first.

EXAMPLE 8   Find the product of $-2[y^3 + 3(y^2 - 2)]$ and $5[y^2 - 2(y + 1)]$.

Solution   We change each expression into polynomial form

$$\begin{array}{ll} -2[y^3 + 3(y^2 - 2)] & 5[y^2 - 2(y + 1)] \\ -2(y^3 + 3y^2 - 6) & 5(y^2 - 2y - 2) \\ -2y^3 - 6y^2 + 12 & 5y^2 - 10y - 10 \end{array}$$

and then do the multiplication:

$$
\begin{array}{r}
-2y^3 - 6y^2 + 12 \\
5y^2 - 10y - 10 \\
\hline
-10y^5 - 30y^4 \qquad\qquad + 60y^2 \\
+ 20y^4 + 60y^3 \qquad\qquad - 120y \\
+ 20y^3 + 60y^2 \qquad\qquad - 120 \\
\hline
-10y^5 - 10y^4 + 80y^3 + 120y^2 - 120y - 120
\end{array}
$$

The following examples show how to multiply expressions that are not polynomials.

EXAMPLE 9   Find the product of $x^{-2} + y$ and $x^2 - y^{-2}$.

Solution   We multiply each term of the second expression by each term of the first expression, and then we simplify.

$$(x^{-2} + y)(x^2 - y^{-2}) = x^{-2}x^2 - x^{-2}y^{-2} + yx^2 - yy^{-2}$$
$$= x^{-2+2} - x^{-2}y^{-2} + yx^2 - y^{1+(-2)}$$
$$= x^0 - \frac{1}{x^2 y^2} + x^2 y - y^{-1}$$
$$= 1 - \frac{1}{x^2 y^2} + x^2 y - \frac{1}{y}$$

EXAMPLE 10 Find the product of $x^n + 2x$ and $x^n + 3x^{-n}$.

Solution We use the FOIL method to multiply each term of the second expression by each term of the first expression, and then we simplify:

$$(x^n + 2x)(x^n + 3x^{-n}) = x^n x^n + x^n(3x^{-n}) + 2x(x^n) + 2x(3x^{-n})$$
$$= x^{n+n} + 3x^{n+(-n)} + 2x^{1+n} + 6xx^{-n}$$
$$= x^{2n} + 3x^0 + 2x^{n+1} + 6x^{1+(-n)}$$
$$= x^{2n} + 3 + 2x^{n+1} + 6x^{1-n} \qquad \blacksquare$$

Orals *Find each product.*

1. $(-2a^2b)(3ab^2)$
2. $(4xy^2)(-2xy)$
3. $3a^2(2a - 1)$
4. $-4n^2(4m - n)$
5. $(x + 1)(2x + 1)$
6. $(3y - 2)(2y + 1)$

EXERCISE 5.3

In Exercises 1–30, find each product.

1. $(2a^2)(-3ab)$
2. $(-3x^2y)(3xy)$
3. $(-3ab^2c)(5ac^2)$
4. $(-2m^2n)(-4mn^3)$
5. $(4a^2b)(-5a^3b^2)(6a^4)$
6. $(2x^2y^3)(4xy^5)(-5y^6)$
7. $(3x^3y^5)(2xy^2)^2$
8. $(a^3b^2c)^3(ab^2c^3)$
9. $(5x^3y^2)^4\left(\dfrac{1}{5}x^{-2}\right)^2$
10. $(4a^{-2}b^{-1})^2(2a^3b^4)^4$
11. $(-5xy^2)(-3xy)^4$
12. $(-2a^2ab^2)^3(-3ab^2b^2)$
13. $[(-2x^3y)(5x^2y^2)]^2$
14. $[(3x^2y^3)(4xy^5)]^3$
15. $3(x + 2)$
16. $-5(a + b)$
17. $-a(a - b)$
18. $y^2(y - 1)$
19. $3x(x^2 + 3x)$
20. $-2x(3x^2 - 2)$
21. $-2x(3x^2 - 3x + 2)$
22. $3a(4a^2 + 3a - 4)$
23. $5a^2b^3(2a^4b - 5a^0b^3)$
24. $-2a^3b(3a^0b^4 - 2a^2b^3)$
25. $7rst(r^2 + s^2 - t^2)$
26. $3x^2yz(x^2 - 2y + 3z^2)$
27. $-4x^2y^3(3x^2 - 4xy + y^2)$
28. $-2x^2y(3x^4y^2 - 2x^2y - 7)$
29. $4m^2n(-3mn)(m + n)$
30. $-3a^2b^3(2b)(3a + b)$

In Exercises 31–60, find each product. If possible, find the product by sight.

31. $(x + 2)(x + 3)$
32. $(y - 3)(y + 4)$
33. $(z - 7)(z - 2)$
34. $(x + 3)(x - 5)$
35. $(2a + 1)(a - 2)$
36. $(3b - 1)(2b - 1)$
37. $(3t - 2)(2t + 3)$
38. $(p + 3)(3p - 4)$
39. $(3y - z)(2y - z)$
40. $(2m + n)(3m + n)$

41. $(2x - 3y)(x + 2y)$

42. $(3y + 2z)(y - 3z)$

43. $(3x + y)(3x - 3y)$

44. $(2x - y)(3x + 2y)$

45. $(4a - 3b)(2a + 5b)$

46. $(3a + 2b)(2a - 7b)$

47. $(x + 2)^2$

48. $(x - 3)^2$

49. $(a - 4)^2$

50. $(y + 5)^2$

51. $(2a + b)^2$

52. $(a - 2b)^2$

53. $(2x - y)^2$

54. $(3m + 4n)(3m + 4n)$

55. $(x + 2)(x - 2)$

56. $(z + 3)(z - 3)$

57. $(a + b)(a - b)$

58. $(p + q)(p - q)$

59. $(2x + 3y)(2x - 3y)$

60. $(3a + 4b)(3a - 4b)$

In Exercises 61–76, find each product.

61. $(x - y)(x^2 + xy + y^2)$

62. $(x + y)(x^2 - xy + y^2)$

63. $(3y + 1)(2y^2 + 3y + 2)$

64. $(a + 2)(3a^2 + 4a - 2)$

65. $(2a - b)(4a^2 + 2ab + b^2)$

66. $(x - 3y)(x^2 + 3xy + 9y^2)$

67. $(a + b + c)(2a - b - 2c)$

68. $(x - 2y - 3z)(3x + 2y + z)$

69. $(x + 2y + 3z)^2$

70. $(3x - 2y - z)^2$

71. $(2x - 1)[2x^2 - 3(x + 2)]$

72. $(x + 1)^2[x^2 - 2(x + 2)]$

73. $(a + b)(a - b)(a - 3b)$

74. $(x - y)(x + 2y)(x - 2y)$

75. $[x + (2y - 1)][x - (2y - 1)]$

76. $[x + (2a - b)]^2$

In Exercises 77–84, find each product. Write all answers without negative exponents.

77. $x^3(2x^2 + x^{-2})$

78. $x^{-4}(2x^{-3} - 5x^2)$

79. $x^3y^{-6}z^{-2}(3x^{-2}y^2z - x^3y^{-4})$

80. $ab^{-2}c^{-3}(a^{-4}bc^3 + a^{-3}b^4c^3)$

81. $(x^{-1} + y)(x^{-1} - y)$

82. $(x^{-1} - y)(x^{-1} - y)$

83. $(2x^{-3} + y^3)(2x^3 - y^{-3})$

84. $(5x^{-4} - 4y^2)(5x^2 - 4y^{-4})$

In Exercises 85–94, find each product. Consider n to be a whole number.

85. $x^n(x^{2n} - x^n)$

86. $a^{2n}(a^n + a^{2n})$

87. $(x^n + 1)(x^n - 1)$

88. $(x^n - a^n)(x^n + a^n)$

89. $(x^n - y^n)(x^n - y^{-n})$

90. $(x^n + y^n)(x^n + y^{-n})$

91. $(x^{2n} + y^{2n})(x^{2n} - y^{2n})$

92. $(a^{3n} - b^{3n})(a^{3n} + b^{3n})$

93. $(x^n + y^n)(x^n + 1)$

94. $(1 - x^n)(x^{-n} - 1)$

In Exercises 95–108, simplify each expression.

95. $3x(2x + 4) - 3x^2$

96. $2y - 3y(y^2 + 4)$

97. $3pq - p(p - q)$

98. $-4rs(r - 2) + 4rs$

99. $2m(m - n) - (m + n)(m - 2n)$

100. $-3y(2y + z) + (2y - z)(3y + 2z)$

101. $(x + 3)(x - 3) + (2x - 1)(x + 2)$

102. $(2b + 3)(b - 1) - (b + 2)(3b - 1)$

103. $(3x - 4)^2 - (2x + 3)^2$

104. $(3y + 1)^2 + (2y - 4)^2$

105. $3(x - 3y)^2 + 2(3x + y)^2$

106. $2(x - y^2)^2 - 3(y^2 + 2x)^2$

107. $5(2y - z)^2 + 4(y + 2z)^2$

108. $3(x + 2z)^2 - 2(2x - z)^2$

In Exercises 109–112, use a calculator to find each product.

109. $(3.21x - 7.85)(2.87x + 4.59)$

110. $(7.44y + 56.7)(-2.1y - 67.3)$

111. $(-17.3y + 4.35)^2$

112. $(-0.31x + 29.3)(-81x - 0.2)$

113. Refer to Illustration 1.
   a. Find the area of the large square.
   b. Find the area of square I.
   c. Find the area of rectangle II.
   d. Find the area of rectangle III.
   e. Find the area of rectangle IV.
   Use the answers to the preceding questions to show that $(x - y)^2 = x^2 - 2xy + y^2$.

114. Refer to Illustration 2.
   a. Find the area of rectangle $ABCD$.
   b. Find the area of rectangle I.
   c. Find the area of rectangle II.
   Use the answers to the preceding questions to show that $(x + y)(x - y) = x^2 - y^2$.

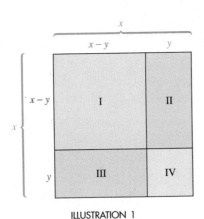

ILLUSTRATION 1

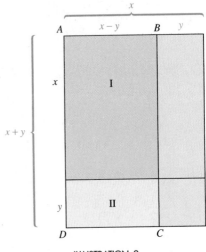

ILLUSTRATION 2

Writing Exercises   *Write a paragraph using your own words.*

1. Explain how the FOIL method is based on the distributive property.

2. Explain how to multiply two trinomials.

Something to Think About

1. The numbers 0.35×10^7 and 1.96×10^7 both involve the same power of 10. Find their sum.

2. Without converting to standard notation, find the sum: $1.435 \times 10^8 + 2.11 \times 10^7$. (*Hint:* The first number in the previous exercise is not in scientific notation.)

Review Exercises   *Let $a = -2$ and $b = 4$. Find the absolute value of each expression.*

1. $|3a - b|$

2. $|ab - b^2|$

3. $-|a^2b - b^0|$

4. $\left| \dfrac{a^3b^2 + ab}{2(ab)^2 - a^3} \right|$

5. A woman bought 200 shares of ABC Company, valued at $125 per share, and 350 shares of WD Company, valued at $75 per share. Then ABC rose $1\frac{1}{2}$ points, and WD fell $1\frac{1}{2}$ points. Find the current value of her portfolio.

6. One light year is approximately 5,870,000,000,000 miles. Write this number in scientific notation.

5.4 The Greatest Common Factor and Factoring by Grouping

■ PRIME-FACTORED FORM OF A NATURAL NUMBER ■ FACTORING OUT THE GREATEST COMMON FACTOR ■ AN APPLICATION OF FACTORING ■ FACTORING BY GROUPING ■ FORMULAS

In this section, we will discuss how to break apart polynomial products. The process of finding the individual factors of a known product is called *factoring*.

■ PRIME-FACTORED FORM OF A NATURAL NUMBER

If one number a divides a second number b, a is called a **factor** of b. For example, because 3 divides 24, it is a factor of 24. Since each of the numbers

1, 2, 3, 4, 6, 8, 12, and 24

divide 24, each number is a factor of 24.

To factor a natural number means to write it as a product of other natural numbers. If each factor is a prime number, the natural number is said to be written in **prime-factored form**.

EXAMPLE 1   Find the prime-factored form of   **a.** 60,   **b.** 84,   and   **c.** 180.

Solution   **a.** $60 = 6 \cdot 10$

$= 2 \cdot 3 \cdot 2 \cdot 5$

$= 2^2 \cdot 3 \cdot 5$

b. $84 = 4 \cdot 21$

$= 2 \cdot 2 \cdot 3 \cdot 7$

$= 2^2 \cdot 3 \cdot 7$

c. $180 = 10 \cdot 18$

$= 2 \cdot 5 \cdot 3 \cdot 6$

$= 2 \cdot 5 \cdot 3 \cdot 3 \cdot 2$

$= 2^2 \cdot 3 \cdot 3 \cdot 5$ ■

The largest natural number that divides 60, 84, and 180 is called the **greatest common factor** or **greatest common divisor** of these three numbers. Because 60, 84, and 180 all have two factors of 2 and one factor of 3, the greatest common factor of these three numbers is $2^2 \cdot 3 = 12$. We note that

$$\frac{60}{12} = 5, \qquad \frac{84}{12} = 7, \qquad \text{and} \qquad \frac{180}{12} = 15$$

There is no natural number greater than 12 that divides 60, 84, and 180.
Algebraic monomials also have greatest common factors.

EXAMPLE 2 Find the greatest common factor of $6a^2b^3c$, $9a^3b^2c$, and $18a^4c^3$.

Solution We begin by factoring each monomial.

$$6a^2b^3c = 3 \cdot 2 \cdot a \cdot a \cdot b \cdot b \cdot b \cdot c$$
$$9a^3b^2c = 3 \cdot 3 \cdot a \cdot a \cdot a \cdot b \cdot b \cdot c$$
$$18a^4c^3 = 2 \cdot 3 \cdot 3 \cdot a \cdot a \cdot a \cdot a \cdot c \cdot c \cdot c$$

Since each monomial has one factor of 3, two factors of a, and one factor of c in common, their greatest common factor is

$$3^1 \cdot a^2 \cdot c^1 = 3a^2c$$ ■

To find the greatest common factor of several monomials, we follow these steps.

> **Steps for Finding the Greatest Common Factor**
> 1. Find the prime-factored form of each monomial.
> 2. List the prime factors and variable factors that are common to each monomial.
> 3. Find the product of the factors found in Step 2, with each factor raised to the smallest power that occurs in any one monomial.

■ FACTORING OUT THE GREATEST COMMON FACTOR

We have seen that the distributive property provides a method for multiplying a polynomial by a monomial. For example,

$$2x^3y^3(3x^2 - 4y^3) = 2x^3y^3 \cdot 3x^2 - 2x^3y^3 \cdot 4y^3$$
$$= 6x^5y^3 - 8x^3y^6$$

If the product of a multiplication is $6x^5y^3 - 8x^3y^6$, we can use the distributive property to find the individual factors.

$$6x^5y^3 - 8x^3y^6 = 2x^3y^3 \cdot 3x^2 - 2x^3y^3 \cdot 4y^3$$
$$= 2x^3y^3(3x^2 - 4y^3)$$

Since $2x^3y^3$ is the greatest common factor of the terms of $6x^5y^3 - 8x^3y^6$, this process is called **factoring out the greatest common factor**. An expression is in factored form when it is written as the product of several quantities.

EXAMPLE 3 Factor $25a^3b + 15ab^3$.

Solution We begin by factoring each monomial:

$$25a^3b = 5 \cdot 5 \cdot a \cdot a \cdot a \cdot b$$
$$15ab^3 = 5 \cdot 3 \cdot a \cdot b \cdot b \cdot b$$

Since each term has at least one factor of 5, one factor of a, and one factor of b in common, and there are no other common factors, $5ab$ is the greatest common factor of the two terms. We can use the distributive property to factor out $5ab$.

$$25a^3b + 15ab^3 = 5ab \cdot 5a^2 + 5ab \cdot 3b^2$$
$$= 5ab(5a^2 + 3b^2)$$ ∎

EXAMPLE 4   Factor $3xy^2z^3 + 6xyz^3 - 3xz^2$.

Solution   We begin by factoring each monomial:

$$3xy^2z^3 = 3 \cdot x \cdot y \cdot y \cdot z \cdot z \cdot z$$
$$6xyz^3 = 3 \cdot 2 \cdot x \cdot y \cdot z \cdot z \cdot z$$
$$-3xz^2 = -3 \cdot x \cdot z \cdot z$$

Since each term has one factor of 3, one factor of x, and two factors of z in common, and because there are no other common factors, $3xz^2$ is the greatest common factor of the three terms. We can use the distributive property to factor out the $3xz^2$.

$$3xy^2z^3 + 6xyz^3 - 3xz^2 = 3xz^2 \cdot y^2z + 3xz^2 \cdot 2yz - 3xz^2 \cdot 1$$
$$= 3xz^2(y^2z + 2yz - 1)$$

WARNING!   The last term $-3xz^2$ of the given trinomial has an understood coefficient of -1. When the $3xz^2$ is factored out, remember to write the -1. ∎

A polynomial that cannot be factored is called a **prime polynomial** or an **irreducible polynomial**.

EXAMPLE 5   Factor $3x^2 + 4y + 7$.

Solution   We factor each monomial:

$$3x^2 = 3 \cdot x \cdot x \qquad 4y = 4 \cdot y \qquad 7 = 7$$

Since there are no common factors other than 1, this polynomial cannot be factored. It is a prime polynomial. ∎

EXAMPLE 6   Factor the negative of the greatest common factor from $-6u^2v^3 + 8u^3v^2$.

Solution   Because the greatest common factor of the two terms is $2u^2v^2$, the negative of the greatest common factor is $-2u^2v^2$. To factor out $-2u^2v^2$, we proceed as follows:

$$-6u^2v^3 + 8u^3v^2 = -2u^2v^2 \cdot 3v + 2u^2v^2 \cdot 4u$$
$$= -2u^2v^2 \cdot 3v - (-2u^2v^2)4u$$
$$= -2u^2v^2(3v - 4u)$$ ∎

We can also factor out a common factor with a variable exponent.

EXAMPLE 7   Factor x^{2n} from $x^{4n} + x^{3n} + x^{2n}$.

Solution   We can write the trinomial in the form $x^{2n} \cdot x^{2n} + x^{2n} \cdot x^n + x^{2n} \cdot 1$ and factor out x^{2n}.

$$x^{4n} + x^{3n} + x^{2n} = x^{2n} \cdot x^{2n} + x^{2n} \cdot x^n + x^{2n} \cdot 1$$
$$= x^{2n}(x^{2n} + x^n + 1) \qquad \blacksquare$$

EXAMPLE 8   Factor $a^{-2}b^{-2}$ from $a^{-2}b - a^3b^{-2}$.

Solution   We write $a^{-2}b - a^3b^{-2}$ in the form $a^{-2}b^{-2} \cdot b^3 - a^{-2}b^{-2} \cdot a^5$ and factor out $a^{-2}b^{-2}$.

$$a^{-2}b - a^3b^{-2} = a^{-2}b^{-2} \cdot b^3 - a^{-2}b^{-2} \cdot a^5$$
$$= a^{-2}b^{-2}(b^3 - a^5) \qquad \blacksquare$$

A common factor can have more than one term. For example, in the expression

$$x(a + b) + y(a + b)$$

the binomial $a + b$ is a factor of both terms. We can factor it out to get

$$x(a + b) + y(a + b) = (a + b)x + (a + b)y \qquad \text{Use the commutative property of multiplication.}$$
$$= (a + b)(x + y)$$

EXAMPLE 9   Factor $a(x - y + z) - b(x - y + z) + 3(x - y + z)$.

Solution   We can factor out the greatest common factor of $(x - y + z)$.

$$a(x - y + z) - b(x - y + z) + 3(x - y + z) = (x - y + z)a - (x - y + z)b + (x - y + z)3$$
$$= (x - y + z)(a - b + 3) \qquad \blacksquare$$

■ AN APPLICATION OF FACTORING

EXAMPLE 10   The polynomial $3x^4 + 2x^3 + 5x^2 + 7x + 1$ can be written as

$$3 \cdot x \cdot x \cdot x \cdot x + 2 \cdot x \cdot x \cdot x + 5 \cdot x \cdot x + 7 \cdot x + 1$$

to illustrate that it involves 10 multiplications and 4 additions. Since computers take more time to do multiplications than additions, a programmer wants to write the polynomial in a way that will require fewer multiplications. Write the polynomial so that it contains only four multiplications.

Solution   Factor the common x from the first four terms of the polynomial to get

$$3x^4 + 2x^3 + 5x^2 + 7x + 1 = x(3x^3 + 2x^2 + 5x + 7) + 1$$

Now factor the common x from the first three terms of the polynomial in the parentheses to get

$$3x^4 + 2x^3 + 5x^2 + 7x + 1 = x\left[x(3x^2 + 2x + 5) + 7\right] + 1$$

Again, factor an x from the first two terms of the polynomial within the set of parentheses to get

1. $3x^4 + 2x^3 + 5x^2 + 7x + 1 = x\left\{x\left[x(3x + 2) + 5\right] + 7\right\} + 1$

To emphasize that there are now only four multiplications, we rewrite the right-hand side of Equation 1 as

$$x \cdot \left\{x \cdot \left[x \cdot (3 \cdot x + 2) + 5\right] + 7\right\} + 1$$

The right-hand side of Equation 1 is called the *nested form* of the polynomial. ■

■ FACTORING BY GROUPING

Suppose we wish to factor

$$ac + ad + bc + bd$$

Although there is no factor common to all four terms, there is a common factor of a in the first two terms and a common factor of b in the last two terms. We can factor out these common factors to get

$$ac + ad + bc + bd = a(c + d) + b(c + d)$$

We can now factor out the common factor of $c + d$ on the right-hand side:

$$ac + ad + bc + bd = (c + d)(a + b)$$

The grouping in this type of problem is not always unique. For example, if we write the expression $ac + ad + bc + bd$ in the form

$$ac + bc + ad + bd$$

and factor c from the first two terms and d from the last two terms, we obtain

$$ac + bc + ad + bd = c(a + b) + d(a + b)$$
$$= (a + b)(c + d)$$

The method used in the previous example is called **factoring by grouping**.

EXAMPLE 11   Factor $3ax^2 + 3bx^2 + a + 5bx + 5ax + b$.

Solution   Although there is no factor common to all six terms, $3x^2$ can be factored out of the first two terms and $5x$ can be factored out of the fourth and fifth terms to get

$$3ax^2 + 3bx^2 + a + 5bx + 5ax + b = 3x^2(a + b) + a + 5x(b + a) + b$$

This result can be written in the form

$$3ax^2 + 3bx^2 + a + 5bx + 5ax + b = 3x^2(a + b) + 5x(a + b) + (a + b)$$

Since $a + b$ is common to all three terms, it can be factored out to get

$$3ax^2 + 3bx^2 + a + 5bx + 5ax + b = (a + b)(3x^2 + 5x + 1)$$ ■

To factor an expression, it is often necessary to factor more than once, as the following example illustrates.

EXAMPLE 12 Factor $3x^3y - 4x^2y^2 - 6x^2y + 8xy^2$.

Solution We begin by factoring out the common factor of xy.

$$3x^3y - 4x^2y^2 - 6x^2y + 8xy^2 = xy(3x^2 - 4xy - 6x + 8y)$$

We can now factor $3x^2 - 4xy - 6x + 8y$ by grouping:

$$3x^3y - 4x^2y^2 - 6x^2y + 8xy^2 = xy(3x^2 - 4xy - 6x + 8y)$$
$$= xy[x(3x - 4y) - 2(3x - 4y)] \qquad \text{Factor } x \text{ from } 3x^2 - 4xy \text{ and } -2 \text{ from } -6x + 8y.$$

$$= xy(3x - 4y)(x - 2) \qquad \text{Factor out } 3x - 4y.$$

Because no more factoring can be done, the factorization is complete. ∎

WARNING! Whenever you factor an expression, always factor it completely. Each factor of a completely factored expression will be prime.

■ FORMULAS

Factoring is often required to solve a literal equation for one of its variables.

EXAMPLE 13 **Simple interest** The formula $A = p + prt$ gives the amount of money in a savings account at the end of a specific time. Solve the formula for p.

Solution We must isolate p on one side of the equation. First, we notice that every term on the right-hand side of the equation has a factor of p, and that p does not occur on the left-hand side. So we factor p from the right-hand side of the equation and solve for p.

$$A = p + prt$$
$$A = p(1 + rt) \qquad \text{Factor out } p.$$
$$\frac{A}{1 + rt} = p \qquad \text{Divide both sides by } 1 + rt.$$
$$p = \frac{A}{1 + rt}$$

∎

EXAMPLE 14 **Electronics** The formula $r_1r_2 = rr_2 + rr_1$ is used in electronics to relate the combined resistance, r, of two resistors wired in parallel. The variable r_1 represents the resistance of the first resistor, and the variable r_2 represents the resistance of the second. Solve the equation for r_2.

Solution   To isolate r_2 on one side of the equation, we get all terms involving r_2 on the left-hand side of the equation and all terms not involving r_2 on the right-hand side. We then proceed as follows:

$$r_1 r_2 = rr_2 + rr_1$$

$$r_1 r_2 - rr_2 = rr_1 \qquad \text{Subtract } rr_2 \text{ from both sides.}$$

$$r_2(r_1 - r) = rr_1 \qquad \text{Factor out } r_2 \text{ on the left-hand side.}$$

$$r_2 = \frac{rr_1}{r_1 - r} \qquad \text{Divide both sides by } r_1 - r.$$

Orals   *Factor each expression.*

1. $3x^2 - x$

2. $7t^3 + 14t^2$

3. $-3a^2 - 6a$

4. $-4x^2 + 12x$

5. $3(a + b) + x(a + b)$

6. $a(m - n) - b(m - n)$

EXERCISE 5.4

In Exercises 1–8, find the prime-factored form of each number.

1. 6

2. 10

3. 135

4. 98

5. 128

6. 357

7. 325

8. 288

In Exercises 9–16, find the greatest common factor of each set of monomials.

9. 36, 48

10. 45, 75

11. 42, 36, 98

12. 16, 40, 60

13. $4a^2b, 8a^3c$

14. $6x^3y^2z, 9xyz^2$

15. $18x^4y^3z^2, -12xy^2z^3$

16. $6x^2y^3, 24xy^3, 40x^2y^2z^3$

In Exercises 17–20, complete each factorization.

17. $3a - 12 = 3(a - \quad)$

18. $5t + 25 = 5(t + \quad)$

19. $8z^2 + 2z = 2z(4z + \quad)$

20. $9t^3 - 3t^2 = 3t^2(3t - \quad)$

In Exercises 21–42, factor each expression, if possible.

21. $2x + 8$

22. $3y - 9$

23. $2x^2 - 6x$

24. $3y^3 + 3y^2$

25. $5xy + 12ab^2$

26. $7x^2 + 14x$

27. $15x^2y - 10x^2y^2$

28. $11m^3n^2 - 12x^2y$

29. $63x^3y^2 + 81x^2y^4$

30. $33a^3b^4c - 16xyz$

31. $14r^2s^3 + 15t^6$

32. $13ab^2c^3 - 26a^3b^2c$

33. $27z^3 + 12z^2 + 3z$

34. $25t^6 - 10t^3 + 5t^2$

35. $24s^3 - 12s^2t + 6st^2$

36. $18y^2z^2 + 12y^2z^3 - 24y^4z^3$

37. $45x^{10}y^3 - 63x^7y^7 + 81x^{10}y^{10}$

38. $48u^6v^6 - 16u^4v^4 - 3u^6v^3$

39. $25x^3 - 14y^3 + 36x^3y^3$

40. $9m^4n^3p^2 + 18m^2n^3p^4 - 27m^3n^4p$

41. $24a^3b^5 + 32a^5b^3 - 64a^5b^5c^5$

42. $32a^4 + 9b^2 + 5a^4b^2$

In Exercises 43–52, factor out the negative of the greatest common factor.

43. $-3a - 6$

44. $-6b + 12$

45. $-3x^2 - x$

46. $-4a^3 + a^2$

47. $-6x^2 - 3xy$

48. $-15y^3 + 25y^2$

49. $-18a^2b - 12ab^2$

50. $-21t^5 + 28t^3$

51. $-63u^3v^6z^9 + 28u^2v^7z^2 - 21u^3v^3z^4$

52. $-56x^4y^3z^2 - 72x^3y^4z^5 + 80xy^2z^3$

In Exercises 53–66, factor out the designated common factor.

53. x^2 from $x^{n+2} + x^{n+3}$

54. y^3 from $y^{n+3} + y^{n+5}$

55. y^n from $2y^{n+2} - 3y^{n+3}$

56. x^n from $4x^{n+3} - 5x^{n+5}$

57. $2t^{2n}$ from $2t^{4n} + 4t^{3n} - 8t^{2n}$

58. $9p^{3n}$ from $36p^{3n} - 27p^{6n} + 18p^{9n}$

59. x^{-2} from $x^4 - 5x^6$

60. y^{-4} from $7y^4 + y$

61. t^{-3} from $t^5 + 4t^{-6}$

62. p^{-5} from $6p^3 - p^{-2}$

63. $4y^{-2n}$ from $8y^{2n} + 12 + 16y^{-2n}$

64. $7x^{-3n}$ from $21x^{6n} + 7x^{3n} + 14$

65. $r^{-3n}s^2$ from $r^{9n}s^4 - 3r^{3n}s^3 - 2s^2$

66. $t^{-4n}s$ from $t^{8n}s^3 - s^2 - t^{-3n}s$

In Exercises 67–78, factor each expression.

67. $4(x + y) + t(x + y)$

68. $5(a - b) - t(a - b)$

69. $(a - b)r - (a - b)s$

70. $(x + y)u + (x + y)v$

71. $3(m + n + p) + x(m + n + p)$

72. $x(x - y - z) + y(x - y - z)$

73. $(x + y)(x + y) + z(x + y)$

74. $(a - b)^2 + (a - b)$

75. $(u + v)^2 - (u + v)$

76. $a(x - y) - (x - y)^2$

77. $-a(x + y) + b(x + y)$

78. $-bx(a - b) - cx(a - b)$

In Exercises 79–80, write each polynomial in nested form.

79. $2x^3 + 5x^2 - 2x + 8$

80. $4x^4 + 5x^2 - 3x + 9$

In Exercises 81–92, factor by grouping.

81. $ax + bx + ay + by$

82. $ar - br + as - bs$

83. $x^2 + yx + 2x + 2y$

84. $2c + 2d - cd - d^2$

85. $3c - cd + 3d - c^2$

86. $x^2 + 4y - xy - 4x$

87. $a^2 - 4b + ab - 4a$

88. $7u + v^2 - 7v - uv$

89. $ax + bx - a - b$

90. $x^2y - ax - xy + a$

91. $x^2 + xy + xz + xy + y^2 + zy$

92. $ab - b^2 - bc + ac - bc - c^2$

In Exercises 93–98, factor by grouping. Factor out all common monomials first.

93. $mpx + mqx + npx + nqx$

94. $abd - abe + acd - ace$

95. $x^2y + xy^2 + 2xyz + xy^2 + y^3 + 2y^2z$

96. $a^3 - 2a^2b + a^2c - a^2b + 2ab^2 - abc$

97. $2n^4p - 2n^2 - n^3p^2 + np + 2mn^3p - 2mn$

98. $a^2c^3 + ac^2 + a^3c^2 - 2a^2bc^2 - 2bc^2 + c^3$

If the greatest common factor of several terms is 1, the terms are called relatively prime. *In Exercises 99–106, tell whether the terms in each set are relatively prime.*

99. 14, 45

100. 24, 63, 112

101. 60, 28, 36

102. 55, 49, 78

103. $12x^2y, 5ab^3, 35x^2b^3$

104. $18uv, 25rs, 12rsuv$

105. $9(a - b), 16(a + b), 25(a + b + c)$

106. $44(x + y - z), 99(x - y + z), 121(x + y + z)$

In Exercises 107–118, solve for the indicated variable.

107. $r_1r_2 = rr_2 + rr_1$ for r_1

108. $r_1r_2 = rr_2 + rr_1$ for r

109. $d_1d_2 = fd_2 + fd_1$ for f

110. $d_1d_2 = fd_2 + fd_1$ for d_1

111. $b^2x^2 + a^2y^2 = a^2b^2$ for a^2

112. $b^2x^2 + a^2y^2 = a^2b^2$ for b^2

113. $S(1 - r) = a - lr$ for r

114. $Sn = (n - 2)180°$ for n

115. $H(a + b) = 2ab$ for a

116. $H(a + b) = 2ab$ for b

117. $x = \dfrac{2y + 1}{3y - 4}$ for y

118. $x = \dfrac{5y - 3}{4y + 7}$ for y

Writing Exercises

1. Explain how to find the greatest common factor of two natural numbers.

2. Explain how to recognize whether a number is prime.

Something to Think About

1. Pick two natural numbers. Divide their product by their greatest common factor. The result is called the **least common multiple** of the two numbers you picked. Why?

2. The number 6 is called a **perfect number** because the sum of all the divisors of 6 is twice 6: $1 + 2 + 3 + 6 = 12$. Verify that 28 is also a perfect number.

Review Exercises   *Do each multiplication.*

1. $(a + 4)(a - 4)$

2. $(2b + 3)(2b - 3)$

3. $(4r^2 + 3s)(4r^2 - 3s)$

4. $(5a + 2b^3)(5a - 2b^3)$

5. $(m + 4)(m^2 - 4m + 16)$

6. $(p - q)(p^2 + pq + q^2)$

5.5   The Difference of Two Squares; The Sum and Difference of Two Cubes

■ THE DIFFERENCE OF TWO SQUARES   ■ THE SUM AND DIFFERENCE OF TWO CUBES   ■ FACTORING BY GROUPING

To factor the difference of two squares, it is helpful to know the first twenty integers that are **perfect squares**.

1, 4, 9, 16, 25, 36, 49, 64, 81, 100, 121, 144, 169, 196, 225, 256, 289, 324, 361, 400

Expressions like $x^6y^4z^2$ are also perfect squares, because they can be written as the square of another polynomial.

$$x^6y^4z^2 = (x^3y^2z)^2$$

Every exponential expression where each factor has an even-numbered exponent is a perfect square.

■ THE DIFFERENCE OF TWO SQUARES

In Section 5.3, we developed the special product formula

1. $(x + y)(x - y) = x^2 - y^2$

The binomial $x^2 - y^2$ is called the **difference of two squares**, because x^2 represents the square of x, y^2 represents the square of y, and $x^2 - y^2$ represents the difference of these squares.

Equation 1 can be written in reverse order to give a formula for factoring the difference of two squares.

> **Factoring the Difference of Two Squares**
>
> $$x^2 - y^2 = (x + y)(x - y)$$

If we think of the difference of two squares as the square of a **First** quantity minus the square of a **Last** quantity, we have the formula

$$F^2 - L^2 = (F + L)(F - L)$$

and we say: *To factor the square of a **First** quantity minus the square of a **Last** quantity, we multiply the **First** plus the **Last** by the **First** minus the **Last**.*

EXAMPLE 1   Factor $49x^2 - 16$.

Solution   We can write $49x^2 - 16$ in the form $(7x)^2 - (4)^2$ and use the formula for factoring the difference of two squares:

$$\mathbf{F^2 - L^2 = (F + L)(F - L)}$$
$$(7x)^2 - 4^2 = (7x + 4)(7x - 4) \qquad \text{Substitute } 7x \text{ for F and 4 for L.}$$

We can verify this result by multiplying $7x + 4$ and $7x - 4$.

$$(7x + 4)(7x - 4) = 49x^2 - 28x + 28x - 16$$
$$= 49x^2 - 16$$

 WARNING! Expressions such as $(7x)^2 + (4)^2$ that represent the sum of two squares cannot be factored in the real number system. The binomial $49x^2 + 16$ is a prime binomial.

EXAMPLE 2   Factor $64a^4 - 25b^2$.

Solution   We can write $64a^4 - 25b^2$ in the form $(8a^2)^2 - (5b)^2$ and use the formula for factoring the difference of two squares.

$$F^2 - L^2 = (F + L)(F - L)$$
$$(8a^2)^2 - (5b)^2 = (8a^2 + 5b)(8a^2 - 5b) \qquad \text{Substitute } 8a^2 \text{ for F and } 5b \text{ for L.}$$

Verify by multiplication that $(8a^2 + 5b)(8a^2 - 5b) = 64a^4 - 25b^2$. ∎

EXAMPLE 3   Factor $x^4 - 1$.

Solution   Because the binomial is the difference of the squares of x^2 and 1, it factors into the sum of x^2 and 1, and the difference of x^2 and 1.

$$x^4 - 1 = (x^2)^2 - (1)^2$$
$$= (x^2 + 1)(x^2 - 1)$$

The factor $x^2 + 1$ is the sum of two squares and is prime. However, the factor $x^2 - 1$ is the difference of two squares and can be factored as $(x + 1)(x - 1)$.

$$x^4 - 1 = (x^2 + 1)(x^2 - 1)$$
$$= (x^2 + 1)(x + 1)(x - 1)$$ ∎

EXAMPLE 4   Factor $(x + y)^4 - z^4$.

Solution   This expression is the difference of two squares that can be factored:

$$(x + y)^4 - z^4 = [(x + y)^2]^2 - (z^2)^2$$
$$= [(x + y)^2 + z^2][(x + y)^2 - z^2]$$

The factor $(x + y)^2 + z^2$ is the sum of two squares and is prime. However, the factor $(x + y)^2 - z^2$ is the difference of two squares and can be factored as $(x + y + z)(x + y - z)$.

$$(x + y)^4 - z^4 = [(x + y)^2 + z^2][(x + y)^2 - z^2]$$
$$= [(x + y)^2 + z^2](x + y + z)(x + y - z)$$ ∎

When possible, we will always factor out a common factor before factoring the difference of two squares. The factoring process is easier when all common factors are factored out first.

EXAMPLE 5   Factor $2x^4y - 32y$.

Solution ·

$$
\begin{aligned}
2x^4y - 32y &= 2y(x^4 - 16) & \text{Factor out } 2y. \\
&= 2y(x^2 + 4)(x^2 - 4) & \text{Factor } x^4 - 16. \\
&= 2y(x^2 + 4)(x + 2)(x - 2) & \text{Factor } x^2 - 4.
\end{aligned}
$$

■

■ THE SUM AND DIFFERENCE OF TWO CUBES

The number 64 is called a **perfect cube**, because $4^3 = 64$. To factor the sum or difference of two cubes, it is helpful to know the first ten perfect cubes:

1, 8, 27, 64, 125, 216, 343, 512, 729, 1000

Expressions like $x^9y^6z^3$ are also perfect cubes, because they can be written as the cube of another polynomial:

$$x^9y^6z^3 = (x^3y^2z)^3$$

To find formulas for factoring the sum or difference of two cubes, we use the following product formulas.

2. $(x + y)(x^2 - xy + y^2) = x^3 + y^3$

3. $(x - y)(x^2 + xy + y^2) = x^3 - y^3$

To verify Equation 2, we multiply $x^2 - xy + y^2$ by $x + y$ to verify that the product is $x^3 + y^3$.

$$
\begin{aligned}
(x + y)(x^2 - xy + y^2) &= (x + y)x^2 - (x + y)xy + (x + y)y^2 \\
&= x \cdot x^2 + y \cdot x^2 - x \cdot xy - y \cdot xy + x \cdot y^2 + y \cdot y^2 \\
&= x^3 + x^2y - x^2y - xy^2 + xy^2 + y^3 \\
&= x^3 + y^3
\end{aligned}
$$

Equation 3 can also be verified by multiplication.

If we write Equations 2 and 3 in reverse order, we have the formulas for factoring the sum and difference of two cubes.

Sum and Difference of Two Cubes

$$x^3 + y^3 = (x + y)(x^2 - xy + y^2)$$
$$x^3 - y^3 = (x - y)(x^2 + xy + y^2)$$

If we think of the sum of two cubes as the sum of the cube of a **F**irst quantity plus the cube of a **L**ast quantity, we have the formula

$$F^3 + L^3 = (F + L)(F^2 - FL + L^2)$$

*To factor the cube of a **F**irst quantity plus the cube of a **L**ast quantity, we multiply the sum of the **F**irst and **L**ast by*

- *the **F**irst squared*
- *minus the **F**irst times the **L**ast*
- *plus the **L**ast squared.*

The formula for the difference of two cubes is

$$F^3 - L^3 = (F - L)(F^2 + FL + L^2)$$

*To factor the cube of a **F**irst quantity minus the cube of a **L**ast quantity, we multiply the difference of the **F**irst and **L**ast by*

- *the **F**irst squared*
- *plus the **F**irst times the **L**ast*
- *plus the **L**ast squared.*

EXAMPLE 6   Factor $a^3 + 8$.

Solution   Since $a^3 + 8$ can be written as $a^3 + 2^3$, we have the sum of two cubes, which factors as follows:

$$F^3 + L^3 = (F + L)(F^2 - FL + L^2)$$
$$a^3 + 2^3 = (a + 2)(a^2 - a2 + 2^2) \qquad \text{Substitute } a \text{ for F and 2 for L.}$$
$$= (a + 2)(a^2 - 2a + 4)$$

Thus, $a^3 + 8 = (a + 2)(a^2 - 2a + 4)$. Verify this result by multiplication.   ■

EXAMPLE 7   Factor $27a^3 - 64b^3$.

Solution   Since $27a^3 - 64b^3$ can be written as $(3a)^3 - (4b)^3$, we have the difference of two cubes, which factors as follows:

$$F^3 - L^3 = (F - L)(F^2 + FL + L^2)$$
$$(3a)^3 - (4b)^3 = (3a - 4b)[(3a)^2 + (3a)(4b) + (4b)^2] \qquad \text{Substitute } 3a \text{ for F and } 4b \text{ for L.}$$
$$= (3a - 4b)(9a^2 + 12ab + 16b^2)$$

Thus, $27a^3 - 64b^3 = (3a - 4b)(9a^2 + 12ab + 16b^2)$. Check by multiplication.   ■

EXAMPLE 8   Factor $a^3 - (c + d)^3$.

Solution
$$a^3 - (c + d)^3 = [a - (c + d)][a^2 + a(c + d) + (c + d)^2]$$
$$= (a - c - d)(a^2 + ac + ad + c^2 + 2cd + d^2) \qquad ■$$

EXAMPLE 9   Factor $x^6 - 64$.

Solution   This expression can be factored as either the difference of two squares or the difference of two cubes.

$$(x^3)^2 - 8^2 \qquad \text{or} \qquad (x^2)^3 - 4^3$$

However, it is much easier to factor the difference of two squares first.

$$x^6 - 64 = (x^3)^2 - 8^2$$
$$= (x^3 + 8)(x^3 - 8)$$

Each of these factors further, however, for one is the sum of two cubes and the other is the difference of two cubes:

$$x^6 - 64 = (x + 2)(x^2 - 2x + 4)(x - 2)(x^2 + 2x + 4)$$

Try to factor $x^6 - 64$ as the difference of two cubes first and see what difficulty you encounter. ■

EXAMPLE 10 Factor $2a^5 + 128a^2$.

Solution We first factor out the common monomial factor of $2a^2$ to obtain

$$2a^5 + 128a^2 = 2a^2(a^3 + 64)$$

Then we factor $a^3 + 64$ as the sum of two cubes to obtain

$$2a^5 + 128a^2 = 2a^2(a + 4)(a^2 - 4a + 16)$$ ■

EXAMPLE 11 Factor $16r^{6m} - 54t^{3n}$.

Solution $16r^{6m} - 54t^{3n} = 2(8r^{6m} - 27t^{3n})$ Factor out a 2.

$$= 2[(2r^{2m})^3 - (3t^n)^3]$$ Write $8r^{6m}$ as $(2r^{2m})^3$ and $27t^{3n}$ as $(3t^n)^3$.

$$= 2[(2r^{2m} - 3t^n)(4r^{4m} + 6r^{2m}t^n + 9t^{2n})]$$ Factor $(2r^{2m})^3 - (3t^n)^3$. ■

■ FACTORING BY GROUPING

EXAMPLE 12 Factor $x^2 - y^2 + x - y$.

Solution If we group the first two terms and factor the difference of two squares, we have

$$x^2 - y^2 + x - y = (x + y)(x - y) + (x - y)$$ Factor $x^2 - y^2$.

$$= (x - y)(x + y + 1)$$ Factor out $x - y$. ■

Orals *Factor each expression, if possible.*

1. $x^2 - 1$ **2.** $a^4 - 16$

3. $x^3 + 1$ **4.** $a^3 - 8$

5. $2x^2 - 8$ **6.** $x^4 + 25$

EXERCISE 5.5

In Exercises 1–20, factor each expression, if possible.

1. $x^2 - 4$ **2.** $y^2 - 9$

3. $9y^2 - 64$ **4.** $16x^4 - 81y^2$

5. $x^2 + 25$

6. $144a^2 - b^4$

7. $625a^2 - 169b^4$

8. $4y^2 + 9z^4$

9. $81a^4 - 49b^2$

10. $64r^6 - 121s^2$

11. $36x^4y^2 - 49z^4$

12. $100a^2b^4c^6 - 225d^8$

13. $(x + y)^2 - z^2$

14. $a^2 - (b - c)^2$

15. $(a - b)^2 - c^2$

16. $(m + n)^2 - p^4$

17. $x^4 - y^4$

18. $16a^4 - 81b^4$

19. $256x^4y^4 - z^8$

20. $225a^4 - 16b^8c^{12}$

In Exercises 21–28, factor each expression.

21. $2x^2 - 288$

22. $8x^2 - 72$

23. $2x^3 - 32x$

24. $3x^3 - 243x$

25. $5x^3 - 125x$

26. $6x^4 - 216x^2$

27. $r^2s^2t^2 - t^2x^4y^2$

28. $16a^4b^3c^4 - 64a^2bc^6$

In Exercises 29–38, factor each expression.

29. $r^3 + s^3$

30. $t^3 - v^3$

31. $x^3 - 8y^3$

32. $27a^3 + b^3$

33. $64a^3 - 125b^6$

34. $8x^6 + 125y^3$

35. $125x^3y^6 + 216z^9$

36. $1000a^6 - 343b^3c^6$

37. $x^6 + y^6$

38. $x^9 + y^9$

In Exercises 39–48, factor each expression.

39. $5x^3 + 625$

40. $2x^3 - 128$

41. $4x^5 - 256x^2$

42. $2x^6 + 54x^3$

43. $128u^2v^3 - 2t^3u^2$

44. $56rs^2t^3 + 7rs^2v^6$

45. $(a + b)x^3 + 27(a + b)$

46. $(c - d)r^3 - (c - d)s^3$

47. $6a^3b^3 - 6z^3$

48. $18x^3y^3 + 18c^3d^3$

In Exercises 49–58, factor each expression. Assume that m and n are natural numbers.

49. $x^{2m} - y^{4n}$

50. $a^{4m} - b^{8n}$

51. $100a^{4m} - 81b^{2n}$

52. $25x^{8m} - 36y^{4n}$

53. $x^{3n} - 8$

54. $a^{3m} + 64$

55. $a^{3m} + b^{3n}$

56. $x^{6m} - y^{3n}$

57. $2x^{6m} + 16y^{3m}$

58. $24 + 3c^{3m}$

In Exercises 59–64, factor each expression by grouping.

59. $a^2 - b^2 + a + b$

60. $x^2 - y^2 - x - y$

61. $a^2 - b^2 + 2a - 2b$

62. $m^2 - n^2 + 3m + 3n$

63. $2x + y + 4x^2 - y^2$

64. $m - 2n + m^2 - 4n^2$

Writing Exercises   *Write a paragraph using your own words.*

1. Describe the pattern used to factor the difference of two squares.

2. Describe the patterns used to factor the sum and the difference of two cubes.

Something to Think About   **1.** Factor $x^{32} - y^{32}$.

2. Find the error in this proof:

$x = y$	Let $x = y$.
$x^2 = xy$	Multiply both sides by x.
$x^2 - y^2 = xy - y^2$	Subtract y^2 from both sides.
$(x + y)(x - y) = y(x - y)$	Factor both sides.
$\dfrac{(x + y)(x - y)}{x - y} = \dfrac{y(x - y)}{x - y}$	Divide both sides by $x - y$.
$x + y = y$	$\dfrac{x - y}{x - y} = 1$.
$y + y = y$	Substitute y for x.
$2y = y$	Combine like terms.
$\dfrac{2y}{y} = \dfrac{y}{y}$	Divide both sides by y.
$2 = 1$	$\dfrac{y}{y} = 1$.

Review Exercises   *Do each multiplication.*

1. $(x + 1)(x + 1)$

2. $(2m - 3)(m - 2)$

3. $(2m + n)(2m + n)$

4. $(3m - 2n)(3m - 2n)$

5. $(a + 4)(a + 3)$

6. $(3b + 2)(2b - 5)$

7. $(4r - 3s)(2r - s)$

8. $(5a - 2b)(3a + 4b)$

5.6 Factoring Trinomials

■ FACTORING TRINOMIALS WITH LEAD COEFFICIENTS OF 1 ■ FACTORING TRINOMIALS WITH LEAD COEFFICIENTS OTHER THAN 1 ■ TEST FOR FACTORABILITY ■ USING SUBSTITUTION TO FACTOR TRINOMIALS ■ FACTORING BY GROUPING ■ USING GROUPING TO FACTOR TRINOMIALS

Many trinomials can be factored by using the following special product formulas.

1. $(x + y)(x + y) = x^2 + 2xy + y^2$

2. $(x - y)(x - y) = x^2 - 2xy + y^2$

To factor the perfect square trinomial $x^2 + 6x + 9$, for example, we note that the trinomial can be written in the form $x^2 + 2(3)x + 3^2$. If $y = 3$, this form matches the right-hand side of Equation 1. Thus, $x^2 + 6x + 9$ factors as

$$x^2 + 6x + 9 = x^2 + 2(3)x + 3^2$$
$$= (x + 3)(x + 3)$$

This result can be verified by multiplication:

$$(x + 3)(x + 3) = x^2 + 3x + 3x + 9$$
$$= x^2 + 6x + 9$$

To factor the perfect square trinomial $x^2 - 4xz + 4z^2$, we note that the trinomial can be written in the form $x^2 - 2x(2z) + (2z)^2$. If $y = 2z$, this form matches the right-hand side of Equation 2. Thus, $x^2 - 4xz + 4z^2$ factors as

$$x^2 - 4xz + 4z^2 = x^2 - 2x(2z) + (2z)^2$$
$$= (x - 2z)(x - 2z)$$

This result can also be verified by multiplication.

We begin our discussion of these *general trinomials* by considering trinomials with lead coefficients (the coefficient of the squared term) of 1.

■ FACTORING TRINOMIALS WITH LEAD COEFFICIENTS OF 1

Since the product of two binomials is often a trinomial, we expect that many trinomials will factor as two binomials. For example, to factor $x^2 + 7x + 12$, we must find two binomials $x + a$ and $x + b$ such that

$$x^2 + 7x + 12 = (x + a)(x + b)$$

where $ab = 12$ and $ax + bx = 7x$

To find the numbers a and b, we list the possible factorizations of 12 and find the one in which the sum of the factors is 7.

The one to choose
↓

$$12(1) \qquad 6(2) \qquad 4(3) \qquad -12(-1) \qquad -6(-2) \qquad -4(-3)$$

Thus, $a = 4$, $b = 3$, and

$$x^2 + 7x + 12 = (x + a)(x + b)$$

3. $\quad x^2 + 7x + 12 = (x + 4)(x + 3)$ Substitute 4 for a and 3 for b.

This factorization can be verified by multiplying $x + 4$ and $x + 3$ and observing that the product is $x^2 + 7x + 12$.

Because of the commutative property of multiplication, the order of the factors in Equation 3 is not important.

To factor trinomials with lead coefficients of 1, we follow these steps.

Factoring Trinomials with Lead Coefficients of 1

1. Write the trinomial in descending powers of one variable.

2. List the factorizations of the third term of the trinomial.

3. Pick the factorization in which the sum of the factors is the coefficient of the middle term.

EXAMPLE 1   Factor $x^2 - 6x + 8$.

Solution   Since this trinomial is already written in descending powers of x, we can move to Step 2 and list the possible factorizations of the third term, which is 8.

The one to choose
↓

$$8(1) \qquad 4(2) \qquad -8(-1) \qquad -4(-2)$$

In this trinomial, the coefficient of the middle term is -6. The only factorization in which the sum of the factors is -6 is $-4(-2)$. Thus, $a = -4$, $b = -2$, and

$$x^2 - 6x + 8 = (x + a)(x + b)$$
$$= (x - 4)(x - 2) \qquad \text{Substitute } -4 \text{ for } a \text{ and } -2 \text{ for } b.$$

We can verify this result by multiplication:

$$(x - 4)(x - 2) = x^2 - 2x - 4x + 8$$
$$= x^2 - 6x + 8 \qquad\qquad ■$$

EXAMPLE 2   Factor $-x + x^2 - 12$.

Solution   We begin by writing the trinomial in descending powers of x:

$$-x + x^2 - 12 = x^2 - x - 12$$

The possible factorizations of the third term are

The one to choose
↓

$$12(-1) \qquad 6(-2) \qquad 4(-3) \qquad 1(-12) \qquad 2(-6) \qquad 3(-4)$$

In this trinomial, the coefficient of the middle term is -1. The only factorization in which the sum of the factors is -1 is $3(-4)$. Thus, $a = 3$, $b = -4$, and

$$-x + x^2 - 12 = (x + a)(x + b)$$
$$= (x + 3)(x - 4) \qquad \text{Substitute } 3 \text{ for } a \text{ and } -4 \text{ for } b. \qquad ■$$

EXAMPLE 3   Factor $30x - 4xy - 2xy^2$.

Solution   We begin by writing the trinomial in descending powers of y:

$$30x - 4xy - 2xy^2 = -2xy^2 - 4xy + 30x$$

Each term in this trinomial has a common monomial factor of $-2x$, which should be factored out.

$$30x - 4xy - 2xy^2 = -2x(y^2 + 2y - 15)$$

To factor $y^2 + 2y - 15$, we list the factors of -15 and find the pair whose sum is 2.

The one to choose
↓

$$15(-1) \qquad 5(-3) \qquad 1(-15) \qquad 3(-5)$$

The only factorization in which the sum of the factors is 2 (the coefficient of the middle term of $y^2 + 2y - 15$) is $5(-3)$. Thus, $a = 5$, $b = -3$, and

$$30x - 4xy - 2xy^2 = -2x(y^2 + 2y - 15)$$
$$= -2x(y + 5)(y - 3) \qquad \blacksquare$$

WARNING!   In Example 3, be sure to include all factors in the final answer. It is a common error to forget to write the $-2x$.

■ FACTORING TRINOMIALS WITH LEAD COEFFICIENTS OTHER THAN 1

There are more combinations to consider when factoring trinomials with lead coefficients other than 1. To factor $5x^2 + 7x + 2$, for example, we must find two binomials of the form $ax + b$ and $cx + d$ such that

$$5x^2 + 7x + 2 = (ax + b)(cx + d)$$

Since the first term of the trinomial $5x^2 + 7x + 2$ is $5x^2$, the first terms of the binomial factors must be $5x$ and x.

$$\overbrace{5x^2 + 7x + 2 = (5x + b)(x}^{5x^2} + d)$$

Since the product of the last terms must be 2, and the sum of the products of the outer and inner terms must be $7x$, we must find two numbers whose product is 2 that will give a middle term of $7x$.

$$\overbrace{5x^2 + 7x + 2 = (5x + \underbrace{b)(x + d)}}^{2}$$
$$\text{O} + \text{I} = 7x$$

Because both $2(1)$ and $(-2)(-1)$ give a product of 2, there are four possible combinations to consider:

$$(5x + 2)(x + 1) \qquad (5x - 2)(x - 1)$$
$$(5x + 1)(x + 2) \qquad (5x - 1)(x - 2)$$

Of these possibilities, only the first one gives the proper middle term of $7x$. Thus,

4. $5x^2 + 7x + 2 = (5x + 2)(x + 1)$

We can verify this result by multiplication:

$$(5x + 2)(x + 1) = 5x^2 + 5x + 2x + 2$$
$$= 5x^2 + 7x + 2$$

■ TEST FOR FACTORABILITY

If a trinomial has the form $ax^2 + bx + c$, with integer coefficients and $a \neq 0$, we can test to see whether it is factorable. If the value of $b^2 - 4ac$ is a perfect square, the trinomial can be factored using only integers. If the value is not a perfect square, the trinomial cannot be factored using only integers.

For example, $5x^2 + 7x + 2$ is a trinomial of the form $ax^2 + bx + c$ with

$$a = 5, \qquad b = 7, \qquad \text{and} \qquad c = 2$$

For this trinomial, the value of $b^2 - 4ac$ is

$$b^2 - 4ac = 7^2 - 4(5)(2) \qquad \text{Substitute 7 for } b, 5 \text{ for } a, \text{ and 2 for } c.$$
$$= 49 - 40$$
$$= 9$$

Since 9 is a perfect square, the trinomial is factorable. Its factorization is shown in Equation 4.

Test for Factorability

A trinomial of the form $ax^2 + bx + c$, with integer coefficients and $a \neq 0$, will factor into two binomials with integer coefficients if the value of

$$b^2 - 4ac$$

is a perfect square. If $b^2 - 4ac = 0$, the factors will be the same.

EXAMPLE 4   Factor $3p^2 - 4p - 4$.

Solution   In this trinomial, $a = 3$, $b = -4$, and $c = -4$. To see whether it factors, we evaluate $b^2 - 4ac$.

$$b^2 - 4ac = (-4)^2 - 4(3)(-4) \qquad \text{Substitute } -4 \text{ for } b, 3 \text{ for } a, \text{ and } -4 \text{ for } c.$$
$$= 16 + 48$$
$$= 64$$

Since 64 is a perfect square, the trinomial is factorable.

To factor the trinomial, we note that the first terms of the binomial factors must be $3p$ and p to give the first term of $3p^2$.

$$3p^2 - 4p - 4 = \overbrace{(3p + ?)(p + ?)}^{3p^2}$$

The product of the last terms must be -4, and the sum of the products of the outer terms and the inner terms must be $-4p$.

$$3p^2 - 4p - 4 = \underbrace{\overbrace{(3p + ?)(p + ?)}^{-4}}_{O + I = -4p}$$

Because $1(-4)$, $-1(4)$, and $-2(2)$ all give a product of -4, there are six possible combinations to consider:

$$(3p + 1)(p - 4) \qquad (3p - 4)(p + 1)$$
$$(3p - 1)(p + 4) \qquad (3p + 4)(p - 1)$$
$$(3p - 2)(p + 2) \qquad (3p + 2)(p - 2)$$

Of these possibilities, only the last gives the required middle term of $-4p$. Thus,

$$3p^2 - 4p - 4 = (3p + 2)(p - 2) \qquad \blacksquare$$

| EXAMPLE 5 | Factor $4t^2 - 3t - 5$, if possible. |

Solution In this trinomial, $a = 4$, $b = -3$, and $c = -5$. To see whether the trinomial is factorable, we evaluate $b^2 - 4ac$ by substituting the values of a, b, and c.

$$b^2 - 4ac = (-3)^2 - 4(4)(-5)$$
$$= 9 + 80$$
$$= 89$$

Since 89 is not a perfect square, the trinomial is not factorable using only integer coefficients. ■

It is not easy to give specific rules for factoring general trinomials, because some guesswork is often necessary. However, the following hints are helpful.

Factoring a General Trinomial

1. Write the trinomial in descending powers of one variable.

2. Test the trinomial for factorability.

3. Factor out any greatest common factor (including -1 if that is necessary to make the coefficient of the first term positive).

4. When the sign of the first term of a trinomial is $+$ and the sign of the third term is $+$, the signs between the terms of each binomial factor are the same as the sign of the middle term of the trinomial.

   When the sign of the first term is $+$ and the sign of the third term is $-$, the signs between the terms of the binomial are opposite.

5. Try various combinations of first terms and last terms until you find the one that works.

6. Check the factorization by multiplication.

| EXAMPLE 6 | Factor $24y + 10xy - 6x^2y$. |

Solution We write the trinomial in descending powers of x and factor out the common factor of $-2y$:

$$24y + 10xy - 6x^2y = -6x^2y + 10xy + 24y$$
$$= -2y(3x^2 - 5x - 12)$$

In the trinomial $3x^2 - 5x - 12$, $a = 3$, $b = -5$, and $c = -12$. Thus,

$$b^2 - 4ac = (-5)^2 - 4(3)(-12)$$
$$= 25 + 144$$
$$= 169$$

Since 169 is a perfect square, the trinomial will factor.

Because the sign of the third term of $3x^2 - 5x - 12$ is $-$ and the sign of the first term is positive, the signs between the binomial factors will be opposite. Because the first term is $3x^2$, the first terms of the binomial factors must be $3x$ and x.

$$\overbrace{}^{3x^2}$$
$$-2y(3x^2 - 5x - 12) = -2y(3x \qquad)(x \qquad)$$

The product of the last terms must be -12, and the sum of the products of the outer terms and the products of the inner terms must be $-5x$.

$$-2y(3x^2 - 5x - 12) = -2y(3x \quad ?)(x \quad ?)$$

$$\overset{-12}{\overbrace{}}$$

$$O + I = -5x$$

Because $1(-12)$, $2(-6)$, $3(-4)$, $12(-1)$, $6(-2)$, and $4(-3)$ all give a product of -12, there are 12 possible combinations to consider.

$$(3x + 1)(x - 12) \qquad (3x - 12)(x + 1)$$
$$(3x + 2)(x - 6) \qquad (3x - 6)(x + 2)$$
$$(3x + 3)(x - 4) \qquad (3x - 4)(x + 3)$$
$$(3x + 12)(x - 1) \qquad (3x - 1)(x + 12)$$
$$(3x + 6)(x - 2) \qquad (3x - 2)(x + 6)$$

The one to choose $\rightarrow (3x + 4)(x - 3) \qquad (3x - 3)(x + 4)$

The six combinations printed in color cannot work, because one of the factors has a common factor. This implies that $3x^2 - 5x - 12$ would have a common factor, which it doesn't.

After mentally trying the remaining combinations, we find that only $(3x + 4)(x - 3)$ gives the proper middle term of $-5x$. Thus,

$$24y + 10xy - 6x^2y = -2y(3x^2 - 5x - 12)$$
$$= -2y(3x + 4)(x - 3)$$

Verify this result by multiplication.   ■

EXAMPLE 7   Factor $6y + 13x^2y + 6x^4y$.

Solution   We write the trinomial in descending powers of x and factor out the common factor of y to obtain

$$6y + 13x^2y + 6x^4y = 6x^4y + 13x^2y + 6y$$
$$= y(6x^4 + 13x^2 + 6)$$

A test of the trinomial $6x^4 + 13x^2 + 6$ will show that it will factor. Because the coefficients of the first and last terms are positive and the sign of the middle term is positive, the signs between the terms in each binomial will be $+$.

Since the first term of the trinomial is $6x^4$, the first terms of the binomial factors must be $2x^2$ and $3x^2$, or perhaps x^2 and $6x^2$.

Since the product of the last terms of the binomial factors must be 6, we must find two numbers whose product is 6 that will lead to a middle term of $13x^2$. After trying some combinations, we find the one that works.

$$6y + 13x^2y + 6x^4y = y(6x^4 + 13x^2 + 6)$$
$$= y(2x^2 + 3)(3x^2 + 2)$$

Verify this result by multiplication.   ■

EXAMPLE 8

Factor $x^{2n} + x^n - 2$.

Solution Since the first term is x^{2n}, the first terms of the binomial factors must be x^n and x^n.

$$x^{2n} + x^n - 2 = (x^n \quad)(x^n \quad)$$

Since the third term of the trinomial is -2, the last terms of the binomial factors must have opposite signs, have a product of -2, and lead to a middle term of x^n. The only combination that works is

$$x^{2n} + x^n - 2 = (x^n + 2)(x^n - 1)$$

Verify this result by multiplication. ∎

■ USING SUBSTITUTION TO FACTOR TRINOMIALS

EXAMPLE 9

Factor the trinomial $(x + y)^2 + 7(x + y) + 12$.

Solution We rewrite the trinomial $(x + y)^2 + 7(x + y) + 12$ as $z^2 + 7z + 12$, where $z = x + y$. The trinomial $z^2 + 7z + 12$ factors as $(z + 4)(z + 3)$.

To find the factorization of $(x + y)^2 + 7(x + y) + 12$, we substitute $x + y$ for z in the expression $(z + 4)(z + 3)$ to obtain

$$z^2 + 7z + 12 = (z + 4)(z + 3)$$
$$(x + y)^2 + 7(x + y) + 12 = (x + y + 4)(x + y + 3)$$ ∎

■ FACTORING BY GROUPING

EXAMPLE 10

Factor $x^2 + 6x + 9 - z^2$.

Solution We group the first three terms together and factor the trinomial to get

$$x^2 + 6x + 9 - z^2 = (x + 3)(x + 3) - z^2$$
$$= (x + 3)^2 - z^2$$

We can now factor the difference of two squares to get

$$x^2 + 6x + 9 - z^2 = (x + 3 + z)(x + 3 - z)$$ ∎

■ USING GROUPING TO FACTOR TRINOMIALS

The method of factoring by grouping can be used to help factor trinomials of the form $ax^2 + bx + c$. For example, to factor the trinomial $6x^2 + 7x - 3$, we proceed as follows:

1. First determine the product ac: $6(-3) = -18$. This number is called the *key number*.

2. Find two factors of the key number -18 whose sum is $b = 7$:

$$9(-2) = -18 \quad \text{and} \quad 9 + (-2) = 7$$

3. Use the factors 9 and -2 as coefficients of terms to be placed between $6x$ and -3:

$$6x^2 + 7x - 3 = 6x^2 + 9x - 2x - 3$$

4. Factor by grouping:

$$6x^2 + 9x - 2x - 3 = 3x(2x + 3) - (2x + 3)$$
$$= (2x + 3)(3x - 1) \qquad \text{Factor out } 2x + 3.$$

We can verify this factorization by multiplication.

EXAMPLE 11 Factor $10x^2 + 13x - 3$.

Solution Since $a = 10$ and $c = -3$ in this trinomial, $ac = -30$. We now find two factors of -30 whose sum is $+13$. Two such factors are 15 and -2. We use these factors as coefficients of two terms to be placed between $10x^2$ and -3:

$$10x^2 + 15x - 2x - 3$$

Finally, we factor $10x^2 + 15x - 2x - 3$ by grouping.

$$5x(2x + 3) - (2x + 3) = (2x + 3)(5x - 1)$$

Thus, $10x^2 + 13x - 3 = (2x + 3)(5x - 1)$. ■

Orals *Factor each expression.*

1. $x^2 + 3x + 2$ **2.** $x^2 + 5x + 4$
3. $x^2 - 5x + 6$ **4.** $x^2 - 3x - 4$
5. $2x^2 + 3x + 1$ **6.** $3x^2 + 4x + 1$

EXERCISE 5.6

In Exercises 1–10, use a special product formula to factor each perfect square trinomial.

1. $x^2 + 2x + 1$ **2.** $y^2 - 2y + 1$
3. $a^2 - 18a + 81$ **4.** $b^2 + 12b + 36$
5. $4y^2 + 4y + 1$ **6.** $9x^2 + 6x + 1$
7. $9b^2 - 12b + 4$ **8.** $4a^2 - 12a + 9$
9. $9z^2 + 24z + 16$ **10.** $16z^2 - 24z + 9$

In Exercises 11–14, test each trinomial for factorability.

11. $x^2 + 5x - 6$ **12.** $6x^2 - 7x - 3$
13. $2b^2 + 5b + 4$ **14.** $8x^2 + 2x - 5$

In Exercises 15–22, test each trinomial for factorability and factor it, if possible.

15. $x^2 + 5x + 6$

16. $y^2 + 7y + 6$

17. $x^2 - 7x + 10$

18. $c^2 - 7c + 12$

19. $b^2 + 8b + 18$

20. $x^2 - 12x + 35$

21. $y^2 - 4y - 21$

22. $x^2 + 4x - 28$

In Exercises 23–34, factor each trinomial. If the coefficient of the first term is negative, begin by factoring out -1.

23. $3x^2 + 12x - 63$

24. $2y^2 + 4y - 48$

25. $a^2b^2 - 13ab^2 + 22b^2$

26. $a^2b^2x^2 - 18a^2b^2x + 81a^2b^2$

27. $b^2x^2 - 12bx^2 + 35x^2$

28. $c^3x^2 + 11c^3x - 42c^3$

29. $-a^2 + 4a + 32$

30. $-x^2 - 2x + 15$

31. $-3x^2 + 15x - 18$

32. $-2y^2 - 16y + 40$

33. $-4x^2 + 4x + 80$

34. $-5a^2 + 40a - 75$

In Exercises 35–72, factor each trinomial. Factor out all common monomials first (including a -1 if the first term is negative). If a trinomial is prime, so indicate.

35. $6y^2 + 7y + 2$

36. $6x^2 - 11x + 3$

37. $8a^2 + 6a - 9$

38. $15b^2 + 4b - 4$

39. $6x^2 - 5x - 4$

40. $18y^2 - 3y - 10$

41. $5x^2 + 4x + 1$

42. $6z^2 + 17z + 12$

43. $8x^2 - 10x + 3$

44. $4a^2 + 20a + 3$

45. $6z^2 + 7z - 20$

46. $7x^2 - 23x + 6$

47. $a^2 - 3ab - 4b^2$

48. $b^2 + 2bc - 80c^2$

49. $2y^2 + yt - 6t^2$

50. $3x^2 - 10xy - 8y^2$

51. $3x^3 - 10x^2 + 3x$

52. $6y^2 + 7y + 2$

53. $-3a^2 + ab + 2b^2$

54. $-2x^2 + 3xy + 5y^2$

55. $9t^2 + 3t - 2$

56. $3t^3 - 3t^2 + t$

57. $9x^2 - 12x + 4$

58. $4a^2 + 28a + 49$

59. $-4x^2 - 9 + 12x$

60. $6x + 4 + 9x^2$

61. $15x^2 + 2 - 13x$

62. $-90x^2 + 2 - 8x$

63. $5a^2 + 45b^2 - 30ab$

64. $x^2 + 324y^2 - 36xy$

65. $8x^2z + 6xyz + 9y^2z$

66. $x^3 - 60xy^2 + 7x^2y$

67. $15x^2 + 74x - 5$

68. $15x^2 - 7x - 30$

69. $21x^4 - 10x^3 - 16x^2$

70. $16x^3 - 50x^2 + 36x$

71. $6x^2y^2 - 17xyz + 12z^2$

72. $6u^2v^2 - uvz + 12z^2$

In Exercises 73–82, factor each trinomial.

73. $x^4 + 8x^2 + 15$

74. $x^4 + 11x^2 + 24$

75. $y^4 - 13y^2 + 30$

76. $y^4 - 13y^2 + 42$

77. $a^4 - 13a^2 + 36$

78. $b^4 - 17b^2 + 16$

79. $z^4 - z^2 - 12$

80. $c^4 - 8c^2 - 9$

81. $4x^3 + x^6 + 3$

82. $a^6 - 2 + a^3$

In Exercises 83–90, factor each expression. Assume that n is a natural number.

83. $x^{2n} + 2x^n + 1$

84. $x^{4n} - 2x^{2n} + 1$

85. $2a^{6n} - 3a^{3n} - 2$

86. $b^{2n} - b^n - 6$

87. $x^{4n} + 2x^{2n}y^{2n} + y^{4n}$

88. $y^{6n} + 2y^{3n}z + z^2$

89. $6x^{2n} + 7x^n - 3$

90. $12y^{4n} + 10y^{2n} + 2$

In Exercises 91–96, factor each expression.

91. $(x + 1)^2 + 2(x + 1) + 1$

92. $(a + b)^2 - 2(a + b) + 1$

93. $(a + b)^2 - 2(a + b) - 24$

94. $(x - y)^2 + 3(x - y) - 10$

95. $6(x + y)^2 - 7(x + y) - 20$

96. $2(x - z)^2 + 9(x - z) + 4$

In Exercises 97–106, factor each expression.

97. $x^2 + 4x + 4 - y^2$

98. $x^2 - 6x + 9 - 4y^2$

99. $x^2 + 2x + 1 - 9z^2$

100. $x^2 + 10x + 25 - 16z^2$

101. $c^2 - 4a^2 + 4ab - b^2$

102. $4c^2 - a^2 - 6ab - 9b^2$

103. $a^2 - b^2 + 8a + 16$

104. $a^2 + 14a - 25b^2 + 49$

105. $4x^2 - z^2 + 4xy + y^2$

106. $x^2 - 4xy - 4z^2 + 4y^2$

In Exercises 107–114, use grouping to help factor each trinomial.

107. $a^2 - 17a + 16$

108. $b^2 - 4b - 21$

109. $2u^2 + 5u + 3$

110. $6y^2 + 5y - 6$

111. $20r^2 - 7rs - 6s^2$

112. $6s^2 + st - 12t^2$

113. $20u^2 + 19uv + 3v^2$

114. $12m^2 + mn - 6n^2$

Writing Exercises *Write a paragraph using your own words.*

1. Explain how you would factor -1 from a trinomial.

2. Explain how you would test the polynomial $ax^2 + bx + c$ for factorability.

Something to Think About

1. Because it is the difference of two squares, $x^2 - q^2$ always factors. Does the test for factorability predict this?

2. The polynomial $ax^2 + ax + a$ factors, because a is a common factor. Does the test for factorability predict this? If not, is there something wrong with the test? Explain.

Review Exercises *Solve each equation.*

1. $\dfrac{2 + x}{11} = 3$

2. $\dfrac{3y - 12}{2} = 9$

3. $\dfrac{2}{3}(5t - 3) = 38$

4. $3(p + 2) = 4p$

5. $11r + 6(3 - r) = 3$

6. $2q^2 - 9 = q(q + 3) + q^2$

5.7 Summary of Factoring Techniques

In this section, we will discuss ways to approach a randomly chosen factoring problem. For example, suppose we wish to factor the trinomial

$$x^2y^2z^3 + 7xy^2z^3 + 6y^2z^3$$

We begin by attempting to identify the problem type. The first to look for is **factoring out a common monomial**. Because the trinomial has a common monomial factor of y^2z^3, we factor it out:

$$x^2y^2z^3 + 7xy^2z^3 + 6y^2z^3 = y^2z^3(x^2 + 7x + 6)$$

We note that $x^2 + 7x + 6$ is a trinomial that can be factored as $(x + 6)(x + 1)$. Thus,

$$x^2y^2z^3 + 7xy^2z^3 + 6y^2z^3 = y^2z^3(x^2 + 7x + 6)$$
$$= y^2z^3(x + 6)(x + 1)$$

To identify the type of factoring problem, we follow these steps.

Strategy for Identifying the Type of Factoring Problem

1. Factor out all common monomial factors.
2. If an expression has two terms, check to see if the problem type is
   a. **The difference of two squares**: $(x^2 - y^2) = (x + y)(x - y)$
   b. **The sum of two cubes**: $(x^3 + y^3) = (x + y)(x^2 - xy + y^2)$
   c. **The difference of two cubes**: $(x^3 - y^3) = (x - y)(x^2 + xy + y^2)$
3. If an expression has three terms, attempt to factor the trinomial as a **general trinomial**.
4. If an expression has four or more terms, try factoring by **grouping**.
5. Continue until each individual factor is prime.
6. Check the results by multiplying.

EXAMPLE 1 Factor $48a^4c^3 - 3b^4c^3$.

Solution We begin by factoring out the common monomial factor of $3c^3$:

$$48a^4c^3 - 3b^4c^3 = 3c^3(16a^4 - b^4)$$

Since the expression $16a^4 - b^4$ has two terms, we check to see if it is the difference of two squares, which it is. As the difference of two squares, it factors as $(4a^2 + b^2)(4a^2 - b^2)$. Thus,

$$48a^4c^3 - 3b^4c^3 = 3c^3(\mathbf{16a^4 - b^4})$$
$$= 3c^3(\mathbf{4a^2 + b^2})(\mathbf{4a^2 - b^2})$$

The binomial $4a^2 + b^2$ is the sum of two squares and is prime. However, the binomial $4a^2 - b^2$ is the difference of two squares and factors as $(2a + b)(2a - b)$. Thus,

$$48a^4c^3 - 3b^4c^3 = 3c^3(16a^4 - b^4)$$
$$= 3c^3(4a^2 + b^2)(\mathbf{4a^2 - b^2})$$
$$= 3c^3(4a^2 + b^2)(\mathbf{2a + b})(\mathbf{2a - b})$$

Since each of the individual factors is prime, the factorization is complete. ■

EXAMPLE 2   Factor $x^5y + x^2y^4 - x^3y^3 - y^6$.

Solution   We begin by factoring out the common monomial factor of y:

$$x^5y + x^2y^4 - x^3y^3 - y^6 = y(x^5 + x^2y^3 - x^3y^2 - y^5)$$

Because the expression $x^5 + x^2y^3 - x^3y^2 - y^5$ has four terms, we try factoring by grouping to obtain

$$x^5y + x^2y^4 - x^3y^3 - y^6 = y(x^5 + x^2y^3 - x^3y^2 - y^5)$$
$$= y[x^2(\mathbf{x^3 + y^3}) - y^2(\mathbf{x^3 + y^3})]$$
$$= y(\mathbf{x^3 + y^3})(x^2 - y^2) \qquad \text{Factor out } x^3 + y^3.$$

Finally, we factor $x^3 + y^3$ (the sum of two cubes) and $x^2 - y^2$ (the difference of two squares) to obtain

$$x^5y + x^2y^4 - x^3y^3 - y^6 = y(x + y)(x^2 - xy + y^2)(x + y)(x - y)$$

Because each of the individual factors is prime, the factorization is complete. ■

EXAMPLE 3   Factor $p^{-4} - p^{-2} - 6$.

Solution   We factor this expression as if it were a trinomial:

$$p^{-4} - p^{-2} - 6 = (p^{-2} - 3)(p^{-2} + 2)$$
$$= \left(\frac{1}{p^2} - 3\right)\left(\frac{1}{p^2} + 2\right)$$

■

EXAMPLE 4   Factor $x^3 + 5x^2 + 6x + x^2y + 5xy + 6y$.

Solution   Since there are more than three terms, we try factoring by grouping.

We can factor x from the first three terms and y from the last three terms and proceed as follows:

$$x^3 + 5x^2 + 6x + x^2y + 5xy + 6y = x(x^2 + 5x + 6) + y(x^2 + 5x + 6)$$

$$= (x^2 + 5x + 6)(x + y) \qquad \text{Factor out } x^2 + 5x + 6.$$

$$= (x + 3)(x + 2)(x + y) \qquad \text{Factor } x^2 + 5x + 6.$$ ∎

EXAMPLE 5   Factor $x^4 + 2x^3 + x^2 + x + 1$.

Solution   Since there are more than three terms, we try factoring by grouping.
We can factor x^2 from the first three terms and proceed as follows:

$$x^4 + 2x^3 + x^2 + x + 1 = x^2(x^2 + 2x + 1) + (x + 1)$$

$$= x^2(x + 1)(x + 1) + (x + 1) \qquad \text{Factor } x^2 + 2x + 1.$$

$$= (x + 1)[x^2(x + 1) + 1] \qquad \text{Factor out } x + 1.$$

$$= (x + 1)(x^3 + x^2 + 1) \qquad \begin{array}{l}\text{Remove the inner} \\ \text{parentheses.}\end{array}$$ ∎

Orals   *Factor each expression.*

1. $x^2 - y^2$ **2.** $2x^3 - 4x^4$

3. $x^2 + 4x + 4$ **4.** $x^2 - 5x + 6$

5. $x^3 - 8$ **6.** $x^3 + 8$

EXERCISE 5.7

In Exercises 1–54, factor each polynomial, if possible.

1. $x^2 + 8x + 16$ **2.** $20 + 11x - 3x^2$

3. $8x^3y^3 - 27$ **4.** $3x^2y + 6xy^2 - 12xy$

5. $xy - ty + xs - ts$ **6.** $bc + b + cd + d$

7. $25x^2 - 16y^2$ **8.** $27x^9 - y^3$

9. $12x^2 + 52x + 35$ **10.** $12x^2 + 14x - 6$

11. $6x^2 - 14x + 8$ **12.** $12x^2 - 12$

13. $56x^2 - 15x + 1$ **14.** $7x^2 - 57x + 8$

15. $4x^2y^2 + 4xy^2 + y^2$ **16.** $100z^2 - 81t^2$

17. $x^3 + (a^2y)^3$ **18.** $4x^2y^2z^2 - 26x^2y^2z^3$

19. $2x^3 - 54$ **20.** $4(xy)^3 + 256$

21. $ae + bf + af + be$ **22.** $a^2x^2 + b^2y^2 + b^2x^2 + a^2y^2$

23. $2(x + y)^2 + (x + y) - 3$ **24.** $(x - y)^3 + 125$

25. $625x^4 - 256y^4$ **26.** $2(a - b)^2 + 5(a - b) + 3$

27. $36x^4 - 36$ **28.** $6x^2 - 63 - 13x$

29. $2x^6 + 2y^6$ **30.** $x^4 - x^4y^4$

31. $a^4 - 13a^2 + 36$ **32.** $x^4 - 17x^2 + 16$

33. $x^2 + 6x + 9 - y^2$ **34.** $x^2 + 10x + 25 - y^8$

35. $4x^2 + 4x + 1 - 4y^2$ **36.** $9x^2 - 6x + 1 - 25y^2$

37. $z^2 + 8z + 16 - 16y^2$

38. $x^2 + 10x + 25 - 25y^2$

39. $x^5 + x^2 - x^3 - 1$

40. $x^5 - x^2 - 4x^3 + 4$

41. $x^5 - 9x^3 + 8x^2 - 72$

42. $x^5 - 4x^3 - 8x^2 + 32$

43. $2x^5z - 2x^2y^3z - 2x^3y^2z + 2y^5z$

44. $x^2y^3 - 4x^2y - 9y^3 + 36y$

45. $x^{2m} - x^m - 6$

46. $a^{2n} - b^{2n}$

47. $a^{3n} - b^{3n}$

48. $x^{3m} + y^{3m}$

49. $x^{-2} + 2x^{-1} + 1$

50. $4a^{-2} - 12a^{-1} + 9$

51. $6x^{-2} - 5x^{-1} - 6$

52. $x^{-4} - y^{-4}$

53. $a^4 + 3a^3 + 2a^2 + a + 2$

54. $x^4 + x^3 - 2x^2 + x - 1$

Writing Exercises *Write a paragraph using your own words.*

1. What is your strategy for factoring a polynomial?

2. Explain how you can know that your factorization is correct.

Something to Think About

1. If you have the choice of factoring a polynomial as the difference of two squares or as the difference of two cubes, which would you do first? Why?

2. Can several polynomials have a greatest common factor? Find the GCF of $2x^2 + 7x + 3$ and $x^2 - 2x - 15$.

3. Factor $x^4 + x^2 + 1$.
(*Hint:* Add and subtract x^2.)

4. Factor $x^4 + 7x^2 + 16$.
(*Hint:* Add and subtract x^2.)

Review Exercises *Do the operations.*

1. $(3a^2 + 4a - 2) + (4a^2 - 3a - 5)$

2. $(-4b^2 - 3b - 2) - (3b^2 - 2b + 5)$

3. $5(2y^2 - 3y + 3) - 2(3y^2 - 2y + 6)$

4. $4(3x^2 + 3x + 3) + 3(x^2 - 3x - 4)$

5. $(m + 4)(m - 2)$

6. $(3p + 4q)(2p - 3q)$

5.8   Solving Equations by Factoring

■ SOLVING QUADRATIC EQUATIONS ■ GRAPHING DEVICES ■ SOLVING HIGHER-DEGREE POLYNOMIAL EQUATIONS ■ GRAPHING DEVICES ■ PROBLEM SOLVING

We have previously solved linear (first-degree) equations. In this section, we solve certain equations of higher degree.

■ SOLVING QUADRATIC EQUATIONS

An equation such as $3x^2 + 4x - 7 = 0$ or $-5y^2 + 3y + 8 = 0$ is called a **quadratic** (or **second-degree**) equation.

Quadratic Equations

A **quadratic equation** is any equation that can be written in the form

$$ax^2 + bx + c = 0$$

where a, b, and c are real numbers and $a \neq 0$.

Many quadratic equations can be solved by factoring and then using the **zero-factor theorem**.

Zero-Factor Theorem

If a and b are real numbers, then

If $ab = 0$, then $a = 0$ or $b = 0$.

The zero-factor theorem states that *if the product of two or more numbers is 0, then at least one of the numbers must be 0.*

To solve the quadratic equation

$$x^2 + 5x + 6 = 0$$

we factor its left-hand side to obtain

$$(x + 3)(x + 2) = 0$$

Since the product of $x + 3$ and $x + 2$ is 0, then at least one of the factors is 0. We can set each factor equal to 0 and solve each resulting linear equation for x:

$$
\begin{aligned}
x + 3 &= 0 &\quad \text{or} \quad& x + 2 = 0 \\
x &= -3 &\quad|\quad& x = -2
\end{aligned}
$$

To check these solutions, we substitute -3 and -2 for x in the equation and verify that each number satisfies the equation.

$$
\begin{array}{ccc}
x^2 + 5x + 6 = 0 & \text{or} & x^2 + 5x + 6 = 0 \\
(-3)^2 + 5(-3) + 6 \stackrel{?}{=} 0 & & (-2)^2 + 5(-2) + 6 \stackrel{?}{=} 0 \\
9 - 15 + 6 \stackrel{?}{=} 0 & & 4 - 10 + 6 \stackrel{?}{=} 0 \\
0 = 0 & & 0 = 0
\end{array}
$$

Both -3 and -2 are solutions, because they both satisfy the equation.

If c is 0 in an equation of the form $ax^2 + bx + c = 0$, the equation can always be solved by factoring.

EXAMPLE 1   Solve the equation $3x^2 + 6x = 0$.

Solution   To solve the equation, we factor the left-hand side, set each factor equal to 0, and solve each resulting equation for x.

$$3x^2 + 6x = 0$$
$$3x(x + 2) = 0 \qquad \text{Factor out the common factor of } 3x.$$

$$3x = 0 \quad \text{or} \quad x + 2 = 0$$
$$x = 0 \qquad \qquad x = -2$$

Verify that both solutions check.   ∎

 WARNING!   In Example 1, do not divide both sides by $3x$, or you will lose the solution $x = 0$.

EXAMPLE 2   Solve the equation $x^2 - 16 = 0$.

Solution   To solve the equation, we factor the difference of two squares on the left-hand side, set each factor equal to 0, and solve each resulting equation.

$$x^2 - 16 = 0$$
$$(x + 4)(x - 4) = 0$$

$$x + 4 = 0 \quad \text{or} \quad x - 4 = 0$$
$$x = -4 \qquad \qquad x = 4$$

Verify that both solutions check.   ∎

Many equations that do not appear to be quadratic can be put into quadratic form ($ax^2 + bx + c = 0$) and then solved by factoring.

EXAMPLE 3   Solve the equation $x = \dfrac{6}{5} - \dfrac{6}{5}x^2$.

Solution   We write the equation in quadratic form and then solve by factoring.

$$x = \frac{6}{5} - \frac{6}{5}x^2$$

$$5x = 6 - 6x^2 \qquad \text{Multiply both sides by 5.}$$

$$6x^2 + 5x - 6 = 0 \qquad \text{Add } 6x^2 \text{ to both sides and subtract 6 from both sides.}$$

$$(3x - 2)(2x + 3) = 0 \qquad \text{Factor the trinomial.}$$

$$3x - 2 = 0 \quad \text{or} \quad 2x + 3 = 0 \qquad \text{Set each factor equal to 0.}$$
$$3x = 2 \qquad \qquad 2x = -3$$
$$x = \frac{2}{3} \qquad \qquad x = -\frac{3}{2}$$

Verify that both solutions check.   ∎

WARNING! To solve a quadratic equation by factoring, be sure to set the quadratic polynomial equal to 0 before factoring and using the zero-factor theorem. Do not make the error

$$6x^2 + 5x = 6$$
$$x(6x + 5) = 6$$
$$x = 6 \quad \text{or} \quad 6x + 5 = 6$$
$$x = \frac{1}{6}$$

Neither solution checks.

■ GRAPHING DEVICES

To solve a quadratic equation such as $x^2 + 4x - 5 = 0$ with a graphing device, we can use settings of $[-10, 10]$ for x and $[-10, 10]$ for y and graph the quadratic function $y = x^2 + 4x - 5$, as shown in Figure 5-4(a). We can then trace to find the x-coordinates of the x-intercepts of the parabola. See Figures 5-4(b) and 5-4(c). For better results, we can zoom in to find good approximations for the numbers x that make $y = 0$. They are the approximate solutions of the equation.

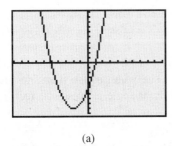
(a)

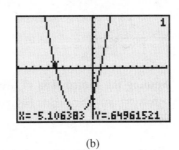
(b)

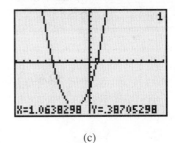
(c)

FIGURE 5-4

■ SOLVING HIGHER-DEGREE POLYNOMIAL EQUATIONS

We can solve many polynomial equations with degree greater than 2 by factoring.

EXAMPLE 4 Solve the equation $6x^3 - x^2 - 2x = 0$.

Solution We factor an x from the third-degree polynomial on the left-hand side and proceed as follows:

$$6x^3 - x^2 - 2x = 0$$
$$x(6x^2 - x - 2) = 0 \qquad \text{Factor out an } x.$$
$$x(3x - 2)(2x + 1) = 0 \qquad \text{Factor } 6x^2 - x - 2.$$
$$x = 0 \quad \text{or} \quad 3x - 2 = 0 \quad \text{or} \quad 2x + 1 = 0 \qquad \text{Set each factor equal to 0.}$$
$$x = \frac{2}{3} \qquad x = -\frac{1}{2}$$

Verify that all three solutions check.

EXAMPLE 5   Solve the equation $x^4 - 5x^2 + 4 = 0$.

Solution   We factor the trinomial on the left-hand side and proceed as follows:

$$x^4 - 5x^2 + 4 = 0$$
$$(x^2 - 1)(x^2 - 4) = 0$$

$(x + 1)(x - 1)(x + 2)(x - 2) = 0$ Factor $x^2 - 1$ and $x^2 - 4$.

$x + 1 = 0$ or $x - 1 = 0$ or $x + 2 = 0$ or $x - 2 = 0$ Set each factor equal to 0.

$x = -1$ | $x = 1$ | $x = -2$ | $x = 2$

Verify that all four solutions check.   ■

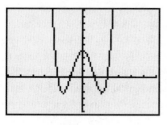

FIGURE 5-5

■ GRAPHING DEVICES

To solve the equation $x^4 - 5x^2 + 4 = 0$ with a graphing device, we can use settings of $[-6, 6]$ for x and $[-5, 10]$ for y and graph the polynomial function $y = x^4 - 5x^2 + 4$, as shown in Figure 5-5. We can then read the values of x that make $y = 0$. They are $x = -2, -1, 1,$ and 2. If the x-coordinates of the x-intercepts were not obvious, we could trace to get their values.

■ PROBLEM SOLVING

EXAMPLE 6   **Finding the dimensions of a truss** The width of the triangular truss in Figure 5-6 is 3 times its height. The area of the triangle is 96 square feet. Find its width and height.

Analyze the Problem   We can let x be the positive number that represents the height of the truss. Then $3x$ represents its width. We can substitute x for h, $3x$ for b, and 96 for A in the formula for the area of a triangle and solve for x.

Form and Solve an Equation

$$A = \frac{1}{2}bh$$

$$96 = \frac{1}{2}(3x)x$$

$192 = 3x^2$ Multiply both sides by 2.

$64 = x^2$ Divide both sides by 3.

$0 = x^2 - 64$ Subtract 64 from both sides.

$0 = (x + 8)(x - 8)$ Factor the difference of two squares.

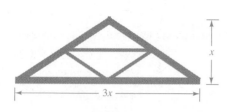

FIGURE 5-6

$x + 8 = 0$ or $x - 8 = 0$

$x = -8$ | $x = 8$

State the Conclusion   Since the height of a triangle cannot be negative, we must discard the negative solution. Thus, the height of the truss is 8 feet, and its width is 3(8), or 24 feet. ·

Check the Result The area of a triangle with a width (base) of 24 feet and a height of 8 feet is 96 square feet:

$$A = \frac{1}{2}bh = \frac{1}{2}(24)(8) = 12(8) = 96$$

The solution checks. ■

EXAMPLE 7 **Ballistics** If the initial velocity of an object thrown straight up into the air is 176 feet per second, when will the object strike the ground?

Analyze the Problem The height of an object thrown straight up into the air with an initial velocity of v feet per second is given by the formula

$$h = vt - 16t^2$$

The height h is in feet, and t represents the number of seconds since the object was released.

Form and Solve an Equation When the object hits the ground, its height will be 0. Thus, we set h equal to 0, set v equal to 176, and solve for t.

$$h = vt - 16t^2$$
$$0 = 176t - 16t^2$$
$$0 = 16t(11 - t) \qquad \text{Factor out } 16t.$$
$$16t = 0 \quad \text{or} \quad 11 - t = 0 \qquad \text{Set each factor equal to 0.}$$
$$t = 0 \quad | \qquad t = 11$$

State the Conclusion When $t = 0$, the object's height above the ground is 0 feet, because it has not been released. When $t = 11$, the height is again 0 feet, and the object has returned to the ground. The solution is 11 seconds.

Check the Result Verify that $h = 0$ when $t = 11$. ■

Orals *Solve each equation.*

1. $(x - 2)(x - 3) = 0$　　　　　　2. $(x + 4)(x - 2) = 0$
3. $(x - 2)(x - 3)(x + 1) = 0$
4. $(x + 3)(x + 2)(x - 5)(x - 6) = 0$

EXERCISE 5.8

In Exercises 1–24, solve each equation.

1. $4x^2 + 8x = 0$　　　　　　　　2. $x^2 - 9 = 0$
3. $y^2 - 16 = 0$　　　　　　　　4. $5y^2 - 10y = 0$
5. $x^2 + x = 0$　　　　　　　　　6. $x^2 - 3x = 0$
7. $5y^2 - 25y = 0$　　　　　　　8. $y^2 - 36 = 0$

9. $z^2 + 8z + 15 = 0$

10. $w^2 + 7w + 12 = 0$

11. $y^2 - 7y + 6 = 0$

12. $n^2 - 5n + 6 = 0$

13. $y^2 - 7y + 12 = 0$

14. $x^2 - 3x + 2 = 0$

15. $x^2 + 6x + 8 = 0$

16. $x^2 + 9x + 20 = 0$

17. $3m^2 + 10m + 3 = 0$

18. $2r^2 + 5r + 3 = 0$

19. $2y^2 - 5y + 2 = 0$

20. $2x^2 - 3x + 1 = 0$

21. $2x^2 - x - 1 = 0$

22. $2x^2 - 3x - 5 = 0$

23. $3s^2 - 5s - 2 = 0$

24. $8t^2 + 10t - 3 = 0$

In Exercises 25–36, write each equation in quadratic form and solve it by factoring.

25. $x(x - 6) + 9 = 0$

26. $x^2 + 8(x + 2) = 0$

27. $8a^2 = 3 - 10a$

28. $5z^2 = 6 - 13z$

29. $b(6b - 7) = 10$

30. $2y(4y + 3) = 9$

31. $\dfrac{3a^2}{2} = \dfrac{1}{2} - a$

32. $x^2 = \dfrac{1}{2}(x + 1)$

33. $x^2 + 1 = \dfrac{5}{2}x$

34. $\dfrac{3}{5}(x^2 - 4) = -\dfrac{9}{5}x$

35. $x\left(3x + \dfrac{22}{5}\right) = 1$

36. $x\left(\dfrac{x}{11} - \dfrac{1}{7}\right) = \dfrac{6}{77}$

In Exercises 37–48, solve each equation.

37. $x^3 + x^2 = 0$

38. $2x^4 + 8x^3 = 0$

39. $y^3 - 49y = 0$

40. $2z^3 - 200z = 0$

41. $x^3 - 4x^2 - 21x = 0$

42. $x^3 + 8x^2 - 9x = 0$

43. $z^4 - 13z^2 + 36 = 0$

44. $y^4 - 10y^2 + 9 = 0$

45. $3a(a^2 + 5a) = -18a$

46. $7t^3 = 2t\left(t + \dfrac{5}{2}\right)$

47. $\dfrac{x^2(6x + 37)}{35} = x$

48. $x^2 = -\dfrac{4x^3(3x + 5)}{3}$

In Exercises 49–54, use grouping to help solve each equation.

49. $x^3 + 3x^2 - x - 3 = 0$

50. $x^3 - x^2 - 4x + 4 = 0$

51. $2r^3 + 3r^2 - 18r - 27 = 0$

52. $3s^3 - 2s^2 - 3s + 2 = 0$

53. $3y^3 + y^2 = 4(3y + 1)$

54. $w^3 + 16 = w(w + 16)$

In Exercises 55–74, solve each problem.

55. Integer problem The product of two consecutive even integers is 288. Find the integers.

57. Integer problem The sum of the squares of two consecutive positive integers is 85. Find the integers.

56. Integer problem The product of two consecutive odd integers is 143. Find the integers.

58. Integer problem The sum of the squares of three consecutive positive integers is 77. Find the integers.

59. Geometry   Find the perimeter of the rectangle in Illustration 1.

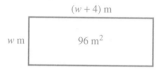

ILLUSTRATION 1

60. Geometry   One side of a rectangle is three times longer than another. If its area is 147 square centimeters, find its dimensions.

61. Geometry   Find the dimensions of the rectangle shown in Illustration 2, given that its area is 375 square feet.

ILLUSTRATION 2

62. Geometry   Find the height of the triangle shown in Illustration 3, given that its area is 162 square centimeters.

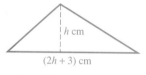

ILLUSTRATION 3

63. Fine art   An artist intends to paint a 60-square-foot mural on the large wall shown in Illustration 4. Find the dimensions of the mural if the artist leaves a border of uniform width around it.

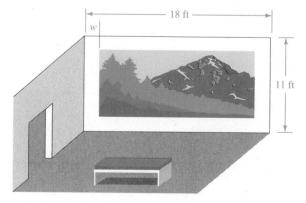

ILLUSTRATION 4

64. Gardening   A woman plans to use one-fourth of her 48-foot-by-100-foot rectangular backyard to plant a

garden. Find the perimeter of the garden if the length is to be 40 feet greater than the width.

65. Architecture   The rectangular room shown in Illustration 5 is twice as long as it is wide. It is divided into two rectangular parts by a partition, positioned as shown. If the larger part of the room contains 560 square feet, find the dimensions of the entire room.

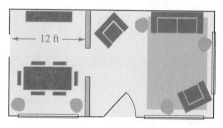

ILLUSTRATION 5

66. Perimeter of a square   If the length of one side of a square is increased by 4 inches, the area of the square becomes 9 times greater than the original area. Find the perimeter of the original square.

67. Time of flight   After how many seconds will an object hit the ground if it was thrown straight up with an initial velocity of 160 feet per second?

68. Time of flight   After how many seconds will an object hit the ground if it was thrown straight up with an initial velocity of 208 feet per second?

69. Ballistics   The muzzle velocity of a cannon is 480 feet per second. If a cannonball is fired vertically, at what times will it be at a height of 3344 feet?

70. Ballistics   A slingshot can provide an initial velocity of 128 feet per second. At what times will a stone, shot vertically upward, be 192 feet above the ground?

71. Winter recreation   The length of the rectangular ice-skating rink in Illustration 6 is 20 meters greater than twice its width. Find the width.

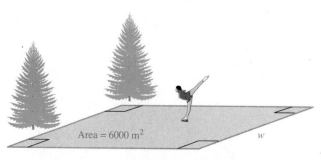

ILLUSTRATION 6

72. Carpentry A 285-square-foot room is 4 feet longer than it is wide. What length of crown molding is needed to trim the perimeter of the ceiling?

73. Designing a swimming pool Building code requires that the rectangular swimming pool in Illustration 7 be surrounded by a uniform-width walkway of at least 516 square feet. The length of the pool is 10 feet less than twice the width. How wide should the border be?

74. House construction The formula for the area of a trapezoid is $A = \frac{h(B+b)}{2}$. The area of the trapezoidal truss in Illustration 8 is 44 square feet. Find the height of the truss if the shorter base is the same as the height.

Area = 1500 ft²

w ft

ILLUSTRATION 7

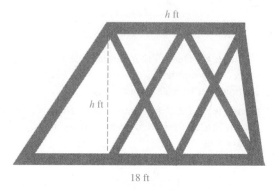

h ft

h ft

18 ft

ILLUSTRATION 8

In Exercises 75–78, use a graphing device to find the solutions of each equation, if possible. Give the answers to the nearest hundredth.

75. $x^2 - 4x + 7 = 0$

76. $2x^2 - 7x + 4 = 0$

77. $-3x^3 - 2x^2 + 5 = 0$

78. $-2x^3 - 3x - 5 = 0$

Writing Exercises *Write a paragraph using your own words.*

1. Describe the steps for solving an application problem.

2. Explain how to check the solution of an application problem.

Something to Think About *Find a quadratic equation with the given roots.*

1. 3, 5

2. −2, 6

3. 0, −5

4. $\frac{1}{2}, \frac{1}{3}$

Review Exercises

1. List the prime numbers less than 10.

2. List the composite numbers between 7 and 17.

3. The formula for the volume of a sphere is $V = \frac{4}{3}\pi r^3$. Find the volume when $r = 21.23$ centimeters. Round the answer to the nearest hundredth.

4. The formula for the volume of a cone is $V = \frac{1}{3}\pi r^2 h$. Find the volume when $r = 12.33$ meters and $h = 14.7$ meters. Round the answer to the nearest hundredth.

■ ■ ■ ■ ■ ■ ■ ■ ■ PROBLEMS AND PROJECTS

1. Explain why the following trick works.
  1. Think of a number.
  2. Add your age.
  3. Double the result.
  4. Subtract 50.
  5. Divide by 2.
  6. Subtract the number you started with.
  7. Add 25.
  8. The result will always be your age.

2. Find the difference of the volumes of the two cubes shown in Illustration 1.

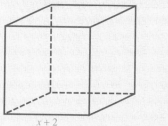 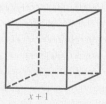

$x + 2$

$x + 1$

ILLUSTRATION 1

3. As a research technician in a medical laboratory, you have received a shipment of yearlings. Some appear to be sick and others healthy, so you need to quarantine the two separate groups. You are going to enclose a rectangular field with a fence and divide it down the middle with a double fence. What is the maximum area that can be enclosed with 2400 feet of fencing?

4. Suspicious Flight Fuel Distributors has a delivery truck that holds 256 gallons of high-octane fuel to service small planes. Each plane uses the same amount of fuel. After each plane is fueled, the truck is refilled with low-octane fuel. For the fuel to work properly in a plane, there must be at least 40% high-octane. After four planes have been serviced, there are only 81 gallons of high-octane left in the truck's mixture. How many planes can SFFD service adding low-octane fuel before more high-octane needs to be added?

PROJECT 1

Mac operates a bait shop on the Snarly River. The shop is located between two sharp bends in the river, and the area around Mac's shop has recently become a popular hiking and camping area. Mac has decided to produce some maps of the area for the use of the visitors. But although he knows the region well, he has very little idea of the actual distances from one location to another. What he knows is:

1. Big Falls, a beautiful waterfall on Snarly River, is due east of Mac's.

2. Grandview Heights, a fabulous rock climbing site, is due west of Mac's, right on the river.

3. Foster's General Store, the only sizable camping and climbing outfitter in the area, is located on the river some distance west and north of Mac's.

Mac hired an aerial photographer to take pictures of the area and found some surprising results. If Mac's bait shop is treated as the origin of a coordinate system, with the y-axis running north-south and the x-axis running east-west, then on the domain $-4 \leq x \leq 4$ (where the units are miles), the river follows the curve

$$P(x) = \frac{1}{4}(x^3 - x^2 - 6x)$$

(continued)

■ ■ ■ ■ ■ ■ ■ ■ ■ ■ **PROBLEMS AND PROJECTS** *(continued)*

The aerial photograph is shown in Illustration 2.

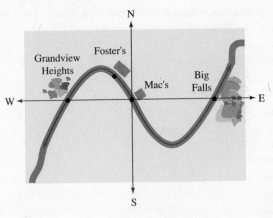

ILLUSTRATION 2

a. Mac would like to include on his maps the exact locations (relative to Mac's) of Big Falls and Grandview Heights. Find these for him, and explain to Mac why your answers must be correct.

b. Mac and Foster have determined that Foster's General Store is 0.7 miles west of the bait shop. Since it is on the river, it is a bit north as well. They decide that to promote business, they will join together to clear a few campsites in the region bordered by the straight-line paths that run between Mac's and Foster's, Mac's and Grandview Heights, and Foster's and Grandview Heights. If they clear 1 campsite for each 40 full acres of area, how many campsites can they put in? (*Hint:* A square mile contains 640 acres.)

c. A path runs in a straight line directly southeast (along the line $y = -x$) from Mac's to the river. How far east and how far south has a hiker on this trail traveled when he or she reaches the river?

PROJECT 2 The rate at which fluid flows through a cylindrical pipe, or any cylinder-shaped tube (an artery, for instance), is

$$\text{Velocity of fluid flow} = V = \frac{p}{nL}(R^2 - r^2)$$

where p is the difference in pressure between the two ends of the tube, L is the length of the tube, R is the radius of the tube, and n is the *viscosity constant*, a measure of the thickness of a fluid. Since the variable r represents the distance from the center of the tube, $0 \leq r \leq R$. (See Illustration 3.) In most situations, p, L, R, and n are constants, so V is a function of r.

$$V(r) = \frac{p}{nL}(R^2 - r^2)$$

Cross section of tube

R

Fluid that is a distance r from center of tube

r

Wall of tube

ILLUSTRATION 3

It can be shown that the velocity of a fluid moving in the tube depends on how far from the center (or how close to the wall of the tube) the fluid is.

a. Consider a pipe with radius 5 centimeters and length 60 centimeters. Suppose that $p = 15$ and $n = 0.001$ (water has a viscosity of approximately 0.001). Find the velocity of the fluid at the center of the pipe. The units of measurement for V will be centimeters per second.

Find the velocity of the fluid when it is halfway between the center and the wall of the pipe. What percent of the velocity at the center of the pipe does this represent? Where in the pipe is the fluid flowing with a velocity of 4000 centimeters per second?

b. Suppose that the situation is the same, but the fluid is now machinery oil, with a viscosity of 0.15. Answer the same questions as in part **a**, except find where in the pipe the oil flows with a velocity of 15 cm per second. Note that the oil travels at a much slower speed than water.

c. Medical doctors use various methods to increase the rate of blood flow through an artery. The patient may take a drug that "thins the blood" (lowers its viscosity) or a drug that dilates the artery, or the patient may undergo angioplasty, a surgical procedure that widens the canal through which the blood passes. Explain why each of these increases the velocity, V, at a given distance r from the center of the artery.

■ Chapter Summary

Key Words

algebraic terms (5.1)	FOIL method (5.3)	polynomial (5.1)
binomial (5.1)	greatest common divisor (5.4)	polynomial function (5.1)
degree of a polynomial (5.1)	greatest common factor (5.4)	prime-factored form (5.4)
difference of two cubes (5.5)	irreducible polynomial (5.4)	prime polynomial (5.4)
difference of two squares (5.5)	like terms (5.2)	quadratic equations (5.8)
factor (5.4)	monomial (5.1)	similar terms (5.2)
factoring by grouping (5.4)	perfect cubes (5.5)	sum of two cubes (5.5)
factoring out the greatest common factor (5.4)	perfect squares (5.5)	trinomial (5.1)
	perfect square trinomials (5.3)	zero-factor theorem (5.8)

Key Ideas

(5.1) A polynomial in x is the sum of one or more terms of the form ax^n, where a is a real number and n is a whole number.

A polynomial in several variables (say, x, y, and z) is a sum of one or more terms of the form $ax^m y^n z^p$, where a is a real number and m, n, and p are whole numbers.

The **degree of a polynomial** is the degree of the term with highest degree contained within the polynomial.

If $P(x)$ is a polynomial in x, then $P(r)$ is the value of the polynomial at $x = r$.

(5.2) To add like terms, add their numerical coefficients and use the same variables with the same exponents.

To add polynomials, add their like terms.

To subtract polynomials, add the negative of the polynomial that is to be subtracted from the other polynomial.

(5.3) To multiply monomials, multiply their numerical factors and multiply their variable factors.

To multiply a polynomial by a monomial, multiply each term of the polynomial by the monomial.

To multiply polynomials, multiply each term of one polynomial by each term of the other polynomial.

(5.4) Always factor out common monomial factors as the first step in a factoring problem.

Use the distributive property to factor out common monomial factors.

If an expression has four or more terms, try to factor the expression by grouping.

(5.5) **Difference of the squares of two quantities:**
$x^2 - y^2 = (x + y)(x - y)$
Sum of two cubes: $x^3 + y^3 = (x + y)(x^2 - xy + y^2)$

Difference of two cubes:
$x^3 - y^3 = (x - y)(x^2 + xy + y^2)$

(5.6) **Special product formulas:**
$x^2 + 2xy + y^2 = (x + y)(x + y)$
$x^2 - 2xy + y^2 = (x - y)(x - y)$

Test for factorability: A trinomial of the form $ax^2 + bx + c$ will factor with integer coefficients if $b^2 - 4ac$ is a perfect square.

(5.7) Use these steps to factor a random expression:

1. Factor out all common monomial factors.

2. If an expression has two terms, check to see if the problem type is
   a. The difference of two squares:
   $x^2 - y^2 = (x + y)(x - y)$
   b. The sum of two cubes:
   $x^3 + y^3 = (x + y)(x^2 - xy + y^2)$
   c. The difference of two cubes:
   $x^3 - y^3 = (x - y)(x^2 + xy + y^2)$

3. If an expression has three terms, attempt to factor the trinomial as a **general trinomial**.

4. If an expression has four or more terms, try factoring by **grouping**.

5. Continue until each individual factor is prime.

6. Check the results by multiplying.

(5.8) **Zero-factor theorem:** If $xy = 0$, then $x = 0$ or $y = 0$.

■ Chapter 5 Review Exercises

In Review Exercises 1–4, $P(x) = -x^2 + 4x + 6$. Find each value.

1. $P(0)$ **2.** $P(1)$ **3.** $P(-t)$ **4.** $P(z)$

5. Give the degree of $P(x) = 3x^5 + 4x^3 + 2$

6. Give the degree of $9x^2y + 13x^3y^2 + 8x^4y^4$

In Review Exercises 7–10, simplify each expression.

7. $(3x^2 + 4x + 9) - (2x^2 - 2x + 7) + (4x^2 - 3x - 2)$

8. $(4x^3 + 4x^2 + 7) - (-2x^3 - x - 2) + (-5x^3 - 3x^2)$

9. $(2x^2 - 5x + 9) - (x^2 - 3) - (-3x^2 + 4x - 7)$

10. $2(7x^3 - 6x^2 + 4x - 3) - 3(7x^3 + 6x^2 + 4x - 3)$

In Review Exercises 11–18, find each product.

11. $(8a^2b^2)(-2abc)$

12. $(-3xy^2z)(2xz^3)$

13. $2xy^2(x^3y - 4xy^5)$

14. $a^2b(a^2 + 2ab + b^2)$

15. $(8x - 5)(2x + 3)$

16. $(3x^2 + 2)(2x - 4)$

17. $(5x^2 - 4x + 5)(3x^2 - 2x + 10)$

18. $(3x^2 + x - 2)(x^2 - x + 2)$

In Review Exercises 19–72, factor each expression.

19. $4x + 8$

20. $3x^2 - 6x$

21. $5x^2y^3 - 10xy^2$

22. $7a^4b^2 + 49a^3b$

23. $-8x^2y^3z^4 - 12x^4y^3z^2$

24. $12a^6b^4c^2 + 15a^2b^4c^6$

25. $27x^3y^3z^3 + 81x^4y^5z^2 - 90x^2y^3z^7$

26. $-36a^5b^4c^2 + 60a^7b^5c^3 - 24a^2b^3c^7$

27. Factor x^n from $x^{2n} + x^n$.

28. Factor y^{2n} from $y^{2n} - y^{4n}$.

29. Factor x^{-2} from $x^{-4} - x^{-2}$.

30. Factor a^{-3} from $a^6 + 1$.

31. $5x^2(x + y)^3 - 15x^3(x + y)^4$

32. $-49a^3b^2(a - b)^4 + 63a^2b^4(a - b)^3$

33. $xy + 2y + 4x + 8$

34. $ac + bc + 3a + 3b$

35. $x^4 + 4y + 4x^2 + x^2y$

36. $a^5 + b^2c + a^2c + a^3b^2$

37. $z^2 - 16$

38. $y^2 - 121$

39. $x^2y^4 - 64z^6$

40. $a^2b^2 + c^2$

41. $(x + z)^2 - t^2$

42. $c^2 - (a + b)^2$

43. $2x^4 - 98$

44. $3x^6 - 300x^2$

45. $x^3 + 343$

46. $a^3 - 125$

47. $8y^3 - 512$

48. $4x^3y + 108yz^3$

49. $y^2 + 21y + 20$

50. $z^2 - 11z + 30$

51. $-x^2 - 3x + 28$

52. $y^2 - 5y - 24$

53. $4a^2 - 5a + 1$

54. $3b^2 + 2b + 1$

55. $7x^2 + x + 2$

56. $-15x^2 + 14x + 8$

57. $y^3 + y^2 - 2y$

58. $2a^4 + 4a^3 - 6a^2$

59. $-3x^2 - 9x - 6$

60. $8x^2 - 4x - 24$

61. $15x^2 - 57xy - 12y^2$

62. $30x^2 + 65xy + 10y^2$

63. $24x^2 - 23xy - 12y^2$

64. $14x^2 + 13xy - 12y^2$

65. $x^3 + 5x^2 - 6x$

66. $3x^2y - 12xy - 63y$

67. $z^2 - 4 + zx - 2x$

68. $x^2 + 2x + 1 - p^2$

69. $x^2 + 4x + 4 - 4p^4$

70. $y^2 + 3y + 2 + 2x + xy$

71. $x^{2m} + 2x^m - 3$

72. $x^{-2} - x^{-1} - 2$

In Review Exercises 73–74, solve for the indicated variable.

73. $S = 2wh + 2wl + 2lh$ for h

74. $S = 2wh + 2wl + 2lh$ for l

In Review Exercises 75–80, solve each equation.

75. $4x^2 - 3x = 0$

76. $x^2 - 36 = 0$

77. $12x^2 + 4x - 5 = 0$

78. $7y^2 - 37y + 10 = 0$

79. $t^2(15t - 2) = 8t$

80. $3u^3 = u(19u + 14)$

81. Volume   The volume, V, of the rectangular solid in Illustration 1 is given by the formula $V = lwh$, where l is its length, w is its width, and h is its height. If the volume is 840 cubic centimeters, the length is 12 centimeters, and the width exceeds the height by 3 centimeters, find the height.

82. Volume of a pyramid   The volume of the pyramid in Illustration 2 is given by the formula $V = \frac{Bh}{3}$, where B is the area of its base and h is its height. The volume of the pyramid is 1020 cubic meters. Find the dimensions of its rectangular base if one edge of the base is 3 meters longer than the other, and the height of the pyramid is 9 meters.

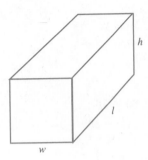

ILLUSTRATION 1

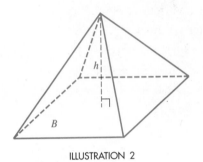

ILLUSTRATION 2

■ Chapter 5 Test

In Problems 1–2, find the degree of each polynomial.

1. $3x^5 - 4x^5 - 3x^2 - 5$

2. $3x^5y^3 - x^8y^2 + 2x^9y^4 - 3x^2y^5 + 4$

In Problems 3–4, let $P(x) = -3x^2 + 2x - 1$ and find each value.

3. $P(2)$

4. $P(-1)$

In Problems 5–12, do the operations.

5. $(2y^2 + 4y + 3) + (3y^2 - 3y - 4)$

6. $(-3u^2 + 2u - 7) - (u^2 + 7)$

7. $3(2a^2 - 4a + 2) - 4(-a^2 - 3a - 4)$

8. $-2(2x^2 - 2) + 3(x^2 + 5x - 2)$

9. $(3x^3y^2z)(-2xy^{-1}z^3)$

10. $-5a^2b(3ab^3 - 2ab^4)$

11. $(z + 4)(z - 4)$

12. $(3x - 2)(4x + 3)$

In Problems 13–16, factor each polynomial.

13. $3xy^2 + 6x^2y$

14. $12a^3b^2c - 3a^2b^2c^2 + 6abc^3$

15. Factor y^n from $x^2y^{n+2} + y^n$.

16. Factor b^n from $a^nb^n - ab^{-n}$.

In Problems 17–18, factor each expression.

17. $(u - v)r + (u - v)s$

18. $ax - xy + ay - y^2$

In Problems 19–24, factor each polynomial.

19. $x^2 - 49$

20. $2x^2 - 32$

21. $4y^4 - 64$

22. $b^3 + 125$

23. $b^3 - 27$

24. $3u^3 - 24$

In Problems 25–30, factor each trinomial.

25. $a^2 - 5a - 6$

26. $6b^2 + b - 2$

27. $6u^2 + 9u - 6$

28. $20r^2 - 15r - 5$

29. $x^{2n} + 2x^n + 1$

30. $x^2 + 6x + 9 - y^2$

31. Solve for r: $r_1r_2 - r_2r = r_1r$

32. Solve for x: $x^2 - 5x - 6 = 0$

33. Integer problem The product of two consecutive positive integers is 156. Find their sum.

34. Preformed concrete The slab of concrete in Illustration 1 is twice as long as it is wide. The area in which it is placed includes a 1-foot-wide border of 70 square feet. Find the dimensions of the slab.

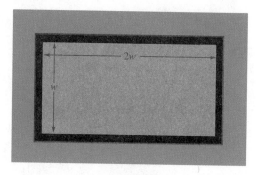

ILLUSTRATION 1

Rational Expressions

Mechanical Engineer

MATHEMATICS IN THE
WORKPLACE

Mechanical engineers design and develop power-producing machines such as internal combustion engines, steam and gas turbines, and jet rocket engines, as well as power-using machines such as refrigeration and air-conditioning equipment, machine tools, printing presses, and steel rolling mills. Many mechanical engineers do research, test, and design work. Others work in maintenance, technical sales, and production operations. Some teach in colleges and universities or work as consultants.

SAMPLE APPLICATION ■ The stiffness of the shaft shown in Illustration 1 is given by the formula

$$k = \dfrac{1}{\dfrac{1}{k_1} + \dfrac{1}{k_2}}$$

ILLUSTRATION 1

where k_1 and k_2 are the individual stiffnesses of each section. If the stiffness, k_2, of Section 2 is 4,200,000 in. lb/rad, and design specifications require that the overall stiffness, k, of the entire shaft be 1,900,000 in. lb/rad, what must the stiffness, k_1, of Section 1 be?
See Exercise 61 in Exercise 6.5.

Rational expressions are algebraic fractions such as

$$\frac{3x}{x - 7}, \qquad \frac{5m + n}{8m + 16}, \qquad \text{and} \qquad \frac{a^3 + 2a^2 + 7}{2a^2 - 5a + 4}$$

that indicate the quotient of two polynomials. Since division by 0 is undefined, the value of any polynomial occurring in a denominator cannot be 0. For example, x cannot be 7 in the rational expression

$$\frac{3x}{x - 7}$$

because if $x = 7$, the denominator would be 0. In the rational expression

$$\frac{5m + n}{8m + 16}$$

m cannot be -2, because the denominator is 0 when $m = -2$.

6.1 Rational Functions and Simplifying Rational Expressions

■ RATIONAL FUNCTIONS ■ GRAPHING DEVICES ■ FINDING THE DOMAIN OF A RATIONAL FUNCTION ■ GRAPHING DEVICES ■ SIMPLIFYING FRACTIONS ■ GRAPHING DEVICES ■ SIMPLIFYING FRACTIONS BY FACTORING OUT −1 ■ ACCENT ON STATISTICS

■ RATIONAL FUNCTIONS

Rational expressions often define functions. For example, if the cost of subscribing to an on-line computer information network is $6 per month plus $1.50 per hour of access time, the average (mean) hourly cost of the service is the total monthly cost, divided by the number of hours of access time:

$$\bar{c} = \frac{C}{n} = \frac{1.50n + 6}{n} \qquad \text{$\bar{c}$ is the mean hourly cost, C is the total monthly cost, and n is the number of hours the service is used.}$$

Thus, the function

1. $\quad \bar{c} = f(n) = \dfrac{1.50n + 6}{n} \qquad (n > 0)$

gives the mean hourly cost of using the information network for n hours per month. Since $n > 0$, the domain of this function is the interval $(0, \infty)$.

■ GRAPHING DEVICES

If we use a graphing device with settings of $[0, 10]$ for x and $[0, 10]$ for y to graph the function $f(n) = \frac{1.50n + 6}{n}$, we get the graph show in Figure 6-1. Note that the graph of the function passes the vertical line test, as expected.

From the graph, we can see that the mean hourly cost decreases as the number of hours of access time increases. Since the cost of each extra hour of access time is $1.50, the mean hourly cost can approach $1.50 but never drop below it. Thus, the graph of the function approaches the line $y = 1.5$ as n increases without bound. Since a graph approaches a line as the number of hours of usage becomes infinitely large, we call the line an **asymptote**. The line $y = 1.5$ is a **horizontal asymptote** of the graph.

As n gets smaller and approaches 0, the graph approaches the y-axis but never touches it. The y-axis is a **vertical asymptote** of the graph.

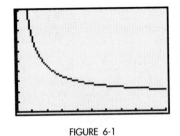

FIGURE 6-1

(b)

EXAMPLE 1   Find the mean hourly cost when the network described above is used for **a.** 3 hours   and   **b.** 10.4 hours.

Solution   **a.** To find the mean hourly cost for 3 hours of access time, we substitute 3 for n in Equation 1 and simplify:

$$\bar{c} = f(3) = \frac{1.50(3) + 6}{3} = 3.5$$

The mean hourly cost for 3 hours of access time is $3.50.
Check this result by graphing Equation 1 with a graphing device, and then using the trace feature.

b. To find the mean hourly cost for 10.4 hours of access time, we substitute 10.4 for n in Equation 1 and simplify:

$$\bar{c} = f(10.4) = \frac{1.50(10.4) + 6}{10.4} = 2.076923$$

The mean hourly cost for 10.4 hours of access time is approximately $2.08.
Check this result with a graphing device.   ■

■ FINDING THE DOMAIN OF A RATIONAL FUNCTION

Since division by 0 is undefined, any values that make the denominator 0 in a rational function must be excluded from the domain of the function.

EXAMPLE 2   Find the domain of the function $f(x) = \dfrac{3x + 2}{x^2 - 7x + 12}$.

Solution   From the set of real numbers, we must exclude any values of x that make the denominator 0. To find these values, we set $x^2 - 7x + 12$ equal to 0 and solve for x.

$$x^2 - 7x + 12 = 0$$
$$(x - 4)(x - 3) = 0 \qquad \text{Factor } x^2 - 7x + 12 = 0.$$
$$x - 4 = 0 \quad \text{or} \quad x - 3 = 0 \qquad \text{Set each factor equal to 0.}$$
$$x = 4 \quad | \quad x = 3 \qquad \text{Solve each linear equation.}$$

The domain of the function is $\{x \mid x$ is a real number, and $x \neq 3$ and $x \neq 4\}$. In interval notation, the domain is $(-\infty, 3) \cup (3, 4) \cup (4, \infty)$.   ■

■ GRAPHING DEVICES

EXAMPLE 3   Use a graphing device to find the domain and range of the function
$$f(x) = \frac{2x + 1}{x - 1}.$$

Solution   If we use settings of $[-10, 10]$ for x and $[-10, 10]$ for y, we will get the graph shown in Figure 6-2.
From the figure, we can see that

- As x approaches 1 from the left, the values of y decrease, and the graph approaches the vertical line $x = 1$.

- As x approaches 1 from the right, the values of y increase, and the graph approaches the vertical line $x = 1$.

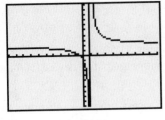

FIGURE 6-2

The line $x = 1$ is a vertical asymptote. Although the vertical line in the graph appears to be the graph of $x = 1$, it is not. Graphing calculators draw graphs by connecting dots whose x-coordinates are close together. Often when two such points straddle a vertical asymptote and their y-coordinates are far apart, the calculator draws a line between them anyway, producing what appears to be the vertical asymptote shown in the figure. If you set your calculator to dot mode instead of connected mode, the vertical line will not appear.

From the figure, we can also see that

- As x increases to the right of 1, the values of y decrease and approach the value $y = 2$.
- As x decreases to the left of 1, the values of y increase and approach the value $y = 2$.

If we were to draw the line $y = 2$, it would be a horizontal asymptote. Graphing calculators do not draw lines that appear to be horizontal asymptotes.

From the graph, we can see that every real number x, except 1, gives a value of y. Thus, the domain of the function is $(-\infty, 1) \cup (1, \infty)$. We can also see that y can be any value except 2. Thus, the range of the function is $(-\infty, 2) \cup (2, \infty)$. ∎

■ SIMPLIFYING FRACTIONS

Since rational expressions are the fractions of algebra, the familiar rules for arithmetic fractions apply.

> **Properties of Fractions**
> If there are no divisions by 0, then
>
> 1. $\dfrac{a}{b} = \dfrac{c}{d}$ if and only if $ad = bc$
>
> 2. $\dfrac{a}{1} = a$ and $\dfrac{a}{a} = 1$ 3. $\dfrac{ak}{bk} = \dfrac{a}{b}$
>
> 4. $\dfrac{a}{b} \cdot \dfrac{c}{d} = \dfrac{ac}{bd}$ and $\dfrac{a}{b} \div \dfrac{c}{d} = \dfrac{a}{b} \cdot \dfrac{d}{c} = \dfrac{ad}{bc}$
>
> 5. $\dfrac{a}{b} + \dfrac{c}{b} = \dfrac{a+c}{b}$ and $\dfrac{a}{b} - \dfrac{c}{b} = \dfrac{a-c}{b}$ 6. $-\dfrac{a}{b} = \dfrac{-a}{b} = \dfrac{a}{-b}$

Property 3 of fractions is true because

$$\frac{ak}{bk} = \frac{a}{b} \cdot \frac{k}{k} = \frac{a}{b} \cdot 1 = \frac{a}{b} \quad (b \neq 0, k \neq 0)$$

Remember that any number times 1 is the number.

To simplify fractions, we will use Property 3, which enables us to divide out factors that are common to the numerator and the denominator.

(b)

EXAMPLE 4   Simplify: **a.** $\dfrac{10k}{25k^2}$ and **b.** $\dfrac{-8y^3z^5}{6y^4z^3}$.

Solution   We factor each numerator and denominator and divide out all common factors:

a. $\dfrac{10k}{25k^2} = \dfrac{5 \cdot 2 \cdot k}{5 \cdot 5 \cdot k \cdot k}$

$= \dfrac{\overset{1}{\cancel{5}} \cdot 2 \cdot \overset{1}{\cancel{k}}}{\underset{1}{\cancel{5}} \cdot 5 \cdot \underset{1}{\cancel{k}} \cdot k}$

$= \dfrac{2}{5k}$

b. $\dfrac{-8y^3z^5}{6y^4z^3} = \dfrac{-2 \cdot 4 \cdot y \cdot y \cdot y \cdot z \cdot z \cdot z \cdot z \cdot z}{2 \cdot 3 \cdot y \cdot y \cdot y \cdot y \cdot z \cdot z \cdot z}$

$= \dfrac{-\overset{1}{\cancel{2}} \cdot 4 \cdot \overset{1}{\cancel{y}} \cdot \overset{1}{\cancel{y}} \cdot \overset{1}{\cancel{y}} \cdot \overset{1}{\cancel{z}} \cdot \overset{1}{\cancel{z}} \cdot \overset{1}{\cancel{z}} \cdot z \cdot z}{\underset{1}{\cancel{2}} \cdot 3 \cdot \underset{1}{\cancel{y}} \cdot \underset{1}{\cancel{y}} \cdot \underset{1}{\cancel{y}} \cdot y \cdot \underset{1}{\cancel{z}} \cdot \underset{1}{\cancel{z}} \cdot \underset{1}{\cancel{z}}}$

$= -\dfrac{4z^2}{3y}$ ■

The fractions in Example 4 can also be simplified by using the rules of exponents:

$\dfrac{10k}{25k^2} = \dfrac{5 \cdot 2}{5 \cdot 5}k^{1-2}$

$= \dfrac{2}{5} \cdot k^{-1}$

$= \dfrac{2}{5} \cdot \dfrac{1}{k}$

$= \dfrac{2}{5k}$

$\dfrac{-8y^3z^5}{6y^4z^3} = \dfrac{-2 \cdot 4}{2 \cdot 3}y^{3-4}z^{5-3}$

$= \dfrac{-4}{3} \cdot y^{-1}z^2$

$= -\dfrac{4}{3} \cdot \dfrac{1}{y} \cdot \dfrac{z^2}{1}$

$= -\dfrac{4z^2}{3y}$

EXAMPLE 5   Simplify: $\dfrac{x^2 - 16}{x + 4}$.

Solution $\dfrac{x^2 - 16}{x + 4} = \dfrac{\overset{1}{\cancel{(x + 4)}}(x - 4)}{1\underset{1}{\cancel{(x + 4)}}}$ Factor $x^2 - 16$, and use the fact that $\dfrac{x + 4}{x + 4} = 1$.

$= \dfrac{x - 4}{1}$

$= x - 4$ ■

■ GRAPHING DEVICES

To show that the simplification in Example 5 is correct, we can graph the functions $f(x) = \dfrac{x^2 - 16}{x + 4}$ (Figure 6-3(a)) and $g(x) = x - 4$ (Figure 6-3(b)). Except for the point where $x = -4$, the graphs are the same. The point where $x = -4$ is excluded from the graph of $f(x) = \dfrac{x^2 - 16}{x + 4}$, because -4 is not in the domain of f. However, graphing devices do not show that this point is excluded.

The point where $x = -4$ is included in the graph of $g(x) = x - 4$, because -4 is in the domain of g.

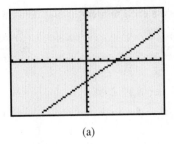   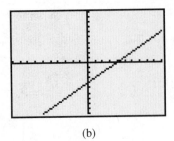

(a)   (b)

FIGURE 6-3

EXAMPLE 6   Simplify: $\dfrac{2x^2 + 11x + 12}{3x^2 + 11x - 4}$.

Solution

$$\frac{2x^2 + 11x + 12}{3x^2 + 11x - 4} = \frac{\overset{1}{(2x + 3)(\cancel{x + 4})}}{\underset{1}{(3x - 1)(\cancel{x + 4})}}$$

Factor and divide out all factors common to the numerator and the denominator: $\dfrac{x + 4}{x + 4} = 1$.

$$= \frac{2x + 3}{3x - 1}$$

WARNING!   Do not divide out the x's in the fraction $\frac{2x + 3}{3x - 1}$. The x in the numerator is a factor of the first term only. It is not a factor of the entire numerator. Likewise, the x in the denominator is a factor of the first term only. It is not a factor of the entire denominator.   ∎

■ SIMPLIFYING FRACTIONS BY FACTORING OUT −1

To simplify $\frac{b-a}{a-b}$ $(a \neq b)$, we factor -1 from the numerator and divide out any factors common to both the numerator and the denominator:

$$\frac{b - a}{a - b} = \frac{-a + b}{a - b} \quad (a \neq b)$$

$$= \frac{-(\overset{1}{\cancel{a - b}})}{(\underset{1}{\cancel{a - b}})} \qquad \frac{a - b}{a - b} = 1.$$

$$= \frac{-1}{1}$$

$$= -1$$

In general, we have the following theorem.

Theorem
The quotient of any nonzero quantity and its negative (or opposite) is -1.

EXAMPLE 7   Simplify the fraction $\dfrac{3x^2 - 10xy - 8y^2}{4y^2 - xy}$.

Solution   We factor the numerator and denominator and apply the previous theorem:

$$\frac{3x^2 - 10xy - 8y^2}{4y^2 - xy} = \frac{(3x + 2y)\overset{-1}{\cancel{(x - 4y)}}}{y\underset{1}{\cancel{(4y - x)}}}$$ Because $x - 4y$ and $4y - x$ are negatives, their quotient is -1.

$$= \frac{-(3x + 2y)}{y}$$

$$= \frac{-3x - 2y}{y}$$ ∎

Many fractions we will encounter are already in simplified form. For example, to attempt to simplify the fraction

$$\frac{x^2 + xa + 2x + 2a}{x^2 + x - 6}$$

we factor the numerator and denominator and attempt to divide out any common factors that exist:

$$\frac{x^2 + xa + 2x + 2a}{x^2 + x - 6} = \frac{x(x + a) + 2(x + a)}{(x - 2)(x + 3)} = \frac{(x + a)(x + 2)}{(x - 2)(x + 3)}$$

Since there are no common factors in the numerator and denominator, the fraction cannot be simplified.

EXAMPLE 8   Simplify: $\dfrac{(x^2 + 2x)(x^2 + 2x - 3)}{(x^2 + x - 2)(x^2 + 3x)}$.

Solution   We factor the polynomials in the numerator and denominator and divide out all common factors:

$$\frac{(x^2 + 2x)(x^2 + 2x - 3)}{(x^2 + x - 2)(x^2 + 3x)} = \frac{\overset{1}{\cancel{x}}\,\overset{1}{\cancel{(x + 2)}}\,\overset{1}{\cancel{(x + 3)}}\,\overset{1}{\cancel{(x - 1)}}}{\underset{1}{\cancel{(x + 2)}}\,\underset{1}{\cancel{(x - 1)}}\,\underset{1}{\cancel{x}}\,\underset{1}{\cancel{(x + 3)}}}$$

$$= 1$$ ∎

WARNING!   Remember that only factors that are common to the entire numerator and the entire denominator can be divided out.

Terms common to both the numerator and denominator cannot be divided out. For example, consider this correct simplification:

$$\frac{3 + 7}{3} = \frac{10}{3}$$

It is incorrect to divide out the common term of 3 in the above simplification, because doing so gives an incorrect answer:

$$\frac{3+7}{3} = \frac{\overset{1}{\cancel{3}}+7}{\underset{1}{\cancel{3}}} = \frac{1+7}{1} = 8$$

The 3's in the fraction

$$\frac{5+3(2)}{3(4)}$$

cannot be divided out, because the 3 in the numerator is a factor of the second term only. To be divided out, the 3 must be a factor of the entire numerator.

It is not correct to divide out the y in the fraction

$$\frac{x^2y + 6x}{y}$$

because y is not a factor of the entire numerator.

ACCENT ON
STATISTICS

Probability

If we toss a coin, it can land in two ways, either heads or tails. The two outcomes, heads and tails, are equally likely; if we were to toss the same coin many times, we would get heads about half of the time. We say that the *probability* of getting heads on a single toss of a coin is $\frac{1}{2}$.

If records show that out of 100 days with today's weather conditions, 30 have received rain, the weather service reports, "Today, there is a $\frac{30}{100}$ or 30% probability of rain."

Activities such as tossing a coin, rolling a die, drawing a card, and predicting rain are called **experiments.** For any experiment, a list of all possible outcomes is called a **sample space.** For example, the sample space, S, for the experiment of tossing two coins is the set

$$S = \{(H, H), (H, T), (T, H), (T, T)\} \qquad \text{There are 4 possible outcomes.}$$

where the ordered pair (H, T) represents the outcome "heads on the first coin and tails on the second."

An **event** is a subset of the sample space of an experiment. For example, if E is the event "getting at least one heads" in the experiment of tossing two coins, then

$$E = \{(H, H), (H, T), (T, H)\} \qquad \text{There are 3 ways of getting at least one heads.}$$

Because the outcome of getting at least one heads can occur in 3 out of 4 possible ways, we say that the probability of E is $\frac{3}{4}$, and we write

$$P(E) = P(\text{at least one heads}) = \frac{3}{4}$$

> ### Probability of an Event
>
> If the sample space of an experiment has n distinct and equally likely out-comes and E is an event that occurs in s of those ways, then the **probability of E** is
>
> $$P(E) = \frac{s}{n}$$

EXAMPLE 9 Find the probability of the event "rolling a sum of 7 on one roll of two dice."

Solution We first find the sample space of the experiment "rolling two dice a single time." If we use ordered-pair notation and let the first number in each ordered pair be the result on the first die and the second number be the result on the second die, the sample space contains the following elements:

$$(1, 1), (1, 2), (1, 3), (1, 4), (1, 5), (1, 6)$$
$$(2, 1), (2, 2), (2, 3), (2, 4), (2, 5), (2, 6)$$
$$(3, 1), (3, 2), (3, 3), (3, 4), (3, 5), (3, 6)$$
$$(4, 1), (4, 2), (4, 3), (4, 4), (4, 5), (4, 6)$$
$$(5, 1), (5, 2), (5, 3), (5, 4), (5, 5), (5, 6)$$
$$(6, 1), (6, 2), (6, 3), (6, 4), (6, 5), (6, 6)$$

Since there are 6 ordered pairs whose numbers give a sum of 7 out of a total of 36 equally likely outcomes, we have

$$P(E) = P(\text{rolling a 7}) = \frac{s}{n} = \frac{6}{36} = \frac{1}{6}$$

Orals *Evaluate the function* $f(x) = \dfrac{2x - 3}{x}$ *when*

1. $x = 1$ **2.** $x = 3$

Give the domain of each function.

3. $f(x) = \dfrac{x + 7}{2x - 4}$ **4.** $f(x) = \dfrac{3x - 4}{x^2 - 9}$

Simplify each fraction.

5. $\dfrac{25}{30}$ **6.** $\dfrac{x^2}{xy}$ **7.** $\dfrac{2x - 4}{x - 2}$ **8.** $\dfrac{x - 2}{2 - x}$

EXERCISE 6.1

In Exercises 1–4, the time, t, it takes to travel 600 miles is a function of the mean rate of speed, r: $t = f(r) = \frac{600}{r}$. Find t for the given values of r.

1. 30 mph   **2.** 40 mph   **3.** 50 mph   **4.** 60 mph

In Exercises 5–8, suppose the cost (in dollars) of removing p% of the pollution in a river is given by the function
$c = f(p) = \frac{50,000p}{100 - p}$ *($0 \leq p \leq 100$). Find the cost of removing each percent of pollution.*

5. 10%   **6.** 30%   **7.** 50%   **8.** 80%

In Exercises 9–14, a service club wants to publish a directory of its members. Some investigation shows that the cost of typesetting and photography will be $700, and the cost of printing each directory will be $1.25.

9. Find a function that gives the total cost c of printing x directories.

10. Find a function that gives the mean cost per directory $\bar{c}$ of printing x directories.

11. Find the total cost of printing 500 directories.

12. Find the mean cost per directory if 500 directories are printed.

13. Find the mean cost per directory if 1000 directories are printed.

14. Find the mean cost per directory if 2000 directories are printed.

In Exercises 15–20, an electric company charges $7.50 per month plus 9¢ for each kilowatt hour (kwh) of electricity used.

15. Find a function that gives the total cost c of n kwh of electricity.

16. Find a function that gives the mean cost per kwh, $\bar{c}$, when using n kwh.

17. Find the total cost for using 775 kwh.

18. Find the mean cost per kwh when 775 kwh are used.

19. Find the mean cost per kwh when 1000 kwh are used.

20. Find the mean cost per kwh when 1200 kwh are used.

In Exercises 21–22, the function $f(t) = \dfrac{t^2 + 2t}{2t + 2}$ gives the number of days it would take two construction crews,
working together, to frame a house that crew 1 (working alone) could complete in t days and crew 2 (working alone)
could complete in t + 2 days.

21. If crew 1 could frame a certain house in 15 days, how long would it take both crews working together?

22. If crew 2 could frame a certain house in 20 days, how long would it take both crews working together?

In Exercises 23–24, the function $f(t) = \dfrac{t^2 + 3t}{2t + 3}$ gives the number of hours it would take two pipes, working together, to
fill a pool that the larger pipe (working alone) could fill in t hours and the smaller pipe (working alone) could fill in
t + 3 hours.

23. If the smaller pipe could fill a pool in 7 hours, how long would it take both pipes to fill the pool?

24. If the larger pipe could fill a pool in 8 hours, how long would it take both pipes to fill the pool?

In Exercises 25–28, use a graphing device to graph each rational function. From the graph, determine its domain.

25. $f(x) = \dfrac{x}{x - 2}$

26. $f(x) = \dfrac{x + 2}{x}$

27. $f(x) = \dfrac{x + 1}{x^2 - 4}$

28. $f(x) = \dfrac{x - 2}{x^2 - 3x - 4}$

In Exercises 29–90, simplify each fraction when possible.

29. $\dfrac{12}{18}$

30. $\dfrac{25}{55}$

31. $-\dfrac{112}{36}$

32. $-\dfrac{49}{21}$

33. $\dfrac{288}{312}$

34. $\dfrac{144}{72}$

35. $-\dfrac{244}{74}$

36. $-\dfrac{512}{236}$

37. $\dfrac{12x^3}{3x}$

38. $-\dfrac{15a^2}{25a^3}$

39. $\dfrac{-24x^3y^4}{18x^4y^3}$

40. $\dfrac{15a^5b^4}{21b^3c^2}$

41. $\dfrac{(3x^3)^2}{9x^4}$

42. $\dfrac{8(x^2y^3)^3}{2(xy^2)^2}$

43. $-\dfrac{11x(x-y)}{22(x-y)}$

44. $\dfrac{x(x-2)^2}{(x-2)^3}$

45. $\dfrac{9y^2(y-z)}{21y(y-z)^2}$

46. $\dfrac{-3ab^2(a-b)}{9ab(b-a)}$

47. $\dfrac{(a-b)(c-d)}{(c-d)(a-b)}$

48. $\dfrac{(p+q)(p-r)}{(r-p)(p+q)}$

49. $\dfrac{x+y}{x^2-y^2}$

50. $\dfrac{x-y}{x^2-y^2}$

51. $\dfrac{5x-10}{x^2-4x+4}$

52. $\dfrac{y-xy}{xy-x}$

53. $\dfrac{12-3x^2}{x^2-x-2}$

54. $\dfrac{x^2+2x-15}{x^2-25}$

55. $\dfrac{3x+6y}{x+2y}$

56. $\dfrac{x^2+y^2}{x+y}$

57. $\dfrac{x^3+8}{x^2-2x+4}$

58. $\dfrac{x^2+3x+9}{x^3-27}$

59. $\dfrac{x^2+2x+1}{x^2+4x+3}$

60. $\dfrac{6x^2+x-2}{8x^2+2x-3}$

61. $\dfrac{3m-6n}{3n-6m}$

62. $\dfrac{ax+by+ay+bx}{a^2-b^2}$

63. $\dfrac{4x^2+24x+32}{16x^2+8x-48}$

64. $\dfrac{a^2-4}{a^3-8}$

65. $\dfrac{3x^2-3y^2}{x^2+2y+2x+yx}$

66. $\dfrac{x^2+x-30}{x^2-x-20}$

67. $\dfrac{4x^2+8x+3}{6+x-2x^2}$

68. $\dfrac{6x^2+13x+6}{6-5x-6x^2}$

69. $\dfrac{a^3+27}{4a^2-36}$

70. $\dfrac{a-b}{b^2-a^2}$

71. $\dfrac{2x^2-3x-9}{2x^2+3x-9}$

72. $\dfrac{6x^2-7x-5}{2x^2+5x+2}$

73. $\dfrac{(m+n)^3}{m^2+2mn+n^2}$

74. $\dfrac{x^3-27}{3x^2-8x-3}$

75. $\dfrac{m^3-mn^2}{mn^2+m^2n-2m^3}$

76. $\dfrac{p^3+p^2q-2pq^2}{pq^2+p^2q-2p^3}$

77. $\dfrac{x^4-y^4}{(x^2+2xy+y^2)(x^2+y^2)}$

78. $\dfrac{-4x-4+3x^2}{4x^2-2-7x}$

79. $\dfrac{4a^2-9b^2}{2a^2-ab-6b^2}$

80. $\dfrac{x^2+2xy}{x+2y+x^2-4y^2}$

81. $\dfrac{x-y}{x^3-y^3-x+y}$

82. $\dfrac{2x^2+2x-12}{x^3+3x^2-4x-12}$

83. $\dfrac{x^6-y^6}{x^4+x^2y^2+y^4}$

84. $\dfrac{6xy-4x-9y+6}{6y^2-13y+6}$

85. $\dfrac{(x^2-1)(x+1)}{(x^2-2x+1)^2}$

86. $\dfrac{(x^2+2x+1)(x^2-2x+1)}{(x^2-1)^2}$

87. $\dfrac{(2x^2+3xy+y^2)(3a+b)}{(x+y)(2xy+2bx+y^2+by)}$

88. $\dfrac{(x-1)(6ax+9x+4a+6)}{(3x+2)(2ax-2a+3x-3)}$

89. $\dfrac{(a^2 + 2a + ab + 2b)(c^2 + c - dc - d)}{(a^2 - 4)(ad + bd - ac - bc)}$

90. $\dfrac{(mp + mq - np - nq)(pm + pn - qm - qn)}{(qm - qn - pm + pn)(pm + pn + qm + qn)}$

In Exercises 91–94, assume that you roll a die one time. Find the probability of each event.

91. Rolling a 2

92. Rolling a 3 or a 4

93. Rolling a 10

94. Rolling an even number

In Exercises 95–98, assume that you draw one card from a standard deck containing 52 cards. Find the probability of each event.

95. Drawing a black card

96. Drawing a jack

97. Drawing an ace

98. Drawing a 4 or a 5

99. Winning a lottery Find the probability of winning a lottery with a single ticket if 6,000,000 tickets are sold and each ticket has the same chance of being the winning number.

100. Winning a raffle Find the probability of winning a raffle if 375 tickets are sold, each with the same chance of being the winning ticket, and you purchased 3 tickets.

101. Probability of parents having all girls Find the probability that a family with three children have all girls.

102. Probability of parents having two girls Find the probability that a family with three children have exactly two girls.

In Exercises 103–104, interpret each statement.

103. The probability of a man being 20 feet tall is 0.

104. The probability of dying is 1.

105. Explain why all probabilities, p, are in the interval $0 \le p \le 1$.

Writing Exercises *Write a paragraph using your own words.*

1. Explain how to simplify a rational expression.

2. Explain how to recognize that a rational expression is in lowest terms.

Something to Think About

1. A student compares his answer of $\frac{a - 3b}{2b - a}$ to an answer of $\frac{3b - a}{a - 2b}$. Are the two answers the same?

2. Is this work correct? Explain:
$$\frac{3x^2 + 6}{3y} = \frac{\cancel{3}x^2 + 6}{\cancel{3}y} = \frac{x^2 + 6}{y}$$

3. In which parts can you divide out the 4's?

   a. $\dfrac{4x}{4y}$ **b.** $\dfrac{4x}{x + 4}$

   c. $\dfrac{4 + x}{4 + y}$ **d.** $\dfrac{4x}{4 + 4y}$

4. In which parts can you divide out the 3's?

   a. $\dfrac{3x + 3y}{3z}$ **b.** $\dfrac{3(x + y)}{3x + y}$

   c. $\dfrac{3x + 3y}{3a - 3b}$ **d.** $\dfrac{x + 3}{3y}$

Review Exercises *Factor each expression.*

1. $3x^2 - 9x$

2. $6t^2 - 5t - 6$

3. $27x^6 + 64y^3$

4. $x^2 + ax + 2x + 2a$

6.2 Multiplying and Dividing Rational Expressions

■ MULTIPLYING RATIONAL EXPRESSIONS ■ GRAPHING DEVICES ■ DIVIDING RATIONAL EXPRESSIONS ■ MIXED OPERATIONS

Property 4 of fractions provides the rules for multiplying and dividing fractions.

Property 4 of Fractions
If no denominators are 0, then

$$\frac{a}{b} \cdot \frac{c}{d} = \frac{a \cdot c}{b \cdot d} = \frac{ac}{bd} \quad \text{and} \quad \frac{a}{b} \div \frac{c}{d} = \frac{a}{b} \cdot \frac{d}{c} = \frac{ad}{bc}$$

To multiply two fractions, we multiply the numerators and multiply the denominators.

$$\frac{3}{5} \cdot \frac{2}{7} = \frac{3 \cdot 2}{5 \cdot 7} \qquad \frac{4}{7} \cdot \frac{5}{8} = \frac{4 \cdot 5}{7 \cdot 8} = \frac{\overset{1}{\cancel{2}} \cdot \overset{1}{\cancel{2}} \cdot 5}{7 \cdot \underset{1}{\cancel{2}} \cdot \underset{1}{\cancel{2}} \cdot 2} \qquad \frac{2}{2} = 1.$$

$$= \frac{6}{35} \qquad\qquad\qquad = \frac{5}{14}$$

■ MULTIPLYING RATIONAL EXPRESSIONS

The same rule applies to algebraic fractions. If $t \neq 0$, then

$$\frac{x^2 y}{t} \cdot \frac{xy^3}{t^3} = \frac{x^2 y \cdot xy^3}{tt^3} = \frac{x^2 x \cdot yy^3}{t^4} = \frac{x^3 y^4}{t^4}$$

EXAMPLE 1 Find the product of $\dfrac{x^2 - 6x + 9}{x}$ and $\dfrac{x^2}{x - 3}$.

Solution We multiply the numerators and multiply the denominators. Then we simplify the resulting fraction.

$$\frac{x^2 - 6x + 9}{x} \cdot \frac{x^2}{x - 3} = \frac{(x^2 - 6x + 9)x^2}{x(x - 3)} \qquad \text{Multiply the numerators and multiply the denominators.}$$

$$= \frac{(x - 3)(x - 3)xx}{x(x - 3)} \qquad \text{Factor the numerator.}$$

$$= \frac{\overset{1}{\cancel{(x - 3)}}(x - 3)\overset{1}{\cancel{x}}x}{\underset{1}{\cancel{x}}\underset{1}{\cancel{(x - 3)}}} \qquad \frac{x-3}{x-3} = 1 \text{ and } \frac{x}{x} = 1.$$

$$= \frac{x(x - 3)}{1}$$

$$= x(x - 3) \qquad\qquad\qquad\qquad ■$$

■ GRAPHING DEVICES

We can check the simplification in Example 1 by graphing the rational functions $f(x) = \left(\frac{x^2 - 6x + 9}{x}\right)\left(\frac{x^2}{x - 3}\right)$, shown in Figure 6-4(a), and $g(x) = x(x - 3)$, shown in Figure 6-4(b), and observing that the graphs are the same, except that 0 and 3 are not included in the domain of the first function. (Recall that a graphing device does not accurately show that 0 and 3 are not included in the domain of $f(x)$.)

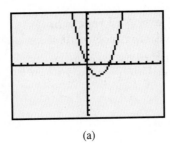   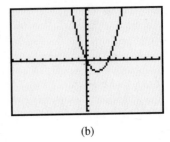

(a)   (b)

FIGURE 6-4

EXAMPLE 2   Multiply: $\dfrac{x^2 - x - 6}{x^2 - 4} \cdot \dfrac{x^2 + x - 6}{x^2 - 9}$.

Solution $\dfrac{x^2 - x - 6}{x^2 - 4} \cdot \dfrac{x^2 + x - 6}{x^2 - 9} = \dfrac{(x^2 - x - 6)(x^2 + x - 6)}{(x^2 - 4)(x^2 - 9)}$ Multiply the numerators and multiply the denominators.

$= \dfrac{(x - 3)(x + 2)(x + 3)(x - 2)}{(x + 2)(x - 2)(x + 3)(x - 3)}$ Factor the polynomials.

$= \dfrac{\overset{1}{\cancel{(x - 3)}}\overset{1}{\cancel{(x + 2)}}\overset{1}{\cancel{(x + 3)}}\overset{1}{\cancel{(x - 2)}}}{\underset{1}{\cancel{(x + 2)}}\underset{1}{\cancel{(x - 2)}}\underset{1}{\cancel{(x + 3)}}\underset{1}{\cancel{(x - 3)}}}$ $\frac{x-3}{x-3} = 1, \frac{x+2}{x+2} = 1,$ $\frac{x+3}{x+3} = 1,$ and $\frac{x-2}{x-2} = 1.$

$= \dfrac{1}{1}$

$= 1$ ■

   WARNING!   Note that when all factors divide out, the result is 1, not 0.

EXAMPLE 3   Multiply: $\dfrac{6x^2 + 5x - 4}{2x^2 + 5x + 3} \cdot \dfrac{8x^2 + 6x - 9}{12x^2 + 7x - 12}$.

Solution

$\dfrac{6x^2 + 5x - 4}{2x^2 + 5x + 3} \cdot \dfrac{8x^2 + 6x - 9}{12x^2 + 7x - 12} = \dfrac{(6x^2 + 5x - 4)(8x^2 + 6x - 9)}{(2x^2 + 5x + 3)(12x^2 + 7x - 12)}$ Multiply the numerators and multiply the denominators.

$= \dfrac{(3x + 4)(2x - 1)(4x - 3)(2x + 3)}{(2x + 3)(x + 1)(3x + 4)(4x - 3)}$ Factor the polynomials.

$$= \frac{\overset{1}{(3x+4)}\overset{1}{(2x-1)}\overset{1}{(4x-3)}(2x+3)}{(2x+3)(x+1)(3x+4)(4x-3)} \qquad \frac{3x+4}{3x+4}=1, \frac{4x-3}{4x-3}=1,$$
$$\frac{2x+3}{2x+3}=1.$$

$$= \frac{2x-1}{x+1}$$

■

EXAMPLE 4 Multiply: $(2x - x^2) \cdot \dfrac{x}{x^2 - 5x + 6}$.

Solution $(2x - x^2) \cdot \dfrac{x}{x^2 - 5x + 6} = \dfrac{2x - x^2}{1} \cdot \dfrac{x}{x^2 - 5x + 6}$ Write $2x - x^2$ as $\dfrac{2x - x^2}{1}$.

$$= \frac{(2x - x^2)x}{1(x^2 - 5x + 6)}$$ Multiply the fractions.

$$= \frac{\overset{-1}{x(2-x)}x}{\underset{1}{(x-2)}(x-3)}$$ Factor the numerator and the denominator and note that the quotient of any nonzero quantity and its negative is -1.

$$= \frac{-x^2}{x - 3}$$

Because $\dfrac{-a}{b} = -\dfrac{a}{b} = \dfrac{a}{-b}$, the negative sign from the numerator can be written in front of the fraction or in the denominator. For this reason, the final result can be written as

$$-\frac{x^2}{x - 3}, \qquad \frac{x^2}{-(x - 3)}, \qquad \text{or} \qquad \frac{x^2}{3 - x}$$

■

In Examples 1–4, we would obtain the same answers by factoring first and dividing out the common factors before multiplying.

■ DIVIDING RATIONAL EXPRESSIONS

Property 4 of fractions also provides the rule for dividing fractions:

$$\frac{a}{b} \div \frac{c}{d} = \frac{a}{b} \cdot \frac{d}{c}$$

We can prove this rule as follows:

$$\frac{a}{b} \div \frac{c}{d} = \frac{\dfrac{a}{b}}{\dfrac{c}{d}} = \frac{\dfrac{a}{b}}{\dfrac{c}{d}} \cdot 1 = \frac{\dfrac{a}{b}}{\dfrac{c}{d}} \cdot \frac{\dfrac{d}{c}}{\dfrac{d}{c}} = \frac{\dfrac{a}{b} \cdot \dfrac{d}{c}}{\dfrac{c}{d} \cdot \dfrac{d}{c}} = \frac{\dfrac{a}{b} \cdot \dfrac{d}{c}}{\dfrac{cd}{cd}} = \frac{\dfrac{a}{b} \cdot \dfrac{d}{c}}{1} = \frac{a}{b} \cdot \frac{d}{c}$$

This shows that *to divide fractions, we can invert the divisor and multiply.*

$$\frac{3}{5} \div \frac{2}{7} = \frac{3}{5} \cdot \frac{7}{2}$$

$$= \frac{3 \cdot 7}{5 \cdot 2}$$

$$= \frac{21}{10}$$

$$\frac{4}{7} \div \frac{2}{21} = \frac{4}{7} \cdot \frac{21}{2}$$

$$= \frac{4 \cdot 21}{7 \cdot 2}$$

$$= \frac{\overset{1}{\cancel{2}} \cdot 2 \cdot 3 \cdot \overset{1}{\cancel{7}}}{\underset{1}{\cancel{7}} \cdot \underset{1}{\cancel{2}}}$$

$$= 6$$

The same rule applies to algebraic fractions.

$$\frac{x^2}{y^3 z^2} \div \frac{x^2}{yz^3} = \frac{x^2}{y^3 z^2} \cdot \frac{yz^3}{x^2}$$ Invert the divisor and multiply.

$$= \frac{x^2 yz^3}{x^2 y^3 z^2}$$ Multiply the numerators and the denominators.

$$= x^{2-2} y^{1-3} z^{3-2}$$ To divide exponential expressions with the same base, keep the base and subtract the exponents.

$$= x^0 y^{-2} z^1$$

$$= 1 \cdot y^{-2} \cdot z$$ $x^0 = 1.$

$$= \frac{z}{y^2}$$ $y^{-2} = \dfrac{1}{y^2}.$

EXAMPLE 5   Divide: $\dfrac{x^3 + 8}{x + 1} \div \dfrac{x^2 - 2x + 4}{2x^2 - 2}.$

Solution   We invert the divisor and multiply.

$$\frac{x^3 + 8}{x + 1} \div \frac{x^2 - 2x + 4}{2x^2 - 2} = \frac{x^3 + 8}{x + 1} \cdot \frac{2x^2 - 2}{x^2 - 2x + 4}$$

$$= \frac{(x^3 + 8)(2x^2 - 2)}{(x + 1)(x^2 - 2x + 4)}$$

$$= \frac{(x + 2)(\cancel{x^2 - 2x + 4})2(\cancel{x + 1})(x - 1)}{(\cancel{x + 1})(\cancel{x^2 - 2x + 4})}$$ $\dfrac{x^2 - 2x + 4}{x^2 - 2x + 4} = 1,$

$\dfrac{x + 1}{x + 1} = 1.$

$$= 2(x + 2)(x - 1)$$

EXAMPLE 6   Divide: $\dfrac{x^2 - 4}{x - 1} \div (x - 2).$

Solution $\dfrac{x^2 - 4}{x - 1} \div (x - 2) = \dfrac{x^2 - 4}{x - 1} \div \dfrac{x - 2}{1}$ Write $x - 2$ as a fraction with a denominator of 1.

$$= \frac{x^2 - 4}{x - 1} \cdot \frac{1}{x - 2}$$ Invert the divisor and multiply.

$$= \frac{x^2 - 4}{(x - 1)(x - 2)} \qquad \text{Multiply the numerators and the denominators.}$$

$$= \frac{\overset{1}{(x + 2)\cancel{(x - 2)}}}{\underset{1}{(x - 1)\cancel{(x - 2)}}} \qquad \text{Factor } x^2 - 4.$$

$$= \frac{x + 2}{x - 1} \qquad\qquad \frac{x - 2}{x - 2} = 1. \qquad\qquad ■$$

■ MIXED OPERATIONS

EXAMPLE 7   Simplify: $\dfrac{x^2 + 2x - 3}{6x^2 + 5x + 1} \div \dfrac{2x^2 - 2}{2x^2 - 5x - 3} \cdot \dfrac{6x^2 + 4x - 2}{x^2 - 2x - 3}.$

Solution   Since multiplications and divisions are done in order from left to right, we change the division to a multiplication.

$$\left(\frac{x^2 + 2x - 3}{6x^2 + 5x + 1} \div \frac{2x^2 - 2}{2x^2 - 5x - 3} \right) \frac{6x^2 + 4x - 2}{x^2 - 2x - 3} = \left(\frac{x^2 + 2x - 3}{6x^2 + 5x + 1} \cdot \frac{2x^2 - 5x - 3}{2x^2 - 2} \right) \frac{6x^2 + 4x - 2}{x^2 - 2x - 3}$$

Then we multiply the fractions and simplify the result.

$$\left(\frac{x^2 + 2x - 3}{6x^2 + 5x + 1} \div \frac{2x^2 - 2}{2x^2 - 5x - 3} \right) \frac{6x^2 + 4x - 2}{x^2 - 2x - 3} = \frac{(x^2 + 2x - 3)(2x^2 - 5x - 3)(6x^2 + 4x - 2)}{(6x^2 + 5x + 1)(2x^2 - 2)(x^2 - 2x - 3)}$$

$$= \frac{\overset{1}{(x + 3)}\overset{1}{\cancel{(x - 1)}}\overset{1}{\cancel{(2x + 1)}}\overset{1}{\cancel{(x - 3)}}\cancel{2}(3x - 1)\overset{1}{\cancel{(x + 1)}}}{(3x + 1)\underset{1}{\cancel{(2x + 1)}}\cancel{2}(x + 1)\underset{1}{\cancel{(x - 1)}}\underset{1}{\cancel{(x - 3)}}\underset{1}{\cancel{(x + 1)}}}$$

$$= \frac{(x + 3)(3x - 1)}{(3x + 1)(x + 1)} \qquad\qquad ■$$

Orals   *Multiply the fractions and simplify, if possible.*

1. $\dfrac{3}{2} \cdot \dfrac{3}{4}$ 　　　　　　**2.** $\dfrac{3x}{7} \cdot \dfrac{7}{6x}$ 　　　　　　**3.** $\dfrac{x - 2}{y} \cdot \dfrac{y}{x + 2}$

Divide the fractions and simplify, if possible.

4. $\dfrac{3}{4} \div \dfrac{4}{3}$ 　　　　　　**5.** $\dfrac{5a}{b} \div \dfrac{a}{b}$ 　　　　　　**6.** $\dfrac{x^2 y}{ab} \div \dfrac{2x^2}{ba}$

EXERCISE 6.2

In Exercises 1–54, do the operations and simplify.

1. $\dfrac{3}{4} \cdot \dfrac{5}{3} \cdot \dfrac{8}{7}$ 　　　**2.** $-\dfrac{5}{6} \cdot \dfrac{3}{7} \cdot \dfrac{14}{25}$ 　　　**3.** $-\dfrac{6}{11} \div \dfrac{36}{55}$ 　　　**4.** $\dfrac{17}{12} \div \dfrac{34}{3}$

5. $\dfrac{x^2 y^2}{cd} \cdot \dfrac{c^{-2} d^2}{x}$

6. $\dfrac{a^{-2} b^2}{x^{-1} y} \cdot \dfrac{a^4 b^4}{x^2 y^3}$

7. $\dfrac{-x^2 y^{-2}}{x^{-1} y^{-3}} \div \dfrac{x^{-3} y^2}{x^4 y^{-1}}$

8. $\dfrac{(a^3)^2}{b^{-1}} \div \dfrac{(a^3)^{-2}}{b^{-1}}$

9. $\dfrac{x^2 + 2x + 1}{x} \cdot \dfrac{x^2 - x}{x^2 - 1}$

10. $\dfrac{a + 6}{a^2 - 16} \cdot \dfrac{3a - 12}{3a + 18}$

11. $\dfrac{2x^2 - x - 3}{x^2 - 1} \cdot \dfrac{x^2 + x - 2}{2x^2 + x - 6}$

12. $\dfrac{9x^2 + 3x - 20}{3x^2 - 7x + 4} \cdot \dfrac{3x^2 - 5x + 2}{9x^2 + 18x + 5}$

13. $\dfrac{x^2 - 16}{x^2 - 25} \div \dfrac{x + 4}{x - 5}$

14. $\dfrac{a^2 - 9}{a^2 - 49} \div \dfrac{a + 3}{a + 7}$

15. $\dfrac{a^2 + 2a - 35}{12x} \div \dfrac{ax - 3x}{a^2 + 4a - 21}$

16. $\dfrac{x^2 - 4}{2b - bx} \div \dfrac{x^2 + 4x + 4}{2b + bx}$

17. $\dfrac{3t^2 - t - 2}{6t^2 - 5t - 6} \cdot \dfrac{4t^2 - 9}{2t^2 + 5t + 3}$

18. $\dfrac{2p^2 - 5p - 3}{p^2 - 9} \cdot \dfrac{2p^2 + 5p - 3}{2p^2 + 5p + 2}$

19. $\dfrac{3n^2 + 5n - 2}{12n^2 - 13n + 3} \div \dfrac{n^2 + 3n + 2}{4n^2 + 5n - 6}$

20. $\dfrac{8y^2 - 14y - 15}{6y^2 - 11y - 10} \div \dfrac{4y^2 - 9y - 9}{3y^2 - 7y - 6}$

21. $(x + 1) \cdot \dfrac{1}{x^2 + 2x + 1}$

22. $\dfrac{x^2 - 4}{x} \div (x + 2)$

23. $(x^2 - x - 2) \cdot \dfrac{x^2 + 3x + 2}{x^2 - 4}$

24. $(2x^2 - 9x - 5) \cdot \dfrac{x}{2x^2 + x}$

25. $(2x^2 - 15x + 25) \div \dfrac{2x^2 - 3x - 5}{x + 1}$

26. $(x^2 - 6x + 9) \div \dfrac{x^2 - 9}{x + 3}$

27. $\dfrac{x^3 + y^3}{x^3 - y^3} \div \dfrac{x^2 - xy + y^2}{x^2 + xy + y^2}$

28. $\dfrac{x^2 - 6x + 9}{4 - x^2} \div \dfrac{x^2 - 9}{x^2 - 8x + 12}$

29. $\dfrac{m^2 - n^2}{2x^2 + 3x - 2} \cdot \dfrac{2x^2 + 5x - 3}{n^2 - m^2}$

30. $\dfrac{x^2 - y^2}{2x^2 + 2xy + x + y} \cdot \dfrac{2x^2 - 5x - 3}{yx - 3y - x^2 + 3x}$

31. $\dfrac{ax + ay + bx + by}{x^3 - 27} \cdot \dfrac{x^2 + 3x + 9}{xc + xd + yc + yd}$

32. $\dfrac{x^2 + 3x + yx + 3y}{x^2 - 9} \cdot \dfrac{x - 3}{x + 3}$

33. $\dfrac{x^2 - x - 6}{x^2 - 4} \cdot \dfrac{x^2 - x - 2}{9 - x^2}$

34. $\dfrac{2x^2 - 7x - 4}{20 - x - x^2} \div \dfrac{2x^2 - 9x - 5}{x^2 - 25}$

35. $\dfrac{2x^2 + 3xy + y^2}{y^2 - x^2} \div \dfrac{6x^2 + 5xy + y^2}{2x^2 - xy - y^2}$

36. $\dfrac{p^3 - q^3}{q^2 - p^2} \cdot \dfrac{q^2 + pq}{p^3 + p^2 q + pq^2}$

37. $\dfrac{3x^2 y^2}{6x^3 y} \cdot \dfrac{-4x^7 y^{-2}}{18x^{-2} y} \div \dfrac{36x}{18y^{-2}}$

38. $\dfrac{9ab^3}{7xy} \cdot \dfrac{14xy^2}{27z^3} \div \dfrac{18a^2 b^2 x}{3z^2}$

39. $(4x + 12) \cdot \dfrac{x^2}{2x - 6} \div \dfrac{2}{x - 3}$

40. $(4x^2 - 9) \div \dfrac{2x^2 + 5x + 3}{x + 2} \div (2x - 3)$

41. $\dfrac{2x^2 - 2x - 4}{x^2 + 2x - 8} \cdot \dfrac{3x^2 + 15x}{x + 1} \div \dfrac{4x^2 - 100}{x^2 - x - 20}$

42. $\dfrac{6a^2 - 7a - 3}{a^2 - 1} \div \dfrac{4a^2 - 12a + 9}{a^2 - 1} \cdot \dfrac{2a^2 - a - 3}{3a^2 - 2a - 1}$

43. $\dfrac{2t^2 + 5t + 2}{t^2 - 4t + 16} \div \dfrac{t + 2}{t^3 + 64} \div \dfrac{2t^3 + 9t^2 + 4t}{t + 1}$

44. $\dfrac{a^6 - b^6}{a^4 - a^3 b} \cdot \dfrac{a^3}{a^4 + a^2 b^2 + b^4} \div \dfrac{1}{a}$

45. $\dfrac{x^4 - 3x^2 - 4}{x^4 - 1} \cdot \dfrac{x^2 + 3x + 2}{x^2 + 4x + 4}$

46. $\dfrac{x^3 + 2x^2 + 4x + 8}{y^2 - 1} \cdot \dfrac{y^2 + 2y + 1}{x^4 - 16}$

47. $(x^2 - x - 6) \div (x - 3) \div (x - 2)$

48. $(x^2 - x - 6) \div [(x - 3) \div (x - 2)]$

49. $\dfrac{3x^2 - 2x}{3x + 2} \div (3x - 2) \div \dfrac{3x}{3x - 3}$

50. $(2x^2 - 3x - 2) \div \dfrac{2x^2 - x - 1}{x - 2} \div (x - 1)$

51. $\dfrac{2x^2 + 5x - 3}{x^2 + 2x - 3} \div \left(\dfrac{x^2 + 2x - 35}{x^2 - 6x + 5} \div \dfrac{x^2 - 9x + 14}{2x^2 - 5x + 2} \right)$

52. $\dfrac{x^2 - 4}{x^2 - x - 6} \div \left(\dfrac{x^2 - x - 2}{x^2 - 8x + 15} \cdot \dfrac{x^2 - 3x - 10}{x^2 + 3x + 2} \right)$

53. $\dfrac{x^2 - x - 12}{x^2 + x - 2} \div \dfrac{x^2 - 6x + 8}{x^2 - 3x - 10} \cdot \dfrac{x^2 - 3x + 2}{x^2 - 2x - 15}$

54. $\dfrac{4x^2 - 10x + 6}{x^4 - 3x^3} \div \dfrac{2x - 3}{2x^3} \cdot \dfrac{x - 3}{2x - 2}$

Writing Exercises *Write a paragraph using your own words.*

1. Explain how to multiply two rational expressions.

2. Explain how to divide one rational expression by another.

Something to Think About *Insert either a multiplication or a division symbol in each box to make a true statement.*

1. $\dfrac{x^2}{y} \; \boxed{} \; \dfrac{x}{y^2} \; \boxed{} \; \dfrac{x^2}{y^2} = \dfrac{x^3}{y}$

2. $\dfrac{x^2}{y} \; \boxed{} \; \dfrac{x}{y^2} \; \boxed{} \; \dfrac{x^2}{y^2} = \dfrac{y^3}{x}$

Review Exercises *Do each operation.*

1. $-2a^2(3a^3 - a^2)$

2. $(2t - 1)^2$

3. $(m^n + 2)(m^n - 2)$

4. $(3b^{-n} + c)(b^{-n} - c)$

6.3 Adding and Subtracting Rational Expressions

■ ADDING AND SUBTRACTING RATIONAL EXPRESSIONS WITH LIKE DENOMINATORS ■ GRAPHING DEVICES ■ ADDING AND SUBTRACTING RATIONAL EXPRESSIONS WITH UNLIKE DENOMINATORS ■ FINDING THE LEAST COMMON DENOMINATOR ■ MIXED OPERATIONS

Fractions are added and subtracted according to the following rules.

Property 5 of Fractions

If there are no divisions by 0, then

$$\frac{a}{b} + \frac{c}{b} = \frac{a + c}{b} \qquad \text{and} \qquad \frac{a}{b} - \frac{c}{b} = \frac{a - c}{b}$$

■ ADDING AND SUBTRACTING RATIONAL EXPRESSIONS WITH LIKE DENOMINATORS

When we add or subtract fractions with like denominators, *we add or subtract the numerators and keep the same denominator.* Whenever possible, we should always simplify the result.

EXAMPLE 1 Simplify: **a.** $\dfrac{17}{22} + \dfrac{13}{22}$, **b.** $\dfrac{3}{2x} + \dfrac{7}{2x}$, and **c.** $\dfrac{4x}{x+2} - \dfrac{7x}{x+2}$.

Solution **a.** $\dfrac{17}{22} + \dfrac{13}{22} = \dfrac{17 + 13}{22}$

$$= \dfrac{30}{22}$$

$$= \dfrac{15 \cdot \cancel{2}}{11 \cdot \cancel{2}}$$

$$= \dfrac{15}{11}$$

b. $\dfrac{3}{2x} + \dfrac{7}{2x} = \dfrac{3 + 7}{2x}$

$$= \dfrac{10}{2x}$$

$$= \dfrac{\cancel{2} \cdot 5}{\cancel{2} \cdot x}$$

$$= \dfrac{5}{x}$$

c. $\dfrac{4x}{x+2} - \dfrac{7x}{x+2} = \dfrac{4x - 7x}{x+2}$

$$= \dfrac{-3x}{x+2}$$ ■

■ GRAPHING DEVICES

We can check the subtraction in part **c** of Example 1 by graphing the rational functions $f(x) = \frac{4x}{x+2} - \frac{7x}{x+2}$, shown in Figure 6-5(a), and $g(x) = \frac{-3x}{x+2}$, shown in Figure 6-5(b), and observing that the graphs are the same. Note that -2 is not in the domain of either function.

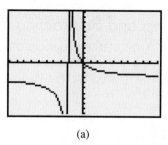 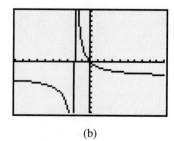

(a)   (b)

FIGURE 6-5

■ ADDING AND SUBTRACTING RATIONAL EXPRESSIONS WITH UNLIKE DENOMINATORS

To add or subtract fractions with unlike denominators, we first convert them to fractions with the same denominator. When the denominators of the fractions are opposites, we can multiply one of the fractions by 1, written in the form $\frac{-1}{-1}$, to get a common denominator.

EXAMPLE 2   Add: $\dfrac{x}{x-y} + \dfrac{y}{y-x}$.

Solution $\dfrac{x}{x-y} + \dfrac{y}{y-x} = \dfrac{x}{x-y} + \left(\dfrac{-1}{-1}\right)\dfrac{y}{y-x}$ $\dfrac{-1}{-1} = 1$, and multiplying a fraction by 1 does not change its value.

$= \dfrac{x}{x-y} + \dfrac{-y}{-y+x}$ Multiply.

$= \dfrac{x}{x-y} + \dfrac{-y}{x-y}$ $-y + x = x - y$.

$= \dfrac{x-y}{x-y}$ Add the numerators and keep the common denominator.

$= 1$ $\dfrac{x-y}{x-y} = 1$.   ■

When the denominators of two or more fractions are different, we often have to multiply one or more fractions by 1, written in some appropriate form, to get a common denominator.

(b)   **EXAMPLE 3**   Simplify: **a.** $\dfrac{2}{3} + \dfrac{3}{2}$ and **b.** $\dfrac{3}{x} + \dfrac{4}{y}$.

Solution   **a.** $\dfrac{2}{3} + \dfrac{3}{2} = \dfrac{2}{3} \cdot \mathbf{1} + \dfrac{3}{2} \cdot \mathbf{1}$ Multiply each fraction by 1.

$= \dfrac{2}{3} \cdot \dfrac{2}{2} + \dfrac{3}{2} \cdot \dfrac{3}{3}$ $\dfrac{2}{2} = 1$ and $\dfrac{3}{3} = 1$.

$= \dfrac{2 \cdot 2}{3 \cdot 2} + \dfrac{3 \cdot 3}{2 \cdot 3}$ Multiply the fractions by multiplying their numerators and denominators.

$= \dfrac{4}{6} + \dfrac{9}{6}$ Do the multiplications.

$= \dfrac{13}{6}$ Since the fractions have the same denominator, add the numerators and keep the denominator.

b. $\dfrac{3}{x} + \dfrac{4}{y} = \dfrac{3}{x} \cdot \mathbf{1} + \dfrac{4}{y} \cdot \mathbf{1}$ Multiply each fraction by 1.

$= \dfrac{3}{x} \cdot \dfrac{y}{y} + \dfrac{4}{y} \cdot \dfrac{x}{x}$ $\dfrac{y}{y} = 1 \; (y \neq 0)$ and $\dfrac{x}{x} = 1 \; (x \neq 0)$.

$= \dfrac{3y}{xy} + \dfrac{4x}{xy}$ Multiply the fractions by multiplying their numerators and denominators.

$= \dfrac{3y + 4x}{xy}$ Since the fractions have the same denominator, add the numerators and keep the denominator.   ■

(a)

EXAMPLE 4   Simplify: **a.** $3 + \dfrac{7}{x-2}$ and   **b.** $\dfrac{4x}{x+2} - \dfrac{7x}{x-2}$.

Solution   **a.** $3 + \dfrac{7}{x-2} = \dfrac{3}{1} + \dfrac{7}{x-2}$ $3 = \dfrac{3}{1}$.

$\qquad\qquad\qquad = \dfrac{3(x-2)}{1(x-2)} + \dfrac{7}{x-2}$ $\dfrac{x-2}{x-2} = 1$, and multiplying 3 by 1 does not change its value.

$\qquad\qquad\qquad = \dfrac{3(x-2)}{x-2} + \dfrac{7}{x-2}$ Simplify.

$\qquad\qquad\qquad = \dfrac{3x-6+7}{x-2}$ Since the denominators are the same, add the numerators and keep the same denominator.

$\qquad\qquad\qquad = \dfrac{3x+1}{x-2}$ Simplify.

b. $\dfrac{4x}{x+2} - \dfrac{7x}{x-2} = \dfrac{4x(x-2)}{(x+2)(x-2)} - \dfrac{(x+2)7x}{(x+2)(x-2)}$ $\dfrac{x-2}{x-2} = 1$ and $\dfrac{x+2}{x+2} = 1$.

$\qquad\qquad\qquad = \dfrac{4x^2-8x}{(x+2)(x-2)} - \dfrac{7x^2+14x}{(x+2)(x-2)}$ Remove parentheses in the numerators.

$\qquad\qquad\qquad = \dfrac{(4x^2-8x)-(7x^2+14x)}{(x+2)(x-2)}$ Subtract the numerators and keep the same denominator.

$\qquad\qquad\qquad = \dfrac{4x^2-8x-7x^2-14x}{(x+2)(x-2)}$ Use the distributive property to remove parentheses in the numerator.

$\qquad\qquad\qquad = \dfrac{-3x^2-22x}{(x+2)(x-2)}$ Simplify the numerator.

⚠ **WARNING!**   The $-$ sign between the fractions in Step 2 of part **b** applies to both terms of the binomial $7x^2 + 14x$.   ∎

■ FINDING THE LEAST COMMON DENOMINATOR

When adding fractions with unlike denominators, it is easiest if we change the fractions into fractions having the smallest common denominator possible, called the **least** (or lowest) **common denominator (LCD)**.

Suppose we have the fractions $\frac{1}{12}$, $\frac{1}{20}$, and $\frac{1}{35}$. To find the LCD of these fractions, we first find the prime factorizations of each denominator.

$$12 = 4 \cdot 3 = 2^2 \cdot 3$$

$$20 = 4 \cdot 5 = 2^2 \cdot 5$$

$$35 = 5 \cdot 7$$

Since the LCD is the smallest number that can be divided by 12, 20, and 35, it must contain factors of 2^2, 3, 5, and 7.

$$\text{LCD} = 2^2 \cdot 3 \cdot 5 \cdot 7 = 420$$

To find the least common denominator of several fractions, we follow these steps.

> **Finding the LCD**
> 1. Factor the denominator of each fraction.
> 2. List the different factors of each denominator.
> 3. Write each factor found in Step 2 to the highest power that occurs in any one denominator.
> 4. The LCD is the product of the factors in Step 3.

EXAMPLE 5 Find the LCD of the fractions $\dfrac{1}{x^2 + 7x + 6}$, $\dfrac{3}{x^2 - 36}$, and $\dfrac{5}{x^2 + 12x + 36}$.

Solution We factor each polynomial denominator:

$$x^2 + 7x + 6 = (x + 6)(x + 1)$$
$$x^2 - 36 = (x + 6)(x - 6)$$
$$x^2 + 12x + 36 = (x + 6)(x + 6) = (x + 6)^2$$

and list the individual factors:

$$x + 6, \quad x + 1, \quad \text{and} \quad x - 6$$

To find the LCD of the three fractions, we use the highest power of each of these factors:

$$\text{LCD} = (x + 6)^2(x + 1)(x - 6)$$

EXAMPLE 6 Simplify: $\dfrac{x}{x^2 - 2x + 1} + \dfrac{3}{x^2 - 1}$.

Solution We factor each denominator and find the LCD:

$$x^2 - 2x + 1 = (x - 1)(x - 1) = (x - 1)^2$$
$$x^2 - 1 = (x + 1)(x - 1)$$

The LCD is $(x - 1)^2(x + 1)$.

We now write each fraction with its denominator in factored form and convert the fractions to fractions with a common denominator of $(x - 1)^2(x + 1)$. Finally, we add the fractions.

$$\frac{x}{x^2 - 2x + 1} + \frac{3}{x^2 - 1} = \frac{x}{(x - 1)(x - 1)} + \frac{3}{(x + 1)(x - 1)}$$

$$= \frac{x(x + 1)}{(x - 1)(x - 1)(x + 1)} + \frac{3(x - 1)}{(x + 1)(x - 1)(x - 1)}$$

$$= \frac{x^2 + x + 3x - 3}{(x - 1)(x - 1)(x + 1)}$$

$$= \frac{x^2 + 4x - 3}{(x - 1)^2(x + 1)} \qquad \text{This result does not simplify.}$$

EXAMPLE 7   Simplify: $\dfrac{3x}{x-1} - \dfrac{2x^2 + 3x - 2}{(x+1)(x-1)}$.

Solution   We write each fraction in a form having the LCD of $(x+1)(x-1)$, remove the resulting parentheses in the first numerator, do the subtraction, and simplify.

1. $\dfrac{3x}{x-1} - \dfrac{2x^2 + 3x - 2}{(x+1)(x-1)} = \dfrac{(x+1)3x}{(x+1)(x-1)} - \dfrac{2x^2 + 3x - 2}{(x+1)(x-1)}$

$\qquad = \dfrac{3x^2 + 3x}{(x+1)(x-1)} - \dfrac{2x^2 + 3x - 2}{(x+1)(x-1)}$

$\qquad = \dfrac{3x^2 + 3x - (2x^2 + 3x - 2)}{(x+1)(x-1)}$

$\qquad = \dfrac{3x^2 + 3x - 2x^2 - 3x + 2}{(x+1)(x-1)}$

$\qquad = \dfrac{x^2 + 2}{(x+1)(x-1)}$

WARNING!   The $-$ sign between the fractions in Equation 1 affects every term of the numerator $2x^2 + 3x - 2$. Whenever we subtract one fraction from another, we must remember to subtract each term of the numerator in the second fraction. ■

■ **MIXED OPERATIONS**

EXAMPLE 8   Simplify: $\dfrac{2x}{x^2 - 4} - \dfrac{1}{x^2 - 3x + 2} + \dfrac{x+1}{x^2 + x - 2}$.

Solution   We factor each denominator to find the LCD, which is

$$\text{LCD} = (x+2)(x-2)(x-1)$$

We then write each fraction as a fraction with the LCD as its denominator and do the subtraction and addition.

$\dfrac{2x}{x^2 - 4} - \dfrac{1}{x^2 - 3x + 2} + \dfrac{x+1}{x^2 + x - 2}$

$= \dfrac{2x}{(x-2)(x+2)} - \dfrac{1}{(x-2)(x-1)} + \dfrac{x+1}{(x-1)(x+2)}$

$= \dfrac{2x(x-1)}{(x-2)(x+2)(x-1)} - \dfrac{1(x+2)}{(x-2)(x-1)(x+2)} + \dfrac{(x+1)(x-2)}{(x-1)(x+2)(x-2)}$

$= \dfrac{2x(x-1) - 1(x+2) + (x+1)(x-2)}{(x+2)(x-2)(x-1)}$

$= \dfrac{2x^2 - 2x - x - 2 + x^2 - x - 2}{(x+2)(x-2)(x-1)}$

$= \dfrac{3x^2 - 4x - 4}{(x+2)(x-2)(x-1)}$

Here, the final result does simplify.

$$\frac{2x}{x^2-4} - \frac{1}{x^2-3x+2} + \frac{x+1}{x^2+x-2} = \frac{3x^2-4x-4}{(x+2)(x-2)(x-1)}$$

$$= \frac{(3x+2)(x-2)}{(x+2)(x-2)(x-1)} \qquad \text{Factor the numerator.}$$

$$= \frac{3x+2}{(x+2)(x-1)} \qquad \frac{x-2}{x-2} = 1. \ \blacksquare$$

EXAMPLE 9   Simplify: $\left(\dfrac{x^2}{x-2} + \dfrac{4}{2-x} \right)^2$.

Solution   To follow order of operations, we do the addition within the parentheses first. Since the denominators are negatives of one another, we can write the fractions as fractions with a common denominator by multiplying both the numerator and denominator of $\frac{4}{2-x}$ by -1. We can then add the fractions, simplify, and square the result.

$$\left(\frac{x^2}{x-2} + \frac{4}{2-x} \right)^2 = \left[\frac{x^2}{x-2} + \frac{(-1)4}{(-1)(2-x)} \right]^2$$

$$= \left[\frac{x^2}{x-2} + \frac{-4}{x-2} \right]^2$$

$$= \left[\frac{x^2-4}{x-2} \right]^2$$

$$= \left[\frac{(x+2)(x-2)}{x-2} \right]^2$$

$$= (x+2)^2 \qquad \frac{x-2}{x-2} = 1.$$

$$= x^2+4x+4 \qquad \blacksquare$$

Orals   *Add or subtract the fractions and simplify the result, if possible.*

1. $\dfrac{x}{2} + \dfrac{x}{2}$ **2.** $\dfrac{3a}{4} - \dfrac{a}{4}$ **3.** $\dfrac{x}{x+2} + \dfrac{2}{x+2}$

4. $\dfrac{2a}{a+4} - \dfrac{a-4}{a+4}$ **5.** $\dfrac{2x}{3} + \dfrac{x}{2}$ **6.** $\dfrac{5}{x} - \dfrac{3}{y}$

EXERCISE 6.3

In Exercises 1–16, do the operations and simplify the result when possible.

1. $\dfrac{3}{4} + \dfrac{7}{4}$ **2.** $\dfrac{5}{11} + \dfrac{2}{11}$ **3.** $\dfrac{10}{33} - \dfrac{21}{33}$ **4.** $\dfrac{8}{15} - \dfrac{2}{15}$

5. $\dfrac{3}{4y} + \dfrac{8}{4y}$ **6.** $\dfrac{5}{3z^2} - \dfrac{6}{3z^2}$ **7.** $\dfrac{3}{a+b} - \dfrac{a}{a+b}$ **8.** $\dfrac{x}{x+4} + \dfrac{5}{x+4}$

9. $\dfrac{3x}{2x+2} + \dfrac{x+4}{2x+2}$ **10.** $\dfrac{4y}{y-4} - \dfrac{16}{y-4}$ **11.** $\dfrac{3x}{x-3} - \dfrac{9}{x-3}$ **12.** $\dfrac{9x}{x-y} - \dfrac{9y}{x-y}$

13. $\dfrac{5x}{x+1} + \dfrac{3}{x+1} - \dfrac{2x}{x+1}$

14. $\dfrac{4}{a+4} - \dfrac{2a}{a+4} + \dfrac{3a}{a+4}$

15. $\dfrac{3(x^2+x)}{x^2-5x+6} + \dfrac{-3(x^2-x)}{x^2-5x+6}$

16. $\dfrac{2x+4}{x^2+13x+12} - \dfrac{x+3}{x^2+13x+12}$

In Exercises 17–24, the denominators of several fractions are given. Find the LCD.

17. 8, 12, 18

18. 10, 15, 28

19. $x^2 + 3x,\ x^2 - 9$

20. $3y^2 - 6y,\ 3y(y-4)$

21. $x^3 + 27,\ x^2 + 6x + 9$

22. $x^3 - 8,\ x^2 - 4x + 4$

23. $2x^2 + 5x + 3,\ 4x^2 + 12x + 9,\ x^2 + 2x + 1$

24. $2x^2 + 5x + 3,\ 4x^2 + 12x + 9,\ 4x + 6$

In Exercises 25–86, do the operations and simplify the result when possible.

25. $\dfrac{1}{2} + \dfrac{1}{3}$

26. $\dfrac{5}{6} + \dfrac{2}{7}$

27. $\dfrac{7}{15} - \dfrac{17}{25}$

28. $\dfrac{8}{9} - \dfrac{5}{12}$

29. $\dfrac{a}{2} + \dfrac{2a}{5}$

30. $\dfrac{b}{6} + \dfrac{3a}{4}$

31. $\dfrac{3a}{2} - \dfrac{4b}{7}$

32. $\dfrac{2m}{3} - \dfrac{4n}{5}$

33. $\dfrac{3}{4x} + \dfrac{2}{3x}$

34. $\dfrac{2}{5a} + \dfrac{3}{2b}$

35. $\dfrac{3a}{2b} - \dfrac{2b}{3a}$

36. $\dfrac{5m}{2n} - \dfrac{3n}{4m}$

37. $\dfrac{a+b}{3} + \dfrac{a-b}{7}$

38. $\dfrac{x-y}{2} + \dfrac{x+y}{3}$

39. $\dfrac{3}{x+2} + \dfrac{5}{x-4}$

40. $\dfrac{2}{a+4} - \dfrac{6}{a+3}$

41. $\dfrac{x+2}{x+5} - \dfrac{x-3}{x+7}$

42. $\dfrac{7}{x+3} + \dfrac{4x}{x+6}$

43. $x + \dfrac{1}{x}$

44. $2 - \dfrac{1}{x+1}$

45. $\dfrac{x+8}{x-3} - \dfrac{x-14}{3-x}$

46. $\dfrac{3-x}{2-x} + \dfrac{x-1}{x-2}$

47. $\dfrac{2a+1}{3a+2} - \dfrac{a-4}{2-3a}$

48. $\dfrac{4}{x-2} + \dfrac{5}{4-x^2}$

49. $\dfrac{x}{x^2+5x+6} + \dfrac{x}{x^2-4}$

50. $\dfrac{x}{3x^2-2x-1} + \dfrac{4}{3x^2+10x+3}$

51. $\dfrac{4}{x^2-2x-3} - \dfrac{x}{3x^2-7x-6}$

52. $\dfrac{2a}{a^2-2a-8} + \dfrac{3}{a^2-5a+4}$

53. $\dfrac{8}{x^2-9} + \dfrac{2}{x-3} - \dfrac{6}{x}$

54. $\dfrac{x}{x^2-4} - \dfrac{x}{x+2} + \dfrac{2}{x}$

55. $\dfrac{x}{x+1} - \dfrac{x}{1-x^2} + \dfrac{1}{x}$

56. $\dfrac{y}{y-2} - \dfrac{2}{y+2} - \dfrac{-8}{4-y^2}$

57. $2x + 3 + \dfrac{1}{x+1}$

58. $x + 1 + \dfrac{1}{x-1}$

59. $1 + x - \dfrac{x}{x-5}$

60. $2 - x + \dfrac{3}{x-9}$

61. $\dfrac{3x}{x-1} - 2x - x^2$

62. $\dfrac{23}{x-1} + 4x - 5x^2$

63. $\dfrac{y+4}{y^2+7y+12} - \dfrac{y-4}{y+3} + \dfrac{47}{y+4}$

64. $\dfrac{x+3}{2x^2-5x+2} - \dfrac{3x-1}{x^2-x-2}$

65. $\dfrac{3}{x+1} - \dfrac{2}{x-1} + \dfrac{x+3}{x^2-1}$

66. $\dfrac{2}{x-2} + \dfrac{3}{x+2} - \dfrac{x-1}{x^2-4}$

67. $\dfrac{x-2}{x^2-3x} + \dfrac{2x-1}{x^2+3x} - \dfrac{2}{x^2-9}$

68. $\dfrac{2}{x-1} - \dfrac{2x}{x^2-1} - \dfrac{x}{x^2+2x+1}$

69. $\dfrac{5}{x^2-25} - \dfrac{3}{2x^2-9x-5} + 1$

70. $\dfrac{3x}{2x-1} + \dfrac{x+1}{3x+2} + \dfrac{2x}{6x^3+x^2-2x}$

71. $\dfrac{3x}{x-3} + \dfrac{4}{x-2} - \dfrac{5x}{x^3-5x^2+6x}$

72. $\dfrac{2x-1}{x^2+x-6} - \dfrac{3x-5}{x^2-2x-15} + \dfrac{2x-3}{x^2-7x+10}$

73. $2 + \dfrac{4a}{a^2-1} - \dfrac{2}{a+1}$

74. $\dfrac{a}{a-1} - \dfrac{a+1}{2a-2} + a$

75. $\dfrac{x+5}{2x^2-2} + \dfrac{x}{2x+2} - \dfrac{3}{x-1}$

76. $\dfrac{a}{2-a} + \dfrac{3}{a-2} - \dfrac{3a-2}{a^2-4}$

77. $\dfrac{a}{a-b} + \dfrac{b}{a+b} + \dfrac{a^2+b^2}{b^2-a^2}$

78. $\dfrac{1}{x+y} - \dfrac{1}{x-y} - \dfrac{2x}{y^2-x^2}$

79. $\dfrac{7n^2}{m-n} + \dfrac{3m}{n-m} - \dfrac{3m^2-n}{m^2-2mn+n^2}$

80. $\dfrac{3b}{2a-b} + \dfrac{2a-1}{b-2a} - \dfrac{3a^2+b}{b^2-4ab+4a^2}$

81. $\dfrac{m+1}{m^2+2m+1} + \dfrac{m-1}{m^2-2m+1} + \dfrac{2}{m^2-1}$

(*Hint:* Think about this before finding the LCD.)

82. $\dfrac{a+2}{a^2+3a+2} + \dfrac{a-1}{a^2-1} + \dfrac{3}{a+1}$

83. $\left(\dfrac{1}{x-1} + \dfrac{1}{1-x}\right)^2$

84. $\left(\dfrac{1}{a-1} - \dfrac{1}{1-a}\right)^2$

85. $\left(\dfrac{x}{x-3} + \dfrac{3}{3-x}\right)^3$

86. $\left(\dfrac{2y}{y+4} + \dfrac{8}{y+4}\right)^3$

87. Show that $\dfrac{a}{b} + \dfrac{c}{d} = \dfrac{ad+bc}{bd}$.

88. Show that $\dfrac{a}{b} - \dfrac{c}{d} = \dfrac{ad-bc}{bd}$.

Writing Exercises   *Write a paragraph using your own words.*

1. Explain how to find the least common denominator.

2. Explain how to add two fractions.

Something to Think About   **1.** Find the error:

$$\dfrac{8x+2}{5} - \dfrac{3x+8}{5}$$

$$= \dfrac{8x+2-3x+8}{5}$$

$$= \dfrac{5x+10}{5}$$

$$= x+2$$

2. Find the error:

$$\dfrac{(x+y)^2}{2} + \dfrac{(x-y)^2}{3}$$

$$= \dfrac{3 \cdot (x+y)^2}{3 \cdot 2} + \dfrac{2 \cdot (x-y)^2}{2 \cdot 3}$$

$$= \dfrac{3x^2+3y^2+2x^2-2y^2}{6}$$

$$= \dfrac{5x^2+y^2}{6}$$

Review Exercises *Graph each interval on a number line.*

1. $(-2, 4] \cup [-1, 5)$

2. $(-1, 4) \cap (2, 5]$

Solve each formula for the indicated letter.

3. $P = 2l + 2w$; for w

4. $S = \dfrac{a - lr}{1 - r}$; for a

6.4 Complex Fractions

■ SIMPLIFYING COMPLEX FRACTIONS ■ GRAPHING DEVICES

A **complex fraction** is one that has a fraction in its numerator or its denominator or both. Examples of complex fractions are

$$\dfrac{\dfrac{3a}{b}}{\dfrac{6ac}{b^2}}, \qquad \dfrac{\dfrac{2}{x} + 1}{3 + x}, \quad \text{and} \quad \dfrac{\dfrac{1}{x} + \dfrac{1}{y}}{\dfrac{1}{x} - \dfrac{1}{y}}$$

■ SIMPLIFYING COMPLEX FRACTIONS

We can use two methods to simplify the complex fraction

$$\dfrac{\dfrac{3a}{b}}{\dfrac{6ac}{b^2}}$$

In one method, we eliminate the fractions in the numerator and denominator by writing the complex fraction as a division and then using the division rule for fractions:

$$\dfrac{\dfrac{3a}{b}}{\dfrac{6ac}{b^2}} = \dfrac{3a}{b} \div \dfrac{6ac}{b^2}$$

$$= \dfrac{3a}{b} \cdot \dfrac{b^2}{6ac} \qquad \text{Invert the divisor and multiply.}$$

$$= \dfrac{b}{2c} \qquad \text{Multiply the fractions and simplify.}$$

In another method, we eliminate the fractions in the numerator and denominator by multiplying the fraction by 1, written in the form $\frac{b^2}{b^2}$. We use $\frac{b^2}{b^2}$ because b^2 is the LCD of $\frac{3a}{b}$ and $\frac{6ac}{b^2}$.

$$\dfrac{\dfrac{3a}{b}}{\dfrac{6ac}{b^2}} = \dfrac{\dfrac{3a}{b} \cdot b^2}{\dfrac{6ac}{b^2} \cdot b^2}$$

$$= \frac{\dfrac{3ab^2}{b}}{\dfrac{6acb^2}{b^2}}$$

$$= \frac{3ab}{6ac} \qquad \text{Simplify the fractions in the numerator and the denominator.}$$

$$= \frac{b}{2c} \qquad \text{Divide out the common factor of } 3a.$$

With either method, the result is the same.

EXAMPLE 1 Simplify the complex fraction $\dfrac{\dfrac{2}{x} + 1}{3 + x}$.

Method 1 We add the fractions in the numerator and proceed as follows:

$$\frac{\dfrac{2}{x} + 1}{3 + x} = \frac{\dfrac{2}{x} + \dfrac{x}{x}}{\dfrac{3 + x}{1}} \qquad \text{Write 1 as } \frac{x}{x}, \text{ and } 3 + x \text{ as } \frac{3 + x}{1}.$$

$$= \frac{\dfrac{2 + x}{x}}{\dfrac{3 + x}{1}} \qquad \text{Add } \frac{2}{x} \text{ and } \frac{x}{x} \text{ to get } \frac{2 + x}{x}.$$

$$= \frac{2 + x}{x} \div \frac{3 + x}{1} \qquad \text{Write the complex fraction as a division.}$$

$$= \frac{2 + x}{x} \cdot \frac{1}{3 + x} \qquad \text{Invert the divisor and multiply.}$$

$$= \frac{2 + x}{x^2 + 3x} \qquad \text{After noting that there are no common factors, multiply the numerators and multiply the denominators.}$$

Method 2 To eliminate the denominator of x, we multiply the numerator and the denominator by x, the LCD of $\frac{2}{x} + 1$ and $\frac{3 + x}{1}$.

$$\frac{\dfrac{2}{x} + 1}{3 + x} = \frac{x\left(\dfrac{2}{x} + 1\right)}{x(3 + x)}$$

$$= \frac{2 + x}{x^2 + 3x} \qquad \text{Use the distributive property to remove parentheses and simplify.} \qquad \blacksquare$$

■ GRAPHING DEVICES

As before, we can check the simplification in Example 1 by graphing the functions $f(x) = \dfrac{\dfrac{2}{x} + 1}{3 + x}$, shown in Figure 6-6(a), and $g(x) = \dfrac{2 + x}{x^2 + 3x}$, shown in Figure 6-6(b), and observing that the graphs are the same. Each graph has window settings of $[-5, 3]$ for x and $[-6, 6]$ for y.

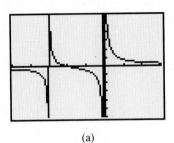 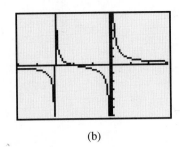

(a)   (b)

FIGURE 6-6

EXAMPLE 2   Simplify the complex fraction $\dfrac{\dfrac{1}{x} + \dfrac{1}{y}}{\dfrac{1}{x} - \dfrac{1}{y}}$.

Method 1   We add the fractions in the numerator and those in the denominator and proceed as follows:

$$\dfrac{\dfrac{1}{x} + \dfrac{1}{y}}{\dfrac{1}{x} - \dfrac{1}{y}} = \dfrac{\dfrac{1y}{xy} + \dfrac{x1}{xy}}{\dfrac{1y}{xy} - \dfrac{x1}{xy}}$$

$$= \dfrac{\dfrac{y + x}{xy}}{\dfrac{y - x}{xy}} \qquad \text{Add the fractions in the numerator and subtract the fractions in the denominator.}$$

$$= \dfrac{y + x}{xy} \div \dfrac{y - x}{xy} \qquad \text{Write the complex fraction as a division.}$$

$$= \dfrac{y + x}{xy} \cdot \dfrac{xy}{y - x} \qquad \text{Invert the divisor and multiply.}$$

$$= \dfrac{xy(y + x)}{xy(y - x)} \qquad \text{Multiply the numerators and multiply the denominators.}$$

$$= \dfrac{y + x}{y - x} \qquad \tfrac{xy}{xy} = 1.$$

Method 2   We multiply the numerator and the denominator by xy (the least common denominator of the fractions appearing in the complex fraction) and simplify.

$$\dfrac{\dfrac{1}{x} + \dfrac{1}{y}}{\dfrac{1}{x} - \dfrac{1}{y}} = \dfrac{xy\left(\dfrac{1}{x} + \dfrac{1}{y}\right)}{xy\left(\dfrac{1}{x} - \dfrac{1}{y}\right)}$$

$$= \dfrac{\dfrac{xy}{x} + \dfrac{xy}{y}}{\dfrac{xy}{x} - \dfrac{xy}{y}} \qquad \text{Use the distributive property to remove parentheses.}$$

$$= \dfrac{y + x}{y - x} \qquad \text{Simplify each fraction.}$$

■

EXAMPLE 3   Simplify $\dfrac{x^{-1} + y^{-1}}{x^{-2} - y^{-2}}$.

Method 1   We proceed as follows:

$$\frac{x^{-1} + y^{-1}}{x^{-2} - y^{-2}} = \frac{\dfrac{1}{x} + \dfrac{1}{y}}{\dfrac{1}{x^2} - \dfrac{1}{y^2}}$$

Write the fraction without using negative exponents.

$$= \frac{\dfrac{y}{xy} + \dfrac{x}{xy}}{\dfrac{y^2}{x^2y^2} - \dfrac{x^2}{x^2y^2}}$$

Get a common denominator in the numerator and the denominator.

$$= \frac{\dfrac{y + x}{xy}}{\dfrac{y^2 - x^2}{x^2y^2}}$$

Add the fractions in the numerator and the denominator.

$$= \frac{y + x}{xy} \div \frac{y^2 - x^2}{x^2y^2}$$

Write the fraction as a division.

$$= \frac{y + x}{xy} \cdot \frac{xxyy}{(y - x)(y + x)}$$

Invert and factor $y^2 - x^2$ in the divisor.

$$= \frac{(y + x)xxyy}{xy(y - x)(y + x)}$$

Multiply the numerators and the denominators.

$$= \frac{xy}{y - x}$$

Divide out the common factors of x, y, and $y + x$ in the numerator and the denominator.

Method 2   We multiply both numerator and denominator by x^2y^2, the LCD of the fractions in the problem, and proceed as follows:

$$\frac{x^{-1} + y^{-1}}{x^{-2} - y^{-2}} = \frac{\dfrac{1}{x} + \dfrac{1}{y}}{\dfrac{1}{x^2} - \dfrac{1}{y^2}}$$

Write the fraction without negative exponents.

$$= \frac{x^2y^2\left(\dfrac{1}{x} + \dfrac{1}{y}\right)}{x^2y^2\left(\dfrac{1}{x^2} - \dfrac{1}{y^2}\right)}$$

Multiply numerator and denominator by x^2y^2.

$$= \frac{xy^2 + x^2y}{y^2 - x^2}$$

Use the distributive property to remove parentheses.

$$= \frac{xy(y + x)}{(y + x)(y - x)}$$

Factor the numerator and the denominator.

$$= \frac{xy}{y - x}$$

$\frac{y + x}{y + x} = 1$.

∎

WARNING! $x^{-1} + y^{-1}$ means $\frac{1}{x} + \frac{1}{y}$, and $(x + y)^{-1}$ means $\frac{1}{x+y}$. Thus,

$$x^{-1} + y^{-1} \ne (x + y)^{-1}$$

EXAMPLE 4 Simplify the fraction $\dfrac{\dfrac{2x}{1 - \dfrac{1}{x}} + 3}{3 - \dfrac{2}{x}}$.

Solution We begin by multiplying the numerator and the denominator of the fraction

$$\dfrac{2x}{1 - \dfrac{1}{x}}$$

by x to eliminate the complex fraction in the numerator of the given fraction.

$$\dfrac{\dfrac{2x}{1 - \dfrac{1}{x}} + 3}{3 - \dfrac{2}{x}} = \dfrac{\dfrac{x(2x)}{x\left(1 - \dfrac{1}{x}\right)} + 3}{3 - \dfrac{2}{x}}$$

$$= \dfrac{\dfrac{2x^2}{x - 1} + 3}{3 - \dfrac{2}{x}}$$

We then multiply the numerator and denominator of the previous fraction by $x(x - 1)$, the LCD of $\frac{2x^2}{x-1}$, 3, and $\frac{2}{x}$, and simplify:

$$\dfrac{\dfrac{2x}{1 - \dfrac{1}{x}} + 3}{3 - \dfrac{2}{x}} = \dfrac{x(x - 1)\left(\dfrac{2x^2}{x - 1} + 3\right)}{x(x - 1)\left(3 - \dfrac{2}{x}\right)}$$

$$= \dfrac{2x^3 + 3x(x - 1)}{3x(x - 1) - 2(x - 1)}$$

$$= \dfrac{2x^3 + 3x^2 - 3x}{3x^2 - 5x + 2}$$

This result does not simplify. ∎

Orals   *Simplify each complex fraction.*

1. $\dfrac{\frac{3}{4}}{\frac{5}{4}}$

2. $\dfrac{\frac{a}{b}}{\frac{d}{b}}$

3. $\dfrac{\frac{3}{4}}{\frac{3}{8}}$

4. $\dfrac{\frac{x+y}{x}}{\frac{x-y}{x}}$

5. $\dfrac{\frac{x}{y}-1}{\frac{x}{y}}$

6. $\dfrac{1+\frac{a}{b}}{\frac{a}{b}}$

EXERCISE 6.4

In Exercises 1–56, simplify each complex fraction.

1. $\dfrac{\frac{1}{2}}{\frac{3}{4}}$

2. $-\dfrac{\frac{3}{4}}{\frac{1}{2}}$

3. $\dfrac{-\frac{2}{3}}{\frac{6}{9}}$

4. $\dfrac{\frac{11}{18}}{\frac{22}{27}}$

5. $\dfrac{\frac{1}{2}+\frac{1}{3}}{\frac{1}{4}}$

6. $\dfrac{\frac{1}{4}-\frac{1}{5}}{\frac{1}{3}}$

7. $\dfrac{\frac{1}{2}-\frac{2}{3}}{\frac{2}{3}+\frac{1}{2}}$

8. $\dfrac{\frac{2}{3}+\frac{4}{5}}{\frac{2}{5}-\frac{1}{3}}$

9. $\dfrac{\frac{4x}{y}}{\frac{6xz}{y^2}}$

10. $\dfrac{\frac{5t^4}{9x}}{\frac{2t}{18x}}$

11. $\dfrac{5ab^2}{\frac{ab}{25}}$

12. $\dfrac{\frac{6a^2b}{4t}}{3a^2b^2}$

13. $\dfrac{\frac{x-y}{xy}}{\frac{y-x}{x}}$

14. $\dfrac{\frac{x^2+5x+6}{3xy}}{\frac{x^2-9}{6xy}}$

15. $\dfrac{\frac{1}{x}-\frac{1}{y}}{xy}$

16. $\dfrac{xy}{\frac{1}{x}-\frac{1}{y}}$

17. $\dfrac{\frac{1}{a}+\frac{1}{b}}{\frac{1}{a}}$

18. $\dfrac{\frac{1}{b}}{\frac{1}{a}-\frac{1}{b}}$

19. $\dfrac{1+\frac{x}{y}}{1-\frac{x}{y}}$

20. $\dfrac{\frac{x}{y}+1}{1-\frac{x}{y}}$

21. $\dfrac{\frac{y}{x}-\frac{x}{y}}{\frac{1}{x}+\frac{1}{y}}$

22. $\dfrac{\frac{y}{x}-\frac{x}{y}}{\frac{1}{y}-\frac{1}{x}}$

23. $\dfrac{\frac{1}{a}-\frac{1}{b}}{\frac{a}{b}-\frac{b}{a}}$

24. $\dfrac{\frac{1}{a}+\frac{1}{b}}{\frac{a}{b}-\frac{b}{a}}$

25. $\dfrac{x + 1 - \dfrac{6}{x}}{\dfrac{1}{x}}$

26. $\dfrac{x - 1 - \dfrac{2}{x}}{\dfrac{x}{3}}$

27. $\dfrac{5xy}{1 + \dfrac{1}{xy}}$

28. $\dfrac{3a}{a + \dfrac{1}{a}}$

29. $\dfrac{a - 4 + \dfrac{1}{a}}{-\dfrac{1}{a} - a + 4}$

30. $\dfrac{a + 1 + \dfrac{1}{a^2}}{\dfrac{1}{a^2} + a - 1}$

31. $\dfrac{1 + \dfrac{6}{x} + \dfrac{8}{x^2}}{1 + \dfrac{1}{x} - \dfrac{12}{x^2}}$

32. $\dfrac{1 - x - \dfrac{2}{x}}{\dfrac{6}{x^2} + \dfrac{1}{x} - 1}$

33. $\dfrac{\dfrac{1}{a + 1} + 1}{\dfrac{3}{a - 1} + 1}$

34. $\dfrac{2 + \dfrac{3}{x + 1}}{\dfrac{1}{x} + x + x^2}$

35. $\dfrac{x^{-1} + y^{-1}}{x}$

36. $\dfrac{x^{-1} - y^{-1}}{y}$

37. $\dfrac{y}{x^{-1} - y^{-1}}$

38. $\dfrac{x^{-1} + y^{-1}}{(x + y)^{-1}}$

39. $\dfrac{x^{-1} + y^{-1}}{x^{-1} - y^{-1}}$

40. $\dfrac{(x + y)^{-1}}{x^{-1} + y^{-1}}$

41. $\dfrac{x + y}{x^{-1} + y^{-1}}$

42. $\dfrac{x - y}{x^{-1} - y^{-1}}$

43. $\dfrac{x - y^{-2}}{y - x^{-2}}$

44. $\dfrac{x^{-2} - y^{-2}}{x^{-1} - y^{-1}}$

45. $\dfrac{1 + \dfrac{a}{b}}{1 - \dfrac{a}{1 - \dfrac{a}{b}}}$

46. $\dfrac{1 + \dfrac{2}{1 + \dfrac{a}{b}}}{1 - \dfrac{a}{b}}$

47. $\dfrac{x - \dfrac{1}{x}}{1 + \dfrac{1}{\dfrac{1}{x}}}$

48. $\dfrac{\dfrac{a^2 + 3a + 4}{ab}}{2 + \dfrac{3 + a}{\dfrac{2}{a}}}$

49. $\dfrac{b}{b + \dfrac{2}{2 + \dfrac{1}{2}}}$

50. $\dfrac{2y}{y - \dfrac{y}{3 - \dfrac{1}{2}}}$

51. $a + \dfrac{a}{1 + \dfrac{a}{a + 1}}$

52. $b + \dfrac{b}{1 - \dfrac{b + 1}{b}}$

53. $\dfrac{x - \dfrac{1}{1 - \dfrac{x}{2}}}{\dfrac{3}{x + \dfrac{2}{3}} - x}$

54. $\dfrac{\dfrac{2x}{x - \dfrac{1}{x}} - \dfrac{1}{x}}{2x + \dfrac{2x}{1 - \dfrac{1}{x}}}$

55. $\dfrac{2x + \dfrac{1}{2 - \dfrac{x}{2}}}{\dfrac{4}{\dfrac{x}{2} - 2} - x}$

56. $\dfrac{3x - \dfrac{1}{3 - \dfrac{x}{2}}}{\dfrac{3}{\dfrac{x}{2} - 3} + x}$

In Exercises 57–58, factor each denominator and simplify the complex fraction.

57. $\dfrac{\dfrac{1}{x^2 + 3x + 2} + \dfrac{1}{x^2 + x - 2}}{\dfrac{3x}{x^2 - 1} - \dfrac{x}{x + 2}}$

58. $\dfrac{\dfrac{1}{x^2 - 1} - \dfrac{2}{x^2 + 4x + 3}}{\dfrac{2}{x^2 + 2x - 3} + \dfrac{1}{x + 3}}$

59. Engineering The stiffness k of the shaft shown in Illustration 1 is given by the formula

$$k = \dfrac{1}{\dfrac{1}{k_1} + \dfrac{1}{k_2}}$$

Section 1   Section 2

ILLUSTRATION 1

where k_1 and k_2 are the individual stiffnesses of each section. Simplify the complex fraction.

60. Transportation If a car travels a distance d_1 at a speed s_1 and then travels a distance d_2 at a speed s_2, the average (mean) speed is given by the formula

$$\bar{s} = \dfrac{d_1 + d_2}{\dfrac{d_1}{s_1} + \dfrac{d_2}{s_2}}$$

Simplify the complex fraction.

Writing Exercises *Write a paragraph using your own words.*

1. Two methods can be used to simplify a complex fraction. Explain one of them.

2. Explain the other method of simplifying a complex fraction.

Something to Think About

1. Simplify: $(x^{-1}y^{-1})(x^{-1} + y^{-1})^{-1}$.

2. Simplify: $[(x^{-1} + 1)^{-1} + 1]^{-1}$.

Review Exercises *Solve each equation.*

1. $\dfrac{8(a - 5)}{3} = 2(a - 4)$

2. $\dfrac{3t^2}{5} + \dfrac{7t}{10} = \dfrac{3t + 6}{5}$

3. $a^4 - 13a^2 + 36 = 0$

4. $|2x - 1| = 9$

6.5 Equations Containing Rational Expressions

■ SOLVING RATIONAL EQUATIONS ■ GRAPHING DEVICES ■ FORMULAS ■ PROBLEM SOLVING

If an equation contains one or more rational expressions, it is called a **rational equation**. Some examples of rational equations are

$$\dfrac{3}{5} + \dfrac{7}{x + 2} = 2, \qquad \dfrac{x + 3}{x - 3} = \dfrac{2}{x^2 - 4}, \qquad \text{and} \qquad \dfrac{-x^2 + 10}{x^2 - 1} + \dfrac{3x}{x - 1} = \dfrac{2x}{x + 1}$$

■ SOLVING RATIONAL EQUATIONS

To solve rational equations, we can multiply both sides of the equation by the least common denominator of the fractions in the equation to clear the equation of fractions.

EXAMPLE 1 Solve the rational equation $\dfrac{3}{5} + \dfrac{7}{x+2} = 2$.

Solution We start by noting that x cannot be -2, because this would give a 0 in the denominator of the fraction $\frac{7}{x+2}$. If $x \neq -2$, we can multiply both sides of the equation by $5(x + 2)$ and simplify to get

$$5(x+2)\left(\frac{3}{5} + \frac{7}{x+2}\right) = 5(x+2)2$$

$$5(x+2)\left(\frac{3}{5}\right) + 5(x+2)\left(\frac{7}{x+2}\right) = 5(x+2)2 \qquad \text{Use the distributive property on the left-hand side.}$$

$$3(x+2) + 5(7) = 10(x+2) \qquad \text{Simplify.}$$

$$3x + 6 + 35 = 10x + 20 \qquad \text{Use the distributive property and simplify.}$$

$$3x + 41 = 10x + 20 \qquad \text{Simplify.}$$

$$-7x = -21 \qquad \text{Add } -10x \text{ and } -41 \text{ to both sides.}$$

$$x = 3 \qquad \text{Divide both sides by } -7.$$

Check: To verify that 3 satisfies the equation, we substitute 3 for x in the original equation and simplify:

$$\frac{3}{5} + \frac{7}{x+2} = 2$$

$$\frac{3}{5} + \frac{7}{3+2} \stackrel{?}{=} 2$$

$$\frac{3}{5} + \frac{7}{5} \stackrel{?}{=} 2$$

$$\frac{10}{5} \stackrel{?}{=} 2$$

$$2 = 2 \qquad \blacksquare$$

■ GRAPHING DEVICES

To use a graphing device to approximate the solution of the equation $\frac{3}{5} + \frac{7}{x+2} = 2$, we graph the functions $f(x) = \frac{3}{5} + \frac{7}{x+2}$ and $g(x) = 2$. If we use win-

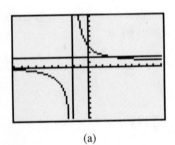

(a)

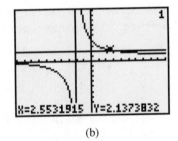

X=2.5531915   Y=2.1373832

(b)

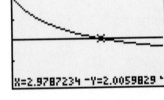

X=2.9787234 -Y=2.0059829

(c)

FIGURE 6-7

dow settings of $[-10, 10]$ for x and $[-10, 10]$ for y, we will obtain the graph shown in Figure 6-7(a). If we trace and move the cursor close to the intersection point of the two graphs, we will get the approximate value of x shown in Figure 6-7(b). If we zoom and trace again, we can get the results shown in Figure 6-7(c). For better results, we can do repeated zooms. Algebra will show that the exact solution is 3.

EXAMPLE 2   Solve the equation $\dfrac{-x^2 + 10}{x^2 - 1} + \dfrac{3x}{x - 1} = \dfrac{2x}{x + 1}$.

Solution   We start by noting that x cannot be 1 or -1, because this would give a 0 in the denominator of a fraction. If $x \neq 1$ and $x \neq -1$, we can clear the equation of fractions by multiplying both sides by the LCD of the three fractions and proceed as follows:

$$\frac{-x^2 + 10}{x^2 - 1} + \frac{3x}{x - 1} = \frac{2x}{x + 1}$$

$$\frac{-x^2 + 10}{(x + 1)(x - 1)} + \frac{3x}{x - 1} = \frac{2x}{x + 1} \qquad \text{Factor } x^2 - 1.$$

$$\frac{(x + 1)(x - 1)(-x^2 + 10)}{(x + 1)(x - 1)} + \frac{3x(x + 1)(x - 1)}{x - 1} = \frac{2x(x + 1)(x - 1)}{x + 1} \qquad \begin{array}{l}\text{Multiply both sides by} \\ (x + 1)(x - 1).\end{array}$$

$$-x^2 + 10 + 3x(x + 1) = 2x(x - 1) \qquad \begin{array}{l}\text{Divide out common} \\ \text{factors.}\end{array}$$

$$-x^2 + 10 + 3x^2 + 3x = 2x^2 - 2x \qquad \text{Remove parentheses.}$$

$$2x^2 + 10 + 3x = 2x^2 - 2x \qquad \text{Combine like terms.}$$

$$10 + 3x = -2x \qquad \text{Add } -2x^2 \text{ to both sides.}$$

$$10 + 5x = 0 \qquad \text{Add } 2x \text{ to both sides.}$$

$$5x = -10 \qquad \text{Add } -10 \text{ to both sides.}$$

$$x = -2 \qquad \text{Divide both sides by 5.}$$

Verify that -2 is a solution of the original equation. The solution set is $\{-2\}$. ∎

When we multiply both sides of an equation by a quantity that contains a variable, we can get apparent solutions that are false. This happens when we multiply both sides of an equation by 0 and get a solution that gives a 0 in the denominator of a fraction. We must exclude from the solution set of an equation any value that makes the denominator of a fraction 0.

EXAMPLE 3   Solve the equation $\dfrac{2(x + 1)}{x - 3} = \dfrac{x + 5}{x - 3}$.

Solution   We start by noting that x cannot be 3, because this would give a 0 in the denominator of a fraction. If $x \neq 3$, we can clear the equation of fractions by multiplying both sides by $x - 3$.

$$\frac{2(x + 1)}{x - 3} = \frac{x + 5}{x - 3}$$

$$(x - 3)\frac{2(x + 1)}{x - 3} = (x - 3)\frac{x + 5}{x - 3}$$ Multiply both sides by $x - 3$.

$$2(x + 1) = x + 5$$ Simplify.

$$2x + 2 = x + 5$$ Remove parentheses.

$$x + 2 = 5$$ Add $-x$ to both sides.

$$x = 3$$ Add -2 to both sides.

Since x cannot be 3, the 3 must be discarded. This equation has no solutions. Its solution set is the empty set, $\emptyset$.   ∎

EXAMPLE 4   Solve the equation $\dfrac{x + 1}{5} - 2 = -\dfrac{4}{x}$.

Solution   We start by noting that x cannot be 0, because this would give a 0 in the denominator of a fraction. If $x \neq 0$, we can clear the equation of fractions by multiplying both sides by $5x$. We then proceed as follows:

$$\frac{x + 1}{5} - 2 = -\frac{4}{x}$$

$$5x\left(\frac{x + 1}{5} - 2\right) = 5x\left(-\frac{4}{x}\right)$$ Multiply both sides by $5x$.

$$x(x + 1) - 10x = -20$$ Remove parentheses and simplify.

$$x^2 + x - 10x = -20$$ Remove parentheses.

$$x^2 - 9x + 20 = 0$$ Combine like terms and add 20 to both sides.

$$(x - 5)(x - 4) = 0$$ Factor $x^2 - 9x + 20$.

$$x - 5 = 0 \quad \text{or} \quad x - 4 = 0$$ Set each factor equal to 0.

$$x = 5 \quad | \quad x = 4$$

Since 4 and 5 both satisfy the original equation, the solution set is $\{4, 5\}$.   ∎

■ **FORMULAS**

Many formulas must be cleared of fractions before we can solve them for specific variables.

EXAMPLE 5   Solve the formula $\dfrac{1}{r} = \dfrac{1}{r_1} + \dfrac{1}{r_2}$ for r.

Solution   We proceed as follows:

$$\frac{1}{r} = \frac{1}{r_1} + \frac{1}{r_2}$$

$$\frac{rr_1r_2}{r} = \frac{rr_1r_2}{r_1} + \frac{rr_1r_2}{r_2}$$ Multiply both sides by rr_1r_2.

$$r_1r_2 = rr_2 + rr_1$$

Simplify each fraction.

$$r_1r_2 = r(r_2 + r_1)$$

Factor out r on the right-hand side.

$$r = \frac{r_1r_2}{r_2 + r_1}$$

Divide both sides by $r_2 + r_1$ and use the symmetric property of equality. ■

■ PROBLEM SOLVING

Many applications lead to rational equations.

EXAMPLE 6

Drywalling a house A contractor knows that his best crew can drywall a house in 4 days and that his second crew can drywall the same house in 5 days. One day must be allowed for the plaster coat to dry. If the contractor uses both crews, can the house be ready for painting in 4 days?

Analyze the Problem

Because 1 day is necessary for drying, the drywallers must complete their work in 3 days. Since the first crew can drywall the house in 4 days, it can do $\frac{1}{4}$ of the job in 1 day. Since the second crew can drywall the house in 5 days, it can do $\frac{1}{5}$ of the job in 1 day. If it takes x days for both crews to finish the house, together they can do $\frac{1}{x}$ of the job in 1 day. The amount of work the first crew can do in 1 day plus the amount of work the second crew can do in 1 day equals the amount of work both crews can do in 1 day working together.

Form an Equation

If x represents the number of days it takes for both crews to drywall the house, we can form the equation

What crew 1 can do in 1 day	$+$	what crew 2 can do in 1 day	$=$	what they can do together in 1 day.

Solve the Equation

$$\frac{1}{4} \qquad + \qquad \frac{1}{5} \qquad = \qquad \frac{1}{x}$$

$$20x\left(\frac{1}{4} + \frac{1}{5}\right) = 20x\left(\frac{1}{x}\right)$$

Multiply both sides by $20x$.

$$5x + 4x = 20$$

Remove parentheses and simplify.

$$9x = 20$$

Combine like terms.

$$x = \frac{20}{9}$$

Divide both sides by 9.

State the Conclusion

Since it will take only $2\frac{2}{9}$ days for both crews to drywall the house, and it takes 1 day for drying, it will be ready for painting in $3\frac{2}{9}$ days, which is less than 4 days.

Check the Result

Check the solution. ■

EXAMPLE 7

Uniform motion problem A man drove 200 miles to a convention. Because of road construction, his average speed on the return trip was 10 miles per hour less than his average speed going to the convention. If the return trip took 1 hour longer, how fast did he drive in each direction?

Analyze the Problem Because the distance traveled is given by the formula.

$$d = rt \quad (d \text{ is distance, } r \text{ is the rate of speed, and } t \text{ is time})$$

the formula for time is

$$t = \frac{d}{r}$$

We can organize the given information in the chart shown in Figure 6-8.

	Rate	·	Time	=	Distance
Going	r		$\dfrac{200}{r}$		200
Returning	$r - 10$		$\dfrac{200}{r - 10}$		200

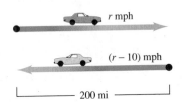

FIGURE 6-8

Form an Equation Let r represent the average rate of speed going to the meeting. Then $r - 10$ represents the average rate of speed on the return trip.

Because the return trip took 1 hour longer, we can form the following equation:

The time it took to travel to the convention	+	1	=	the time it took to return.
$\dfrac{200}{r}$	+	1	=	$\dfrac{200}{r - 10}$

Solve the Equation We can solve the equation as follows:

$$r(r - 10)\left(\frac{200}{r} + 1\right) = r(r - 10)\left(\frac{200}{r - 10}\right) \quad \text{Multiply both sides by } r(r - 10).$$

$$200(r - 10) + r(r - 10) = 200r \quad \text{Remove parentheses and simplify.}$$

$$200r - 2000 + r^2 - 10r = 200r \quad \text{Remove parentheses.}$$

$$r^2 - 10r - 2000 = 0 \quad \text{Add } -200r \text{ to both sides.}$$

$$(r - 50)(r + 40) = 0 \quad \text{Factor } r^2 - 10r - 2000.$$

$$r - 50 = 0 \quad \text{or} \quad r + 40 = 0 \quad \text{Set each factor equal to 0.}$$

$$r = 50 \quad | \quad r = -40$$

State the Conclusion We must exclude the solution of -40, because a speed cannot be negative. Thus, the man averaged 50 miles per hour going to the convention, and he averaged $50 - 10$ or 40 miles per hour returning.

Check the Result   At 50 miles per hour, the 200-mile trip took 4 hours. At 40 miles per hour, the return trip took 5 hours, which is 1 hour longer.   ■

EXAMPLE 8   **Upstream/downstream problem**   The Forest City Queen can make a 9-mile trip down the Rock River and the 9-mile trip back in a total of 1.6 hours. If the riverboat travels 12 miles per hour in still water, find the speed of the current in the Rock River.

Analyze the Problem   We can let c represent the speed of the current. Since the boat travels 12 mph and a current of c mph pushes the boat while it is going downstream, the speed of the boat going downstream is $(12 + c)$ mph. On the return trip, the current pushes against the boat, and its speed is $(12 - c)$ mph. Since $t = \frac{d}{r}$ (time $= \frac{\text{distance}}{\text{rate}}$), the time required for the downstream leg of the trip is $\frac{9}{12+c}$ hours, and the time required for the upstream leg of the trip is $\frac{9}{12-c}$ hours.

We can organize this information in the chart shown in Figure 6-9.

	Rate	·	Time	=	Distance
Going downstream	$12 + c$		$\dfrac{9}{12 + c}$		9
Going upstream	$12 - c$		$\dfrac{9}{12 - c}$		9

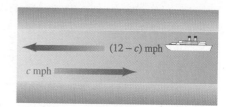

FIGURE 6-9

Furthermore, we know that the total time required for the round trip is 1.6 hours.

Form an Equation   If c represents the speed of the current in the Rock River, then $12 + c$ represents the speed of the boat going downstream, and $12 - c$ represents the speed of the boat going upstream.

Since the time going downstream is $\frac{9}{12+c}$ hours, the time going upstream is $\frac{9}{12-c}$ hours, and total time is 1.6 hours $\left(\frac{8}{5} \text{ hours}\right)$, we have

The time it takes to travel downstream	+	the time it takes to travel upstream	=	the total time for the round trip.

Solve the Equation
$$\frac{9}{12 + c} \quad + \quad \frac{9}{12 - c} \quad = \quad \frac{8}{5}$$

We can multiply both sides of the previous equation by $5(12 + c)(12 - c)$ to clear the fractions and proceed as follows:

$$\frac{5(12 + c)(12 - c)9}{12 + c} + \frac{5(12 + c)(12 - c)9}{12 - c} = \frac{5(12 + c)(12 - c)8}{5}$$

$45(12 - c) + 45(12 + c) = 8(12 + c)(12 - c)$	$\frac{12+c}{12+c} = 1$, and $\frac{12-c}{12-c} = 1$.
$540 - 45c + 540 + 45c = 8(144 - c^2)$	Multiply.
$1080 = 1152 - 8c^2$	Combine like terms and multiply.
$8c^2 - 72 = 0$	Add $8c^2$ and -1152 to both sides.
$c^2 - 9 = 0$	Divide both sides by 8.
$(c + 3)(c - 3) = 0$	Factor $c^2 - 9$.
$c + 3 = 0 \quad \text{or} \quad c - 3 = 0$	Set each factor equal to 0.
$c = -3 \quad | \quad c = 3$	

State the Conclusion Since the current cannot be negative, the apparent solution -3 must be discarded. The current in the Rock River is 3 mph.

Check the result. ∎

Orals *Solve each equation.*

1. $\dfrac{4}{x} = 2$

2. $\dfrac{9}{y} = 3$

3. $\dfrac{4}{p} + \dfrac{5}{p} = 9$

4. $\dfrac{5}{r} - \dfrac{2}{r} = 1$

5. $\dfrac{4}{y} - \dfrac{1}{y} = 3$

6. $\dfrac{8}{t} - \dfrac{2}{t} = 3$

EXERCISE 6.5

In Exercises 1–30, solve each equation. If an equation has no solution, so indicate.

1. $\dfrac{1}{4} + \dfrac{9}{x} = 1$

2. $\dfrac{1}{3} - \dfrac{10}{x} = -3$

3. $\dfrac{34}{x} - \dfrac{3}{2} = -\dfrac{13}{20}$

4. $\dfrac{1}{2} + \dfrac{7}{x} = 2 + \dfrac{1}{x}$

5. $\dfrac{3}{y} + \dfrac{7}{2y} = 13$

6. $\dfrac{2}{x} + \dfrac{1}{2} = \dfrac{7}{2x}$

7. $\dfrac{x + 1}{x} - \dfrac{x - 1}{x} = 0$

8. $\dfrac{2}{x} + \dfrac{1}{2} = \dfrac{9}{4x} - \dfrac{1}{2x}$

9. $\dfrac{7}{5x} - \dfrac{1}{2} = \dfrac{5}{6x} + \dfrac{1}{3}$

10. $\dfrac{x - 3}{x - 1} - \dfrac{2x - 4}{x - 1} = 0$

11. $\dfrac{3 - 5y}{2 + y} = \dfrac{3 + 5y}{2 - y}$

12. $\dfrac{x}{x - 2} = 1 + \dfrac{1}{x - 3}$

13. $\dfrac{a + 2}{a + 1} = \dfrac{a - 4}{a - 3}$

14. $\dfrac{z + 2}{z + 8} - \dfrac{z - 3}{z - 2} = 0$

15. $\dfrac{x + 2}{x + 3} - 1 = \dfrac{1}{3 - 2x - x^2}$

16. $\dfrac{x - 3}{x - 2} - \dfrac{1}{x} = \dfrac{x - 3}{x}$

17. $\dfrac{x}{x + 2} = 1 - \dfrac{3x + 2}{x^2 + 4x + 4}$

18. $\dfrac{3 + 2a}{a^2 + 6 + 5a} + \dfrac{2 - 5a}{a^2 - 4} = \dfrac{2 - 3a}{a^2 - 6 + a}$

19. $\dfrac{2}{x - 2} + \dfrac{1}{x + 1} = \dfrac{1}{x^2 - x - 2}$

20. $\dfrac{5}{y - 1} + \dfrac{3}{y - 3} = \dfrac{8}{y - 2}$

21. $\dfrac{a-1}{a+3} - \dfrac{1-2a}{3-a} = \dfrac{2-a}{a-3}$

22. $\dfrac{5}{2z^2+z-3} - \dfrac{2}{2z+3} = \dfrac{z+1}{z-1} - 1$

23. $\dfrac{5}{x+4} + \dfrac{1}{x+4} = x-1$

24. $\dfrac{2}{x-1} + \dfrac{x-2}{3} = \dfrac{4}{x-1}$

25. $\dfrac{3}{x+1} - \dfrac{x-2}{2} = \dfrac{x-2}{x+1}$

26. $\dfrac{x-4}{x-3} + \dfrac{x-2}{x-3} = x-3$

27. $\dfrac{2}{x-3} + \dfrac{3}{4} = \dfrac{17}{2x}$

28. $\dfrac{30}{y-2} + \dfrac{24}{y-5} = 13$

29. $\dfrac{x+4}{x+7} - \dfrac{x}{x+3} = \dfrac{3}{8}$

30. $\dfrac{5}{x+4} - \dfrac{1}{3} = \dfrac{x-1}{x}$

In Exercises 31–36, solve each formula for the indicated variable.

31. $\dfrac{1}{p} + \dfrac{1}{q} = \dfrac{1}{f}$ for f

32. $\dfrac{1}{p} + \dfrac{1}{q} = \dfrac{1}{f}$ for p

33. $S = \dfrac{a-lr}{1-r}$ for r

34. $H = \dfrac{2ab}{a+b}$ for a

35. $\dfrac{1}{R} = \dfrac{1}{r_1} + \dfrac{1}{r_2} + \dfrac{1}{r_3}$ for R

36. $\dfrac{1}{R} = \dfrac{1}{r_1} + \dfrac{1}{r_2} + \dfrac{1}{r_3}$ for r_1

In Exercises 37–61, solve each problem.

37. Focal length The design of a camera lens uses the equation $\dfrac{1}{f} = \dfrac{1}{s_1} + \dfrac{1}{s_2}$ which relates the focal length, f, of a lens to the image distance, s_1, and the object distance, s_2. Find the focal length of the lens in Illustration 1. (*Hint:* Convert feet to inches.)

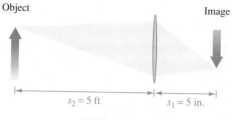

Object   Image

$s_2 = 5$ ft $s_1 = 5$ in.

ILLUSTRATION 1

38. Lens maker's formula The focal length, f, of a lens is given by the lens maker's formula,

$$\frac{1}{f} = 0.6\left(\frac{1}{r_1} + \frac{1}{r_2}\right)$$

where f is the focal length of the lens and r_1 and r_2 are the radii of the two circular surfaces. Find the focal length of the lens in Illustration 2.

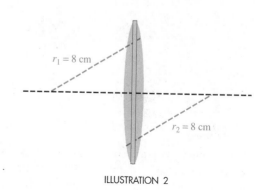

$r_1 = 8$ cm

$r_2 = 8$ cm

ILLUSTRATION 2

39. House painting If one painter can paint a house in 5 days and another painter can paint the same house in 3 days, how long will it take them to paint the house if they work together?

40. Reading proof A proofreader can read 50 pages in 3 hours, and a second proofreader can read 50 pages in 1 hour. If they both work on a 250-page book, can they meet a six-hour deadline?

41. Storing corn In 10 minutes, a conveyor belt can move 1000 bushels of corn into the storage bin shown in Illustration 3. A smaller belt can move 1000 bushels to the storage bin in 14 minutes. If both belts are used, how long it will take to move 1000 bushels to the storage bin?

ILLUSTRATION 3

42. **Roofing a house**   One roofing crew can finish a 2800-square-foot roof in 12 hours, and another crew can do the job in 10 hours. If they work together, can they finish before a predicted rain in 5 hours?

43. **Draining a swimming pool**   A drain can empty the swimming pool shown in Illustration 4 in 3 days. A second drain can empty the pool in 2 days. How long will it take to empty the pool if both drains are used?

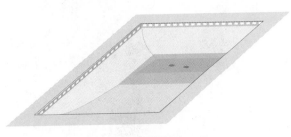

ILLUSTRATION 4

44. **Filling a pool**   A pipe can fill a pool in 9 hours. If a second pipe is also used, the pool can be filled in 3 hours. How long would it take the second pipe alone to fill the pool?

45. **Filling a pond**   One pipe can fill a pond in 3 weeks, and a second pipe can fill the pond in 5 weeks. However, evaporation and seepage can empty the pond in 10 weeks. If both pipes are used, how long will it take to fill the pond?

46. **House cleaning**   Sally can clean the house in 6 hours, and her father can clean the house in 4 hours. Sally's younger brother, Dennis, can completely mess up the house in 8 hours. If Sally and her father clean and Dennis plays, how long will it take to clean the house?

47. **Touring the countryside**   A man bicycles 5 mph faster than he can walk. He bicycles 24 miles and walks back along the same route in 11 hours. How fast does he walk?

48. **Finding rates**   Two trains made the same 315-mile run. Since one train traveled 10 mph faster than the other, it arrived 2 hours earlier. Find the speed of each train.

49. **Train travel**   A train traveled 120 miles from Freeport to Chicago and returned the same distance in a total time of 5 hours. If the train traveled 20 miles per hour slower on the return trip, how fast did the train travel in each direction?

50. **Time on the road**   A car traveled from Rockford to Chicago in 3 hours less time than it took a second car to travel from Rockford to St. Louis. If the cars traveled at the same average speed, how long was the first driver on the road? (See Illustration 5.)

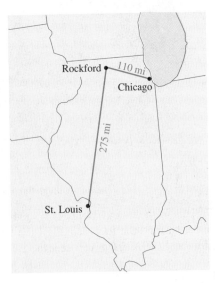

ILLUSTRATION 5

51. **Driving a boat**   A boy can drive a motorboat 45 miles down the Rock River in the same amount of time that he can drive 27 miles upstream. Find the speed of the current if the speed of the boat is 12 miles per hour in still water.

52. **Rowing a boat**   A woman who can row 3 miles per hour in still water rows 10 miles downstream on the Eagle River and returns upstream in a total of 12 hours. Find the speed of the current.

53. **Aviation**   A plane that can fly 340 mph in still air can fly 200 miles downwind in the same amount of time that it can fly 140 miles upwind. Find the velocity of the wind.

54. **Aviation**   A plane can fly 650 miles with the wind in the same amount of time it can fly 475 miles against the wind. If the wind speed is 40 mph, find the speed of the plane in still air.

55. Number puzzle If three times a certain integer is added to four times its reciprocal, the result is 8. Find the integer.

56. Number puzzle If three times a number is subtracted from four times its reciprocal, the result is 11. Find the number. (There are two possibilities.)

57. Discount buying A repairman purchased several washing-machine motors for a total of $224. When the unit cost decreased by $4, he was able to buy one extra motor for the same total price. How many motors did he buy originally? (*Hint:* Cost = unit price times quantity.)

58. Price increase An appliance store manager bought several microwave ovens for a total of $1800. When her unit cost increased by $25, she was able to buy one less oven for the same total price. How many ovens did she buy originally?

59. Stretching a vacation A student saved $1200 for a trip to Europe. By cutting $20 from her daily expenses, she was able to stay three extra days. How long had she originally planned to be gone?

60. Planting a windscreen A homeowner wants to plant a windscreen of several trees, spaced equally in a 120-foot row. A landscape architect recommends that he plant two more trees and space them two feet closer to each other. How many trees was the homeowner originally planning to use? (*Hint:* A row of n trees has $n - 1$ spaces between; a row of $n + 2$ trees has $n + 2 - 1$ spaces between.)

61. Engineering The stiffness of the shaft shown in Illustration 6 is given by the formula

$$k = \frac{1}{\dfrac{1}{k_1} + \dfrac{1}{k_2}}$$

ILLUSTRATION 6

where k_1 and k_2 are the individual stiffnesses of each section. If the stiffness, k_2, of Section 2 is 4,200,000 in. lb/rad, and design specifications require that the overall stiffness, k, of the entire shaft be 1,900,000 in. lb/rad, what must the stiffness, k_1, of Section 1 be?

Writing Exercises   *Write a paragraph using your own words.*

1. Why is it necessary to check the solutions of a rational equation?

2. Explain the steps you would use to solve a rational equation.

Something to Think About

1. Find a rational equation that has an apparent solution of 3 that will not satisfy the equation.

2. Solve: $(x - 1)^{-1} - x^{-1} = 6^{-1}$.

Review Exercises   *Simplify each expression. Write each answer without using negative exponents.*

1. $(m^2 n^{-3})^{-2}$

2. $\dfrac{a^{-1}}{a^{-1} + 1}$

3. $\dfrac{a^0 + 2a^0 - 3a^0}{(a - b)^0}$

4. $(4x^{-2} + 3)(2x - 4)$

6.6 Proportion and Variation

■ SOLVING PROPORTIONS ■ SIMILAR TRIANGLES ■ DIRECT VARIATION ■ INVERSE VARIATION
■ JOINT VARIATION ■ COMBINED VARIATION

The quotient of two numbers is often called a **ratio**. For example, the fraction $\frac{2}{3}$ can be read as "the ratio of 2 to 3." An equation indicating that two ratios are equal is called a **proportion**. Two examples of proportions are

$$\frac{1}{4} = \frac{2}{8} \qquad \text{and} \qquad \frac{4}{7} = \frac{12}{21}$$

In the proportion $\frac{a}{b} = \frac{c}{d}$, a and d are called the **extremes** of the proportion, and b and c are called the **means**.

To develop a fundamental property of proportions, we suppose that

$$\frac{a}{b} = \frac{c}{d}$$

is a proportion and multiply both sides by bd to obtain

$$bd\left(\frac{a}{b}\right) = bd\left(\frac{c}{d}\right).$$

$$\frac{\cancel{b}da}{\cancel{b}} = \frac{b\cancel{d}c}{\cancel{d}}$$

$$ad = bc$$

Thus, if $\frac{a}{b} = \frac{c}{d}$, then $ad = bc$. In a proportion, the *product of the extremes equals the product of the means*.

■ SOLVING PROPORTIONS

EXAMPLE 1 Solve the proportion $\dfrac{x+1}{x} = \dfrac{x}{x+2}$ $(x \neq 0, x \neq -2)$.

Solution

$$\frac{x+1}{x} = \frac{x}{x+2}$$

$(x+1)(x+2) = x \cdot x$ The product of the extremes equals the product of the means.

$x^2 + 3x + 2 = x^2$ Multiply.

$3x + 2 = 0$ Subtract x^2 from both sides.

$x = -\dfrac{2}{3}$ Subtract 2 from both sides and divide by 3.

Thus, $x = -\dfrac{2}{3}$.   ■

EXAMPLE 2 Solve the proportion $\dfrac{5a+2}{2a} = \dfrac{18}{a+4}$ $(a \neq 0, a \neq -4)$.

Solution

$$\frac{5a+2}{2a} = \frac{18}{a+4}$$

$(5a+2)(a+4) = 2a(18)$ The product of the extremes equals the product of the means.

$5a^2 + 22a + 8 = 36a$ Multiply the polynomials.

$5a^2 - 14a + 8 = 0$ Subtract $36a$ from both sides.

$(5a - 4)(a - 2) = 0$ Factor $5a^2 - 14a + 8$.

$$5a - 4 = 0 \quad \text{or} \quad a - 2 = 0 \qquad \text{Set each factor equal to 0.}$$
$$5a = 4 \qquad\qquad\qquad a = 2 \qquad \text{Solve each linear equation.}$$
$$a = \frac{4}{5}$$

Thus, $a = \dfrac{4}{5}$ or $a = 2$. ∎

EXAMPLE 3

Grocery shopping   If 7 pears cost $2.73, how much will 11 pears cost?

Solution   We can let c represent the cost of 11 pears. The ratio of the numbers of pears is the same as the ratio of their costs. We express this relationship as a proportion and find c.

$$\frac{7}{11} = \frac{2.73}{c}$$

$$7c = 11(2.73) \qquad \text{The product of the extremes is equal to the product of the means.}$$

$$7c = 30.03 \qquad\qquad \text{Multiply.}$$

$$c = \frac{30.03}{7} \qquad\qquad \text{Divide both sides by 7.}$$

$$c = 4.29 \qquad\qquad \text{Simplify.}$$

Eleven pears will cost $4.29. ∎

■ SIMILAR TRIANGLES

If two triangles have the same shape, they are called **similar triangles**. The following theorem points out an important fact about similar triangles.

> **Theorem**
> If two triangles are similar, then all pairs of corresponding sides are in proportion.

   This theorem often enables us to measure sides of triangles indirectly. For example, on a sunny day we can find the height of a tree and stay safely on the ground.

EXAMPLE 4

On a sunny day, a tree casts a shadow of 29 feet at the same time a vertical yardstick casts a shadow of 2.5 feet. Find the height of the tree.

Solution   Refer to Figure 6–10, which shows the triangles determined by the tree and its shadow and the yardstick and its shadow. Because the triangles have the same shape, they are similar, and the measures of their corresponding sides are in proportion. If we let h represent the height of the tree, we can find h by setting up and solving the following proportion.

$$\frac{h}{3} = \frac{29}{2.5}$$

$2.5h = 3(29)$ In a proportion, the product of the extremes is equal to the product of the means.

$2.5h = 87$ Simplify.

$h = 34.8$ Divide both sides by 2.5.

The tree is about 35 feet tall.

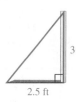

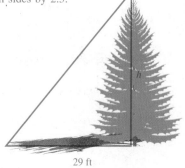

3 ft

2.5 ft

29 ft

FIGURE 6-10   ■

■ DIRECT VARIATION

We consider the formula

$$C = \pi D$$

for the circumference of a circle where C is the circumference, D is the diameter, and $\pi \approx 3.14159$. If we double the diameter of a circle, we determine another circle with a larger circumference C_1 such that

$$C_1 = \pi(2D) = 2\pi D = 2C$$

Thus, doubling the diameter results in doubling the circumference. Likewise, if we triple the diameter, we triple the circumference.

In this formula, we say that the variables C and D *vary directly*, or that they are *directly proportional*. This is because as one variable gets larger, so does the other, and in a predictable way. In this example, the constant π is called the *constant of variation* or the *constant of proportionality*.

Direct Variation

The words "y varies directly with x" or "y is directly proportional to x" mean that $y = kx$ for some constant k.

k is called the **constant of variation** or the **constant of proportionality**.

An example of direct variation is Hooke's law from physics. Hooke's law states that the distance a spring will stretch varies directly with the force that is applied to it.

If d represents a distance and f represents a force, Hooke's law is expressed mathematically as

$$d = kf$$

where k is the constant of variation. If the spring stretches 10 inches when a weight of 6 pounds is attached, k can be found as follows:

$$d = kf$$
$$10 = k(6) \qquad \text{Substitute 10 for } d \text{ and 6 for } f.$$
$$\frac{5}{3} = k \qquad \text{Divide both sides by 6 and simplify.}$$

To find the force required to stretch the spring a distance of 35 inches, we can solve the equation $d = kf$ for f, with $d = 35$ and $k = \frac{5}{3}$.

$$d = kf$$
$$35 = \frac{5}{3}f \qquad \text{Substitute 35 for } d \text{ and } \tfrac{5}{3} \text{ for } k.$$
$$105 = 5f \qquad \text{Multiply both sides by 3.}$$
$$21 = f \qquad \text{Divide both sides by 5.}$$

The force required to stretch the spring a distance of 35 inches is 21 pounds.

EXAMPLE 5 The distance traveled in a given time is directly proportional to the speed. If a car travels 70 miles at 30 miles per hour, how far could it travel in the same time at 45 miles per hour?

Solution The words "Distance is directly proportional to speed" can be expressed by the equation

1. $d = ks$

where d is distance, k is the constant of variation, and s is the speed. To find k, we substitute 70 for d and 30 for s, and solve for k.

$$d = ks$$
$$70 = k(30)$$
$$k = \frac{7}{3}$$

To find the distance traveled at 45 miles per hour, we substitute $\frac{7}{3}$ for k and 45 for s in Equation 1 and simplify.

$$d = ks$$
$$d = \frac{7}{3}(45)$$
$$= 105$$

In the time it took to go 70 miles at 30 miles per hour, the car could travel 105 miles at 45 miles per hour. ■

■ INVERSE VARIATION

In the formula $w = \frac{12}{l}$, w gets smaller as l gets larger, and w gets larger as l gets smaller. Since these variables vary in opposite directions in a predictable way, we say that the variables *vary inversely*, or that they are *inversely proportional*. The constant 12 is the constant of variation.

> **Inverse Variation**
>
> The words "*y* varies inversely with *x*," or "*y* is inversely proportional to *x*" mean that $y = \frac{k}{x}$ for some constant *k*.
>
> *k* is called the **constant of variation**.

Because of gravity, an object in space is attracted to the earth. The force of this attraction varies inversely with the square of the distance from the object to the center of the earth.

If *f* represents the force and *d* represents the distance, this information can be expressed by the equation

$$f = \frac{k}{d^2}$$

If we know that an object 4000 miles from the center of the earth is attracted to the earth with a force of 90 pounds, we can find *k*.

$$f = \frac{k}{d^2}$$

$$90 = \frac{k}{4000^2} \qquad \text{Substitute 90 for } f \text{ and 4000 for } d.$$

$$k = 90(4000^2)$$

$$= 1.44 \times 10^9$$

To find the force of attraction when the object is 5000 miles from the center of the earth, we proceed as follows:

$$f = \frac{k}{d^2}$$

$$f = \frac{1.44 \times 10^9}{5000^2} \qquad \text{Substitute } 1.44 \times 10^9 \text{ for } k \text{ and 5000 for } d.$$

$$= 57.6$$

The object will be attracted to earth with a force of 57.6 pounds when it is 5000 miles from the earth's center.

EXAMPLE 6   The intensity *I* of light received from a light source varies inversely with the square of the distance from the light source. If the intensity from a light source 4 feet from an object is 8 candelas, find the intensity at a distance of 2 feet.

Solution   The words "Intensity varies inversely with the square of the distance *d*" can be expressed by the equation

$$I = \frac{k}{d^2}$$

To find *k*, we substitute 8 for *I* and 4 for *d* and solve for *k*.

$$I = \frac{k}{d^2}$$

$$8 = \frac{k}{4^2}$$

$$128 = k$$

To find the intensity when the object is 2 feet from the light source, we substitute 2 for d and 128 for k and simplify.

$$I = \frac{k}{d^2}$$

$$I = \frac{128}{2^2}$$

$$= 32$$

The intensity at 2 feet is 32 candelas. ■

■ JOINT VARIATION

There are times when one variable varies with the product of several variables. For example, the area of a triangle varies directly with the product of its base and height:

$$A = \frac{1}{2}bh$$

Such variation is called *joint variation*.

> **Joint Variation**
> If one variable varies directly with the product of two or more variables, the relationship is called **joint variation**. If y varies jointly with x and z, then $y = kxz$.
>
> The constant k is called the **constant of variation**.

EXAMPLE 7   The volume V of a cone varies jointly with its height h and the area of its base B. Express this relationship as an equation.

Solution   The words "V varies jointly with h and B" mean that V varies directly as the product of h and B. Thus,

$$V = khB$$

The relationship can also be read as "V is directly proportional to the product of h and B." ■

■ COMBINED VARIATION

Many applied problems involve a combination of direct and inverse variation. Such variation is called **combined variation**.

EXAMPLE 8   The time it takes to build a highway varies directly with the length of the road, but inversely with the number of workers. If it takes 100 workers 4 weeks to build 2 miles of highway, how long will it take 80 workers to build 10 miles of highway?

Solution   We can let t represent the time in weeks, l represent the length in miles, and w represent the number of workers. The relationship between these variables can be expressed by the equation

$$t = \frac{kl}{w}$$

We substitute 4 for t, 100 for w, and 2 for l to find k:

$$4 = \frac{k(2)}{100}$$

$400 = 2k$ Multiply both sides by 100.

$200 = k$ Divide both sides by 2.

We now substitute 80 for w, 10 for l, and 200 for k in the equation $t = \frac{kl}{w}$ and simplify:

$$t = \frac{kl}{w}$$

$$t = \frac{200(10)}{80}$$

$$= 25$$

It will take 25 weeks for 80 workers to build 10 miles of highway.   ∎

Orals   *Solve each proportion.*

1. $\dfrac{x}{2} = \dfrac{3}{6}$ **2.** $\dfrac{3}{x} = \dfrac{4}{12}$ **3.** $\dfrac{5}{7} = \dfrac{2}{x}$

Express each sentence with a formula.

4. a varies directly with b.   **5.** a varies inversely with b.

6. a varies jointly with b and c.   **7.** a varies directly with b but inversely with c.

EXERCISE 6.6

In Exercises 1–16, solve each proportion for the variable, if possible.

1. $\dfrac{x}{5} = \dfrac{15}{25}$ **2.** $\dfrac{4}{y} = \dfrac{6}{27}$ **3.** $\dfrac{r-2}{3} = \dfrac{r}{5}$ **4.** $\dfrac{x+1}{x-1} = \dfrac{6}{4}$

5. $\dfrac{3}{n} = \dfrac{2}{n+1}$ **6.** $\dfrac{4}{x+3} = \dfrac{3}{5}$ **7.** $\dfrac{5}{5z+3} = \dfrac{2z}{2z^2+6}$ **8.** $\dfrac{9t+6}{t(t+3)} = \dfrac{7}{t+3}$

9. $\dfrac{2}{c} = \dfrac{c-3}{2}$ **10.** $\dfrac{y}{4} = \dfrac{4}{y}$ **11.** $\dfrac{2}{3x} = \dfrac{6x}{36}$ **12.** $\dfrac{2}{x+6} = \dfrac{-2x}{5}$

13. $\dfrac{2(x+3)}{3} = \dfrac{4(x-4)}{5}$ **14.** $\dfrac{x+4}{5} = \dfrac{3(x-2)}{3}$ **15.** $\dfrac{1}{x+3} = \dfrac{-2x}{x+5}$ **16.** $\dfrac{x-1}{x+1} = \dfrac{2}{3x}$

In Exercises 17–26, set up and solve the required proportion.

17. Selling shirts Sport shirts are on sale. How much will 5 shirts cost? (See Illustration 1.)

ILLUSTRATION 1

18. Increasing a recipe A recipe for spaghetti sauce requires four 16-ounce bottles of ketchup to make two gallons of sauce. How many bottles of ketchup are needed to make 10 gallons of sauce?

19. Gas consumption A car gets 42 miles per gallon of gas. How much gas is needed to drive 315 miles?

20. Model railroading An HO-scale model railroad engine is 9 inches long. The HO scale is 87 feet to 1 foot. How long is a real engine?

21. Dollhouses Standard dollhouse scale is 1 inch to 1 foot. Heidi's dollhouse is 32 inches wide. How wide would it be if it were a real house?

22. Staffing A school board has determined that there should be 3 teachers for every 50 students. How many teachers are needed for an enrollment of 2700 students?

23. Drafting In a scale drawing, a 280-foot antenna tower is drawn 7 inches high. The building next to it is drawn 2 inches high. How tall is the actual building?

24. Mixing fuel The instructions on a can of oil intended to be added to lawnmower gasoline read as shown in the table.

Recommended	Gasoline	Oil
50 to 1	6 gal	16 oz

Are these instructions correct? (*Hint:* There are 128 ounces in 1 gallon.)

25. Recommended dosage The recommended child's dose of the sedative hydroxine is 0.006 gram per kilogram of body mass. Find the dosage for a 30-kg child.

26. Body mass Proper dosage of the antibiotic cephalexin in children is 0.025 gram per kilogram of body mass. Find the mass of a child receiving a $1\frac{1}{8}$ gram dose.

In Exercises 27–32, use similar triangles to help solve each problem.

27. Height of a tree A tree casts a shadow of 28 feet at the same time a 6-foot man casts a shadow of 4 feet. (See Illustration 2.) Find the height of the tree.

28. Height of a flagpole A man places a mirror on the ground and sees the reflection of the top of a flagpole, as in Illustration 3. The two triangles in the illustration are similar. Find the height, *h*, of the flagpole.

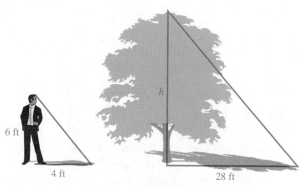

ILLUSTRATION 2

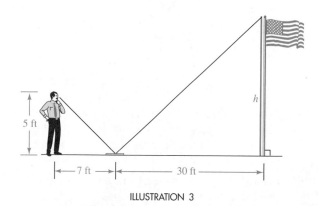

ILLUSTRATION 3

29. Width of a river Use the dimensions in Illustration 4 to find w, the width of the river. The two triangles in the illustration are similar.

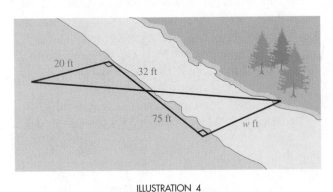

ILLUSTRATION 4

30. Flight path An airplane ascends 150 feet as it flies a horizontal distance of 1000 feet. How much altitude

will it gain as it flies a horizontal distance of 1 mile? (See Illustration 5.) (*Hint:* 5280 feet = 1 mile.)

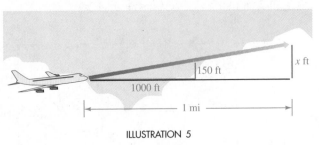

ILLUSTRATION 5

31. Flight path An airplane descends 1350 feet as it flies a horizontal distance of 1 mile. How much altitude is lost as it flies a horizontal distance of 5 miles?

32. Ski runs A ski course with $\frac{1}{2}$ mile of run falls 100 feet in every 300 feet of run. Find the height of the hill.

In Exercises 33–42, express each sentence as a formula.

33. A varies directly with the square of p.

34. z varies inversely with the cube of t.

35. v varies inversely with the cube of r.

36. r varies directly with the square of s.

37. B varies jointly with m and n.

38. C varies jointly with x, y, and z.

39. P varies directly with the square of a, and inversely with the cube of j.

40. M varies inversely with the cube of n, and jointly with x and the square of z.

41. The force of attraction between two masses m_1 and m_2 varies directly with the product of m_1 and m_2 and inversely with the square of the distance between them.

42. The force of wind on a vertical surface varies jointly with the area of the surface and the square of the velocity of the wind.

In Exercises 43–50, express each formula in words. In each formula, k is the constant of variation.

43. $L = kmn$

44. $P = \dfrac{km}{n}$

45. $E = kab^2$

46. $U = krs^2t$

47. $X = \dfrac{kx^2}{y^2}$

48. $Z = \dfrac{kw}{xy}$

49. $R = \dfrac{kL}{d^2}$

50. $e = \dfrac{kPL}{A}$

51. Area of a circle The area of a circle varies directly with the square of its radius, and the constant of variation is π. Find the area of a circle with a radius of 6 inches.

52. Falling objects An object in free fall travels a distance s that is directly proportional to the square of the time t. If an object falls 1024 feet

in 8 seconds, how far will it fall in 10 seconds?

53. Finding distance The distance that a car can travel is directly proportional to the number of gallons of gasoline it consumes. If a car can go 288 miles on 12 gallons of gasoline, how far can it go on a full tank of 18 gallons?

54. Farming A farmer's harvest in bushels varies directly with the number of acres planted. If 8 acres can produce 144 bushels, how many acres are required to produce 1152 bushels?

55. Farming The length of time that a given number of bushels of corn will last when feeding cattle varies inversely with the number of animals. If x bushels will feed 25 cows for 10 days, how long will the feed last for 10 cows?

56. Geometry For a fixed area, the length of a rectangle is inversely proportional to its width. A rectangle has a width of 18 feet and a length of 12 feet. If the length is increased to 16 feet, find the width.

57. Gas pressure Under a constant temperature, the volume occupied by a gas is inversely proportional to the pressure applied. If the gas occupies a volume of 20 cubic inches under a pressure of 6 pounds per square inch, find the volume when the gas is subjected to a pressure of 10 pounds per square inch.

58. Value of a car The value of a car usually varies inversely with its age. If a car is worth $7000 when it is 3 years old, how much will it be worth when it is 7 years old?

59. Organ pipes The frequency of vibration of air in an organ pipe is inversely proportional to the length of the pipe. (See Illustration 6.) If a pipe 2 feet long vibrates 256 times per second, how many times per second will a 6-foot pipe vibrate?

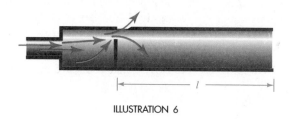

ILLUSTRATION 6

60. Geometry The area of a rectangle varies jointly with its length and width. If both the length and the width are tripled, by what factor is the area multiplied?

61. Geometry The volume of a rectangular solid varies jointly with its length, width, and height. If the length is doubled, the width is tripled, and the height is doubled, by what factor is the volume multiplied?

62. Costs of a trucking company The costs incurred by a trucking company vary jointly with the number of trucks in service and the number of hours they are used. When 4 trucks are used for 6 hours each, the costs are $1800. Find the costs of using 10 trucks, each for 12 hours.

63. Storing oil The number of gallons of oil that can be stored in a cylindrical tank varies jointly with the height of the tank and the square of the radius of its base. The constant of proportionality is 23.5. Find the number of gallons that can be stored in the cylindrical tank in Illustration 7.

ILLUSTRATION 7

64. Finding the constant of variation A quantity l varies jointly with x and y and inversely with z. If the value of l is 30 when $x = 15$, $y = 5$, and $z = 10$, find k.

65. Electronics The voltage (in volts) measured across a resistor is directly proportional to the current (in amperes) flowing through the resistor. The constant of variation is the **resistance** (in ohms). If 6 volts is measured across a resistor carrying a current of 2 amperes, find the resistance.

66. Electronics The power (in watts) lost in a resistor (in the form of heat) is directly proportional to the square of the current (in amperes) passing through it. The constant of proportionality is the resistance (in ohms). What power is lost in a 5-ohm resistor carrying a 3-ampere current?

67. Building construction The deflection of a beam is inversely proportional to its width and the cube of its depth. If the deflection of a 4-inch-by-4-inch beam is 1.1 inches, find the deflection of a 2-inch-by-8-inch beam positioned as in Illustration 8.

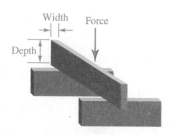

ILLUSTRATION 8

68. Building construction Find the deflection of the beam in Exercise 67 when the beam is positioned as in Illustration 9.

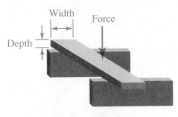

ILLUSTRATION 9

69. Gas pressure The pressure of a certain amount of gas is directly proportional to the temperature (measured in degrees Kelvin) and inversely proportional to the volume. A sample of gas at a pressure of 1 atmosphere occupies a volume of 1 cubic meter at a temperature of 273 Kelvin. When heated, the gas expands to twice its volume, but the pressure remains constant. To what temperature is it heated?

70. Tension A yo-yo, twirled at the end of a string, is kept in its circular path by the tension of the string. The tension, T, is directly proportional to the square of the speed, s, and inversely proportional to the radius, r, of the circle. In Illustration 10, the tension is 32 pounds when the speed is 8 feet/second and the radius is 6 feet. Find the tension when the speed is 4 feet/second and the radius is 3 feet.

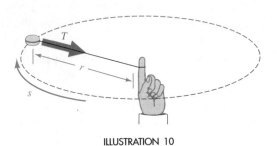

ILLUSTRATION 10

Writing Exercises *Write a paragraph using your own words.*

1. Explain the terms *means* and *extremes*.

2. Distinguish between a *ratio* and a *proportion*.

3. Explain the term *joint variation*.

4. Explain why the equation $\frac{y}{x} = k$ indicates that y varies directly with x.

Something to Think About

1. As temperature increases on the Fahrenheit scale, it also increases on the Celsius scale. Is this direct variation? Explain.

2. As the cost of a purchase (less than \$5) increases, the amount of change received from a five-dollar bill decreases. Is this inverse variation? Explain.

3. Is a proportion useful for solving this problem? Explain.
   A water bill for 1000 gallons was \$15, and a bill for 2000 gallons was \$25. Find the bill for 3000 gallons.

4. How would you solve the problem in Exercise 3?

Review Exercises *Simplify each expression.*

1. $(x^2x^3)^2$

2. $\left(\dfrac{a^3a^5}{a^{-2}}\right)^3$

3. $\dfrac{b^0 - 2b^0}{b^0}$

4. $\left(\dfrac{2r^{-2}r^{-3}}{4r^{-5}}\right)^{-3}$

5. Write 35,000 in scientific notation.

6. Write 0.00035 in scientific notation.

7. Write 2.5×10^{-3} in standard notation.

8. Write 2.5×10^4 in standard notation.

6.7 Dividing Polynomials

■ DIVIDING A MONOMIAL BY A MONOMIAL ■ DIVIDING A POLYNOMIAL BY A MONOMIAL
■ DIVIDING A POLYNOMIAL BY A POLYNOMIAL ■ THE CASE OF THE MISSING TERMS

■ DIVIDING A MONOMIAL BY A MONOMIAL

EXAMPLE 1   Simplify the expression $(3a^2b^3) \div (2a^3b)$.

Method 1   We write the expression as a fraction and divide out all common factors:

$$\frac{3a^2b^3}{2a^3b} = \frac{3aabbb}{2aaab}$$

$$= \frac{3\cancel{a}\cancel{a}bbb}{2\cancel{a}\cancel{a}a\cancel{b}}$$

$$= \frac{3b^2}{2a}$$

Method 2   We write the expression as a fraction and use the rules of exponents:

$$\frac{3a^2b^3}{2a^3b} = \frac{3}{2}a^{2-3}b^{3-1}$$

$$= \frac{3}{2}a^{-1}b^2$$

$$= \frac{3}{2}\left(\frac{1}{a}\right)\frac{b^2}{1}$$

$$= \frac{3b^2}{2a}$$

■ DIVIDING A POLYNOMIAL BY A MONOMIAL

EXAMPLE 2   Divide $4x^3y^2 + 3xy^5 - 12xy$ by $3x^2y^3$.

Solution   We write the expression as a fraction and then as the sum of three separate fractions:

$$\frac{4x^3y^2 + 3xy^5 - 12xy}{3x^2y^3} = \frac{4x^3y^2}{3x^2y^3} + \frac{3xy^5}{3x^2y^3} + \frac{-12xy}{3x^2y^3}$$

We then pick one of the methods used in Example 1 and simplify each of the three fractions on the right-hand side of the equal sign to get

$$\frac{4x^3y^2 + 3xy^5 - 12xy}{3x^2y^3} = \frac{4x}{3y} + \frac{y^2}{x} + \frac{-4}{xy^2}$$

$$= \frac{4x}{3y} + \frac{y^2}{x} - \frac{4}{xy^2}$$

■ DIVIDING A POLYNOMIAL BY A POLYNOMIAL

EXAMPLE 3   Divide $x^2 + 7x + 12$ by $x + 4$.

Solution   There is a repeating series of steps to follow when the divisor is not a monomial. To divide a polynomial by a polynomial, we write the division in long division form and proceed as follows:

$$x + 4\overline{)x^2 + 7x + 12} \quad \text{quotient } x$$

What number multiplied by x will give x^2?
$\dfrac{x^2}{x} = x$. Place x in the quotient.

$$\begin{array}{r} x \\ x + 4\overline{)x^2 + 7x + 12} \\ \underline{x^2 + 4x} \\ 3x + 12 \end{array}$$

Multiply each term in the divisor by x to get $x^2 + 4x$, subtract $x^2 + 4x$ from $x^2 + 7x$, and bring down the 12.

$$\begin{array}{r} x + 3 \\ x + 4\overline{)x^2 + 7x + 12} \\ \underline{x^2 + 4x} \\ 3x + 12 \end{array}$$

What number multiplied by x will give $3x$?
$\dfrac{3x}{x} = 3$. Place $+3$ in the quotient.

$$\begin{array}{r} x + 3 \\ x + 4\overline{)x^2 + 7x + 12} \\ \underline{x^2 + 4x} \\ 3x + 12 \\ \underline{3x + 12} \\ 0 \end{array}$$

Multiply each term in the divisor by 3 to get $3x + 12$, subtract $3x + 12$ from $3x + 12$ to get 0.

The division process stops when the result of the subtraction is either a constant or a polynomial with a degree that is less than the degree of the divisor. In this problem, the quotient is $x + 3$, and the remainder is 0.

We can check the quotient by multiplying the divisor by the quotient. The product should be the dividend. The quotient checks, because

$$\overbrace{\text{divisor} \cdot \text{quotient}}^{} = \overbrace{\text{dividend}}^{}$$
$$\overbrace{(x + 4)}\overbrace{(x + 3)} = \overbrace{x^2 + 7x + 12}$$ ■

EXAMPLE 4   Divide $2a^3 + 9a^2 + 5a - 6$ by $2a + 3$.

Solution

$$\begin{array}{r} a^2 \\ 2a + 3\overline{)2a^3 + 9a^2 + 5a - 6} \end{array}$$

What number multiplied by $2a$ will give $2a^3$?
$\dfrac{2a^3}{2a} = a^2$. Place a^2 in the quotient.

$$\begin{array}{r} a^2 \\ 2a + 3\overline{)2a^3 + 9a^2 + 5a - 6} \\ \underline{2a^3 + 3a^2} \\ 6a^2 + 5a \end{array}$$

Multiply each term in the divisor by a^2 to get $2a^3 + 3a^2$, subtract $2a^3 + 3a^2$ from $2a^3 + 9a^2$, and bring down the $5a$.

$$\begin{array}{r} a^2 + 3a \\ 2a + 3\overline{)2a^3 + 9a^2 + 5a - 6} \\ \underline{2a^3 + 3a^2} \\ 6a^2 + 5a \end{array}$$

What number multiplied by $2a$ will give $6a^2$?
$\dfrac{6a^2}{2a} = 3a$. Place $+ 3a$ in the quotient.

$$\begin{array}{r} a^2 + 3a \\ 2a + 3\overline{)2a^3 + 9a^2 + 5a - 6} \\ \underline{2a^3 + 3a^2} \\ 6a^2 + 5a \\ \underline{6a^2 + 9a} \\ -4a - 6 \end{array}$$

Multiply each term in the divisor by $3a$ to get $6a^2 + 9a$, subtract $6a^2 + 9a$ from $6a^2 + 5a$, and bring down -6.

$$\begin{array}{r} a^2 + 3a \;\; - 2 \\ 2a + 3\overline{)2a^3 + 9a^2 + 5a - 6} \\ \underline{2a^3 + 3a^2} \\ 6a^2 + 5a \\ \underline{6a^2 + 9a} \\ -4a - 6 \end{array}$$

What number multiplied by $2a$ will give $-4a$?
$$\frac{-4a}{2a} = -2.$$
Place -2 in the quotient.

$$\begin{array}{r} a^2 + 3a \;\; - 2 \\ 2a + 3\overline{)2a^3 + 9a^2 + 5a - 6} \\ \underline{2a^3 + 3a^2} \\ 6a^2 + 5a \\ \underline{6a^2 + 9a} \\ -4a - 6 \\ \underline{-4a - 6} \\ 0 \end{array}$$

Multiply each term in the divisor by -2 to get $-4a - 6$; subtract $-4a - 6$ from $-4a - 6$ to get 0.

The remainder is 0, and the quotient is $a^2 + 3a - 2$. We can check the quotient by verifying that

$$\underbrace{\text{divisor}}\; \cdot \;\underbrace{\text{quotient}} \;=\; \underbrace{\text{dividend}}$$
$$\underbrace{(2a + 3)}\underbrace{(a^2 + 3a - 2)} = \underbrace{2a^3 + 9a^2 + 5a - 6}$$ ∎

EXAMPLE 5   Divide $3x^3 + 2x^2 - 3x + 8$ by $x - 2$.

Solution
$$\begin{array}{r} 3x^2 + 8x \;\; + 13 \\ x - 2\overline{)3x^3 + 2x^2 - \;\; 3x + \;\; 8} \\ \underline{3x^3 - 6x^2} \\ 8x^2 - \;\; 3x \\ \underline{8x^2 - 16x} \\ 13x + \;\; 8 \\ \underline{13x - 26} \\ 34 \end{array}$$

This division gives a quotient of $3x^2 + 8x + 13$ and a remainder of 34. It is common to form a fraction with the remainder as the numerator and the divisor as the denominator and to write the result as

$$3x^2 + 8x + 13 + \frac{34}{x - 2}$$

To check this result, we verify that

$$(x - 2)\left(3x^2 + 8x + 13 + \frac{34}{x - 2}\right) = 3x^3 + 2x^2 - 3x + 8$$ ∎

EXAMPLE 6 Divide $-9x + 8x^3 + 10x^2 - 9$ by $3 + 2x$.

Solution The division process works best when the polynomials in the dividend and the divisor are written in descending powers of x. We can use the commutative property of addition to rearrange the terms. Then the division is routine:

$$
\begin{array}{r}
4x^2 - x - 3 \\
2x + 3 \overline{)8x^3 + 10x^2 - 9x - 9} \\
\underline{8x^3 + 12x^2} \\
-2x^2 - 9x \\
\underline{-2x^2 - 3x} \\
-6x - 9 \\
\underline{-6x - 9} \\
0
\end{array}
$$

Thus,

$$
\frac{-9x + 8x^3 + 10x^2 - 9}{3 + 2x} = 4x^2 - x - 3
$$

■ THE CASE OF THE MISSING TERMS

EXAMPLE 7 Divide $8x^3 + 1$ by $2x + 1$.

Solution When we write the terms in the dividend in descending powers of x, we see that the terms involving x^2 and x are missing. To make the division convenient, we must either include the terms $0x^2$ and $0x$ in the dividend or leave spaces for them. After this adjustment, the division is routine.

$$
\begin{array}{r}
4x^2 - 2x + 1 \\
2x + 1 \overline{)8x^3 + 0x^2 + 0x + 1} \\
\underline{8x^3 + 4x^2} \\
-4x^2 + 0x \\
\underline{-4x^2 - 2x} \\
+2x + 1 \\
\underline{+2x + 1} \\
0
\end{array}
$$

Thus,

$$
\frac{8x^3 + 1}{2x + 1} = 4x^2 - 2x + 1
$$

EXAMPLE 8 Divide $-17x^2 + 5x + x^4 + 2$ by $x^2 - 1 + 4x$.

Solution We write the problem with the divisor and the dividend in descending powers of x. After leaving space for the missing term in the dividend, we proceed as follows:

$$
\begin{array}{r}
x^2 - 4x \\
x^2 + 4x - 1 \overline{)x^4 - 17x^2 + 5x + 2} \\
\underline{x^4 + 4x^3 - x^2} \\
-4x^3 - 16x^2 + 5x \\
\underline{-4x^3 - 16x^2 + 4x} \\
x + 2
\end{array}
$$

This division gives a quotient of $x^2 - 4x$ and a remainder of $x + 2$. Thus,

$$\frac{-17x^2 + 5x + x^4 + 2}{x^2 - 1 + 4x} = x^2 - 4x + \frac{x + 2}{x^2 + 4x - 1}$$ ∎

Orals   *Divide.*

1. $\dfrac{6x^2y^2}{2xy}$ **2.** $\dfrac{4ab^2 + 8a^2b}{2ab}$

3. $\dfrac{x^2 + 2x + 1}{x + 1}$ **4.** $\dfrac{x^2 - 4}{x - 2}$

EXERCISE 6.7

In Exercises 1–18, do each division. Write all answers without using negative exponents.

1. $\dfrac{4x^2y^3}{8x^5y^2}$ **2.** $\dfrac{25x^4y^7}{5xy^9}$ **3.** $\dfrac{33a^{-2}b^2}{44a^2b^{-2}}$ **4.** $\dfrac{-63a^4b^{-3}}{81a^{-3}b^3}$

5. $\dfrac{45x^{-2}y^{-3}t^0}{-63x^{-1}y^4t^2}$ **6.** $\dfrac{112a^0b^2c^{-3}}{48a^4b^0c^4}$

7. $\dfrac{-65a^{2n}b^nc^{3n}}{-15a^nb^{-n}c}$ **8.** $\dfrac{-32x^{-3n}y^{-2n}z}{40x^{-2}y^{-n}z^{n+1}}$

9. $\dfrac{4x^2 - x^3}{6x}$ **10.** $\dfrac{5y^4 + 45y^3}{15y^2}$

11. $\dfrac{4x^2y^3 + x^3y^2}{8xy}$ **12.** $\dfrac{3a^3y^2 - 18a^4y^3}{27a^2y^2}$

13. $\dfrac{24x^6y^7 - 12x^5y^{12} + 36xy}{48x^2y^3}$ **14.** $\dfrac{9x^4y^3 + 18x^2y - 27xy^4}{9x^3y^3}$

15. $\dfrac{3a^{-2}b^3 - 6a^2b^{-3} + 9a^{-2}}{12a^{-1}b}$ **16.** $\dfrac{4x^3y^{-2} + 8x^{-2}y^2 - 12y^4}{12x^{-1}y^{-1}}$

17. $\dfrac{x^ny^n - 3x^{2n}y^{2n} + 6x^{3n}y^{3n}}{x^ny^n}$ **18.** $\dfrac{2a^n - 3a^nb^{2n} - 6b^{4n}}{a^nb^{n-1}}$

In Exercises 19–54, find each quotient.

19. $\dfrac{x^2 + 5x + 6}{x + 3}$ **20.** $\dfrac{x^2 - 5x + 6}{x - 3}$

21. $\dfrac{x^2 + 10x + 21}{x + 3}$ **22.** $\dfrac{x^2 + 10x + 21}{x + 7}$

23. $\dfrac{6x^2 - x - 12}{2x + 3}$ **24.** $\dfrac{6x^2 - x - 12}{2x - 3}$

25. $\dfrac{3x^3 - 2x^2 + x + 6}{x - 1}$ **26.** $\dfrac{4a^3 + a^2 - 3a + 7}{a + 1}$

27. $\dfrac{6x^3 + 11x^2 - x - 2}{3x - 2}$ **28.** $\dfrac{6x^3 + 11x^2 - x + 10}{2x + 3}$

29. $\dfrac{6x^3 - x^2 - 6x - 9}{2x - 3}$ **30.** $\dfrac{16x^3 + 16x^2 - 9x - 5}{4x + 5}$

31. $\dfrac{2a + 1 + a^2}{a + 1}$

32. $\dfrac{a - 15 + 6a^2}{2a - 3}$

33. $\dfrac{6y - 4 + 10y^2}{5y - 2}$

34. $\dfrac{-10xy + x^2 + 16y^2}{x - 2y}$

35. $\dfrac{-18x + 12 + 6x^2}{x - 1}$

36. $\dfrac{27x + 23x^2 + 6x^3}{2x + 3}$

37. $\dfrac{-9x^2 + 8x + 9x^3 - 4}{3x - 2}$

38. $\dfrac{6x^2 + 8x^3 - 13x + 3}{4x - 3}$

39. $\dfrac{13x + 16x^4 + 3x^2 + 3}{4x + 3}$

40. $\dfrac{3x^2 + 9x^3 + 4x + 4}{3x + 2}$

41. $\dfrac{a^3 + 1}{a - 1}$

42. $\dfrac{27a^3 - 8b^3}{3a - 2b}$

43. $\dfrac{15a^3 - 29a^2 + 16}{3a - 4}$

44. $\dfrac{4x^3 - 12x^2 + 17x - 12}{2x - 3}$

45. $y - 2\overline{)-24y + 24 + 6y^2}$

46. $3 - a\overline{)21a - a^2 - 54}$

47. $2x + y\overline{)32x^5 + y^5}$

48. $3x - y\overline{)81x^4 - y^4}$

49. $x^2 - 2\overline{)x^6 - x^4 + 2x^2 - 8}$

50. $x^2 + 3\overline{)x^6 + 2x^4 - 6x^2 - 9}$

51. $\dfrac{x^4 + 2x^3 + 4x^2 + 3x + 2}{x^2 + x + 2}$

52. $\dfrac{2x^4 + 3x^3 + 3x^2 - 5x - 3}{2x^2 - x - 1}$

53. $\dfrac{x^3 + 3x + 5x^2 + 6 + x^4}{x^2 + 3}$

54. $\dfrac{x^5 + 3x + 2}{x^3 + 1 + 2x}$

In Exercises 55–56, use a calculator to find each quotient.

55. $x - 2\overline{)9.8x^2 - 3.2x - 69.3}$

56. $2.5x - 3.7\overline{)-22.25x^2 - 38.9x - 16.65}$

Writing Exercises   *Write a paragraph using your own words.*

1. Explain how to divide a monomial by a monomial.

2. Explain how to check the result of a division problem.

Something to Think About   **1.** Since 6 is a factor of 24, 6 divides 24 with no remainder. Decide whether $2x - 3$ is a factor of $10x^2 - x - 21$.

2. Is $x - 1$ a factor of $x^5 - 1$?

Review Exercises   *Remove parentheses and simplify.*

1. $2(x^2 + 4x - 1) + 3(2x^2 - 2x + 2)$

2. $3(2a^2 - 3a + 2) - 4(2a^2 + 4a - 7)$

3. $-2(3y^3 - 2y + 7) - 3(y^2 + 2y - 4) + 4(y^3 + 2y - 1)$

4. $3(4y^3 + 3y - 2) + 2(3y^2 - y + 3) - 5(2y^3 - y^2 - 2)$

6.8 Synthetic Division (Optional)

■ THE REMAINDER THEOREM ■ THE FACTOR THEOREM ■ GRAPHING DEVICES

There is a shortcut method, called **synthetic division**, that we can use to divide a polynomial by a binomial of the form $x - r$. To see how this method works, we consider the division of $4x^3 - 5x^2 - 11x + 20$ by $x - 2$.

$$
\begin{array}{r}
4x^2 + 3x - 5 \\
x - 2 \overline{)4x^3 - 5x^2 - 11x + 20} \\
\underline{4x^3 - 8x^2} \\
3x^2 - 11x \\
\underline{3x^2 - 6x} \\
-5x + 20 \\
\underline{-5x + 10} \\
10 \quad \text{(remainder)}
\end{array}
$$

$$
\begin{array}{r}
4 \quad 3 \quad -5 \\
1 - 2 \overline{)4 \quad -5 \quad -11 \quad 20} \\
\underline{4 \quad -8} \\
3 \quad -11 \\
\underline{3 \quad -6} \\
-5 \quad 20 \\
\underline{-5 \quad 10} \\
10 \quad \text{(remainder)}
\end{array}
$$

On the left is the familiar long-division process, and on the right is the same division with the variables and their exponents removed. The various powers of x can be remembered without actually writing them, because the exponents of the terms in the divisor, dividend, and quotient were written in descending order.

We can further shorten the version on the right. The numbers printed in color need not be written, because they are duplicates of the numbers immediately above them. Thus, we can write the division in the following form:

$$
\begin{array}{r}
4 \quad 3 \quad -5 \\
1 - 2 \overline{)4 \quad -5 \quad -11 \quad 20} \\
\underline{-8} \\
3 \\
\underline{-6} \\
-5 \\
\underline{10} \\
10
\end{array}
$$

We can shorten the process still further by compressing the work vertically and eliminating the 1 (the coefficient of x in the divisor):

$$
\begin{array}{r}
4 \quad 3 \quad -5 \\
-2 \overline{)4 \quad -5 \quad -11 \quad 20} \\
\underline{-8 \quad -6 \quad 10} \\
3 \quad -5 \quad 10
\end{array}
$$

There is no reason why the quotient represented by the numbers 4   3 -5, must appear above the long division. If we write the 4 on the bottom line, the bottom line gives the coefficients of the quotient, and it also gives the remainder. The entire top line can be eliminated. The division now appears as follows:

$$
\begin{array}{r}
\underline{-2|} \quad 4 \quad -5 \quad -11 \quad 20 \\
\underline{-8 \quad -6 \quad 10} \\
4 \quad 3 \quad -5 \quad 10
\end{array}
$$

The bottom line was obtained by subtracting the middle line from the top line. If we were to replace the -2 in the divisor by $+2$, the division process would reverse the signs of every entry in the middle line. Then the bottom line could be obtained by addition. Thus, we have this final form of the synthetic division.

$$\underline{+2\rvert\quad} \begin{array}{rrrr} 4 & -5 & -11 & 20 \\ & 8 & 6 & -10 \\ \hline 4 & 3 & -5 & 10 \end{array}$$

The coefficients of the dividend.

The coefficients of the quotient and the remainder.

Thus,

$$\frac{4x^3 - 5x^2 - 11x + 20}{x - 2} = 4x^2 + 3x - 5 + \frac{10}{x - 2}$$

EXAMPLE 1 Use synthetic division to divide $6x^2 + 5x - 2$ by $x - 5$.

Solution We begin by writing the coefficients of the dividend, and the 5 from the divisor, in the following form:

$$\underline{5\rvert\quad 6 \quad 5 \quad -2}$$

Then we follow these steps:

$$\underline{5\rvert\quad 6 \quad 5 \quad -2} \\ \ \ 6$$

Begin by bringing down the 6.

$$\underline{5\rvert\quad \begin{array}{rrr} 6 & 5 & -2 \\ & 30 & \end{array}} \\ \ \ 6$$

Multiply 5 by 6 to get 30.

$$\underline{5\rvert\quad \begin{array}{rrr} 6 & 5 & -2 \\ & 30 & \end{array}} \\ \ \ 6 \quad 35$$

Add 5 and 30 to get 35.

$$\underline{5\rvert\quad \begin{array}{rrr} 6 & 5 & -2 \\ & 30 & 175 \end{array}} \\ \ \ 6 \quad 35$$

Multiply 5 and 35 to get 175.

$$\underline{5\rvert\quad \begin{array}{rrr} 6 & 5 & -2 \\ & 30 & 175 \end{array}} \\ \ \ 6 \quad 35 \quad 173$$

Add −2 and 175 to get 173.

The numbers 6 and 35 represent the quotient: $6x + 35$. The number 173 is the remainder. Thus,

$$\frac{6x^2 + 5x - 2}{x - 5} = 6x + 35 + \frac{173}{x - 5}$$

EXAMPLE 2 Use synthetic division to divide $5x^3 + x^2 - 3$ by $x - 2$.

Solution We begin by writing

$$\underline{2|} \quad 5 \quad 1 \quad 0 \quad -3 \qquad \text{Write 0 for the coefficient of } x, \text{ the missing term.}$$

and complete the division as follows:

$$
\begin{array}{c|cccc}
2 & 5 & 1 & 0 & -3 \\
  &   & 10 &   &   \\
\hline
  & 5 & 11 &   &
\end{array}
\qquad
\begin{array}{c|cccc}
2 & 5 & 1 & 0 & -3 \\
  &   & 10 & 22 &   \\
\hline
  & 5 & 11 & 22 &
\end{array}
\qquad
\begin{array}{c|cccc}
2 & 5 & 1 & 0 & -3 \\
  &   & 10 & 22 & 44 \\
\hline
  & 5 & 11 & 22 & 41
\end{array}
$$

Thus,

$$\frac{5x^3 + x^2 - 3}{x - 2} = 5x^2 + 11x + 22 + \frac{41}{x - 2}$$

EXAMPLE 3 Use synthetic division to divide $5x^2 + 6x^3 + 2 - 4x$ by $x + 2$.

Solution First, we write the dividend with the exponents in descending order:

$$6x^3 + 5x^2 - 4x + 2$$

Since the divisor is $x + 2$, we write it in $x - r$ form: $x - (-2)$. Using synthetic division, we begin by writing

$$\underline{-2|} \quad 6 \quad 5 \quad -4 \quad 2$$

and complete the division:

$$
\begin{array}{c|cccc}
-2 & 6 & 5 & -4 & 2 \\
   &   & -12 & 14 & -20 \\
\hline
   & 6 & -7 & 10 & -18
\end{array}
$$

Thus,

$$\frac{5x^2 + 6x^3 + 2 - 4x}{x + 2} = 6x^2 - 7x + 10 + \frac{-18}{x + 2}$$

■ THE REMAINDER THEOREM

Synthetic division is important in mathematics because of the *remainder theorem*.

> **Remainder Theorem**
> If a polynomial $P(x)$ is divided by $x - r$, the remainder is $P(r)$.

We illustrate the remainder theorem in the next example.

EXAMPLE 4 Let $P(x) = 2x^3 - 3x^2 - 2x + 1$. Find **a.** $P(3)$ and **b.** the remainder when $P(x)$ is divided by $x - 3$.

Solution   **a.** $P(3) = 2(3)^3 - 3(3)^2 - 2(3) + 1$ Substitute 3 for x.

$= 2(27) - 3(9) - 6 + 1$

$= 54 - 27 - 6 + 1$

$= 22$

b. We use synthetic division to find the remainder when $P(x) = 2x^3 - 3x^2 - 2x + 1$ is divided by $x - 3$.

$$
\begin{array}{r|rrrr}
3 & 2 & -3 & -2 & 1 \\
  &   & 6 & 9 & 21 \\
\hline
  & 2 & 3 & 7 & 22
\end{array}
$$

The remainder is 22.

The results of parts **a** and **b** show that when $P(x)$ is divided by $x - 3$, the remainder is $P(3)$.   ■

It is often easier to find $P(r)$ by using synthetic division than by substituting r for x in $P(x)$. This is especially true if r is a decimal.

■ THE FACTOR THEOREM

Recall that if two quantities are multiplied, each is called a *factor* of the product. Thus, $x - 2$ is one factor of $6x - 12$, because $6(x - 2) = 6x - 12$. A theorem, called the *factor theorem*, tells us how to find one factor of a polynomial if the remainder of a certain division is 0.

Factor Theorem
If $P(x)$ is a polynomial in x, then

$P(r) = 0$ if and only if $x - r$ is a factor of $P(x)$.

If $P(x)$ is a polynomial in x and if $P(r) = 0$, then r is called a **zero** of the polynomial. A zero of a polynomial in x is any value of x that makes the value of the polynomial equal to 0.

EXAMPLE 5   Let $P(x) = 3x^3 - 5x^2 + 3x - 10$. Show that   **a.** $P(2) = 0$ and   **b.** $x - 2$ is a factor of $P(x)$.

Solution   **a.** Use the remainder theorem to evaluate $P(2)$ by dividing $P(x) = 3x^3 - 5x^2 + 3x - 10$ by $x - 2$.

$$
\begin{array}{r|rrrr}
2 & 3 & -5 & 3 & -10 \\
  &   & 6 & 2 & 10 \\
\hline
  & 3 & 1 & 5 & 0
\end{array}
$$

The remainder in this division is 0. By the remainder theorem, the remainder is $P(2)$. Thus, $P(2) = 0$, and 2 is a zero of the polynomial.

b. The remainder is 0, and the numbers 3, 1, and 5 in the synthetic division in part **a** represent the coefficients of quotient $3x^2 + x + 5$. Thus,

$$\underbrace{(x - 2)}_{\text{divisor}} \cdot \underbrace{(3x^2 + x + 5)}_{\text{quotient}} + \underbrace{0}_{\text{remainder}} = \underbrace{3x^3 - 5x^2 + 3x - 10}_{\text{the dividend, } P(x)}$$

or

$$(x - 2)(3x^2 + x + 5) = 3x^3 - 5x^2 + 3x - 10$$

Thus, $x - 2$ is a factor of $P(x)$. Note that $3x^3 - 5x^2 + 3x - 10$ does not factor easily using the technique discussed in the previous chapter. ■

The result of Example 5 is true because the remainder, $P(2)$, is 0. If the remainder had not been 0, then $x - 2$ would not have been a factor of $P(x)$.

■ GRAPHING DEVICES

We can use a graphing device to approximate the real zeros of a polynomial function $f(x)$. For example, to find the real zeros of $f(x) = 2x^3 - 6x^2 + 7x - 21$, we graph the function as in Figure 6–11.

It is clear from the figure that the function f has a zero at $x = 3$. To verify this, we calculate $f(3)$.

$$f(3) = 2(3)^3 - 6(3)^2 + 7(3) - 21$$
$$= 2(27) - 6(9) + 21 - 21$$
$$= 0$$

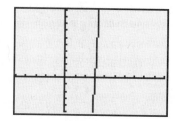

FIGURE 6-11

From the factor theorem, we know that $x - 3$ is a factor of the polynomial. To find the other factor, we can synthetically divide by 3.

$$\begin{array}{r|rrrr}
3 & 2 & -6 & 7 & -21 \\
  &   & 6 & 0 & 21 \\
\hline
  & 2 & 0 & 7 & 0
\end{array}$$

Thus, $f(x) = (x - 3)(2x^2 + 7)$. Since $2x^2 + 7$ cannot be factored over the real numbers, we can conclude that 3 is the only real zero of the polynomial function.

Orals   *Find the remainder in each division.*

1. $(x^2 + 2x + 1) \div (x - 2)$ **2.** $(x^2 - 4) \div (x + 1)$

Tell whether $x - 2$ is a factor of each polynomial.

3. $x^3 - 2x^2 + x - 2$ **4.** $x^3 + 4x^2 - 1$

EXERCISE 6.8

In Exercises 1–14, use synthetic division to do each division.

1. $(x^2 + x - 2) \div (x - 1)$ **2.** $(x^2 + x - 6) \div (x - 2)$

3. $(x^2 - 7x + 12) \div (x - 4)$ **4.** $(x^2 - 6x + 5) \div (x - 5)$

5. $(x^2 + 8 + 6x) \div (x + 4)$

6. $(x^2 - 15 - 2x) \div (x + 3)$

7. $(x^2 - 5x + 14) \div (x + 2)$

8. $(x^2 + 13x + 42) \div (x + 6)$

9. $(3x^3 - 10x^2 + 5x - 6) \div (x - 3)$

10. $(2x^3 - 9x^2 + 10x - 3) \div (x - 3)$

11. $(2x^3 - 5x - 6) \div (x - 2)$

12. $(4x^3 + 5x^2 - 1) \div (x + 2)$

13. $(5x^2 + 6x^3 + 4) \div (x + 1)$

14. $(4 - 3x^2 + x) \div (x - 4)$

In Exercises 15–20, use a calculator and synthetic division to do each division.

15. $(7.2x^2 - 2.1x + 0.5) \div (x - 0.2)$

16. $(8.1x^2 + 3.2x - 5.7) \div (x - 0.4)$

17. $(2.7x^2 + x - 5.2) \div (x + 1.7)$

18. $(1.3x^2 - 0.5x - 2.3) \div (x + 2.5)$

19. $(9x^3 - 25) \div (x + 57)$

20. $(0.5x^3 + x) \div (x - 2.3)$

In Exercises 21–28, let $P(x) = 2x^3 - 4x^2 + 2x - 1$. Evaluate $P(x)$ by substituting the given value of x into the polynomial and simplifying. Then evaluate the polynomial by using the remainder theorem and synthetic division.

21. $P(1)$

22. $P(2)$

23. $P(-2)$

24. $P(-1)$

25. $P(3)$

26. $P(-4)$

27. $P(0)$

28. $P(4)$

In Exercises 29–36, let $Q(x) = x^4 - 3x^3 + 2x^2 + x - 3$. Evaluate $Q(x)$ by substituting the given value of x into the polynomial and simplifying. Then evaluate the polynomial by using the remainder theorem and synthetic division.

29. $Q(-1)$

30. $Q(1)$

31. $Q(2)$

32. $Q(-2)$

33. $Q(3)$

34. $Q(0)$

35. $Q(-3)$

36. $Q(-4)$

In Exercises 37–44, use the remainder theorem and synthetic division to find $P(r)$.

37. $P(x) = x^3 - 4x^2 + x - 2; r = 2$

38. $P(x) = x^3 - 3x^2 + x + 1; r = 1$

39. $P(x) = 2x^3 + x + 2; r = 3$

40. $P(x) = x^3 + x^2 + 1; r = -2$

41. $P(x) = x^4 - 2x^3 + x^2 - 3x + 2; r = -2$

42. $P(x) = x^5 + 3x^4 - x^2 + 1; r = -1$

43. $P(x) = 3x^5 + 1; r = -\dfrac{1}{2}$

44. $P(x) = 5x^7 - 7x^4 + x^2 + 1; r = 2$

In Exercises 45–48, use the factor theorem and tell whether the first expression is a factor of $P(x)$.

45. $x - 3; P(x) = x^3 - 3x^2 + 5x - 15$

46. $x + 1; P(x) = x^3 + 2x^2 - 2x - 3$
(*Hint:* Write $x + 1$ as $x - (-1)$.)

47. $x + 2; P(x) = 3x^2 - 7x + 4$
(*Hint:* Write $x + 2$ as $x - (-2)$.)

48. $x; P(x) = 7x^3 - 5x^2 - 8x$
(*Hint:* $x = x - 0$.)

In Exercises 49–50, use a calculator to work each problem.

49. Find 2^6 by using synthetic division to evaluate the polynomial $P(x) = x^6$ at $x = 2$. Then check the answer by evaluating 2^6 with a calculator.

50. Find $(-3)^5$ by using synthetic division to evaluate the polynomial $P(x) = x^5$ at $x = -3$. Then check the answer by evaluating $(-3)^5$ with a calculator.

Writing Exercises   *Write a paragraph using your own words.*

1. If you are given $P(x)$, explain how to use synthetic division to calculate $P(a)$.

2. Explain the factor theorem.

Something to Think About   *Suppose that* $P(x) = x^{100} - x^{99} + x^{98} - x^{97} + \cdots + x^2 - x + 1.$

1. Find the remainder when $P(x)$ is divided by $x - 1$.

2. Find the remainder when $P(x)$ is divided by $x + 1$.

Review Exercises   *Let* $f(x) = 3x^2 + 2x - 1$ *and find each value.*

1. $f(1)$ **2.** $f(-2)$ **3.** $f(2a)$ **4.** $f(-t)$

Remove parentheses and simplify.

5. $2(x^2 + 4x - 1) + 3(2x^2 - 2x + 2)$

6. $-2(3y^3 - 2y + 7) - 3(y^2 + 2y - 4) + 4(y^3 + 2y - 1)$

■ ■ ■ ■ ■ ■ ■ ■ ■ ■ **PROBLEMS AND PROJECTS**

1. A rectangle has an area of 64 square meters. Find the dimensions that will give a perimeter of 40 meters.

2. You are an engineer who is about to bid on constructing a new school for the local public school district. From past records, you have found that one of your work crews can build $28\frac{4}{7}\%$ of the school in 4 months, and another crew can build $33\frac{1}{3}\%$ of the school in 7 months. How many months would you expect it to take to build the school if both crews worked together on the job?

3. As an archaeologist, you find a clay tablet during an excavation in Mesopotamia. Upon translation, the inscription describes a caravan that a man and some other merchants had formed at a total cost of 60 talents of gold. In the inscription, the man lamented the fact that if three more merchants had joined the caravan for the same total cost, his cost would have been one talent less. How many merchants formed the caravan, and what did each pay?

4. After changing from electric steel casting to gas steel casting, PDP Enterprises found that the cost in dollars per cubic foot of gas, $f(r)$, to run the gas steel casting furnaces at a rate r (in cubic feet per hour) was given by the function

$$f(r) = \frac{r^2}{50} + \frac{135}{r}$$

What is the most economical rate at which to operate the gas furnaces, between 5 cubic feet per hour and 20 cubic feet per hour?

PROJECT 1   Suppose that a motorist usually makes a certain 80-mile trip at an average speed of 50 miles per hour.

a. How much time does the driver save by increasing her rate by 5 miles per hour? By 10 miles per hour? By 15 miles per hour? Give all answers to the nearest minute.

b. Consider your work in part **a**, and then find a formula that will tell how much time is saved if the motorist travels x mph faster than 50 mph. That is, find an expression involving x that will give the time saved by traveling $(50 + x)$ mph instead of 50 mph.

(continued)

■ ■ ■ ■ ■ ■ ■ ■ ■ ■ **PROBLEMS AND PROJECTS** *(continued)*

c. Find a formula that will give the time saved on a trip of *d* miles by traveling at an average rate that is *x* mph faster than a usual speed of *y* mph. When you have this formula, simplify it into a nice, compact form.

d. Test your formula by doing part **a** over again using the formula. The answers you get should agree with those found earlier.

e. Use your formula to solve the following problem. Every holiday season, Kurt and Ellen travel to visit their relatives, a distance of 980 miles. Under normal circumstances, they can average 60 mph during the trip. However, improved roads will enable them to travel 4 mph faster this year. How much time (to the nearest minute) will they save on this year's trip? How much faster than normal (to the nearest tenth of a mph) would Kurt and Ellen have to travel to save two hours?

PROJECT 2 By careful cross-fertilization among a number of species of corn, Professor Greenthumb has succeeded in creating a miracle hybrid corn plant. Under ideal conditions, the plant will produce twice as much corn as an ordinary hybrid. However, the new hybrid is very sensitive to the amount of sun and water it receives. If conditions are not ideal, the corn yield diminishes.

To determine what yield the plants will have, daily measurements record the amount of sun the plants receive (the sun index, S) and the amount of water they receive (W, measured in millimeters). Then the daily yield diminution is calculated using the formula

$$\text{Daily yield diminution} = \frac{S^3 + W^3 - (S + W)^2}{(S + W)^3}$$

At the end of a 100 day growing season, the daily yield diminutions are added together to make the total diminution (D). The yield for the year is then determined using the formula

$$\text{Annual yield} = 1 - .01D$$

The resulting decimal is the percent of maximum yield that the corn plants will achieve. Remember that maximum yield for these plants is twice the yield of an ordinary corn plant.

As Greenthumb's research assistant, you have been asked to handle a few questions that have come up regarding her research.

a. First, show that if $S = W = 2$, the daily yield diminution is 0. These would be the optimal conditions for Greenthumb's plants.

b. Now suppose that for each day of the 100-day growing season, the sun index is 8, and 6 millimeters of rain fall on the plants. Find the daily yield diminution. To the nearest tenth of a percent, what percentage of the yield of normal plants will the special hybrid plants produce?

c. Show that when the daily yield diminution is .5 for each of the 100 days in the growing season, the annual yield will be .5 (exactly the yield of ordinary corn plants). Now suppose that through the use of an irrigation system, you arrange for the corn to receive 10 millimeters of rain each day. What would the sun index have to be each day to give an annual yield of .5? To answer this question, first simplify the daily yield diminution formula. Do this *be-fore* you substitute any numbers into the formula.

d. Another assistant is also working with Greenthumb's formula, but he is getting different results. Something is wrong; for the situation given in part **b**, he finds that the daily yield diminution for one day is 34. He simplified Greenthumb's formula first, then substituted in the values for S and W. He shows you the following work. Find all of his mistakes and explain what he did wrong in making each of them.

$$\text{Daily yield diminution} = \frac{S^3 + W^3 - (S + W)^2}{(S + W)^3}$$

$$= \frac{S^3 + W^3 - S^2 + W^2}{(S + W)^3}$$

$$= \frac{(S^3 + W^3) - (S^2 - W^2)}{(S + W)^3}$$

$$= \frac{(S + W) \cdot (S^2 + SW + W^2) - (S + W) \cdot (S - W)}{(S + W)^3}$$

$$= \frac{(S^2 + SW + W^2) - S - W}{(S + W)^2}$$

$$= \frac{S^2 + SW + W^2 - S - W}{S^2 + W^2}$$

$$= SW - S - W$$

So for $S = 8$ and $W = 6$, he gets 34 as the daily yield diminution.

Chapter Summary

Key Words

Key Ideas

(6.1) Division by 0 is undefined.

$\dfrac{ak}{bk} = \dfrac{a}{b}$, provided that $b \neq 0$ and $k \neq 0$.

To simplify a fraction, factor the numerator and denominator and divide out all factors common to both the numerator and denominator.

(6.2) $\dfrac{a}{b} \cdot \dfrac{c}{d} = \dfrac{ac}{bd}$ ($b \neq 0$ and $d \neq 0$)

$\dfrac{a}{b} \div \dfrac{c}{d} = \dfrac{a}{b} \cdot \dfrac{d}{c}$ ($b \neq 0$, $d \neq 0$, and $c \neq 0$)

(6.3) $\dfrac{a}{b} + \dfrac{c}{b} = \dfrac{a+c}{b}$ $(b \neq 0)$

$\dfrac{a}{b} - \dfrac{c}{b} = \dfrac{a-c}{b}$ $(b \neq 0)$

Fractions must have common denominators before they can be added or subtracted.

To find the least common denominator (LCD) of two fractions, factor each denominator and use each factor the greatest number of times that it appears in any one denominator. The product of these factors is the LCD of the fractions.

(6.4) A fraction that has a fraction within its numerator or its denominator is called a **complex fraction**.

(6.5) Multiplying both sides of an equation by a quantity that contains a variable can lead to false solutions. All possible solutions of a rational equation must be checked.

(6.6) In a proportion, the product of the extremes is equal to the product of the means.

If $y = kx$ (k is a constant), then x and y vary directly.

If $y = \frac{k}{x}$ (k is a constant), then x and y vary inversely.

If $y = kxz$ (k is a constant), then y varies jointly with x and z.

The expression $y = \frac{kx}{z}$ (k is a constant) represents combined variation, with y and x varying directly and y and z varying inversely.

(6.7) To find the quotient of two monomials, express the quotient as a fraction and use the rules of exponents to simplify.

(6.8) Synthetic division can be used to divide polynomials by binomials of the form $x - r$.

The remainder theorem. If a polynomial $P(x)$ is divided by $x - r$, the remainder is $P(r)$.

The factor theorem. If a polynomial $P(x)$ is divided by $x - r$, then $P(r) = 0$, if and only if $x - r$ is a factor of $P(x)$.

■ Chapter 6 Review Exercises

In Review Exercises 1–8, simplify each fraction.

1. $\dfrac{248x^2y}{576xy^2}$

2. $\dfrac{212m^3n}{588m^2n^3}$

3. $\dfrac{x^2 - 49}{x^2 + 14x + 49}$

4. $\dfrac{x^2 + 6x + 36}{x^3 - 216}$

5. $\dfrac{x^2 - 2x + 4}{2x^3 + 16}$

6. $\dfrac{x - y}{y - x}$

7. $\dfrac{2m - 2n}{n - m}$

8. $\dfrac{ac - ad + bc - bd}{d^2 - c^2}$

In Review Exercises 9–10, find each probability.

9. Rolling an 11 on one roll of two dice.

10. Three children, all boys.

In Review Exercises 11–24, do the operations and simplify.

11. $\dfrac{x^2 + 4x + 4}{x^2 - x - 6} \cdot \dfrac{x^2 - 9}{x^2 + 5x + 6}$

12. $\dfrac{x^3 - 64}{x^2 + 4x + 16} \div \dfrac{x^2 - 16}{x + 4}$

13. $\dfrac{5y}{x - y} - \dfrac{3}{x - y}$

14. $\dfrac{3x - 1}{x^2 + 2} + \dfrac{3(x - 2)}{x^2 + 2}$

15. $\dfrac{3}{x + 2} + \dfrac{2}{x + 3}$

16. $\dfrac{4x}{x - 4} - \dfrac{3}{x + 3}$

17. $\dfrac{x^2 + 3x + 2}{x^2 - x - 6} \cdot \dfrac{3x^2 - 3x}{x^2 - 3x - 4} \div \dfrac{x^2 + 3x + 2}{x^2 - 2x - 8}$

18. $\dfrac{x^2 - x - 6}{x^2 - 3x - 10} \div \dfrac{x^2 - x}{x^2 - 5x} \cdot \dfrac{x^2 - 4x + 3}{x^2 - 6x + 9}$

19. $\dfrac{2x}{x+1} + \dfrac{3x}{x+2} + \dfrac{4x}{x^2+3x+2}$

20. $\dfrac{5x}{x-3} + \dfrac{5}{x^2-5x+6} + \dfrac{x+3}{x-2}$

21. $\dfrac{3(x+2)}{x^2-1} - \dfrac{2}{x+1} + \dfrac{4(x+3)}{x^2-2x+1}$

22. $\dfrac{x}{x^2+4x+4} + \dfrac{2x}{x^2-4} - \dfrac{x^2-4}{x-2}$

23. $\dfrac{x+2}{x^2-9} - \dfrac{x^2+6x+9}{x^2+5x+6} + \dfrac{3x}{x-3}$

24. $\dfrac{-2(3+x)}{x^2+6x+9} + \dfrac{3(x+2)}{x^2-6x+9} - \dfrac{1}{x^2-9}$

In Review Exercises 25–34, simplify each complex fraction.

25. $\dfrac{\dfrac{3}{x} - \dfrac{2}{y}}{xy}$

26. $\dfrac{\dfrac{1}{x} + \dfrac{2}{y}}{\dfrac{2}{x} - \dfrac{1}{y}}$

27. $\dfrac{2x + 3 + \dfrac{1}{x}}{x + 2 + \dfrac{1}{x}}$

28. $\dfrac{6x + 13 + \dfrac{6}{x}}{6x + 5 - \dfrac{6}{x}}$

29. $\dfrac{1 + \dfrac{3}{x}}{x + 3}$

30. $\dfrac{1 - \dfrac{1}{x} - \dfrac{2}{x^2}}{1 + \dfrac{4}{x} + \dfrac{3}{x^2}}$

31. $\dfrac{(x-y)^{-2}}{x^{-2} - y^{-2}}$

32. $\dfrac{x^{-1} + 1}{x + 1}$

33. $\dfrac{x^{-2} + 1}{x^2 + 1}$

34. $\dfrac{x^{-1} - y^{-1}}{x^{-1} + y^{-1}}$

In Review Exercises 35–38, solve each equation, if possible.

35. $\dfrac{4}{x} - \dfrac{1}{10} = \dfrac{7}{2x}$

36. $\dfrac{2}{x+5} - \dfrac{1}{6} = \dfrac{1}{x+4}$

37. $\dfrac{2(x-5)}{x-2} = \dfrac{6x+12}{4-x^2}$

38. $\dfrac{7}{x+9} - \dfrac{x+2}{2} = \dfrac{x+4}{x+9}$

In Review Exercises 39–40, solve each formula for the indicated variable.

39. $\dfrac{x^2}{a^2} - \dfrac{y^2}{b^2} = 1$ for y^2

40. $H = \dfrac{2ab}{a+b}$ for b

In Review Exercises 41–44, solve each problem.

41. Trip length   Heavy traffic reduced Jim's usual average speed by 10 miles per hour, which lengthened his 200-mile trip by 1 hour. Find his usual average speed.

42. Flying speed   On a 600-mile trip, a pilot can save 30 minutes by increasing her usual speed by 40 miles per hour. Find her usual speed.

43. Draining a tank If one outlet pipe can drain a tank in 24 hours and another pipe can drain the tank in 36 hours, how long will it take for both pipes to drain the tank?

44. Siding a house Two men have estimated that they can side a house in 8 days. If one of them, who could have sided the house alone in 14 days, gets sick, how long will it take the other man to side the house alone?

In Review Exercises 45–46, solve each proportion.

45. $\dfrac{x+1}{8} = \dfrac{4x-2}{24}$

46. $\dfrac{1}{x+6} = \dfrac{x+10}{12}$

47. Assume that x varies directly with y. If $x = 12$ when $y = 2$, find the value of x when $y = 12$.

48. Assume that x varies inversely with y. If $x = 24$ when $y = 3$, find y when $x = 12$.

49. Assume that x varies jointly with y and z. Find the constant of variation if $x = 24$ when $y = 3$ and $z = 4$.

50. Assume that x varies directly with t and inversely with y. Find the constant of variation if $x = 2$ when $t = 8$ and $y = 64$.

In Review Exercises 51–54, do each division.

51. $\dfrac{-5x^6y^3}{10x^3y^6}$

52. $\dfrac{30x^3y^2 - 15x^2y - 10xy^2}{-10xy}$

53. $(3x^2 + 13xy - 10y^2) \div (3x - 2y)$

54. $(2x^3 + 7x^2 + 3 + 4x) \div (2x + 3)$

(Optional) In Review Exercises 55–56, use the factor theorem and synthetic division to decide whether the first expression is a factor of $P(x)$.

55. $x - 5$; $P(x) = x^3 - 3x^2 - 8x - 10$

56. $x + 5$; $P(x) = x^3 + 4x^2 - 5x + 5$
 (*Hint*: Write $x + 5$ as $x - (-5)$.)

■ Chapter 6 Test

In Problems 1–4, simplify each fraction.

1. $\dfrac{-12x^2y^3z^2}{18x^3y^4z^2}$

2. $\dfrac{2x+4}{x^2-4}$

3. $\dfrac{3y-6z}{2z-y}$

4. $\dfrac{2x^2+7x+3}{4x+12}$

In Problems 5–6, find each probability.

5. Tossing a coin three times and getting no heads.

6. Drawing an ace or king from an ordinary card deck.

In Problems 7–16, do the operations and simplify, if necessary. Write all answers without negative exponents.

7. $\dfrac{x^2y^{-2}}{x^3z^2} \cdot \dfrac{x^2z^4}{y^2z}$

8. $\dfrac{(x+1)(x+2)}{10} \cdot \dfrac{5}{x+2}$

9. $\dfrac{u^2+5u+6}{u^2-4} \cdot \dfrac{u^2-5u+6}{u^2-9}$

10. $\dfrac{x^3+y^3}{4} \div \dfrac{x^2-xy+y^2}{2x+2y}$

11. $\dfrac{xu+2u+3x+6}{u^2-9} \cdot \dfrac{2u-6}{x^2+3x+2}$

12. $\dfrac{a^2+7a+12}{a+3} \div \dfrac{16-a^2}{a-4}$

13. $\dfrac{3t}{t+3} + \dfrac{9}{t+3}$

14. $\dfrac{3w}{w-5} + \dfrac{w+10}{5-w}$

15. $\dfrac{2}{r} + \dfrac{r}{s}$

16. $\dfrac{x+2}{x+1} - \dfrac{x+1}{x+2}$

In Problems 17–18, simplify each complex fraction.

17. $\dfrac{\dfrac{2u^2w^3}{v^2}}{\dfrac{4uw^4}{uv}}$

18. $\dfrac{\dfrac{x}{y} + \dfrac{1}{2}}{\dfrac{x}{2} - \dfrac{1}{y}}$

In Problems 19–20, solve each equation.

19. $\dfrac{2}{x-1} + \dfrac{5}{x+2} = \dfrac{11}{x+2}$

20. $\dfrac{u-2}{u-3} + 3 = u + \dfrac{u-4}{3-u}$

In Problems 21–22, solve each formula for the indicated variable.

21. $\dfrac{x^2}{a^2} + \dfrac{y^2}{b^2} = 1$ for a^2

22. $\dfrac{1}{r} = \dfrac{1}{r_1} + \dfrac{1}{r_2}$ for r_2

In Problems 23–24, solve each problem.

23. Sailing time   A boat sails a distance of 440 nautical miles. If the boat had averaged 11 nautical miles more each day, the trip would have required 2 fewer days. How long did the trip take?

24. Investing   A student can earn $300 interest annually by investing in a bank certificate of deposit at a certain interest rate. If she were to receive an annual interest rate that is 4% higher, she could receive the same annual interest by investing $2000 less. How much would she be investing at each rate? (*Hint:* 4% is 0.04.)

25. Solve the proportion $\dfrac{3}{x-2} = \dfrac{x+3}{2x}$.

26. V varies inversely with t. If $V = 55$ when $t = 20$, find t when $V = 75$.

27. Divide: $\dfrac{18x^2y^3 - 12x^3y^2 + 9xy}{-3xy^4}$.

28. Divide: $(6x^3 + 5x^2 - 2) \div (2x - 1)$.

29. Find the remainder: $\dfrac{x^3 - 4x^2 + 5x + 3}{x+1}$.

30. Optional   Use synthetic division to find the remainder when $4x^3 + 3x^2 + 2x - 1$ is divided by $x - 2$.

$$\underline{2\,|}\quad 4 \quad 3 \quad 2 \quad -1$$

Cumulative Review Exercises (Chapters 1–6)

In Exercises 1-4, simplify each expression.

1. $a^3b^2a^5b^2$

2. $\dfrac{a^3b^6}{a^7b^2}$

3. $\left(\dfrac{2a^2}{3b^4}\right)^{-4}$

4. $\left(\dfrac{x^{-2}y^3}{x^2x^3y^4}\right)^{-3}$

In Exercises 5-6, write each number in standard notation.

5. 4.25×10^4

6. 7.12×10^{-4}

In Exercises 7–8, solve each equation.

7. $\dfrac{a+2}{5} - \dfrac{8}{5} = 4a - \dfrac{a+9}{2}$

8. $\dfrac{3x-4}{6} - \dfrac{x-2}{2} = \dfrac{-2x-3}{3}$

In Exercises 9–12, find the slope of the line with the given properties.

9. Passing through $P(-2, 5)$ and $Q(4, 10)$

10. Has an equation of $3x + 4y = 13$

11. Parallel to a line with equation of $y = 3x + 2$

12. Perpendicular to a line with equation of $y = 3x + 2$

In Exercises 13–16, let $f(x) = x^2 - 2x$ and find each value.

13. $f(0)$

14. $f(-2)$

15. $f\left(\frac{2}{5}\right)$

16. $f(t - 1)$

17. Express as a formula: y varies directly with the product of x and z, but inversely with r.

18. Does the graph represent a function?

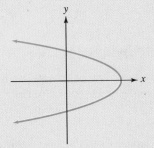

In Exercises 19–20, solve each inequality and graph the solution set.

19. $x - 2 \le 3x + 1 \le 5x - 4$

20. $\left| \dfrac{3a}{5} - 2 \right| + 1 \ge \dfrac{6}{5}$

21. Is $3 + x + x^2$ a monomial, a binomial, or a trinomial?

22. Find the degree of $3 + x^2y + 17x^3y^4$.

23. If $f(x) = -3x^3 + x - 4$, find $f(-2)$.

24. Graph $y = f(x) = 2x^2 - 3$.

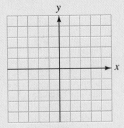

In Exercises 25–28, do the operations and simplify.

25. $(3x^2 - 2x + 7) + (-2x^2 + 2x + 5) + (3x^2 - 4x + 2)$

26. $(-5x^2 + 3x + 4) - (-2x^2 + 3x + 7)$

27. $(3x + 4)(2x - 5)$

28. $(2x^n - 1)(x^n + 2)$

In Exercises 29–40, factor each expression.

29. $3r^2s^3 - 6rs^4$

30. $5(x - y) - a(x - y)$

31. $xu + yv + xv + yu$

32. $81x^4 - 16y^4$

33. $8x^3 - 27y^6$

34. $6x^2 + 5x - 6$

35. $9x^2 - 30x + 25$

36. $15x^2 - x - 6$

37. $27a^3 + 8b^3$

38. $6x^2 + x - 35$

39. $x^2 + 10x + 25 - y^4$

40. $y^2 - x^2 + 4x - 4$

In Exercises 41–42, solve each equation by factoring.

41. $x^3 - 4x = 0$

42. $6x^2 + 7 = -23x$

In Exercises 43–46, simplify each expression.

43. $\dfrac{2x^2y + xy - 6y}{3x^2y + 5xy - 2y}$

44. $\dfrac{x^2 - 4}{x^2 + 9x + 20} \div \dfrac{x^2 + 5x + 6}{x^2 + 4x - 5} \cdot \dfrac{x^2 + 3x - 4}{(x - 1)^2}$

45. $\dfrac{2}{x + y} + \dfrac{3}{x - y} - \dfrac{x - 3y}{x^2 - y^2}$

46. $\dfrac{\dfrac{a}{b} + b}{a - \dfrac{b}{a}}$

In Exercises 47–48, solve each equation.

47. $\dfrac{5x - 3}{x + 2} = \dfrac{5x + 3}{x - 2}$

48. $\dfrac{3}{x - 2} + \dfrac{x^2}{(x + 3)(x - 2)} = \dfrac{x + 4}{x + 3}$

In Exercises 49–50, do the operations.

49. $(x^2 + 9x + 20) \div (x + 5)$

50. $(2x^2 + 4x - x^3 + 3) \div (x - 1)$

Rational Exponents and Radicals

7

Photographer

Photographers use cameras and film to portray people, objects, places, and events. Some specialize in scientific, medical, or engineering photography and provide illustrations and documentation for publications and research reports. Others specialize in portrait, fashion, or industrial photography and provide the pictures for catalogs and other publications. Photojournalists capture newsworthy events, people, and places; their work is seen in newspapers and magazines, as well as on television.

SAMPLE APPLICATION ■ Many camera lenses have an adjustable opening called the *aperture*, which controls the amount of light passing through the lens. Various lenses—wide-angle, close-up, and telephoto—are distinguished by their *focal length*. The *f-number* of a lens is its *focal length* divided by the diameter of its circular aperture:

$$f\text{-number} = \frac{f}{d} \qquad f \text{ is the focal length, and } d \text{ is the diameter of the aperture.}$$

A lens with a focal length of 12 centimeters and an aperture with a diameter of 6 centimeters has an *f*-number of $\frac{12}{6}$ and is called an *f*/2 lens. If the area of the aperture is reduced to admit half as much light, the *f*-number of the lens will change. Find the new *f*-number.
See Section 7.4, Example 5.

To square a number, we raise it to the second power. For example,

- Since $0^2 = 0 \cdot 0 = 0$, the square of 0 is 0.
- Since $4^2 = 4 \cdot 4 = 16$, the square of 4 is 16.
- Since $(-4)^2 = (-4)(-4) = 16$, the square of -4 is also 16.
- Since $(7xy)^2 = (7xy)(7xy) = 49x^2y^2$, the square of $7xy$ is $49x^2y^2$.
- Since $(-7xy)^2 = (-7xy)(-7xy) = 49x^2y^2$, the square of $-7xy$ is also $49x^2y^2$.

In this chapter, we will reverse the squaring process and find square roots of numbers. We will also discuss how to find other roots of numbers.

7.1 Radical Expressions

■ SQUARE ROOTS ■ SQUARE ROOTS OF EXPRESSIONS WITH VARIABLES ■ THE SQUARE ROOT FUNCTION ■ GRAPHING DEVICES ■ CUBE ROOTS ■ NTH ROOTS ■ ACCENT ON STATISTICS

■ SQUARE ROOTS

In application problems, we must often find what number can be squared to obtain a second number a. If such a number can be found, it is called a *square root of a*. For example,

- 0 is a square root of 0, because $0^2 = 0$.
- 4 is a square root of 16, because $4^2 = 16$.
- -4 is a square root of 16, because $(-4)^2 = 16$.
- $7xy$ is a square root of $49x^2y^2$, because $(7xy)^2 = 49x^2y^2$.
- $-7xy$ is a square root of $49x^2y^2$, because $(-7xy)^2 = 49x^2y^2$.

The preceding examples illustrate the following definition.

> **Square Root of a**
> The number b is a **square root of** a when $b^2 = a$.

All positive numbers have two real-number square roots, one that is positive and one that is negative. The only number with a single square root is 0, whose square root is 0.

EXAMPLE 1   Find the two square roots of 121.

Solution   The two square roots of 121 are 11 and -11, because

$$11^2 = 121 \quad \text{and} \quad (-11)^2 = 121$$ ■

In the following definition, the symbol $\sqrt{}$ is called the **radical sign**, and the number x under the radical sign is called the **radicand**.

> **Principal Square Root**
> If $x > 0$, the **principal square root of** x is the positive square root of x, denoted as $\sqrt{x}$.
>
> The principal square root of 0 is 0: $\sqrt{0} = 0$.

By definition, the principal square root of a positive number is always positive. Although 5 and -5 are both square roots of 25, only 5 is the principal square root. The radical $\sqrt{25}$ represents 5. The radical $-\sqrt{25}$ represents -5.

(b, c, g)   EXAMPLE 2

a. $\sqrt{1} = 1$ **b.** $\sqrt{81} = 9$ **c.** $-\sqrt{81} = -9$

d. $-\sqrt{225} = -15$ **e.** $\sqrt{\dfrac{1}{4}} = \dfrac{1}{2}$ **f.** $-\sqrt{\dfrac{16}{121}} = -\dfrac{4}{11}$

g. $\sqrt{0.04} = 0.2$ **h.** $-\sqrt{0.0009} = -0.03$ ∎

Numbers such as 4, 9, 16, 49, and 1600 are *integer squares*, because each number is the square of an integer. The square root of every integer square is a rational number.

$$\sqrt{4} = 2, \quad \sqrt{9} = 3, \quad \sqrt{16} = 4, \quad \sqrt{49} = 7, \quad \sqrt{1600} = 40$$

The square roots of many positive integers are not rational numbers. For example, $\sqrt{11}$ is an *irrational number*. To find an approximate value of $\sqrt{11}$, we enter 11 into a calculator and press the $\sqrt{}$ key.

$$\sqrt{11} \approx 3.3166247 \qquad \text{Read} \approx \text{as "is approximately equal to."}$$

Square roots of negative numbers are not real numbers. For example, $\sqrt{-9}$ is not a real number, because no real number squared equals -9. Square roots of negative numbers come from a set called the *imaginary numbers*, which we will discuss in the next chapter.

■ SQUARE ROOTS OF EXPRESSIONS WITH VARIABLES

If $x \neq 0$, the positive number x^2 has x and $-x$ for its two square roots, because $(x)^2 = x^2$ and $(-x)^2 = x^2$. To denote the positive square root of $\sqrt{x^2}$, we must know whether x is positive or negative.

If $x > 0$, we can write

$$\sqrt{x^2} = x \qquad \sqrt{x^2} \text{ represents the positive square root of } x^2, \text{ which is } x.$$

If x is negative, then $-x > 0$, and we can write

$$\sqrt{x^2} = -x \qquad \sqrt{x^2} \text{ represents the positive square root of } x^2, \text{ which is } -x.$$

If we don't know whether x is positive or negative, we can use absolute value symbols to guarantee that $\sqrt{x^2}$ is positive.

> **Definition of $\sqrt{x^2}$**
> If x can be any real number, then
> $$\sqrt{x^2} = |x|$$

 (a, c)   EXAMPLE 3

If x can be any real number, then we have

a. $\sqrt{16x^2} = \sqrt{(4x)^2}$ Write $16x^2$ as $(4x)^2$.

$\phantom{\sqrt{16x^2}} = |4x|$ Because $(|4x|)^2 = 16x^2$. Since x could be negative, the absolute value symbols are necessary.

$\phantom{\sqrt{16x^2}} = 4|x|$ Since 4 is a positive constant in the product $4x$, we can write it outside the absolute value symbols.

b. $\sqrt{(x+4)^2} = |x+4|$

Because $|x+4|^2 = (x+4)^2$. Since x can be any real number, $x+4$ can be negative (for example, when $x = -8$), so the absolute value symbols are necessary.

c. $\sqrt{x^2 + 2x + 1} = \sqrt{(x+1)^2}$ Factor $x^2 + 2x + 1$.

$\qquad\qquad\qquad = |x+1|$

Because $|x+1|^2 = (x+1)^2$. Since x can be any real number, $x+1$ can be negative (for example, when $x = -5$), so the absolute value symbols are necessary.

d. $\sqrt{x^4} = x^2$

Because $(x^2)^2 = x^4$. Since $x^2 \geq 0$, no absolute value symbols are necessary. ∎

■ THE SQUARE ROOT FUNCTION

Since there is only one principal square root for every nonnegative real number x, the equation $y = f(x) = \sqrt{x}$ determines a *square root function*.

EXAMPLE 4   Graph the function $y = f(x) = \sqrt{x}$ and find its domain and range.

Solution   We can make a table of values and plot points to get the graph shown in Figure 7-1(a), or we can use a graphing device with window settings of $[-2, 8]$ for x and $[-2, 8]$ for y to get the graph shown in Figure 7-1(b). Since the equation defines a function, its graph passes the vertical line test.

From either graph, we can see that both the domain and the range are the interval $[0, \infty)$.

$y = f(x) = \sqrt{x}$

x	$f(x)$	$(x, f(x))$
0	0	$(0, 0)$
1	1	$(1, 1)$
4	2	$(4, 2)$
9	3	$(9, 3)$
16	4	$(16, 4)$

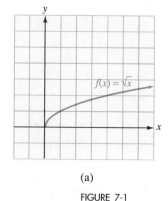

(a)

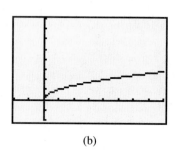

(b)

FIGURE 7-1   ∎

In general, we have the following definition.

> **Square Root Functions**
> A **square root function** is any function of the form
> $$y = f(x) = \sqrt{u} \quad (u \geq 0)$$
> where u is an algebraic expression in x.

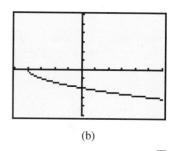

EXAMPLE 5 Graph the function $y = f(x) = -\sqrt{x + 4}$ and find its domain and range.

Solution We can make a table of values and plot points to get the graph shown in Figure 7-2(a), or we can use a graphing device with window settings of $[-5, 6]$ for x and $[-5, 6]$ for y to get the graph shown in Figure 7-2(b).

From either graph, we can see that the domain is the interval $[-4, \infty)$, and that the range is the interval $(-\infty, 0]$.

$y = f(x) = -\sqrt{x + 4}$

x	$f(x)$	$(x, f(x))$
-4	0	$(-4, 0)$
-3	-1	$(-3, -1)$
0	-2	$(0, -2)$
5	-3	$(5, -3)$
12	-4	$(12, -4)$

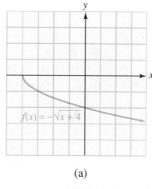

(a)   (b)

FIGURE 7-2 ∎

EXAMPLE 6 **Period of a pendulum** The *period* of a pendulum is the time required for the pendulum to swing back and forth to complete one cycle. (See Figure 7-3.) The period t (in seconds) is a function of the pendulum's length l, which is defined by the formula

$$t = f(l) = 1.11\sqrt{l}$$

Find the period of a pendulum that is 5 feet long.

Solution We substitute 5 for l in the formula and simplify.

$$t = 1.11\sqrt{l}$$
$$t = 1.11\sqrt{5}$$
$$= 2.482035455$$

The period is approximately 2.5 seconds.

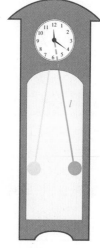

FIGURE 7-3 ∎

■ GRAPHING DEVICES

To solve Example 6 with a graphing device with window settings of $[-2, 10]$ for x and $[-2, 10]$ for y, we graph the function $f(x) = 1.11\sqrt{x}$ as in Figure 7-4(a). We then trace, and move the cursor until we see the coordinates shown in Figure 7-4(b). By zooming in, we can get better results.

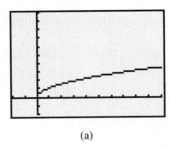

(a)

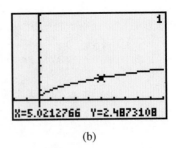

(b)

FIGURE 7-4

■ CUBE ROOTS

The **cube root of x** is any number whose cube is x. For example,

4 is a cube root of 64, because $4^3 = 64$.

$3x^2y$ is a cube root of $27x^6y^3$, because $(3x^2y)^3 = 27x^6y^3$.

$-2y$ is a cube root of $-8y^3$, because $(-2y)^3 = -8y^3$.

Cube Roots

The **cube root of x** is denoted as $\sqrt[3]{x}$, and is defined by

$$\sqrt[3]{x} = y \qquad \text{if} \qquad y^3 = x$$

We note that 64 has two real-number square roots, 8 and -8. However, 64 has only one real-number cube root, 4, because 4 is the only real number whose cube is 64.

Definition of $\sqrt[3]{x^3}$

$$\sqrt[3]{x^3} = x$$

Since all real numbers have exactly one real cube root, some of which are positive and some of which are negative, it is not necessary to use absolute value symbols when simplifying cube roots.

◉◉ (b, d)　　**EXAMPLE 7**

a. $\sqrt[3]{125} = 5$ 　　　 Because $5^3 = 5 \cdot 5 \cdot 5 = 125$.

b. $\sqrt[3]{\dfrac{1}{8}} = \dfrac{1}{2}$ 　　　 Because $\left(\dfrac{1}{2}\right)^3 = \dfrac{1}{2} \cdot \dfrac{1}{2} \cdot \dfrac{1}{2} = \dfrac{1}{8}$.

c. $\sqrt[3]{-27x^3} = -3x$ 　　　 Because $(-3x)^3 = (-3x)(-3x)(-3x) = -27x^3$.

d. $\sqrt[3]{-\dfrac{8a^3}{27b^3}} = -\dfrac{2a}{3b}$ 　　　 Because $\left(-\dfrac{2a}{3b}\right)^3 = \left(-\dfrac{2a}{3b}\right)\left(-\dfrac{2a}{3b}\right)\left(-\dfrac{2a}{3b}\right)$

$$= -\dfrac{8a^3}{27b^3}.$$

e. $\sqrt[3]{0.216x^3y^6} = 0.6xy^2$ 　　　 Because $(0.6xy^2)^3 = (0.6xy^2)(0.6xy^2)(0.6xy^2)$

$$= 0.216x^3y^6.$$ ■

The equation $y = f(x) = \sqrt[3]{x}$ defines a function. From the graphs shown in Figure 7-5, we can see that the domain of the function $f(x) = \sqrt[3]{x}$ is the interval $(-\infty, \infty)$ and the range is also the interval $(-\infty, \infty)$. Note that the graph of $f(x) = \sqrt[3]{x}$ passes the vertical line test.

$y = f(x) = \sqrt[3]{x}$

x	$f(x)$	$(x, f(x))$
-8	-2	$(-8, -2)$
-1	-1	$(-1, -1)$
0	0	$(0, 0)$
1	1	$(1, 1)$
8	2	$(8, 2)$

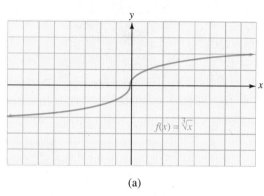

(a)

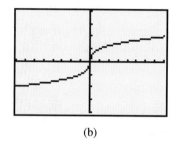

(b)

FIGURE 7-5

■ NTH ROOTS

Just as there are square roots and cube roots, there are fourth roots, fifth roots, sixth roots, and so on.

When n is odd, the radical $\sqrt[n]{x}$ ($n > 1$) represents an **odd root**. Since every real number has just one real nth root when n is odd, we don't need to worry about absolute value symbols when finding odd roots. For example,

$$\sqrt[5]{243} = \sqrt[5]{3^5} = 3 \qquad \text{because} \qquad 3^5 = 243$$

$$\sqrt[7]{-128x^7} = \sqrt[7]{(-2x)^7} = -2x \qquad \text{because} \qquad (-2x)^7 = -128x^7$$

When n is even, the radical $\sqrt[n]{x}$ ($n > 1$) represents an **even root**. In this case, there will be one positive and one negative real nth root. For example, the two real sixth roots of 729 are 3 and -3, because $3^6 = 729$ and $(-3)^6 = 729$. When finding even roots, we often use absolute value symbols to guarantee that the nth root is positive.

$$\sqrt[4]{(-3)^4} = |-3| = 3$$

We could also simplify this as follows: $\sqrt[4]{(-3)^4} = \sqrt[4]{81} = 3$.

$$\sqrt[6]{729x^6} = \sqrt[6]{(3x)^6} = |3x| = 3|x|$$

The absolute value symbols guarantee that the sixth root is positive.

In general, we have the following rules.

> **Rules for $\sqrt[n]{x^n}$**
>
> If x is a real number and $n > 1$, then
>
> If n is an odd natural number, then $\sqrt[n]{x^n} = x$.
>
> If n is an even natural number, then $\sqrt[n]{x^n} = |x|$.

In the radical $\sqrt[n]{x}$, n is called the **index** (or **order**) of the radical. When the index is 2, the radical is a square root, and we usually do not write the index.

WARNING! When n is even ($n > 1$) and $x < 0$, the radical $\sqrt[n]{x}$ is not a real number. For example, $\sqrt[4]{-81}$ is not a real number, because no real number raised to the 4th power is -81.

 (a, b)

EXAMPLE 8

a. $\sqrt[4]{625} = 5$, because $5^4 = 625$ Read $\sqrt[4]{625}$ as "the fourth root of 625."

b. $\sqrt[5]{-32} = -2$, because $(-2)^5 = -32$ Read $\sqrt[5]{-32}$ as "the fifth root of -32."

c. $\sqrt[6]{\dfrac{1}{64}} = \dfrac{1}{2}$, because $\left(\dfrac{1}{2}\right)^6 = \dfrac{1}{64}$ Read $\sqrt[6]{\dfrac{1}{64}}$ as "the sixth root of $\dfrac{1}{64}$."

d. $\sqrt[7]{10^7} = 10$, because $10^7 = 10^7$ Read $\sqrt[7]{10^7}$ as "the seventh root of 10^7." ∎

(b, e)

EXAMPLE 9 Simplify each radical. Assume that x can be any real number.

Solution   **a.** $\sqrt[5]{x^5} = x$ Since n is odd, there is no need for absolute value symbols.

b. $\sqrt[4]{16x^4} = |2x| = 2|x|$ Since n is even and x can be negative, absolute value symbols are necessary to guarantee that the simplified result is positive.

c. $\sqrt[6]{(x+4)^6} = |x+4|$ Since n is even and $x + 4$ can be negative (for example, when $x = -5$), absolute value symbols are necessary.

d. $\sqrt[3]{(x+1)^3} = x + 1$ Since n is odd, there is no need for absolute value symbols.

e. $\sqrt{(x^2 + 4x + 4)^2} = \sqrt{[(x+2)^2]^2}$ Factor $x^2 + 4x + 4$.

$\phantom{\sqrt{(x^2 + 4x + 4)^2}} = \sqrt{(x+2)^4}$

$\phantom{\sqrt{(x^2 + 4x + 4)^2}} = (x+2)^2$ Since $(x+2)^2$ is always positive, no absolute value symbols are necessary. ∎

We summarize the definitions concerning $\sqrt[n]{x}$ as follows.

> **Summary of the Definitions of $\sqrt[n]{x}$**
>
> If n is a natural number greater than 1 and x is a real number, then
>
>   If $x > 0$, then $\sqrt[n]{x}$ is the positive number such that $\left(\sqrt[n]{x}\right)^n = x$.
>
>   If $x = 0$, then $\sqrt[n]{x} = 0$.
>
>   If $x < 0$ $\begin{cases} \text{and } n \text{ is odd, then } \sqrt[n]{x} \text{ is the real number such that } \left(\sqrt[n]{x}\right)^n = x. \\ \text{and } n \text{ is even, then } \sqrt[n]{x} \text{ is not a real number.} \end{cases}$

■ ■ ■ ■ ■ ■ ■ ■ ■ ■   **The Standard Deviation**

ACCENT ON
STATISTICS   In statistics, the *standard deviation* is used to tell which of a set of distributions has the most variability. To see how to compute the standard deviation of a distribution, we consider the distribution 4, 5, 5, 8, 13 and construct the following table.

Original term	Mean of the distribution	Difference (Original term minus mean)	Square of the difference from the mean
4	7	−3	9
5	7	−2	4
5	7	−2	4
8	7	1	1
13	7	6	36

The **standard deviation** of the distribution is the positive square root of the mean of the numbers shown in the rightmost column of the previous table.

$$\text{Standard deviation} = \sqrt{\frac{\text{sum of the squares of the differences from the mean}}{\text{number of differences}}}$$

$$= \sqrt{\frac{9 + 4 + 4 + 1 + 36}{5}}$$

$$= \sqrt{\frac{54}{5}}$$

$$\approx 3.2863353$$

The standard deviation of the given distribution is approximately 3.29.

The symbol for standard deviation is σ, the lowercase Greek letter sigma.

EXAMPLE 10

Which of the following distributions has the most variability?
a. 3, 5, 7, 8, 12   and   **b.** 1, 4, 6, 11.

Solution   We compute the standard deviation of each distribution.

a.

Original term	Mean of the distribution	Difference (Original term minus mean)	Square of the difference from the mean
3	7	−4	16
5	7	−2	4
7	7	0	0
8	7	1	1
12	7	5	25

$$\sigma = \sqrt{\frac{16 + 4 + 0 + 1 + 25}{5}} = \sqrt{\frac{46}{5}} \approx 3.03$$

b.

Original term	Mean of the distribution	Difference (Original term minus mean)	Square of the difference from the mean
1	5.5	−4.5	20.25
4	5.5	−1.5	2.25
6	5.5	0.5	0.25
11	5.5	5.5	30.25

$$\sigma = \sqrt{\frac{20.25 + 2.25 + 0.25 + 30.25}{4}} = \sqrt{\frac{53}{4}} \approx 3.64$$

Since the standard deviation for the second distribution is greater than the standard deviation for the first distribution, the second distribution has the greater variability. ■

Orals *Simplify each radical, if possible.*

1. $\sqrt{9}$ **2.** $-\sqrt{16}$ **3.** $\sqrt[3]{-8}$ **4.** $\sqrt[5]{32}$

5. $\sqrt{64x^2}$ **6.** $\sqrt[3]{-27x^3}$

7. $\sqrt{-3}$ **8.** $\sqrt[4]{(x+1)^8}$

EXERCISE 7.1

In Exercises 1–4, identify the radicand in each expression.

1. $\sqrt{3x^2}$ **2.** $5\sqrt{x}$ **3.** $ab^2\sqrt{a^2+b^3}$ **4.** $\frac{1}{2}x\sqrt{\frac{x}{y}}$

In Exercises 5–20, find each square root, if possible.

5. $\sqrt{121}$ **6.** $\sqrt{144}$ **7.** $-\sqrt{64}$ **8.** $-\sqrt{1}$

9. $\sqrt{\frac{1}{9}}$ **10.** $-\sqrt{\frac{4}{25}}$ **11.** $-\sqrt{\frac{25}{49}}$ **12.** $\sqrt{\frac{49}{81}}$

13. $\sqrt{-25}$ **14.** $\sqrt{0.25}$ **15.** $\sqrt{0.16}$ **16.** $\sqrt{-49}$

17. $\sqrt{(-4)^2}$ **18.** $\sqrt{(-9)^2}$ **19.** $\sqrt{-36}$ **20.** $-\sqrt{-4}$

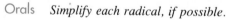 *In Exercises 21–24, use a calculator to find each square root. Give each answer to four decimal places.*

21. $\sqrt{12}$ **22.** $\sqrt{340}$ **23.** $\sqrt{679.25}$ **24.** $\sqrt{0.0063}$

In Exercises 25–32, find each square root. Assume that all variables are unrestricted, and use absolute value symbols when necessary.

25. $\sqrt{4x^2}$ **26.** $\sqrt{16y^4}$ **27.** $\sqrt{(t+5)^2}$ **28.** $\sqrt{(a+6)^2}$

29. $\sqrt{(-5b)^2}$ **30.** $\sqrt{(-8c)^2}$ **31.** $\sqrt{a^2+6a+9}$ **32.** $\sqrt{x^2+10x+25}$

In Exercises 33–36, find each value, given that $f(x) = \sqrt{x-4}$.

33. $f(4)$ **34.** $f(8)$ **35.** $f(20)$ **36.** $f(29)$

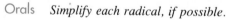 *In Exercises 37–40, find each value, given that $f(x) = \sqrt{x^2+1}$. Give each answer to four decimal places.*

37. $f(4)$ **38.** $f(6)$ **39.** $f(2.35)$ **40.** $f(21.57)$

In Exercises 41–44, graph each square root function and find its domain and range.

41. $f(x) = \sqrt{x+4}$
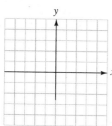

42. $f(x) = -\sqrt{x-2}$
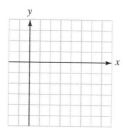

43. $f(x) = 3\sqrt{x}$
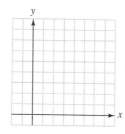

44. $f(x) = -2\sqrt{x}$
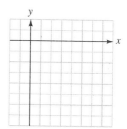

In Exercises 45–50, use a calculator.

45. Radius of a circle The radius r of a circle is given by the formula $r = \sqrt{\frac{A}{\pi}}$, where A is its area. Find the radius of a circle whose area is 9π square units.

46. Diagonal of a baseball diamond The diagonal d of a square is given by the formula $d = \sqrt{2s^2}$, where s is the length of each side. To the nearest hundredth, find the diagonal of the baseball diamond shown in Illustration 1.

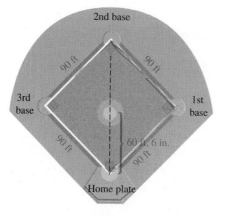

ILLUSTRATION 1

47. Falling objects The time t (in seconds) that it will take for an object to fall a distance of s feet is given

by the formula $t = \sqrt{s}/4$. If a stone is dropped down a 256-foot well, how long will it take it to hit bottom?

48. Law enforcement The police sometimes use the formula $s = k\sqrt{l}$ to estimate the speed s (in mph) of a car involved in an accident. In this formula, l is the length of the skid in feet, and k is a constant related to the condition of the pavement. For wet pavement, $k \approx 3.24$. How fast was a car going if its skid was 400 feet on wet pavement?

49. Electronics When the resistance in a circuit is 18 ohms, the current I (measured in amperes) and the power P (measured in watts) are related by the formula

$$I = \sqrt{\frac{P}{18}}$$

Find the current used by an electrical appliance that is rated at 980 watts.

50. Medicine The approximate pulse rate p (in beats per minute) of an adult who is t inches tall is given by the formula

$$p = \frac{590}{\sqrt{t}}$$

Find the approximate pulse rate of an adult who is 71 inches tall.

In Exercises 51–66, simplify each cube root.

51. $\sqrt[3]{1}$

52. $\sqrt[3]{-8}$

53. $\sqrt[3]{-125}$

54. $\sqrt[3]{512}$

55. $\sqrt[3]{-\dfrac{8}{27}}$

56. $\sqrt[3]{\dfrac{125}{216}}$

57. $\sqrt[3]{0.064}$

58. $\sqrt[3]{0.001}$

59. $\sqrt[3]{8a^3}$

60. $\sqrt[3]{-27x^6}$

61. $\sqrt[3]{-1000p^3q^3}$

62. $\sqrt[3]{343a^6b^3}$

63. $\sqrt[3]{-\dfrac{1}{8}m^6n^3}$

64. $\sqrt[3]{\dfrac{27}{1000}a^6b^6}$

65. $\sqrt[3]{0.008z^9}$

66. $\sqrt[3]{0.064s^9t^6}$

In Exercises 67–78, simplify each radical, if possible.

67. $\sqrt[4]{81}$

68. $\sqrt[6]{64}$

69. $-\sqrt[5]{243}$

70. $-\sqrt[4]{625}$

71. $\sqrt[5]{-32}$

72. $\sqrt[6]{729}$

73. $\sqrt[4]{\dfrac{16}{625}}$

74. $\sqrt[5]{-\dfrac{243}{32}}$

75. $-\sqrt[5]{-\dfrac{1}{32}}$

76. $\sqrt[6]{-729}$

77. $\sqrt[4]{-256}$

78. $-\sqrt[4]{\dfrac{81}{256}}$

In Exercises 79–90, simplify each radical. Assume that all variables are unrestricted, and use absolute value symbols where necessary.

79. $\sqrt[4]{16x^4}$

80. $\sqrt[5]{32a^5}$

81. $\sqrt[3]{8a^3}$

82. $\sqrt[6]{64x^6}$

83. $\sqrt[4]{\dfrac{1}{16}x^4}$

84. $\sqrt[4]{\dfrac{1}{81}x^8}$

85. $\sqrt[4]{x^{12}}$

86. $\sqrt[8]{x^{24}}$

87. $\sqrt[5]{-x^5}$

88. $\sqrt[3]{-x^6}$

89. $\sqrt[3]{-27a^6}$

90. $\sqrt[5]{-32x^5}$

In Exercises 91–94, simplify each radical. Assume that all variables are unrestricted, and use absolute value symbols when necessary.

91. $\sqrt[25]{(x+2)^{25}}$

92. $\sqrt[44]{(x+4)^{44}}$

93. $\sqrt[8]{0.00000001x^{16}y^8}$

94. $\sqrt[5]{0.00032x^{10}y^5}$

95. Find the standard deviation of the following distribution to the nearest hundredth: 2, 5, 5, 6, 7.

96. Find the standard deviation of the following distribution to the nearest hundredth: 3, 6, 7, 9, 11, 12.

97. Statistics In statistics, the formula

$$s_{\bar{x}} = \frac{s}{\sqrt{N}}$$

gives an estimate of the standard error of the mean. Find $s_{\bar{x}}$ to four decimal places when $s = 65$ and $N = 30$.

98. Statistics In statistics, the formula

$$\sigma_{\bar{x}} = \frac{\sigma}{\sqrt{N}}$$

gives the standard deviation of means of samples of size N. Find $\sigma_{\bar{x}}$ to four decimal places when $\sigma = 12.7$ and $N = 32$.

Writing Exercises *Write a paragraph using your own words.*

1. If x is any real number, then $\sqrt{x^2} = x$ is not correct. Explain.

2. If x is any real number, then $\sqrt[3]{x^3} = |x|$ is not correct. Explain.

Something to Think About

1. Is $\sqrt{x^2 - 4x + 4} = x - 2$? What are the exceptions?

2. When is $\sqrt{x^2} \neq x$?

Review Exercises *Simplify each fraction.*

1. $\dfrac{x^2 + 7x + 12}{x^2 - 16}$

2. $\dfrac{a^3 - b^3}{b^2 - a^2}$

Do the operations.

3. $\dfrac{x^2 - x - 6}{x^2 - 2x - 3} \cdot \dfrac{x^2 - 1}{x^2 + x - 2}$

4. $\dfrac{x^2 - 3x - 4}{x^2 - 5x + 6} \div \dfrac{x^2 - 2x - 3}{x^2 - x - 2}$

5. $\dfrac{3}{m + 1} + \dfrac{3m}{m - 1}$

6. $\dfrac{2x + 3}{3x - 1} - \dfrac{x - 4}{2x + 1}$

7.2 Rational Exponents

■ EXPONENTIAL EXPRESSIONS WITH VARIABLES IN THEIR BASES ■ FRACTIONAL EXPONENTS WITH NUMERATORS OTHER THAN 1 ■ NEGATIVE FRACTIONAL EXPONENTS ■ SIMPLIFYING RADICAL EXPRESSIONS

We have seen that positive integer exponents indicate the number of times that a base is to be used as a factor in a product. For example, x^5 means that x is to be used as a factor five times.

$$x^5 = \overbrace{x \cdot x \cdot x \cdot x \cdot x}^{5 \text{ factors of } x}$$

Furthermore, we recall the following properties of exponents from Section 1.3.

Rules of Exponents

If there are no divisions by 0, then for all integers m and n

1. $x^m x^n = x^{m+n}$

2. $(x^m)^n = x^{mn}$

3. $(xy)^n = x^n y^n$

4. $\left(\dfrac{x}{y}\right)^n = \dfrac{x^n}{y^n}$

5. $x^0 = 1 \; (x \neq 0)$

6. $x^{-n} = \dfrac{1}{x^n}$

7. $\dfrac{x^m}{x^n} = x^{m-n}$

8. $\left(\dfrac{x}{y}\right)^{-n} = \left(\dfrac{y}{x}\right)^n$

It is possible to raise many bases to fractional powers. Since we want fractional exponents to obey the same rules as integer exponents, the square of $10^{1/2}$ must be 10, because

$$\left(10^{1/2}\right)^2 = 10^{(1/2)2} \qquad \text{Keep the base and multiply the exponents.}$$
$$= 10^1 \qquad \tfrac{1}{2} \cdot 2 = \tfrac{1}{2} \cdot \tfrac{2}{1} = 1.$$
$$= 10 \qquad 10^1 = 10.$$

However, in the last section, we saw that

$$\left(\sqrt{10}\right)^2 = 10$$

Since $\left(10^{1/2}\right)^2$ and $\left(\sqrt{10}\right)^2$ both equal 10, we define $10^{1/2}$ to be $\sqrt{10}$. Likewise, we define

$$10^{1/3} \text{ to be } \sqrt[3]{10} \qquad \text{and} \qquad 10^{1/4} \text{ to be } \sqrt[4]{10}$$

> **Rational Exponents**
> If n is a natural number greater than 1 and $\sqrt[n]{x}$ is a real number, then
> $$x^{1/n} = \sqrt[n]{x}$$

(a, c, e)   **EXAMPLE 1**

a. $9^{1/2} = \sqrt{9} = 3$

b. $-\left(\dfrac{16}{9}\right)^{1/2} = -\sqrt{\dfrac{16}{9}} = -\dfrac{4}{3}$

c. $(-64)^{1/3} = \sqrt[3]{-64} = -4$

d. $16^{1/4} = \sqrt[4]{16} = 2$

e. $\left(\dfrac{1}{32}\right)^{1/5} = \sqrt[5]{\dfrac{1}{32}} = \dfrac{1}{2}$

f. $0^{1/8} = \sqrt[8]{0} = 0$

g. $-(32x^5)^{1/5} = -\sqrt[5]{32x^5} = -2x$

h. $(xyz)^{1/4} = \sqrt[4]{xyz}$ ■

EXAMPLE 2   Write each radical as an expression with a fractional exponent:

a. $\sqrt[4]{5xyz}$ and

b. $\sqrt[5]{\dfrac{xy^2}{15}}$.

Solution   a. $\sqrt[4]{5xyz} = (5xyz)^{1/4}$

b. $\sqrt[5]{\dfrac{xy^2}{15}} = \left(\dfrac{xy^2}{15}\right)^{1/5}$ ■

■ EXPONENTIAL EXPRESSIONS WITH VARIABLES IN THEIR BASES

As with radicals, when n is odd in the expression $x^{1/n}$ ($n > 1$), there is exactly one real nth root, and we don't have to worry about absolute value symbols.

However, when n is even, there are two real nth roots. Since we want the expression $x^{1/n}$ to represent the positive nth root, we must often use absolute value symbols to guarantee that the simplified result is positive. Thus, if n is even,

$$(x^n)^{1/n} = |x|$$

When n is even and x is negative, the expression $x^{1/n}$ is not a real number.

(a, c, e)   **EXAMPLE 3**   Assume that all variables can be any real number and simplify each expression.

a. $(-27x^3)^{1/3} = -3x$

Because $(-3x)^3 = -27x^3$. Since n is odd, no absolute value symbols are necessary.

b. $(49x^2)^{1/2} = |7x| = 7|x|$

Because $(7|x|)^2 = 49x^2$. Since x can be any real number, $7x$ can be negative. Thus, absolute value symbols are necessary.

c. $(256a^8)^{1/8} = 2|a|$

Because $(2|a|)^8 = 256a^8$. Since a can be any real number, $2a$ can be negative. Thus, absolute value symbols are necessary.

d. $\left[(y+4)^2\right]^{1/2} = |y+4|$

Because $(|y+4|)^2 = (y+4)^2$. Since y can be any real number, $y + 4$ can be negative, and the absolute value symbols are necessary.

e. $(25b^4)^{1/2} = 5b^2$

Because $(5b^2)^2 = 25b^4$. Since $b^2 \geq 0$, no absolute value symbols are necessary.

f. $(-256x^4)^{1/4}$ is not a real number.

Because no real number raised to the 4th power is $-256x^4$. ∎

We summarize the cases as follows.

Summary of the Definitions of $x^{1/n}$

If n is a natural number greater than 1 and x is a real number, then

If $x > 0$, then $x^{1/n}$ is the positive number such that $(x^{1/n})^n = x$.

If $x = 0$, then $x^{1/n} = 0$.

If $x < 0$ $\begin{cases} \text{and } n \text{ is odd, then } x^{1/n} \text{ is the real number such that } (x^{1/n})^n = x. \\ \text{and } n \text{ is even, then } x^{1/n} \text{ is not a real number.} \end{cases}$

■ FRACTIONAL EXPONENTS WITH NUMERATORS OTHER THAN 1

We can extend the definition of $x^{1/n}$ to include fractional exponents with numerators other than 1. For example, since $4^{3/2}$ can be written as $(4^{1/2})^3$, we have

$$4^{3/2} = (4^{1/2})^3 = (\sqrt{4})^3 = 2^3 = 8$$

Thus, we can simplify $4^{3/2}$ by cubing the square root of 4. We can also simplify $4^{3/2}$ by taking the square root of 4 cubed.

$$4^{3/2} = (4^3)^{1/2} = 64^{1/2} = \sqrt{64} = 8$$

In general, we have the following rule.

Changing from Rational Exponents to Radicals

If m and n are positive integers, $x \geq 0$, and $\frac{m}{n}$ is in simplified form, then

$$x^{m/n} = \sqrt[n]{x^m} = (\sqrt[n]{x})^m$$

Because of the previous definition, we can interpret $x^{m/n}$ in two ways:

1. $x^{m/n}$ means the nth root of the mth power of x.

2. $x^{m/n}$ means the mth power of the nth root of x.

EXAMPLE 4   **a.** $27^{2/3} = (\sqrt[3]{27})^2$ or $27^{2/3} = \sqrt[3]{27^2}$

$\qquad\qquad\qquad = 3^2 \qquad\qquad\qquad\qquad = \sqrt[3]{729}$

$\qquad\qquad\qquad = 9 \qquad\qquad\qquad\qquad\quad = 9$

b. $\left(\dfrac{1}{16}\right)^{3/4} = \left(\sqrt[4]{\dfrac{1}{16}}\right)^3$ or $\left(\dfrac{1}{16}\right)^{3/4} = \sqrt[4]{\left(\dfrac{1}{16}\right)^3}$

$$= \left(\dfrac{1}{2}\right)^3 \qquad\qquad = \sqrt[4]{\dfrac{1}{4096}}$$

$$= \dfrac{1}{8} \qquad\qquad\qquad = \dfrac{1}{8}$$

c. $(-8x^3)^{4/3} = \left(\sqrt[3]{-8x^3}\right)^4$ or $(-8x^3)^{4/3} = \sqrt[3]{(-8x^3)^4}$

$$= (-2x)^4 \qquad\qquad = \sqrt[3]{4096x^{12}}$$

$$= 16x^4 \qquad\qquad\quad = 16x^4 \qquad\qquad ■$$

To avoid large numbers, it is usually better to find the root of the base first, as shown in Example 4.

■ NEGATIVE FRACTIONAL EXPONENTS

To be consistent with the definition of negative integer exponents, we define $x^{-m/n}$ as follows.

Definition of $x^{-m/n}$

If m and n are positive integers, $\dfrac{m}{n}$ is in simplified form, and $x^{1/n}$ is a real number, then

$$x^{-m/n} = \dfrac{1}{x^{m/n}} \qquad \text{and} \qquad \dfrac{1}{x^{-m/n}} = x^{m/n} \quad (x \neq 0)$$

 (b)   **EXAMPLE 5**   **a.** $64^{-1/2} = \dfrac{1}{64^{1/2}}$ **b.** $16^{-3/2} = \dfrac{1}{16^{3/2}}$

$$= \dfrac{1}{8} \qquad\qquad\qquad\qquad\qquad = \dfrac{1}{\left(16^{1/2}\right)^3}$$

$$= \dfrac{1}{4^3}$$

$$= \dfrac{1}{64}$$

c. $(-32x^5)^{-2/5} = \dfrac{1}{(-32x^5)^{2/5}}$ $(x \neq 0)$ **d.** $(-16)^{-3/4}$ is not a real number, because $(-16)^{1/4}$ is not a real number.

$$= \dfrac{1}{\left[(-32x^5)^{1/5}\right]^2}$$

$$= \dfrac{1}{(-2x)^2}$$

$$= \dfrac{1}{4x^2} \qquad\qquad\qquad\qquad\qquad\qquad\qquad ■$$

WARNING! By definition, 0^0 is undefined. A base of 0 raised to a negative power is also undefined, since 0^{-2} would equal $\frac{1}{0^2}$, which is undefined because we cannot divide by 0.

We can use the laws of exponents to simplify many expressions with fractional exponents. If all variables represent positive numbers, no absolute value symbols are necessary.

(a, b, c)   EXAMPLE 6   Assume that all variables represent positive numbers. Write all answers without using negative exponents.

a. $5^{2/7}5^{3/7} = 5^{2/7 + 3/7}$ Use the rule $x^m x^n = x^{m+n}$.

$= 5^{5/7}$ Add: $\frac{2}{7} + \frac{3}{7} = \frac{5}{7}$.

b. $\left(5^{2/7}\right)^3 = 5^{(2/7)(3)}$ Use the rule $(x^m)^n = x^{mn}$.

$= 5^{6/7}$ Multiply: $\frac{2}{7}(3) = \frac{6}{7}$.

c. $\left(a^{2/3}b^{1/2}\right)^6 = \left(a^{2/3}\right)^6\left(b^{1/2}\right)^6$ Use the rule $(xy)^n = x^n y^n$.

$= a^{12/3}b^{6/2}$ Use the rule $(x^m)^n = x^{mn}$ twice.

$= a^4 b^3$ Simplify each exponent.

d. $\dfrac{a^{8/3}a^{1/3}}{a^2} = a^{8/3 + 1/3 - 2}$ Use the rules $x^m x^n = x^{m+n}$ and $\dfrac{x^m}{x^n} = x^{m-n}$.

$= a^{8/3 + 1/3 - 6/3}$ $2 = \dfrac{6}{3}$.

$= a^{3/3}$ $\dfrac{8}{3} + \dfrac{1}{3} - \dfrac{6}{3} = \dfrac{3}{3}$.

$= a$ $\dfrac{3}{3} = 1$.   ∎

(b, c)   EXAMPLE 7   Assume that all variables represent positive numbers and do the operations. Write all answers without using negative exponents.

a. $a^{4/5}\left(a^{1/5} + a^{3/5}\right) = a^{4/5}a^{1/5} + a^{4/5}a^{3/5}$ Use the distributive property.

$= a^{4/5 + 1/5} + a^{4/5 + 3/5}$ Use the rule $x^m x^n = x^{m+n}$.

$= a^{5/5} + a^{7/5}$ Simplify the exponents.

$= a + a^{7/5}$ $\frac{5}{5} = 1$.

WARNING! Note that $a + a^{7/5} \neq a^{1 + 7/5}$. The expression $a + a^{7/5}$ cannot be simplified because a and $a^{7/5}$ are not like terms.

b. $x^{1/2}\left(x^{-1/2} + x^{1/2}\right) = x^{1/2}x^{-1/2} + x^{1/2}x^{1/2}$ Use the distributive property.

$= x^{1/2 - 1/2} + x^{1/2 + 1/2}$ Use the rule $x^m x^n = x^{m+n}$.

$= x^0 + x^1$ Simplify.

$= 1 + x$ $x^0 = 1$.

c. $\left(x^{2/3} + 1\right)\left(x^{2/3} - 1\right) = x^{4/3} - x^{2/3} + x^{2/3} - 1$ Use the FOIL method.
$$= x^{4/3} - 1$$ Combine like terms.

d. $\left(x^{1/2} + y^{1/2}\right)^2 = \left(x^{1/2} + y^{1/2}\right)\left(x^{1/2} + y^{1/2}\right)$
$$= x + 2x^{1/2}y^{1/2} + y$$ Use the FOIL method. ∎

■ SIMPLIFYING RADICAL EXPRESSIONS

We can simplify many radical expressions by using the following steps.

> ### Using Fractional Exponents to Simplify Radical Expressions
> 1. Change the radical expression into an exponential expression with rational exponents.
> 2. Simplify the rational exponents.
> 3. Change the exponential expression back into a radical.

EXAMPLE 8   Simplify   **a.** $\sqrt[4]{3^2}$,   **b.** $\sqrt[8]{x^6}$,   and   **c.** $\sqrt[9]{27x^6y^3}$.

Solution   **a.** $\sqrt[4]{3^2} = (3^2)^{1/4}$ Change the radical to an exponential expression.
$$= 3^{2/4}$$ Use the rule $(x^m)^n = x^{mn}$.
$$= 3^{1/2}$$ $\frac{2}{4} = \frac{1}{2}$.
$$= \sqrt{3}$$ Change back to radical notation.

   b. $\sqrt[8]{x^6} = (x^6)^{1/8}$ Change the radical to an exponential expression.
$$= x^{6/8}$$ Use the rule $(x^m)^n = x^{mn}$.
$$= x^{3/4}$$ $\frac{6}{8} = \frac{3}{4}$.
$$= (x^3)^{1/4}$$ $\frac{3}{4} = 3\left(\frac{1}{4}\right)$.
$$= \sqrt[4]{x^3}$$ Change back to radical notation.

   c. $\sqrt[9]{27x^6y^3} = (3^3x^6y^3)^{1/9}$ Write 27 as 3^3 and change the radical to an exponential expression.
$$= 3^{3/9}x^{6/9}y^{3/9}$$ Raise each factor to the $\frac{1}{9}$ power by keeping the bases and multiplying the fractional exponents.
$$= 3^{1/3}x^{2/3}y^{1/3}$$ Simplify each fractional exponent.
$$= (3x^2y)^{1/3}$$ Use the rule $(xy)^n = x^ny^n$ twice.
$$= \sqrt[3]{3x^2y}$$ Change back to radical notation. ∎

Orals   *Simplify each expression.*

1. $4^{1/2}$ 2. $9^{1/2}$ 3. $27^{1/3}$ 4. $1^{1/4}$

5. $4^{3/2}$ 6. $8^{2/3}$ 7. $\left(\dfrac{1}{4}\right)^{1/2}$ 8. $\left(\dfrac{1}{4}\right)^{-1/2}$

9. $(8x^3)^{1/3}$ 10. $(16x^8)^{1/4}$

EXERCISE 7.2

In Exercises 1–8, change each expression into radical notation.

1. $7^{1/3}$
2. $26^{1/2}$
3. $(3x)^{1/4}$
4. $(4ab)^{1/6}$

5. $\left(\dfrac{1}{2}x^3y\right)^{1/4}$
6. $\left(\dfrac{3}{4}a^2b^2\right)^{1/5}$
7. $(x^2 + y^2)^{1/2}$
8. $(x^3 + y^3)^{1/3}$

In Exercises 9–16, change each radical to an exponential expression.

9. $\sqrt{11}$
10. $\sqrt[3]{12}$
11. $\sqrt[4]{3a}$
12. $3\sqrt[5]{a}$

13. $\sqrt[6]{\dfrac{1}{7}abc}$
14. $\sqrt[7]{\dfrac{3}{8}p^2q}$
15. $\sqrt[3]{a^2 - b^2}$
16. $\sqrt{x^2 + y^2}$

In Exercises 17–36, simplify each expression, if possible.

17. $4^{1/2}$
18. $25^{1/2}$
19. $8^{1/3}$
20. $125^{1/3}$

21. $16^{1/4}$
22. $625^{1/4}$
23. $32^{1/5}$
24. $0^{1/5}$

25. $\left(\dfrac{1}{4}\right)^{1/2}$
26. $\left(\dfrac{1}{16}\right)^{1/2}$
27. $\left(\dfrac{1}{8}\right)^{1/3}$
28. $\left(\dfrac{1}{16}\right)^{1/4}$

29. $-16^{1/4}$
30. $-125^{1/3}$
31. $(-27)^{1/3}$
32. $(-125)^{1/3}$

33. $(-64)^{1/2}$
34. $(-243)^{1/5}$
35. $0^{1/3}$
36. $(-216)^{1/2}$

In Exercises 37–44, simplify each expression, if possible. Assume that all variables are unrestricted, and use absolute value symbols when necessary.

37. $(25y^2)^{1/2}$
38. $(-27x^3)^{1/3}$
39. $(16x^4)^{1/4}$
40. $(-16x^4)^{1/2}$

41. $(243x^5)^{1/5}$
42. $[(x + 1)^4]^{1/4}$
43. $(-64x^8)^{1/4}$
44. $[(x + 5)^3]^{1/3}$

In Exercises 45–56, simplify each expression.

45. $36^{3/2}$
46. $27^{2/3}$
47. $81^{3/4}$
48. $100^{3/2}$

49. $144^{3/2}$
50. $1000^{2/3}$
51. $\left(\dfrac{1}{8}\right)^{2/3}$
52. $\left(\dfrac{4}{9}\right)^{3/2}$

53. $(25x^4)^{3/2}$
54. $(27a^3b^3)^{2/3}$
55. $\left(\dfrac{8x^3}{27}\right)^{2/3}$
56. $\left(\dfrac{27}{64y^6}\right)^{2/3}$

In Exercises 57–72, write each expression without using negative exponents. Assume that all variables represent positive numbers.

57. $4^{-1/2}$
58. $8^{-1/3}$
59. $4^{-3/2}$
60. $25^{-5/2}$

61. $(16x^2)^{-3/2}$
62. $(81c^4)^{-3/2}$
63. $(-27y^3)^{-2/3}$
64. $(-8z^9)^{-2/3}$

65. $(-32p^5)^{-2/5}$
66. $(16q^6)^{-5/2}$
67. $\left(\dfrac{1}{4}\right)^{-3/2}$
68. $\left(\dfrac{4}{25}\right)^{-3/2}$

69. $\left(\dfrac{27}{8}\right)^{-4/3}$
70. $\left(\dfrac{25}{49}\right)^{-3/2}$
71. $\left(-\dfrac{8x^3}{27}\right)^{-1/3}$
72. $\left(\dfrac{16}{81y^4}\right)^{-3/4}$

In Exercises 73–84, do the operations. Write answers without negative exponents. Assume that all variables represent positive numbers.

73. $5^{3/7} 5^{2/7}$

74. $4^{2/5} 4^{2/5}$

75. $\left(4^{1/5}\right)^3$

76. $\left(3^{1/3}\right)^5$

77. $\dfrac{9^{4/5}}{9^{3/5}}$

78. $\dfrac{7^{2/3}}{7^{1/2}}$

79. $\dfrac{7^{1/2}}{7^0}$

80. $5^{1/3} 5^{-5/3}$

81. $6^{-2/3} 6^{-4/3}$

82. $\dfrac{3^{4/3} 3^{1/3}}{3^{2/3}}$

83. $\dfrac{2^{5/6} 2^{1/3}}{2^{1/2}}$

84. $\dfrac{5^{1/3} 5^{1/2}}{5^{1/3}}$

In Exercises 85–96, do the operations. Assume that all variables are unrestricted, and write all answers without using negative exponents.

85. $a^{2/3} a^{1/3}$

86. $b^{3/5} b^{1/5}$

87. $\left(a^{2/3}\right)^{1/3}$

88. $\left(t^{4/5}\right)^{10}$

89. $\left(a^{1/2} b^{1/3}\right)^{3/2}$

90. $\left(a^{3/5} b^{3/2}\right)^{2/3}$

91. $\left(mn^{-2/3}\right)^{-3/5}$

92. $\left(r^{-2} s^3\right)^{1/3}$

93. $\dfrac{\left(4x^3 y\right)^{1/2}}{\left(9xy\right)^{1/2}}$

94. $\dfrac{\left(27x^3 y\right)^{1/3}}{\left(8xy^2\right)^{2/3}}$

95. $\left(27x^{-3}\right)^{-1/3}$

96. $\left(16a^{-2}\right)^{-1/2}$

In Exercises 97–108, do the multiplications. Assume that all variables are unrestricted, and write all answers without using negative exponents.

97. $y^{1/3}\left(y^{2/3} + y^{5/3}\right)$

98. $y^{2/5}\left(y^{-2/5} + y^{3/5}\right)$

99. $x^{3/5}\left(x^{7/5} - x^{2/5} + 1\right)$

100. $x^{4/3}\left(x^{2/3} + 3x^{5/3} - 4\right)$

101. $\left(x^{1/2} + 2\right)\left(x^{1/2} - 2\right)$

102. $\left(x^{1/2} + y^{1/2}\right)\left(x^{1/2} - y^{1/2}\right)$

103. $\left(x^{2/3} - x\right)\left(x^{2/3} + x\right)$

104. $\left(x^{1/3} + x^2\right)\left(x^{1/3} - x^2\right)$

105. $\left(x^{2/3} + y^{2/3}\right)^2$

106. $\left(a^{1/2} - b^{2/3}\right)^2$

107. $\left(a^{3/2} - b^{3/2}\right)^2$

108. $\left(x^{-1/2} - x^{1/2}\right)^2$

In Exercises 109–112, use rational exponents to simplify each radical. Assume that all variables represent positive numbers.

109. $\sqrt[6]{p^3}$

110. $\sqrt[8]{q^2}$

111. $\sqrt[4]{25b^2}$

112. $\sqrt[9]{-8x^6}$

Writing Exercises   *Write a paragraph using your own words.*

1. Explain how you would decide whether $a^{1/n}$ is a real number.

2. The expression $\left(a^{1/2} + b^{1/2}\right)^2$ is not equal to $a + b$. Explain.

Something to Think About   **1.** The fraction $\frac{2}{4}$ is equal to $\frac{1}{2}$. Is $16^{2/4}$ equal to $16^{1/2}$? Explain.

2. How would you evaluate an expression with a mixed-number exponent? For example, what is $8^{1\frac{1}{3}}$? What is $25^{2\frac{1}{2}}$? Discuss.

Review Exercises   *Solve each inequality.*

1. $5x - 4 < 11$

2. $2(3t - 5) \geq 8$

3. $\dfrac{4}{5}(r - 3) > \dfrac{2}{3}(r + 2)$

4. $-4 < 2x - 4 \leq 8$

5. How much water must be added to 5 pints of a 20% alcohol solution to dilute it to a 15% solution?

6. A grocer bought some boxes of apples for $70. However, 4 boxes were spoiled. The grocer sold the remaining boxes at a profit of $2 each. How many boxes did the grocer sell if she managed to break even?

7.3 Simplifying and Combining Radical Expressions

■ PROPERTIES OF RADICALS ■ SIMPLIFYING RADICAL EXPRESSIONS ■ ADDING AND SUBTRACTING RADICAL EXPRESSIONS

■ PROPERTIES OF RADICALS

Many properties of exponents have counterparts in radical notation. One such property of radicals involves products. Because $a^{1/n}b^{1/n} = (ab)^{1/n}$, we have

1. $\sqrt[n]{a}\sqrt[n]{b} = \sqrt[n]{ab}$

For example,

$$\sqrt{5}\sqrt{5} = \sqrt{5 \cdot 5} = \sqrt{5^2} = 5$$
$$\sqrt[3]{7x}\sqrt[3]{49x^2} = \sqrt[3]{7x \cdot 7^2x^2} = \sqrt[3]{7^3x^3} = 7x$$
$$\sqrt[4]{2x^3}\sqrt[4]{8x} = \sqrt[4]{2x^3 2^3 x} = \sqrt[4]{2^4 x^4} = 2|x|$$

If we apply the symmetric property of equality to Equation 1, we have the following rule.

> **Multiplication Property of Radicals**
> If $\sqrt[n]{a}$ and $\sqrt[n]{b}$ are real numbers, then
> $$\sqrt[n]{ab} = \sqrt[n]{a}\sqrt[n]{b}$$

As long as all radicals represent real numbers, *the nth root of the product of two numbers is equal to the product of the nth roots of the numbers.*

 WARNING! The multiplication property of radicals applies to the *n*th root of the product of two numbers. There is no such property for sums or differences. For example,

$$\sqrt{9 + 4} \neq \sqrt{9} + \sqrt{4} \qquad \sqrt{9 - 4} \neq \sqrt{9} - \sqrt{4}$$
$$\sqrt{13} \neq 3 + 2 \qquad\qquad \sqrt{5} \neq 3 - 2$$
$$\sqrt{13} \neq 5 \qquad\qquad\qquad \sqrt{5} \neq 1$$

Thus, $\sqrt{a + b} \neq \sqrt{a} + \sqrt{b}$ and $\sqrt{a - b} \neq \sqrt{a} - \sqrt{b}$.

Another property of radicals involves quotients. Because

$$\frac{a^{1/n}}{b^{1/n}} = \left(\frac{a}{b}\right)^{1/n}$$

it follows that

2. $\dfrac{\sqrt[n]{a}}{\sqrt[n]{b}} = \sqrt[n]{\dfrac{a}{b}}$ $(b \ne 0)$

For example,

$$\frac{\sqrt{8x^3}}{\sqrt{2x}} = \sqrt{\frac{8x^3}{2x}} = \sqrt{4x^2} = 2x \quad (x > 0)$$

$$\frac{\sqrt[3]{54x^5}}{\sqrt[3]{2x^2}} = \sqrt[3]{\frac{54x^5}{2x^2}} = \sqrt[3]{27x^3} = 3x$$

If we apply the symmetric property of equality to Equation 2, we have the following rule.

Quotient Property of Radicals

If $\sqrt[n]{a}$ and $\sqrt[n]{b}$ are real numbers, then

$$\sqrt[n]{\frac{a}{b}} = \frac{\sqrt[n]{a}}{\sqrt[n]{b}} \quad (b \ne 0)$$

As long as all radicals represent real numbers, *the nth root of the quotient of two numbers is equal to the quotient of their nth roots.*

■ SIMPLIFYING RADICAL EXPRESSIONS

A radical expression is said to be in simplest form when each of the following statements is true.

Simplified Form of a Radical Expression

A radical expression is in simplest form when

1. No radicals appear in the denominator of a fraction.

2. The radicand contains no fractions or negative numbers.

3. Each factor in the radicand appears to a power that is less than the index of the radical.

(b, c)   **EXAMPLE 1**   Simplify   **a.** $\sqrt{12}$,   **b.** $\sqrt{98}$,   and   **c.** $\sqrt[3]{54}$.

Solution   **a.** Recall that numbers such as 1, 4, 9, 16, 25, and 36 that are squares of integers are *perfect squares*. To simplify $\sqrt{12}$, we first factor 12 so that one factor is the largest perfect square that divides 12. Since 4 is the largest perfect square factor of 12, we write 12 as $4 \cdot 3$, use the multiplication property of radicals, and simplify.

$$\sqrt{12} = \sqrt{4 \cdot 3} \qquad \text{Write 12 as } 4 \cdot 3.$$
$$= \sqrt{4}\sqrt{3} \qquad \sqrt{4 \cdot 3} = \sqrt{4}\sqrt{3}.$$
$$= 2\sqrt{3} \qquad \sqrt{4} = 2.$$

b. The largest perfect square factor of 98 is 49. Thus,

$$\sqrt{98} = \sqrt{49 \cdot 2} \qquad \text{Write 98 as } 49 \cdot 2.$$
$$= \sqrt{49}\sqrt{2} \qquad \sqrt{49 \cdot 2} = \sqrt{49}\sqrt{2}.$$
$$= 7\sqrt{2} \qquad \sqrt{49} = 7.$$

c. Numbers such as 1, 8, 27, 64, 125, and 216 that are cubes of integers are called *perfect cubes*. Since the largest perfect cube factor of 54 is 27, we have

$$\sqrt[3]{54} = \sqrt[3]{27 \cdot 2} \qquad \text{Write 54 as } 27 \cdot 2.$$
$$= \sqrt[3]{27}\sqrt[3]{2} \qquad \sqrt[3]{27 \cdot 2} = \sqrt[3]{27}\sqrt[3]{2}.$$
$$= 3\sqrt[3]{2} \qquad \sqrt[3]{27} = 3.$$ ∎

EXAMPLE 2   Simplify   **a.** $\sqrt{\dfrac{15}{49x^2}}$ $(x > 0)$ and   **b.** $\sqrt[3]{\dfrac{10x^2}{27y^6}}$ $(y \neq 0)$.

Solution   **a.** We can write the square root of the quotient as the quotient of the square roots and simplify the denominator. Since $x > 0$, we have

$$\sqrt{\dfrac{15}{49x^2}} = \dfrac{\sqrt{15}}{\sqrt{49x^2}}$$
$$= \dfrac{\sqrt{15}}{7x}$$

b. We can write the cube root of the quotient as the quotient of two cube roots. Since $y \neq 0$, we have

$$\sqrt[3]{\dfrac{10x^2}{27y^6}} = \dfrac{\sqrt[3]{10x^2}}{\sqrt[3]{27y^6}}$$
$$= \dfrac{\sqrt[3]{10x^2}}{3y^2}$$ ∎

EXAMPLE 3   Simplify each expression. Assume that all variables represent positive numbers.

a. $\sqrt{128a^5}$,   **b.** $\sqrt[3]{24x^5}$,   **c.** $\dfrac{\sqrt{45xy^2}}{\sqrt{5x}}$,   and   **d.** $\dfrac{\sqrt[3]{-432x^5}}{\sqrt[3]{8x}}$

Solution   **a.** We write $128a^5$ as $64a^4 \cdot 2a$, and use the multiplication property of radicals.

$$\sqrt{128a^5} = \sqrt{64a^4 \cdot 2a} \qquad 64a^4 \text{ is the largest perfect square that divides } 128a^5.$$
$$= \sqrt{64a^4}\sqrt{2a} \qquad \text{Use the multiplication property of radicals.}$$
$$= 8a^2\sqrt{2a} \qquad \sqrt{64a^4} = 8a^2.$$

b. We write $24x^5$ as $8x^3 \cdot 3x^2$ and use the multiplication property of radicals.

$$\sqrt[3]{24x^5} = \sqrt[3]{8x^3 \cdot 3x^2} \qquad 8x^3 \text{ is the largest perfect cube that divides } 24x^5.$$
$$= \sqrt[3]{8x^3}\sqrt[3]{3x^2} \qquad \text{Use the multiplication property of radicals.}$$
$$= 2x\sqrt[3]{3x^2} \qquad \sqrt[3]{8x^3} = 2x.$$

c. We can write the quotient of the square roots as the square root of a quotient.

$$\frac{\sqrt{45xy^2}}{\sqrt{5x}} = \sqrt{\frac{45xy^2}{5x}} \qquad \text{Use the quotient property of radicals.}$$

$$= \sqrt{9y^2} \qquad \text{Simplify the fraction.}$$

$$= 3y$$

d. We can write the quotient of the cube roots as the cube root of a quotient.

$$\frac{\sqrt[3]{-432x^5}}{\sqrt[3]{8x}} = \sqrt[3]{\frac{-432x^5}{8x}} \qquad \text{Use the quotient property of radicals.}$$

$$= \sqrt[3]{-54x^4} \qquad \text{Simplify the fraction.}$$

$$= \sqrt[3]{-27x^3 \cdot 2x} \qquad \begin{array}{l}-27x^3 \text{ is the largest perfect cube that} \\ \text{divides } -54x^4.\end{array}$$

$$= \sqrt[3]{-27x^3}\sqrt[3]{2x} \qquad \text{Use the multiplication property of radicals.}$$

$$= -3x\sqrt[3]{2x} \qquad \sqrt[3]{-27x^3} = -3x. \qquad \blacksquare$$

To simplify more complicated radicals, we can use the prime factorization of the radicand to find its perfect square factors. For example, to simplify $\sqrt{3168x^5y^7}$, we first find the prime factorization of $3168x^5y^7$.

$$3168x^5y^7 = 2^5 \cdot 3^2 \cdot 11 \cdot x^5 \cdot y^7$$

Then we have

$$\sqrt{3168x^5y^7} = \sqrt{2^4 \cdot 3^2 \cdot x^4 \cdot y^6 \cdot 2 \cdot 11 \cdot x \cdot y}$$

$$= \sqrt{2^4 \cdot 3^2 \cdot x^4 \cdot y^6}\sqrt{2 \cdot 11 \cdot x \cdot y} \qquad \begin{array}{l}\text{Write each perfect square} \\ \text{under the left radical and} \\ \text{each nonperfect square} \\ \text{under the right radical.}\end{array}$$

$$= 2^2 \cdot 3x^2y^3\sqrt{22xy}$$

$$= 12x^2y^3\sqrt{22xy}$$

■ ADDING AND SUBTRACTING RADICAL EXPRESSIONS

Radical expressions with the same index and the same radicand are called **like** or **similar radicals**. For example, $3\sqrt{2}$ and $2\sqrt{2}$ are like radicals. However,

$3\sqrt{5}$ and $4\sqrt{2}$ are not like radicals, because the radicands are different.

$3\sqrt{5}$ and $2\sqrt[3]{5}$ are not like radicals, because the indexes are different.

We can often combine like radicals. For example, to simplify the expression $3\sqrt{2} + 2\sqrt{2}$, we use the distributive property to factor out $\sqrt{2}$ and simplify.

$$3\sqrt{2} + 2\sqrt{2} = (3 + 2)\sqrt{2}$$

$$= 5\sqrt{2}$$

Radicals with the same index but different radicands often simplify as like radicals. For example, to simplify the expression $\sqrt{27} - \sqrt{12}$, we simplify both radicals and then combine the like radicals.

$$\sqrt{27} - \sqrt{12} = \sqrt{9 \cdot 3} - \sqrt{4 \cdot 3}$$
$$= \sqrt{9}\sqrt{3} - \sqrt{4}\sqrt{3} \qquad \sqrt{ab} = \sqrt{a}\sqrt{b}.$$
$$= 3\sqrt{3} - 2\sqrt{3} \qquad \sqrt{9} = 3 \text{ and } \sqrt{4} = 2.$$
$$= (3 - 2)\sqrt{3} \qquad \text{Factor out } \sqrt{3}.$$
$$= \sqrt{3}$$

As the previous examples suggest, we can use the following rule to add or subtract radicals.

Adding and Subtracting Radicals
To add or subtract radicals, simplify each radical and then combine all like radicals. To combine like radicals, add the coefficients and keep the common radical.

EXAMPLE 4   Simplify $2\sqrt{12} - 3\sqrt{48} + 3\sqrt{3}$.

Solution   We simplify each radical separately and combine like radicals.

$$2\sqrt{12} - 3\sqrt{48} + 3\sqrt{3} = 2\sqrt{4 \cdot 3} - 3\sqrt{16 \cdot 3} + 3\sqrt{3}$$
$$= 2\sqrt{4}\sqrt{3} - 3\sqrt{16}\sqrt{3} + 3\sqrt{3}$$
$$= 2(2)\sqrt{3} - 3(4)\sqrt{3} + 3\sqrt{3}$$
$$= 4\sqrt{3} - 12\sqrt{3} + 3\sqrt{3}$$
$$= (4 - 12 + 3)\sqrt{3}$$
$$= -5\sqrt{3}$$

EXAMPLE 5   Simplify $\sqrt[3]{16} - \sqrt[3]{54} + \sqrt[3]{24}$.

Solution   We simplify each radical separately and combine like radicals:

$$\sqrt[3]{16} - \sqrt[3]{54} + \sqrt[3]{24} = \sqrt[3]{8 \cdot 2} - \sqrt[3]{27 \cdot 2} + \sqrt[3]{8 \cdot 3}$$
$$= \sqrt[3]{8}\sqrt[3]{2} - \sqrt[3]{27}\sqrt[3]{2} + \sqrt[3]{8}\sqrt[3]{3}$$
$$= 2\sqrt[3]{2} - 3\sqrt[3]{2} + 2\sqrt[3]{3}$$
$$= -\sqrt[3]{2} + 2\sqrt[3]{3}$$

 WARNING!   We cannot combine $-\sqrt[3]{2}$ and $2\sqrt[3]{3}$, because the radicals have different radicands.

EXAMPLE 6   Simplify $\sqrt[3]{16x^4} + \sqrt[3]{54x^4} - \sqrt[3]{-128x^4}$ $(x > 0)$.

Solution   We simplify each radical expression separately, factor out $\sqrt[3]{2x}$, and simplify.

$$\sqrt[3]{16x^4} + \sqrt[3]{54x^4} - \sqrt[3]{-128x^4} = \sqrt[3]{8x^3 \cdot 2x} + \sqrt[3]{27x^3 \cdot 2x} - \sqrt[3]{-64x^3 \cdot 2x}$$
$$= \sqrt[3]{8x^3}\sqrt[3]{2x} + \sqrt[3]{27x^3}\sqrt[3]{2x} - \sqrt[3]{-64x^3}\sqrt[3]{2x}$$
$$= 2x\sqrt[3]{2x} + 3x\sqrt[3]{2x} + 4x\sqrt[3]{2x}$$
$$= (2x + 3x + 4x)\sqrt[3]{2x}$$
$$= 9x\sqrt[3]{2x}$$

Orals   *Simplify.*

1. $\sqrt{7}\sqrt{7}$ **2.** $\sqrt[3]{4^2}\sqrt[3]{4}$ **3.** $\dfrac{\sqrt[3]{54}}{\sqrt[3]{2}}$

Simplify each expression. Assume $b \neq 0$.

4. $\sqrt{18}$ **5.** $\sqrt[3]{16}$ **6.** $\sqrt[3]{\dfrac{3x^2}{64b^6}}$

Combine like radicals.

7. $3\sqrt{3} + 4\sqrt{3}$ **8.** $5\sqrt{7} - 2\sqrt{7}$

9. $2\sqrt[3]{9} + 3\sqrt[3]{9}$ **10.** $10\sqrt[5]{4} - 2\sqrt[5]{4}$

EXERCISE 7.3

In Exercises 1–16, simplify each expression. Assume that all variables represent positive numbers.

1. $\sqrt{6}\sqrt{6}$ **2.** $\sqrt{11}\sqrt{11}$ **3.** $\sqrt{t}\sqrt{t}$ **4.** $-\sqrt{z}\sqrt{z}$

5. $\sqrt[3]{5x^2}\sqrt[3]{25x}$ **6.** $\sqrt[4]{25a}\sqrt[4]{25a^3}$ **7.** $\dfrac{\sqrt{500}}{\sqrt{5}}$ **8.** $\dfrac{\sqrt{128}}{\sqrt{2}}$

9. $\dfrac{\sqrt{98x^3}}{\sqrt{2x}}$ **10.** $\dfrac{\sqrt{75y^5}}{\sqrt{3y}}$ **11.** $\dfrac{\sqrt{180ab^4}}{\sqrt{5ab^2}}$ **12.** $\dfrac{\sqrt{112ab^3}}{\sqrt{7ab}}$

13. $\dfrac{\sqrt[3]{48}}{\sqrt[3]{6}}$ **14.** $\dfrac{\sqrt[3]{64}}{\sqrt[3]{8}}$ **15.** $\dfrac{\sqrt[3]{189a^4}}{\sqrt[3]{7a}}$ **16.** $\dfrac{\sqrt[3]{243x^7}}{\sqrt[3]{9x}}$

In Exercises 17–36, simplify each radical.

17. $\sqrt{20}$ **18.** $\sqrt{8}$ **19.** $-\sqrt{200}$ **20.** $-\sqrt{250}$

21. $\sqrt[3]{80}$ **22.** $\sqrt[3]{270}$ **23.** $\sqrt[3]{-81}$ **24.** $\sqrt[3]{-72}$

25. $\sqrt[4]{32}$ **26.** $\sqrt[4]{48}$ **27.** $\sqrt[5]{96}$ **28.** $\sqrt[7]{256}$

29. $\sqrt{\dfrac{7}{9}}$ **30.** $\sqrt{\dfrac{3}{4}}$ **31.** $\sqrt[3]{\dfrac{7}{64}}$ **32.** $\sqrt[3]{\dfrac{4}{125}}$

33. $\sqrt[4]{\dfrac{3}{10,000}}$ **34.** $\sqrt[5]{\dfrac{4}{243}}$ **35.** $\sqrt[5]{\dfrac{3}{32}}$ **36.** $\sqrt[6]{\dfrac{5}{64}}$

In Exercises 37–56, simplify each radical. Assume that all variables represent positive numbers.

37. $\sqrt{50x^2}$ **38.** $\sqrt{75a^2}$ **39.** $\sqrt{32b}$ **40.** $\sqrt{80c}$

41. $-\sqrt{112a^3}$ **42.** $\sqrt{147a^5}$ **43.** $\sqrt{175a^2b^3}$ **44.** $\sqrt{128a^3b^5}$

45. $-\sqrt{300xy}$ **46.** $\sqrt{200x^2y}$ **47.** $\sqrt[3]{-54x^6}$ **48.** $-\sqrt[3]{-81a^3}$

49. $\sqrt[3]{16x^{12}y^3}$ **50.** $\sqrt[3]{40a^3b^6}$ **51.** $\sqrt[4]{32x^{12}y^4}$ **52.** $\sqrt[5]{64x^{10}y^5}$

53. $\sqrt{\dfrac{z^2}{16x^2}}$ **54.** $\sqrt{\dfrac{b^4}{64a^8}}$ **55.** $\sqrt[4]{\dfrac{5x}{16z^4}}$ **56.** $\sqrt[3]{\dfrac{11a^2}{125b^6}}$

In Exercises 57–96, simplify and combine like radicals. All variables represent positive numbers.

57. $4\sqrt{2x} + 6\sqrt{2x}$

58. $6\sqrt[3]{5y} + 3\sqrt[3]{5y}$

59. $8\sqrt[5]{7a^2} - 7\sqrt[5]{7a^2}$

60. $10\sqrt[6]{12xyz} - \sqrt[6]{12xyz}$

61. $\sqrt{3} + \sqrt{27}$

62. $\sqrt{8} + \sqrt{32}$

63. $\sqrt{2} - \sqrt{8}$

64. $\sqrt{20} - \sqrt{125}$

65. $\sqrt{98} - \sqrt{50}$

66. $\sqrt{72} - \sqrt{200}$

67. $3\sqrt{24} + \sqrt{54}$

68. $\sqrt{18} + 2\sqrt{50}$

69. $\sqrt[3]{24} + \sqrt[3]{3}$

70. $\sqrt[3]{16} + \sqrt[3]{128}$

71. $\sqrt[3]{32} - \sqrt[3]{108}$

72. $\sqrt[3]{80} - \sqrt[3]{10,000}$

73. $2\sqrt[3]{125} - 5\sqrt[3]{64}$

74. $3\sqrt[3]{27} + 12\sqrt[3]{216}$

75. $14\sqrt[4]{32} - 15\sqrt[4]{162}$

76. $23\sqrt[4]{768} + \sqrt[4]{48}$

77. $3\sqrt[4]{512} + 2\sqrt[4]{32}$

78. $4\sqrt[4]{243} - \sqrt[4]{48}$

79. $\sqrt{98} - \sqrt{50} - \sqrt{72}$

80. $\sqrt{20} + \sqrt{125} - \sqrt{80}$

81. $\sqrt{18} + \sqrt{300} - \sqrt{243}$

82. $\sqrt{80} - \sqrt{128} + \sqrt{288}$

83. $2\sqrt[3]{16} - \sqrt[3]{54} - 3\sqrt[3]{128}$

84. $\sqrt[4]{48} - \sqrt[4]{243} - \sqrt[4]{768}$

85. $\sqrt{25y^2z} - \sqrt{16y^2z}$

86. $\sqrt{25yz^2} + \sqrt{9yz^2}$

87. $\sqrt{36xy^2} + \sqrt{49xy^2}$

88. $3\sqrt{2x} - \sqrt{8x}$

89. $2\sqrt[3]{64a} + 2\sqrt[3]{8a}$

90. $3\sqrt[4]{x^4y} - 2\sqrt[4]{x^4y}$

91. $\sqrt{y^5} - \sqrt{9y^5} - \sqrt{25y^5}$

92. $\sqrt{8y^7} + \sqrt{32y^7} - \sqrt{2y^7}$

93. $\sqrt[5]{x^6y^2} + \sqrt[5]{32x^6y^2} + \sqrt[5]{x^6y^2}$

94. $\sqrt[3]{xy^4} + \sqrt[3]{8xy^4} - \sqrt[3]{27xy^4}$

95. $\sqrt{x^2 + 2x + 1} + \sqrt{x^2 + 2x + 1}$

96. $\sqrt{4x^2 + 12x + 9} + \sqrt{9x^2 + 6x + 1}$

Writing Exercises   *Write a paragraph using your own words.*

1. Explain how to recognize like radicals.

2. Explain how to combine like radicals.

Something to Think About

1. Can you find any numbers a and b such that $\sqrt{a + b} = \sqrt{a} + \sqrt{b}$?

2. Find the sum:
$\sqrt{3} + \sqrt{3^2} + \sqrt{3^3} + \sqrt{3^4} + \sqrt{3^5}$.

Review Exercises   *Do each operation.*

1. $3x^2y^3(-5x^3y^{-4})$

2. $-2a^2b^{-2}(4a^{-2}b^4 - 2a^2b + 3a^3b^2)$

3. $(3t + 2)^2$

4. $(5r - 3s)(5r + 2s)$

5. $2p - 5\overline{)6p^2 - 7p - 25}$

6. $3m + n\overline{)6m^3 - m^2n + 2mn^2 + n^3}$

7.4 Multiplying and Dividing Radical Expressions

■ MULTIPLYING A MONOMIAL BY A MONOMIAL ■ MULTIPLYING A POLYNOMIAL BY A MONOMIAL
■ MULTIPLYING A POLYNOMIAL BY A POLYNOMIAL ■ PROBLEM SOLVING ■ RATIONALIZING
DENOMINATORS AND NUMERATORS

Radical expressions with the same index can be multiplied and divided.

■ MULTIPLYING A MONOMIAL BY A MONOMIAL

EXAMPLE 1   Multiply $3\sqrt{6}$ by $2\sqrt{3}$.

Solution   We use the commutative and associative properties of multiplication to multiply the coefficients and the radicals separately. We then simplify any radicals in the product, if possible.

$$3\sqrt{6} \cdot 2\sqrt{3} = 3(2)\sqrt{6}\sqrt{3} \qquad \text{Multiply the coefficients and multiply the radicals.}$$
$$= 6\sqrt{18} \qquad 3(2) = 6 \text{ and } \sqrt{6}\sqrt{3} = \sqrt{18}.$$
$$= 6\sqrt{9}\sqrt{2} \qquad \sqrt{18} = \sqrt{9 \cdot 2} = \sqrt{9}\sqrt{2}.$$
$$= 6(3)\sqrt{2} \qquad \sqrt{9} = 3.$$
$$= 18\sqrt{2} \qquad \blacksquare$$

■ MULTIPLYING A POLYNOMIAL BY A MONOMIAL

To multiply a polynomial by a monomial, we use the distributive property to remove parentheses and then simplify each resulting term, if possible.

EXAMPLE 2   Multiply: $3\sqrt{3}(4\sqrt{8} - 5\sqrt{10})$.

Solution $3\sqrt{3}(4\sqrt{8} - 5\sqrt{10}) = 3\sqrt{3} \cdot 4\sqrt{8} - 3\sqrt{3} \cdot 5\sqrt{10}$ Use the distributive property.

$$= 12\sqrt{24} - 15\sqrt{30} \qquad \text{Multiply the coefficients and multiply the radicals.}$$
$$= 12\sqrt{4}\sqrt{6} - 15\sqrt{30} \qquad \sqrt{24} = \sqrt{4}\sqrt{6}.$$
$$= 12(2)\sqrt{6} - 15\sqrt{30}$$
$$= 24\sqrt{6} - 15\sqrt{30} \qquad \blacksquare$$

■ MULTIPLYING A POLYNOMIAL BY A POLYNOMIAL

To multiply a binomial by a binomial, we use the FOIL method.

EXAMPLE 3   Multiply $\left(\sqrt{7} + \sqrt{2}\right)\left(\sqrt{7} - 3\sqrt{2}\right)$.

Solution
$$\left(\sqrt{7} + \sqrt{2}\right)\left(\sqrt{7} - 3\sqrt{2}\right) = \left(\sqrt{7}\right)^2 - 3\sqrt{7}\sqrt{2} + \sqrt{2}\sqrt{7} - 3\sqrt{2}\sqrt{2}$$
$$= 7 - 3\sqrt{14} + \sqrt{14} - 3(2)$$
$$= 7 - 2\sqrt{14} - 6$$
$$= 1 - 2\sqrt{14} \qquad ■$$

EXAMPLE 4   Multiply $\left(\sqrt{3x} - \sqrt{5}\right)\left(\sqrt{2x} + \sqrt{10}\right)$.

Solution
$$\left(\sqrt{3x} - \sqrt{5}\right)\left(\sqrt{2x} + \sqrt{10}\right) = \sqrt{3}\sqrt{2}x^2 + \sqrt{3}\sqrt{10}x - \sqrt{5}\sqrt{2}x - \sqrt{5}\sqrt{10}$$
$$= \sqrt{6}x^2 + \sqrt{30}x - \sqrt{10}x - \sqrt{50}$$
$$= \sqrt{6}x^2 + \sqrt{30}x - \sqrt{10}x - \sqrt{25}\sqrt{2}$$
$$= \sqrt{6}x^2 + \sqrt{30}x - \sqrt{10}x - 5\sqrt{2} \qquad ■$$

WARNING!   It is important to draw radical signs carefully so that they completely cover a radicand, but no more than the radicand. To avoid confusion, we often write an expression such as $\sqrt{30}x$ in the form $x\sqrt{30}$.

■ PROBLEM SOLVING

EXAMPLE 5

FIGURE 7-6

Photography   Many camera lenses (see Figure 7-6) have an adjustable opening called the *aperture*, which controls the amount of light passing through the lens. The *f-number* of a lens is its *focal length* divided by the diameter of its circular aperture:

$$f\text{-number} = \frac{f}{d} \qquad f \text{ is the focal length, and } d \text{ is the diameter of the aperture.}$$

A lens with a focal length of 12 centimeters and an aperture with a diameter of 6 centimeters has an *f*-number of $\frac{12}{6}$ and is an $f/2$ lens. If the area of the aperture is reduced to admit half as much light, the *f*-number of the lens will change. Find the new *f*-number.

Solution   We first find the area of the aperture when its diameter is 6 centimeters.

$$A = \pi r^2 \qquad \text{The formula for the area of a circle.}$$
$$A = \pi(3)^2 \qquad \text{Since a radius is half the diameter, substitute 3 for } r.$$
$$A = 9\pi$$

When the size of the aperture is reduced to admit half as much light, the area of the aperture will be $\frac{9\pi}{2}$ square centimeters. To find the diameter of a circle with this area, we proceed as follows:

$$A = \pi r^2$$ The formula for the area of a circle.

$$\frac{9\pi}{2} = \pi \left(\frac{d}{2}\right)^2$$ Substitute $\frac{9\pi}{2}$ for A and $\frac{d}{2}$ for r.

$$\frac{9\pi}{2} = \frac{\pi d^2}{4}$$ $\left(\frac{d}{2}\right)^2 = \frac{d^2}{4}$.

$$18 = d^2$$ Multiply both sides by 4, and divide both sides by π.

$$d = 3\sqrt{2}$$ $\sqrt{18} = \sqrt{9}\sqrt{2} = 3\sqrt{2}$.

Since the focal length of the lens is still 12 centimeters and the diameter is now $3\sqrt{2}$ centimeters, the new f-number of the lens is

$$f\text{-number} = \frac{f}{d} = \frac{12}{3\sqrt{2}}$$ Substitute 12 for f and $3\sqrt{2}$ for d.

$$\approx 2.828427125$$ Use a calculator and press 3 $\boxed{\times}$ 2 $\boxed{\sqrt{}}$ $\boxed{=}$ $\boxed{1/x}$ $\boxed{\times}$ 12 $\boxed{=}$.

The lens is now an $f/2.8$ lens.   ∎

■ RATIONALIZING DENOMINATORS AND NUMERATORS

To divide radical expressions, we **rationalize the denominator** of a fraction to replace the denominator with a rational number. For example, to divide $\sqrt{70}$ by $\sqrt{3}$, we write the division as the fraction

$$\frac{\sqrt{70}}{\sqrt{3}}$$

To eliminate the radical in the denominator, we multiply both the numerator and the denominator by a number that will give a perfect square under the radical in the denominator. Because $3 \cdot 3 = 9$ is a perfect square, $\sqrt{3}$ is such a number.

$$\frac{\sqrt{70}}{\sqrt{3}} = \frac{\sqrt{70} \cdot \sqrt{3}}{\sqrt{3} \cdot \sqrt{3}}$$ Multiply numerator and denominator by $\sqrt{3}$.

$$= \frac{\sqrt{210}}{3}$$ Multiply the radicals.

Since there is no radical in the denominator and $\sqrt{210}$ cannot be simplified, the expression $\sqrt{210}/3$ is in simplest form, and the division is complete.

EXAMPLE 6 Rationalize each denominator: **a.** $\sqrt{\dfrac{20}{7}}$ and **b.** $\dfrac{4}{\sqrt[3]{2}}$.

Solution **a.** We first write the square root of the quotient as the quotient of two square roots:

$$\sqrt{\frac{20}{7}} = \frac{\sqrt{20}}{\sqrt{7}}$$

We then proceed as follows:

$$\frac{\sqrt{20}}{\sqrt{7}} = \frac{\sqrt{20} \cdot \sqrt{7}}{\sqrt{7} \cdot \sqrt{7}} \qquad \text{Multiply numerator and denominator by } \sqrt{7}.$$

$$= \frac{\sqrt{140}}{7} \qquad \text{Multiply the radicals, and note that } \sqrt{7}\sqrt{7} = 7.$$

$$= \frac{2\sqrt{35}}{7} \qquad \text{Simplify } \sqrt{140}\colon \sqrt{140} = \sqrt{4 \cdot 35} = \sqrt{4}\sqrt{35} = 2\sqrt{35}.$$

b. Since the denominator is a cube root, we multiply the numerator and the denominator by a number that will give a perfect cube under the radical sign. Because $2 \cdot 4 = 8$ is a perfect cube, $\sqrt[3]{4}$ is such a number.

$$\frac{4}{\sqrt[3]{2}} = \frac{4 \cdot \sqrt[3]{4}}{\sqrt[3]{2} \cdot \sqrt[3]{4}} \qquad \text{Multiply numerator and denominator by } \sqrt[3]{4}.$$

$$= \frac{4\sqrt[3]{4}}{\sqrt[3]{8}} \qquad \text{Multiply the radicals in the denominator.}$$

$$= \frac{4\sqrt[3]{4}}{2} \qquad \sqrt[3]{8} = 2.$$

$$= 2\sqrt[3]{4} \qquad \text{Simplify.} \qquad \blacksquare$$

EXAMPLE 7 Rationalize the denominator of $\dfrac{\sqrt[3]{5}}{\sqrt[3]{18}}$.

Solution We multiply the numerator and the denominator by a number that will result in a perfect cube under the radical sign in the denominator.

Since 216 is the smallest perfect cube that is divisible by 18 ($216 \div 18 = 12$), multiplying both the numerator and the denominator by $\sqrt[3]{12}$ will give the smallest possible perfect cube under the radical in the denominator.

$$\frac{\sqrt[3]{5}}{\sqrt[3]{18}} = \frac{\sqrt[3]{5} \cdot \sqrt[3]{12}}{\sqrt[3]{18} \cdot \sqrt[3]{12}} \qquad \text{Multiply numerator and denominator by } \sqrt[3]{12}.$$

$$= \frac{\sqrt[3]{60}}{\sqrt[3]{216}} \qquad \text{Multiply the radicals.}$$

$$= \frac{\sqrt[3]{60}}{6} \qquad \sqrt[3]{216} = 6, \text{ and } \sqrt[3]{60} \text{ does not simplify.} \qquad \blacksquare$$

EXAMPLE 8   Rationalize the denominator of $\dfrac{\sqrt{5xy^2}}{\sqrt{xy^3}}$ (x and y are positive numbers).

Method 1
$$\frac{\sqrt{5xy^2}}{\sqrt{xy^3}} = \sqrt{\frac{5xy^2}{xy^3}}$$
$$= \sqrt{\frac{5}{y}}$$
$$= \frac{\sqrt{5}}{\sqrt{y}}$$
$$= \frac{\sqrt{5}\sqrt{y}}{\sqrt{y}\sqrt{y}}$$
$$= \frac{\sqrt{5y}}{y}$$

Method 2
$$\frac{\sqrt{5xy^2}}{\sqrt{xy^3}} = \sqrt{\frac{5xy^2}{xy^3}}$$
$$= \sqrt{\frac{5}{y}}$$
$$= \sqrt{\frac{5y}{yy}}$$
$$= \frac{\sqrt{5y}}{\sqrt{y^2}}$$
$$= \frac{\sqrt{5y}}{y}$$
∎

To rationalize the denominator of a fraction with square roots in a binomial denominator, we multiply its numerator and denominator by the *conjugate* of its denominator. Conjugate binomials are binomials with the same terms but with opposite signs between their terms.

Conjugate Binomials
The **conjugate** of the binomial $a + b$ is $a - b$, and the conjugate of $a - b$ is $a + b$.

EXAMPLE 9   Rationalize the denominator of $\dfrac{1}{\sqrt{2} + 1}$.

Solution   We multiply the numerator and denominator of the fraction by $\sqrt{2} - 1$, which is the conjugate of the denominator.

$$\frac{1}{\sqrt{2} + 1} = \frac{1(\sqrt{2} - 1)}{(\sqrt{2} + 1)(\sqrt{2} - 1)}$$
$$= \frac{\sqrt{2} - 1}{(\sqrt{2})^2 - 1}$$
$$= \frac{\sqrt{2} - 1}{2 - 1}$$
$$= \sqrt{2} - 1$$

Multiply the numerator and the denominator by the conjugate of the denominator.

$(\sqrt{2} + 1)(\sqrt{2} - 1) =$
$(\sqrt{2})^2 - \sqrt{2} + \sqrt{2} - 1 = (\sqrt{2})^2 - 1$

$(\sqrt{2})^2 = 2$.

$\dfrac{\sqrt{2} - 1}{2 - 1} = \dfrac{\sqrt{2} - 1}{1} = \sqrt{2} - 1$. ∎

EXAMPLE 10   Rationalize the denominator of $\dfrac{\sqrt{x} + \sqrt{2}}{\sqrt{x} - \sqrt{2}}$ ($x > 0$).

Solution   We multiply the numerator and denominator by $\sqrt{x} + \sqrt{2}$, which is the conjugate of $\sqrt{x} - \sqrt{2}$, and simplify.

$$\frac{\sqrt{x} + \sqrt{2}}{\sqrt{x} - \sqrt{2}} = \frac{(\sqrt{x} + \sqrt{2})(\sqrt{x} + \sqrt{2})}{(\sqrt{x} - \sqrt{2})(\sqrt{x} + \sqrt{2})}$$

$$= \frac{x + \sqrt{2x} + \sqrt{2x} + 2}{x - 2} \qquad \text{Use the FOIL method.}$$

$$= \frac{x + 2\sqrt{2x} + 2}{x - 2}$$

■

In calculus, we sometimes have to rationalize the numerator of a fraction by multiplying the numerator and denominator of the fraction by the conjugate of the numerator.

EXAMPLE 11   Rationalize the numerator of $\dfrac{\sqrt{x} - 3}{\sqrt{x}}$ $(x > 0)$.

Solution   We multiply the numerator and denominator by $\sqrt{x} + 3$, which is the conjugate of the numerator.

$$\frac{\sqrt{x} - 3}{\sqrt{x}} = \frac{(\sqrt{x} - 3)(\sqrt{x} + 3)}{\sqrt{x}(\sqrt{x} + 3)}$$

$$= \frac{x + 3\sqrt{x} - 3\sqrt{x} - 9}{x + 3\sqrt{x}}$$

$$= \frac{x - 9}{x + 3\sqrt{x}}$$

Technically, the final expression is not in simplified form. However, this non-simplified form is sometimes desirable in calculus.   ■

Orals   *Multiply and simplify.*

1. $\sqrt{3}\sqrt{3}$

2. $\sqrt[3]{2}\sqrt[3]{2}\sqrt[3]{2}$

3. $\sqrt{3}\sqrt{9}$

4. $\sqrt{a^3b}\sqrt{ab}$

5. $3\sqrt{2}(\sqrt{2} + 1)$

6. $(\sqrt{2} + 1)(\sqrt{2} - 1)$

7. $\dfrac{1}{\sqrt{2}}$

8. $\dfrac{1}{\sqrt{3} - 1}$

EXERCISE 7.4

In Exercises 1–24, do each multiplication and simplify, if possible. All variables represent positive numbers.

1. $\sqrt{2}\sqrt{8}$

2. $\sqrt{3}\sqrt{27}$

3. $\sqrt{5}\sqrt{10}$

4. $\sqrt{7}\sqrt{35}$

5. $2\sqrt{3}\sqrt{6}$

6. $3\sqrt{11}\sqrt{33}$

7. $\sqrt[3]{5}\sqrt[3]{25}$

8. $\sqrt[3]{7}\sqrt[3]{49}$

9. $(3\sqrt[3]{9})(2\sqrt[3]{3})$

10. $(2\sqrt[3]{16})(-\sqrt[3]{4})$

11. $\sqrt[3]{2}\sqrt[3]{12}$

12. $\sqrt[3]{3}\sqrt[3]{18}$

13. $\sqrt{ab^3}\sqrt{ab}$

14. $\sqrt{8x}\sqrt{2x^3y}$

15. $\sqrt{5ab}\sqrt{5a}$

16. $\sqrt{15rs^2}\sqrt{10r}$

17. $\sqrt[3]{5r^2s}\sqrt[3]{2r}$

18. $\sqrt[3]{3xy^2}\sqrt[3]{9x^3}$

19. $\sqrt[3]{a^5b}\sqrt[3]{16ab^5}$

20. $\sqrt[3]{3x^4y}\sqrt[3]{18x}$

21. $\sqrt{x(x + 3)}\sqrt{x^3(x + 3)}$

22. $\sqrt{y^2(x + y)}\sqrt{(x + y)^3}$

23. $\sqrt[3]{6x^2(y + z)^2}\sqrt[3]{18x(y + z)}$

24. $\sqrt[3]{9x^2y(z + 1)^2}\sqrt[3]{6xy^2(z + 1)}$

In Exercises 25–44, do each multiplication and simplify. All variables represent positive numbers.

25. $3\sqrt{5}(4 - \sqrt{5})$

26. $2\sqrt{7}(3\sqrt{7} - 1)$

27. $3\sqrt{2}(4\sqrt{3} + 2\sqrt{7})$

28. $-\sqrt{3}(\sqrt{7} - \sqrt{5})$

29. $-2\sqrt{5x}(4\sqrt{2x} - 3\sqrt{3})$

30. $3\sqrt{7t}(2\sqrt{7t} + 3\sqrt{3t^2})$

31. $(\sqrt{2} + 1)(\sqrt{2} - 3)$

32. $(2\sqrt{3} + 1)(\sqrt{3} - 1)$

33. $(4\sqrt{x} + 3)(2\sqrt{x} - 5)$

34. $(7\sqrt{y} + 2)(3\sqrt{y} - 5)$

35. $(\sqrt{5z} + \sqrt{3})(\sqrt{5z} + \sqrt{3})$

36. $(\sqrt{3p} - \sqrt{2})(\sqrt{3p} + \sqrt{2})$

37. $(\sqrt{3x} - \sqrt{2y})(\sqrt{3x} + \sqrt{2y})$

38. $(\sqrt{3m} + \sqrt{2n})(\sqrt{3m} + \sqrt{2n})$

39. $(2\sqrt{3a} - \sqrt{b})(\sqrt{3a} + 3\sqrt{b})$

40. $(5\sqrt{p} - \sqrt{3q})(\sqrt{p} + 2\sqrt{3q})$

41. $(3\sqrt{2r} - 2)^2$

42. $(2\sqrt{3t} + 5)^2$

43. $-2(\sqrt{3x} + \sqrt{3})^2$

44. $3(\sqrt{5x} - \sqrt{3})^2$

45. Photography In Example 5, we saw that a lens with a focal length of 12 centimeters and an aperture $3\sqrt{2}$ centimeters in diameter is an $f/2.8$ lens. Find the f-number if the area of the aperture is again cut in half.

46. Photography In Exercise 45, we saw that a lens with a focal length of 12 centimeters and an aperture 3 centimeters in diameter is an $f/4$ lens. Find the f-number if the area of the aperture is again cut in half.

In Exercises 47–70, rationalize each denominator. All variables represent positive numbers.

47. $\sqrt{\dfrac{1}{7}}$

48. $\sqrt{\dfrac{5}{3}}$

49. $\sqrt{\dfrac{2}{3}}$

50. $\sqrt{\dfrac{3}{2}}$

51. $\dfrac{\sqrt{5}}{\sqrt{8}}$

52. $\dfrac{\sqrt{3}}{\sqrt{50}}$

53. $\dfrac{\sqrt{8}}{\sqrt{2}}$

54. $\dfrac{\sqrt{27}}{\sqrt{3}}$

55. $\dfrac{1}{\sqrt[3]{2}}$

56. $\dfrac{2}{\sqrt[3]{6}}$

57. $\dfrac{3}{\sqrt[3]{9}}$

58. $\dfrac{2}{\sqrt[3]{a}}$

59. $\dfrac{\sqrt[3]{2}}{\sqrt[3]{9}}$

60. $\dfrac{\sqrt[3]{9}}{\sqrt[3]{54}}$

61. $\dfrac{\sqrt[3]{8x^2y}}{\sqrt{xy}}$

62. $\dfrac{\sqrt{9xy}}{\sqrt{3x^2y}}$

63. $\dfrac{\sqrt{10xy^2}}{\sqrt{2xy^3}}$

64. $\dfrac{\sqrt{5ab^2c}}{\sqrt{10abc}}$

65. $\dfrac{\sqrt[3]{4a^2}}{\sqrt[3]{2ab}}$

66. $\dfrac{\sqrt[3]{9x}}{\sqrt[3]{3xy}}$

67. $\dfrac{1}{\sqrt[4]{4}}$

68. $\dfrac{1}{\sqrt[5]{2}}$

69. $\dfrac{1}{\sqrt[5]{16}}$

70. $\dfrac{4}{\sqrt[4]{32}}$

In Exercises 71–86, do each division by rationalizing the denominator and simplifying. All variables represent positive numbers.

71. $\dfrac{1}{\sqrt{2} - 1}$

72. $\dfrac{3}{\sqrt{3} - 1}$

73. $\dfrac{\sqrt{2}}{\sqrt{5} + 3}$

74. $\dfrac{\sqrt{3}}{\sqrt{3} - 2}$

75. $\dfrac{\sqrt{3} + 1}{\sqrt{3} - 1}$

76. $\dfrac{\sqrt{2} - 1}{\sqrt{2} + 1}$

77. $\dfrac{\sqrt{7} - \sqrt{2}}{\sqrt{2} + \sqrt{7}}$

78. $\dfrac{\sqrt{3} + \sqrt{2}}{\sqrt{3} - \sqrt{2}}$

79. $\dfrac{2}{\sqrt{x}+1}$ **80.** $\dfrac{3}{\sqrt{x}-2}$ **81.** $\dfrac{x}{\sqrt{x}-4}$ **82.** $\dfrac{2x}{\sqrt{x}+1}$

83. $\dfrac{2z-1}{\sqrt{2z}-1}$ **84.** $\dfrac{3t-1}{\sqrt{3t}+1}$

85. $\dfrac{\sqrt{x}-\sqrt{y}}{\sqrt{x}+\sqrt{y}}$ **86.** $\dfrac{\sqrt{x}+\sqrt{y}}{\sqrt{x}-\sqrt{y}}$

In Exercises 87–92, rationalize each numerator. All variables represent positive numbers.

87. $\dfrac{\sqrt{3}+1}{2}$ **88.** $\dfrac{\sqrt{5}-1}{2}$ **89.** $\dfrac{\sqrt{x}+3}{x}$ **90.** $\dfrac{2+\sqrt{x}}{5x}$

91. $\dfrac{\sqrt{x}+\sqrt{y}}{\sqrt{x}}$ **92.** $\dfrac{\sqrt{x}-\sqrt{y}}{\sqrt{x}+\sqrt{y}}$

Writing Exercises *Write a paragraph using your own words.*

1. Explain how to simplify a fraction with a denominator of $\sqrt[3]{3}$.

2. Explain how to simplify a fraction with a denominator of $\sqrt[3]{9}$.

Something to Think About *Assume that x is a rational number.*

1. Change the numerator of $\dfrac{\sqrt{x}-3}{4}$ to a rational number.

2. Rationalize the numerator of $\dfrac{2\sqrt{3x}+4}{\sqrt{3x}-1}$.

Review Exercises *Solve each equation.*

1. $\dfrac{2}{3-a}=1$

2. $5(s-4)=-5(s-4)$

3. $\dfrac{8}{b-2}+\dfrac{3}{2-b}=-\dfrac{1}{b}$

4. $\dfrac{2}{x-2}+\dfrac{1}{x+1}=\dfrac{1}{(x+1)(x-2)}$

7.5 Radical Equations

■ EQUATIONS CONTAINING ONE RADICAL ■ GRAPHING DEVICES ■ EQUATIONS CONTAINING TWO RADICALS ■ GRAPHING DEVICES ■ EQUATIONS CONTAINING THREE RADICALS

To solve equations that contain radicals, we will use the **power rule**.

Power Rule

If x, y, and n are real numbers and $x=y$, then

$$x^n = y^n$$

If we raise both sides of an equation to the same power, the resulting equation may or may not be equivalent to the original equation. For example, if we square both sides of the equation

1. $x = 3$ With a solution set of $\{3\}$.

we obtain the equation

2. $x^2 = 9$ With a solution set of $\{3, -3\}$.

Equations 1 and 2 are not equivalent, because they have different solution sets, and the solution -3 of Equation 2 does not satisfy Equation 1. Because raising both sides of an equation to the same power can produce an equation with roots that don't satisfy the original equation, we must always check each suspected solution in the original equation.

■ EQUATIONS CONTAINING ONE RADICAL

EXAMPLE 1   Solve the equation $\sqrt{x + 3} = 4$.

Solution   To eliminate the radical, we apply the power rule by squaring both sides of the equation and proceed as follows:

$$\sqrt{x + 3} = 4$$
$$\left(\sqrt{x + 3}\right)^2 = (4)^2 \qquad \text{Square both sides.}$$
$$x + 3 = 16$$
$$x = 13 \qquad \text{Subtract 3 from both sides.}$$

We must check the apparent solution of 13 to see whether it satisfies the original equation.

Check: $\sqrt{x + 3} = 4$
$$\sqrt{13 + 3} \overset{?}{=} 4 \qquad \text{Substitute 13 for } x.$$
$$\sqrt{16} \overset{?}{=} 4$$
$$4 = 4$$

Since 13 satisfies the original equation, it is a solution.   ■

To solve an equation with radicals, we follow these steps.

> ### Steps for Solving an Equation with Radicals
> **1.** Isolate one radical expression on one side of the equation.
> **2.** Raise both sides of the equation to the power that is the same as the index of the radical.
> **3.** Solve the resulting equation. If it still contains a radical, go back to Step 1.
> **4.** Check the possible solutions to eliminate the ones that do not satisfy the original equation.

EXAMPLE 2 **Height of a bridge** The distance d (in feet) that an object will fall in t seconds is given by the formula

$$t = \sqrt{\frac{d}{16}}$$

To find the height of a bridge above a river, a man drops a stone into the water (see Figure 7-7). If it takes the stone 3 seconds to hit the water, how far above the river is the bridge?

Solution We substitute 3 for t in the formula and solve for d.

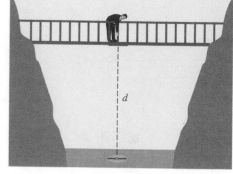

$$t = \sqrt{\frac{d}{16}}$$

$$3 = \sqrt{\frac{d}{16}}$$

$$9 = \frac{d}{16} \qquad \text{Square both sides.}$$

$$144 = d \qquad \text{Multiply both sides by 16.}$$

The bridge is 144 feet above the river.

FIGURE 7-7

EXAMPLE 3 Solve the equation $\sqrt{3x + 1} + 1 = x$.

Solution We first subtract 1 from both sides to isolate the radical. Then, to eliminate the radical, we square both sides of the equation and proceed as follows:

$$\sqrt{3x + 1} + 1 = x$$

$$\sqrt{3x + 1} = x - 1 \qquad \text{Subtract 1 from both sides.}$$

$$\left(\sqrt{3x + 1}\right)^2 = (x - 1)^2 \qquad \text{Square both sides to eliminate the square root.}$$

$$3x + 1 = x^2 - 2x + 1 \qquad (x-1)^2 \neq x^2 - 1. \text{ Instead, } (x-1)^2 = (x-1)(x-1) = x^2 - x - x + 1 = x^2 - 2x + 1.$$

$$0 = x^2 - 5x \qquad \text{Subtract } 3x \text{ and 1 from both sides.}$$

$$0 = x(x - 5) \qquad \text{Factor } x^2 - 5x.$$

$$x = 0 \quad \text{or} \quad x - 5 = 0 \qquad \text{Set each factor equal to 0.}$$

$$x = 0 \quad | \quad x = 5$$

We must check each apparent solution to see whether it satisfies the original equation.

Check:

$$\sqrt{3x + 1} + 1 = x \qquad\qquad \sqrt{3x + 1} + 1 = x$$

$$\sqrt{3(0) + 1} + 1 \overset{?}{=} 0 \qquad\qquad \sqrt{3(5) + 1} + 1 \overset{?}{=} 5$$

$$\sqrt{1} + 1 \overset{?}{=} 0 \qquad\qquad\qquad \sqrt{16} + 1 \overset{?}{=} 5$$

$$2 \neq 0 \qquad\qquad\qquad\qquad\qquad 5 = 5$$

Since the apparent solution 0 does not check, it must be discarded. The only solution of the original equation is 5.

EXAMPLE 4 Solve the equation $\sqrt[3]{x^3 + 7} = x + 1$.

Solution To eliminate the radical, we cube both sides of the equation and proceed as follows:

$$\sqrt[3]{x^3 + 7} = x + 1$$

$$\left(\sqrt[3]{x^3 + 7}\right)^3 = (x + 1)^3 \qquad \text{Cube both sides to eliminate the cube root.}$$

$$x^3 + 7 = x^3 + 3x^2 + 3x + 1$$

$$0 = 3x^2 + 3x - 6 \qquad \text{Subtract } x^3 \text{ and } 7 \text{ from both sides.}$$

$$0 = x^2 + x - 2 \qquad \text{Divide both sides by 3.}$$

$$0 = (x + 2)(x - 1) \qquad \text{Factor the trinomial.}$$

$$x + 2 = 0 \qquad \text{or} \qquad x - 1 = 0$$
$$x = -2 \qquad | \qquad x = 1$$

We check each apparent solution to see whether it satisfies the original equation.

Check:

$$\sqrt[3]{x^3 + 7} = x + 1 \qquad\qquad \sqrt[3]{x^3 + 7} = x + 1$$

$$\sqrt[3]{(-2)^3 + 7} \stackrel{?}{=} -2 + 1 \qquad\qquad \sqrt[3]{1^3 + 7} \stackrel{?}{=} 1 + 1$$

$$\sqrt[3]{-8 + 7} \stackrel{?}{=} -1 \qquad\qquad \sqrt[3]{8} \stackrel{?}{=} 2$$

$$\sqrt[3]{-1} \stackrel{?}{=} -1 \qquad\qquad 2 = 2$$

$$-1 = -1$$

Both solutions satisfy the original equation. ∎

■ GRAPHING DEVICES

To find approximate solutions for $\sqrt{3x + 1} + 1 = x$ with a graphing device, we use window settings of $[-5, 10]$ for x and $[-2, 8]$ for y and graph the functions $f(x) = \sqrt{3x + 1} + 1$ and $g(x) = x$, as in Figure 7-8(a). We then trace to find the approximate x-coordinate of their intersection point, as in Figure 7-8(b). We can zoom to get better results.

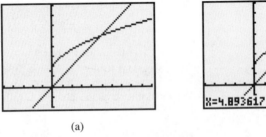

(a)   (b)

FIGURE 7-8

■ EQUATIONS CONTAINING TWO RADICALS

When more than one radical appears in an equation, it is often necessary to apply the power rule more than once.

EXAMPLE 5 Solve the equation $\sqrt{x} + \sqrt{x+2} = 2$.

Solution To remove the radicals, we must square both sides of the equation. This is easier to do if one radical is on each side of the equation. So, we subtract $\sqrt{x}$ from both sides to isolate one radical on one side of the equation.

$$\sqrt{x} + \sqrt{x+2} = 2$$

$$\sqrt{x+2} = 2 - \sqrt{x} \qquad \text{Subtract } \sqrt{x} \text{ from both sides.}$$

$$\left(\sqrt{x+2}\right)^2 = \left(2 - \sqrt{x}\right)^2 \qquad \text{Square both sides to eliminate one square root.}$$

$$x + 2 = 4 - 4\sqrt{x} + x \qquad \begin{aligned}\left(2 - \sqrt{x}\right)\left(2 - \sqrt{x}\right) &= 4 - 2\sqrt{x} - 2\sqrt{x} + x \\ &= 4 - 4\sqrt{x} + x.\end{aligned}$$

$$2 = 4 - 4\sqrt{x} \qquad \text{Subtract } x \text{ from both sides.}$$

$$-2 = -4\sqrt{x} \qquad \text{Subtract 4 from both sides.}$$

$$\frac{1}{2} = \sqrt{x} \qquad \text{Divide both sides by } -4.$$

$$\frac{1}{4} = x \qquad \text{Square both sides.}$$

Check:
$$\sqrt{x} + \sqrt{x+2} = 2$$

$$\sqrt{\frac{1}{4}} + \sqrt{\frac{1}{4} + 2} \stackrel{?}{=} 2$$

$$\frac{1}{2} + \sqrt{\frac{9}{4}} \stackrel{?}{=} 2$$

$$\frac{1}{2} + \frac{3}{2} \stackrel{?}{=} 2$$

$$2 = 2$$

The solution checks. ∎

■ GRAPHING DEVICES

To find approximate solutions for $\sqrt{x} + \sqrt{x+2} = 2$ with a graphing device, we use settings of $[-2, 10]$ for x and $[-2, 8]$ for y and graph the functions $f(x) = \sqrt{x} + \sqrt{x+2}$ and $g(x) = 2$, as in Figure 7-9(a). We then trace to find an approximation of the x-coordinate of their intersection point, as in Figure 7-9(b). We can zoom to get better results.

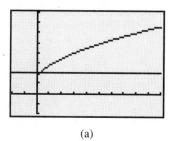

(a)

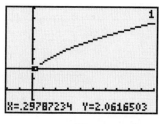

X=.29787234   Y=2.0616503

(b)

FIGURE 7-9

■ EQUATIONS CONTAINING THREE RADICALS

EXAMPLE 6   Solve the equation $\sqrt{x+2} + \sqrt{2x} = \sqrt{18 - x}$.

Solution   In this case, it is impossible to isolate one radical on each side of the equation, so we begin by squaring both sides. Then we proceed as follows.

$$\sqrt{x+2} + \sqrt{2x} = \sqrt{18 - x}$$
$$\left(\sqrt{x+2} + \sqrt{2x}\right)^2 = \left(\sqrt{18-x}\right)^2 \qquad \text{Square both sides to eliminate one square root.}$$

$$x + 2 + 2\sqrt{x+2}\sqrt{2x} + 2x = 18 - x$$
$$2\sqrt{x+2}\sqrt{2x} = 16 - 4x \qquad \text{Subtract } 3x \text{ and } 2 \text{ from both sides.}$$

$$\sqrt{x+2}\sqrt{2x} = 8 - 2x \qquad \text{Divide both sides by 2.}$$
$$\left(\sqrt{x+2}\sqrt{2x}\right)^2 = (8 - 2x)^2 \qquad \text{Square both sides to eliminate the other square roots.}$$

$$(x+2)2x = 64 - 32x + 4x^2$$
$$2x^2 + 4x = 64 - 32x + 4x^2$$
$$0 = 2x^2 - 36x + 64 \qquad \text{Write the equation in quadratic form.}$$

$$0 = x^2 - 18x + 32 \qquad \text{Divide both sides by 2.}$$
$$0 = (x - 16)(x - 2) \qquad \text{Factor the trinomial.}$$

$$x - 16 = 0 \qquad \text{or} \qquad x - 2 = 0 \qquad \text{Set each factor equal to 0.}$$
$$x = 16 \qquad | \qquad x = 2$$

Verify that 2 satisfies the equation, but 16 does not. The only solution is 2.   ■

Orals   *Solve each equation.*

1. $\sqrt{x+2} = 3$ **2.** $\sqrt{x-2} = 1$

3. $\sqrt[3]{x+1} = 1$ **4.** $\sqrt[3]{x-1} = 2$

5. $\sqrt[4]{x-1} = 2$ **6.** $\sqrt[5]{x+1} = 2$

EXERCISE 7.5

In Exercises 1–54, solve each equation. Write all possible solutions and cross out those that do not satisfy the equation.

1. $\sqrt{5x - 6} = 2$ **2.** $\sqrt{7x - 10} = 12$ **3.** $\sqrt{6x + 1} + 2 = 7$ **4.** $\sqrt{6x + 13} - 2 = 5$

5. $2\sqrt{4x + 1} = \sqrt{x + 4}$ **6.** $\sqrt{3(x + 4)} = \sqrt{5x - 12}$

7. $\sqrt[3]{7n - 1} = 3$ **8.** $\sqrt[3]{12m + 4} = 4$

9. $\sqrt[4]{10p + 1} = \sqrt[4]{11p - 7}$ **10.** $\sqrt[4]{10y + 2} = 2\sqrt[4]{2}$

11. $x = \dfrac{\sqrt{12x - 5}}{2}$ **12.** $x = \dfrac{\sqrt{16x - 12}}{2}$

13. $\sqrt{x + 2} = \sqrt{4 - x}$ **14.** $\sqrt{6 - x} = \sqrt{2x + 3}$

15. $2\sqrt{x} = \sqrt{5x - 16}$

16. $3\sqrt{x} = \sqrt{3x + 12}$

17. $r - 9 = \sqrt{2r - 3}$

18. $-s - 3 = 2\sqrt{5 - s}$

19. $\sqrt{-5x + 24} = 6 - x$

20. $\sqrt{-x + 2} = x - 2$

21. $\sqrt{y + 2} = 4 - y$

22. $\sqrt{22y + 86} = y + 9$

23. $\sqrt{x}\sqrt{x + 16} = 15$

24. $\sqrt{x}\sqrt{x + 6} = 4$

25. $\sqrt[3]{x^3 - 7} = x - 1$

26. $\sqrt[3]{x^3 + 56} - 2 = x$

27. $\sqrt[4]{x^4 + 4x^2 - 4} = -x$

28. $\sqrt[4]{8x - 8} + 2 = 0$

29. $\sqrt[4]{12t + 4} + 2 = 0$

30. $u = \sqrt[4]{u^4 - 6u^2 + 24}$

31. $\sqrt{2y + 1} = 1 - 2\sqrt{y}$

32. $\sqrt{u} + 3 = \sqrt{u - 3}$

33. $\sqrt{y + 7} + 3 = \sqrt{y + 4}$

34. $1 + \sqrt{z} = \sqrt{z + 3}$

35. $\sqrt{v} + \sqrt{3} = \sqrt{v + 3}$

36. $\sqrt{x + 2} = \sqrt{x + 4}$

37. $2 + \sqrt{u} = \sqrt{2u + 7}$

38. $5r + 4 = \sqrt{5r + 20} + 4r$

39. $\sqrt{6t + 1} - 3\sqrt{t} = -1$

40. $\sqrt{4s + 1} - \sqrt{6s} = -1$

41. $\sqrt{2x + 5} + \sqrt{x + 2} = 5$

42. $\sqrt{2x + 5} + \sqrt{2x + 1} + 4 = 0$

43. $\sqrt{z - 1} + \sqrt{z + 2} = 3$

44. $\sqrt{16v + 1} + \sqrt{8v + 1} = 12$

45. $\sqrt{x - 5} - \sqrt{x + 3} = 4$

46. $\sqrt{x + 8} - \sqrt{x - 4} = -2$

47. $\sqrt{x + 1} + \sqrt{3x} = \sqrt{5x + 1}$

48. $\sqrt{3x} - \sqrt{x + 1} = \sqrt{x - 2}$

49. $\sqrt{\sqrt{a} + \sqrt{a + 8}} = 2$

50. $\sqrt{\sqrt{2y} - \sqrt{y - 1}} = 1$

51. $\dfrac{6}{\sqrt{x + 5}} = \sqrt{x}$

52. $\dfrac{\sqrt{2x}}{\sqrt{x + 2}} = \sqrt{x - 1}$

53. $\sqrt{x + 2} + \sqrt{2x - 3} = \sqrt{11 - x}$

54. $\sqrt{8 - x} - \sqrt{3x - 8} = \sqrt{x - 4}$

55. Banked curves   A highway curve banked at 8° will accommodate traffic traveling s mph if the radius of the curve is r feet, according to the formula $s = 1.45\sqrt{r}$. If highway engineers expect 65-mph traffic, what radius should they specify? (See Illustration 1.)

56. Horizon distance   The higher a lookout tower is built, the farther an observer can see. (See Illustration 2.) That distance d (called the *horizon distance*, measured in miles) is related to the height h of the observer (measured in feet) by the formula $d = 1.4\sqrt{h}$. How tall must a lookout tower be to see the edge of the forest, 25 miles away?

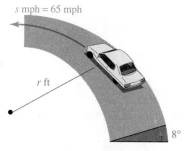

ILLUSTRATION 1

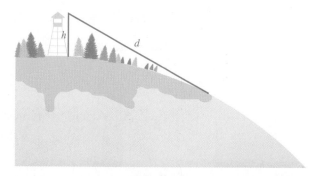

ILLUSTRATION 2

57. Producing power   The power generated by a certain windmill is related to the velocity of the wind by the formula

$$v = \sqrt[3]{\frac{P}{0.02}}$$

where P is the power (in watts) and v is the velocity of the wind (in mph). Find the speed of the wind when the windmill is generating 500 watts of power.

58. Carpentry   During construction, carpenters often brace walls as shown in Illustration 3, where the length of the brace is given by the formula $l = \sqrt{f^2 + h^2}$. If a carpenter nails a 10-foot brace to the wall 6 feet above the floor, how far from the base of the wall should he nail the brace to the floor?

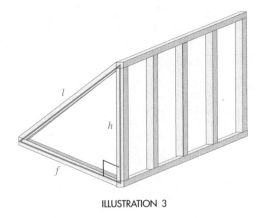

ILLUSTRATION 3

In Exercises 59–60, use a graphing device.

59. Marketing   The number of wrenches that will be produced at a given price can be predicted by the formula $s = \sqrt{5x}$, where s is the supply (in thousands) and x is the price (in dollars). If the demand, d, for wrenches can be predicted by the formula $d = \sqrt{100 - 3x^2}$, find the equilibrium price.

60. Marketing   The number of footballs that will be produced at a given price can be predicted by the formula $s = \sqrt{23x}$, where s is the supply (in thousands) and x is the price (in dollars). If the demand, d, for footballs can be predicted by the formula $d = \sqrt{312 - 2x^2}$, find the equilibrium price.

Writing Exercises   *Write a paragraph using your own words.*

1. If both sides of an equation are raised to the same power, the resulting equation might not be equivalent to the original equation. Explain.

2. Explain why you must check each apparent solution of a radical equation.

Something to Think About   **1.** Solve $\sqrt[3]{2x} = \sqrt{x}$.
(*Hint:* Square and then cube both sides.)

2. Solve $\sqrt[4]{x} = \sqrt{\dfrac{x}{4}}$.

Review Exercises   *If $f(x) = 3x^2 - 4x + 2$, find each quantity.*

1. $f(0)$ **2.** $f(-3)$ **3.** $f(2)$ **4.** $f\left(\dfrac{1}{2}\right)$

7.6 Applications of Radicals

■ THE PYTHAGOREAN THEOREM ■ SOME SPECIAL TRIANGLES ■ THE DISTANCE FORMULA

One of the important results from geometry is the Pythagorean theorem.

■ THE PYTHAGOREAN THEOREM

If we know the lengths of two sides of a right triangle, we can always find the length of the **hypotenuse** (the side opposite the 90° angle) by using the *Pythagorean theorem*.

> **Pythagorean Theorem**
> If a and b are the lengths of two sides of a right triangle and c is the length of the hypotenuse, then
> $$a^2 + b^2 = c^2$$

In words, the Pythagorean theorem says:

> *In any right triangle, the square of the hypotenuse is equal to the sum of the squares of the other two sides.*

Suppose the right triangle shown in Figure 7-10 has sides of length 3 and 4 units. To find the length of the hypotenuse, we use the Pythagorean theorem.

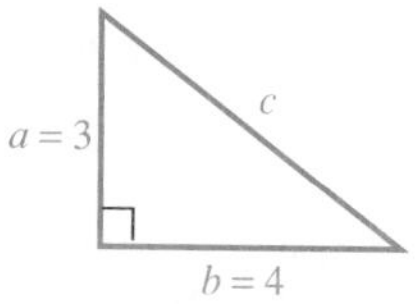

$$a^2 + b^2 = c^2$$
$$3^2 + 4^2 = c^2$$
$$9 + 16 = c^2$$
$$25 = c^2$$
$$\sqrt{25} = \sqrt{c^2} \qquad \text{Take the positive square root of both sides.}$$
$$5 = c$$

FIGURE 7-10

The length of the hypotenuse is 5 units.

EXAMPLE 1

Fighting fires To fight a forest fire, the forestry department plans to clear a rectangular fire break around the fire, as shown in Figure 7-11. Crews are equipped with mobile communications with a 3000-yard range. Can crews at points A and B remain in radio contact?

Solution Points A, B, and C form a right triangle. To find the distance c from point A to point B, we can use the Pythagorean theorem, substituting 2400 for a and 1000 for b and solving for c.

$$a^2 + b^2 = c^2$$
$$2400^2 + 1000^2 = c^2$$
$$5{,}760{,}000 + 1{,}000{,}000 = c^2$$
$$6{,}760{,}000 = c^2$$
$$\sqrt{6{,}760{,}000} = \sqrt{c^2} \qquad \text{Take the positive square root of both sides.}$$
$$2600 = c \qquad \text{Use a calculator to find the square root.}$$

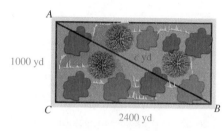

FIGURE 7-11

The two fire crews are 2600 yards apart. Because this distance is less than the range of the radios, the crews can communicate.

■ SOME SPECIAL TRIANGLES

An **isosceles right triangle** is a right triangle with two sides of equal length. If we know the length of one leg of an isosceles right triangle, we can use the Pythagorean theorem to find the length of the hypotenuse. Since the triangle shown in Figure 7-12 is a right triangle, we have

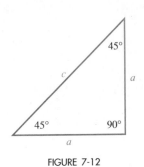

FIGURE 7-12

$$c^2 = a^2 + a^2 \qquad \text{Use the Pythagorean theorem.}$$
$$c^2 = 2a^2 \qquad \text{Combine like terms.}$$
$$c = \sqrt{2a^2} \qquad \text{Take the positive square root of both sides.}$$
$$c = a\sqrt{2} \qquad \sqrt{2a^2} = \sqrt{2}\sqrt{a^2} = a\sqrt{2}.$$

Thus, in an isosceles right triangle, the length of the hypotenuse is the length of one leg times $\sqrt{2}$.

EXAMPLE 2 If one leg of the isosceles right triangle shown in Figure 7-12 is 10 feet long, find the length of the hypotenuse.

Solution Since the length of the hypotenuse is the length of a leg times $\sqrt{2}$, we have

$$c = 10\sqrt{2}$$

The length of the hypotenuse is $10\sqrt{2}$ feet. To two decimal places, the length is 14.14 feet. ■

If the length of the hypotenuse of an isosceles right triangle is known, we can use the Pythagorean theorem to find the length of each leg.

EXAMPLE 3 Find the length of each leg of the isosceles right triangle shown in Figure 7-13.

Solution We use the Pythagorean theorem.

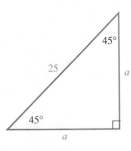

FIGURE 7-13

$$c^2 = a^2 + a^2$$
$$25^2 = 2a^2 \qquad \text{Substitute 25 for } c \text{ and combine like terms.}$$
$$\frac{625}{2} = a^2 \qquad \text{Square 25 and divide both sides by 2.}$$
$$\sqrt{\frac{625}{2}} = a \qquad \text{Take the positive square root of both sides.}$$
$$\sqrt{\frac{625 \cdot 2}{2 \cdot 2}} = a \qquad \text{Multiply the numerator and denominator of the radicand by 2.}$$
$$\frac{25\sqrt{2}}{2} = a \qquad \text{Simplify the expression:}$$

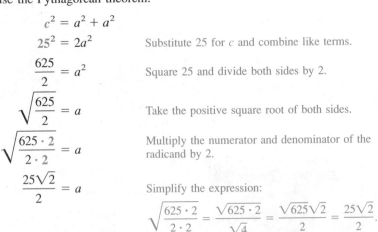

The length of each leg is $\dfrac{25\sqrt{2}}{2}$ units. To two decimal places, the length is 17.68 units. ■

From geometry, we know that an **equilateral triangle** is a triangle with three sides of equal length and three 60° angles. If an *altitude* is drawn upon the base of an equilateral triangle, as shown in Figure 7-14, it bisects the base and divides the triangle into two 30°–60°–90° triangles. We can see that the shortest side of each 30°–60°–90° triangle is a units long. Thus,

The shortest side of a 30°–60°–90° right triangle is half as long as its hypotenuse.

We can find the length of the altitude, h, by using the Pythagorean theorem.

$$a^2 + h^2 = (2a)^2$$
$$a^2 + h^2 = 4a^2 \qquad (2a)^2 = (2a)(2a) = 4a^2.$$
$$h^2 = 3a^2 \qquad \text{Subtract } a^2 \text{ from both sides.}$$
$$h = \sqrt{3a^2} \qquad \text{Take the positive square root of both sides.}$$
$$h = a\sqrt{3} \qquad \sqrt{3a^2} = \sqrt{3}\sqrt{a^2} = a\sqrt{3}.$$

Thus,

The length of the longer leg is the length of the shorter leg times $\sqrt{3}$.

FIGURE 7-14

EXAMPLE 4 Find the length of the hypotenuse and the longer leg of the right triangle shown in Figure 7-15.

Solution Since the shorter leg of a 30°–60°–90° right triangle is half as long as its hypotenuse, the hypotenuse is 12 centimeters long.

Since the length of the longer leg is the length of the shorter leg times $\sqrt{3}$, the longer leg is $6\sqrt{3}$ (or about 10.39) centimeters long.

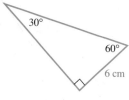

FIGURE 7-15 ∎

EXAMPLE 5 Find the length of each leg of the triangle shown in Figure 7-16.

Solution Since the shorter leg of a 30°–60°–90° right triangle is half as long as its hypotenuse, the shorter leg is $\frac{9}{2}$ centimeters long.

Since the length of the longer leg is the length of the shorter leg times $\sqrt{3}$, the longer leg is $\frac{9}{2}\sqrt{3}$ (or about 7.79) centimeters long.

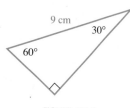

FIGURE 7-16 ∎

■ THE DISTANCE FORMULA

With the *distance formula*, we can find the distance between any two points that are graphed on a rectangular coordinate system.

To find the distance d between points $P(x_1, y_1)$ and $Q(x_2, y_2)$ shown in Figure 7-17, we construct the right triangle PRQ. The distance between P and R is $|x_2 - x_1|$, and the distance between R and Q is $|y_2 - y_1|$. We apply the Pythagorean theorem to the right triangle PRQ to get

$$[d(PQ)]^2 = |x_2 - x_1|^2 + |y_2 - y_1|^2 \qquad \begin{array}{l}\text{Read } d(PQ) \text{ as "the distance between} \\ P \text{ and } Q."\end{array}$$

$$= (x_2 - x_1)^2 + (y_2 - y_1)^2 \qquad \begin{array}{l}\text{Because } |x_2 - x_1|^2 = (x_2 - x_1)^2 \text{ and} \\ |y_2 - y_1|^2 = (y_2 - y_1)^2.\end{array}$$

or

1. $d(PQ) = \sqrt{(x_2 - x_1)^2 + (y_2 - y_1)^2}$

Equation 1 is the *distance formula*. Because it is one of the most important formulas in mathematics, take the time to memorize it.

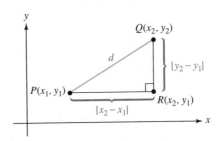

FIGURE 7-17

Distance Formula
The distance between two points $P(x_1, y_1)$ and $Q(x_2, y_2)$ is given by the formula
$$d(PQ) = \sqrt{(x_2 - x_1)^2 + (y_2 - y_1)^2}$$

EXAMPLE 6   Find the distance between points $P(-2, 3)$ and $Q(4, -5)$.

Solution   To find the distance, we can use the distance formula by substituting 4 for x_2, -2 for x_1, -5 for y_2, and 3 for y_1.

$$\begin{aligned} d(PQ) &= \sqrt{(x_2 - x_1)^2 + (y_2 - y_1)^2} \\ &= \sqrt{[4 - (-2)]^2 + (-5 - 3)^2} \\ &= \sqrt{(4 + 2)^2 + (-5 - 3)^2} \\ &= \sqrt{6^2 + (-8)^2} \\ &= \sqrt{36 + 64} \\ &= \sqrt{100} \\ &= 10 \end{aligned}$$

The distance between points P and Q is 10 units.   ∎

EXAMPLE 7   **Building a freeway**   In a city, streets run north and south and avenues run east and west. Streets and avenues are 750 feet apart. The city plans to construct a straight freeway from the intersection of 21st Street and 4th Avenue to the intersection of 111th Street and 60th Avenue. How long will the freeway be?

Solution   We can represent the roads of the city by the coordinate system in Figure 7-18, where the units on each axis represent 750 feet. We represent the end of the freeway at 21st Street and 4th Avenue by the point $(x_1, y_1) = (21, 4)$. The other end is $(x_2, y_2) = (111, 60)$.

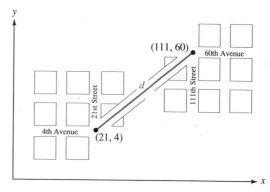

FIGURE 7-18

We can now use the distance formula to find the length of the freeway.

$$d = \sqrt{(x_2 - x_1)^2 + (y_2 - y_1)^2}$$
$$d = \sqrt{(111 - 21)^2 + (60 - 4)^2}$$
$$= \sqrt{8100 + 3136}$$
$$= \sqrt{11{,}236}$$
$$= 106 \qquad\qquad \text{Use a calculator to find the square root.}$$

Because each unit represents 750 feet, the length of the freeway is $106 \cdot 750 = 79{,}500$ feet. Since 5280 feet $= 1$ mile, we can divide 79,500 by 5280 to convert 79,500 feet to 15.056818 miles. The freeway will be about 15 miles long.   ■

EXAMPLE 8   **Bowling**   The velocity, v, of an object after it has fallen d feet is given by the equation $v^2 = 64d$. An inexperienced bowler lofts the ball 4 feet. With what velocity will it strike the alley?

Solution   We find the velocity by substituting 4 for d in the equation $v^2 = 64d$ and solving for v.

$$v^2 = 64d$$
$$v^2 = 64(4)$$
$$v^2 = 256$$
$$v = \sqrt{256} \qquad\qquad \text{Take the square root of both sides. Only}$$
$$\text{the positive square root is meaningful.}$$
$$= 16$$

The ball will strike the alley with a velocity of 16 feet per second.   ■

Orals   *Evaluate each expression.*

1. $\sqrt{25}$ **2.** $\sqrt{100}$ **3.** $\sqrt{169}$

4. $\sqrt{3^2 + 4^2}$ **5.** $\sqrt{8^2 + 6^2}$ **6.** $\sqrt{5^2 + 12^2}$

7. $\sqrt{5^2 - 3^2}$ **8.** $\sqrt{5^2 - 4^2}$ **9.** $\sqrt{169 - 12^2}$

EXERCISE 7.6

In Exercises 1–4, the lengths of two sides of the right triangle ABC shown in Illustration 1 are given. Find the length of the missing side.

1. $a = 6$ ft and $b = 8$ ft

2. $a = 10$ cm and $c = 26$ cm

3. $b = 18$ m and $c = 82$ m

4. $a = 14$ in. and $c = 50$ in.

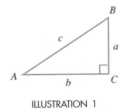

ILLUSTRATION 1

5. Sailing Refer to the sailboat in Illustration 2. How long must a rope be to fasten the top of the mast to the bow?

6. Carpentry The gable end of the roof shown in Illustration 3 is divided in half by a vertical brace. Find the distance from eaves to peak.

ILLUSTRATION 2

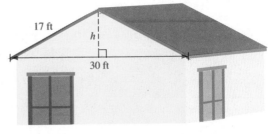

ILLUSTRATION 3

In Exercises 7–10, use a calculator. The baseball diamond shown in Illustration 4 is a square, 90 feet on a side.

7. Baseball How far must a catcher throw the ball to throw out a runner stealing second base?

8. Baseball In baseball, the pitcher's mound is 60 feet, 6 inches from home plate. How far from the mound is second base?

9. Baseball If the third baseman fields a ground ball 10 feet directly behind third base, how far must he throw the ball to throw the batter out at first base?

10. Baseball A shortstop fields a grounder at a point one-third of the way from second base to third base. How far will he have to throw the ball to make an out at first base?

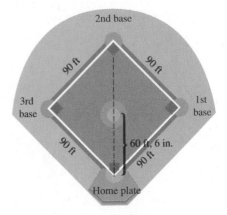

ILLUSTRATION 4

In Exercises 11–18, find the missing lengths in each triangle. Give each answer to two decimal places.

11.

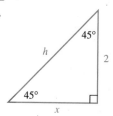

12.

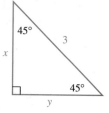

13.

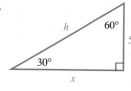

14.

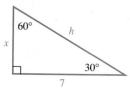

15.

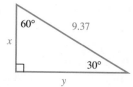

16.

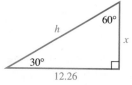

17.

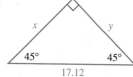

18.

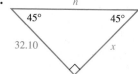

19. Geometry Find the length of the diagonal of one of the faces of the cube shown in Illustration 5.

20. Geometry Find the length of the diagonal of the cube shown in Illustration 5.

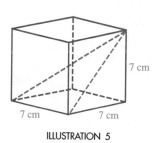

ILLUSTRATION 5

In Exercises 21–30, find the distance between P and Q.

21. $Q(0, 0)$, $P(3, -4)$

22. $Q(0, 0)$, $P(-6, 8)$

23. $P(2, 4)$, $Q(5, 8)$

24. $P(5, 9)$, $Q(8, 13)$

25. $P(-2, -8)$, $Q(3, 4)$

26. $P(-5, -2)$, $Q(7, 3)$

27. $P(6, 8)$, $Q(12, 16)$

28. $P(10, 4)$, $Q(2, -2)$

29. $Q(-3, 5)$, $P(-5, -5)$

30. $Q(2, -3)$, $P(4, -8)$

31. Geometry Show that a triangle with vertices at $(-2, 4)$, $(2, 8)$, and $(6, 4)$ is isosceles.

32. Geometry Show that a triangle with vertices at $(-2, 13)$, $(-8, 9)$, and $(-2, 5)$ is isosceles.

33. Finding the equation of a line Every point on the line CD in Illustration 6 is equidistant from points A and B. Use the distance formula to find the equation of line CD.

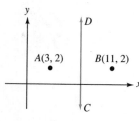

ILLUSTRATION 6

34. Geometry Show that a triangle with vertices at $(2, 3)$, $(-3, 4)$, and $(1, -2)$ is a right triangle. (*Hint:* If the Pythagorean relation holds, then the triangle is a right triangle.)

35. Geometry Find the coordinates of the two points on the x-axis that are $\sqrt{5}$ units from the point $(5, 1)$.

36. Geometry The square in Illustration 7 has an area of 18 square units, and its diagonals lie on the x- and y-axes. Find the coordinates of each corner of the square.

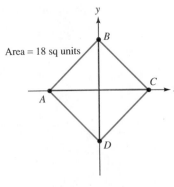

ILLUSTRATION 7

 In Exercises 37–46, use a calculator.

37. Packing a shotgun The diagonal d of a rectangular box with dimensions $a \times b \times c$ is given by

$$d = \sqrt{a^2 + b^2 + c^2}$$

Can a hunter fit a 32-inch shotgun into the shipping carton in Illustration 8?

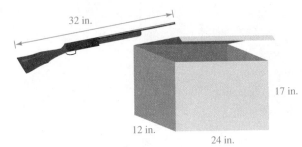

ILLUSTRATION 8

38. Shipping packages A delivery service won't accept a package for shipping if any dimension exceeds 21 inches. An archeologist wants to ship a 36-inch femur. Will it fit in a 3-inch-tall box that has a 21-inch-square base?

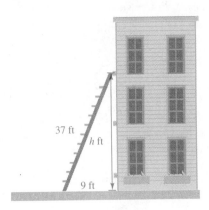

ILLUSTRATION 9

39. Shipping packages Can the archeologist in Exercise 38 ship the femur in a cubical box 21 inches on an edge?

40. Reach of a ladder The base of the 37-foot ladder in Illustration 9 is 9 feet from the wall. Will the top reach a window ledge that is 35 feet above the ground?

41. Telephone service The telephone cable in Illustration 10 runs from A to B to C to D. How much cable is required to run from A to D directly?

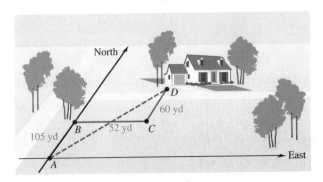

ILLUSTRATION 10

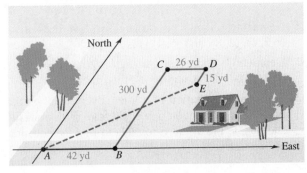

ILLUSTRATION 11

42. Electric service   The power company routes its lines as in Illustration 11. How much wire could be saved by going directly from A to E?

43. Supporting a weight   A weight placed on the tight wire in Illustration 12 pulls the center down 1 foot. By how much is the wire stretched? Round the answer to the nearest hundredth of a foot.

44. Geometry   The side, s, of a square with area A square feet is given by the formula $s = \sqrt{A}$. Find the perimeter of a square with an area of 49 square feet.

45. Volume of a cube   The total surface area, A, of a cube is related to its volume, V, by the formula $A = 6\sqrt[3]{V^2}$. Find the volume of a cube with a surface area of 24 square centimeters.

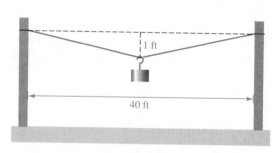

ILLUSTRATION 12

46. Area of many cubes   A grain of table salt is a cube with a volume of approximately 6×10^{-6} cubic inches, and there are about 1.5 million grains of salt in one cup. Find the total surface area of the salt in one cup. (See Exercise 45.)

Writing Exercises   *Write a paragraph using your own words.*

1. State the Pythagorean theorem.

2. Explain the distance formula.

Something to Think About

1. The formula

$$I = \frac{704w}{h^2}$$

(where w is weight in pounds and h is height in inches) can be used to estimate body mass index, I. The scale, shown in Table 1, can be used to judge a person's risk of heart attack. A girl weighing 104 pounds has a body mass index of 25. How tall is she?

Mass Index	Risk
20–26	normal
27–29	higher risk
30 and above	very high risk

TABLE 1

Review Exercises   *Find each product.*

1. $(4x + 2)(3x - 5)$

2. $(3y - 5)(2y + 3)$

3. $(5t + 4s)(3t - 2s)$

4. $(4r - 3)(2r^2 + 3r - 4)$

■ ■ ■ ■ ■ ■ ■ ■ ■ PROBLEMS AND PROJECTS

1. The area of a square tile is 15 square inches. Find the perimeter of the tile. Determine the number of tiles needed to make a tile border 10 feet long if the tiles are to be spaced $\frac{1}{4}$ inch apart.

2. Cardboard boxes in the shape of a cube are to be manufactured so that each one will hold 10 cubic feet of styrofoam pellets. Find the surface area of each box.

3. The area of a triangle can be found by using Heron's formula if the lengths of all three sides are known.

$$A = \sqrt{s(s - a)(s - b)(s - c)},$$

where $s = \dfrac{a + b + c}{2}$

Find the cost of the property shown in Illustration 1 if the owner has set a price of $6600 per acre. (1 acre = 43,560 ft².)

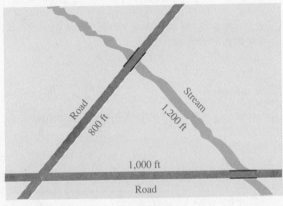

Road
800 ft
Stream
1,200 ft
1,000 ft
Road

ILLUSTRATION 1

4. The back corner of a fenced yard is to be enclosed to form a dog pen. What will it cost to enclose the area shown in Illustration 2? (Assume that new fencing costs $21.45 per foot.) How much will it cost to paint the existing fence if the painter has bid $3.25 per foot?

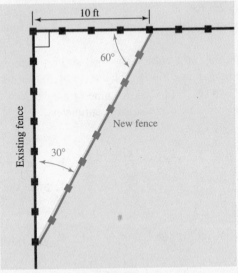

10 ft
60°
30°
Existing fence
New fence

ILLUSTRATION 2

5. Determine the lengths of sides a, b, and c, shown in Illustration 3. Do you notice a pattern? How many triangles will be necessary to produce a side of length 3?

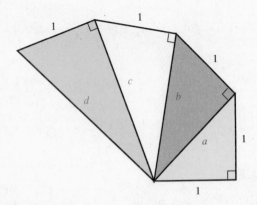

1
1
1
1
c
d
b
a
1
1

ILLUSTRATION 3

6. A sheet metal shop has a surplus of circular sheets of metal. However, the only orders received lately are for square sheets of metal. Express the dimensions of the largest square that can be cut from circular sheets of diameter 2 meters. (See Illustration 4.) If the sheet metal is priced at $3.56 per square meter and the cost to cut each sheet is 50¢, how much money will be lost by converting 169 circular sheets into square sheets?

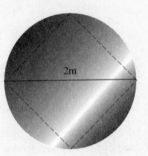

2m

ILLUSTRATION 4

PROJECT 1   Tom and Brian arrange to have a bicycle race. Each leaves his own house at the same time and rides to the other's house, whereupon the winner of the race calls his own house and leaves a message for the loser. A map of the race is shown in Illustration 5. Brian stays on the highway, averaging 21 miles per hour. Tom knows that he and Brian are evenly matched when biking on the highway, so he cuts across country for the first part of his trip, averaging 15 miles per hour. When Tom reaches the highway at point A, he turns right and follows the highway, averaging 21 miles per hour.

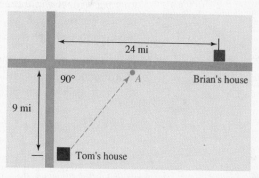

24 mi

90°   A   Brian's house

9 mi

Tom's house

ILLUSTRATION 5

Tom and Brian never meet during the race, and amazingly, the race is a tie. Each of them calls the other at exactly the same moment!

a. How long (to the nearest second) did it take each person to complete the race?

b. How far from the intersection of the two highways is point A? (*Hint:* Set the travel times for Brian and Tom equal to each other. You may find two answers, but only one of them matches all of the information.)

c. Show that if Tom had started straight across country for Brian's house (in order to minimize the distance he had to travel), he would have lost the race. By how much time (to the nearest second) would he have lost? Then show that if Tom had biked across country to a point 9 miles from the intersection of the two highways, he would have won the race. By how much time (to the nearest second) would he have won?

PROJECT 2   You have been hired by the accounts receivable department of Widget Industries. Your boss, I. Tally, the head of accounting, informs you that those who purchase from Widget Industries are billed according to a pricing structure given by the following function:

$$\text{Price per widget in an order of } x \text{ widgets} = p(x) = p_0 - \frac{p_0}{2}\left(1 - \sqrt{1 - \frac{x^2}{k^2}}\right)$$

where x is the number of widgets ordered, p_0 is the standard price of a widget, and k is the maximum number of widgets that can be ordered at one time.

The formula was developed because it allows for changes in the standard price and the maximum order size without requiring that the entire pricing scheme be redone. Note that the total cost of an order of x widgets is $x \cdot p(x)$.

Tally is quite a taskmaster. On your first day, he assigns the following problems to you.

a. The pricing function is designed to benefit those who order the largest number of widgets. To see this, calculate the price (as a percentage of p_0) that a customer would pay for each widget if the customer ordered 100, 500, 800, or 1000 widgets. Assume that $k = 1000$. For what value of x is the cost per widget a minimum, and what is that minimum? Looking at the graph of $p(x)$ (using, say, 8 as p_0) on a graphing calculator will help you see how buying more widgets means that each widget costs less.

b. Tally tells you that in the past the company lost money because someone sent around a memo with a simplified, but incorrect, pricing function. Find the error(s) in the simplification given below.

$$p(x) = p_0 - \frac{p_0}{2}\left(1 - \sqrt{1 - \frac{x^2}{k^2}}\right)$$

$$= p_0 - \frac{p_0}{2}\left[1 - \left(1 - \frac{x}{k}\right)\right]$$

$$= p_0 - \frac{p_0}{2}\left(\frac{x}{k}\right)$$

$$= p_0\left(1 - \frac{x}{2k}\right)$$

Show that $x = 0$ and 1000 are the *only* values of x for which the two formulas agree. (*Hint:* Set the two formulas equal to each other and solve for x.)

Now calculate the amount of money lost by using the incorrect formula, rather than the correct formula, on orders of 100, 500, 800, and 1000 widgets. For these calculations, assume that $k = 1000$ and $p_0 = \$5.60$.

c. Now is your chance to show Tally a thing or two. The pricing function *can* be simplified. Demonstrate to Tally that his pricing function is equivalent to

$$p(x) = \frac{p_0}{2}\left(1 + \frac{\sqrt{k^2 - x^2}}{k}\right)$$

Use this function to help a customer who wants to know how many widgets she must order to receive a 30% discount off the standard price p_0. Again, assume that $k = 1000$ when doing this calculation.

Chapter Summary

Key Words

conjugate binomials (7.4)
cube root of x (7.1)
equilateral triangle (7.6)

even root (7.1)
hypotenuse (7.6)
index of a radical (7.1)

isosceles right triangle (7.6)
like radicals (7.3)
odd root (7.1)

power rule (7.5)

principal square root of x (7.1)

radical sign (7.1)

radicand (7.1)

rationalizing the
denominator (7.4)

similar radicals (7.3)

square root function (7.1)

square root of a (7.1)

standard deviation (7.1)

Key Ideas

(7.1) If n is an even natural number, $\sqrt[n]{a^n} = |a|$.
If n is an odd natural number, $\sqrt[n]{a^n} = a$.

If n is a natural number greater than 1 and x is a real number, then

If $x > 0$, then $\sqrt[n]{x}$ is the positive number such that $\left(\sqrt[n]{x}\right)^n = x$.

If $x = 0$, then $\sqrt[n]{x} = 0$.

If $x < 0$

$\begin{cases} \text{and } n \text{ is odd, then } \sqrt[n]{x} \text{ is the real number such} \\ \quad \text{that } \left(\sqrt[n]{x}\right)^n = x. \\ \text{and } n \text{ is even, then } \sqrt[n]{x} \text{ is not a real number.} \end{cases}$

(7.2) If n $(n > 1)$ is a natural number and $\sqrt[n]{x}$ is a real number, then $x^{1/n} = \sqrt[n]{x}$.

If n is even, $(x^n)^{1/n} = |x|$.

If n is a natural number greater than 1 and x is a real number, then

If $x > 0$, then $x^{1/n}$ is the positive number such that $\left(x^{1/n}\right)^n = x$.

If $x = 0$, then $x^{1/n} = 0$.

If $x < 0$

$\begin{cases} \text{and } n \text{ is odd, then } x^{1/n} \text{ is the real number such} \\ \quad \text{that } \left(x^{1/n}\right)^n = x. \\ \text{and } n \text{ is even, then } x^{1/n} \text{ is not a real number.} \end{cases}$

If m and n are positive integers, $x > 0$, and $\frac{m}{n}$ is in simplest form, then

$$x^{m/n} = \sqrt[n]{x^m} = \left(\sqrt[n]{x}\right)^m$$

$$x^{-m/n} = \frac{1}{x^{m/n}} \quad \text{and} \quad \frac{1}{x^{-m/n}} = x^{m/n} \quad (x \neq 0)$$

(7.3) Properties of radicals: $\sqrt[n]{ab} = \sqrt[n]{a}\sqrt[n]{b}$ and
if $b \neq 0$, $\sqrt[n]{\dfrac{a}{b}} = \dfrac{\sqrt[n]{a}}{\sqrt[n]{b}}$.

Like radicals can be combined by addition and subtraction: $3\sqrt{2} + 5\sqrt{2} = 8\sqrt{2}$.

Radicals that are not similar can often be simplified to radicals that are similar and then combined:
$$\sqrt{2} + \sqrt{8} = \sqrt{2} + \sqrt{4}\sqrt{2} = \sqrt{2} + 2\sqrt{2} = 3\sqrt{2}$$

(7.4) If two radicals have the same index, they can be multiplied: $\sqrt{3x}\sqrt{6x} = \sqrt{18x^2} = 3x\sqrt{2}$ $(x \geq 0)$.

To rationalize the binomial denominator of a fraction with two square roots, multiply both the numerator and the denominator by the conjugate of the binomial in the denominator.

(7.5) The power rule: If $x = y$, then $x^n = y^n$.

Raising both sides of an equation to the same power can lead to apparent solutions that do not satisfy the original equation. Be sure to check all suspected solutions.

(7.6) The Pythagorean theorem: If a and b are the lengths of the legs of a right triangle and c is the length of the hypotenuse, then $a^2 + b^2 = c^2$.

In an isosceles right triangle, the length of the hypotenuse is the length of either leg times $\sqrt{2}$.

The shorter leg of a 30°–60°–90° right triangle is half as long as the hypotenuse.

The length of the longer leg of a 30°–60°–90° right triangle is the length of the shorter leg times $\sqrt{3}$.

The distance formula: $d = \sqrt{(x_2 - x_1)^2 + (y_2 - y_1)^2}$.

■ Chapter 7 Review Exercises

In Review Exercises 1–12, simplify each radical. Assume that x can be any number.

1. $\sqrt{49}$ 　　　　**2.** $-\sqrt{121}$ 　　　　**3.** $-\sqrt{36}$ 　　　　**4.** $\sqrt{225}$

5. $\sqrt[3]{-27}$ 　　　　**6.** $-\sqrt[3]{216}$ 　　　　**7.** $\sqrt[4]{625}$ 　　　　**8.** $\sqrt[5]{-32}$

9. $\sqrt{25x^2}$ 　　　　**10.** $\sqrt{x^2 + 4x + 4}$ 　　　　**11.** $\sqrt[3]{27a^6b^3}$ 　　　　**12.** $\sqrt[4]{256x^8y^4}$

In Review Exercises 13–16, graph each function.

13. $y = f(x) = \sqrt{2x}$

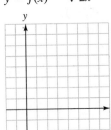

14. $y = f(x) = 2\sqrt{x}$

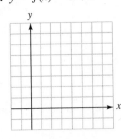

15. $y = f(x) = -2\sqrt{x}$

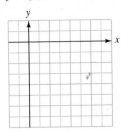

16. $y = f(x) = -\sqrt[3]{x}$

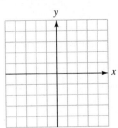

In Review Exercises 17–18, consider the distribution: 4, 8, 12, 16, 20.

17. Find the mean of the distribution.

18. Find the standard deviation.

In Review Exercises 19–34, simplify each expression, if possible. Assume that all variables represent positive numbers.

19. $25^{1/2}$

20. $-36^{1/2}$

21. $9^{3/2}$

22. $16^{3/2}$

23. $(-8)^{1/3}$

24. $-8^{2/3}$

25. $8^{-2/3}$

26. $8^{-1/3}$

27. $-49^{5/2}$

28. $\dfrac{1}{25^{5/2}}$

29. $\left(\dfrac{1}{4}\right)^{-3/2}$

30. $\left(\dfrac{4}{9}\right)^{-3/2}$

31. $(27x^3y)^{1/3}$

32. $(81x^4y^2)^{1/4}$

33. $(25x^3y^4)^{3/2}$

34. $(8u^2v^3)^{-2/3}$

In Review Exercises 35–40, do the multiplications. Assume that all variables represent positive numbers, and write all answers without negative exponents.

35. $5^{1/4}5^{1/2}$

36. $a^{3/7}a^{2/7}$

37. $u^{1/2}\left(u^{1/2} - u^{-1/2}\right)$

38. $v^{2/3}\left(v^{1/3} + v^{4/3}\right)$

39. $\left(x^{1/2} + y^{1/2}\right)^2$

40. $\left(a^{2/3} + b^{2/3}\right)\left(a^{2/3} - b^{2/3}\right)$

In Review Exercises 41–56, simplify each expression. Assume that all variables are unrestricted.

41. $\sqrt[6]{5^2}$

42. $\sqrt[8]{x^4}$

43. $\sqrt[9]{27a^3b^6}$

44. $\sqrt[4]{\dfrac{a^2}{25b^2}}$

45. $\sqrt{240}$

46. $\sqrt[3]{54}$

47. $\sqrt[4]{32}$

48. $\sqrt[5]{96}$

49. $\sqrt{8x^2y}$

50. $\sqrt{18x^4y^2}$

51. $\sqrt[3]{16x^5y^4}$

52. $\sqrt[3]{54x^7y^3}$

53. $\dfrac{\sqrt{32x^3}}{\sqrt{2x}}$

54. $\dfrac{\sqrt[3]{16x^5}}{\sqrt[3]{2x^2}}$

55. $\sqrt[3]{\dfrac{2a^2b}{27x^3}}$

56. $\sqrt{\dfrac{17xy}{64a^4}}$

In Review Exercises 57–64, simplify and combine like radicals. Assume that all variables represent positive numbers.

57. $\sqrt{2} + \sqrt{8}$

58. $\sqrt{20} - \sqrt{5}$

59. $2\sqrt[3]{3} - \sqrt[3]{24}$

60. $\sqrt[4]{32} + 2\sqrt[4]{162}$

61. $2x\sqrt{8} + 2\sqrt{200x^2} + \sqrt{50x^2}$

62. $3\sqrt{27a^3} - 2a\sqrt{3a} + 5\sqrt{75a^3}$

63. $\sqrt[3]{54} - 3\sqrt[3]{16} + 4\sqrt[3]{128}$

64. $2\sqrt[4]{32x^5} + 4\sqrt[4]{162x^5} - 5x\sqrt[4]{512x}$

In Review Exercises 65–78, simplify each expression. Assume that all variables represent positive numbers.

65. $(2\sqrt{5})(3\sqrt{2})$　　**66.** $2\sqrt{6}\sqrt{216}$　　**67.** $\sqrt{9x}\sqrt{x}$　　**68.** $\sqrt[3]{3}\sqrt[3]{9}$

69. $-\sqrt[3]{2x^2}\sqrt[3]{4x}$

70. $-\sqrt[4]{256x^5y^{11}}\sqrt[4]{625x^9y^3}$

71. $\sqrt{2}(\sqrt{8}-3)$

72. $\sqrt{2}(\sqrt{2}+3)$

73. $\sqrt{5}(\sqrt{2}-1)$

74. $\sqrt{3}(\sqrt{3}+\sqrt{2})$

75. $(\sqrt{2}+1)(\sqrt{2}-1)$

76. $(\sqrt{3}+\sqrt{2})(\sqrt{3}+\sqrt{2})$

77. $(\sqrt{x}+\sqrt{y})(\sqrt{x}-\sqrt{y})$

78. $(2\sqrt{u}+3)(3\sqrt{u}-4)$

In Review Exercises 79–86, rationalize each denominator.

79. $\dfrac{1}{\sqrt{3}}$　　**80.** $\dfrac{\sqrt{3}}{\sqrt{5}}$　　**81.** $\dfrac{x}{\sqrt{xy}}$　　**82.** $\dfrac{\sqrt[3]{uv}}{\sqrt[3]{u^5v^7}}$

83. $\dfrac{2}{\sqrt{2}-1}$　　**84.** $\dfrac{\sqrt{2}}{\sqrt{3}-1}$　　**85.** $\dfrac{2x-32}{\sqrt{x}+4}$　　**86.** $\dfrac{\sqrt{a}+1}{\sqrt{a}-1}$

In Review Exercises 87–90, rationalize each numerator.

87. $\dfrac{\sqrt{3}}{5}$　　**88.** $\dfrac{\sqrt[3]{9}}{3}$　　**89.** $\dfrac{3-\sqrt{x}}{2}$　　**90.** $\dfrac{\sqrt{a}-\sqrt{b}}{\sqrt{a}}$

In Review Exercises 91–96, solve each equation.

91. $\sqrt{y+3}=\sqrt{2y-19}$

92. $u=\sqrt{25u-144}$

93. $r=\sqrt{12r-27}$

94. $\sqrt{z+1}+\sqrt{z}=2$

95. $\sqrt{2x+5}-\sqrt{2x}=1$

96. $\sqrt[3]{x^3+8}=x+2$

 In Review Exercises 97–98, find x to two decimal places.

97.

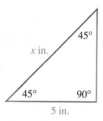

98.

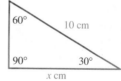

In Review Exercises 99–100, find the distance between points P and Q.

99. $P(0, 0)$, $Q(5, -12)$

100. $P(-2, -7)$, $Q(4, 1)$

In Review Exercises 101–104, recall that the horizon distance d (measured in miles) is related to the height h (measured in feet) of the observer by the formula $d = 1.4\sqrt{h}$.

101. View from a submarine　A submarine's periscope extends 4.7 feet above the surface. How far is the horizon?

102. View from a submarine　How far out of the water must a submarine periscope extend to provide a 4-mile horizon?

103. Sailing   A technique called *tacking* allows a sailboat to make progress into the wind. A sailboat follows the course in Illustration 1. Find *d*, the distance the boat advances into the wind.

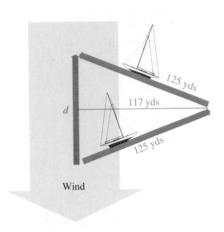

Wind

ILLUSTRATION 1

104. Communications   Some campers 3900 yards from a highway are chatting with truckers on a citizen's band transceiver with an 8900-yard range. Over what length of highway can these conversations take place? (See Illustration 2.)

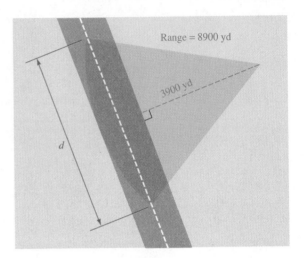

Range = 8900 yd

3900 yd

d

ILLUSTRATION 2

▪ Chapter 7 Test

In Problems 1–4, simplify each expression. Assume that all variables represent positive numbers.

1. $\sqrt{48}$

2. $\sqrt{250x^3y^5}$

3. $\dfrac{\sqrt[3]{24x^{15}y^4}}{\sqrt[3]{y}}$

4. $\sqrt{\dfrac{3a^5}{48a^7}}$

In Problems 5–8, simplify each expression. Assume that the variables are unrestricted.

5. $\sqrt{x^2}$

6. $\sqrt{8xy^2}$

7. $\sqrt[3]{27x^3}$

8. $\sqrt{18x^4y^9}$

In Problems 9–10, consider the distribution 7, 8, 12, 13.

9. Find the mean of the distribution.

10. Find the standard deviation.

In Problems 11–16, simplify each expression. Assume that all variables represent positive numbers, and write answers without using negative exponents.

11. $16^{1/4}$

12. $27^{2/3}$

13. $36^{-3/2}$

14. $\left(-\dfrac{8}{27}\right)^{-2/3}$

15. $\dfrac{2^{5/3}2^{1/6}}{2^{1/2}}$

16. $\dfrac{(8x^3y)^{1/2}(8xy^5)^{1/2}}{(x^3y^6)^{1/3}}$

In Problems 17–20, simplify and combine like radicals. Assume that all variables represent positive numbers.

17. $\sqrt{12} - \sqrt{27}$

18. $2\sqrt[3]{40} - \sqrt[3]{5000} + 4\sqrt[3]{625}$

19. $2\sqrt{48y^5} - 3y\sqrt{12y^3}$

20. $\sqrt[4]{768z^5} + z\sqrt[4]{48z}$

In Problems 21–22, do each operation and simplify, if possible. All variables represent positive numbers.

21. $-2\sqrt{xy}\left(3\sqrt{x} + \sqrt{xy^3}\right)$

22. $\left(3\sqrt{2} + \sqrt{3}\right)\left(2\sqrt{2} - 3\sqrt{3}\right)$

In Problems 23–26, rationalize each denominator.

23. $\dfrac{1}{\sqrt{5}}$

24. $\dfrac{6}{\sqrt[3]{9}}$

25. $\dfrac{-4\sqrt{2}}{\sqrt{5} + 3}$

26. $\dfrac{3t - 1}{\sqrt{3t} - 1}$

In Problems 27–28, rationalize each numerator.

27. $\dfrac{\sqrt{3}}{\sqrt{7}}$

28. $\dfrac{\sqrt{a} + \sqrt{b}}{\sqrt{a} - \sqrt{b}}$

In Problems 29–30, solve and check each equation.

29. $\sqrt[3]{6n + 4} - 4 = 0$

30. $1 - \sqrt{u} = \sqrt{u - 3}$

 In Problems 31–32, find x to two decimal places.

31.

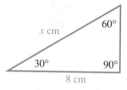

32.

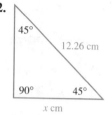

In Problems 33–34, find the distance between points P and Q.

33. $P(6, 8)$, $Q(0, 0)$

34. $P(-2, 5)$, $Q(22, 12)$

 In Exercises 35–36, use a calculator.

35. Shipping crates The diagonal brace on the shipping crate in Illustration 1 is 53 inches. Find the height, h, of the crate.

36. Pendulum The 2-meter pendulum in Illustration 2 rises 0.1 meter at the extremes of its swing. Find the width, w, of the swing.

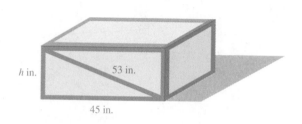

ILLUSTRATION 1

ILLUSTRATION 2

Quadratic Functions, Inequalities, and Algebra of Functions

Chemist

Chemists search for knowledge about substances and put it to practical use. Most chemists work in research and development. In basic research, chemists investigate the properties, composition, and structure of matter and the laws that govern the combination of elements and the reactions of substances. Their research has resulted in the development of a tremendous variety of synthetic materials, of ingredients that have improved other substances, and of processes that help save energy and reduce pollution. In applied research and development, chemists create new products or improve existing ones.

SAMPLE APPLICATION ■ A weak acid (0.1 M concentration) breaks down into free cations (the hydrogen ion, H^+) and anions (A^-). When this acid dissociates, the following equilibrium equation is established.

1. $$\frac{[H^+][A^-]}{[HA]} = 4 \times 10^{-4}$$

where $[H^+]$, the hydrogen ion concentration, is equal to $[A^-]$, the anion concentration. $[HA]$ is the concentration of the undissociated acid itself. Find $[H^+]$ at equilibrium.
See Exercise 75 in Exercise 8.1.

We have discussed how to solve linear equations and certain quadratic equations in which the quadratic expression is factorable. In this chapter, we will discuss more general methods for solving quadratic equations, and we will consider their graphs.

8.1 Completing the Square and the Quadratic Formula

■ THE SQUARE ROOT PROPERTY ■ COMPLETING THE SQUARE ■ SOLVING EQUATIONS BY COMPLETING THE SQUARE ■ THE QUADRATIC FORMULA ■ PROBLEM SOLVING

A *quadratic equation* is an equation of the form $ax^2 + bx + c = 0$, $(a \neq 0)$, where a, b, and c are real numbers. We have discussed how to solve certain quadratic equations by using factoring. For example, to solve $6x^2 - 7x - 3 = 0$, we proceed as follows:

$$6x^2 - 7x - 3 = 0$$
$$(2x - 3)(3x + 1) = 0 \qquad \text{Factor the quadratic trinomial.}$$
$$2x - 3 = 0 \quad \text{or} \quad 3x + 1 = 0 \qquad \text{Set each factor equal to 0.}$$
$$x = \frac{3}{2} \quad \Big| \quad x = -\frac{1}{3} \qquad \text{Solve each linear equation.}$$

However, many quadratic expressions do not factor easily. For example, it would be difficult to solve $2x^2 + 4x + 1 = 0$ by factoring, because $2x^2 + 4x + 1$ cannot be factored using only integers.

■ THE SQUARE ROOT PROPERTY

To develop general methods for solving all quadratic equations, we first consider the equation $x^2 = c$. If $c > 0$, we can find the real solutions of $x^2 = c$ as follows:

$$x^2 = c$$
$$x^2 - c = 0 \qquad \text{Subtract } c \text{ from both sides.}$$
$$x^2 - \left(\sqrt{c}\right)^2 = 0 \qquad c = \left(\sqrt{c}\right)^2.$$
$$\left(x + \sqrt{c}\right)\left(x - \sqrt{c}\right) = 0 \qquad \text{Factor the difference of two squares.}$$
$$x + \sqrt{c} = 0 \qquad \text{or} \qquad x - \sqrt{c} = 0 \qquad \text{Set each factor equal to 0.}$$
$$x = -\sqrt{c} \qquad | \qquad x = \sqrt{c} \qquad \text{Solve each linear equation.}$$

The solutions of $x^2 = c$ are $x = \sqrt{c}$ and $x = -\sqrt{c}$.

Square Root Property
If $c > 0$, the equation $x^2 = c$ has two real solutions. They are
$$x = \sqrt{c} \qquad \text{and} \qquad x = -\sqrt{c}$$

EXAMPLE 1   Solve the equation $x^2 - 12 = 0$.

Solution   We can write the equation as $x^2 = 12$ and use the square root property.

$$x^2 - 12 = 0$$
$$x^2 = 12 \qquad \text{Add 12 to both sides.}$$
$$x = \sqrt{12} \quad \text{or} \quad x = -\sqrt{12} \qquad \text{Use the square root property.}$$
$$x = 2\sqrt{3} \quad | \quad x = -2\sqrt{3} \qquad \sqrt{12} = \sqrt{4}\sqrt{3} = 2\sqrt{3}.$$

Verify that each solution satisfies the equation.   ■

EXAMPLE 2   Solve the equation $(x - 3)^2 = 16$.

Solution
$$(x - 3)^2 = 16$$

$x - 3 = \sqrt{16}$ or $x - 3 = -\sqrt{16}$ Use the square root property.

$x - 3 = 4$ $x - 3 = -4$ $\sqrt{16} = 4$.

$x = 3 + 4$ $x = 3 - 4$ Add 3 to both sides.

$x = 7$ $x = -1$ Simplify.

Verify that each solution satisfies the equation.   ■

■ COMPLETING THE SQUARE

All quadratic equations can be solved by **completing the square**. This method is based on the special products

$$x^2 + 2ax + a^2 = (x + a)^2 \quad \text{and} \quad x^2 - 2ax + a^2 = (x - a)^2$$

The trinomials $x^2 + 2ax + a^2$ and $x^2 - 2ax + a^2$ are both perfect square trinomials, because both factor as the square of a binomial. In each case, the coefficient of the first term is 1, and if we take one-half of the coefficient of x in the middle term and square it, we obtain the third term.

$$\left[\frac{1}{2}(2a)\right]^2 = a^2$$

$$\left[\frac{1}{2}(-2a)\right]^2 = (-a)^2 = a^2$$

EXAMPLE 3   (b,c)   Add a number to each binomial to make it a perfect square trinomial:
a. $x^2 + 10x$,   **b.** $x^2 - 6x$,   and   **c.** $x^2 - 11x$.

Solution   **a.** To make $x^2 + 10x$ a perfect square trinomial, we find one-half of 10 to get 5, square 5 to get 25, and add 25 to $x^2 + 10x$.

$$x^2 + 10x + \left[\frac{1}{2}(10)\right]^2 = x^2 + 10x + (5)^2$$

$$= x^2 + 10x + 25 \qquad \text{Note that } x^2 + 10x + 25 = (x + 5)^2.$$

b. To make $x^2 - 6x$ a perfect square trinomial, we find one-half of -6 to get -3, square -3 to get 9, and add 9 to $x^2 - 6x$.

$$x^2 - 6x + \left[\frac{1}{2}(-6)\right]^2 = x^2 - 6x + (-3)^2$$

$$= x^2 - 6x + 9 \qquad \text{Note that } x^2 - 6x + 9 = (x - 3)^2.$$

c. To make $x^2 - 11x$ a perfect square trinomial, we find one-half of -11 to get $-\frac{11}{2}$, square $-\frac{11}{2}$ to get $\frac{121}{4}$, and add $\frac{121}{4}$ to $x^2 - 11x$.

$$x^2 - 11x + \left[\frac{1}{2}(-11)\right]^2 = x^2 - 11x + \left(-\frac{11}{2}\right)^2$$

$$= x^2 - 11x + \frac{121}{4} \qquad \text{Note that } x^2 - 11x + \frac{121}{4} = \left(x - \frac{11}{2}\right)^2.$$ ■

■ SOLVING EQUATIONS BY COMPLETING THE SQUARE

To solve a quadratic equation of the form $ax^2 + bx + c = 0$ by completing the square, we use the following steps.

> **Completing the Square**
>
> 1. Make sure that the coefficient of x^2 is 1. If it is not, make it 1 by dividing both sides of the equation by the coefficient of x^2.
> 2. If necessary, add a number to both sides of the equation to put the constant term on the right-hand side of the equal sign.
> 3. Complete the square:
>   a. Find one-half of the coefficient of x and square it.
>   b. Add that square to both sides of the equation.
> 4. Factor the perfect square trinomial and combine like terms.
> 5. Solve the resulting equation by using the square root property. If the constant is greater than 0, there will be two solutions.

EXAMPLE 4   Use completing the square to solve the equation $x^2 + 8x + 7 = 0$.

Solution   *Step 1:* In this example, the coefficient of x^2 is 1.

Step 2: We add -7 to both sides to put the constant on the right-hand side of the equal sign:

$$x^2 + 8x + 7 = 0$$
$$x^2 + 8x = -7$$

Step 3: The coefficient of x is 8, one-half of 8 is 4, and $4^2 = 16$. Thus, we add 16 to both sides.

$$x^2 + 8x + \mathbf{16} = \mathbf{16} - 7$$
1. $x^2 + 8x + 16 = 9$ $16 - 7 = 9.$

Step 4: Since the left-hand side of Equation 1 is a perfect square trinomial, we can factor it to get $(x + 4)^2$.

$$x^2 + 8x + 16 = 9$$
2. $(x + 4)^2 = 9$

Step 5: We can solve Equation 2 by using the square root property.

$$(x + 4)^2 = 9$$

$x + 4 = \sqrt{9}$ or $x + 4 = -\sqrt{9}$
$x + 4 = 3$ $x + 4 = -3$
$x = -1$ $x = -7$

Verify that both solutions satisfy the equation.   ■

EXAMPLE 5 Solve $6x^2 + 5x - 6 = 0$.

Solution *Step 1:* To make the coefficient of x^2 equal to 1, we divide both sides of the equation by 6.

$$6x^2 + 5x - 6 = 0$$

$$\frac{6x^2}{6} + \frac{5}{6}x - \frac{6}{6} = \frac{0}{6} \qquad \text{Divide both sides by 6.}$$

$$x^2 + \frac{5}{6}x - 1 = 0 \qquad \text{Simplify.}$$

Step 2: We add 1 to both sides to put the constant on the right-hand side of the equal sign:

$$x^2 + \frac{5}{6}x = 1$$

Step 3: The coefficient of x is $\frac{5}{6}$, one-half of $\frac{5}{6}$ is $\frac{5}{12}$, and $\left(\frac{5}{12}\right)^2 = \frac{25}{144}$. Thus, we add $\frac{25}{144}$ to both sides.

$$x^2 + \frac{5}{6}x + \frac{25}{144} = 1 + \frac{25}{144}$$

3. $x^2 + \frac{5}{6}x + \frac{25}{144} = \frac{169}{144} \qquad 1 + \frac{25}{144} = \frac{144}{144} + \frac{25}{144} = \frac{169}{144}.$

Step 4: Since the left-hand side of Equation 3 is a perfect square trinomial, we can factor it to get $\left(x + \frac{5}{12}\right)^2$.

4. $\left(x + \frac{5}{12}\right)^2 = \frac{169}{144}$

Step 5: We can solve Equation 4 by using the square root property.

$$x + \frac{5}{12} = \sqrt{\frac{169}{144}} \qquad \text{or} \qquad x + \frac{5}{12} = -\sqrt{\frac{169}{144}}$$

$$x + \frac{5}{12} = \frac{13}{12} \qquad\qquad x + \frac{5}{12} = -\frac{13}{12}$$

$$x = -\frac{5}{12} + \frac{13}{12} \qquad\qquad x = -\frac{5}{12} - \frac{13}{12}$$

$$x = \frac{8}{12} \qquad\qquad\qquad x = -\frac{18}{12}$$

$$x = \frac{2}{3} \qquad\qquad\qquad x = -\frac{3}{2}$$

Verify that both solutions satisfy the original equation.

EXAMPLE 6 Solve $2x^2 + 4x + 1 = 0$.

Solution $$2x^2 + 4x + 1 = 0$$

$$x^2 + 2x + \frac{1}{2} = \frac{0}{2} \qquad \text{Divide both sides by 2 to make the coefficient of } x^2 \text{ equal to 1.}$$

$$x^2 + 2x = -\frac{1}{2} \qquad \text{Subtract } \tfrac{1}{2} \text{ from both sides.}$$

$$x^2 + 2x + 1 = 1 - \frac{1}{2} \qquad \begin{array}{l}\text{Square half the coefficient of } x \text{ and add it}\\ \text{to both sides.}\end{array}$$

$$(x + 1)^2 = \frac{1}{2} \qquad \text{Factor and combine like terms.}$$

$$x + 1 = \sqrt{\frac{1}{2}} \qquad \text{or} \qquad x + 1 = -\sqrt{\frac{1}{2}}$$

$$x + 1 = \frac{\sqrt{2}}{2} \qquad\qquad x + 1 = -\frac{\sqrt{2}}{2}$$

$$x = -1 + \frac{\sqrt{2}}{2} \qquad\qquad x = -1 - \frac{\sqrt{2}}{2}$$

$$x = \frac{-2 + \sqrt{2}}{2} \qquad\qquad x = \frac{-2 - \sqrt{2}}{2}$$

Both values check.

■ THE QUADRATIC FORMULA

To develop a formula we can use to solve quadratic equations, we solve the general quadratic equation $ax^2 + bx + c = 0 \quad (a \neq 0)$.

$$ax^2 + bx + c = 0$$

$$\frac{ax^2}{a} + \frac{bx}{a} + \frac{c}{a} = \frac{0}{a} \qquad \begin{array}{l}\text{Since } a \neq 0, \text{ we can divide both sides by}\\ a \text{ to make the coefficient of } x^2 \text{ equal to 1.}\end{array}$$

$$x^2 + \frac{bx}{a} = -\frac{c}{a} \qquad \text{Simplify and subtract } \tfrac{c}{a} \text{ from both sides.}$$

$$x^2 + \frac{b}{a}x + \left(\frac{b}{2a}\right)^2 = \left(\frac{b}{2a}\right)^2 - \frac{c}{a} \qquad \begin{array}{l}\text{Complete the square on } x \text{ and add } \left(\dfrac{b}{2a}\right)^2\\ \text{to both sides.}\end{array}$$

$$x^2 + \frac{b}{a}x + \frac{b^2}{4a^2} = \frac{b^2}{4a^2} - \frac{4ac}{4aa} \qquad \begin{array}{l}\text{Remove parentheses and get a common}\\ \text{denominator on the right-hand side.}\end{array}$$

5. $$\left(x + \frac{b}{2a}\right)^2 = \frac{b^2 - 4ac}{4a^2} \qquad \begin{array}{l}\text{Factor the left-hand side and add the}\\ \text{fractions on the right-hand side.}\end{array}$$

We can solve Equation 5 by using the square root property.

$$x + \frac{b}{2a} = \sqrt{\frac{b^2 - 4ac}{4a^2}} \qquad \text{or} \qquad x + \frac{b}{2a} = -\sqrt{\frac{b^2 - 4ac}{4a^2}}$$

$$x + \frac{b}{2a} = \frac{\sqrt{b^2 - 4ac}}{2a} \qquad\qquad x + \frac{b}{2a} = -\frac{\sqrt{b^2 - 4ac}}{2a}$$

$$x = -\frac{b}{2a} + \frac{\sqrt{b^2 - 4ac}}{2a} \qquad\qquad x = -\frac{b}{2a} - \frac{\sqrt{b^2 - 4ac}}{2a}$$

$$= \frac{-b + \sqrt{b^2 - 4ac}}{2a} \qquad\qquad = \frac{-b - \sqrt{b^2 - 4ac}}{2a}$$

These two solutions give the **quadratic formula**.

The Quadratic Formula

The solutions of $ax^2 + bx + c = 0$ $(a \neq 0)$ are given by the formula

$$x = \frac{-b \pm \sqrt{b^2 - 4ac}}{2a}$$

Read the symbol $\pm$ as "plus or minus."

WARNING! Be sure to draw the fraction bar under both parts of the numerator, and be sure to draw the radical sign exactly over $b^2 - 4ac$. Do not write the quadratic formula as

$$x = -b \pm \frac{\sqrt{b^2 - 4ac}}{2a} \qquad \text{or as} \qquad x = -b \pm \sqrt{\frac{b^2 - 4ac}{2a}}$$

EXAMPLE 7 Use the quadratic formula to solve $2x^2 - 3x - 5 = 0$.

Solution In this equation, $a = 2$, $b = -3$, and $c = -5$.

$$x = \frac{-b \pm \sqrt{b^2 - 4ac}}{2a}$$

$$= \frac{-(-3) \pm \sqrt{(-3)^2 - 4(2)(-5)}}{2(2)}$$

Substitute 2 for a, -3 for b, and -5 for c.

$$= \frac{3 \pm \sqrt{9 + 40}}{4}$$

$$= \frac{3 \pm \sqrt{49}}{4}$$

$$= \frac{3 \pm 7}{4}$$

$$x = \frac{3 + 7}{4} \qquad \text{or} \qquad x = \frac{3 - 7}{4}$$

$$x = \frac{10}{4} \qquad\qquad\qquad x = \frac{-4}{4}$$

$$x = \frac{5}{2} \qquad\qquad\qquad x = -1$$

Verify that both solutions satisfy the original equation.

EXAMPLE 8 Solve the equation $2x^2 + 4x + 1 = 0$.

Solution In this equation, $a = 2$, $b = 4$, and $c = 1$.

$$x = \frac{-b \pm \sqrt{b^2 - 4ac}}{2a}$$

$$= \frac{-4 \pm \sqrt{4^2 - 4(2)(1)}}{2(2)}$$

Substitute 2 for a, 4 for b, and 1 for c.

$$= \frac{-4 \pm \sqrt{16 - 8}}{4}$$

$$= \frac{-4 \pm \sqrt{8}}{4}$$

$$= \frac{-4 \pm 2\sqrt{2}}{4} \qquad \sqrt{8} = \sqrt{4 \cdot 2} = \sqrt{4}\sqrt{2} = 2\sqrt{2}.$$

$$= \frac{-2 \pm \sqrt{2}}{2} \qquad \frac{-4 \pm 2\sqrt{2}}{4} = \frac{2(-2 \pm \sqrt{2})}{4} = \frac{-2 \pm \sqrt{2}}{2}.$$

Thus, $x = \dfrac{-2 + \sqrt{2}}{2}$ or $x = \dfrac{-2 - \sqrt{2}}{2}$. ∎

■ PROBLEM SOLVING

EXAMPLE 9 **Dimensions of a rectangle** Find the dimensions of the rectangle shown in Figure 8-1, given that it is 12 centimeters longer than it is wide.

Solution If we let w represent the width of the rectangle, then $w + 12$ represents its length. Since the area of the rectangle is 253 square centimeters, we can form the equation

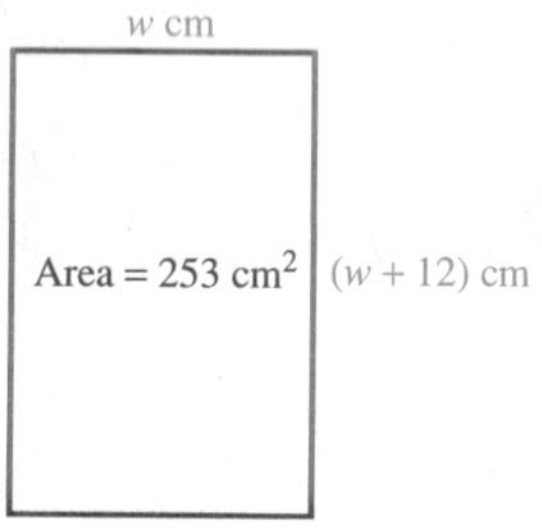

$$w(w + 12) = 253 \qquad \text{Area of a rectangle} = \text{width} \times \text{length.}$$

and solve it as follows:

$$w(w + 12) = 253$$
$$w^2 + 12w = 253 \qquad \text{Use the distributive property to remove parentheses.}$$
$$w^2 + 12w - 253 = 0 \qquad \text{Subtract 253 from both sides.}$$

w cm

Area = 253 cm² | (*w* + 12) cm

FIGURE 8-1

Solution by factoring

$$(w - 11)(w + 23) = 0$$

$$w - 11 = 0 \quad \text{or} \quad w + 23 = 0$$

$$w = 11 \qquad \qquad w = -23$$

Solution by formula

$$w = \frac{-12 \pm \sqrt{12^2 - 4(1)(-253)}}{2(1)}$$

$$= \frac{-12 \pm \sqrt{144 + 1012}}{2}$$

$$= \frac{-12 \pm \sqrt{1156}}{2}$$

$$= \frac{-12 \pm 34}{2}$$

$$w = 11 \quad \text{or} \quad w = -23$$

Because a rectangle cannot have a negative width, we discard -23 as a solution. Since the rectangle is 11 centimeters wide and $(11 + 12)$ centimeters long, its dimensions are 11 centimeters by 23 centimeters.

Check: 23 is 12 more than 11, and the area of a rectangle with dimensions of 23 centimeters by 11 centimeters is 253 square centimeters. ■

Orals   *Solve each equation.*

1. $x^2 = 49$ **2.** $x^2 = 10$

Find the number that must be added to the binomial to make it a perfect square trinomial.

3. $x^2 + 4x$ **4.** $x^2 - 6x$ **5.** $x^2 - 3x$ **6.** $x^2 + 5x$

Identify a, b, and c in each quadratic equation.

7. $3x^2 - 4x + 7 = 0$ **8.** $-2x^2 + x = 5$

EXERCISE 8.1

In Exercises 1–12, use factoring to solve each equation.

1. $6x^2 + 12x = 0$ **2.** $5x^2 + 11x = 0$ **3.** $2y^2 - 50 = 0$

4. $4y^2 - 64 = 0$ **5.** $r^2 + 6r + 8 = 0$ **6.** $x^2 + 9x + 20 = 0$

7. $7x - 6 = x^2$ **8.** $5t - 6 = t^2$ **9.** $2z^2 - 5z + 2 = 0$

10. $2x^2 - x - 1 = 0$ **11.** $6s^2 + 11s - 10 = 0$ **12.** $3x^2 + 10x - 8 = 0$

In Exercises 13–24, use the square root property to solve each equation.

13. $x^2 = 36$ **14.** $x^2 = 144$ **15.** $z^2 = 5$

16. $u^2 = 24$ **17.** $3x^2 - 16 = 0$ **18.** $5x^2 - 49 = 0$

19. $(x + 1)^2 = 1$ **20.** $(x - 1)^2 = 4$ **21.** $(s - 7)^2 - 9 = 0$

22. $(t + 4)^2 = 16$ **23.** $(x + 5)^2 - 3 = 0$ **24.** $(x + 3)^2 - 7 = 0$

In Exercises 25–38, use completing the square to solve each equation.

25. $x^2 + 2x - 8 = 0$ **26.** $x^2 + 6x + 5 = 0$ **27.** $x^2 - 6x + 8 = 0$

28. $x^2 + 8x + 15 = 0$ **29.** $x^2 + 5x + 4 = 0$ **30.** $x^2 - 11x + 30 = 0$

31. $x + 1 = 2x^2$ **32.** $-2 = 2x^2 - 5x$ **33.** $6x^2 + 11x + 3 = 0$

34. $6x^2 + x - 2 = 0$ **35.** $9 - 6r = 8r^2$ **36.** $11w - 10 = 3w^2$

37. $\dfrac{7x + 1}{5} = -x^2$ **38.** $\dfrac{3x^2}{8} = \dfrac{1}{8} - x$

In Exercises 39–50, use the quadratic formula to solve each equation.

39. $x^2 + 3x + 2 = 0$ **40.** $x^2 - 3x + 2 = 0$ **41.** $x^2 + 12x = -36$

42. $y^2 - 18y = -81$ **43.** $5x^2 + 5x + 1 = 0$ **44.** $4w^2 + 6w + 1 = 0$

45. $8u = -4u^2 - 3$ **46.** $4t + 3 = 4t^2$ **47.** $16y^2 + 8y - 3 = 0$

48. $16x^2 + 16x + 3 = 0$ **49.** $\dfrac{x^2}{2} + \dfrac{5}{2}x = -1$ **50.** $-3x = \dfrac{x^2}{2} + 2$

In Exercises 51–52, use the quadratic formula and a scientific calculator to solve each equation. Give all answers to the nearest hundredth.

51. $0.7x^2 - 3.5x - 25 = 0$

52. $-4.5x^2 + 0.2x + 3.75 = 0$

53. Integer problem The product of two consecutive even positive integers is 288. Find the integers. (*Hint:* If one integer is x, the next consecutive even integer is $x + 2$.)

54. Integer problem The product of two consecutive odd negative integers is 143. Find the integers. (*Hint:* If one integer is x, the next consecutive odd integer is $x + 2$.)

55. Integer problem The sum of the squares of two consecutive positive integers is 85. Find the integers. (*Hint:* If one integer is x, the next consecutive positive integer is $x + 1$.)

56. Integer problem The sum of the squares of three consecutive positive integers is 77. Find the integers. (*Hint:* If one integer is x, the next consecutive positive integer is $x + 1$ and the third is $x + 2$.)

57. Dimensions of a rectangle A rectangle is 4 feet longer than it is wide, and its area is 96 square feet. Find its dimensions.

58. Dimensions of a rectangle One side of a rectangle is 3 times as long as another. The area of the rectangle is 147 square meters. Find its dimensions.

59. Side of a square The area of a square is numerically equal to its perimeter. Find the length of one side of the square.

60. Perimeter of a rectangle A rectangle is 2 inches longer than it is wide. Numerically, its area exceeds its perimeter by 11. Find the perimeter.

61. Base of a triangle The height of a triangle is 5 centimeters longer than three times its base. Find the base of the triangle if its area is 6 square centimeters.

62. Height of a triangle The height of a triangle is 4 meters longer than twice its base. Find the height of the triangle if its area is 15 square meters.

63. Finding rates A woman drives her snowmobile 150 miles at the rate of r miles per hour. She could have gone the same distance in 2 hours less time if she had increased her speed by 20 miles per hour. Find r.

64. Finding rates Jeff bicycles 160 miles at the rate of r miles per hour. The same trip would have taken 2 hours longer if he had decreased his speed by 4 miles per hour. Find r.

65. Pricing concert tickets Tickets to a rock concert cost $4, and the projected attendance is 300 persons. It is further projected that for every 10¢ increase in ticket price, the average attendance will decrease by 5. At what ticket price will the nightly receipts be $1248?

66. Setting bus fares A bus company has 3000 passengers daily, paying a 25¢ fare. For each nickel increase in fare, the company estimates that it will lose 80 passengers. What is the smallest increase in fare that will produce $994 in daily revenue?

67. Computing profit The *Gazette*'s profit is $20 per year for each of its 3000 subscribers. Management estimates that the profit per subscriber will increase by 1¢ for each additional subscriber over the current 3000. How many subscribers will bring a total profit of $120,000?

68. Finding interest rates A woman invests $1000 in a mutual fund for which interest is compounded annually at a rate r. After one year, she deposits an additional $2000. After two years, the balance in the account is

$$\$1000(1 + r)^2 + \$2000(1 + r)$$

If this amount is $3368.10, find r.

69. Framing a picture The frame surrounding the picture in Illustration 1 has a constant width. How wide is the frame if the area of the frame equals the area of the picture?

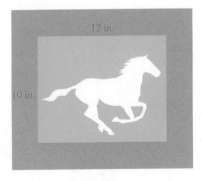

ILLUSTRATION 1

70. Metal fabrication A box with no top is to be made by cutting a 2-inch square from each corner of the square sheet of metal shown in Illustration 2. After bending up the sides, the volume of the box is to be 200 cubic inches. How large should the piece of metal be?

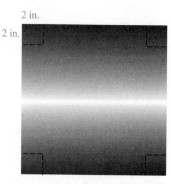

ILLUSTRATION 2

In Exercises 71–74, note that a and b are solutions to the equation $(x - a)(x - b) = 0$.

71. Find a quadratic equation with a solution set of $\{3, 5\}$.

72. Find a quadratic equation with a solution set of $\{-4, 6\}$.

73. Find a third-degree equation with a solution set of $\{2, 3, -4\}$.

74. Find a fourth-degree equation with a solution set of $\{3, -3, 4, -4\}$.

In Exercises 75–76, use a calculator.

75. Chemistry A weak acid (0.1 M concentration) breaks down into free cations (the hydrogen ion, H^+) and anions (A^-). When this acid dissociates, the following equilibrium equation is established.

$$\frac{[H^+][A^-]}{[HA]} = 4 \times 10^{-4}$$

where $[H^+]$, the hydrogen ion concentration, is equal to $[A^-]$, the anion concentration. $[HA]$ is the concentration of the undissociated acid itself. Find $[H^+]$ at equilibrium. (*Hint:* If $H^+ = x$, then $[HA] = 0.1 - x$.)

76. Chemistry A saturated solution of hydrogen sulfide (0.1 M concentration) dissociates into cation H^+ and anion HS^-, where $H^+ = HS^-$. When this solution dissociates, the following equilibrium equation is established.

$$\frac{[H^+][HS^-]}{[HHS]} = 1.0 \times 10^{-7}$$

Find $[H^+]$.
(*Hint:* If $H^+ = x$, then $[HHS] = 0.1 - x$.)

Writing Exercises *Write a paragraph using your own words.*

1. Explain how to complete the square.

2. Tell why a cannot be 0 in the quadratic equation $ax^2 + bx + c = 0$.

Something to Think About

1. What number must be added to $x^2 + \sqrt{3}x$ to make it a perfect square trinomial?

2. Solve $x^2 + \sqrt{3}x - \frac{1}{4} = 0$ by completing the square.

Review Exercises *Solve each equation or inequality.*

1. $\dfrac{t + 9}{2} + \dfrac{t + 2}{5} = \dfrac{8}{5} + 4t$

2. $\dfrac{1 - 5x}{2x} + 4 = \dfrac{x + 3}{x}$

3. $3(t - 3) + 3t - 5 \leq 2(t + 1) + t - 4$

4. $-2(y + 4) - 3y + 3 \geq 3(2y - 3) - y - 5$

Solve for the indicated variable.

5. $Ax + By = C$ for B **6.** $R = \dfrac{kL}{d^2}$ for L

8.2 Graphs of Quadratic Functions

■ GRAPHS OF $f(x) = ax^2$ ■ GRAPHS OF $f(x) = ax^2 + c$ ■ GRAPHS OF $f(x) = a(x - h)^2$ ■ GRAPHS OF $f(x) = a(x - h)^2 + k$ ■ GRAPHS OF $f(x) = ax^2 + bx + c$ ■ GRAPHING DEVICES ■ PROBLEM SOLVING ■ ACCENT ON STATISTICS

The graph in Figure 8-2 shows the height (in relation to time) of a toy rocket launched straight up into the air.

 WARNING! Note that the graph describes the height of the rocket, not the path of the rocket. The rocket goes straight up and comes straight down.

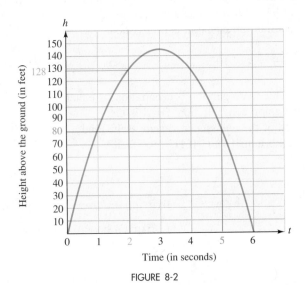

FIGURE 8-2

From the graph, we can see that the height of the rocket 2 seconds after launch is about 128 feet, and the height of the rocket 5 seconds after launch is 80 feet.

The parabola shown in Figure 8-2 is the graph of a *quadratic function*, the topic of this section.

Quadratic Function
A **quadratic function** is a second-degree polynomial function of the form

$$y = f(x) = ax^2 + bx + c \quad (a \neq 0)$$

where a, b, and c are real numbers.

We begin the discussion of graphing quadratic functions by considering the graph of $f(x) = ax^2 + bx + c$, where $b = 0$ and $c = 0$.

■ GRAPHS OF $f(x) = ax^2$

EXAMPLE 1   Graph   **a.** $f(x) = x^2$,   **b.** $g(x) = 3x^2$,   and   **c.** $h(x) = \dfrac{1}{3}x^2$.

Solution   We make a table of ordered pairs that satisfy each equation, plot each point, and join them with a smooth curve, as in Figure 8-3. We note that the graph of $h(x) = \frac{1}{3}x^2$ is wider than the graph of $f(x) = x^2$, and that the graph of $g(x) = 3x^2$ is narrower than the graph of $f(x) = x^2$. In the function $f(x) = ax^2$, the smaller the value of $|a|$, the wider the graph.

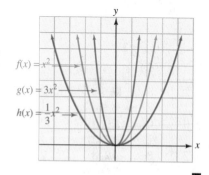

$f(x) = x^2$

x	$f(x)$	$(x, f(x))$
-2	4	$(-2, 4)$
-1	1	$(-1, 1)$
0	0	$(0, 0)$
1	1	$(1, 1)$
2	4	$(2, 4)$

$g(x) = 3x^2$

x	$g(x)$	$(x, g(x))$
-2	12	$(-2, 12)$
-1	3	$(-1, 3)$
0	0	$(0, 0)$
1	3	$(1, 3)$
2	12	$(2, 12)$

$h(x) = \frac{1}{3}x^2$

x	$h(x)$	$(x, h(x))$
-2	$4/3$	$\left(-2, 4/3\right)$
-1	$1/3$	$\left(-1, 1/3\right)$
0	0	$(0, 0)$
1	$1/3$	$\left(1, 1/3\right)$
2	$4/3$	$\left(2, 4/3\right)$

FIGURE 8-3   ■

If we consider the graph of $f(x) = -3x^2$, we see that it opens downward and has the same shape as the graph of $g(x) = 3x^2$.

EXAMPLE 2   Graph $f(x) = -3x^2$.

Solution   We make a table of ordered pairs that satisfy the equation, plot each point, and join them with a smooth curve, as in Figure 8-4.

$f(x) = -3x^2$

x	$f(x)$	$(x, f(x))$
-2	-12	$(-2, -12)$
-1	-3	$(-1, -3)$
0	0	$(0, 0)$
1	-3	$(1, -3)$
2	-12	$(2, -12)$

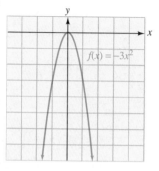

FIGURE 8-4   ■

The graphs of quadratic functions are called **parabolas**. They open upward when $a > 0$ and downward when $a < 0$. The lowest point of a parabola that opens upward, or the highest point of a parabola that opens downward, is called the **vertex** of the parabola. The vertex of the parabola shown in Figure 8-4 is the point $(0, 0)$.

The vertical line, called an **axis of symmetry**, that passes through the vertex divides the parabola into two congruent halves. The axis of symmetry of the parabola shown in Figure 8-4 is the y-axis.

■ GRAPHS OF $f(x) = ax^2 + c$

EXAMPLE 3   Graph   **a.** $f(x) = 2x^2$,   **b.** $g(x) = 2x^2 + 3$,   and   **c.** $h(x) = 2x^2 - 3$.

Solution   We make a table of ordered pairs that satisfy each equation, plot each point, and join them with a smooth curve, as in Figure 8-5. We note that the graph of $g(x) = 2x^2 + 3$ is identical to the graph of $f(x) = 2x^2$, except that it has been shifted 3 units upward. The graph of $h(x) = 2x^2 - 3$ is identical to the graph of $f(x) = 2x^2$, except that it has been shifted 3 units downward.

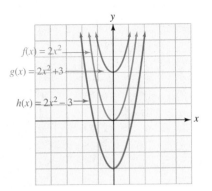

$f(x) = 2x^2$			$g(x) = 2x^2 + 3$			$h(x) = 2x^2 - 3$		
x	$f(x)$	$(x, f(x))$	x	$g(x)$	$(x, g(x))$	x	$h(x)$	$(x, h(x))$
-2	8	$(-2, 8)$	-2	11	$(-2, 11)$	-2	5	$(-2, 5)$
-1	2	$(-1, 2)$	-1	5	$(-1, 5)$	-1	-1	$(-1, -1)$
0	0	$(0, 0)$	0	3	$(0, 3)$	0	-3	$(0, -3)$
1	2	$(1, 2)$	1	5	$(1, 5)$	1	-1	$(1, -1)$
2	8	$(2, 8)$	2	11	$(2, 11)$	2	5	$(2, 5)$

FIGURE 8-5

The results of Example 3 suggest the following facts.

Vertical Shifts of Graphs
If f is a function and k is a positive number, then
- The graph of $y = f(x) + k$ is identical to the graph of $y = f(x)$, except that it is shifted k units upward.
- The graph of $y = f(x) - k$ is identical to the graph of $y = f(x)$, except that it is shifted k units downward.

■ GRAPHS OF $f(x) = a(x - h)^2$

EXAMPLE 4   Graph   **a.** $f(x) = 2x^2$,   **b.** $g(x) = 2(x - 3)^2$,   and   **c.** $h(x) = 2(x + 3)^2$.

Solution   We make a table of ordered pairs that satisfy each equation, plot each point, and join them with a smooth curve, as in Figure 8-6. We note that the graph of $g(x) = 2(x - 3)^2$ is identical to the graph of $f(x) = 2x^2$, except that it has been shifted 3 units to the right. The graph of $h(x) = 2(x + 3)^2$ is identical to the graph of $f(x) = 2x^2$, except that it has been shifted 3 units to the left.

$f(x) = 2x^2$			$g(x) = 2(x - 3)^2$			$h(x) = 2(x + 3)^2$		
x	$f(x)$	$(x, f(x))$	x	$g(x)$	$(x, g(x))$	x	$h(x)$	$(x, h(x))$
-2	8	$(-2, 8)$	1	8	$(1, 8)$	-5	8	$(-5, 8)$
-1	2	$(-1, 2)$	2	2	$(2, 2)$	-4	2	$(-4, 2)$
0	0	$(0, 0)$	3	0	$(3, 0)$	-3	0	$(-3, 0)$
1	2	$(1, 2)$	4	2	$(4, 2)$	-2	2	$(-2, 2)$
2	8	$(2, 8)$	5	8	$(5, 8)$	-1	8	$(-1, 8)$

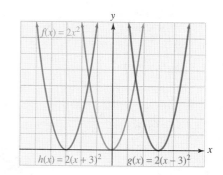

FIGURE 8-6

The results of Example 4 suggest the following facts.

Horizontal Shifts of Graphs
If f is a function and h is a positive number, then

- The graph of $y = f(x - h)$ is identical to the graph of $y = f(x)$, except that it is shifted h units to the right.

- The graph of $y = f(x + h)$ is identical to the graph of $y = f(x)$, except that it is shifted h units to the left.

■ GRAPHS OF $f(x) = a(x - h)^2 + k$

EXAMPLE 5   Graph $f(x) = 2(x - 3)^2 - 4$.

Solution   The graph of $f(x) = 2(x - 3)^2 - 4$ is identical to the graph of $g(x) = 2(x - 3)^2$, except that it has been shifted 4 units downward. The graph of $g(x) = 2(x - 3)^2$ is identical to the graph of $h(x) = 2x^2$, except that it has been shifted 3 units to the right. Thus, to graph $f(x) = 2(x - 3)^2 - 4$, we can graph $h(x) = 2x^2$, shift it 3 units to the right, and then shift it 4 units downward. (See Figure 8-7.)

Note that the vertex of the graph of $f(x) = 2(x - 3)^2 - 4$ is the point $(3, -4)$, and the axis of symmetry is the line $x = 3$.

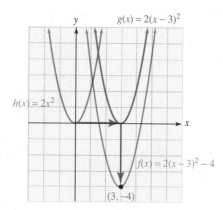

FIGURE 8-7

The results of Example 5 suggest the following theorem.

Vertex and Axis of Symmetry of a Parabola

The graph of the function

$$y = f(x) = a(x - h)^2 + k \quad (a \neq 0)$$

is a parabola with vertex at (h, k). (See Figure 8-8.)

The parabola opens upward when $a > 0$ and downward when $a < 0$. The axis of symmetry is the line $x = h$.

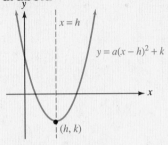

FIGURE 8-8

■ GRAPHS OF $f(x) = ax^2 + bx + c$

To graph functions of the form $f(x) = ax^2 + bx + c$, we use completing the square to write the function in the form $f(x) = a(x - h)^2 + k$.

EXAMPLE 6   Graph $f(x) = 2x^2 - 4x - 1$.

Solution   We complete the square on x to write the function in the form $f(x) = a(x - h)^2 + k$.

$f(x) = 2x^2 - 4x - 1$	
$f(x) = 2(x^2 - 2x) - 1$	Factor 2 from $2x^2 - 4x$.
$f(x) = 2(x^2 - 2x + 1 - 1) - 1$	Add and subtract 1.
$f(x) = 2(x^2 - 2x + 1) - 2 - 1$	$2(-1) = -2$.
1. $\quad f(x) = 2(x - 1)^2 - 3$	Factor $x^2 - 2x + 1$, and combine like terms.

From Equation 1, we can see that the vertex will be at the point $(1, -3)$. We can plot the vertex and a few points on either side of the vertex and draw the graph, which appears in Figure 8-9.

$$f(x) = 2x^2 - 4x - 1$$

x	$f(x)$	$(x, f(x))$
-1	5	$(-1, 5)$
0	-1	$(0, -1)$
1	-3	$(1, -3)$
2	-1	$(2, -1)$
3	5	$(3, 5)$

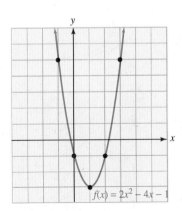

FIGURE 8-9

■ GRAPHING DEVICES

EXAMPLE 7

Graph the function $f(x) = 2x^2 + 6x - 3$ and find the coordinates of the vertex and the axis of symmetry of the parabola.

Solution

If we use a graphing device with window settings of $[-10, 10]$ for x and $[-10, 10]$ for y, we will obtain the graph shown in Figure 8-10(a).

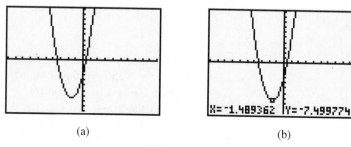

(a)   (b)

FIGURE 8-10

We use the trace feature to move the cursor to the lowest point on the graph, as shown in Figure 8-10(b). By zooming in, we can determine that the vertex is the point $(-1.5, -7.5)$ and that the line $x = -1.5$ is the axis of symmetry. ■

Because it is easy to graph quadratic functions with a graphing device, we can use graphing to find approximate solutions of quadratic equations. For example, the solutions of $0.7x^2 + 2x - 3.5 = 0$ are the numbers x that will make $y = 0$ in the quadratic function $y = f(x) = 0.7x^2 + 2x - 3.5$. To approximate these numbers, we graph the quadratic function and read the x-intercepts from the graph.

We can use the standard window settings of $[-10, 10]$ for x and $[-10, 10]$ for y and graph the function, as in Figure 8-11(a). We then trace to move the cursor to each x-intercept, as in Figures 8-11(b) and 8-11(c). From the graph, we can read the approximate value of the x-coordinate of each x-intercept. For better results, we can zoom in.

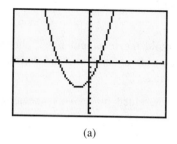 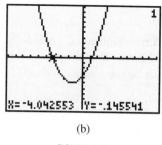 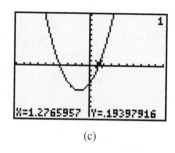

(a)   (b)   (c)

FIGURE 8-11

■ PROBLEM SOLVING

EXAMPLE 8

Ballistics   The ball shown in Figure 8-12(a) is thrown straight up with a velocity of 128 feet per second. The function $s = h(t) = -16t^2 + 128t$ gives the relation between s (the number of feet the ball is above the ground) and t (the time measured in seconds). How high does the ball travel, and when will it hit the ground?

Solution   The graph of $s = -16t^2 + 128t$ is a parabola. Since the coefficient of t^2 is negative, it opens downward, and the maximum height of the ball is given by the s-coordinate of the vertex of the parabola. We can find the coordinates of the vertex by completing the square:

$$s = -16t^2 + 128t$$
$$= -16(t^2 - 8t)$$ Factor out -16.
$$= -16(t^2 - 8t + \mathbf{16} - \mathbf{16})$$ Add and subtract 16.
$$= -16(t^2 - 8t + 16) + 256$$ $(-16)(-16) = 256$.
$$= -16(t - 4)^2 + \mathbf{256}$$ Factor $t^2 - 8t + 16$.

From the result, we can see that the coordinates of the vertex are $(\mathbf{4}, 256)$. Since $t = 4$ and $s = 256$ are the coordinates of the vertex, the ball reaches a maximum height of 256 feet in 4 seconds.

From the graph, we can see that the ball will hit the ground in 8 seconds, because the height is 0 when $t = 8$.

To solve this problem with a graphing device with window settings of $[0, 10]$ for x and $[0, 300]$ for y, we graph the function $h(t) = -16t^2 + 128t$ to get the graph in Figure 8-12(b). By using trace and zoom, we can determine that the ball reaches a height of 256 feet in 4 seconds and that the ball will hit the ground in 8 seconds.

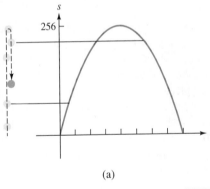

(a)   (b)

FIGURE 8-12

EXAMPLE 9   **Maximizing area**   A man wants to build the rectangular pen shown in Figure 8-13(a) to house his dog. If he uses one side of his barn, find the maximum area that he can enclose with 80 feet of fencing.

Solution   We can let the width of the pen be represented by w. Then the length is represented by $80 - 2w$.

We can find the maximum value of A as follows:

$$A = (80 - 2w)w$$ $A = lw$.
$$= 80w - 2w^2$$ Use the distributive property to remove parentheses.
$$= -2(w^2 - 40w)$$ Factor out -2.
$$= -2(w^2 - 40w + 400 - 400)$$ Add and subtract 400.
$$= -2(w^2 - 40w + 400) + 800$$ $(-2)(-400) = 800$.
$$= -2(w - 20)^2 + 800$$ Factor $w^2 - 40w + 400$.

The coordinates of the vertex of the graph of the quadratic function are (20, 800), and the maximum area is 800 square feet.

To solve this problem with a graphing device with window settings of [0, 50] for x and [0, 1000] for y, we graph the function $s(t) = -2w^2 + 80w$ to get the graph in Figure 8-13(b). By using trace and zoom, we can determine that the maximum area is 800 square feet when the width is 20 feet.

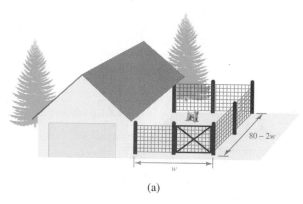

(a)

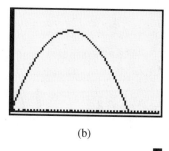

(b)

FIGURE 8-13

The Variance

ACCENT ON
STATISTICS

In statistics, the square of the standard deviation is called the **variance**.

EXAMPLE 10

If p is the probability that a person selected at random has AIDS, then $1 - p$ is the probability that the person does not have AIDS. If 100 people in Minneapolis are randomly sampled, we know from statistics that the variance of the sample will be $100p(1 - p)$. What value of p will maximize the variance?

Solution   The variance is given by the function

$$v(p) = 100p(1 - p) \quad \text{or} \quad v(p) = -100p^2 + 100p$$

Since all probabilities have values between 0 and 1, including 0 and 1, we use window settings of [0, 1] for x when graphing the function $v(p) = -100p^2 + 100p$ on a graphing device. If we also use settings of [0, 30] for y, we will obtain the graph shown in Figure 8-14(a). After using trace and zoom to obtain Figure 8-14(b), we can see that a probability of 0.5 will give the maximum variance.

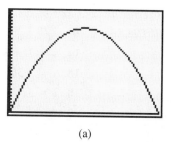

(a)

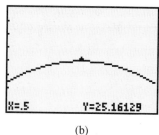

(b)

FIGURE 8-14

Orals   *Tell whether the graph of each equation opens up or down.*

1. $y = -3x^2 + x - 5$ **2.** $y = 4x^2 + 2x - 3$

3. $y = 2(x - 3)^2 - 1$ **4.** $y = -3(x + 2)^2 + 2$

Find the vertex of the parabola determined by each equation.

5. $y = 2(x - 3)^2 - 1$ **6.** $y = -3(x + 2)^2 + 2$

EXERCISE 8.2

In Exercises 1–12, graph each function.

1. $f(x) = x^2$

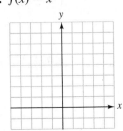

2. $f(x) = -x^2$

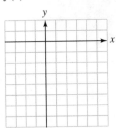

3. $f(x) = x^2 + 2$

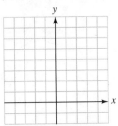

4. $f(x) = x^2 - 3$

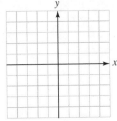

5. $f(x) = -(x - 2)^2$

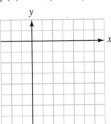

6. $f(x) = (x + 2)^2$

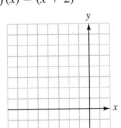

7. $f(x) = (x - 3)^2 + 2$

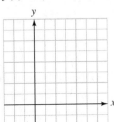

8. $f(x) = (x + 1)^2 - 2$

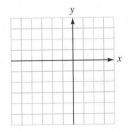

9. $f(x) = x^2 + x - 6$

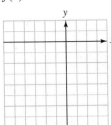

10. $f(x) = x^2 - x - 6$

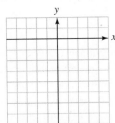

11. $f(x) = 12x^2 + 6x - 6$

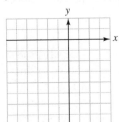

12. $f(x) = -2x^2 + 4x + 3$

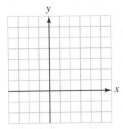

In Exercises 13–24, find the coordinates of the vertex and the axis of symmetry of the graph of each equation. If necessary, complete the square on x to write the equation in the form $y = a(x - h)^2 + k$. **Do not graph the equation.**

13. $y = (x - 1)^2 + 2$

14. $y = 2(x - 2)^2 - 1$

15. $y = 2(x + 3)^2 - 4$

16. $y = -3(x + 1)^2 + 3$

17. $y = -3x^2$

18. $y = 3x^2 - 3$

19. $y = 2x^2 - 4x$

20. $y = 3x^2 + 6x$

21. $y = -4x^2 + 16x + 5$

22. $y = 5x^2 + 20x + 25$

23. $y - 7 = 6x^2 - 5x$

24. $y - 2 = 3x^2 + 4x$

25. The equation $y - 2 = (x - 5)^2$ represents a quadratic function whose graph is a parabola. Find its vertex.

26. Show that $y = ax^2$, where $a \neq 0$, represents a quadratic function whose vertex is at the origin.

In Exercises 27–30, use a graphing device to find the coordinates of the vertex of the graph of each quadratic function.

27. $y = 2x^2 - x + 1$ **28.** $y = x^2 + 5x - 6$ **29.** $y = 7 + x - x^2$ **30.** $y = 2x^2 - 3x + 2$

In Exercises 31–34, use a graphing device to solve each equation.

31. $x^2 + x - 6 = 0$ **32.** $2x^2 - 5x - 3 = 0$ **33.** $0.5x^2 - 0.7x - 3 = 0$ **34.** $2x^2 - 0.5x - 2 = 0$

35. Ballistics   If a ball is thrown straight up with an initial velocity of 48 feet per second, its height s after t seconds is given by the equation $s = 48t - 16t^2$. Find the maximum height attained by the ball and the time it takes for the ball to return to earth.

36. Ballistics   From the top of the building in Illustration 1, a ball is thrown straight up with an initial velocity of 32 feet per second. The equation $s = -16t^2 + 32t + 48$ gives the height s of the ball t seconds after it is thrown. Find the maximum height reached by the ball and the time it takes for the ball to hit the ground.

37. Maximizing area   Find the dimensions of the rectangle of maximum area that can be constructed with 200 feet of fencing. Find the maximum area.

38. Fencing a field   A farmer wants to fence in three sides of a rectangular field (see Illustration 2) with 1000 feet of fencing. The other side of the rectangle will be a river. If the enclosed area is to be maximum, find the dimensions of the field.

ILLUSTRATION 1

ILLUSTRATION 2

In Exercises 39–46, use a graphing device to help solve each problem.

39. Maximizing revenue   The revenue R received for selling x stereos is given by the equation

$$R = -\frac{x^2}{1000} + 10x$$

Find the number of stereos that must be sold to obtain the maximum revenue.

40. Maximizing revenue   In Exercise 39, find the maximum revenue.

41. Maximizing revenue   The revenue received for selling x radios is given by the formula

$$R = -\frac{x^2}{728} + 9x$$

How many radios must be sold to obtain the maximum revenue? Find the maximum revenue.

42. Maximizing revenue The revenue received for selling x stereos is given by the formula

$$R = -\frac{x^2}{5} + 80x - 1000$$

How many stereos must be sold to obtain the maximum revenue? Find the maximum revenue.

43. Maximizing revenue When priced at $30 each, a toy has annual sales of 4000 units. The manufacturer estimates that each $1 increase in cost will decrease sales by 100 units. Find the unit price that will maximize total revenue. (*Hint:* Total revenue = price · number of units sold.)

44. Maximizing revenue When priced at $57, one type of camera has annual sales of 525 units. For each $1 the camera is reduced in price, management expects to sell an additional 75 cameras. Find the unit price that will maximize total revenue. (*Hint:* Total revenue = price · number of units sold.)

45. Finding the variance If p is the probability that a person sampled at random has high blood pressure, then $1 - p$ is the probability that the person doesn't. If 50 people are sampled at random, the variance of the sample will be $50p(1 - p)$. What two probabilities p will give a variance of 9.375?

46. Finding the variance If p is the probability that a person sampled at random smokes, then $1 - p$ is the probability that the person doesn't. If 75 people are sampled at random, the variance of the sample will be $75p(1 - p)$. What two probabilities p will give a variance of 12?

Writing Exercises   *Write a paragraph using your own words.*

1. The graph of $y = ax^2 + bx + c$ $(a \neq 0)$ passes the vertical line test. Explain why this shows that the equation defines a function.

2. The graph of $x = y^2 - 2y$ is a parabola. Explain why its graph does not represent a function.

Something to Think About   **1.** Can you use a graphing device to find solutions of the equation $0 = x^2 + x + 1$? What is the problem? How do you interpret the result?

2. Complete the square on x in the equation $y = ax^2 + bx + c$ and show that the vertex of the parabolic graph is the point with coordinates of $\left(-\dfrac{b}{2a}, c - \dfrac{b^2}{4a}\right)$.

Review Exercises   **1.** Find the value of x.

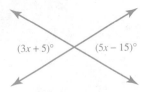

$(3x + 5)°$ $(5x - 15)°$

2. Lines r and s are parallel. Find the value of x.

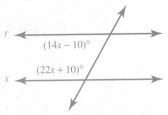

r $(14x - 10)°$
$(22x + 10)°$
s

3. Madison and St. Louis are 385 miles apart. One train leaves Madison and heads toward St. Louis at the rate of 30 miles per hour. Three hours later, a second train leaves Madison, also bound for St. Louis. If the second train travels at the rate of 55 miles per hour, in how many hours will the faster train overtake the slower train?

4. A woman invests $25,000, some at 7% annual interest and the rest at 8%. If the annual income from both investments is $1900, how much is invested at the higher rate?

8.3 Complex Numbers

So far, all of our work with quadratic equations has involved real numbers only. However, the solutions to many quadratic equations are not real numbers.

EXAMPLE 1   Solve the equation $x^2 + x + 1 = 0$.

Solution   Because the quadratic trinomial is prime and cannot be factored, we will use the quadratic formula, with $a = 1$, $b = 1$, and $c = 1$:

$$x = \frac{-b \pm \sqrt{b^2 - 4ac}}{2a}$$

$$= \frac{-1 \pm \sqrt{1^2 - 4(1)(1)}}{2(1)} \qquad \text{Substitute 1 for } a, \text{ 1 for } b, \text{ and 1 for } c.$$

$$= \frac{-1 \pm \sqrt{1 - 4}}{2}$$

$$= \frac{-1 \pm \sqrt{-3}}{2}$$

$$x = \frac{-1 + \sqrt{-3}}{2} \qquad \text{or} \qquad x = \frac{-1 - \sqrt{-3}}{2}$$ ∎

Each solution in Example 1 contains the number $\sqrt{-3}$. Since the square of no real number is -3, $\sqrt{-3}$ cannot be a real number. For years, people believed that numbers such as

$$\sqrt{-1}, \sqrt{-3}, \sqrt{-4}, \text{ and } \sqrt{-9}$$

were nonsense. In the 17th century, René Descartes (1596–1650) called them **imaginary numbers**. Today, imaginary numbers have many important uses, such as describing the behavior of alternating current in electronics.

The imaginary number $\sqrt{-1}$ is often denoted by the letter i:

$$i = \sqrt{-1}$$

Because i represents the square root of -1, it follows that

$$i^2 = -1$$

■ POWERS OF i

The powers of i produce an interesting pattern:

$$i^1 = \sqrt{-1} = i \qquad\qquad i^5 = i^4 i = 1i = i$$
$$i^2 = \left(\sqrt{-1}\right)^2 = -1 \qquad\qquad i^6 = i^4 i^2 = 1(-1) = -1$$
$$i^3 = i^2 i = -1i = -i \qquad\qquad i^7 = i^4 i^3 = 1(-i) = -i$$
$$i^4 = i^2 i^2 = (-1)(-1) = 1 \qquad\qquad i^8 = i^4 i^4 = (1)(1) = 1$$

The pattern continues: $i, -1, -i, 1, \ldots$.

EXAMPLE 2   Simplify i^{29}.

Solution   We note that 29 divided by 4 gives a quotient of 7 and a remainder of 1. Thus, $29 = 4 \cdot 7 + 1$, and

$$i^{29} = i^{4 \cdot 7 + 1} \qquad 4 \cdot 7 + 1 = 29.$$
$$= (i^4)^7 \cdot i^1$$
$$= 1^7 \cdot i \qquad i^4 = 1.$$
$$= i$$

The results of Example 2 illustrate the following theorem.

Powers of *i*
If n is a natural number that has a remainder of r when divided by 4, then
$$i^n = i^r$$
When n is divisible by 4, the remainder r is 0 and $i^0 = 1$.

EXAMPLE 3   Simplify i^{55}.

Solution   We divide 55 by 4 and get a remainder of 3. Therefore,

$$i^{55} = i^3 = -i$$

■ SIMPLIFYING IMAGINARY NUMBERS

If we assume that multiplication of imaginary numbers is commutative and associative, then

$$(2i)^2 = 2^2 i^2$$
$$= 4(-1) \qquad i^2 = -1.$$
$$= -4$$

Since $(2i)^2 = -4$, $2i$ is a square root of -4, and we can write

$$\sqrt{-4} = 2i$$

This result can also be obtained by using the multiplication property of radicals:

$$\sqrt{-4} = \sqrt{4(-1)} = \sqrt{4}\sqrt{-1} = 2i$$

We can use the multiplication property of radicals to simplify any imaginary number. For example,

$$\sqrt{-25} = \sqrt{25(-1)} = \sqrt{25}\sqrt{-1} = 5i$$
$$\sqrt{\frac{-100}{49}} = \sqrt{\frac{100}{49}(-1)} = \frac{\sqrt{100}}{\sqrt{49}}\sqrt{-1} = \frac{10}{7}i$$

These examples illustrate the following rule.

Properties of Radicals

If at least one of a and b is a nonnegative real number, then

$$\sqrt{ab} = \sqrt{a}\sqrt{b} \quad \text{and} \quad \sqrt{\frac{a}{b}} = \frac{\sqrt{a}}{\sqrt{b}} \quad (b \neq 0)$$

 WARNING! If a and b are both negative, then $\sqrt{ab} \neq \sqrt{a}\sqrt{b}$. For example, if $a = -16$ and $b = -4$,

$$\sqrt{(-16)(-4)} = \sqrt{64} = 8 \quad \text{but} \quad \sqrt{-16}\sqrt{-4} = (4i)(2i) = 8i^2$$
$$= 8(-1) = -8$$

■ COMPLEX NUMBERS

The imaginary numbers are a subset of a set of numbers that are called the **complex numbers**.

Complex Numbers

A **complex number** is any number that can be written in the form $a + bi$, where a and b are real numbers and $i = \sqrt{-1}$.

In the complex number $a + bi$, a is called the **real part**, and b is called the **imaginary part**.

If $b = 0$, the complex number $a + bi$ is a real number. If $b \neq 0$ and $a = 0$, the complex number $0 + bi$ (or just bi) is an imaginary number.

Any imaginary number can be expressed in bi form. For example,

$$\sqrt{-1} = i$$
$$\sqrt{-9} = \sqrt{9(-1)} = \sqrt{9}\sqrt{-1} = 3i$$
$$\sqrt{-3} = \sqrt{3(-1)} = \sqrt{3}\sqrt{-1} = \sqrt{3}i$$

 WARNING! The expression $\sqrt{3}i$ is often written as $i\sqrt{3}$ to make it clear that i is not part of the radicand. Do not confuse $\sqrt{3}i$ with $\sqrt{3i}$.

The relationship between the real numbers, the imaginary numbers, and the complex numbers is shown in Figure 8-15.

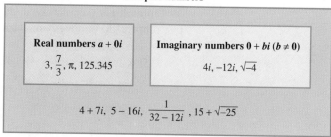

FIGURE 8-15

Equality of Complex Numbers

The complex numbers $a + bi$ and $c + di$ are equal if and only if

$$a = c \quad \text{and} \quad b = d$$

Because of the previous definition, complex numbers are equal when their real parts are equal and their imaginary parts are equal.

(a, c)   **EXAMPLE 4**

a. $2 + 3i = \sqrt{4} + \dfrac{6}{2}i$ because $2 = \sqrt{4}$ and $3 = \dfrac{6}{2}$.

b. $4 - 5i = \dfrac{12}{3} - \sqrt{25}i$ because $4 = \dfrac{12}{3}$ and $-5 = -\sqrt{25}$.

c. $x + yi = 4 + 7i$ if and only if $x = 4$ and $y = 7$. ∎

■ ARITHMETIC OF COMPLEX NUMBERS

Addition of Complex Numbers

Complex numbers are added as if they were binomials:

$$(a + bi) + (c + di) = (a + c) + (b + d)i$$

(b, d)   **EXAMPLE 5**

a. $(8 + 4i) + (12 + 8i) = 8 + 4i + 12 + 8i$
$$= 20 + 12i$$

b. $(7 - 4i) + (9 + 2i) = 7 - 4i + 9 + 2i$
$$= 16 - 2i$$

c. $(-6 + i) - (3 - 4i) = -6 + i - 3 + 4i$
$$= -9 + 5i$$

d. $(2 - 4i) - (-4 + 3i) = 2 - 4i + 4 - 3i$
$$= 6 - 7i$$ ∎

To multiply a complex number by an imaginary number, we use the distributive property to remove parentheses and then simplify. For example,

$$-5i(4 - 8i) = -5i(4) - (-5i)8i \qquad \text{Use the distributive property.}$$
$$= -20i + 40i^2 \qquad \text{Simplify.}$$
$$= -40 - 20i \qquad \text{Remember that } i^2 = -1 \text{, and write the number in } a + bi \text{ form.}$$

To multiply two complex numbers, we use the following definition.

Multiplying Complex Numbers

Complex numbers are multiplied as if they were binomials, with $i^2 = -1$:

$$(a + bi)(c + di) = ac + adi + bci + bdi^2$$
$$= (ac - bd) + (ad + bc)i$$

(b, c)

EXAMPLE 6

a. $(2 + 3i)(3 - 2i) = 6 - 4i + 9i - 6i^2$ Use the FOIL method.

$= 6 + 5i + 6$ $i^2 = -1$, and combine $-4i$ and $9i$.

$= 12 + 5i$

b. $(3 + i)(1 + 2i) = 3 + 6i + i + 2i^2$ Use the FOIL method.

$= 3 + 7i - 2$ $i^2 = -1$, and combine $6i$ and i.

$= 1 + 7i$

c. $(-4 + 2i)(2 + i) = -8 - 4i + 4i + 2i^2$ Use the FOIL method.

$= -8 - 2$ $i^2 = -1$, and combine $-4i$ and $4i$.

$= -10$ ∎

The next two examples show how to write complex numbers in $a + bi$ form. It is common to use $a - bi$ as a substitute for $a + (-b)i$.

EXAMPLE 7

a. $7 = 7 + 0i$

b. $3i = 0 + 3i$

c. $4 - \sqrt{-16} = 4 - \sqrt{-1(16)}$

$= 4 - \sqrt{16}\sqrt{-1}$

$= 4 - 4i$

d. $5 + \sqrt{-11} = 5 + \sqrt{-1(11)}$

$= 5 + \sqrt{11}\sqrt{-1}$

$= 5 + \sqrt{11}i$ ∎

EXAMPLE 8

a. $2i^2 + 4i^3 = 2(-1) + 4(-i)$

$= -2 - 4i$

b. $\dfrac{3}{2i} = \dfrac{3}{2i} \cdot \dfrac{i}{i}$ $\dfrac{i}{i} = 1$.

$= \dfrac{3i}{2i^2}$

$= \dfrac{3i}{2(-1)}$

$= \dfrac{3i}{-2}$

$= 0 - \dfrac{3}{2}i$

c. $-\dfrac{5}{i} = -\dfrac{5}{i} \cdot \dfrac{i^3}{i^3}$ $\dfrac{i^3}{i^3} = 1$.

$= -\dfrac{5(-i)}{1}$

$= 5i$

$= 0 + 5i$

d. $\dfrac{6}{i^3} = \dfrac{6i}{i^3 i}$ $\dfrac{i}{i} = 1$.

$= \dfrac{6i}{i^4}$

$= \dfrac{6i}{1}$

$= 6i$

$= 0 + 6i$ ∎

> **Complex Conjugates**
>
> The complex numbers $a + bi$ and $a - bi$ are called **complex conjugates**.

For example,

$3 + 4i$ and $3 - 4i$ are complex conjugates.

$5 - 7i$ and $5 + 7i$ are complex conjugates.

$8 + 17i$ and $8 - 17i$ are complex conjugates.

EXAMPLE 9   Find the product of $3 + i$ and its complex conjugate.

Solution   The complex conjugate of $3 + i$ is $3 - i$. We can find the product as follows:

$$(3 + i)(3 - i) = 9 - 3i + 3i - i^2 \qquad \text{Use the FOIL method.}$$
$$= 9 - i^2 \qquad \text{Combine like terms.}$$
$$= 9 - (-1) \qquad i^2 = -1.$$
$$= 10 \qquad \blacksquare$$

The product of the complex number $a + bi$ and its complex conjugate $a - bi$ is the real number $a^2 + b^2$, as the following work shows:

$$(a + bi)(a - bi) = a^2 - abi + abi - b^2 i^2 \qquad \text{Use the FOIL method.}$$
$$= a^2 - b^2(-1) \qquad i^2 = -1.$$
$$= a^2 + b^2$$

■ RATIONALIZING DENOMINATORS WITH TWO TERMS

To divide complex numbers, we often have to rationalize a denominator.

EXAMPLE 10   Do the division and write the result in $a + bi$ form: $\dfrac{1}{3 + i}$.

Solution   We can rationalize the denominator by multiplying both the numerator and the denominator of the fraction by the complex conjugate of the denominator.

$$\frac{1}{3 + i} = \frac{1}{3 + i} \cdot \frac{3 - i}{3 - i} \qquad \frac{3 - i}{3 - i} = 1.$$
$$= \frac{3 - i}{9 - 3i + 3i - i^2} \qquad \text{Multiply the numerators and multiply the denominators.}$$
$$= \frac{3 - i}{9 - (-1)} \qquad i^2 = -1.$$
$$= \frac{3 - i}{10}$$
$$= \frac{3}{10} - \frac{1}{10}i \qquad \blacksquare$$

EXAMPLE 11 Write $\dfrac{3 - i}{2 + i}$ in $a + bi$ form.

Solution We multiply both the numerator and the denominator of the fraction by the complex conjugate of the denominator.

$$\frac{3 - i}{2 + i} = \frac{3 - i}{2 + i} \cdot \frac{2 - i}{2 - i} \qquad \frac{2 - i}{2 - i} = 1.$$

$$= \frac{6 - 3i - 2i + i^2}{4 - 2i + 2i - i^2} \qquad \text{Multiply the numerators and multiply the denominators.}$$

$$= \frac{5 - 5i}{4 - (-1)} \qquad i^2 = -1.$$

$$= \frac{5(1 - i)}{5} \qquad \text{Factor out 5 in the numerator.}$$

$$= 1 - i \qquad \text{Simplify.} \qquad \blacksquare$$

EXAMPLE 12 Write $\dfrac{4 + \sqrt{-16}}{2 + \sqrt{-4}}$ in $a + bi$ form.

Solution
$$\frac{4 + \sqrt{-16}}{2 + \sqrt{-4}} = \frac{4 + 4i}{2 + 2i} \qquad \text{Write each number in } a + bi \text{ form.}$$

$$= \frac{\overset{1}{\cancel{2(2 + 2i)}}}{\underset{1}{\cancel{2 + 2i}}} \qquad \text{Factor out 2 in the numerator and simplify.}$$

$$= 2 + 0i \qquad \blacksquare$$

WARNING! To avoid mistakes, always put complex numbers in $a + bi$ form before doing any complex number arithmetic.

■ ABSOLUTE VALUE OF A COMPLEX NUMBER

> **Absolute Value of a Complex Number**
> The **absolute value** of the complex number $a + bi$ is $\sqrt{a^2 + b^2}$. In symbols,
> $$|a + bi| = \sqrt{a^2 + b^2}$$

(b, c)

EXAMPLE 13

a. $|3 + 4i| = \sqrt{3^2 + 4^2}$
$= \sqrt{9 + 16}$
$= \sqrt{25}$
$= 5$

b. $|3 - 4i| = \sqrt{3^2 + (-4)^2}$
$= \sqrt{9 + 16}$
$= \sqrt{25}$
$= 5$

c. $|-5 - 12i| = \sqrt{(-5)^2 + (-12)^2}$
$= \sqrt{25 + 144}$
$= \sqrt{169}$
$= 13$

d. $|a + 0i| = \sqrt{a^2 + 0^2}$
$= \sqrt{a^2}$
$= |a|$

$\blacksquare$

Orals *Simplify each power of i.*

1. i^3 **2.** i^2 **3.** i^4 **4.** i^5

Write each imaginary number in bi form.

5. $\sqrt{-49}$ **6.** $\sqrt{-64}$ **7.** $\sqrt{-100}$ **8.** $\sqrt{-81}$

Find each absolute value.

9. $|4 + 3i|$ **10.** $|5 - 12i|$

EXERCISE 8.3

In Exercises 1–12, solve each equation. Write all roots in bi or a + bi form.

1. $x^2 + 9 = 0$ **2.** $x^2 + 16 = 0$ **3.** $3x^2 = -16$

4. $2x^2 = -25$ **5.** $x^2 + 2x + 2 = 0$ **6.** $x^2 + 3x + 3 = 0$

7. $2x^2 + x + 1 = 0$ **8.** $3x^2 + 2x + 1 = 0$ **9.** $3x^2 - 4x = -2$

10. $2x^2 + 3x = -3$ **11.** $3x^2 - 2x = -3$ **12.** $5x^2 = 2x - 1$

In Exercises 13–20, simplify each expression.

13. i^{21} **14.** i^{19} **15.** i^{27} **16.** i^{22}

17. i^{100} **18.** i^{42} **19.** i^{97} **20.** i^{200}

In Exercises 21–26, tell whether the complex numbers are equal.

21. $3 + 7i, \sqrt{9} + (5 + 2)i$ **22.** $\sqrt{4} + \sqrt{25}i, 2 - (-5)i$ **23.** $8 + 5i, 2^3 + \sqrt{25}i^3$

24. $4 - 7i, -4i^2 + 7i^3$ **25.** $\sqrt{4} + \sqrt{-4}, 2 - 2i$ **26.** $\sqrt{-9} - i, 4i$

In Exercises 27–60, do the operations. Write all answers in a + bi form.

27. $(3 + 4i) + (5 - 6i)$ **28.** $(5 + 3i) - (6 - 9i)$ **29.** $(7 - 3i) - (4 + 2i)$

30. $(8 + 3i) + (-7 - 2i)$ **31.** $(8 + 5i) + (7 + 2i)$ **32.** $(-7 + 9i) - (-2 - 8i)$

33. $(1 + i) - 2i + (5 - 7i)$ **34.** $(-9 + i) - 5i + (2 + 7i)$

35. $(5 + 3i) - (3 - 5i) + \sqrt{-1}$ **36.** $(8 + 7i) - \left(-7 - \sqrt{-64}\right) + (3 - i)$

37. $\left(-8 - \sqrt{3}i\right) - \left(7 - 3\sqrt{3}i\right)$ **38.** $\left(2 + 2\sqrt{2}i\right) + \left(-3 - \sqrt{2}i\right)$

39. $3i(2 - i)$ **40.** $-4i(3 + 4i)$ **41.** $-5i(5 - 5i)$

42. $2i(7 + 2i)$ **43.** $(2 + i)(3 - i)$ **44.** $(4 - i)(2 + i)$

45. $(2 - 4i)(3 + 2i)$ **46.** $(3 - 2i)(4 - 3i)$

47. $\left(2 + \sqrt{2}i\right)\left(3 - \sqrt{2}i\right)$ **48.** $\left(5 + \sqrt{3}i\right)\left(2 - \sqrt{3}i\right)$

49. $\left(8 - \sqrt{-1}\right)\left(-2 - \sqrt{-16}\right)$ **50.** $\left(-1 + \sqrt{-4}\right)\left(2 + \sqrt{-9}\right)$

51. $(2 + i)^2$ **52.** $(3 - 2i)^2$ **53.** $(2 + 3i)^2$

54. $(1 - 3i)^2$ **55.** $i(5 + i)(3 - 2i)$ **56.** $i(-3 - 2i)(1 - 2i)$

57. $(2 + i)(2 - i)(1 + i)$ **58.** $(3 + 2i)(3 - 2i)(i + 1)$

59. $(3 + i)[(3 - 2i) + (2 + i)]$ **60.** $(2 - 3i)[(5 - 2i) - (2i + 1)]$

In Exercises 61–88, write each expression in a + bi form.

61. $\dfrac{1}{i}$

62. $\dfrac{1}{i^3}$

63. $\dfrac{4}{5i^3}$

64. $\dfrac{3}{2i}$

65. $\dfrac{3i}{8\sqrt{-9}}$

66. $\dfrac{5i^3}{2\sqrt{-4}}$

67. $\dfrac{-3}{5i^5}$

68. $\dfrac{-4}{6i^7}$

69. $\dfrac{5}{2-i}$

70. $\dfrac{26}{3-2i}$

71. $\dfrac{13i}{5+i}$

72. $\dfrac{2i}{5+3i}$

73. $\dfrac{-12}{7-\sqrt{-1}}$

74. $\dfrac{4}{3+\sqrt{-1}}$

75. $\dfrac{5i}{6+2i}$

76. $\dfrac{-4i}{2-6i}$

77. $\dfrac{3-2i}{3+2i}$

78. $\dfrac{2+3i}{2-3i}$

79. $\dfrac{3+2i}{3+i}$

80. $\dfrac{2-5i}{2+5i}$

81. $\dfrac{\sqrt{5}-\sqrt{3}i}{\sqrt{5}+\sqrt{3}i}$

82. $\dfrac{\sqrt{3}+\sqrt{2}i}{\sqrt{3}-\sqrt{2}i}$

83. $\left(\dfrac{i}{3+2i}\right)^2$

84. $\left(\dfrac{5+i}{2+i}\right)^2$

85. $\dfrac{i(3-i)}{3+i}$

86. $\dfrac{5+3i}{i(3-5i)}$

87. $\dfrac{(2-5i)-(5-2i)}{5-i}$

88. $\dfrac{5i}{(5+2i)+(2+i)}$

In Exercises 89–96, find each value.

89. $|6+8i|$

90. $|12+5i|$

91. $|12-5i|$

92. $|3-4i|$

93. $|5+7i|$

94. $|6-5i|$

95. $\left|\dfrac{3}{5}-\dfrac{4}{5}i\right|$

96. $\left|\dfrac{5}{13}+\dfrac{12}{13}i\right|$

97. Show that $1-5i$ is a solution of $x^2-2x+26=0$.

98. Show that $3-2i$ is a solution of $x^2-6x+13=0$.

99. Show that i is a solution of $x^4-3x^2-4=0$.

100. Show that i is *not* a solution of $x^2+x+1=0$.

Writing Exercises

Write a paragraph using your own words.

1. Tell how to decide whether two complex numbers are equal.

2. Define the complex conjugate of a complex number.

Something to Think About

1. Rationalize the numerator of $\dfrac{3-i}{2}$.

2. Rationalize the numerator of $\dfrac{2+3i}{2-3i}$.

Review Exercises

Do each operation.

1. $\dfrac{x^2-x-6}{9-x^2}\cdot\dfrac{x^2+x-6}{x^2-4}$

2. $\dfrac{3x+4}{x-2}+\dfrac{x-4}{x+2}$

3. Wind speed A plane that can fly 200 miles per hour in still air makes a 330-mile flight with a tail wind and returns, flying into the same wind. Find the speed of the wind if the total flying time is $3\frac{1}{3}$ hours.

4. Finding rates A student drove a distance of 135 miles at average speed of 50 mph. How much faster would he have to drive on the return trip to save 30 minutes of driving time?

8.4 The Discriminant and Equations That Can Be Written in Quadratic Form

■ EQUATIONS THAT CAN BE WRITTEN IN QUADRATIC FORM   ■ SOLUTIONS OF A QUADRATIC EQUATION

We can predict the type of solutions of a particular quadratic equation without solving it. To see how, we suppose that the coefficients a, b, and c in the equation $ax^2 + bx + c = 0$ $(a \neq 0)$ are real numbers. Then the solutions of the equation are given by the quadratic formula

$$x = \frac{-b \pm \sqrt{b^2 - 4ac}}{2a} \quad (a \neq 0)$$

If $b^2 - 4ac \geq 0$, the solutions are real numbers. If $b^2 - 4ac < 0$, the solutions are nonreal complex numbers. Thus, the value of $b^2 - 4ac$, called the **discriminant**, determines the type of solutions for a particular quadratic equation.

The Discriminant
If a, b, and c are real numbers and

If $b^2 - 4ac$ is . . .	the solutions are . . .
positive,	real numbers and unequal.
0,	real numbers and equal.
negative,	nonreal complex numbers and complex conjugates.

If a, b, and c are rational numbers and

If $b^2 - 4ac$ is . . .	the solutions are . . .
a perfect square greater than 0,	rational numbers and unequal.
positive and not a perfect square,	irrational numbers and unequal.

EXAMPLE 1   Determine the type of solutions for the equation $x^2 + x + 1 = 0$.

Solution   We calculate the discriminant:

$$b^2 - 4ac = 1^2 - 4(1)(1) \qquad a = 1, b = 1, \text{ and } c = 1.$$
$$= -3$$

Since $b^2 - 4ac < 0$, the solutions are nonreal complex conjugates.   ■

EXAMPLE 2   Determine the type of solutions for the equation $3x^2 + 5x + 2 = 0$.

Solution   We calculate the discriminant:

$$b^2 - 4ac = 5^2 - 4(3)(2) \qquad a = 3, b = 5, \text{ and } c = 2.$$
$$= 25 - 24$$
$$= 1$$

Since $b^2 - 4ac > 0$ and $b^2 - 4ac$ is a perfect square, the solutions are rational and unequal. ∎

EXAMPLE 3   What value of k will make the solutions of the equation $kx^2 - 12x + 9 = 0$ equal?

Solution   We calculate the discriminant:

$$b^2 - 4ac = (-12)^2 - 4(k)(9) \qquad a = k, b = -12, \text{ and } c = 9.$$
$$= 144 - 36k$$
$$= -36k + 144$$

Since the solutions are to be equal, we let $-36k + 144 = 0$ and solve for k.

$$-36k + 144 = 0$$
$$-36k = -144 \qquad \text{Subtract 144 from both sides.}$$
$$k = 4 \qquad \text{Divide both sides by } -36.$$

If $k = 4$, the solutions will be equal. Verify this by solving $4x^2 - 12x + 9 = 0$ and showing that the solutions are equal. ∎

■ EQUATIONS THAT CAN BE WRITTEN IN QUADRATIC FORM

Many nonquadratic equations can be put into quadratic form and then solved with the techniques used for solving quadratic equations. For example, we can solve $x^4 - 5x^2 + 4 = 0$ as follows:

$$x^4 - 5x^2 + 4 = 0$$
$$(x^2)^2 - 5(x^2) + 4 = 0$$
$$y^2 - 5y + 4 = 0 \qquad \text{Let } y = x^2.$$
$$(y - 4)(y - 1) = 0 \qquad \text{Factor } y^2 - 5y + 4.$$
$$y - 4 = 0 \quad \text{or} \quad y - 1 = 0 \qquad \text{Set each factor equal to 0.}$$
$$y = 4 \quad | \quad y = 1$$

Since $x^2 = y$, it follows that $x^2 = 4$ or $x^2 = 1$. Thus,

$$x^2 = 4 \qquad \text{or} \qquad x^2 = 1$$
$$x = 2 \quad \text{or} \quad x = -2 \quad | \quad x = 1 \quad \text{or} \quad x = -1$$

This equation has four solutions: 1, -1, 2, and -2. Verify that each one satisfies the original equation. Note that this equation can be solved by factoring.

EXAMPLE 4   Solve the equation $x - 7x^{1/2} + 12 = 0$.

Solution   If y^2 is substituted for x and y is substituted for $x^{1/2}$, the equation

$$x - 7x^{1/2} + 12 = 0$$

becomes a quadratic equation that can be solved by factoring:

$$y^2 - 7y + 12 = 0 \qquad \text{Substitute } y^2 \text{ for } x \text{ and } y \text{ for } x^{1/2}.$$
$$(y - 3)(y - 4) = 0 \qquad \text{Factor } y^2 - 7y + 12 = 0.$$
$$y - 3 = 0 \quad \text{or} \quad y - 4 = 0 \qquad \text{Set each factor equal to 0.}$$
$$y = 3 \quad | \quad y = 4$$

Because $x = y^2$, it follows that

$$x = 3^2 \quad \text{or} \quad x = 4^2$$
$$= 9 \quad | \quad = 16$$

Verify that both solutions satisfy the original equation.   ■

EXAMPLE 5   Solve the equation $\dfrac{24}{x} + \dfrac{12}{x + 1} = 11$.

Solution   Since the denominator of a fraction cannot be 0, x cannot be 0 or -1. If either 0 or -1 appears as a suspected solution, it must be discarded.

$$\frac{24}{x} + \frac{12}{x + 1} = 11$$

$$x(x + 1)\left(\frac{24}{x} + \frac{12}{x + 1}\right) = x(x + 1)11 \qquad \text{Multiply both sides by } x(x + 1).$$

$$24(x + 1) + 12x = (x^2 + x)11 \qquad \text{Simplify.}$$

$$24x + 24 + 12x = 11x^2 + 11x \qquad \text{Use the distributive property to remove parentheses.}$$

$$36x + 24 = 11x^2 + 11x \qquad \text{Combine like terms.}$$

$$0 = 11x^2 - 25x - 24 \qquad \text{Subtract } 36x \text{ and } 24 \text{ from both sides.}$$

$$0 = (11x + 8)(x - 3) \qquad \text{Factor } 11x^2 - 25x - 24.$$

$$11x + 8 = 0 \qquad \text{or} \qquad x - 3 = 0 \qquad \text{Set each factor equal to 0,}$$

$$x = -\frac{8}{11} \qquad \bigg| \qquad x = 3$$

Verify that both $-\frac{8}{11}$ and 3 satisfy the original equation.   ■

EXAMPLE 6   Solve the formula $s = 16t^2 - 32$ for t.

Solution   We proceed as follows:

$$s = 16t^2 - 32$$

$$s + 32 = 16t^2 \qquad \text{Add 32 to both sides.}$$

$$\frac{s + 32}{16} = t^2 \qquad \text{Divide both sides by 16.}$$

$$t^2 = \frac{s + 32}{16} \qquad \text{Apply the symmetric property of equality.}$$

$$t = \pm\sqrt{\frac{s + 32}{16}} \qquad \text{Apply the square root property.}$$

$$t = \frac{\pm\sqrt{s + 32}}{\sqrt{16}}$$

$$t = \frac{\pm\sqrt{s + 32}}{4}$$

■

■ SOLUTIONS OF A QUADRATIC EQUATION

Solutions of a Quadratic Equation

If r_1 and r_2 are the solutions of the quadratic equation $ax^2 + bx + c = 0$, with $a \neq 0$, then

$$r_1 + r_2 = -\frac{b}{a} \quad \text{and} \quad r_1 r_2 = \frac{c}{a}$$

Proof   To prove this property, we note that the solutions to the equation are given by the quadratic formula

$$r_1 = \frac{-b + \sqrt{b^2 - 4ac}}{2a} \quad \text{and} \quad r_2 = \frac{-b - \sqrt{b^2 - 4ac}}{2a}$$

Thus,

$$r_1 + r_2 = \frac{-b + \sqrt{b^2 - 4ac}}{2a} + \frac{-b - \sqrt{b^2 - 4ac}}{2a}$$

$$= \frac{-b + \sqrt{b^2 - 4ac} - b - \sqrt{b^2 - 4ac}}{2a} \qquad \text{Keep the denominator and add the numerators.}$$

$$= -\frac{2b}{2a} \qquad \text{Combine like terms.}$$

$$= -\frac{b}{a}$$

and

$$r_1 r_2 = \frac{-b + \sqrt{b^2 - 4ac}}{2a} \cdot \frac{-b - \sqrt{b^2 - 4ac}}{2a}$$

$$= \frac{b^2 - (b^2 - 4ac)}{4a^2} \qquad \text{Multiply the numerators and multiply the denominators, and combine like terms.}$$

$$= \frac{b^2 - b^2 + 4ac}{4a^2}$$

$$= \frac{4ac}{4a^2} \qquad \text{Combine } b^2 \text{ and } -b^2.$$

$$= \frac{c}{a} \qquad \qquad \qquad \qquad \qquad \square$$

It can also be shown that if

$$r_1 + r_2 = -\frac{b}{a} \quad \text{and} \quad r_1 r_2 = \frac{c}{a}$$

then r_1 and r_2 are solutions of $ax^2 + bx + c = 0$. We can use this fact to check the solutions of quadratic equations.

EXAMPLE 7   Show that $\frac{3}{2}$ and $-\frac{1}{3}$ are solutions of the quadratic equation $6x^2 - 7x - 3 = 0$.

Solution   Since $a = 6$, $b = -7$, and $c = -3$, we have

$$-\frac{b}{a} = -\frac{-7}{6} = \frac{7}{6} \quad \text{and} \quad \frac{c}{a} = \frac{-3}{6} = -\frac{1}{2}$$

Since $\frac{3}{2} + \left(-\frac{1}{3}\right) = \frac{7}{6}$ and $\left(\frac{3}{2}\right)\left(-\frac{1}{3}\right) = -\frac{1}{2}$, these numbers are solutions. Solve the equation to see that the roots are $\frac{3}{2}$ and $-\frac{1}{3}$.   ■

Orals   *Find $b^2 - 4ac$ when*

1. $a = 1, b = 1, c = 1$ **2.** $a = 2, b = 1, c = 1$

Determine the type of solutions for

3. $x^2 - 4x + 1 = 0$ **4.** $8x^2 - x + 2 = 0$

Are the following numbers solutions of $x^2 - 7x + 6 = 0$?

5. $1, 5$ **6.** $1, 6$

EXERCISE 8.4

In Exercises 1–8, use the discriminant to determine what type of solutions exist for each quadratic equation. **Do not solve the equation.**

1. $4x^2 - 4x + 1 = 0$ **2.** $6x^2 - 5x - 6 = 0$

3. $5x^2 + x + 2 = 0$ **4.** $3x^2 + 10x - 2 = 0$

5. $2x^2 = 4x - 1$ **6.** $9x^2 = 12x - 4$

7. $x(2x - 3) = 20$ **8.** $x(x - 3) = -10$

In Exercises 9–16, find the value(s) of k that will make the solutions of each given quadratic equation equal.

9. $x^2 + kx + 9 = 0$ **10.** $kx^2 - 12x + 4 = 0$

11. $9x^2 + 4 = -kx$ **12.** $9x^2 - kx + 25 = 0$

13. $(k - 1)x^2 + (k - 1)x + 1 = 0$ **14.** $(k + 3)x^2 + 2kx + 4 = 0$

15. $(k + 4)x^2 + 2kx + 9 = 0$ **16.** $(k + 15)x^2 + (k - 30)x + 4 = 0$

17. Use the discriminant to determine whether the solutions of $1492x^2 + 1776x - 1984 = 0$ are real numbers.

18. Use the discriminant to determine whether the solutions of $1776x^2 - 1492x + 1984 = 0$ are real numbers.

19. Determine k such that the solutions of $3x^2 + 4x = k$ are nonreal complex numbers.

20. Determine k such that the solutions of $kx^2 - 4x = 7$ are nonreal complex numbers.

In Exercises 21–48, solve each equation.

21. $x^4 - 17x^2 + 16 = 0$ **22.** $x^4 - 10x^2 + 9 = 0$

23. $x^4 - 3x^2 = -2$ **24.** $x^4 - 29x^2 = -100$

25. $x^4 = 6x^2 - 5$ **26.** $x^4 = 8x^2 - 7$

27. $2x^4 - 10x^2 = -8$

28. $2x^4 + 24 = 26x^2$

29. $2x + x^{1/2} - 3 = 0$

30. $2x - x^{1/2} - 1 = 0$

31. $3x + 5x^{1/2} + 2 = 0$

32. $3x - 4x^{1/2} + 1 = 0$

33. $x^{2/3} + 5x^{1/3} + 6 = 0$

34. $x^{2/3} - 7x^{1/3} + 12 = 0$

35. $x^{2/3} - 2x^{1/3} - 3 = 0$

36. $x^{2/3} + 4x^{1/3} - 5 = 0$

37. $x + 5 + \dfrac{4}{x} = 0$

38. $x - 4 + \dfrac{3}{x} = 0$

39. $x + 1 = \dfrac{20}{x}$

40. $x + \dfrac{15}{x} = 8$

41. $\dfrac{1}{x-1} + \dfrac{3}{x+1} = 2$

42. $\dfrac{6}{x-2} - \dfrac{12}{x-1} = -1$

43. $\dfrac{1}{x+2} + \dfrac{24}{x+3} = 13$

44. $\dfrac{3}{x} + \dfrac{4}{x+1} = 2$

45. $x^{-4} - 2x^{-2} + 1 = 0$

46. $4x^{-4} + 1 = 5x^{-2}$

47. $x + \dfrac{2}{x-2} = 0$

48. $x + \dfrac{x+5}{x-3} = 0$

In Exercises 49–56, solve each equation for the indicated variable.

49. $x^2 + y^2 = r^2$ for x

50. $x^2 + y^2 = r^2$ for y

51. $I = \dfrac{k}{d^2}$ for d

52. $V = \dfrac{1}{3}\pi r^2 h$ for r

53. $xy^2 + 3xy + 7 = 0$ for y

54. $kx = ay - x^2$ for x

55. $\sigma = \sqrt{\dfrac{\Sigma x^2}{N} - \mu^2}$ for μ^2

56. $\sigma = \sqrt{\dfrac{\Sigma x^2}{N} - \mu^2}$ for N

In Exercises 57–64, solve each quadratic equation and verify that the sum of the solutions is $-\frac{b}{a}$ and that the product of the solutions is $\frac{c}{a}$.

57. $12x^2 - 5x - 2 = 0$

58. $8x^2 - 2x - 3 = 0$

59. $2x^2 + 5x + 1 = 0$

60. $3x^2 + 9x + 1 = 0$

61. $3x^2 - 2x + 4 = 0$

62. $2x^2 - x + 4 = 0$

63. $x^2 + 2x + 5 = 0$

64. $x^2 - 4x + 13 = 0$

Writing Exercises   *Write a paragraph using your own words.*

1. Describe how to predict what type of solutions the equation $3x^2 - 4x + 5 = 0$ will have.

2. How is the discriminant related to the quadratic formula?

Something to Think About   **1.** Can a quadratic equation with integer coefficients have one real and one complex solution? Why?

2. Can a quadratic equation with complex coefficients have one real and one complex solution? Why?

Review Exercises   *Solve each equation.*

1. $\dfrac{1}{4} + \dfrac{1}{t} = \dfrac{1}{2t}$

2. $\dfrac{p-3}{3p} + \dfrac{1}{2p} = \dfrac{1}{4}$

3. Find the slope of the line passing through $P(-2, -4)$ and $Q(3, 5)$.

4. Write the equation of the line passing through $P(-2, -4)$ and $Q(3, 5)$ in general form.

8.5 Quadratic and Other Nonlinear Inequalities

■ SOLVING QUADRATIC INEQUALITIES ■ SOLVING OTHER INEQUALITIES ■ GRAPHING DEVICES
■ GRAPHS OF NONLINEAR INEQUALITIES IN TWO VARIABLES

■ SOLVING QUADRATIC INEQUALITIES

To solve the inequality $x^2 + x - 6 < 0$, we must find the values of x that make the inequality true. To find these values, we can factor the trinomial to obtain

$$(x + 3)(x - 2) < 0$$

Since the product of $x + 3$ and $x - 2$ is less than 0, their values must be opposite in sign. To find the intervals where this is true, we keep track of their signs by constructing the chart in Figure 8-16. The chart shows that

- The binomial $x - 2 = 0$ when $x = 2$, the binomial is positive when $x > 2$, and the binomial is negative when $x < 2$.
- The binomial $x + 3 = 0$ when $x = -3$, the binomial is positive when $x > -3$, and the binomial is negative when $x < -3$.

The only place where the values of the binomial are opposite in sign is in the interval $(-3, 2)$. Therefore,

$$-3 < x < 2$$

The graph of the solution set is shown on the number line in Figure 8-16.

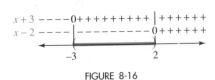

FIGURE 8-16

EXAMPLE 1   Solve the inequality $x^2 + 2x - 3 \geq 0$.

Solution   We factor the quadratic trinomial to get $(x - 1)(x + 3)$ and construct a sign chart as in Figure 8-17.

- The binomial $x - 1 = 0$ when $x = 1$, is positive when $x > 1$, and is negative when $x < 1$.
- The binomial $x + 3 = 0$ when $x = -3$, is positive when $x > -3$, and is negative when $x < -3$.

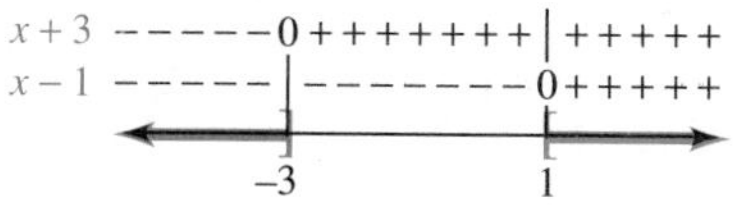

The product of $x - 1$ and $x + 3$ will be greater than 0 when the signs of the binomial factors are the same. This occurs in the intervals $(-\infty, -3)$ or $(1, \infty)$. The numbers -3 and 1 are also included because they make the product equal to 0. Thus, the solution set is

$$(-\infty, -3] \cup [1, \infty) \qquad \text{or} \qquad x \le -3 \quad \text{or} \quad x \ge 1$$

The graph of the solution set is shown on the number line in Figure 8-17.   ■

$x + 3$ $-----0+++++++|+++++$
$x - 1$ $-----|-------0+++++$

$-3 \qquad 1$

FIGURE 8-17

■ SOLVING OTHER INEQUALITIES

Making a sign chart is useful for solving many inequalities that are neither linear nor quadratic.

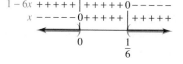

EXAMPLE 2   Solve the inequality $\dfrac{1}{x} < 6$.

Solution   We subtract 6 from both sides to make the right-hand side equal to 0. We then find a common denominator and add the fractions:

$$\frac{1}{x} < 6$$

$$\frac{1}{x} - 6 < 0 \qquad \text{Subtract 6 from both sides.}$$

$$\frac{1}{x} - \frac{6x}{x} < 0 \qquad \text{Get a common denominator.}$$

$$\frac{1 - 6x}{x} < 0 \qquad \text{Keep the denominator and subtract the numerators.}$$

We now make a sign chart as in Figure 8-18.

- The denominator x is 0 when $x = 0$, is positive when $x > 0$, and is negative when $x < 0$.

- The numerator $1 - 6x$ is 0 when $x = \frac{1}{6}$, is positive when $x < \frac{1}{6}$, and is negative when $x > \frac{1}{6}$.

The fraction $\frac{1 - 6x}{x}$ will be less than 0 when the numerator and denominator are opposite in sign. This occurs in the interval

$$\left(-\infty, 0\right) \cup \left(\frac{1}{6}, \infty\right) \qquad \text{or} \qquad x < 0 \quad \text{or} \quad x > \frac{1}{6}$$

The graph of this interval is shown in Figure 8-18.   ■

$1 - 6x$ $+++++|+++++0-----$
x $-----0+++++|+++++$

$0 \qquad \frac{1}{6}$

FIGURE 8-18

WARNING!   Since we don't know if x is positive, 0, or negative, multiplying both sides of the inequality $\frac{1}{x} < 6$ by x is a three-case situation:

- If $x > 0$, then $1 < 6x$.

- If $x = 0$, then the fraction $\frac{1}{x}$ is undefined.

- If $x < 0$, then $1 > 6x$.

If you multiply both sides by x and solve the linear inequality $1 < 6x$, you are only considering one case and will get only part of the answer.

EXAMPLE 3 Solve the inequality $\dfrac{x^2 - 3x + 2}{x - 3} \geq 0$.

Solution We write the fraction with the numerator in factored form.

$$\frac{(x - 2)(x - 1)}{x - 3} \geq 0$$

To keep track of the signs of the three binomials, we construct the sign chart shown in Figure 8-19. The fraction will be positive in the intervals where all factors are positive, or where two factors are negative. The numbers 1 and 2 are included because they make the numerator (and thus the fraction) equal to 0. The number 3 is not included because it gives a 0 in the denominator.

The solution is the interval $[1, 2] \cup (3, \infty)$. The graph appears in Figure 8-19. ∎

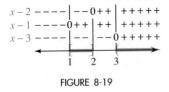

FIGURE 8-19

EXAMPLE 4 Solve the inequality $\dfrac{3}{x - 1} < \dfrac{2}{x}$.

Solution We subtract $\frac{2}{x}$ from both sides to get 0 on the right-hand side and then proceed as follows:

$$\frac{3}{x - 1} < \frac{2}{x}$$

$$\frac{3}{x - 1} - \frac{2}{x} < 0 \qquad \text{Subtract } \tfrac{2}{x} \text{ from both sides.}$$

$$\frac{3x}{(x - 1)x} - \frac{2(x - 1)}{x(x - 1)} < 0 \qquad \text{Get a common denominator.}$$

$$\frac{3x - 2x + 2}{x(x - 1)} < 0 \qquad \text{Keep the denominator and subtract the numerators.}$$

$$\frac{x + 2}{x(x - 1)} < 0 \qquad \text{Combine like terms.}$$

We can keep track of the signs of the three polynomials with the sign chart shown in Figure 8-20. The fraction will be negative in the intervals with either one or three negative factors. The numbers 0 and 1 are not included because they give a 0 in the denominator, and the number −2 is not included because it does not satisfy the inequality.

The solution is the interval $(-\infty, -2) \cup (0, 1)$, as shown in Figure 8-20. ∎

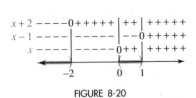

FIGURE 8-20

■ GRAPHING DEVICES

To approximate the solutions of the inequality $x^2 + 2x - 3 \geq 0$ (Example 1) by graphing, we can use window settings of $[-10, 10]$ for x and $[-10, 10]$ for y and graph the quadratic function $y = x^2 + 2x - 3$, as in Figure 8-21. The solution of the inequality will be those numbers x for which the graph of $y = x^2 + 2x - 3$ lies above or on the x-axis. We can trace to find that this interval is $(-\infty, -3] \cup [1, \infty)$.

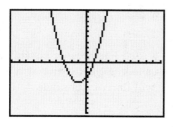

FIGURE 8-21

To approximate the solutions of $\dfrac{3}{x-1} < \dfrac{2}{x}$ (Example 4), we first write the inequality in the form

$$\frac{x+2}{x(x-1)} < 0$$

use window settings of $[-10, 10]$ for x and $[-10, 10]$ for y, and graph the function $y = \frac{x+2}{x(x-1)}$, as in Figure 8-22(a). The solution of the inequality will be those numbers x for which the graph lies below the x-axis.

We can trace to see that the graph is below the x-axis when x is less than -2. Since it is hard to see the graph in the interval $0 < x < 1$, we redraw the graph using settings of $[-1, 2]$ for x and $[-25, 10]$ for y. (See Figure 8-22(b).)

We can now see that the graph is below the x-axis in the interval $(0, 1)$. Thus, the solution to the inequality is the union of two intervals:

$$(-\infty, -2) \cup (0, 1)$$

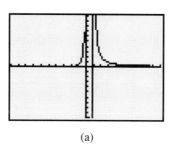

(a)   (b)

FIGURE 8-22

■ GRAPHS OF NONLINEAR INEQUALITIES IN TWO VARIABLES

We now consider the graphs of nonlinear inequalities in two variables.

EXAMPLE 5   Graph the inequality $y < -x^2 + 4$.

Solution   The graph of $y = -x^2 + 4$ is the parabolic boundary separating the region representing $y < -x^2 + 4$ and the region representing $y > -x^2 + 4$.

We graph the quadratic function $y = -x^2 + 4$ as a broken parabola, because equality is not permitted. Since the coordinates of the origin satisfy the inequality $y < -x^2 + 4$, the point $(0, 0)$ is in the graph. The complete graph is shown in Figure 8-23.

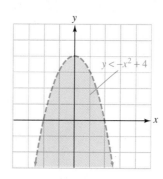

FIGURE 8-23

EXAMPLE 6   Graph the inequality $x \leq |y|$.

Solution   We first graph $x = |y|$ as in Figure 8-24(a), using a solid line because equality is permitted. Because the origin is on the graph, we cannot use the origin as a test point. However, any another point, such as $(1, 0)$, will do. We substitute 1 for x and 0 for y into the inequality to get

$$x \leq |y|$$
$$1 \leq |0|$$
$$1 \leq 0$$

Since $1 \leq 0$ is a false statement, the point $(1, 0)$ does not satisfy the inequality and is not part of the graph. Thus, the graph of $x \leq |y|$ is to the left of the boundary. The complete graph is shown in Figure 8-24(b).

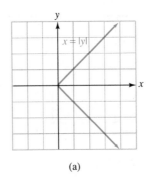
(a)

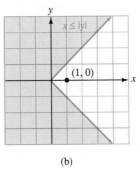
(b)

FIGURE 8-24   ∎

Orals   *Find the interval where $x - 2$ is*

1. 0   **2.** positive   **3.** negative

Find the interval where $x + 3$ is

4. 0   **5.** positive   **6.** negative

Multiply both sides of the inequality $\frac{1}{x} < 2$ by x when x is

7. positive   **8.** negative

EXERCISE 8.5

In Exercises 1–40, solve each inequality. Give each result in interval notation and graph the solution set.

1. $x^2 - 5x + 4 < 0$ **2.** $x^2 - 3x - 4 > 0$

3. $x^2 - 8x + 15 > 0$ **4.** $x^2 + 2x - 8 < 0$

5. $x^2 + x - 12 \leq 0$ **6.** $x^2 + 7x + 12 \geq 0$

7. $x^2 + 2x \geq 15$

8. $x^2 - 8x \leq -15$

9. $x^2 + 8x < -16$

10. $x^2 + 6x > -9$

11. $x^2 \geq 9$

12. $x^2 \geq 16$

13. $2x^2 - 50 < 0$

14. $3x^2 - 243 < 0$

15. $\dfrac{1}{x} < 2$

16. $\dfrac{1}{x} > 3$

17. $\dfrac{4}{x} \geq 2$

18. $-\dfrac{6}{x} < 12$

19. $-\dfrac{5}{x} < 3$

20. $\dfrac{4}{x} \geq 8$

21. $\dfrac{x^2 - x - 12}{x - 1} < 0$

22. $\dfrac{x^2 + x - 6}{x - 4} \geq 0$

23. $\dfrac{x^2 + x - 20}{x + 2} \geq 0$

24. $\dfrac{x^2 - 10x + 25}{x + 5} < 0$

25. $\dfrac{x^2 - 4x + 4}{x + 4} < 0$

26. $\dfrac{2x^2 - 5x + 2}{x + 2} > 0$

27. $\dfrac{6x^2 - 5x + 1}{2x + 1} > 0$

28. $\dfrac{6x^2 + 11x + 3}{3x - 1} < 0$

29. $\dfrac{3}{x - 2} < \dfrac{4}{x}$

30. $\dfrac{-6}{x + 1} \geq \dfrac{1}{x}$

31. $\dfrac{-5}{x + 2} \geq \dfrac{4}{2 - x}$

32. $\dfrac{-6}{x - 3} < \dfrac{5}{3 - x}$

33. $\dfrac{7}{x - 3} \geq \dfrac{2}{x + 4}$

34. $\dfrac{-5}{x - 4} < \dfrac{3}{x + 1}$

35. $\dfrac{x}{x + 4} \leq \dfrac{1}{x + 1}$

36. $\dfrac{x}{x + 9} \geq \dfrac{1}{x + 1}$

37. $\dfrac{x}{x + 16} > \dfrac{1}{x + 1}$

38. $\dfrac{x}{x + 25} < \dfrac{1}{x + 1}$

39. $(x + 2)^2 > 0$

40. $(x - 3)^2 < 0$

In Exercises 41–44, use a graphing device to solve each inequality. Give the answer in interval notation.

41. $x^2 - 2x - 3 < 0$ **42.** $x^2 + x - 6 > 0$ **43.** $\dfrac{x + 3}{x - 2} > 0$ **44.** $\dfrac{3}{x} < 2$

In Exercises 45–56, graph each inequality.

45. $y < x^2 + 1$

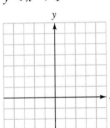

46. $y > x^2 - 3$

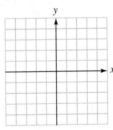

47. $y \leq x^2 + 5x + 6$

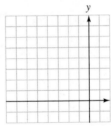

48. $y \geq x^2 + 5x + 4$

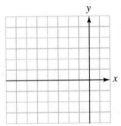

49. $y \geq (x - 1)^2$

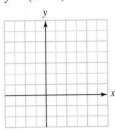

50. $y \leq (x + 2)^2$

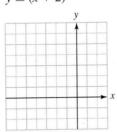

51. $-x^2 - y + 6 > -x$

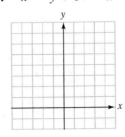

52. $y > (x + 3)(x - 2)$

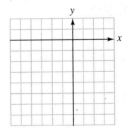

53. $y < |x + 4|$

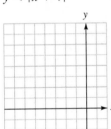

54. $y \geq |x - 3|$

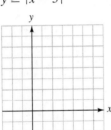

55. $y \leq -|x| + 2$

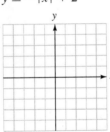

56. $y < |x| - 2$

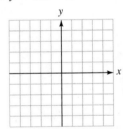

Writing Exercises   *Write a paragraph using your own words.*

1. Explain why $(x - 4)(x + 5)$ will be positive only when the signs of $x - 4$ and $x + 5$ are the same.

2. Tell how to find the graph of $y \geq x^2$.

Something to Think About

1. Under what conditions will the fraction $\frac{(x - 1)(x + 4)}{(x + 2)(x + 1)}$ be positive?

2. Under what conditions will the fraction $\frac{(x - 1)(x + 4)}{(x + 2)(x + 1)}$ be negative?

Review Exercises   *Write each expression as an equation.*

1. x varies directly with y.

2. y varies inversely with t.

3. t varies jointly with x and y.

4. d varies directly with t but inversely with u^2.

Find the slope of the graph of each equation.

5. $y = 3x - 4$

6. $\dfrac{2x - y}{5} = 8$

8.6 Algebra and Composition of Functions

■ ALGEBRA OF FUNCTIONS ■ COMPOSITION OF FUNCTIONS ■ IDENTITY FUNCTION ■ PROBLEM SOLVING

We now consider how functions can be added, subtracted, multiplied, and divided.

■ ALGEBRA OF FUNCTIONS

Operations on Functions
If the domains and ranges of functions f and g are subsets of the real numbers, then

The **sum** of f and g, denoted as $f + g$, is defined by
$$(f + g)(x) = f(x) + g(x)$$
The **difference** of f and g, denoted as $f - g$, is defined by
$$(f - g)(x) = f(x) - g(x)$$
The **product** of f and g, denoted as $f \cdot g$, is defined by
$$(f \cdot g)(x) = f(x)g(x)$$
The **quotient** of f and g, denoted as f/g, is defined by
$$(f/g)(x) = \frac{f(x)}{g(x)} \quad (g(x) \neq 0)$$

The domain of each of these functions is the set of real numbers x that are in the domain of both f and g. In the case of the quotient, there is the further restriction that $g(x) \neq 0$.

EXAMPLE 1   Let $f(x) = 2x^2 + 1$ and $g(x) = 5x - 3$. Find each of the following functions and its domain: **a.** $f + g$ and   **b.** $f - g$.

Solution   **a.** $(f + g)(x) = f(x) + g(x)$
$$= (2x^2 + 1) + (5x - 3)$$
$$= 2x^2 + 5x - 2$$

The domain of $f + g$ is the set of real numbers that are in the domain of both f and g. Since the domain of both f and g is the interval $(-\infty, \infty)$, the domain of $f + g$ is also the interval $(-\infty, \infty)$.

b. $(f - g)(x) = f(x) - g(x)$
$$= (2x^2 + 1) - (5x - 3)$$
$$= 2x^2 + 1 - 5x + 3 \qquad \text{Remove parentheses.}$$
$$= 2x^2 - 5x + 4 \qquad \text{Combine like terms.}$$

Since the domain of both f and g is the interval $(-\infty, \infty)$, the domain of $f - g$ is also the interval $(-\infty, \infty)$.   ■

EXAMPLE 2   Let $f(x) = 2x^2 + 1$ and $g(x) = 5x - 3$. Find each of the following functions and its domain: **a.** $f \cdot g$ and   **b.** f/g.

Solution   **a.** $(f \cdot g)(x) = f(x)g(x)$

$$= (2x^2 + 1)(5x - 3)$$

$$= 10x^3 - 6x^2 + 5x - 3 \qquad \text{Multiply the binomials.}$$

The domain $f \cdot g$ is the set of real numbers that are in the domain of both f and g. Since the domain of both f and g is the interval $(-\infty, \infty)$, the domain of $f \cdot g$ is also the interval $(-\infty, \infty)$.

b. $\left(f/g\right)(x) = \dfrac{f(x)}{g(x)}$

$$= \dfrac{2x^2 + 1}{5x - 3}$$

Since the denominator of the fraction cannot be 0, $x \neq \frac{3}{5}$. The domain of f/g is the interval $\left(-\infty, \frac{3}{5}\right) \cup \left(\frac{3}{5}, \infty\right)$.   ■

■ COMPOSITION OF FUNCTIONS

Often one quantity is a function of a second quantity that depends, in turn, on a third quantity. For example, the cost of a car trip is a function of the gasoline consumed. The amount of gasoline consumed, in turn, is a function of the number of miles driven. Such chains of dependence can be analyzed mathematically as **compositions of functions**.

Suppose that $y = f(x)$ and $y = g(x)$ define two functions. Any number x in the domain of g will produce the corresponding value $g(x)$ in the range of g. If $g(x)$ is in the domain of function f, then $g(x)$ can be substituted into f, and a corresponding value $f(g(x))$ will be determined. This two-step process defines a new function, called a **composite function**, denoted by $f \circ g$. (See Figure 8-25.)

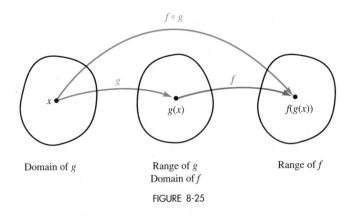

Domain of g Range of g Range of f
                     Domain of f

FIGURE 8-25

Composite Function

The **composite function** $f \circ g$ is defined by

$$(f \circ g)(x) = f(g(x))$$

For example, if $f(x) = 4x$ and $g(x) = 3x + 2$, then

$$(f \circ g)(x) = f(g(x)) \qquad \text{or} \qquad (g \circ f)(x) = g(f(x))$$
$$= f(3x + 2) \qquad\qquad\qquad = g(4x)$$
$$= 4(3x + 2) \qquad\qquad\qquad = 3(4x) + 2$$
$$= 12x + 8 \qquad\qquad\qquad\quad = 12x + 2$$

 WARNING! Note that in the previous example, $(f \circ g)(x) \neq (g \circ f)(x)$. This shows that the composition of functions is not commutative.

We have seen that a function can be represented by a machine: We put in a number from the domain, and a number from the range comes out. For example, if we put the number 2 into the machine shown in Figure 8-26(a), the number $f(2) = 5(2) - 2 = 8$ comes out. In general, if we put x into the machine shown in Figure 8-26(b), the value $f(x)$ comes out.

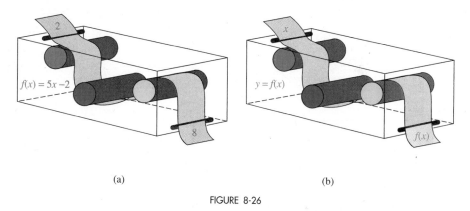

(a)   (b)

FIGURE 8-26

The function machines shown in Figure 8-27 illustrate the composition $f \circ g$. When we put a number x into the function g, $g(x)$ comes out. The value $g(x)$ goes

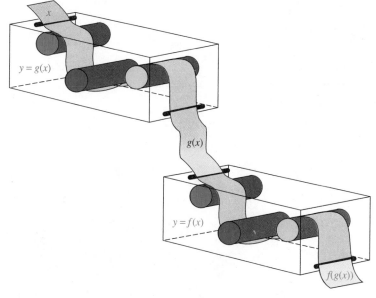

FIGURE 8-27

into function f, which transforms $g(x)$ into $f(g(x))$. If the function machines for g and f were connected to make a single machine, that machine would be named $f \circ g$.

To be in the domain of the composite function $f \circ g$, a number x has to be in the domain of g. Also, the output of g must be in the domain of f. Thus, the domain of $f \circ g$ consists of those numbers x that are in the domain of g, and for which $g(x)$ is in the domain of f.

EXAMPLE 3   Let $f(x) = 2x + 1$ and $g(x) = x - 4$. Find   **a.** $(f \circ g)(9)$,   **b.** $(f \circ g)(x)$,   and   **c.** $(g \circ f)(-2)$.

Solution   **a.** $(f \circ g)(9)$ means $f(g(9))$. In Figure 8-28(a), function g receives the number 9, subtracts 4, and releases the number $g(x) = 5$. The 5 then goes into the f function, which doubles 5 and adds 1. The final result, 11, is the output of the composite function $f \circ g$:

$$(f \circ g)(9) = f(g(9)) = f(5) = 2(5) + 1 = 11$$

b. $(f \circ g)(x)$ means $f(g(x))$. In Figure 8-28(a), function g receives the number x, subtracts 4, and releases the number $x - 4$. The $x - 4$ then goes into the f function, which doubles $x - 4$ and adds 1. The final result, $2x - 7$, is the output of the composite function $f \circ g$:

$$(f \circ g)(x) = f(g(x)) = f(x - 4) = 2(x - 4) + 1 = 2x - 7$$

c. $(g \circ f)(-2)$ means $g(f(-2))$. In Figure 8-28(b), function f receives the number -2, doubles it and adds 1, and releases -3 into the g function. Function g subtracts 4 from -3 and releases a final result of -7. Thus,

$$(g \circ f)(-2) = g(f(-2)) = g(-3) = -3 - 4 = -7$$

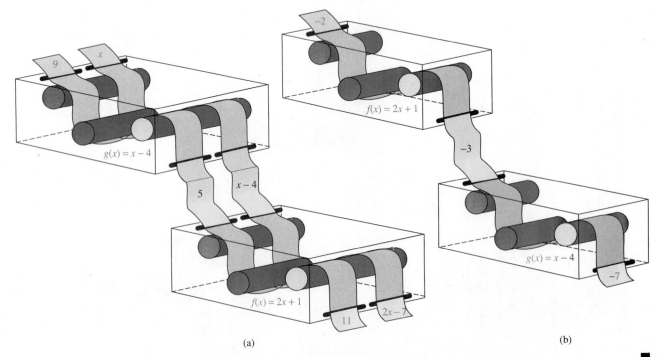

(a)   (b)

FIGURE 8-28

■ IDENTITY FUNCTION

The **identity function** is defined by the equation $I(x) = x$. Under this function, the value that corresponds to any real number x is x itself. If f is any function, the composition of f with the identity function is just the function f:

$$(f \circ I)(x) = (I \circ f)(x) = f(x)$$

EXAMPLE 4 Let f be any function and I be the identity function, $I(x) = x$. Show that
a. $(f \circ I)(x) = f(x)$ and **b.** $(I \circ f)(x) = f(x)$.

Solution **a.** $(f \circ I)(x)$ means $f(I(x))$. Because $I(x) = x$, we have

$$(f \circ I)(x) = f(I(x)) = f(x)$$

b. $(I \circ f)(x)$ means $I(f(x))$. Because I passes any number through unchanged, we have $I(f(x)) = f(x)$ and

$$(I \circ f)(x) = I(f(x)) = f(x) \qquad ■$$

■ PROBLEM SOLVING

EXAMPLE 5 **Temperature change** A laboratory sample is removed from a cooler at a temperature of 15° Fahrenheit. Technicians are warming the sample at a controlled rate of 3° F per hour. Express the sample's Celsius temperature as a function of the time, t, since it was removed from refrigeration.

Solution The temperature of the sample is 15° F when $t = 0$. Because it warms 3° F per hour after that, it warms $3t°$, and the Fahrenheit temperature is given by the function

$$F(t) = 3t + 15$$

The Celsius temperature, C, is a function of this Fahrenheit temperature, F, given by the formula

$$C(F) = \frac{5}{9}(F - 32)$$

To express the sample's Celsius temperature as a function of time, we find the composition function

$$(C \circ F)(t) = C(F(t))$$

$$= \frac{5}{9}(F(t) - 32) \qquad \text{Substitute } F(t) \text{ for } F \text{ in } \tfrac{5}{9}(F - 32).$$

$$= \frac{5}{9}[(3t + 15) - 32] \qquad \text{Substitute } 3t + 15 \text{ for } F(t).$$

$$= \frac{5}{9}(3t - 17) \qquad \text{Simplify.} \qquad ■$$

Orals   *If $f(x) = 2x$, $g(x) = 3x$, and $h(x) = 4x$, find*

1. $f + g$ **2.** $h - g$ **3.** $f \cdot h$

4. g/f **5.** h/f **6.** $g \cdot h$

If $f(x) = x + 1$ and $g(x) = 2x$, find

7. $(f \circ g)(x)$ **8.** $(g \circ f)(x)$

EXERCISE 8.6

In Exercises 1–8, $f(x) = 3x$ and $g(x) = 4x$. Find each function and its domain.

1. $f + g$ **2.** $f - g$ **3.** $f \cdot g$ **4.** f/g

5. $g - f$ **6.** $g + f$ **7.** g/f **8.** $g \cdot f$

In Exercises 9–16, $f(x) = 2x + 1$ and $g(x) = x - 3$. Find each function and its domain.

9. $f + g$ **10.** $f - g$ **11.** $f \cdot g$ **12.** f/g

13. $g - f$ **14.** $g + f$ **15.** g/f **16.** $g \cdot f$

In Exercises 17–20, $f(x) = 3x - 2$ and $g(x) = 2x^2 + 1$. Find each function and its domain.

17. $f - g$ **18.** $f + g$ **19.** f/g **20.** $f \cdot g$

In Exercises 21–24, $f(x) = x^2 - 1$ and $g(x) = x^2 - 4$. Find each function and its domain.

21. $f - g$ **22.** $f + g$ **23.** g/f **24.** $g \cdot f$

In Exercises 25–36, $f(x) = 2x + 1$ and $g(x) = x^2 - 1$. Find each value.

25. $(f \circ g)(2)$ **26.** $(g \circ f)(2)$ **27.** $(g \circ f)(-3)$ **28.** $(f \circ g)(-3)$

29. $(f \circ g)(0)$ **30.** $(g \circ f)(0)$ **31.** $(f \circ g)\left(\dfrac{1}{2}\right)$ **32.** $(g \circ f)\left(\dfrac{1}{3}\right)$

33. $(f \circ g)(x)$ **34.** $(g \circ f)(x)$ **35.** $(g \circ f)(2x)$ **36.** $(f \circ g)(2x)$

In Exercises 37–44, $f(x) = 3x - 2$ and $g(x) = x^2 + x$. Find each value.

37. $(f \circ g)(4)$ **38.** $(g \circ f)(4)$ **39.** $(g \circ f)(-3)$ **40.** $(f \circ g)(-3)$

41. $(g \circ f)(0)$ **42.** $(f \circ g)(0)$ **43.** $(g \circ f)(x)$ **44.** $(f \circ g)(x)$

45. If $f(x) = x + 1$ and $g(x) = 2x - 5$, show that $(f \circ g)(x) \neq (g \circ f)(x)$.

46. If $f(x) = x^2 + 1$ and $g(x) = 3x^2 - 2$, show that $(f \circ g)(x) \neq (g \circ f)(x)$.

47. If $f(x) = x^2 + 2x - 3$, find $f(a)$, $f(h)$, and $f(a + h)$. Then show that $f(a + h) \neq f(a) + f(h)$.

48. If $g(x) = 2x^2 + 10$, find $g(a)$, $g(h)$, and $g(a + h)$. Then show that $g(a + h) \neq g(a) + g(h)$.

*In Exercises 49–52, the fraction $\dfrac{f(x + h) - f(x)}{h}$ is called the **difference quotient**.*

49. If $f(x) = x^2 + 2$, find $\dfrac{f(x + h) - f(x)}{h}$.

50. If $f(x) = x^2 - x$, find $\dfrac{f(x + h) - f(x)}{h}$.

51. If $f(x) = x^3 - 1$, find $\dfrac{f(x + h) - f(x)}{h}$.

52. If $f(x) = x^3 + 2$, find $\dfrac{f(x + h) - f(x)}{h}$.

53. Alloys A molten alloy must be cooled slowly to control crystalization. When removed from the furnace, its temperature is 2700° F, and it will be cooled at 200° per hour. Express the Celsius temperature as a function of the number of hours, t, since cooling began.

54. Weather forecasting A high pressure area promises increasingly warmer weather for the next 48 hours. The temperature is now 34° Celsius and is expected to rise 1° every 6 hours. Express the Fahrenheit temperature as a function of the number of hours from now. $\left(Hint: F = \frac{9}{5}C + 32.\right)$

Writing Exercises *Write a paragraph using your own words.*

1. Explain how to find the domain of f/g.

2. Explain how to find the domain of $(f \circ g)(x)$.

Something to Think About

1. Is composition of functions associative? Choose functions f, g, and h, and determine whether $[f \circ (g \circ h)](x) = [(f \circ g) \circ h](x)$.

2. Choose functions f, g, and h and determine whether $f \circ (g + h) = f \circ g + f \circ h$.

Review Exercises *Simplify each expression.*

1. $\dfrac{3x^2 + x - 14}{4 - x^2}$

2. $\dfrac{2x^3 + 14x^2}{3 + 2x - x^2} \cdot \dfrac{x^2 - 3x}{x}$

3. $\dfrac{8 + 2x - x^2}{12 + x - 3x^2} \div \dfrac{3x^2 + 5x - 2}{3x - 1}$

4. $\dfrac{x - 1}{1 + \dfrac{x}{x - 2}}$

8.7 Inverses of Functions

■ INVERSES OF FUNCTIONS ■ ONE-TO-ONE FUNCTIONS ■ THE HORIZONTAL LINE TEST

The linear function defined by $C = \frac{5}{9}(F - 32)$ gives a formula to convert degrees Fahrenheit to degrees Celsius. If we substitute a Fahrenheit reading into the formula, a Celsius reading comes out. For example, if we substitute 41° for F, we obtain a Celsius reading of 5°:

$$C = \frac{5}{9}(F - 32)$$

$$= \frac{5}{9}(41 - 32)$$

$$= \frac{5}{9}(9)$$

$$= 5$$

If we want to find a Fahrenheit reading from a Celsius reading, we need a formula into which we can substitute a Celsius reading and have a Fahrenheit reading come out. Such a formula is $F = \frac{9}{5}C + 32$, which takes the Celsius reading of $5°$ and turns it back into a Fahrenheit reading of $41°$.

$$F = \frac{9}{5}C + 32$$

$$= \frac{9}{5}(5) + 32$$

$$= 41$$

The functions defined by these two formulas do opposite things. The first turns $41°$ F into $5°$ Celsius, and the second turns $5°$ Celsius back into $41°$ F. Such functions are called **inverse functions**.

■ INVERSES OF FUNCTIONS

If R is a function determined by the set of ordered pairs

$R = \{(1, 10), (2, 20), (3, 30)\}$

then R turns the number 1 into 10, turns 2 into 20, and turns 3 into 30. Since the inverse of R, denoted as R^{-1}, must turn 10 back into 1, turn 20 back into 2, and turn 30 back into 3, it must be the set of ordered pairs

$R^{-1} = \{(10, 1), (20, 2), (30, 3)\}$

To form the inverse of the function R, we simply interchange the coordinates of each ordered pair that determines R.

WARNING!   The -1 in the notation R^{-1} is not an exponent. It refers to the inverse of R. The symbol R^{-1} is read as "the inverse of R" or just "R inverse."

The domain of R and the range of R^{-1} is $\{1, 2, 3\}$. The range of R and the domain of R^{-1} is $\{10, 20, 30\}$.

The function R can be illustrated by the diagram in Figure 8-29(a). The function R is represented by arrows drawn from each number in the domain to each corresponding value of y in the range.

R^{-1} is illustrated in Figure 8-29(b). In this figure, the direction of the arrows has been reversed. Because each arrow in the figure points to exactly one value of y, R^{-1} is a function.

Inverse Relations
If R is any function and R^{-1} is the relation obtained from R by interchanging the components of each ordered pair of R, then R^{-1} is called the **inverse relation of R**.

The domain of R^{-1} is the range of R, and the range of R^{-1} is the domain of R.

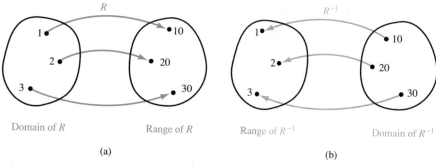

Domain of R	Range of R	Range of R^{-1}	Domain of R^{-1}

(a)   (b)

FIGURE 8-29

For example, the linear equation $y = 4x + 2$ determines the linear function R consisting of infinitely many ordered pairs (x, y). If $x = 0$, the corresponding value of y is

$$4(0) + 2 = 2$$

A partial listing of ordered pairs in R is

$$R = \left\{ (0, 2), (1, 6), \left(\frac{1}{4}, 3 \right), (-3, -10), \ldots \right\}$$

To form a partial listing of R^{-1}, we interchange the x- and y-coordinates of the elements of R to obtain

$$R^{-1} = \left\{ (2, 0), (6, 1), \left(3, \frac{1}{4} \right), (-10, -3), \ldots \right\}$$

To generate other elements of R^{-1} conveniently, and to decide whether R^{-1} is a function, we need an equation. To find an equation for R^{-1}, we proceed as in Example 1.

EXAMPLE 1   Find R^{-1}, the inverse relation of $y = 4x + 2$, and tell whether the inverse is a function.

Solution   To find the inverse relation of $y = 4x + 2$, we interchange the variables x and y to obtain

1. $x = 4y + 2$

To decide whether the inverse relation is a function, we solve Equation 1 for y:

$$x = 4y + 2$$
$$x - 2 = 4y \qquad \text{To isolate the } 4y \text{ term, subtract 2 from both sides.}$$

2. $y = \dfrac{x - 2}{4}$ To isolate y, divide both sides by 4 and use the symmetric property.

Because each number x that is substituted into Equation 2 gives a single value y, the inverse relation R^{-1} is a function. ∎

In Example 1, we found that the inverse relation of the function $y = 4x + 2$ is the function $y = \frac{x-2}{4}$. In function notation, this inverse function can be denoted as

$$f^{-1}(x) = \frac{x - 2}{4} \qquad \text{Read as "} f \text{ inverse of } x \text{ is } \tfrac{x-2}{4}.\text{"}$$

WARNING! The symbol $f^{-1}(x)$ means "the inverse of x." Remember that $f^{-1}(x) \neq \frac{1}{f(x)}$.

To see an important relationship between a function and its inverse, we substitute some number x, such as $x = 3$, into the function $f(x) = 4x + 2$ of Example 1. The corresponding value of y produced is

$$y = f(3)$$
$$= 4(3) + 2$$
$$= 14$$

If we substitute 14 into the inverse function, f^{-1}, the corresponding value of y that is produced is

$$y = f^{-1}(14)$$
$$= \frac{14 - 2}{4}$$
$$= 3$$

Thus, the function f turns 3 into 14, and the inverse function f^{-1} turns 14 back into 3. In general, the composition of a function and its inverse is the identity function.

To prove that $f(x) = 4x + 2$ and $f^{-1}(x) = \frac{x-2}{4}$ are inverse functions, we must show that their composition (in both directions) is the identity function:

$$(f \circ f^{-1})(x) = f(f^{-1}(x)) \qquad\qquad (f^{-1} \circ f)(x) = f^{-1}(f(x))$$

$$= f\left(\frac{x-2}{4}\right) \qquad\qquad\qquad = f^{-1}(4x+2)$$

$$= 4\left(\frac{x-2}{4}\right) + 2 \qquad\qquad = \frac{4x+2-2}{4}$$

$$= x - 2 + 2 \qquad\qquad\qquad = \frac{4x}{4}$$

$$= x \qquad\qquad\qquad\qquad\qquad = x$$

Thus, $(f \circ f^{-1})(x) = (f^{-1} \circ f)(x) = x$, which is the identity function $I(x)$.

Steps for Finding the Inverse of a Relation

1. If a relation is given as a set of ordered pairs, interchange the x- and y-values.
2. If a relation is given as an equation,
   a. Interchange the variables x and y.
   b. Solve the resulting equation for y. This equation defines the inverse relation.

EXAMPLE 2 The set of all pairs (x, y) determined by the equation $3x + 2y = 6$ is a function. Find its inverse function and graph both functions on a single coordinate system.

Solution To find the inverse function of $3x + 2y = 6$, we interchange x and y to obtain

$$3y + 2x = 6$$

and then solve the equation for y.

$$3y + 2x = 6$$

$$3y = -2x + 6 \qquad \text{Subtract } 2x \text{ from both sides.}$$

$$y = -\frac{2}{3}x + 2 \qquad \text{Divide both sides by 3.}$$

Thus, $y = f^{-1}(x) = -\frac{2}{3}x + 2$.

The graphs of the equations $3x + 2y = 6$ and $y = f^{-1}(x) = -\frac{2}{3}x + 2$ appear in Figure 8-30.

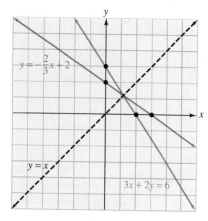

FIGURE 8-30　■

In Example 2, the graph of the equation $3x + 2y = 6$ and $y = f^{-1}(x) = -\frac{2}{3}x + 2$ are symmetric about the line $y = x$. This is always the case, because when the coordinates (a, b) satisfy an equation, the coordinates (b, a) will satisfy its inverse.

In each example so far, the inverse of a function has been another function. This is not always true, as the following example will show.

EXAMPLE 3　Find the inverse relation of the function determined by $y = x^2$.

Solution

$$y = x^2$$

$$x = y^2 \qquad \text{Interchange } x \text{ and } y.$$

$$y = \pm\sqrt{x} \qquad \text{Use the symmetric property of equality and the square root property.}$$

When the inverse relation $y = \pm\sqrt{x}$ is graphed as in Figure 8-31, we see that the graph does not pass the vertical line test. Thus, it is not a function.

The graph of $y = x^2$ is also shown in the figure. As expected, the graphs of $y = x^2$ and $y = \pm\sqrt{x}$ are symmetric about the line $y = x$.

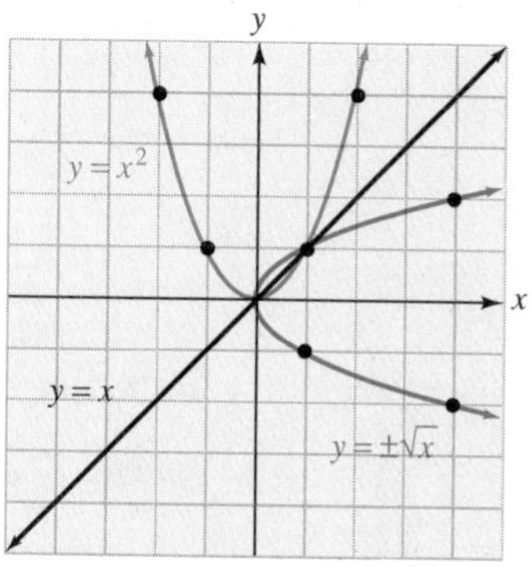

FIGURE 8-31

■

■ ONE-TO-ONE FUNCTIONS

We have seen that a function f can be depicted by a diagram like the one shown in Figure 8-32(a), where the function f is represented by arrows drawn from each number x in the domain to its corresponding value y in the range.

We can visualize the inverse of f by reversing the direction of the arrows in Figure 8-32(a) to produce the diagram shown in Figure 8-32(b). In both parts of Figure 8-32, each arrow points to a single value. Thus, both diagrams represent functions, and each function is the inverse of the other.

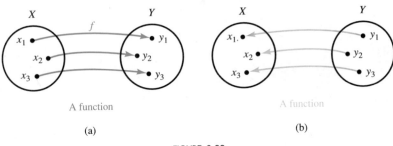

FIGURE 8-32

The correspondence shown in Figure 8-33(a) is a function from set X to set Y. However, when the arrows are reversed, as in Figure 8-33(b), the correspondence is not a function from set Y to set X, because to a single number y there correspond several values of x.

The diagrams in Figures 8-32 and 8-33 suggest these facts:

- If each y corresponds to a single number x, as in Figure 8-32(a), then the correspondence represents a function when the arrows are reversed.

- However, if one y corresponds to several numbers x, as in Figure 8-33(a), then the inverse is not a function.

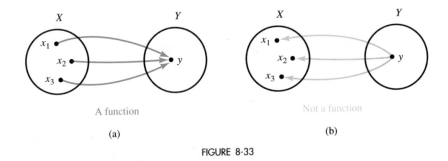

FIGURE 8-33

One-to-One Function
A function is called **one-to-one** if and only if each value of y in the range corresponds to only one number x in the domain.

Because of the definition, the inverse of any one-to-one function is also a function.

■ THE HORIZONTAL LINE TEST

The **horizontal line test** can be used to decide whether a graph of a function represents a one-to-one function. If any horizontal line that intersects the graph of a function does so only once, the function is one-to-one. Otherwise, the function is not one-to-one.

EXAMPLE 4 Use the horizontal line test to decide whether the functions defined by
a. $y = x^2 - 4$ and **b.** $y = x^3$ are one-to-one.

Solution **a.** Graph the equation as in Figure 8-34(a). Because many horizontal lines that intersect the graph do so twice, some values of y correspond to two numbers x. Thus, the function is not one-to-one.

b. Graph the equation as in Figure 8-34(b). Because each horizontal line that intersects the graph does so exactly once, each value of y corresponds to only one number x. Thus, the function defined by $y = x^3$ is one-to-one.

$y = x^2 - 4$

x	y	(x, y)
-3	5	$(-3, 5)$
-2	0	$(-2, 0)$
-1	-3	$(-1, -3)$
0	-4	$(0, -4)$
1	-3	$(1, -3)$
2	0	$(2, 0)$
3	5	$(3, 5)$

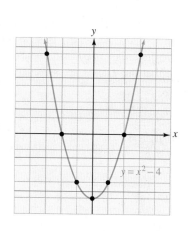

(a)

$y = x^3$

x	y	(x, y)
-2	-8	$(-2, -8)$
-1	-1	$(-1, -1)$
0	0	$(0, 0)$
1	1	$(1, 1)$
2	8	$(2, 8)$

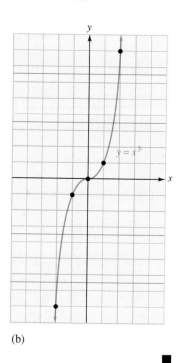

(b)

FIGURE 8-34 ■

  WARNING! Make sure to use the vertical line test to determine whether the graph of a relation is a function. If it is, use the horizontal line test to determine whether the function is one-to-one.

EXAMPLE 5 Express the inverse of the function $y = f(x) = x^3$.

Solution To find the inverse of the function, we proceed as follows:

$$y = x^3$$
$$x = y^3 \qquad \text{Interchange the variables } x \text{ and } y.$$
$$\sqrt[3]{x} = y \qquad \text{Take the cube root of both sides.}$$

We note that to each number x there corresponds a single real cube root. Thus, the equation $y = \sqrt[3]{x}$ represents a function. In $f^{-1}(x)$ notation, we have

$$y = f^{-1}(x) = \sqrt[3]{x}$$ ■

If a function is not one-to-one, it is often possible to construct a one-to-one function by restricting its domain.

EXAMPLE 6   Find the inverse of the function defined by $y = x^2$ with $x \geq 0$. Then tell whether the inverse relation is a function. Graph the function and its inverse on a single set of coordinate axes.

Solution   The inverse of the function $y = x^2$ with $x \geq 0$ is

$$x = y^2 \quad \text{with} \quad y \geq 0 \qquad \text{Interchange the variables } x \text{ and } y.$$

This equation can be written in the form

$$y = \pm\sqrt{x} \quad \text{with} \quad y \geq 0$$

Since $y \geq 0$, we choose the $+$ sign in the previous equation so that each number x gives only one value of y: $y = \sqrt{x}$. Thus, the inverse relation is a function.

The graphs of the two functions appear in Figure 8-35. The line $y = x$ is included so that we can see that the graphs are symmetric about the line $y = x$.

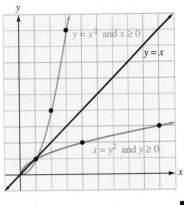

$y = x^2$ and $x \geq 0$		
x	y	(x, y)
0	0	(0, 0)
1	1	(1, 1)
2	4	(2, 4)
3	9	(3, 9)

$x = y^2$ and $y \geq 0$		
x	y	(x, y)
0	0	(0, 0)
1	1	(1, 1)
4	2	(4, 2)
9	3	(9, 3)

FIGURE 8-35   ■

Orals   *Find the inverse relation of each set of ordered pairs.*

1. $\{(1, 2), (2, 3), (5, 10)\}$ **2.** $\{(1, 1), (2, 8), (4, 64)\}$

Find the inverse function of each linear function.

3. $y = \dfrac{1}{2}x$ **4.** $y = 2x$

Tell whether each function is one-to-one.

5. $y = x^2 - 2$ **6.** $y = x^3$

EXERCISE 8.7

In Exercises 1–6, find the inverse relation of each set of ordered pairs (x, y) and tell whether the inverse relation is a function.

1. {(3, 2), (2, 1), (1, 0)}

2. {(4, 1), (5, 1), (6, 1), (7, 1)}

3. {(1, 2), (2, 3), (1, 3), (1, 5)}

4. {(−1, −1), (0, 0), (1, 1), (2, 2)}

5. {(1, 1), (2, 4), (3, 9), (4, 16)}

6. {(1, 1), (2, 1), (3, 1), (4, 1)}

In Exercises 7–14, find the inverse of the relation determined by the given equation and tell whether that inverse relation is a function. If the inverse relation is a function, express it in the form $y = f^{-1}(x)$.

7. $y = 3x + 1$

8. $y + 1 = 5x$

9. $x + 4 = 5y$

10. $x = 3y + 1$

11. $y = \dfrac{x - 4}{5}$

12. $y = \dfrac{2x + 6}{3}$

13. $4x - 5y = 20$

14. $3x + 5y = 15$

In Exercises 15–24, find the inverse of each linear function. Then graph both the function and its inverse on a single coordinate system. Find the equation of the line of symmetry.

15. $y = 4x + 3$

16. $x = 3y - 1$

17. $x = \dfrac{y - 2}{3}$

18. $y = \dfrac{x + 3}{4}$

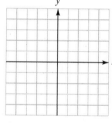   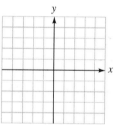 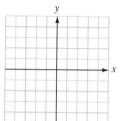

19. $3x - y = 5$

20. $2x + 3y = 9$

21. $3(x + y) = 2x + 4$

22. $-4(y - 1) + x = 2$

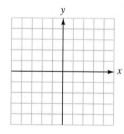 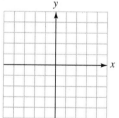 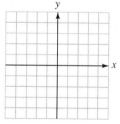  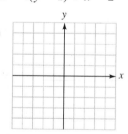

23. $3x = 2(1 - y)$

24. $2\left(y + \dfrac{3}{2}\right) = -x$

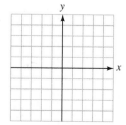 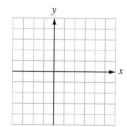

In Exercises 25–34, find the inverse of the relation determined by each equation. Tell whether the inverse relation is a function.

25. $y = x^2 + 4$

26. $x = y^2 - 2$

27. $x = y^2 - 4$

28. $y = x^2 + 5$

29. $y = x^3$

30. $xy = 4$

31. $y = \pm\sqrt{x}$

32. $y = \sqrt[3]{x}$

33. $x = \sqrt{y}$

34. $4y^2 = x - 3$

In Exercises 35–36, show that the inverse of the function determined by each equation is also a function. Express it using $f^{-1}(x)$ notation.

35. $y = 2x^3 - 3$

36. $y = \dfrac{3}{x^3} - 1$

In Exercises 37–40, graph each equation and its inverse on one set of coordinate axes. Find the axis of symmetry.

37. $y = x^2 + 1$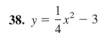

38. $y = \dfrac{1}{4}x^2 - 3$

39. $y = \sqrt{x}$

40. $y = |x|$

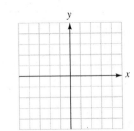
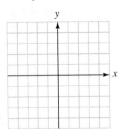
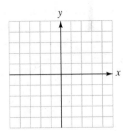
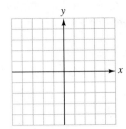

In Exercises 41–50, graph the function represented by each equation. Use the horizontal line test to decide whether each is one-to-one.

41. $y = 3x + 2$

42. $y = 5 - 3x$

43. $y = \dfrac{x + 5}{2}$

44. $y = \dfrac{5 - x}{2}$

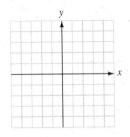
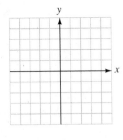
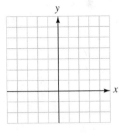
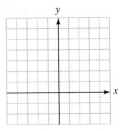

45. $y = 3x^2 + 2$

46. $y = 5 - x^2$

47. $y = \sqrt[3]{x}$

48. $y = \sqrt{x}$

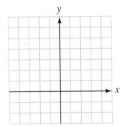
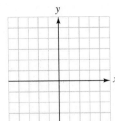
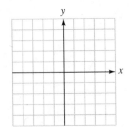
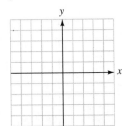

49. $y = x^3 - x$

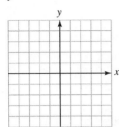

50. $y = -x^4 + x^2$

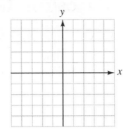

Writing Exercises   *Write a paragraph using your own words.*

1. Explain the purpose of the vertical line test.

2. Explain the purpose of the horizontal line test.

Something to Think About   **1.** Find the inverse of $y = \dfrac{x + 1}{x - 1}$.

2. Using the functions of Question 1, show that $(f \circ f^{-1})(x) = x$.

Review Exercises   *Write each complex number in $a + bi$ form or find each value.*

1. $3 - \sqrt{-64}$

2. $(2 - 3i) + (4 + 5i)$

3. $(3 + 4i)(2 - 3i)$

4. $\dfrac{6 + 7i}{3 - 4i}$

5. $|6 - 8i|$

6. $\left| \dfrac{2 + i}{3 - i} \right|$

■ ■ ■ ■ ■ ■ ■ ■ ■ ■ PROBLEMS AND PROJECTS

1. The accounting department of the CM Cookie Company projects a cost function of

$$C(q) = 1.02q^2 - 86q + 3500$$

where q is in dozens of cookies baked. Find the number of dozens of cookies that will give the minimum cost. Then find the minimum cost.

2. The marketing director for the Fun Toy Company has noticed that when the Big Blaster gun is priced at $18, weekly sales are 3600 guns. However, when the price is increased by $2, weekly sales decrease by 200 guns. Assuming a linear sales curve, find the revenue function. Then find the unit price that will maximize revenue.

3. The accounting department of the Sunshine Carwash uses a cost function of $C(t) = 54t + 390$ and a revenue function of $R(t) = 75t + 250$, where t is the number of hours the carwash is in operation each week. Find the profit function for the company.

4. The accounting department of the Ultra-Max Water Blaster Company uses a revenue function of $R(x) = 37.5x + 247$, where x is $x(t) = 10t + 15$ and t is the number of hours worked. The cost function is $C(t) = 24t + 190$. Find the profit function in terms of the hours worked.

(continued)

■ ■ ■ ■ ■ ■ ■ ■ ■ ■ **PROBLEMS AND PROJECTS** *(continued)*

PROJECT 1

Ballistics is the study of how projectiles fly. The general formula for the height above the ground of an object thrown straight up or straight down is given by the function

$$h(t) = -16t^2 + v_0 t + h_0$$

where h is the object's height (in feet) above the ground t seconds after it is thrown. The initial velocity v_0 is the velocity with which the object is thrown, measured in feet per second. The initial height h_0 is the object's height (in feet) above the ground when it is thrown. (If $v_0 > 0$, the object is thrown upward; if $v_0 < 0$, the object is thrown downward.)

This formula takes into account the force of gravity, but it disregards the force of air resistance. It is much more accurate for a smooth, dense ball than for a crumpled piece of paper.

One of the most popular acts of the Bungling Brothers Circus is the Amazing Glendo and his cannonball catching act. A cannon fires a ball vertically into the air; Glendo, standing on a platform above the cannon, uses his catlike reflexes to catch the ball as it passes by on its way toward the roof of the big top. As the balls fly past, they are within Glendo's reach only during a two-foot interval of their upward path.

As an investigator for the company that insures the circus, you have been asked to find answers to the following questions. The answers will determine whether or not Bungling Brothers' insurance policy will be renewed.

a. In the first part of the act, cannonballs are fired from the end of a six-foot cannon with an initial velocity of 80 feet per second. Glendo catches one ball between 40 and 42 feet above the ground. Then he lowers his platform and catches another ball between 25 and 27 feet above the ground.

   i. Show that if Glendo missed a cannonball, it would hit the roof of the 56-foot tall big top. How long would it take for a ball to hit the big top? To prevent this from happening, a special net near the roof catches and holds any missed cannonballs.

   ii. Find (to the nearest thousandth of a second) how long the cannon balls are within Glendo's reach for each of his catches. Which catch is easier? Why does your answer make sense? Your company is willing to insure against injuries to Glendo if he has at least .025 second to make each catch. Should the insurance be offered?

b. For Glendo's grand finale, the special net at the roof of the big top is removed, making Glendo's catch more significant to the people in the audience, who worry that if Glendo misses, the tent will collapse around them. To make it even more dramatic, Glendo's arms are tied to restrict his reach to a one-foot interval of the ball's flight, and he stands on a platform just under the peak of the big top, so that his catch is made at the very last instant (between 54 and 55 feet above the ground). For this part of the act, however, Glendo has the cannon charged with less gunpowder, so that the muzzle velocity of the cannon is 56 feet per second. Show work to prove that Glendo's big finale is in fact his *easiest* catch, and that even if he misses, the big top is never in any danger of collapsing, so insurance should be offered against injury to the audience.

PROJECT 2   The center of Sterlington is the intersection of Main Street (running east–west) and DueNorth Road (running north–south). The recreation area for the towns-people is Robin Park, a few blocks from there. The park is bounded on the south by Main Street and on every other side by Parabolic Boulevard, named for its distinctive shape. In fact, if Main Street and DueNorth Road were used as the axes of a rectangular coordinate system, Parabolic Boulevard would have the equation $y = -(x - 4)^2 + 5$, where each unit on the axes is 100 yards.

The city council has recently begun to consider whether or not to put two walkways through the park. (See Illustration 1.) The walkways would run from two points on Main Street and converge at the northernmost point of the park, dividing the area of the park exactly into thirds.

The city council is pleased with the esthetics of this arrangement but needs to know two important facts.

a. For planning purposes, they need to know exactly where on Main Street the walkways would begin.

b. In order to budget for the construction, they need to know how long the walk-ways will be.

Provide answers for the city council, along with explanations and work to show that your answers are correct. You will need to use the formula shown in Illustration 2, due to Archimedes (287–212 B.C.), for the area under a parabola but above a line perpendicular to the axis of symmetry of the parabola.

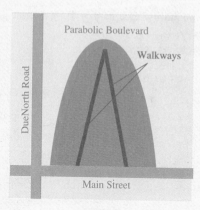

ILLUSTRATION 1

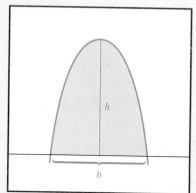

Shaded area $= \frac{2}{3} \cdot b \cdot h$

ILLUSTRATION 2

Chapter Summary

Key Words

absolute value of a complex number (8.3)

axis of symmetry (8.2)

completing the square (8.1)

complex conjugates (8.3)

complex numbers (8.3)

composite functions (8.6)

discriminant (8.4)

horizontal line test (8.7)

identity function (8.6)

imaginary numbers (8.3)

imaginary part of a complex number (8.3)

inverse functions (8.7)

one-to-one functions (8.7)

parabola (8.2)
quadratic formula (8.1)
quadratic function (8.2)

real part of a complex
number (8.3)

variance (8.2)
vertex of a parabola (8.2)

Key Ideas

(8.1) The square root property: If $c > 0$, the equation $x^2 = c$ has two real solutions. They are

$$x = \sqrt{c} \quad \text{and} \quad x = -\sqrt{c}$$

The **quadratic formula**:

$$x = \frac{-b \pm \sqrt{b^2 - 4ac}}{2a} \quad (a \neq 0)$$

(8.2) If f is a function and h and k are positive numbers, then

- The graph of $y = f(x) + k$ is identical to the graph of $y = f(x)$, except that it is shifted k units upward.
- The graph of $y = f(x) - k$ is identical to the graph of $y = f(x)$, except that it is shifted k units downward.
- The graph of $y = f(x - h)$ is identical to the graph of $y = f(x)$, except that it is shifted h units to the right.
- The graph of $y = f(x + h)$ is identical to the graph of $y = f(x)$, except that it is shifted h units to the left.

If $a \neq 0$, the graph of the equation $y = a(x - h)^2 + k$ is a parabola with vertex at (h, k). It opens upward when $a > 0$ and downward when $a < 0$.

(8.3) Properties of complex numbers: If a, b, c, and d are real numbers and $i^2 = -1$, then

$$a + bi = c + di \text{ if and only if } a = c \text{ and } b = d$$
$$(a + bi) + (c + di) = (a + c) + (b + d)i$$
$$(a + bi)(c + di) = (ac - bd) + (ad + bc)i$$
$$|a + bi| = \sqrt{a^2 + b^2}$$

(8.4) The discriminant:

If $b^2 - 4ac > 0$, the solutions of $ax^2 + bx + c = 0$ are unequal real numbers.
If $b^2 - 4ac = 0$, the solutions of $ax^2 + bx + c = 0$ are equal real numbers.
If $b^2 - 4ac < 0$, the solutions of $ax^2 + bx + c = 0$ are complex conjugates.

If r_1 and r_2 are solutions of $ax^2 + bx + c = 0$, then

$$r_1 + r_2 = -\frac{b}{a} \quad \text{and} \quad r_1 r_2 = \frac{c}{a}$$

(8.5) To find the solution of a quadratic inequality in one variable, make a sign chart.

To solve inequalities with rational expressions, get 0 on the right-hand side, add the fractions, and then factor the numerator and the denominator. Then use a sign chart.

To graph an inequality such as $y > 4x - 3$, first graph the equation $y = 4x - 3$. Then determine the half-plane that represents the graph of $y > 4x - 3$.

(8.6) Operations with functions:

$$(f + g)(x) = f(x) + g(x) \qquad (f - g)(x) = f(x) - g(x)$$

$$(f \cdot g)(x) = f(x)g(x) \qquad (f/g)(x) = \frac{f(x)}{g(x)} \quad \begin{array}{l} \text{provided} \\ (g(x) \neq 0) \end{array}$$

$$(f \circ g)(x) = f(g(x))$$

(8.7) To find the inverse of a relation, interchange the positions of variables x and y and solve for y.

The horizontal line test: If any horizontal line that intersects the graph of a function does so only once, the function is one-to-one.

■ Chapter 8 Review Exercises

In Review Exercises 1–4, solve each equation by factoring or by using the square root property.

1. $12x^2 + x - 6 = 0$

2. $6x^2 + 17x + 5 = 0$

3. $15x^2 + 2x - 8 = 0$

4. $(x + 2)^2 = 36$

In Review Exercises 5–6, solve each equation by completing the square.

5. $x^2 + 6x + 8 = 0$

6. $2x^2 - 9x + 7 = 0$

In Review Exercises 7–10, solve each equation by using the quadratic formula.

7. $x^2 - 8x - 9 = 0$

8. $x^2 - 10x = 0$

9. $2x^2 + 13x - 7 = 0$

10. $3x^2 - 20x - 7 = 0$

11. Dimensions of a rectangle A rectangle is 2 centimeters longer than it is wide. If both the length and width are doubled, its area is increased by 72 square centimeters. Find the dimensions of the original rectangle.

12. Dimensions of a rectangle A rectangle is 1 foot longer than it is wide. If the length is tripled and the width is doubled, its area is increased by 30 square feet. Find the dimensions of the original rectangle.

13. Ballistics If a rocket is launched straight up into the air with an initial velocity of 112 feet per second, its height after t seconds is given by the formula $h = 112t - 16t^2$, where h represents the height of the object in feet. After launch, how long will it be before it hits the ground?

14. Ballistics What is the maximum height of the rocket discussed in Exercise 13?

In Review Exercises 15–18, graph each equation and give the coordinates of the vertex of the resulting parabola.

15. $y = 2x^2 - 3$

16. $y = -2x^2 - 1$

17. $y = -4(x - 2)^2 + 1$

18. $y = 5(x + 1)^2 - 6$

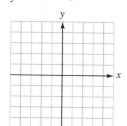
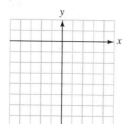
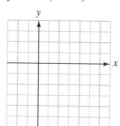
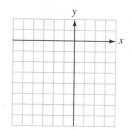

In Review Exercises 19–36, do the operations and give all answers in $a + bi$ form.

19. $(5 + 4i) + (7 - 12i)$

20. $(-6 - 40i) - (-8 + 28i)$

21. $\left(-32 + \sqrt{-144}\right) - \left(64 + \sqrt{-81}\right)$

22. $\left(-8 + \sqrt{-8}\right) + \left(6 - \sqrt{-32}\right)$

23. $(2 - 7i)(-3 + 4i)$

24. $(-5 + 6i)(2 + i)$

25. $\left(5 - \sqrt{-27}\right)\left(-6 + \sqrt{-12}\right)$

26. $\left(2 + \sqrt{-128}\right)\left(3 - \sqrt{-98}\right)$

27. $\dfrac{3}{4i}$

28. $\dfrac{-2}{5i^3}$

29. $\dfrac{6}{2 + i}$

30. $\dfrac{7}{3 - i}$

31. $\dfrac{4 + i}{4 - i}$

32. $\dfrac{3 - i}{3 + i}$

33. $\dfrac{3}{5 + \sqrt{-4}}$

34. $\dfrac{2}{3 - \sqrt{-9}}$

35. $|9 + 12i|$

36. $|24 - 10i|$

In Review Exercises 37–38, use the discriminant to determine what type of solutions exist for each equation.

37. $3x^2 + 4x - 3 = 0$

38. $4x^2 - 5x + 7 = 0$

39. Find the values of k that will make the solutions of $(k - 8)x^2 + (k + 16)x = -49$ equal.

40. Find the values of k such that the solutions of $3x^2 + 4x = k + 1$ are real numbers.

In Review Exercises 41–44, solve each equation.

41. $x - 13x^{1/2} + 12 = 0$

42. $a^{2/3} + a^{1/3} - 6 = 0$

43. $\dfrac{1}{x + 1} - \dfrac{1}{x} = -\dfrac{1}{x + 1}$

44. $\dfrac{6}{x + 2} + \dfrac{6}{x + 1} = 5$

45. Find the sum of the solutions of the equation $3x^2 - 14x + 3 = 0$.

46. Find the product of the solutions of the equation $3x^2 - 14x + 3 = 0$.

In Review Exercises 47–50, solve each inequality. Give each result in interval notation and graph the solution set.

47. $x^2 + 2x - 35 > 0$

48. $x^2 + 7x - 18 < 0$

49. $\dfrac{3}{x} \leq 5$

50. $\dfrac{2x^2 - x - 28}{x - 1} > 0$

In Review Exercises 51–54, use a graphing device to solve each inequality. Compare the results with Review Exercises 47–50.

51. $x^2 + 2x - 35 > 0$

52. $x^2 + 7x - 18 < 0$

53. $\dfrac{3}{x} \leq 5$

54. $\dfrac{2x^2 - x - 28}{x - 1} > 0$

In Review Exercises 55–56, graph each inequality.

55. $y < \dfrac{1}{2}x^2 - 1$

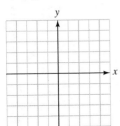

56. $y \geq -|x|$

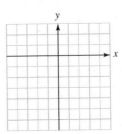

In Review Exercises 57–64, $f(x) = 2x$ and $g(x) = x + 1$. Find each function or value.

57. $f + g$

58. $f - g$

59. $f \cdot g$

60. f/g

61. $(f \circ g)(2)$

62. $(g \circ f)(-1)$

63. $(f \circ g)(x)$

64. $(g \circ f)(x)$

In Review Exercises 65–68, find the inverse of each function.

65. $y = 6x - 3$

66. $y = 4x + 5$

67. $y = 2x^2 - 1$ $(x \geq 0)$

68. $y = |x|$

In Review Exercises 69–72, graph each equation and use the vertical and horizontal line tests to decide whether the equation determines a one-to-one function.

69. $y = 2(x - 3)$

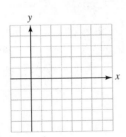

70. $y = x(2x - 3)$

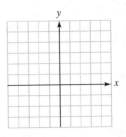

71. $y = -3(x - 2)^2 + 5$

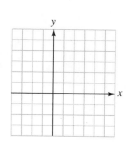

72. $x = \dfrac{1}{2}y^2 - 1$

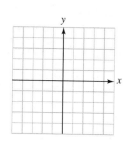

◼ Chapter 8 Test

In Problems 1–2, solve each equation.

1. $x^2 + 3x - 18 = 0$

2. $x(6x + 19) = -15$

In Problems 3–4, determine what number must be added to each binomial to make it a perfect square.

3. $x^2 + 24x$

4. $x^2 - 50x$

In Problems 5–6, solve each equation.

5. $x^2 + 4x + 1 = 0$

6. $x^2 - 5x - 3 = 0$

In Problems 7–12, do the operations. Give all answers in $a + bi$ form.

7. $(2 + 4i) + (-3 + 7i)$

8. $\left(3 - \sqrt{-9}\right) - \left(-1 + \sqrt{-16}\right)$

9. $2i(3 - 4i)$

10. $(3 + 2i)(-4 - i)$

11. $\dfrac{1}{\sqrt{2}i}$

12. $\dfrac{2 + i}{3 - i}$

13. Determine whether the solutions of $3x^2 + 5x + 17 = 0$ are real or nonreal.

14. For what value(s) of k are the solutions of $4x^2 - 2kx + k - 1 = 0$ equal?

15. One leg of a right triangle is 14 inches longer than the other, and the hypotenuse is 26 inches. Find the length of the shortest leg.

16. Solve the equation $2y - 3y^{1/2} + 1 = 0$.

17. Graph the equation $f(x) = \frac{1}{2}x^2 - 4$ and give the coordinates of its vertex.

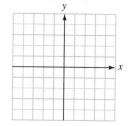

18. Graph the inequality $y \le -x^2 + 3$.

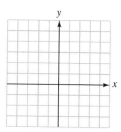

In Problems 19–20, solve each inequality and graph the solution set.

19. $x^2 - 2x - 8 > 0$

20. $\dfrac{x - 2}{x + 3} \le 0$

In Problems 21–24, $f(x) = 4x$ and $g(x) = x - 1$. Find each function.

21. $g + f$ **22.** $f - g$ **23.** $g \cdot f$ **24.** g/f

In Problems 25–28, $f(x) = 4x$ and $g(x) = x - 1$. Find each value.

25. $(g \circ f)(1)$ **26.** $(f \circ g)(0)$ **27.** $(f \circ g)(-1)$ **28.** $(g \circ f)(-2)$

In Problems 29–30, $f(x) = 4x$ and $g(x) = x - 1$. Find each function.

29. $(f \circ g)(x)$ **30.** $(g \circ f)(x)$

In Problems 31–32, find the inverse of each function.

31. $3x + 2y = 12$ **32.** $y = 3x^2 + 4 \ (x \le 0)$

■ **Cumulative Review Exercises (Chapters 1–8)**

In Exercises 1–2, find the domain and range of each function.

1. $y = f(x) = 2x^2 - 3$ **2.** $y = f(x) = -|x - 4|$

In Exercises 3–4, write the equation of the line with the given properties.

3. $m = 3$, passing through $(-2, -4)$ **4.** parallel to the graph of $2x + 3y = 6$ and passing through $(0, -2)$

In Exercises 5–6, do each operation.

5. $(2a^2 + 4a - 7) - 2(3a^2 - 4a)$ **6.** $(3x + 2)(2x - 3)$

In Exercises 7–8, factor each expression.

7. $x^4 - 16y^4$ **8.** $15x^2 - 2x - 8$

In Exercises 9–10, solve each equation.

9. $x^2 - 5x - 6 = 0$ **10.** $6a^3 - 2a = a^2$

11. Divide and simplify: $\dfrac{x^2 - 2x - 3}{x^2 + x - 2} \div \dfrac{x^2 - 9}{x^2 + 5x + 6}$.   **12.** Add and simplify: $\dfrac{x}{x + 4} + \dfrac{2x}{x - 1}$.

In Exercises 13–18, simplify each expression. Assume that all variables represent positive numbers.

13. $\sqrt{48t^3}$ **14.** $\sqrt[3]{-27x^3}$ **15.** $\sqrt[3]{\dfrac{128x^4}{2x}}$

16. $64^{2/3}$ **17.** $\dfrac{y^{2/3} y^{5/3}}{y^{1/3}}$ **18.** $\dfrac{x^{5/3} x^{1/2}}{x^{3/4}}$

In Exercises 19–20, graph each function and give the domain and the range.

19. $f(x) = \sqrt{x - 2}$

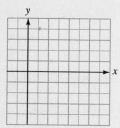

20. $f(x) = -\sqrt{x + 2}$

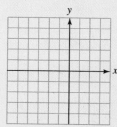

In Exercises 21–22, do the operations.

21. $\left(x^{2/3} - x^{1/3}\right)\left(x^{2/3} + x^{1/3}\right)$

22. $\left(x^{-1/2} + x^{1/2}\right)^2$

In Exercises 23–28, simplify each statement.

23. $\sqrt{50} - \sqrt{8} + \sqrt{32}$

24. $-3\sqrt[4]{32} - 2\sqrt[4]{162} + 5\sqrt[4]{48}$

25. $3\sqrt{2}\left(2\sqrt{3} - 4\sqrt{12}\right)$

26. $\dfrac{5}{\sqrt[3]{x}}$

27. $\dfrac{\sqrt{x} + 2}{\sqrt{x} - 1}$

28. $\sqrt[6]{x^3 y^3}$

In Exercises 29–30, solve each equation.

29. $5\sqrt{x + 2} = x + 8$

30. $\sqrt{x} + \sqrt{x + 2} = 2$

31. Find the length of the hypotenuse of the right triangle shown in Illustration 1.

32. Find the length of the hypotenuse of the right triangle shown in Illustration 2.

ILLUSTRATION 1

ILLUSTRATION 2

33. Find the distance between $P(-2, 6)$ and $Q(4, 14)$.

34. What number must be added to $x^2 + 6x$ to make a trinomial square?

35. Use the method of completing the square to solve the equation $2x^2 + x - 3 = 0$.

36. Use the quadratic formula to solve the equation $3x^2 + 4x - 1 = 0$.

37. Graph $y = f(x) = \frac{1}{2}x^2 + 5$ and find the coordinates of its vertex.

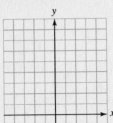

38. Graph the inequality $y \leq -x^2 + 3$.

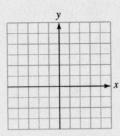

In Exercises 39–46, express each expression as a real number or as a complex number in a + bi form.

39. $(3 + 5i) + (4 - 3i)$

40. $(7 - 4i) - (12 + 3i)$

41. $(2 - 3i)(2 + 3i)$

42. $(3 + i)(3 - 3i)$

43. $(3 - 2i) - (4 + i)^2$

44. $\dfrac{5}{3 - i}$

45. $|3 + 2i|$

46. $|5 - 6i|$

47. For what values of k will the solutions of $2x^2 + 4x = k$ be equal?

48. Solve $a - 7a^{1/2} + 12 = 0$.

In Exercises 49–50, solve each inequality and graph the solution set on the number line.

49. $x^2 - x - 6 > 0$

50. $x^2 - x - 6 \leq 0$

In Exercises 51–54, $f(x) = 3x^2 + 2$ and $g(x) = 2x - 1$. Find each value or composite function.

51. $f(-1)$ **52.** $(g \circ f)(2)$ **53.** $(f \circ g)(x)$ **54.** $(g \circ f)(x)$

In Exercises 55–56, find the inverse of each function.

55. $y = f(x) = 3x + 2$

56. $y = f(x) = x^3 + 4$

Exponential and Logarithmic Functions

9

Medical Laboratory Worker

Medical laboratory workers include three levels of personnel: medical technologists, technicians, and assistants. They perform laboratory tests on specimens taken from patients by other health professionals, such as physicians.

Employment of medical laboratory workers is expected to expand at a rate faster than the average for all occupations through the year 2005, as physicians continue to use laboratory tests in routine physical checkups and in the diagnosis and treatment of disease.

SAMPLE APPLICATION ■ If a medium is inoculated with a bacterial culture that contains 1000 cells per milliliter, how many generations will pass by the time the culture has grown to a population of 1 million cells per milliliter? See Example 11 in Section 9.5.

In this chapter, we will discuss two functions that are important in many applications of mathematics. **Exponential functions** are used to compute compound interest, model population growth, and find radioactive decay. **Logarithmic functions** are used to measure the acidity of solutions, drug dosage, the gain of an amplifier, the intensity of earthquakes, and safe noise levels in factories.

9.1 Exponential Functions

■ IRRATIONAL EXPONENTS ■ EXPONENTIAL FUNCTIONS ■ GRAPHING EXPONENTIAL FUNCTIONS
■ COMPOUND INTEREST ■ GRAPHING DEVICES ■ BASE-*e* EXPONENTIAL FUNCTIONS
■ CONTINUOUS COMPOUND INTEREST ■ MALTHUSIAN POPULATION GROWTH

The graph in Figure 9-1 shows the balance in a bank account in which $5000 was invested in 1990 at 8% annual interest, compounded monthly. The graph shows that in the year 2005, the value of the account will be approximately $17,000, and in the year 2030, its value will be approximately $121,000.

The curve shown in Figure 9-1 is the graph of a function called an *exponential function.*

Value of $5000 invested at 8% compounded monthly

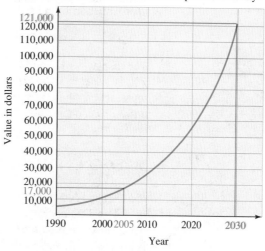

FIGURE 9-1

■ IRRATIONAL EXPONENTS

We have previously discussed expressions of the form b^x, where x is a rational number. For example,

$4^{1/2}$ means "the square root of 4."

$2^{1.25} = 2^{5/4}$ means "the fourth root of 2^5."

$5^{-2/3} = \dfrac{1}{5^{2/3}}$ means "the reciprocal of the cube root of 5^2."

To give meaning to b^x, where x is an irrational number, we consider the expression

$3^{\sqrt{2}}$ where $\sqrt{2}$ is the irrational number $1.414213562373\ldots$

Because $1 < \sqrt{2} < 2$, it can be shown that

$3^1 < 3^{\sqrt{2}} < 3^2$

Because $1.4 < \sqrt{2} < 1.5$, it can be shown that

$3^{1.4} < 3^{\sqrt{2}} < 3^{1.5}$

As the colored exponents on each side of the following list get closer to $\sqrt{2}$, the value of $3^{\sqrt{2}}$ gets squeezed into a smaller and smaller interval:

$$3^1 = 3 < 3^{\sqrt{2}} < 9 = 3^2$$
$$3^{1.4} \approx 4.656 < 3^{\sqrt{2}} < 5.196 \approx 3^{1.5}$$
$$3^{1.41} \approx 4.7070 < 3^{\sqrt{2}} < 4.7590 \approx 3^{1.42}$$
$$3^{1.414} \approx 4.727695 < 3^{\sqrt{2}} < 4.732892 \approx 3^{1.415}$$

There is exactly one real number that is greater than all of the increasing numbers on the left-hand side of the previous list and less than all of the decreasing numbers on the right-hand side. We define this number to be $3^{\sqrt{2}}$. We can get a good approximation by pressing these keys on a scientific calculator:

$$3 \quad \boxed{y^x} \quad 2 \quad \boxed{\sqrt{x}} \quad \boxed{=} \qquad \text{You may have to press} \quad \boxed{\text{2nd}} \quad \text{before} \quad \boxed{\sqrt{x}}.$$

The display will show 4.728804388. Thus, to four decimal places, $3^{\sqrt{2}} = 4.7288$.

In general, if $b > 0$ and x is any real number, the exponential expression b^x represents a single positive number. It can be shown that all of the familiar properties of exponents hold for irrational exponents as well.

■ EXPONENTIAL FUNCTIONS

If $b > 0$ and $b \neq 1$, the function $f(x) = b^x$ is an **exponential function**. Since x can be any real number, the domain of an exponential function is the set of real numbers. Since the base b of an exponential function is always positive, the value $f(x)$ is always positive, and the range is the set of positive numbers.

Since $b \neq 1$, an exponential function cannot be the constant function $f(x) = 1^x$, in which $f(x) = 1$ for every real number x.

> **Exponential Function with Base b**
> An **exponential function with base b** is defined by the equation
>
> $$y = f(x) = b^x \qquad \text{where } b > 0, b \neq 1, \text{ and } x \text{ is a real number}$$
>
> The **domain of an exponential function** is the set of real numbers (the interval $(-\infty, \infty)$), and the **range** is the set of positive real numbers (the interval $(0, \infty)$).

■ GRAPHING EXPONENTIAL FUNCTIONS

Because the domain and range of the function $f(x) = b^x$ are sets of real numbers, we can graph exponential functions on a rectangular coordinate system.

 (a)

EXAMPLE 1   Graph the functions   **a.** $f(x) = 2^x$ and   **b.** $g(x) = 4^x$.

Solution   **a.** To graph the function $f(x) = 2^x$, we find several points (x, y) whose coordinates satisfy the equation, plot those points, and join them with a smooth curve. The graph is shown in Figure 9-2(a).

   b. To graph the function $g(x) = 4^x$, we proceed as in part **a**. The graph is shown in Figure 9-2(b).

   By looking at the graphs, we see that the domain of each function is the interval $(-\infty, \infty)$, and the range is the interval $(0, \infty)$.

   Note that as x decreases, the values of each function decrease and approach 0. Thus, the x-axis is a horizontal asymptote for each graph.

   Also note that the graph of $f(x) = 2^x$ passes through the point $(1, 2)$, the graph of $g(x) = 4^x$ passes through the point $(1, 4)$, and each graph passes through the point $(0, 1)$.

$$f(x) = 2^x$$

x	$f(x)$	$(x, f(x))$
-1	$\frac{1}{2}$	$\left(-1, \frac{1}{2}\right)$
0	1	$(0, 1)$
1	2	$(1, 2)$
2	4	$(2, 4)$
3	8	$(3, 8)$

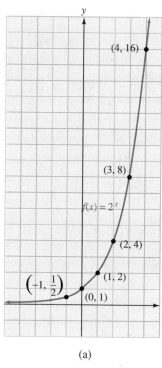

$$g(x) = 4^x$$

x	$g(x)$	$(x, g(x))$
-1	$\frac{1}{4}$	$\left(-1, \frac{1}{4}\right)$
0	1	$(0, 1)$
1	4	$(1, 4)$
2	16	$(2, 16)$

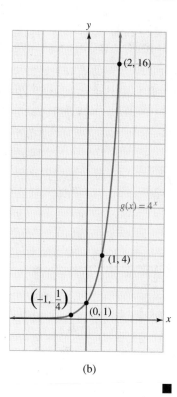

(a)

(b)

FIGURE 9-2

Since the values of $f(x)$ and $g(x)$ in Example 1 increase as the values of x increase, the functions $f(x) = 2^x$ and $g(x) = 4^x$ are called **increasing functions**. When $b > 1$ in the function $f(x) = b^x$, the larger the value of b, the steeper the curve.

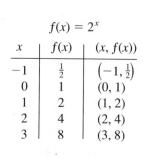 (a)

EXAMPLE 2 Graph the functions $f(x) = \left(\frac{1}{2}\right)^x$ and $g(x) = \left(\frac{1}{4}\right)^x$.

Solution We find and plot pairs (x, y) that satisfy each equation. The graph of $f(x) = \left(\frac{1}{2}\right)^x$ appears in Figure 9-3(a), and the graph of $g(x) = \left(\frac{1}{4}\right)^x$ appears in Figure 9-3(b).

$$f(x) = \left(\frac{1}{2}\right)^x$$

x	$f(x)$	$(x, f(x))$
-1	2	$(-1, 2)$
0	1	$(0, 1)$
1	$\frac{1}{2}$	$\left(1, \frac{1}{2}\right)$

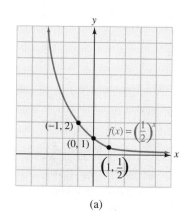

$$g(x) = \left(\frac{1}{4}\right)^x$$

x	$g(x)$	$(x, g(x))$
-1	4	$(-1, 4)$
0	1	$(0, 1)$
1	$\frac{1}{4}$	$\left(1, \frac{1}{4}\right)$

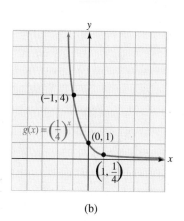

(a)

(b)

FIGURE 9-3

By looking at the graphs, we see that the domain of each function is the interval $(-\infty, \infty)$, and the range is the interval $(0, \infty)$.

In this case, as x increases, the function values decrease and approach 0. The x-axis is again a horizontal asymptote for each graph. Note that the graph of $f(x) = \left(\frac{1}{2}\right)^x$ passes through the point $\left(1, \frac{1}{2}\right)$, the graph of $g(x) = \left(\frac{1}{4}\right)^x$ passes through the point $\left(1, \frac{1}{4}\right)$, and each graph passes through the point $(0, 1)$.   ■

In each graph in Example 2, the values of y decrease as the values of x increase. Thus, the functions $f(x) = \left(\frac{1}{2}\right)^x$ and $g(x) = \left(\frac{1}{4}\right)^x$ are called **decreasing functions**. If $0 < b < 1$, the smaller the value of b, the steeper the curve.

Examples 1 and 2 illustrate the following fact.

Points on the Graph of $f(x) = b^x$

The graph of the exponential function defined by $y = f(x) = b^x$ passes through the points $(0, 1)$ and $(1, b)$.

An exponential function with base b is either increasing (for $b > 1$) or decreasing (for $0 < b < 1$). Since distinct real numbers x will determine distinct values b^x, exponential functions are one-to-one.

Exponential Function

An exponential function defined by

$$y = f(x) = b^x \qquad \text{where } b > 0 \text{ and } b \neq 1$$

is one-to-one. Thus,

1. If $b^r = b^s$, then $r = s$.
2. If $r \neq s$, then $b^r \neq b^s$.

EXAMPLE 3   On the same set of axes, graph   **a.** $f(x) = \left(\frac{2}{3}\right)^x$ and   **b.** $g(x) = \left(\frac{3}{2}\right)^x$.

Solution   **a.** The graph of $f(x) = \left(\frac{2}{3}\right)^x$ passes through the points $(0, 1)$ and $\left(1, \frac{2}{3}\right)$. Since $\frac{2}{3} < 1$, the function is decreasing. The graph appears in Figure 9-4.

b. The graph of $g(x) = \left(\frac{3}{2}\right)^x$ passes through the points $(0, 1)$ and $\left(1, \frac{3}{2}\right)$. Since $\frac{3}{2} > 1$, the function is increasing. The graph also appears in Figure 9-4.

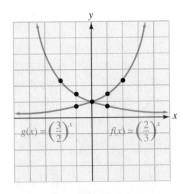

$f(x) = \left(\frac{2}{3}\right)^x$

x	$f(x)$	$(x, f(x))$
-2	$\frac{9}{4}$	$\left(-2, \frac{9}{4}\right)$
-1	$\frac{3}{2}$	$\left(-1, \frac{3}{2}\right)$
0	1	$(0, 1)$
1	$\frac{2}{3}$	$\left(1, \frac{2}{3}\right)$
2	$\frac{4}{9}$	$\left(2, \frac{4}{9}\right)$

$g(x) = \left(\frac{3}{2}\right)^x$

x	$g(x)$	$(x, g(x))$
-2	$\frac{4}{9}$	$\left(-2, \frac{4}{9}\right)$
-1	$\frac{2}{3}$	$\left(-1, \frac{2}{3}\right)$
0	1	$(0, 1)$
1	$\frac{3}{2}$	$\left(1, \frac{3}{2}\right)$
2	$\frac{9}{4}$	$\left(2, \frac{9}{4}\right)$

FIGURE 9-4

Since both graphs pass the horizontal line test, each function is one-to-one. ∎

■ COMPOUND INTEREST

If we deposit P dollars in an account paying an annual interest rate r, we can find the amount A in the account at the end of t years by using the formula $A = P + Prt$, or $A = P(1 + rt)$.

Suppose that we deposit \$500 in such an account. Then $P = 500$, and after six months $\left(\frac{1}{2} \text{ year}\right)$, the amount will be

$$A = 500(1 + rt) = 500\left(1 + r \cdot \frac{1}{2}\right) = 500\left(1 + \frac{r}{2}\right)$$

The account will begin the second six-month period with a value of $\$500\left(1 + \frac{r}{2}\right)$. After the second six-month period, the amount will be

$$A = \left[500\left(1 + \frac{r}{2}\right)\right]\left(1 + r \cdot \frac{1}{2}\right) = 500\left(1 + \frac{r}{2}\right)\left(1 + \frac{r}{2}\right) = 500\left(1 + \frac{r}{2}\right)^2$$

At the end of 10 years, or 20 six-month periods, the amount will be

$$A = 500\left(1 + \frac{r}{2}\right)^{20}$$

In the preceding example, the earned interest is deposited back in the account and also earns interest. When this is the case, we say that the account is earning **compound interest**. The general formula for compound interest is as follows.

> **Formula for Compound Interest**
> If \$$P$ is deposited in an account, and interest is paid k times a year at an annual rate r, the amount \$$A$ in the account after t years is given by
> $$A = P\left(1 + \frac{r}{k}\right)^{kt}$$

EXAMPLE 4

Saving for college To save for college, parents invest \$12,000 for their newborn child in a mutual fund that should average 10% annual interest. If the quarterly dividends are reinvested, how much will be available in 18 years?

Solution We substitute 12,000 for P, 0.10 for r, and 18 for t into the formula for compound interest and find A. Since interest is paid quarterly, $k = 4$.

$$A = P\left(1 + \frac{r}{k}\right)^{kt}$$

$$A = 12,000\left(1 + \frac{0.10}{4}\right)^{4(18)}$$

$$= 12,000(1 + 0.025)^{72}$$

$$= 12,000(1.025)^{72}$$

$$= 71,006.74$$

Use a calculator, and press these keys:

1.025 y^x 72 = × 12,000 = .

In 18 years, the account will be worth \$71,006.74. ∎

In business applications, the initial amount of money deposited is sometimes called the *present value* (*PV*). The amount to which the money will grow is called the *future value* (*FV*). The interest rate used for each compounding period is the *periodic interest rate* (*i*), and the number of times interest is compounded is the *number of compounding periods* (*n*). Using these definitions, an alternate formula for the compound interest formula is as follows.

Formula for Compound Interest

$$FV = PV(1 + i)^n$$

This alternate formula appears on business calculators. To use this formula to solve Example 4, we proceed as follows:

$$FV = PV(1 + i)^n$$
$$FV = 12{,}000(1 + 0.025)^{72} \qquad i = \tfrac{0.10}{4} = 0.025 \text{ and } n = 4(18) = 72.$$
$$= 71{,}006.74$$

■ GRAPHING DEVICES

EXAMPLE 5

Suppose $1 is deposited in an account earning 6% annual interest, compounded monthly. Use a graphing device to estimate how much will be in the account in 100 years.

Solution

We can substitute 1 for P, 0.06 for r, and 12 for k into the formula

$$A = P\left(1 + \frac{r}{k}\right)^{kt}$$

and simplify to get

$$A = (1.005)^{12t}$$

We now use a graphing device to graph $A = (1.005)^{12t}$ using window settings of [0, 120] for t and [0, 400] for A, obtaining the graph shown in Figure 9-5. We can then trace and zoom to estimate that $1 grows to approximately $397 in 100 years.

From the graph, we can see that the money grows slowly in the early years and rapidly in the later years. ■

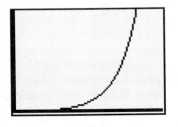

FIGURE 9-5

■ BASE-*e* EXPONENTIAL FUNCTIONS

We have seen that if $P is deposited in an account and interest is paid k times a year at an annual rate r, the amount $A in the account after t years is given by

$$A = P\left(1 + \frac{r}{k}\right)^{kt}$$

If we substitute $1 for P, 100% for r, and 1 year for t into the formula, we have

$$A = 1\left(1 + \frac{1}{k}\right)^{k(1)} = \left(1 + \frac{1}{k}\right)^{k} \qquad 100\% = 1.$$

To find the value of $\left(1 + \frac{1}{k}\right)^k$ when interest is compounded daily ($k = 365$), we press these keys on a scientific calculator:

$$1 \;\boxed{\div}\; 365 \;\boxed{=}\; \boxed{+}\; 1 \;\boxed{=}\; \boxed{y^x}\; 365 \;\boxed{=}$$

The display will read 2.71456748.

Table 1 shows this value of $\left(1 + \frac{1}{k}\right)^k$ and several others for specific values of k.

Number of times interest is compounded each year (k)	$\left(1 + \dfrac{1}{k}\right)^k$
1	2
2	2.25
4	2.44140625 . . .
12	2.61303529 . . .
365	2.71456748 . . .
1000	2.71692393 . . .
100,000	2.71826823 . . .
1,000,000	2.71828046 . . .

TABLE 1

Table 1 suggests that as k increases, the value of $\left(1 + \frac{1}{k}\right)^k$ approaches a fixed number. This does happen, and the fixed number is called e, which has the following value.

$$e = 2.718281828459 \ldots$$

The function defined by the equation $y = f(x) = e^x$ is called the **exponential function**. Its graph is shown in Figure 9-6.

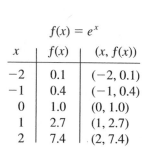

$$f(x) = e^x$$

x	$f(x)$	$(x, f(x))$
-2	0.1	$(-2, 0.1)$
-1	0.4	$(-1, 0.4)$
0	1.0	$(0, 1.0)$
1	2.7	$(1, 2.7)$
2	7.4	$(2, 7.4)$

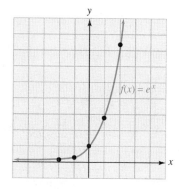

FIGURE 9-6

■ CONTINUOUS COMPOUND INTEREST

When banks compound interest quarterly, the account balance increases four times each year. In many situations, however, quantities grow continuously. For example, the population of a town does not increase abruptly every quarter but grows steadily all year long.

To develop a formula for computing **continuous compound interest**, we recall the formula for compound interest and proceed as follows:

$$A = P\left(1 + \frac{r}{k}\right)^{kt}$$ The formula for compound interest.

$$A = P\left(1 + \frac{r}{rn}\right)^{rnt}$$ Since k and r are positive constants, we can let $k = rn$, where n is some other positive constant.

$$A = P\left[\left(1 + \frac{r}{rn}\right)^{n}\right]^{rt}$$ Use the property $a^{mn} = (a^m)^n$.

$$A = P\left[\left(1 + \frac{1}{n}\right)^{n}\right]^{rt}$$ Simplify $\frac{r}{rn}$.

Since we have continuous compound interest, n is infinitely large, and we can replace $\left(1 + \frac{1}{n}\right)^n$ with e to get

$$A = Pe^{rt}$$

Formula for Exponential Growth

If a quantity P increases or decreases at an annual rate r compounded continuously, then the amount A after t years is given by

$$A = Pe^{rt}$$

If time is measured in years, r is called the *annual growth rate*. If r is negative, the "growth" represents a decrease.

To compute the amount to which $12,000 will grow if invested for 18 years at 10% annual interest, compounded continuously, we substitute 12,000 for P, 0.10 for r, and 18 for t in the formula for exponential growth:

$$A = Pe^{rt}$$
$$A = 12,000e^{0.10(18)}$$
$$= 12,000e^{1.8}$$
$$= 72,595.77 \qquad \text{Use a calculator and press these keys:}$$
$$1.8 \boxed{e^x} \boxed{\times} 12,000 \boxed{=} .$$

After 18 years, the account will contain $72,595.77. This is $1589.03 more than the result in Example 4, where interest was compounded quarterly.

■ MALTHUSIAN POPULATION GROWTH

In the **Malthusian model for population growth**, the future population of a colony is related to the present population by the formula for exponential growth.

EXAMPLE 6   **City planning**   The population of a city is currently 15,000, but changing economic conditions are causing the population to decrease 2% each year. If this trend continues, find the population in 30 years.

Solution   Since the population is decreasing 2% each year, the annual growth rate is -2%, or -0.02. We can substitute -0.02 for r, 30 for t, and 15,000 for P in the formula for exponential growth and find A.

$$A = Pe^{rt}$$
$$A = \mathbf{15,000}e^{-0.02\,(30)}$$
$$= 15,000e^{-0.6}$$
$$= 8232.17$$

In 30 years, city planners expect a population of approximately 8232 persons. ∎

Orals   *If $x = 2$, evaluate each expression.*

1. 2^x **2.** 5^x **3.** $2(3^x)$ **4.** 3^{x-1}

If $x = -2$, evaluate each expression.

5. 2^x **6.** 5^x **7.** $2(3^x)$ **8.** 3^{x-1}

EXERCISE 9.1

In Exercises 1–4, find each value to four decimal places.

1. $2^{\sqrt{2}}$ **2.** $7^{\sqrt{2}}$ **3.** $5^{\sqrt{5}}$ **4.** $6^{\sqrt{3}}$

In Exercises 5–12, graph each exponential function. Check your work with a graphing calculator.

5. $f(x) = 3^x$ **6.** $f(x) = 4^x$ **7.** $f(x) = 5^x$ **8.** $f(x) = 6^x$

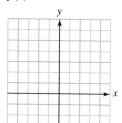 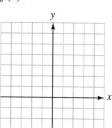 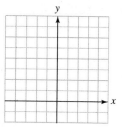 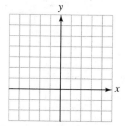

9. $f(x) = \left(\dfrac{1}{3}\right)^x$ **10.** $f(x) = \left(\dfrac{1}{4}\right)^x$ **11.** $f(x) = \left(\dfrac{1}{5}\right)^x$ **12.** $f(x) = \left(\dfrac{1}{6}\right)^x$

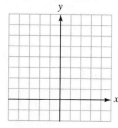 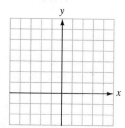 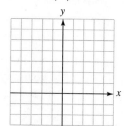 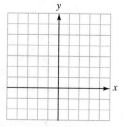

In Exercises 13–16, use a graphing device to do each experiment. Describe what you find out.

13. Graph $f(x) = k + 2^x$ for many values of k.

14. Graph $f(x) = 2^{x+k}$ for many values of k.

15. Graph $f(x) = k2^x$ for many values of k.

16. Graph $f(x) = 2^{kx}$ for many values of k.

In Exercises 17–20, use what you learned in Exercises 13–16 to graph each function.

17. $f(x) = -2 + 3^x$

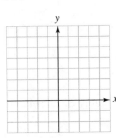

18. $f(x) = 3^{x-1}$

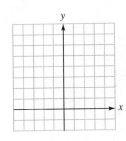

19. $f(x) = 3^{-x}$

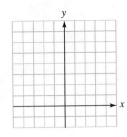

20. $f(x) = \dfrac{1}{3}(3^x)$

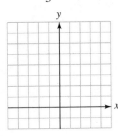

In Exercises 21–28, find the value of b, if any, that would cause the graph of $y = b^x$ to look like the graph indicated.

21.

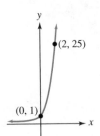

22.

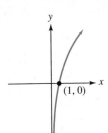

23.

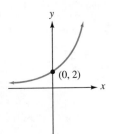

24.

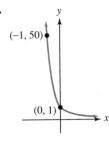

25.

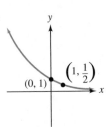

26.

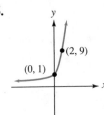

27.

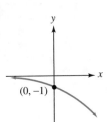

28.

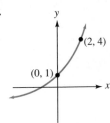

In Exercises 29–30, assume there are no deposits or withdrawals.

29. Compound interest An initial deposit of $10,000 earns 8% interest, compounded quarterly. How much will be in the account after 10 years?

30. Compound interest An initial deposit of $10,000 earns 8% interest, compounded monthly. How much will be in the account after 10 years?

In Exercises 31–46, use a calculator to help solve each problem.

31. Comparing interest rates How much more interest could $1000 earn in 5 years, compounded quarterly, if the annual interest rate were $5\frac{1}{2}\%$ instead of 5%?

32. Comparing savings plans Which institution in Illustration 1 provides the better investment?

33. Compound interest If $1 had been invested on July 4, 1776 at 5% interest, compounded annually, what would it be worth on July 4, 2076?

34. Frequency of compounding $10,000 is invested in each of two accounts, both paying 6% annual interest. In the first account, interest compounds quarterly, and in the second account, interest compounds daily. Find the difference in value between the accounts after 20 years.

35. Fish population A population of fish is growing according to the Malthusian model. How many fish will there be in 10 years if the annual growth rate is 3% and the initial population is 2700 fish?

Fidelity Savings & Loan

earn 5.25%
compounded monthly

Union Trust

Money Market account
paying 5.35%
compounded annually

ILLUSTRATION 1

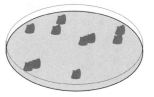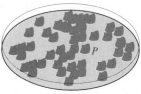

12:00 PM   4:00 PM

ILLUSTRATION 2

36. Town population The population of a town is 1350. The town is expected to grow according to the Malthusian model, with an annual growth rate of 6%. Find the population of the town in 20 years.

37. World population The population of the world is approximately 6 billion. If the population is growing according to the Malthusian model with an annual growth rate of 1.8%, what will be the population of the world in 30 years?

38. World population The population of the world is approximately 6 billion. If the population is growing according to the Malthusian model with an annual growth rate of 1.8%, what will be the population of the world in 60 years?

39. Bacterial cultures A colony of 6 million bacteria is growing in a culture medium. (See Illustration 2.) The population P after t hours is given by the formula $P = (6 \times 10^6)(2.3)^t$. Find the population after 4 hours.

40. Discharging a battery The charge remaining in a battery decreases as the battery discharges. The charge C (in coulombs) after t days is given by the formula $C = (3 \times 10^{-4})(0.7)^t$. Find the charge after 5 days.

41. Radioactive decay A radioactive material decays according to the formula $A = A_0\left(\frac{2}{3}\right)^t$, where A_0 is the initial amount present and t is measured in years. Find the amount present in 5 years.

42. Town population The population of North Rivers is decreasing exponentially according to the formula $P = 3745(0.93)^t$, where t is measured in years from the present date. Find the population in 6 years, 9 months.

43. Depreciation A camping trailer originally purchased for $4570 is continuously losing value at the rate of 6% per year. Find its value when it is $6\frac{1}{2}$ years old.

44. Depreciation A boat purchased for $7500 has been continuously decreasing in value at the rate of 2% each year. It is now 8 years, 3 months old. Find its value.

45. Salvage value A small business purchased a computer for $4700. It is expected that its value each year will be 75% of its value in the preceding year. If the business disposes of the computer after 5 years, find its salvage value (the value after 5 years). (*Hint:* Do not use continuous compounding.)

46. Louisiana Purchase In 1803, the United States acquired the Louisiana Purchase from France. The country doubled its territory by adding 827,000 square miles of land for $15 million. If the land has appreciated at the rate of 6% each year, what would one square mile of land be worth in 1996? (*Hint:* Do not use continuous compounding.)

Writing Exercises   *Write a paragraph using your own words.*

1. If world population is increasing exponentially, why is there cause for concern?

2. How do the graphs of $y = b^x$ differ when $b > 1$ and $0 < b < 1$?

Something to Think About   **1.** In the definition of the exponential function, b could not equal 0. Why not?

2. In the definition of the exponential function, b could not be negative. Why not?

Review Exercises   *In Illustration 1, lines r and s are parallel.*

1. Find x.

2. Find the measure of angle 1.

3. Find the measure of angle 2.

4. Find the measure of angle 3.

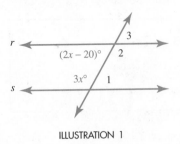

ILLUSTRATION 1

9.2 Logarithmic Functions

■ COMMON LOGARITHMS ■ NATURAL LOGARITHMS ■ GRAPHING LOGARITHMIC FUNCTIONS

Since an exponential function $y = f(x) = b^x$ is one-to-one, it has an inverse function that is defined by $x = b^y$. To express this inverse function in the form $y = f^{-1}(x)$, we must solve the equation $x = b^y$ for y. To do this, we need the following definition.

> **Logarithmic Function**
> If $b > 0$ and $b \neq 1$, the **logarithmic function with base b** is defined by
>
> $$y = f(x) = \log_b x \qquad \text{if and only if} \qquad x = b^y$$
>
> The **domain of the logarithmic function** is the set of positive real numbers, which is the interval $(0, \infty)$.
> The **range of the logarithmic function** is the set of real numbers, which is the interval $(-\infty, \infty)$.

Since the function $f(x) = \log_b x$ is the inverse of the one-to-one exponential function $f(x) = b^x$, the logarithmic function is a one-to-one function.

WARNING!   Because the domain of the logarithmic function is restricted to the set of positive numbers, it is impossible to find the logarithm of 0, the logarithm of a negative number, or the logarithm of a nonreal number.

The previous definition guarantees that any pair (x, y) that satisfies the equation $y = \log_b x$ also satisfies the equation $x = b^y$. Thus,

$$\log_7 1 = 0 \qquad \text{because} \qquad 1 = 7^0$$
$$\log_5 25 = 2 \qquad \text{because} \qquad 25 = 5^2$$
$$\log_5 \frac{1}{25} = -2 \qquad \text{because} \qquad \frac{1}{25} = 5^{-2}$$

$$\log_{16} 4 = \frac{1}{2} \qquad \text{because} \qquad 4 = 16^{1/2}$$

$$\log_2 \frac{1}{8} = -3 \qquad \text{because} \qquad \frac{1}{8} = 2^{-3}$$

$$\log_b x = y \qquad \text{when} \qquad x = b^y$$

In each of the previous examples, the logarithm of a number is an exponent, and furthermore,

$\log_b x$ **is the exponent to which** b **is raised to get** x.

To say this with an equation, we write

$$b^{\log_b x} = x$$

EXAMPLE 1   Find y in each equation: **a.** $\log_6 1 = y$,   **b.** $\log_3 27 = y$,   and   **c.** $\log_5 \frac{1}{5} = y$.

Solution   **a.** We can change the equation $\log_6 1 = y$ into the equivalent exponential equation $1 = 6^y$. Since $1 = 6^0$, it follows that $y = 0$. Thus,

$$\log_6 1 = 0$$

b. $\log_3 27 = y$ is equivalent to $27 = 3^y$. Since $27 = 3^3$, it follows that $3^y = 3^3$, and $y = 3$. Thus,

$$\log_3 27 = 3$$

c. $\log_5 \frac{1}{5} = y$ is equivalent to $\frac{1}{5} = 5^y$. Since $\frac{1}{5} = 5^{-1}$, it follows that $5^y = 5^{-1}$, and $y = -1$. Thus,

$$\log_5 \frac{1}{5} = -1$$

■

EXAMPLE 2   Find the value of x in each equation: **a.** $\log_3 81 = x$,   **b.** $\log_x 125 = 3$,   and **c.** $\log_4 x = 3$.

Solution   **a.** $\log_3 81 = x$ is equivalent to $3^x = 81$. Because $3^4 = 81$, it follows that $3^x = 3^4$, and $x = 4$.

b. $\log_x 125 = 3$ is equivalent to $x^3 = 125$. Because $5^3 = 125$, it follows that $x^3 = 5^3$, and $x = 5$.

c. $\log_4 x = 3$ is equivalent to $4^3 = x$. Because $4^3 = 64$, it follows that $x = 64$.

■

EXAMPLE 3   Find the value of x in each equation: **a.** $\log_{1/3} x = 2$,   **b.** $\log_{1/3} x = -2$,   and **c.** $\log_{1/3} \frac{1}{27} = x$.

Solution   **a.** $\log_{1/3} x = 2$ is equivalent to $\left(\frac{1}{3}\right)^2 = x$. Thus, $x = \frac{1}{9}$.

b. $\log_{1/3} x = -2$ is equivalent to $\left(\frac{1}{3}\right)^{-2} = x$. Thus,

$$x = \left(\frac{1}{3}\right)^{-2} = 3^2 = 9$$

c. $\log_{1/3} \dfrac{1}{27} = x$ is equivalent to $\left(\dfrac{1}{3}\right)^x = \dfrac{1}{27}$. Because $\left(\dfrac{1}{3}\right)^3 = \dfrac{1}{27}$, it follows that $x = 3$. ∎

■ COMMON LOGARITHMS

For computational purposes, base-10 logarithms are convenient. For this reason, base-10 logarithms are called **common logarithms**. When the base b is not indicated in the notation $\log x$, we assume that $b = 10$:

$$\log x \quad \text{means} \quad \log_{10} x$$

Because base-10 logarithms appear often in mathematics and science, it is helpful to become familiar with the following base-10 logarithms:

$$\log_{10} \frac{1}{100} = -2 \qquad \text{because} \qquad 10^{-2} = \frac{1}{100}$$

$$\log_{10} \frac{1}{10} = -1 \qquad \text{because} \qquad 10^{-1} = \frac{1}{10}$$

$$\log_{10} 1 = 0 \qquad \text{because} \qquad 10^0 = 1$$

$$\log_{10} 10 = 1 \qquad \text{because} \qquad 10^1 = 10$$

$$\log_{10} 100 = 2 \qquad \text{because} \qquad 10^2 = 100$$

$$\log_{10} 1000 = 3 \qquad \text{because} \qquad 10^3 = 1000$$

In general, we have

$$\log_{10} 10^x = x$$

Base-10 logarithms are easy to find with a calculator. For example, to find $\log 6.35$, we press these keys on a scientific calculator:

6.35 LOG   You may have to press a 2nd function key before LOG .

The display will read 0.802773725. To four decimal places,

$$\log 6.35 = 0.8028$$

EXAMPLE 4   Find x in the equation $\log x = 0.3568$.

Solution   The equation $\log x = 0.3568$ is equivalent to the equation $10^{0.3568} = x$. To find x with a scientific calculator, we press these keys:

10 y^x .3568 =

The display will read 2.274049951. To four decimal places,

$$x = 2.2740$$

If your calculator has a 10^x key, enter .3568 and press it to get the same result. (You may have to press a 2nd function key first.) ∎

■ NATURAL LOGARITHMS

Because the number e appears often in mathematical models of natural events, base-e logarithms are called **natural logarithms**. Natural logarithms are usually denoted by the symbol $\ln x$, rather than $\log_e x$:

$$\ln x \quad \text{means} \quad \log_e x$$

We can also find natural logarithms with a calculator. For example, to find $\ln 6.35$ with a scientific calculator, we press these keys:

6.35 LN   (You may have to press a 2nd function key first.)

The display will read 1.848454813. To four decimal places,

$$\ln 6.35 = 1.8485$$

EXAMPLE 5   Find x in each equation: **a.** $\ln x = 3.441$ and   **b.** $\ln x = \log 6.7$.

Solution   **a.** The equation $\ln x = 3.441$ is equivalent to the equation $e^{3.441} = x$. To use a scientific calculator to find x, we press these keys:

3.441 e^x (You may have to press a 2nd function key first.)

The display will read 31.21816073. To four decimal places,

$$x = 31.2182$$

b. The equation $\ln x = \log 6.7$ is equivalent to the equation $e^{\log 6.7} = x$. To use a scientific calculator to find x, we press these keys:

6.7 LOG e^x (You may have to press a 2nd function key first.)

The display will read 2.284334655. To four decimal places,

$$x = 2.2843$$ ■

■ GRAPHING LOGARITHMIC FUNCTIONS

To graph the logarithmic function $f(x) = \log_2 x$, we find and plot several points with coordinates (x, y) that satisfy the equation $x = 2^y$. After joining them with a smooth curve, we have the graph that is shown in Figure 9-7(a).

To graph $g(x) = \log_{1/2} x$, we find and plot several points with coordinates (x, y) that satisfy the equation $x = \left(\frac{1}{2}\right)^y$. After joining them with a smooth curve, we have the graph shown in Figure 9-7(b).

We can see from the graphs that the domain of each function is the interval $(0, \infty)$, and the range is the interval $(-\infty, \infty)$.

$f(x) = \log_2 x$ or $x = 2^y$

x	$f(x)$	$(x, f(x))$
$\frac{1}{4}$	-2	$\left(\frac{1}{4}, -2\right)$
$\frac{1}{2}$	-1	$\left(\frac{1}{2}, -1\right)$
1	0	$(1, 0)$
2	1	$(2, 1)$
4	2	$(4, 2)$
8	3	$(8, 3)$

$g(x) = \log_{1/2} x$ or $x = \left(\frac{1}{2}\right)^y$

x	$g(x)$	$(x, g(x))$
8	-3	$(8, -3)$
4	-2	$(4, -2)$
2	-1	$(2, -1)$
1	0	$(1, 0)$
$\frac{1}{2}$	1	$\left(\frac{1}{2}, 1\right)$
$\frac{1}{4}$	2	$\left(\frac{1}{4}, 2\right)$

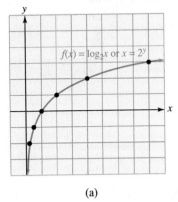

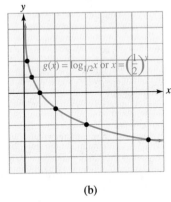

(a) (b)

FIGURE 9-7

EXAMPLE 6 Graph the functions **a.** $f(x) = \log x$ and **b.** $g(x) = \ln x$.

Solution **a.** The equation $y = \log x$ is equivalent to the equation $x = 10^y$. To get the graph of $f(x) = \log x$, we can plot points that satisfy the equation $x = 10^y$ and join them with a smooth curve, as in Figure 9-8(a).

$f(x) = \log x$

x	$f(x)$	$(x, f(x))$
$\frac{1}{10}$	-1	$\left(\frac{1}{10}, -1\right)$
1	0	$(1, 0)$
10	1	$(10, 1)$
100	2	$(100, 2)$

$g(x) = \ln x$

x	$g(x)$	$(x, g(x))$
$\frac{1}{e} \approx 0.4$	-1	$(0.4, -1)$
1	0	$(1, 0)$
$e \approx 2.7$	1	$(2.7, 1)$
$e^2 \approx 7.4$	2	$(7.4, 2)$

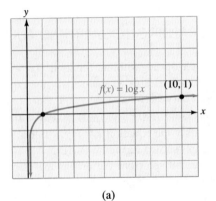

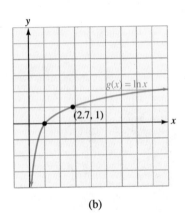

(a) (b)

FIGURE 9-8

b. The equation $y = \ln x$ is equivalent to the equation $x = e^y$. To get the graph of $g(x) = \ln x$, we can plot points that satisfy the equation $x = e^y$ and join them with a smooth curve, as in Figure 9-8(b).

We can see from the graphs that the domain of each function is the interval $(0, \infty)$, and the range is the interval $(-\infty, \infty)$. ∎

The previous examples suggest that the graphs of all logarithmic functions are similar to those shown in Figure 9-9. If $b > 1$, the logarithmic function is increasing, as in Figure 9-9(a). If $0 < b < 1$, the logarithmic function is decreasing, as in Figure 9-9(b). Each graph of $y = \log_b x$ passes through the points $(1, 0)$ and $(b, 1)$ and has the y-axis as an asymptote.

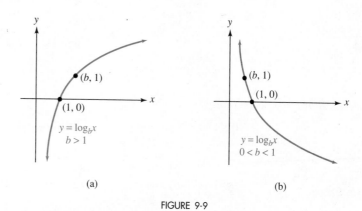

FIGURE 9-9

Since the exponential and logarithmic functions are inverses of each other, their graphs are symmetric about the line $y = x$. The graphs of $f(x) = \log_b x$ and $g(x) = b^x$ are shown in Figure 9-10(a) when $b > 0$, and in Figure 9-10(b) when $0 < b < 1$.

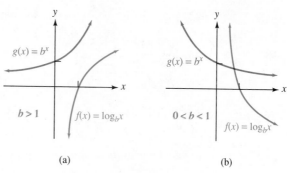

FIGURE 9-10

Orals   *Find the value of x in each equation.*

1. $\log_2 8 = x$ **2.** $\log_3 9 = x$ **3.** $\log_x 125 = 3$

4. $\log_x 8 = 3$ **5.** $\log_4 16 = x$ **6.** $\log_x 32 = 5$

7. $\log_{1/2} x = 2$ **8.** $\log_9 3 = x$ **9.** $\log_x \frac{1}{4} = -2$

EXERCISE 9.2

In Exercises 1–8, write each equation in exponential form.

1. $\log_3 81 = 4$

2. $\log_7 7 = 1$

3. $\log_{1/2} \dfrac{1}{8} = 3$

4. $\log_{1/5} 1 = 0$

5. $\log_4 \dfrac{1}{64} = -3$

6. $\log_6 \dfrac{1}{36} = -2$

7. $\log_{1/2} \dfrac{1}{8} = 3$

8. $\log_{1/5} 1 = 0$

In Exercises 9–16, write each equation in logarithmic form.

9. $8^2 = 64$

10. $10^3 = 1000$

11. $4^{-2} = \dfrac{1}{16}$

12. $3^{-4} = \dfrac{1}{81}$

13. $\left(\dfrac{1}{2}\right)^{-5} = 32$

14. $\left(\dfrac{1}{3}\right)^{-3} = 27$

15. $x^y = z$

16. $m^n = p$

In Exercises 17–56, find each value of x.

17. $\log_2 8 = x$

18. $\log_3 9 = x$

19. $\log_4 64 = x$

20. $\log_6 216 = x$

21. $\log_{1/2} \dfrac{1}{8} = x$

22. $\log_{1/3} \dfrac{1}{81} = x$

23. $\log_9 3 = x$

24. $\log_{125} 5 = x$

25. $\log_{1/2} 8 = x$

26. $\log_{1/2} 16 = x$

27. $\log_8 x = 2$

28. $\log_7 x = 0$

29. $\log_7 x = 1$

30. $\log_2 x = 3$

31. $\log_{25} x = \dfrac{1}{2}$

32. $\log_4 x = \dfrac{1}{2}$

33. $\log_5 x = -2$

34. $\log_3 x = -4$

35. $\log_{36} x = -\dfrac{1}{2}$

36. $\log_{27} x = -\dfrac{1}{3}$

37. $\log_{100} \dfrac{1}{1000} = x$

38. $\log_{5/2} \dfrac{4}{25} = x$

39. $\log_{27} 9 = x$

40. $\log_{12} x = 0$

41. $\log_x 5^3 = 3$

42. $\log_x 5 = 1$

43. $\log_x \dfrac{9}{4} = 2$

44. $\log_x \dfrac{\sqrt{3}}{3} = \dfrac{1}{2}$

45. $\log_x \dfrac{1}{64} = -3$

46. $\log_x \dfrac{1}{100} = -2$

47. $\log_{2\sqrt{2}} x = 2$

48. $\log_4 8 = x$

49. $2^{\log_2 5} = x$

50. $3^{\log_3 4} = x$

51. $x^{\log_4 6} = 6$

52. $x^{\log_3 8} = 8$

53. $\log 10^3 = x$

54. $\log 10^{-2} = x$

55. $10^{\log x} = 100$

56. $10^{\log x} = \dfrac{1}{10}$

In Exercises 57–68, use a calculator to find each value, if possible. Give answers to four decimal places.

57. $\log 3.25$

58. $\log 0.57$

59. $\log 0.00467$

60. $\log 375.876$

61. $\ln 0.93$

62. $\ln 7.39$

63. $\ln 37.896$

64. $\ln 0.00465$

65. $\log (\ln 1.7)$

66. $\ln (\log 9.8)$

67. $\ln (\log 0.1)$

68. $\log (\ln 0.01)$

In Exercises 69–76, use a calculator to find each value of y, if possible. Give answers to two decimal places.

69. $\log y = 1.4023$

70. $\ln y = 2.6490$

71. $\ln y = 4.24$

72. $\log y = 0.926$

73. $\log y = -3.71$

74. $\ln y = -0.28$

75. $\log y = \ln 8$

76. $\ln y = \log 7$

In Exercises 77–80, graph each function.

77. $f(x) = \log_3 x$

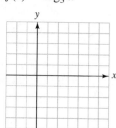

78. $f(x) = \log_{1/3} x$

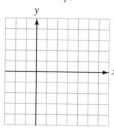

79. $f(x) = \log_{1/2} x$

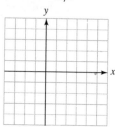

80. $f(x) = \log_4 x$

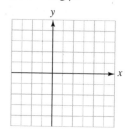

In Exercises 81–84, graph each pair of inverse functions on a single coordinate system.

81. $f(x) = 2^x$,
$g(x) = \log_2 x$

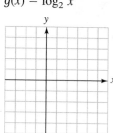

82. $f(x) = \left(\dfrac{1}{2}\right)^x$,
$g(x) = \log_{1/2} x$

83. $f(x) = \left(\dfrac{1}{4}\right)^x$,
$g(x) = \log_{1/4} x$

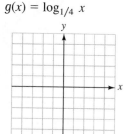

84. $f(x) = 4^x$,
$g(x) = \log_4 x$

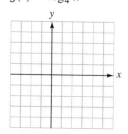

In Exercises 85–92, find the value of b, if any, that would cause the graph of $f(x) = \log_b x$ to look like the graph indicated.

85.

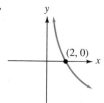

86.

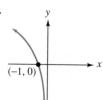

87.

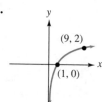

88.

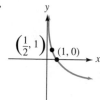

89.

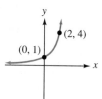

90.

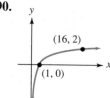

91.

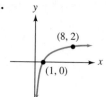

92.

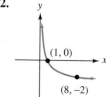

In Exercises 93–96, use a graphing device to do each experiment. Describe what you find.

93. Graph $f(x) = k + \log x$ for many values of k.

94. Graph $f(x) = \log (x + k)$ for many values of k.

95. Graph $f(x) = k \log x$ for many values of k.

96. Graph $f(x) = \log kx$ for many values of k.

Writing Exercises *Write a paragraph using your own words.*

1. Describe the appearance of the graph of $f(x) = \log_b x$ when $0 < b < 1$ and when $b > 1$.

2. Explain why it is impossible to find the logarithm of a negative number.

Something to Think About

1. Find the value of $(1 + p)^{1/p}$ for $p = 1, \frac{1}{10}, \frac{1}{100}, \frac{1}{1000},$ and $\frac{1}{10,000}$. What do you find?

2. Add the first 3 terms, the first 4 terms, the first 5 terms, and the first 6 terms of the list

$$1, 1, \frac{1}{2}, \frac{1}{2 \cdot 3}, \frac{1}{2 \cdot 3 \cdot 4}, \frac{1}{2 \cdot 3 \cdot 4 \cdot 5}$$

What do you find?

Review Exercises *Solve each equation.*

1. $\sqrt[3]{6x + 4} = 4$

2. $\sqrt{3x - 4} = \sqrt{-7x + 2}$

3. $\sqrt{a + 1} - 1 = 3a$

4. $3 - \sqrt{t - 3} = \sqrt{t}$

9.3 Properties of Logarithms

■ THE CHANGE-OF-BASE FORMULA

Since logarithms are exponents, the laws of exponents have counterparts in logarithmic notation. We will use properties of exponents to develop many of the properties of logarithms.

Properties of Logarithms
If b is a positive number and $b \neq 1$, then

1. $\log_b 1 = 0$
2. $\log_b b = 1$
3. $\log_b b^x = x$
4. $b^{\log_b x} = x \quad (x > 0)$

Properties 1 through 4 follow directly from the definition of logarithms:

1. $\log_b 1 = 0$ because $b^0 = 1$.
2. $\log_b b = 1$ because $b^1 = b$.
3. $\log_b b^x = x$ because $b^x = b^x$.
4. $b^{\log_b x} = x$ because $\log_b x$ is the exponent to which b is raised to get x.

EXAMPLE 1 Simplify each expression: **a.** $\log_5 1$, **b.** $\log_3 3$, **c.** $\log_7 7^3$, and **d.** $b^{\log_b 7}$.

Solution **a.** By Property 1, $\log_5 1 = 0$ because $5^0 = 1$.

b. By Property 2, $\log_3 3 = 1$ because $3^1 = 3$.

c. By Property 3, $\log_7 7^3 = 3$ because $7^3 = 7^3$.

d. By Property 4, $b^{\log_b 7} = 7$ because $\log_b 7$ is the power to which b is raised to get 7. ■

Properties 3 and 4 show that the compositions of the exponential and logarithmic functions with the same base (in both directions) are the identity function. This is expected, because the exponential and logarithmic functions are inverse functions.

More Properties of Logarithms

If M, N, and b are positive numbers and $b \neq 1$, then

5. Logarithm of a product: $\log_b MN = \log_b M + \log_b N$

6. Logarithm of a quotient: $\log_b \dfrac{M}{N} = \log_b M - \log_b N$

7. Logarithm of a power: $\log_b M^p = p \log_b M$

8. If $\log_b x = \log_b y$, then $x = y$.

Proof To prove the rule for the logarithm of a product, we let $x = \log_b M$ and $y = \log_b N$. Because of the definition of logarithm, these equations can be written in the form

$$M = b^x \qquad \text{and} \qquad N = b^y$$

Then $MN = b^x b^y$, and the product rule of exponents gives

$$MN = b^{x+y}$$

Using the definition of logarithms gives

$$\log_b MN = x + y$$

Substituting the values of x and y completes the proof:

$$\log_b MN = \log_b M + \log_b N \qquad\qquad \square$$

WARNING! The rule for the logarithm of a product asserts that the logarithm of the product of two numbers is equal to the sum of their logarithms. The logarithm of a sum or a difference usually does not simplify. In general,

$$\log_b (M + N) \neq \log_b M + \log_b N \qquad \text{and} \qquad \log_b (M - N) \neq \log_b M - \log_b N$$

Proof To prove the rule for the logarithm of a quotient, we again let $x = \log_b M$ and $y = \log_b N$. These equations can be written as

$$M = b^x \qquad \text{and} \qquad N = b^y$$

Then $\dfrac{M}{N} = \dfrac{b^x}{b^y}$, and the quotient rule of exponents gives

$$\frac{M}{N} = b^{x-y}$$

Using the definition of logarithms gives

$$\log_b \frac{M}{N} = x - y$$

Substituting the values of x and y completes the proof:

$$\log_b \frac{M}{N} = \log_b M - \log_b N \qquad\qquad \square$$

WARNING! The rule for the logarithm of a quotient asserts that the logarithm of the quotient of two numbers is equal to the difference of their logarithms. The logarithm of a quotient is not the quotient of the logarithms:

$$\log_b \frac{M}{N} \neq \frac{\log_b M}{\log_b N}$$

Proof To prove the rule for the logarithm of a power, we let $x = \log_b M$ and proceed as follows:

$M = b^x$	Write $x = \log_b M$ in exponential form.
$(M)^p = (b^x)^p$	Raise both sides to the pth power.
$M^p = b^{px}$	Remove parentheses.
$\log_b M^p = px$	Use the definition of logarithm.

Substituting the value for x completes the proof:

$$\log_b M^p = p \log_b M \qquad\qquad \square$$

Property 8 follows from the fact that the logarithmic function is a one-to-one function.

EXAMPLE 2 Use a calculator to verify the rule for the logarithm of a quotient by showing that

$$\ln \frac{25.37}{37.25} = \ln 25.37 - \ln 37.25$$

Solution We can find the left- and right-hand sides of the equation separately and compare the results. To use a scientific calculator to find $\ln \frac{25.37}{37.25}$, we press these keys:

25.37 $\div$ 37.25 $=$ LN

The display will read -0.384084571.
To find $\ln 25.37 - \ln 37.25$, we press these keys:

25.37 LN $-$ 37.25 LN $=$

The display will read -0.384084571. Since the left- and right-hand sides are equal, the equation is verified. ■

The rules for the logarithm of a product, quotient, or power can be used to expand or condense many logarithmic expressions.

(b, c)

EXAMPLE 3 Assume that x, y, z, and b are positive numbers ($b \neq 1$). Use the properties of logarithms to write each expression in terms of the logarithms of x, y, and z.

a. $\log_b xyz$, **b.** $\log_b \dfrac{x}{yz}$ and **c.** $\log_b \dfrac{y^2 \sqrt{x}}{z}$

Solution
a. $\log_b xyz = \log_b (xy)z$	Use the associative property of multiplication.
$= \log_b (xy) + \log_b z$	Use the product rule of logarithms.
$= \log_b x + \log_b y + \log_b z$	Use the product rule of logarithms.

b. $\log_b \dfrac{x}{yz} = \log_b x - \log_b (yz)$

Use the quotient rule of logarithms.

$= \log_b x - (\log_b y + \log_b z)$

Use the product rule of logarithms.

$= \log_b x - \log_b y - \log_b z$

Remove parentheses.

c. $\log_b \dfrac{y^2 \sqrt{x}}{z} = \log_b y^2 \sqrt{x} - \log_b z$

Use the quotient rule of logarithms.

$= \log_b y^2 + \log_b \sqrt{x} - \log_b z$

Use the product rule of logarithms.

$= \log_b y^2 + \log_b x^{1/2} - \log_b z$

Write $\sqrt{x}$ as $x^{1/2}$.

$= 2 \log_b y + \dfrac{1}{2} \log_b x - \log_b z$

Use the power rule of logarithms twice. ∎

EXAMPLE 4   Assume that x, y, z, and b are positive numbers ($b \neq 1$). Use the properties of logarithms to write each expression as the logarithm of a single quantity.
a. $2 \log_b x + \frac{1}{3} \log_b y$ and   **b.** $\frac{1}{2} \log_b (x - 2) - \log_b y + 3 \log_b z$

Solution   **a.** $2 \log_b x + \dfrac{1}{3} \log_b y = \log_b x^2 + \log_b y^{1/3}$

Use the power rule of logarithms.

$= \log_b (x^2 y^{1/3})$

Use the product rule of logarithms.

$= \log_b (x^2 \sqrt[3]{y})$

Change $y^{1/3}$ to $\sqrt[3]{y}$.

b. $\dfrac{1}{2} \log_b (x - 2) - \log_b y + 3 \log_b z$

$= \log_b (x - 2)^{1/2} - \log_b y + \log_b z^3$

Use the power rule of logarithms.

$= \log_b \dfrac{(x - 2)^{1/2}}{y} + \log_b z^3$

Use the quotient rule of logarithms.

$= \log_b \dfrac{z^3 \sqrt{x - 2}}{y}$

Use the product rule of logarithms. ∎

(b, d)   **EXAMPLE 5**   Given that $\log 2 \approx 0.3010$ and $\log 3 \approx 0.4771$, find approximations for **a.** $\log 6$, **b.** $\log 9$,   **c.** $\log 5$,   **d.** $\log \sqrt{5}$,   and   **e.** $\log 1.5$. **Do not use a calculator**.

Solution   **a.** $\log 6 = \log (2 \cdot 3)$

Factor 6 as $2 \cdot 3$.

$= \log 2 + \log 3$

Use the product rule of logarithms.

$\approx 0.3010 + 0.4771$

Substitute the value of each logarithm.

≈ 0.7781

b. $\log 9 = \log (3^2)$

Write 9 as 3^2.

$= 2 \log 3$

Use the power rule of logarithms.

$\approx 2(0.4771)$

Substitute the value of the base-10 logarithm of 3.

≈ 0.9542

c. $\log 5 = \log \left(\dfrac{10}{2} \right)$ Write 5 as $\frac{10}{2}$.

$\qquad\quad = \log 10 - \log 2$ Use the quotient rule of logarithms.

$\qquad\quad = 1 - \log 2$ $\log 10 = 1$, because $10^1 = 1$.

$\qquad\quad \approx 1 - 0.3010$ Substitute the value of the base-10 logarithm of 2.

$\qquad\quad \approx 0.6990$

d. $\log \sqrt{5} = \log (5^{1/2})$ Write $\sqrt{5}$ as $5^{1/2}$.

$\qquad\quad = \dfrac{1}{2} \log 5$ Use the power rule of logarithms.

$\qquad\quad \approx \dfrac{1}{2}(0.6990)$ Use the answer from part **c.**

$\qquad\quad \approx 0.3495$

e. $\log 1.5 = \log \left(\dfrac{3}{2} \right)$ Write 1.5 as $\frac{3}{2}$.

$\qquad\quad = \log 3 - \log 2$ Use the quotient rule of logarithms.

$\qquad\quad \approx 0.4771 - 0.3010$ Substitute the values of the base-10 logarithm of 2 and the base-10 logarithm of 3.

$\qquad\quad \approx 0.1761$ ■

■ THE CHANGE-OF-BASE FORMULA

If we know the base-a logarithm of a number, we can find its logarithm to some other base b by using the **change-of-base formula.**

Change-of-Base Formula
If a, b, and x are positive numbers, $a \neq 1$, and $b \neq 1$, then
$$\log_b x = \frac{\log_a x}{\log_a b}$$

Proof   We begin with the equation $\log_b x = y$ and proceed as follows:

1. $\log_b x = y$

$\qquad\quad b^y = x$ Change the equation into logarithmic form.

$\qquad\quad \log_a b^y = \log_a x$ Take the base-a logarithm of both sides.

$\qquad\quad y \log_a b = \log_a x$ Use the power rule of logarithms.

$\qquad\quad y = \dfrac{\log_a x}{\log_a b}$ Divide both sides by $\log_a b$.

$\qquad\quad \log_b x = \dfrac{\log_a x}{\log_a b}$ Refer to Equation 1 and substitute $\log_b x$ for y.   □

If we know logarithms to some base a (for example, $a = 10$), we can find the logarithm of x to a new base b. *We simply divide the base-a logarithm of x by the base-a logarithm of b.*

EXAMPLE 6   Use the change-of-base formula to find $\log_3 5$.

Solution   We substitute 3 for b, 5 for x, and 10 for a (because we can then use a calculator) into the change-of-base formula:

$$\log_b x = \frac{\log_a x}{\log_a b}$$

$$\log_3 5 = \frac{\log_{10} 5}{\log_{10} 3}$$

$$\approx 1.464973521$$

To four decimal places, $\log_3 5 = 1.4650$.   ■

Orals   *Find the value of* x.

1. $\log_3 9 = x$ **2.** $\log_x 5 = 1$ **3.** $\log_7 x = 7$ **4.** $\log_2 x = -2$

5. $\log_4 x = \dfrac{1}{2}$ **6.** $\log_x 4 = 2$ **7.** $\log_3 \dfrac{1}{9} = x$ **8.** $\log_{\sqrt{3}} x = 2$

EXERCISE 9.3

In Exercises 1–8, simplify each expression.

1. $\log_7 1$ **2.** $\log_5 5$ **3.** $\log_3 3^7$ **4.** $5^{\log_5 8}$

5. $8^{\log_8 10}$ **6.** $\log_4 4^2$ **7.** $\log_9 9$ **8.** $\log_3 1$

In Exercises 9–14, use a calculator to verify each equation.

9. $\log\,[(2.5)(3.7)] = \log 2.5 + \log 3.7$

10. $\ln \dfrac{11.3}{6.1} = \ln 11.3 - \ln 6.1$

11. $\ln\,(2.25)^4 = 4 \ln 2.25$

12. $\log 45.37 = \dfrac{\ln 45.37}{\ln 10}$

13. $\log \sqrt{24.3} = \dfrac{1}{2} \log 24.3$

14. $\ln 8.75 = \dfrac{\log 8.75}{\log e}$

In Exercises 15–26, assume that x, y, z, *and* b ($b \neq 1$) *are positive numbers. Use the properties of logarithms to write each expression in terms of the logarithms of* x, y, *and* z.

15. $\log_b xyz$

16. $\log_b 4xz$

17. $\log_b \dfrac{2x}{y}$

18. $\log_b \dfrac{x}{yz}$

19. $\log_b x^3 y^2$

20. $\log_b xy^2 z^3$

21. $\log_b (xy)^{1/2}$

22. $\log_b x^3 y^{1/2}$

23. $\log_b x\sqrt{z}$

24. $\log_b \sqrt{xy}$

25. $\log_b \dfrac{\sqrt[3]{x}}{\sqrt[4]{yz}}$

26. $\log_b \sqrt[4]{\dfrac{x^3 y^2}{z^4}}$

In Exercises 27–34, assume that x, y, z, and b (b ≠ 1) are positive numbers. Use the properties of logarithms to write each expression as the logarithm of a single quantity.

27. $\log_b (x + 1) - \log_b x$

28. $\log_b x + \log_b (x + 2) - \log_b 8$

29. $2 \log_b x + \dfrac{1}{2} \log_b y$

30. $-2 \log_b x - 3 \log_b y + \log_b z$

31. $-3 \log_b x - 2 \log_b y + \dfrac{1}{2} \log_b z$

32. $3 \log_b (x + 1) - 2 \log_b (x + 2) + \log_b x$

33. $\log_b \left(\dfrac{x}{z} + x \right) - \log_b \left(\dfrac{y}{z} + y \right)$

34. $\log_b (xy + y^2) - \log_b (xz + yz) + \log_b z$

In Exercises 35–54, tell whether the given statement is true. If a statement is false, explain why.

35. $\log_b 0 = 1$

36. $\log_b (x + y) \neq \log_b x + \log_b y$

37. $\log_b xy = (\log_b x)(\log_b y)$

38. $\log_b ab = \log_b a + 1$

39. $\log_7 7^7 = 7$

40. $7^{\log_7 7} = 7$

41. $\dfrac{\log_b A}{\log_b B} = \log_b A - \log_b B$

42. $\log_b (A - B) = \dfrac{\log_b A}{\log_b B}$

43. $3 \log_b \sqrt[3]{a} = \log_b a$

44. $\dfrac{1}{3} \log_b a^3 = \log_b a$

45. $\log_b \dfrac{1}{a} = -\log_b a$

46. $\log_b 2 = \log_2 b$

47. If $\log_b b = c$, then $\log_b a = c$.

48. If $\log_a b = c$, then $\log_b a = \dfrac{1}{c}$.

49. $\log_b (-x) = -\log_b x$

50. If $\log_b a = c$, then $\log_b a^p = pc$.

51. $\log_b \dfrac{1}{5} = -\log_b 5$

52. $\log_{4/3} y = -\log_{3/4} y$

53. $\log_b y + \log_{1/b} y = 0$

54. $\log_{10} 10^3 = 3(10^{\log_{10} 3})$

In Exercises 55–66, assume that log 4 = 0.6021, log 7 = 0.8451, and log 9 = 0.9542. Use these values and the properties of logarithms to find each value.

55. $\log 28$

56. $\log \dfrac{7}{4}$

57. $\log 2.25$

58. $\log 36$

59. $\log \dfrac{63}{4}$

60. $\log \dfrac{4}{63}$

61. $\log 252$

62. $\log 49$

63. $\log 112$

64. $\log 324$

65. $\log \dfrac{144}{49}$

66. $\log \dfrac{324}{63}$

In Exercises 67–74, use the change-of-base formula to find each logarithm. Give each answer to four decimal places.

67. $\log_3 7$

68. $\log_7 3$

69. $\log_{1/3} 3$

70. $\log_{1/2} 6$

71. $\log_3 8$

72. $\log_5 10$

73. $\log_{\sqrt{2}} \sqrt{5}$

74. $\log_\pi e$

Writing Exercises   *Write a paragraph using your own words.*

1. Explain why ln(log 0.9) is undefined.

2. Explain why $\log_b (\ln 1)$ is undefined.

Something to Think About

1. Show that $\ln(e^x) = x$.

2. If $\log_b 3x = 1 + \log_b x$, find b.

3. Show that $\log_{b^2} x = \frac{1}{2} \log_b x$.

4. Show that $e^{x \ln a} = a^x$.

Review Exercises   *Consider the line that passes through $P(-2, 3)$ and $Q(4, -4)$.*

1. Find the slope of line PQ.

2. Find the distance PQ.

3. Find the midpoint of segment PQ.

4. Write the equation of line PQ.

9.4   Applications of Logarithms

■ APPLICATIONS OF BASE-10 LOGARITHMS ■ APPLICATIONS OF BASE-e LOGARITHMS

In this section, we will consider many applications of logarithmic functions.

■ APPLICATIONS OF BASE-10 LOGARITHMS

In chemistry, common logarithms are used to express the acidity of solutions. The more acidic a solution, the greater the concentration of hydrogen ions. This concentration is indicated indirectly by the *pH scale*, or *hydrogen ion index*. The pH of a solution is defined by the following equation.

> **pH of a Solution**
>
> If [H$^+$] is the hydrogen ion concentration in gram-ions per liter, then
>
> $$pH = -\log [H^+]$$

EXAMPLE 1

Finding pH   Find the pH of pure water, which has a hydrogen ion concentration of 10^{-7} gram-ions per liter.

Solution   Since pure water has approximately 10^{-7} gram-ions per liter, its pH is

$$pH = -\log 10^{-7}$$
$$= -(-7) \log 10 \qquad \text{Use the power rule of logarithms.}$$
$$= -(-7) \cdot 1 \qquad \quad \log 10 = 1.$$
$$= 7$$

The pH of pure water is 7.

EXAMPLE 2   **Finding hydrogen ion concentration**   Find the hydrogen ion concentration of sea-water if its pH is 8.5.

Solution   To find its hydrogen ion concentration, we solve the following equation for $[H^+]$.

$$8.5 = -\log [H^+]$$
$$-8.5 = \log [H^+] \qquad \text{Multiply both sides by } -1.$$
$$[H^+] = 10^{-8.5} \qquad \text{Change the equation to exponential form.}$$

We can use a calculator to find that

$$[H^+] \approx 3.2 \times 10^{-9} \text{ gram-ions per liter} \qquad\blacksquare$$

Common logarithms are used in electrical engineering to express the voltage gain (or loss) of an electronic device such as an amplifier. The unit of gain (or loss), called the *decibel*, is defined by a logarithmic relation.

> **Decibel Voltage Gain**
> If E_O is the output voltage of a device and E_I is the input voltage, the decibel voltage gain is given by
> $$\text{db gain} = 20 \log \frac{E_O}{E_I}$$

EXAMPLE 3   **Finding db gain**   If the input to an amplifier is 0.5 volt and the output is 40 volts, find the decibel voltage gain of the amplifier.

Solution   We can find the decibel voltage gain by substituting 0.5 for E_I and 40 for E_O in the formula for db gain:

$$\text{db gain} = 20 \log \frac{E_O}{E_I}$$
$$\text{db gain} = 20 \log \frac{40}{0.5}$$
$$= 20 \log 80$$
$$\approx 38 \qquad \text{Use a calculator.}$$

The amplifier provides a 38-decibel voltage gain. $\blacksquare$

In seismology, common logarithms are used to measure the intensity of earthquakes on the *Richter scale*. The intensity of an earthquake is given by the following logarithmic function.

> **Richter Scale**
> If R is the intensity of an earthquake, A is the amplitude (measured in micrometers), and P is the period (the time of one oscillation of the earth's surface measured in seconds), then
> $$R = \log \frac{A}{P}$$

EXAMPLE 4   **Measuring an earthquake**   Find the measure on the Richter scale of an earthquake with an amplitude of 10,000 micrometers (1 centimeter) and a period of 0.1 second.

Solution   We substitute 10,000 for A and 0.1 for P in the Richter scale formula and simplify:

$$R = \log \frac{A}{P}$$

$$R = \log \frac{10,000}{0.1}$$

$$= \log 100,000$$

$$= \log 10^5$$

$$= 5 \log 10 \qquad \text{Use power rule of logarithms.}$$

$$= 5 \qquad \log 10 = 1.$$

The earthquake measures 5 on the Richter scale.   ■

■ APPLICATIONS OF BASE-*e* LOGARITHMS

If a population grows exponentially at a certain annual rate, the time required for the population to double is called the *doubling time*. It is given by the following formula.

> **Formula for Doubling Time**
> If r is the annual rate and t is the time required for a population to double, then
>
> $$t = \frac{\ln 2}{r}$$

You will be asked to prove the formula for doubling time in the exercises.

EXAMPLE 5   **Doubling time**   The population of the earth is growing at the rate of approximately 2% per year. If this rate continues, how long will it take the population to double?

Solution   Because the population is growing at the rate of 2% per year, we substitute 0.02 for r in the formula for doubling time and simplify.

$$t = \frac{\ln 2}{r}$$

$$t = \frac{\ln 2}{0.02}$$

$$\approx 34.65735903$$

At the current rate of growth, the population of the earth will double in about $34\frac{1}{2}$ years.   ■

The formula for doubling time can also be used to find the length of time it will take money to double when interest is compounded continuously.

EXAMPLE 6

Doubling time   How long will it take $1000 to double if it is invested at an annual rate of 8%, compounded continuously?

Solution   We substitute 0.08 for r and simplify:

$$t = \frac{\ln 2}{r}$$

$$t = \frac{\ln 2}{0.08}$$

$$\approx 8.664339757 \qquad \text{Use a calculator.}$$

It will take approximately $8\frac{2}{3}$ years for the money to double.   ∎

In physiology, experiments suggest that the relationship between the loudness and the intensity of sound is a logarithmic one, known as the *Weber–Fechner law*.

Weber–Fechner Law
If L is the apparent loudness of a sound, I is the actual intensity, and k is a constant, then

$$L = k \ln I$$

EXAMPLE 7

Weber–Fechner law   Find the increase in intensity that will cause the apparent loudness of a sound to double.

Solution   If the original loudness L_O is caused by an actual intensity I_O, then

1. $L_O = k \ln I_O$

To double the apparent loudness, we multiply both sides of Equation 1 by 2 and use the power rule of logarithms:

$$2L_O = 2k \ln I_O$$
$$= k \ln (I_O)^2$$

To double the apparent loudness of a sound, the intensity must be squared.   ∎

Orals   *Give the following formulas.*

1. pH

2. db gain

3. Richter scale

4. Doubling time

5. Weber–Fechner law

EXERCISE 9.4

In Exercises 1–26, solve each problem.

1. pH of a solution   Find the pH of a solution with a hydrogen ion concentration of 1.7×10^{-5} gram-ions per liter.

2. Hydrogen ion concentration   Find the hydrogen ion concentration of a saturated solution of calcium hydroxide whose pH is 13.2.

3. Aquariums   To test for safe pH levels in a fresh-water aquarium, a test strip is compared with the scale shown in Illustration 1. Find the corresponding safe range in the hydrogen ion concentration.

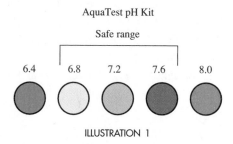

ILLUSTRATION 1

4. pH of sour pickles   The hydrogen ion concentration of sour pickles is 6.31×10^{-4}. Find the pH.

5. Finding input voltage   The db gain of an amplifier is 29. Find the input voltage when the output voltage is 20 volts.

6. Finding output voltage   The db gain of an amplifier is 35. Find the output voltage when the input voltage is 0.05 volt.

7. db gain of an amplifier   Find the db gain of the amplifier shown in Illustration 2.

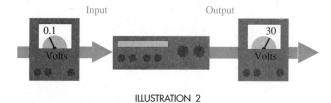

ILLUSTRATION 2

8. db gain of an amplifier   An amplifier produces an output of 80 volts when driven by an input of 0.12 volts. Find the amplifier's db gain.

9. Earthquakes   An earthquake has an amplitude of 5000 micrometers and a period of 0.2 second. Find its measure on the Richter scale.

10. Earthquakes   Find the period of an earthquake with an amplitude of 80,000 micrometers that measures 6 on the Richter scale.

11. Earthquakes   An earthquake with a period of $\frac{1}{4}$ second measures 4 on the Richter scale. Find its amplitude.

12. Earthquakes   By what factor must the amplitude of an earthquake change to increase its severity by 1 point on the Richter scale? Assume that the period remains constant.

13. Population growth   How long will it take the population of River City to double? (See Illustration 3.)

River City
A growing community

• 6 parks   • 12% annual growth
• 10 churches   • low crime rate

ILLUSTRATION 3

14. Doubling money   How long will it take $1000 to double if it is invested at an annual rate of 5%, compounded continuously?

15. Population growth   A population growing at an annual rate r will triple in a time t given by the formula

$$t = \frac{\ln 3}{r}$$

How long will it take the population of a town growing at the rate of 12% per year to triple?

16. Tripling money   Find the length of time needed for $25,000 to triple if invested at 6% annual interest, compounded continuously.

17. Change in loudness   If the intensity of a sound is doubled, find the apparent change in loudness.

18. Change in intensity   If the intensity of a sound is tripled, find the apparent change in loudness.

19. Change in loudness   What change in intensity of sound will cause an apparent tripling of the loudness?

20. Change in intensity   What increase in the intensity of a sound will cause the apparent loudness to be multiplied by 4?

21. Depreciation   In business, equipment is often depreciated using the double declining-balance method. In this method, a piece of equipment with a life expectancy of N years, costing $\$C$, will depreciate to a value of $\$V$ in n years, where n is given by the formula

$$n = \frac{\log V - \log C}{\log\left(1 - \dfrac{2}{N}\right)}$$

A computer that cost $\$37,000$ has a life expectancy of 5 years. If it has depreciated to a value of $\$8000$, how old is it?

22. Depreciation   A typewriter worth $\$470$ when new had a life expectancy of 12 years. If it is now worth $\$189$, how old is it? (See Exercise 21.)

23. Time for money to grow   If $\$P$ is invested at the end of each year in an annuity earning annual interest at rate r, then the amount in the account will be $\$A$ after n years, where

$$n = \frac{\log\left[\dfrac{Ar}{P} + 1\right]}{\log(1 + r)}$$

If $\$1000$ is invested each year in an annuity earning 12% annual interest, how long will it take for the account to be worth $\$20,000$?

24. Time for money to grow tenfold   If $\$5000$ is invested each year in an annuity earning 8% annual interest, how long will it take for the account to be worth $\$50,000$? (See Exercise 23.)

25. Use the formula $P = P_0 e^{rt}$ to verify that P will be twice P_0 when $t = \frac{\ln 2}{r}$.

26. Use the formula $P = P_0 e^{rt}$ to verify that P will be three times as large as P_0 when $t = \frac{\ln 3}{r}$.

Writing Exercises   *Write a paragraph using your own words.*

1. Explain why an earthquake measuring 7 on the Richter scale is much worse than an earthquake measuring 6.

2. The time it takes money to double at an annual rate r, compounded continuously, is given by the formula $(\ln 2)/r$. Explain why money doubles more quickly the higher the rate.

Something to Think About   **1.** Find a formula to find how long it will take money to quadruple.

2. Graph the *logistic function*

$$y = \frac{1}{1 + e^{-2x}}$$ and discuss its graph.

Review Exercises   *Let $f(x) = 3x - 2$ and $g(x) = x^2 + 3$. Find each function or value.*

1. $f + g$

2. $f - g$

3. $f \cdot g$

4. g/f

5. $(g \circ f)(-2)$

6. $(f \circ g)(x)$

9.5 Exponential and Logarithmic Equations

■ SOLVING EXPONENTIAL EQUATIONS ■ SOLVING EXPONENTIAL EQUATIONS WITH A GRAPHING DEVICE ■ SOLVING LOGARITHMIC EQUATIONS ■ SOLVING LOGARITHMIC EQUATIONS WITH A GRAPHING DEVICE ■ APPLICATIONS OF EXPONENTIAL AND LOGARITHMIC EQUATIONS

An **exponential equation** is an equation that contains a variable in one of its exponents. Some examples of exponential equations are

$$3^x = 5, \qquad 6^{x-3} = 2^x, \qquad \text{and} \qquad 3^{2x+1} - 10(3^x) + 3 = 0$$

A **logarithmic equation** is an equation with a logarithmic expression that contains a variable. Some examples of logarithmic equations are

$$\log 2x = 25, \qquad \ln x - \ln (x - 12) = 24, \qquad \text{and} \qquad \log x = \log \frac{1}{x} + 4$$

In this section, we will learn how to solve many of these equations.

■ SOLVING EXPONENTIAL EQUATIONS

EXAMPLE 1   Solve the exponential equation $3^x = 5$.

Solution   Since logarithms of equal numbers are equal, we can take the common logarithm of each side of the equation. The power rule of logarithms then provides a way of moving the variable x from its position as an exponent to a position as a coefficient.

$$3^x = 5$$
$$\log 3^x = \log 5 \qquad \text{Take the common logarithm of each side.}$$
$$x \log 3 = \log 5 \qquad \text{Use the power rule of logarithms.}$$
1. $$\qquad x = \frac{\log 5}{\log 3} \qquad \text{Divide both sides by } \log 3.$$
$$\approx 1.464973521 \qquad \text{Use a calculator.}$$

To four decimal places, $x = 1.4650$. ■

  WARNING!   A careless reading of Equation 1 leads to a common error. The right-hand side of Equation 1 calls for a division, not a subtraction.

$$\frac{\log 5}{\log 3} \quad \text{means} \quad (\log 5) \div (\log 3)$$

It is the expression $\log \frac{5}{3}$ that means $\log 5 - \log 3$.

EXAMPLE 2   Solve the exponential equation $6^{x-3} = 2^x$.

Solution

$$6^{x-3} = 2^x$$

$\log 6^{x-3} = \log 2^x$ Take the common logarithm of each side.

$(x - 3) \log 6 = x \log 2$ Use the power rule of logarithms.

$x \log 6 - 3 \log 6 = x \log 2$ Use the distributive property.

$x \log 6 - x \log 2 = 3 \log 6$ Add 3 log 6 and subtract x log 2 from both sides.

$x(\log 6 - \log 2) = 3 \log 6$ Factor out x on the left-hand side.

$$x = \frac{3 \log 6}{\log 6 - \log 2}$$ Divide both sides by log 6 − log 2.

$x \approx 4.892789261$ Use a calculator.   ■

EXAMPLE 3   Solve the exponential equation $2^{x^2+2x} = \dfrac{1}{2}$.

Solution   Since $\frac{1}{2} = 2^{-1}$, we can write the equation in the form

$$2^{x^2+2x} = 2^{-1}$$

Since equal quantities with equal bases have equal exponents, we have

$x^2 + 2x = -1$

$x^2 + 2x + 1 = 0$ Add 1 to both sides.

$(x + 1)(x + 1) = 0$ Factor the trinomial.

$x + 1 = 0$ or $x + 1 = 0$ Set each factor equal to 0.

$x = -1$ | $x = -1$

Verify that -1 satisfies the equation.   ■

■ SOLVING EXPONENTIAL EQUATIONS WITH A GRAPHING DEVICE

EXAMPLE 4   Use a graphing device to approximate the solutions of $2^{x^2+2x} = \dfrac{1}{2}$. (See Example 3.)

Solution   We can subtract $\frac{1}{2}$ from both sides of the equation to get

$$2^{x^2+2x} - \frac{1}{2} = 0$$

and graph the corresponding function

$$y = 2^{x^2+2x} - \frac{1}{2}$$

using window settings of $[-4, 4]$ for x and $[-2, 6]$ for y, obtaining the graph shown in Figure 9-11(a).

Since the solutions of the equation are its x-intercepts, we can approximate the solutions by zooming in on the values of the x-intercepts, as in Figure 9-11(b). Since $x = -1$ is the only x-intercept, -1 is the only solution. In this case, we have found an exact solution.

We could also solve this equation by graphing

$$y = 2^{x^2 + 2x} \quad \text{and} \quad y = \frac{1}{2}$$

and finding the coordinates of the intersection point.

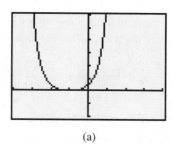
(a)

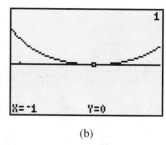
(b)

FIGURE 9-11

■ SOLVING LOGARITHMIC EQUATIONS

EXAMPLE 5   Solve the logarithmic equation $\log x + \log (x - 3) = 1$.

Solution

$\log x + \log (x - 3) = 1$	
$\log x(x - 3) = 1$	Use the product rule of logarithms.
$x(x - 3) = 10^1$	Use the definition of logarithms to change the equation to exponential form.
$x^2 - 3x - 10 = 0$	Remove parentheses and subtract 10 from both sides.
$(x + 2)(x - 5) = 0$	Factor the trinomial.
$x + 2 = 0 \quad \text{or} \quad x - 5 = 0$	Set each factor equal to 0.
$x = -2 \quad | \quad x = 5$	

Check: The number -2 is not a solution, because when we substitute -2 for x, it results in the logarithm of a negative number. We will check the remaining number, 5.

$\log x + \log (x - 3) = 1$	
$\log \mathbf{5} + \log (\mathbf{5} - 3) \stackrel{?}{=} 1$	Substitute 5 for x.
$\log 5 + \log (2) \stackrel{?}{=} 1$	
$\log 10 \stackrel{?}{=} 1$	Use the product rule of logarithms.
$1 = 1$	$\log 10 = 1$.

Since 5 satisfies the equation, it is a solution.

EXAMPLE 6 Solve the logarithmic equation $\log_b (3x + 2) - \log_b (2x - 3) = 0$.

Solution

$$\log_b (3x + 2) - \log_b (2x - 3) = 0$$

$$\log_b (3x + 2) = \log_b (2x - 3) \qquad \text{Add } \log_b (2x - 3) \text{ to both sides.}$$

$$3x + 2 = 2x - 3 \qquad \text{If } \log_b r = \log_b s, \text{ then } r = s.$$

$$x = -5 \qquad \text{Subtract } 2x \text{ and } 2 \text{ from both sides.}$$

Check:
$$\log_b (3x + 2) - \log_b (2x - 3) = 0$$
$$\log_b [3(-5) + 2] - \log_b [2(-5) - 3] \stackrel{?}{=} 0$$
$$\log_b (-13) - \log_b (-13) \stackrel{?}{=} 0$$

Since the logarithm of a negative number does not exist, the apparent solution of -5 must be discarded. This equation has no solutions. ■

EXAMPLE 7 Solve the logarithmic equation $\dfrac{\log (5x - 6)}{\log x} = 2 \ (x \neq 1)$.

Solution We can multiply both sides of the equation by $\log x$ to get

$$\log (5x - 6) = 2 \log x$$

and apply the power rule of logarithms to get

$$\log (5x - 6) = \log x^2$$

By Property 8 of logarithms, $5x - 6 = x^2$, because they have equal logarithms. Thus,

$$5x - 6 = x^2$$

$$0 = x^2 - 5x + 6 \qquad \text{Add 6 and subtract } 5x \text{ from both sides.}$$

$$0 = (x - 3)(x - 2) \qquad \text{Factor } x^2 - 5x + 6.$$

$$x - 3 = 0 \quad \text{or} \quad x - 2 = 0 \qquad \text{Set each factor equal to 0.}$$

$$x = 3 \quad | \quad x = 2$$

Verify that both 2 and 3 satisfy the equation. ■

■ SOLVING LOGARITHMIC EQUATIONS WITH A GRAPHING DEVICE

EXAMPLE 8 Use a graphing device to approximate the solutions of $\log x + \log (x - 3) = 1$. (See Example 5.)

Solution We can subtract 1 from both sides of the equation to get

$$\log x + \log (x - 3) - 1 = 0$$

and graph the corresponding function

$$y = \log x + \log (x - 3) - 1$$

using window settings of $[0, 20]$ for x and $[-2, 2]$ for y, obtaining the graph shown in Figure 9-12. Since the solution of the equation is the x-intercept, we can find the solution by zooming in on the value of the x-intercept. The solution is $x = 5$.

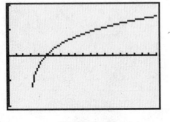

FIGURE 9-12

We could also solve this equation by graphing

$$y = \log x + \log (x - 3) \qquad \text{and} \qquad y = 1$$

and finding the coordinates of the intersection point. ■

■ APPLICATIONS OF EXPONENTIAL AND LOGARITHMIC EQUATIONS

Experiments have determined the time it takes for half of a sample of a given radioactive material to decompose. This time is a constant, called the material's **half-life**.

When living organisms die, the oxygen/carbon dioxide cycle common to all living things ceases and carbon-14, a radioactive isotope with a half-life of approximately 5700 years, is no longer absorbed. By measuring the amount of carbon-14 present in an ancient object, archaeologists can estimate the object's age by using the radioactive decay formula.

> **Radioactive Decay Formula**
> If A is the amount of radioactive material present at time t, A_0 was the amount present at $t = 0$, and h is the material's half-life, then
> $$A = A_0 2^{-t/h}$$

EXAMPLE 9   **Carbon-14 dating**   How old is a wooden statue that retains only one-third of its original carbon-14 content?

Solution   To find the time t when $A = \frac{1}{3}A_0$, we substitute $\dfrac{A_0}{3}$ for A and 5700 for h into the radioactive decay formula and solve for t:

$$A = A_0 2^{-t/h}$$

$$\frac{A_0}{3} = A_0 2^{-t/5700}$$

$$1 = 3(2^{-t/5700}) \qquad \text{Divide both sides by } A_0 \text{ and multiply both sides by 3.}$$

$$\log 1 = \log 3(2^{-t/5700}) \qquad \text{Take the common logarithm of each side.}$$

$$0 = \log 3 + \log 2^{-t/5700} \qquad \log 1 = 0, \text{ and use the product rule of logarithms.}$$

$$-\log 3 = -\frac{t}{5700} \log 2 \qquad \text{Subtract } \log 3 \text{ from both sides and use the power rule of logarithms.}$$

$$5700\left(\frac{\log 3}{\log 2}\right) = t \qquad \text{Multiply both sides by } -\frac{5700}{\log 2}.$$

$$t \approx 9034.286254 \qquad \text{Use a calculator.}$$

The wooden statue is approximately 9000 years old. ■

When there is sufficient food and space, populations of living organisms tend to increase exponentially according to the Malthusian growth model.

Malthusian Growth Model
If P is the population at some time t, P_0 is the initial population at $t = 0$, and k depends on the rate of growth, then

$$P = P_0 e^{kt}$$

EXAMPLE 10   **Population growth**   The bacteria in a laboratory culture increased from an initial population of 500 to 1500 in 3 hours. How long will it take for the population to reach 10,000?

Solution   We substitute 500 for P_0, 1500 for P, and 3 for t and simplify to find k:

$$P = P_0 e^{kt}$$
$$1500 = 500(e^{k3}) \qquad \text{Substitute 1500 for } P, \text{ 500 for } P_0, \text{ and 3 for } t.$$
$$3 = e^{3k} \qquad \text{Divide both sides by 500.}$$
$$3k = \ln 3 \qquad \text{Change the equation from exponential to logarithmic form.}$$
$$k = \frac{\ln 3}{3} \qquad \text{Divide both sides by 3.}$$

To find when the population will reach 10,000, we substitute 10,000 for P, 500 for P_0, and $\frac{\ln 3}{3}$ for k in the equation $P = P_0 e^{kt}$ and solve for t:

$$P = P_0 e^{kt}$$
$$10{,}000 = 500 e^{\left[(\ln 3)/3 \right] t}$$
$$20 = e^{\left[(\ln 3)/3 \right] t} \qquad \text{Divide both sides by 500.}$$
$$\left(\frac{\ln 3}{3} \right) t = \ln 20 \qquad \text{Change the equation to logarithmic form.}$$
$$t = \frac{3 \ln 20}{\ln 3} \qquad \text{Multiply both sides by } \frac{3}{\ln 3}.$$
$$\approx 8.180499084 \qquad \text{Use a calculator.}$$

The culture will reach 10,000 bacteria in a little more than 8 hours. ∎

EXAMPLE 11   **Generation time**   If a medium is inoculated with a bacterial culture that contains 1000 cells per milliliter, how many generations will pass by the time the culture has grown to a population of 1 million cells per milliliter?

Solution   During bacterial reproduction, the time required for a population to double is called the *generation time*. If b bacteria are introduced into a medium, then after the generation time of the organism has elapsed, there are $2b$ cells. After another generation, there are $2(2b)$, or $4b$ cells, and so on. After n generations, the number of cells present will be

1. $B = b \cdot 2^n$

To find the number of generations that have passed while the population grows from b bacteria to B bacteria, we solve Equation 1 for n.

$$\log B = \log (b \cdot 2^n) \qquad \text{Take the common logarithm of both sides.}$$

$$\log B = \log b + n \log 2 \qquad \text{Apply the product and power rules of logarithms.}$$

$$\log B - \log b = n \log 2 \qquad \text{Subtract } \log b \text{ from both sides.}$$

$$n = \frac{1}{\log 2}(\log B - \log b) \qquad \text{Multiply both sides by } \tfrac{1}{\log 2}.$$

2. $$n = \frac{1}{\log 2}\left(\log \frac{B}{b}\right) \qquad \text{Use the quotient rule of logarithms.}$$

Equation 2 is a formula that gives the number of generations that will pass as the population grows from b bacteria to B bacteria.

To find the number of generations that have passed while a population of 1000 cells per milliliter has grown to a population of 1 million cells per milliliter, we substitute 1000 for b and 1,000,000 for B in Equation 2 and solve for n.

$$n = \frac{1}{\log 2}\log \frac{1,000,000}{1000}$$

$$= \frac{1}{\log 2}\log 1000 \qquad \text{Simplify.}$$

$$= 3.321928095(3) \qquad \tfrac{1}{\log 2} \approx 3.321928095 \text{ and } \log 1000 = 3.$$

$$= 9.965784285$$

Approximately 10 generations will have passed. ∎

Orals *Solve each equation for x. Do not simplify answers.*

1. $3^x = 5$

2. $5^x = 3$

3. $2^{-x} = 7$

4. $6^{-x} = 1$

5. $\log 2x = \log (x + 2)$

6. $\log 2x = 0$

7. $\log x^4 = 4$

8. $\log \sqrt{x} = \dfrac{1}{2}$

EXERCISE 9.5

In Exercises 1–20, solve each exponential equation. If an answer is not exact, give the answer to four decimal places.

1. $4^x = 5$

2. $7^x = 12$

3. $13^{x-1} = 2$

4. $5^{x+1} = 3$

5. $2^{x+1} = 3^x$

6. $5^{x-3} = 3^{2x}$

7. $2^x = 3^x$

8. $3^{2x} = 4^x$

9. $7^{x^2} = 10$

10. $8^{x^2} = 11$

11. $8^{x^2} = 9^x$

12. $5^{x^2} = 2^{5x}$

13. $2^{x^2-2x} = 8$

14. $3^{x^2-3x} = 81$

15. $3^{x^2+4x} = \dfrac{1}{81}$

16. $7^{x^2+3x} = \dfrac{1}{49}$

17. $4^{x+2} - 4^x = 15$ (*Hint:* $4^{x+2} = 4^x 4^2$.)

18. $3^{x+3} + 3^x = 84$ (*Hint:* $3^{x+3} = 3^x 3^3$.)

19. $2(3^x) = 6^{2x}$

20. $2(3^{x+1}) = 3(2^{x-1})$

In Exercises 21–24, use a graphing device to solve each equation. Give all answers to the nearest tenth.

21. $2^{x+1} = 7$

22. $3^{x-1} = 2^x$

23. $4(2^{x^2}) = 8^{3x}$

24. $3^x - 10 = 3^{-x}$

In Exercises 25–54, solve each logarithmic equation. Check all solutions.

25. $\log 2x = \log 4$

26. $\log 3x = \log 9$

27. $\log (3x + 1) = \log (x + 7)$

28. $\log (x^2 + 4x) = \log (x^2 + 16)$

29. $\log (3 - 2x) - \log (x + 24) = 0$

30. $\log (3x + 5) - \log (2x + 6) = 0$

31. $\log \dfrac{4x + 1}{2x + 9} = 0$

32. $\log \dfrac{2 - 5x}{2(x + 8)} = 0$

33. $\log x^2 = 2$

34. $\log x^3 = 3$

35. $\log x + \log (x - 48) = 2$

36. $\log x + \log (x + 9) = 1$

37. $\log x + \log (x - 15) = 2$

38. $\log x + \log (x + 21) = 2$

39. $\log (x + 90) = 3 - \log x$

40. $\log (x - 90) = 3 - \log x$

41. $\log (x - 6) - \log (x - 2) = \log \dfrac{5}{x}$

42. $\log (3 - 2x) - \log (x + 9) = 0$

43. $\log x^2 = (\log x)^2$

44. $\log (\log x) = 1$

45. $\dfrac{\log (3x - 4)}{\log x} = 2$

46. $\dfrac{\log (8x - 7)}{\log x} = 2$

47. $\dfrac{\log (5x + 6)}{2} = \log x$

48. $\dfrac{1}{2} \log (4x + 5) = \log x$

49. $\log_3 x = \log_3 \left(\dfrac{1}{x} \right) + 4$

50. $\log_5 (7 + x) + \log_5 (8 - x) - \log_5 2 = 2$

51. $2 \log_2 x = 3 + \log_2 (x - 2)$

52. $2 \log_3 x - \log_3 (x - 4) = 2 + \log_3 2$

53. $\log (7y + 1) = 2 \log (y + 3) - \log 2$

54. $2 \log (y + 2) = \log (y + 2) - \log 12$

In Exercises 55–58, use a graphing device to solve each equation. If an answer is not exact, give all answers to the nearest tenth.

55. $\log x + \log (x - 15) = 2$

56. $\log x + \log (x + 3) = 1$

57. $\ln (2x + 5) - \ln 3 = \ln (x - 1)$

58. $2 \log (x^2 + 4x) = 1$

In Exercises 59–78, solve each problem.

59. Tritium decay The half-life of tritium is 12.4 years. How long will it take for 25% of a sample of tritium to decompose?

60. Radioactive decay In two years, 20% of a radioactive element decays. Find its half-life.

61. Thorium decay An isotope of thorium, ^{227}Th, has a half-life of 18.4 days. How long will it take for 80% of the sample to decompose?

62. Lead decay An isotope of lead, ^{201}Pb, has a half-life of 8.4 hours. How many hours ago was there 30% more of the substance?

63. Carbon-14 dating The bone fragment shown in Illustration 1 contains 60% of the carbon-14 that it is assumed to have had initially. How old is it?

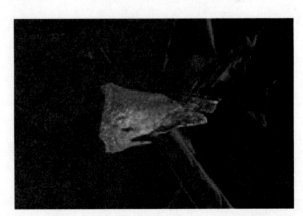

ILLUSTRATION 1
(Archaeological Consulting/Gary Breschini & Trudy Haverstat)

64. Carbon-14 dating Only 10% of the carbon-14 in a small wooden bowl remains. How old is the bowl?

65. Compound interest If $500 is deposited in an account paying 8.5% annual interest, compounded semiannually, how long will it take for the account to increase to $800?

66. Continuous compound interest In Exercise 65, how long will it take if the interest is compounded continuously?

67. Compound interest If $1300 is deposited in a savings account paying 9% interest, compounded quarterly, how long will it take the account to increase to $2100?

68. Compound interest A sum of $5000 deposited in an account grows to $7000 in 5 years. Assuming annual compounding, what interest rate is being paid?

69. Rule of seventy A rule of thumb for finding how long it takes an investment to double is called the **rule of seventy**. To apply the rule, divide 70 by the interest rate written as a percent. At 5%, doubling requires $\frac{70}{5} = 14$ years to double an investment. At 7%, it takes $\frac{70}{7} = 10$ years. Explain why this formula works.

70. Bacterial growth A bacterial culture grows according to the formula

$$P = P_0 a^t$$

If it takes 5 days for the culture to triple in size, how long will it take to double in size?

71. Rodent control The rodent population in a city is currently estimated at 30,000. If it is expected to double every 5 years, when will the population reach 1 million?

72. Population growth The population of a city is expected to triple every 15 years. When can the city planners expect the present population of 140 persons to double?

73. Bacterial culture A bacterial culture doubles in size every 24 hours. By how much will it have increased in 36 hours?

74. Oceanography The intensity I of a light a distance x meters beneath the surface of a lake decreases exponentially. From Illustration 2, find the depth at which the intensity will be 20%.

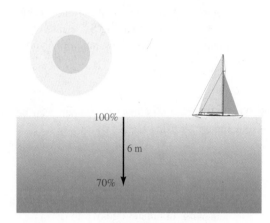

ILLUSTRATION 2

75. Medicine If a medium is inoculated with a bacterial culture containing 500 cells per milliliter, how many generations will have passed by the time the culture contains 5×10^6 cells per milliliter?

76. Medicine If a medium is inoculated with a bacterial culture containing 800 cells per milliliter, how many generations will have passed by the time the culture contains 6×10^7 cells per milliliter?

77. Newton's law of cooling Water initially at 100°C is left to cool in a room at temperature 60°C. After 3 minutes, the water temperature is 90°. If the water temperature T is a function of time t given by $T = 60 + 40e^{kt}$, find k.

78. Newton's law of cooling Refer to Exercise 77 and find the time for the water temperature to reach 70°C.

Writing Exercises *Write a paragraph using your own words.*

1. Explain how to solve the equation $2^x = 7$.

2. Explain how to solve the equation $x^2 = 7$.

Something to Think About

1. Without solving the following equation, find the values of x that cannot be a solution:

$$\log (x - 3) - \log (x^2 + 2) = 0$$

2. Solve the equation $x^{\log x} = 10,000$.

Review Exercises   *Solve each equation.*

1. $5x^2 - 25x = 0$ **2.** $4y^2 - 25 = 0$

3. $3p^2 + 10p = 8$ **4.** $4t^2 + 1 = -6t$

■ ■ ■ ■ ■ ■ ■ ■ ■ ■   **PROBLEMS AND PROJECTS**

1. You receive $1500 interest at the end of the first year from an account paying 8.5% interest, compounded quarterly. How much did you originally deposit?

2. A deer population is declining because of overcrowded feeding grounds. If the decline fits the Malthusian model, how long will it take for a herd of 400 deer to decline to a herd of 150 deer? Assume a decay rate of 4.7% per month.

3. The half-life of a radioactive element can be calculated by using the formula

$$\text{Half-life} = \frac{\ln \frac{1}{2}}{k}$$

where k is the decay rate. Find the half-life of an element with an annual decay rate of $k = -0.12\%$.

4. Find the half-life of strontium, which has a daily decay rate of -24.8%. (See Problem 3.)

PROJECT 1   When an object moves through air, it encounters air resistance. So far, all ballistics problems in this text have ignored air resistance. We now consider the case where an object's fall is affected by air resistance.

At relatively low velocities ($v < 200$ feet per second), the force resisting an object's motion is a constant multiple of the object's velocity:

Resisting force $= f_r = bv$

where b is a constant that depends on the size, shape, and texture of the object, and has units of kilograms per second. This is known as *Stokes' law of resistance*.

In a vacuum, the downward velocity of an object dropped with an initial velocity of 0 feet per second is

$$v(t) = 32t \qquad \text{(no air resistance)}$$

t seconds after is is released. However, with air resistance, the velocity is given by the formula

$$v(t) = \frac{32m}{b}\left(1 - e^{-\frac{b}{m}t}\right)$$

where m is the object's mass (in kilograms). There is also a formula for the distance an object falls (in feet) during the first t seconds after release with air resistance:

$$d(t) = \frac{32m}{b}t - \frac{32m^2}{b^2}\left(1 - e^{-\frac{b}{m}t}\right)$$

Without air resistance, the formula would be

$$d(t) = 16t^2$$

a. Fearless Freda, a renowned skydiving daredevil, performs a practice dive from a hot-air balloon with an altitude of 5000 feet. With her parachute on, Freda has a mass of 75 kg, so that $b = 15$ kg/sec. How far (to the nearest foot) will Freda fall in 5 seconds? Compare this with the answer you get by disregarding air resistance.

b. What downward velocity (to the nearest ft/sec) does Freda have after she has fallen for 2 seconds? For 5 seconds? Compare these answers with the answers you get by disregarding air resistance.

c. Find Freda's downward velocity after falling for 20, 22, and 25 seconds (without air resistance, Freda would hit the ground in less than 18 seconds). Note that Freda's velocity increases only slightly. This is because for a large enough velocity, the air resistance force almost counteracts the force of gravity; after Freda has been falling for a few seconds, her velocity becomes nearly constant. The constant velocity that a falling object approaches is called the *terminal velocity*.

$$\text{Terminal velocity} = \frac{32m}{b}$$

Find Freda's terminal velocity for her practice dive.

d. In Freda's show, she dives from a hot-air balloon with an altitude of only 550 feet, and pulls her ripcord when her velocity is 100 feet per second. (She can't tell her speed, but she knows how long it takes to reach that speed.) It takes a fall of 80 more feet for the chute to open fully, but then the chute increases the air resistance force, making $b = 80$. After that, Freda's velocity approaches the terminal velocity of an object with this new b value.

To the nearest hundredth of a second, how long should Freda fall before she pulls the ripcord? To the nearest foot, how close is she to the ground when she pulls the ripcord? How close to the ground is she when the chute takes full effect? At what velocity will Freda hit the ground?

PROJECT 2   If an object at temperature T_0 is surrounded by a constant temperature T_s (for instance, an oven or a large amount of fluid that has a constant temperature), the temperature of the object will change with time t according to the formula

$$T(t) = T_s + (T_0 - T_s)e^{-kt}$$

This is *Newton's law of cooling and warming*. The number k is a constant that depends on how well the object absorbs and dispels heat.

In the course of brewing "yo ho! grog," the dread pirates of Hancock Isle have learned that it is important that their rather disgusting, soupy mash be heated slowly to allow all of the ingredients a chance to add their particular offensiveness to the mixture. However, after the mixture has simmered for several hours, it is equally important that the grog be cooled very quickly, so that it retains its potency. The kegs of grog are then stored in a cool spring.

(continued)

■ ■ ■ ■ ■ ■ ■ ■ ■ ■ **PROBLEMS AND PROJECTS**   *(continued)*

By trial and error, the pirates have learned that by placing the mash pot into a tub of boiling water (100°C), they can heat the mash in the correct amount of time. They have also learned that they can cool the grog to the temperature of the spring by placing it in ice caves for 1 hour.

With a thermometer, you find that the pirates heat the mash from 20°C to 95°C and then cool the grog from 95°C to 7°C. Calculate how long the pirates cook the mash, and how cold the ice caves are. Assume that $k = 0.5$, and t is measured in hours.

■ Chapter Summary

Key Words

change-of-base formula (9.3)

common logarithm (9.2)

compound interest (9.1)

continuous compound interest (9.1)

decreasing functions (9.1)

domain of an exponential function (9.1)

domain of a logarithmic function (9.2)

e (9.1)

exponential equation (9.5)

exponential function (9.1)

exponential growth (9.1)

half-life (9.5)

increasing function (9.1)

logarithmic equation (9.5)

logarithmic function (9.2)

Malthusian population growth (9.1)

natural logarithm (9.2)

radioactive decay (9.5)

range of an exponential function (9.1)

range of a logarithmic function (9.2)

Key Ideas

(9.1) The exponential function $y = b^x$, where $b > 0$, $b \neq 1$, and x is a real number, is one-to-one. Its domain is the set·of real numbers, and its range is the set of positive numbers.

$e = 2.718281828 \ldots$

The exponential function $y = e^x$ is one-to-one. Its domain is the set of real numbers, and its range is the set of positive numbers.

Formula for compound interest: $A = P\left(1 + \dfrac{r}{k}\right)^{kt}$ or $FV = PV(1 + i)^n$

Formula for continuous compound interest: $A = Pe^{rt}$

(9.2) The logarithmic function $y = \log_b x$, where $b > 0$, $b \neq 1$, and x is a positive number, is one-to-one. Its domain is the set of positive numbers, and its range is the set of real numbers.

The equation $y = \log_b x$ is equivalent to the equation $x = b^y$.

Logarithms of negative numbers do not exist.

The functions defined by $y = \log_b x$ and $y = b^x$ are inverse functions.

Common logarithms are base-10 logarithms.

Natural logarithms are base-e logarithms.

(9.3) Properties of logarithms: If M, N, and b are positive numbers and $b \neq 1$, then

1. $\log_b 1 = 0$

2. $\log_b b = 1$

3. $\log_b b^x = x$

4. $b^{\log_b x} = x$ $(x > 0)$

5. $\log_b MN = \log_b M + \log_b N$

6. $\log_b \dfrac{M}{N} = \log_b M - \log_b N$

7. $\log_b M^p = p \log_b M$

8. If $\log_b x = \log_b y$, then $x = y$.

The change-of-base formula: $\log_b y = \dfrac{\log_a y}{\log_a b}$

(9.4) **pH of a solution**: $\text{pH} = -\log [\text{H}^+]$

db gain: db gain $= 20 \log \dfrac{E_O}{E_I}$

Richter scale: $R = \log \dfrac{A}{P}$

Doubling time: $t = \dfrac{\ln 2}{r}$

Weber–Fechner law: $L = k \ln I$

(9.5) **Radioactive decay formula**: $A = A_0 2^{-t/h}$

Malthusian growth: $P = P_0 e^{kt}$

Chapter 9 Review Exercises

In Review Exercises 1–4, graph each function.

1. $f(x) = \left(\dfrac{6}{5}\right)^x$

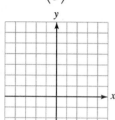

2. $f(x) = e^x$

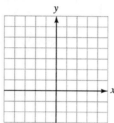

3. $f(x) = \log x$

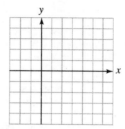

4. $f(x) = \ln x$

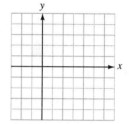

In Review Exercises 5–8, graph each pair of functions on one set of coordinate axes.

5. $f(x) = \left(\dfrac{1}{3}\right)^x$ and
   $g(x) = \log_{1/3} x$

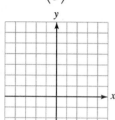

6. $f(x) = \left(\dfrac{2}{5}\right)^x$ and
   $g(x) = \log_{2/5} x$

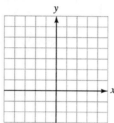

7. $f(x) = 4^x$ and
   $g(x) = \log_4 x$

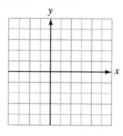

8. $f(x) = 3^x$ and
   $g(x) = \log_3 x$

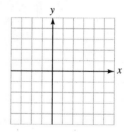

In Review Exercises 9–20, find each value.

9. $\log_3 9$

10. $\log_9 \dfrac{1}{3}$

11. $\log_\pi 1$

12. $\log_5 0.04$

13. $\log_a \sqrt{a}$

14. $\log_a \sqrt[3]{a}$

15. $\ln e^4$

16. $\ln 1$

17. $10^{\log_{10} 7}$

18. $e^{\ln 3}$

19. $\log_b b^4$

20. $\ln e^9$

In Review Exercises 21–44, find x.

21. $\log_2 x = 3$

22. $\log_3 x = -2$

23. $\log_x 9 = 2$

24. $\log_x 0.125 = -3$

25. $\log_7 7 = x$

26. $\log_3 \sqrt{3} = x$

27. $\log_8 \sqrt{2} = x$

28. $\log_6 36 = x$

29. $\log_{1/3} 9 = x$

30. $\log_{1/2} 1 = x$

31. $\log_x 3 = \dfrac{1}{3}$

32. $\log_x 25 = -2$

33. $\log_2 x = 5$

34. $\log_{\sqrt{3}} x = 4$

35. $\log_{\sqrt{3}} x = 6$

36. $\log_{0.1} 10 = x$

37. $\log_x 2 = -\dfrac{1}{3}$

38. $\log_x 32 = 5$

39. $\log_{0.25} x = -1$

40. $\log_{0.125} x = -\dfrac{1}{3}$

41. $\log_{\sqrt{2}} 32 = x$

42. $\log_{\sqrt{5}} x = -4$

43. $\log_{\sqrt{3}} 9\sqrt{3} = x$

44. $\log_{\sqrt{5}} 5\sqrt{5} = x$

In Review Exercises 45–46, write each expression in terms of the logarithms of x, y, and z.

45. $\log_b \dfrac{x^2 y^3}{z^4}$

46. $\log_b \sqrt{\dfrac{x}{yz^2}}$

In Review Exercises 47–48, write each expression as the logarithm of a single quantity.

47. $3 \log_b x - 5 \log_b y + 7 \log_b z$

48. $\dfrac{1}{2} \log_b x + 3 \log_b y - 7 \log_b z$

In Review Exercises 49–52, assume that $\log a = 0.6$, $\log b = 0.36$, *and* $\log c = 2.4$. *Find each value.*

49. $\log abc$

50. $\log a^2 b$

51. $\log \dfrac{ac}{b}$

52. $\log \dfrac{a^2}{c^3 b^2}$

In Review Exercises 53–66, solve for x, where possible.

53. $3^x = 7$

54. $5^{x+2} = 625$

55. $2^x = 3^{x-1}$

56. $2^{x^2+4x} = \dfrac{1}{8}$

57. $\log x + \log (29 - x) = 2$

58. $\log_2 x + \log_2 (x - 2) = 3$

59. $\log_2 (2 - x) + \log_2 (-x) = 3$

60. $\dfrac{\log (7x - 12)}{\log x} = 2$

61. $\log x + \log (x - 5) = \log 6$

62. $\log 3 - \log (x - 1) = -1$

63. $e^{x \ln 2} = 9$

64. $\ln x = \ln (x - 1)$

65. $\ln x = \ln (x - 1) + 1$

66. $\ln x = \log_{10} x$
(*Hint:* Use the change-of-base formula.)

In Review Exercises 67–70, solve each problem.

67. Carbon-14 dating A wooden statue excavated in Egypt has a carbon-14 content that is two-thirds of that found in living wood. If the half-life of carbon-14 is 5700 years, how old is the statue?

68. Hydrogen ion concentration The pH of grapefruit juice is approximately 3.1. Find its hydrogen ion concentration.

69. Radioactive decay One-third of a radioactive material decays in 20 years. Find its half-life.

70. Formula for pH Some chemistry textbooks define the pH of a solution with the formula

$$pH = \log_{10} \dfrac{1}{[H^+]}$$

Show that this definition is equivalent to the one given in this book.

Chapter 9 Test

In Questions 1–2, graph each function.

1. $f(x) = 2^x + 1$

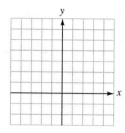

2. $f(x) = 2^{-x}$

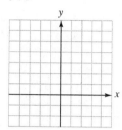

In Questions 3–4, solve each problem.

3. A radioactive material decays according to the formula $A = A_0(2)^{-t}$. How much of a 3-gram sample will be left in 6 years?

4. An initial deposit of \$1000 earns 6% interest, compounded twice a year. How much will be in the account in one year?

5. Graph the function $f(x) = e^x$.

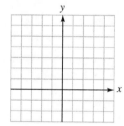

6. An account contains \$2000 and has been earning 8% interest, compounded continuously. How much will be in the account in 10 years?

In Questions 7–12, find x.

7. $\log_4 16 = x$

8. $\log_x 81 = 4$

9. $\log_3 x = -3$

10. $\log_x 100 = 2$

11. $\log_{3/2} \dfrac{9}{4} = x$

12. $\log_{2/3} x = -3$

In Questions 13–14, graph each function.

13. $f(x) = -\log_3 x$

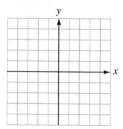

14. $f(x) = \ln x$

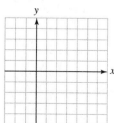

In Questions 15–16, write each expression in terms of the logarithms of a, b, and c.

15. $\log a^2 bc^3$

16. $\ln \sqrt{\dfrac{a}{b^2 c}}$

In Questions 17–18, write each expression as a logarithm of a single quantity.

17. $\dfrac{1}{2} \log (a + 2) + \log b - 3 \log c$

18. $\dfrac{1}{3}(\log a - 2 \log b) - \log c$

In Questions 19–20, assume that $\log 2 = 0.3010$ and $\log 3 = 0.4771$. Find each value. **Do not use a calculator.**

19. $\log 24$

20. $\log \dfrac{8}{3}$

In Questions 21–22, use the change-of-base formula to find each logarithm. Do not attempt to simplify the answer.

21. $\log_7 3$

22. $\log_\pi e$

In Questions 23–26, tell whether each statement is true. If a statement is not true, explain why.

23. $\log_a ab = 1 + \log_a b$

24. $\dfrac{\log a}{\log b} = \log a - \log b$

25. $\log a^{-3} = \dfrac{1}{3 \log a}$

26. $\ln (-x) = -\ln x$

27. Find the pH of a solution with a hydrogen ion concentration of 3.7×10^{-7}. (*Hint:* pH $= -\log [H^+]$.)

28. Find the db gain of an amplifier when $E_O = 60$ volts and $E_I = 0.3$ volt. (*Hint:* db gain $= 20 \log (E_O/E_I)$.)

In Questions 29–30, solve each equation. Do not simplify the logarithms.

29. $5^x = 3$

30. $3^{x-1} = 100^x$

In Questions 31–32, solve each equation.

31. $\log (5x + 2) = \log (2x + 5)$

32. $\log x + \log (x - 9) = 1$

Miscellaneous Topics

10

■ ■ ■ ■ ■ ■ ■ ■ ■ **Actuary**

Why do young persons pay more for automobile insurance than older persons? How much should an insurance policy cost? How much should an organization contribute each year to its pension fund? Answers to these and similar questions are provided by actuaries, who design insurance and pension plans and follow their performance to make sure that the plans are maintained on a sound financial basis.

SAMPLE APPLICATION ■ An *annuity* is a sequence of equal payments made periodically over a length of time. The sum of the payments and the interest earned during the *term* of the annuity is called the *amount* of the annuity.

After a sales clerk works six months, her employer will begin an annuity for her and will contribute $500 semiannually to a fund that pays 8% annual interest. After she has been employed for two years, what will be the amount of her annuity?
See Example 7 in Section 10.7.

The graphs of second-degree equations in x and y represent figures that have interested people since the time of the ancient Greeks. The equations of these graphs were studied carefully in the 17th century by René Descartes (1596–1650) and Blaise Pascal (1623–1662).

Descartes discovered that graphs of second-degree equations fall into one of several categories: a pair of lines, a point, a circle, a parabola, an ellipse, a hyperbola, or no graph at all. Because all of these graphs can be formed by the intersection of a plane and a right-circular cone, they are called **conic sections**. (See Figure 10-1.)

Conic sections have many applications. For example, everyone knows the importance of circular wheels and gears, pizza cutters, and ferris wheels.

Parabolas can be rotated to generate dish-shaped surfaces called *paraboloids*. Any light or sound placed at the *focus* of the paraboloid is reflected outward in parallel paths, as shown in Figure 10-2(a). This property makes parabolic surfaces ideal for flashlight and headlight reflectors. It also makes parabolic surfaces good antennas, because signals captured by such antennas are concentrated at the focus. Parabolic mirrors are capable of concentrating the rays of the sun at a single point and thereby generating tremendous heat. This property is used in the design of certain solar furnaces.

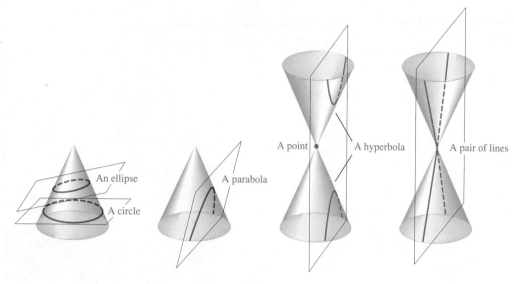

FIGURE 10-1

Any object thrown upward and outward travels in a parabolic path, as shown in Figure 10-2(b). In architecture, many arches are parabolic in shape because this gives strength. Cables that support suspension bridges hang in the form of a parabola. (See Figure 10-2(c).)

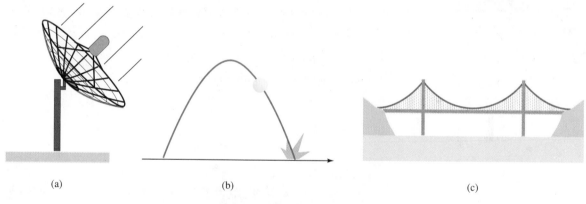

(a)   (b)   (c)

FIGURE 10-2

Ellipses have optical and acoustical properties that are useful in architecture and engineering. For example, many arches are portions of an ellipse, because the shape is pleasing to the eye. (See Figure 10-3(a).) The planets and some comets have elliptical orbits. (See Figure 10-3(b).) Gears are often cut into elliptical shapes to provide nonuniform motion. (See Figure 10-3(c).)

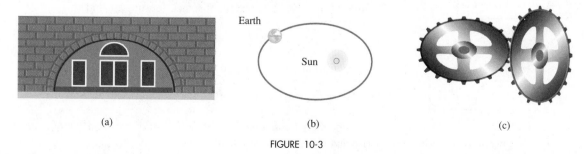

(a)   (b)   (c)

FIGURE 10-3

Hyperbolas serve as the basis of a navigational system known as LORAN (LOng RAnge Navigation). (See Figure 10-4.) They are also used to find the source of a distress signal, are the basis for the design of hypoid gears, and describe the orbits of some comets.

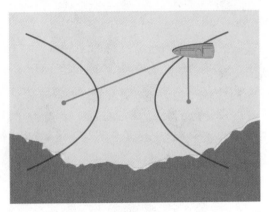

FIGURE 10-4

10.1 The Circle and the Parabola

■ THE CIRCLE ■ GRAPHING DEVICES ■ PROBLEM SOLVING ■ THE PARABOLA ■ PROBLEM SOLVING

The best known conic section is the circle.

■ THE CIRCLE

> **The Circle**
> A **circle** is the set of all points in a plane that are a fixed distance from a point called its *center*.
> The fixed distance is the *radius* of the circle.

To develop the general equation of a circle, we must write the equation of a circle with a radius of r and with center at some point $C(h, k)$, as in Figure 10-5. This task is equivalent to finding all points $P(x, y)$ such that the length of line segment CP is r. We can use the distance formula to find r.

$$r = \sqrt{(x - h)^2 + (y - k)^2}$$

We then square both sides to obtain

1. $r^2 = (x - h)^2 + (y - k)^2$

Equation 1 is called the *standard form of the equation of a circle* with a radius of r and center at the point with coordinates (h, k).

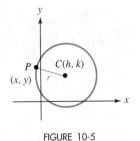

FIGURE 10-5

Standard Equation of a Circle with Center at (*h, k*)

Any equation that can be written in the form

$$(x - h)^2 + (y - k)^2 = r^2$$

has a graph that is a circle with radius r and center at point (h, k).

If $r = 0$, the graph reduces to a single point called a *point circle*. If $r^2 < 0$, then a circle does not exist. If both coordinates of the center are 0, then the center of the circle is the origin.

Standard Equation of a Circle with Center at (0, 0)

Any equation that can be written in the form

$$x^2 + y^2 = r^2$$

has a graph that is a circle with radius r and center at the origin.

EXAMPLE 1 Graph the equation $x^2 + y^2 = 25$.

Solution Because this equation can be written in the form $x^2 + y^2 = r^2$, its graph is a circle with center at the origin. Since $r^2 = 25 = 5^2$, the circle has a radius of 5. The graph appears in Figure 10-6.

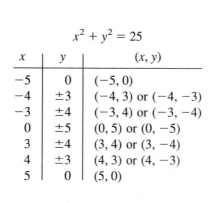

$$x^2 + y^2 = 25$$

x	y	(x, y)
-5	0	$(-5, 0)$
-4	± 3	$(-4, 3)$ or $(-4, -3)$
-3	± 4	$(-3, 4)$ or $(-3, -4)$
0	± 5	$(0, 5)$ or $(0, -5)$
3	± 4	$(3, 4)$ or $(3, -4)$
4	± 3	$(4, 3)$ or $(4, -3)$
5	0	$(5, 0)$

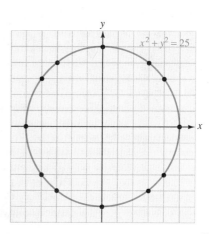

FIGURE 10-6

EXAMPLE 2 Find the equation of the circle with radius 5 and center at $C(3, 2)$.

Solution We substitute 5 for r, 3 for h, and 2 for k in standard form and simplify.

$$(x - h)^2 + (y - k)^2 = r^2$$
$$(x - 3)^2 + (y - 2)^2 = 5^2$$
$$x^2 - 6x + 9 + y^2 - 4y + 4 = 25$$
$$x^2 + y^2 - 6x - 4y - 12 = 0$$

The equation of the circle is $x^2 + y^2 - 6x - 4y - 12 = 0$.

EXAMPLE 3   Graph the circle $x^2 + y^2 - 4x + 2y = 20$.

Solution   Since the equation is not in standard form, the coordinates of the center and the length of the radius are not obvious. To put the equation in standard form, we complete the square on both x and y as follows:

$$x^2 + y^2 - 4x + 2y = 20$$
$$x^2 - 4x + y^2 + 2y = 20$$
$$x^2 - 4x + 4 + y^2 + 2y + 1 = 20 + 4 + 1 \qquad \text{To complete the square on } x$$
$$\text{and } y, \text{ add 4 and 1 to both sides.}$$
$$(x - 2)^2 + (y + 1)^2 = 25 \qquad \text{Factor } x^2 - 4x + 4 \text{ and}$$
$$y^2 + 2y + 1.$$
$$(x - 2)^2 + [y - (-1)]^2 = 5^2$$

The radius of the circle is 5, and the coordinates of its center are $h = 2$ and $k = -1$. We plot the center of the circle and draw a circle with a radius of 5 units, as shown in Figure 10-7. ∎

$$x^2 + y^2 - 4x + 2y = 20$$

FIGURE 10-7

■ GRAPHING DEVICES

Since the graphs of circles do not pass the vertical line test, their equations do not represent functions. Most graphing devices are limited to graphing functions, so it is usually not easy to use such a device to graph a circle. For example, to graph the circle described by $(x - 1)^2 + (y - 2)^2 = 4$, we must split the equation into two functions and graph each one separately. We begin by solving the equation for y.

$$(x - 1)^2 + (y - 2)^2 = 4$$
$$(y - 2)^2 = 4 - (x - 1)^2 \qquad \text{Subtract } (x - 1)^2 \text{ from both sides.}$$
$$y - 2 = \pm\sqrt{4 - (x - 1)^2} \qquad \text{Take the square root of both sides.}$$
$$y = 2 \pm \sqrt{4 - (x - 1)^2} \qquad \text{Add 2 to both sides.}$$

This equation defines two functions. If we use window settings of $[-3, 5]$ for x and $[-3, 5]$ for y and graph the two functions

$$y = 2 + \sqrt{4 - (x - 1)^2} \qquad \text{and} \qquad y = 2 - \sqrt{4 - (x - 1)^2}$$

we get the distorted circle shown in Figure 10-8(a). To get a better circle, most graphing devices have a squaring feature that gives an equal unit distance on both the x- and y-axes. After using the squaring feature, we get the circle shown in Figure 10-8(b).

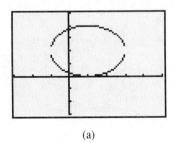

(a)

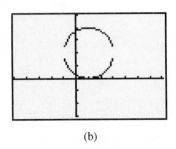

(b)

FIGURE 10-8

■ PROBLEM SOLVING

EXAMPLE 4

Radio translators The effective broadcast area of a television station is bounded by the circle $x^2 + y^2 = 3600$, where x and y are measured in miles. A translator station picks up the signal and retransmits it from the center of a circular area bounded by $(x + 30)^2 + (y - 40)^2 = 1600$. Find the location of the translator and the greatest distance from the main transmitter that the signal can be received.

Solution The coverage of the television station is bounded by $x^2 + y^2 = 60^2$, a circle centered at the origin with a radius of 60 miles, as shown in Figure 10-9. Because the translator is at the center of the circle $(x + 30)^2 + (y - 40)^2 = 1600$, it is located at $(-30, 40)$, a point 30 miles west and 40 miles north of the television station. The radius of the translator's coverage is $\sqrt{1600}$, or 40 miles.

As shown in Figure 10-9, the greatest distance of reception is the sum of A, the distance of the translator from the television station, and 40 miles, the radius of the translator's coverage.

To find A, we use the distance formula to find the distance between $(x_1, y_1) = (-30, 40)$ and the origin, $(x_2, y_2) = (0, 0)$.

$$A = \sqrt{(x_1 - x_2)^2 + (y_1 - y_2)^2}$$
$$A = \sqrt{(-30 - 0)^2 + (40 - 0)^2}$$
$$= \sqrt{(-30)^2 + 40^2}$$
$$= \sqrt{2500}$$
$$= 50$$

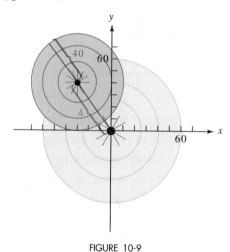

FIGURE 10-9

The translator is located 50 miles from the television station, and it broadcasts the signal an additional 40 miles. The greatest reception distance is $50 + 40$, or 90 miles. ■

■ THE PARABOLA

We have seen that equations of the form $y = a(x - h)^2 + k$, with $a \neq 0$, represent parabolas with the vertex at the point (h, k). They open upward when $a > 0$ and downward when $a < 0$.

Equations of the form $x = a(y - k)^2 + h$ also represent parabolas with vertex at point (h, k). However, they open to the right when $a > 0$ and to the left when $a < 0$. Parabolas that open to the right or left do not represent functions, because their graphs do not pass the vertical line test.

Several types of parabolas are summarized in the following chart. (In all cases, $a > 0$.)

Equations of Parabolas ($a > 0$)

Parabola opening	Vertex at origin	Vertex at (h, k)
Up	$y = ax^2$	$y = a(x - h)^2 + k$
Down	$y = -ax^2$	$y = -a(x - h)^2 + k$
Right	$x = ay^2$	$x = a(y - k)^2 + h$
Left	$x = -ay^2$	$x = -a(y - k)^2 + h$

EXAMPLE 5 Graph the equations **a.** $x = \dfrac{1}{2}y^2$ and **b.** $x = -2(y - 2)^2 + 3$.

Solution **a.** We make a table of ordered pairs that satisfy the equation, plot each pair, and draw the parabola as in Figure 10-10(a). Because the equation is of the form $x = ay^2$ with $a > 0$, the parabola opens to the right and has its vertex at the origin.

b. We make a table of ordered pairs that satisfy the equation, plot each pair, and draw the parabola as in Figure 10-10(b). Because the equation is of the form $x = -a(y - k)^2 + h$, the parabola opens to the left and has its vertex at the point with coordinates $(3, 2)$.

$x = \frac{1}{2}y^2$

x	y	(x, y)
0	0	$(0, 0)$
2	2	$(2, 2)$
2	-2	$(2, -2)$
8	4	$(8, 4)$
8	-4	$(8, -4)$

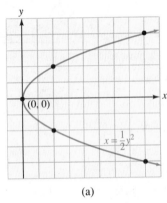

(a)

$x = -2(y - 2)^2 + 3$

x	y	(x, y)
-5	0	$(-5, 0)$
1	1	$(1, 1)$
3	2	$(3, 2)$
1	3	$(1, 3)$
-5	4	$(-5, 4)$

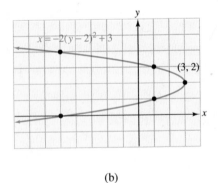

(b)

FIGURE 10-10

EXAMPLE 6 Graph the function $f(x) = -2x^2 + 12x - 15$.

Solution Because the equation is not in standard form, the coordinates of its vertex are not obvious. To write the equation in standard form, we complete the square on x.

$$f(x) = -2x^2 + 12x - 15$$
$$f(x) = -2(x^2 - 6x) - 15 \qquad \text{Factor out } -2 \text{ from } -2x^2 + 12x.$$
$$f(x) = -2(x^2 - 6x + 9) - 15 + 18 \qquad \text{Subtract and add 18: } (-2)(9) = -18.$$
$$f(x) = -2(x - 3)^2 + 3$$

Because the equation is written in the form $y = -a(x - h)^2 + k$, we can see that the parabola opens downward and has its vertex at $(3, 3)$. The graph of the function is shown in Figure 10-11.

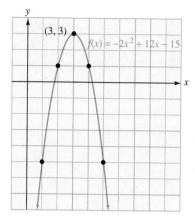

$$f(x) = -2x^2 + 12x - 15$$

x	$f(x)$	$(x, f(x))$
1	-5	$(1, -5)$
2	1	$(2, 1)$
3	3	$(3, 3)$
4	1	$(4, 1)$
5	-5	$(5, -5)$

FIGURE 10-11   ■

■ PROBLEM SOLVING

EXAMPLE 7   **Gateway Arch**   The shape of the Gateway Arch in St. Louis is approximately a parabola, as shown in Figure 10-12(a). How high is the arch 100 feet from its foundation?

Solution   We place the parabola in a coordinate system as in Figure 10-12(b), with ground level on the x-axis and the vertex of the parabola at the point $(h, k) = (0, 630)$. The equation of this downward-opening parabola has the form

$$y = -a(x - h)^2 + k$$
$$y = -a(x - 0)^2 + \mathbf{630} \qquad \text{Substitute } h = 0 \text{ and } k = 630.$$
$$y = -ax^2 + 630 \qquad \text{Simplify.}$$

Because the Gateway Arch is 630 feet wide at its base, the parabola passes through the point $\left(\frac{630}{2}, 0\right)$, or $(315, 0)$. To find a in the equation of the parabola, we substitute $x = 315$ and $y = 0$ and proceed as follows:

$$y = -ax^2 + 630$$
$$0 = -a(\mathbf{315})^2 + 630$$
$$\frac{-630}{315^2} = -a$$
$$\frac{2}{315} = a$$

The equation of the parabola that approximates the shape of the Gateway Arch is

$$y = -\frac{2}{315}x^2 + 630$$

To find the height of the arch at a point 100 feet from its foundation, we substitute $315 - 100$, or 215, for x into the equation of the parabola and solve for y.

$$y = -\frac{2}{315}x^2 + 630$$

$$y = -\frac{2}{315}(\mathbf{215})^2 + 630$$

$$= 336.5079365 \qquad \text{Use a calculator.}$$

At a point 100 feet from the foundation, the height of the arch is about 337 feet.

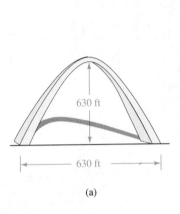

(a)

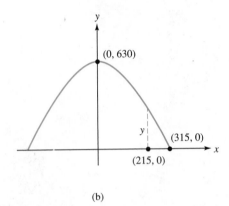

(b)

FIGURE 10-12

Orals   *Find the center and the radius of each circle.*

1. $x^2 + y^2 = 144$ **2.** $x^2 + y^2 = 121$

3. $(x - 2)^2 + y^2 = 16$ **4.** $x^2 + (y + 1)^2 = 9$

Tell whether each parabola opens up or down or left or right.

5. $y = -3x^2 - 2$ **6.** $y = 7x^2 - 5$

7. $x = -3y^2$ **8.** $x = (y - 3)^2$

EXERCISE 10.1

In Exercises 1–10, graph each equation.

1. $x^2 + y^2 = 9$ **2.** $x^2 + y^2 = 16$ **3.** $(x - 2)^2 + y^2 = 9$ **4.** $x^2 + (y - 3)^2 = 4$

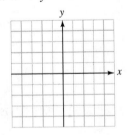   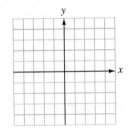   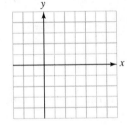   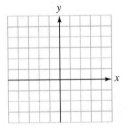

5. $(x - 2)^2 + (y - 4)^2 = 4$ **6.** $(x - 3)^2 + (y - 2)^2 = 4$ **7.** $(x + 3)^2 + (y - 1)^2 = 16$

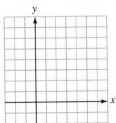   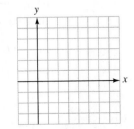   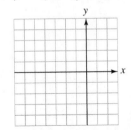

8. $(x - 1)^2 + (y + 4)^2 = 9$

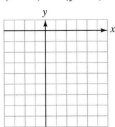

9. $x^2 + (y + 3)^2 = 1$

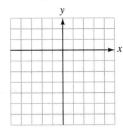

10. $(x + 4)^2 + y^2 = 1$

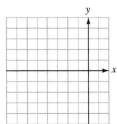

In Exercises 11–14. use a graphing device to graph each equation.

11. $3x^2 + 3y^2 = 16$ **12.** $2x^2 + 2y^2 = 9$ **13.** $(x + 1)^2 + y^2 = 16$ **14.** $x^2 + (y - 2)^2 = 4$

In Exercises 15–22, write the equation of the circle with the following properties.

15. Center at the origin; radius 1

16. Center at the origin; radius 4

17. Center at $(6, 8)$; radius 5

18. Center at $(5, 3)$; radius 2

19. Center at $(-2, 6)$; radius 12

20. Center at $(5, -4)$; radius 6

21. Center at the origin; diameter of $4\sqrt{2}$

22. Center at the origin; diameter of $8\sqrt{3}$

In Exercises 23–30, graph each circle. Give the coordinates of the center.

23. $x^2 + y^2 + 2x - 8 = 0$

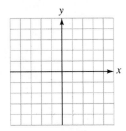

24. $x^2 + y^2 - 4y = 12$

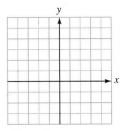

25. $9x^2 + 9y^2 - 12y = 5$

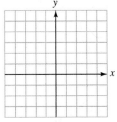

26. $4x^2 + 4y^2 + 4y = 15$

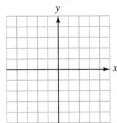

27. $x^2 + y^2 - 2x + 4y = -1$

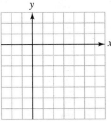

28. $x^2 + y^2 + 4x + 2y = 4$

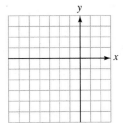

29. $x^2 + y^2 + 6x - 4y = -12$

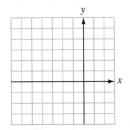

30. $x^2 + y^2 + 8x + 2y = -13$

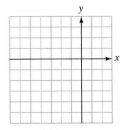

In Exercises 31–44, find the vertex of each parabola and graph it.

31. $x = y^2$

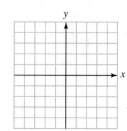

32. $x = -y^2 + 1$

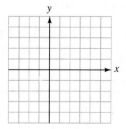

33. $x = -\dfrac{1}{4}y^2$

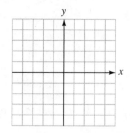

34. $x = 4y^2$

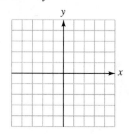

35. $y = x^2 + 4x + 5$

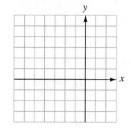

36. $y = -x^2 - 2x + 3$

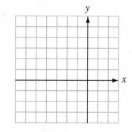

37. $y = -x^2 - x + 1$

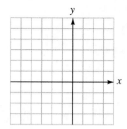

38. $x = \dfrac{1}{2}y^2 + 2y$

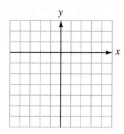

39. $y^2 + 4x - 6y = -1$

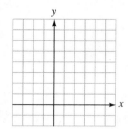

40. $x^2 - 2y - 2x = -7$

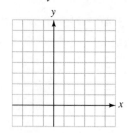

41. $y = 2(x - 1)^2 + 3$

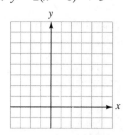

42. $y = -2(x + 1)^2 + 2$

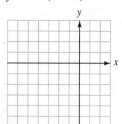

43. $x = -3(y + 2)^2 - 2$

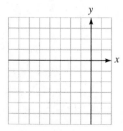

44. $x = 2(y - 3)^2 - 4$

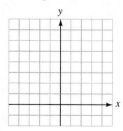

In Exercises 45–48, use a graphing device to graph each equation.

45. $x = 2y^2$ **46.** $x = y^2 - 4$ **47.** $x^2 - 2x + y = 6$ **48.** $x = -2(y - 1)^2 + 2$

49. Meshing gears   For design purposes, the large gear in Illustration 1 is the circle $x^2 + y^2 = 16$. The smaller gear is a circle centered at $(7, 0)$ and tangent to the larger circle. Find the equation of the smaller gear.

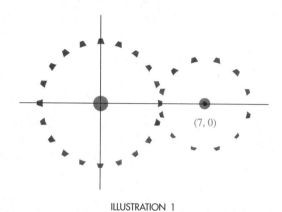

ILLUSTRATION 1

50. Width of a walkway   The walkway in Illustration 2 is bounded by the two circles $x^2 + y^2 = 2500$ and $(x - 10)^2 + y^2 = 900$, measured in feet. Find the largest and the smallest width of the walkway.

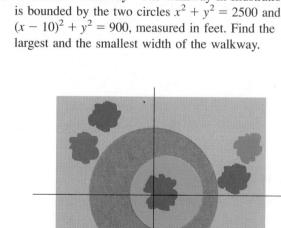

ILLUSTRATION 2

51. Broadcast ranges   Radio stations applying for licensing may not use the same frequency if their broadcast areas overlap. One station's coverage is bounded by $x^2 + y^2 - 8x - 20y + 16 = 0$, and the other's by $x^2 + y^2 + 2x + 4y - 11 = 0$. May they be licensed for the same frequency?

52. Highway curves   Highway design engineers want to join two sections of highway with a curve that is one-quarter of a circle, as in Illustration 3. The equation of the circle is $x^2 + y^2 - 16x - 20y + 155 = 0$, where distances are measured in kilometers. Find the locations (relative to the center of town) of the intersections of the highway with State and with Main.

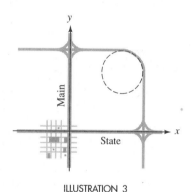

ILLUSTRATION 3

53. Flight of a projectile   The cannonball in Illustration 4 follows the parabolic trajectory $y = 30x - x^2$. Where does it land?

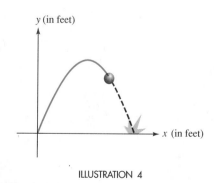

ILLUSTRATION 4

54. Flight of a projectile   In Exercise 53, how high does the cannonball get?

55. Orbit of a comet If the orbit of the comet shown in Illustration 5 is given by the equation $2y^2 - 9x = 18$, how far is it from the sun at the vertex of the orbit? Distances are in astronomical units (AU).

56. Satellite antenna The cross section of the satellite antenna in Illustration 6 is a parabola given by the equation $y = \frac{1}{16}x^2$, with distances measured in feet. If the dish is 8 feet wide, how deep is it?

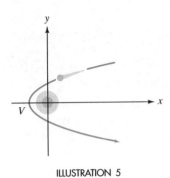

ILLUSTRATION 5

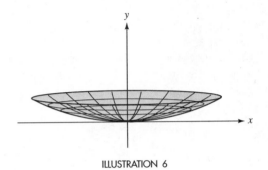

ILLUSTRATION 6

Writing Exercises   *Write a paragraph using your own words.*

1. Explain how to decide from its equation whether the graph of a parabola opens up, down, right, or left.

2. From the equation of a circle, explain how to determine the radius and the coordinates of the center.

Something to Think About

1. From the values of a, h, and k, explain how to determine the number of x-intercepts of the graph of $y = a(x - h)^2 + k$.

2. Under what conditions will the graph of $x = a(y - k)^2 + h$ have no y-intercepts?

Review Exercises   *Solve each equation.*

1. $|3x - 4| = 11$

2. $\left| \dfrac{4 - 3x}{5} \right| = 12$

3. $|3x + 4| = |5x - 2|$

4. $|6 - 4x| = |x + 2|$

10.2 The Ellipse and the Hyperbola

■ THE ELLIPSE ■ GRAPHING DEVICES ■ PROBLEM SOLVING ■ THE HYPERBOLA ■ PROBLEM SOLVING

■ THE ELLIPSE

The Ellipse
An **ellipse** is the set of all points P in the plane the sum of whose distances from two fixed points is a constant. See Figure 10-13, in which $d_1 + d_2$ is a constant.

Each of the two points is called a **focus**. Midway between the foci is the *center* of the ellipse.

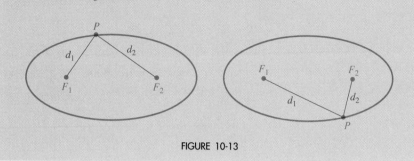

FIGURE 10-13

Using the previous definition, we can construct an ellipse by placing two thumbtacks fairly close together, as in Figure 10-14. We then tie each end of a piece of string to a thumbtack, catch the loop with the point of a pencil, and while keeping the string taut, draw the ellipse.

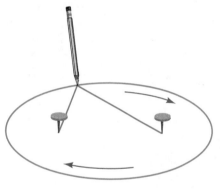

FIGURE 10-14

The graph of the equation

$$\frac{x^2}{36} + \frac{y^2}{9} = 1 \qquad \text{or} \qquad \frac{x^2}{6^2} + \frac{y^2}{3^2} = 1$$

is an ellipse. To graph it, we make a table of ordered pairs that satisfy the equation, plot each pair, and join them with the smooth curve shown in Figure 10-15.

$$\frac{x^2}{6^2} + \frac{y^2}{3^2} = 25$$

x	y	(x, y)
-6	0	$(-6, 0)$
-4	± 2.2	$(-4, 2.2)$ or $(-4, -2.2)$
-2	± 2.8	$(-2, 2.8)$ or $(-2, -2.8)$
0	± 3	$(0, 3)$ or $(0, -3)$
2	± 2.8	$(2, 2.8)$ or $(2, -2.8)$
4	± 2.2	$(4, 2.2)$ or $(4, -2.2)$
6	0	$(6, 0)$

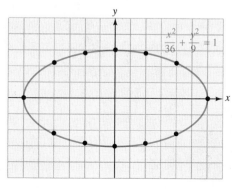

FIGURE 10-15

The center of the ellipse is the origin. It intersects the x-axis at points $(6, 0)$ and $(-6, 0)$, and it intersects the y-axis at the points $(0, 3)$ and $(0, -3)$.

The previous example suggests that the graph of the equation

$$\frac{x^2}{a^2} + \frac{y^2}{b^2} = 1$$

is an ellipse centered at the origin. To find the x-intercepts of the graph, we can let $y = 0$ and solve for x.

$$\frac{x^2}{a^2} + \frac{0^2}{b^2} = 1$$

$$\frac{x^2}{a^2} + 0 = 1$$

$$x^2 = a^2$$

$$x = a \qquad \text{or} \qquad x = -a$$

The x-intercepts of the graph are $(a, 0)$ and $(-a, 0)$.

To find the y-intercepts of the graph, we can let $x = 0$ and solve for y.

$$\frac{0^2}{a^2} + \frac{y^2}{b^2} = 1$$

$$0 + \frac{y^2}{b^2} = 1$$

$$y^2 = b^2$$

$$y = b \qquad \text{or} \qquad y = -b$$

The y-intercepts of the graph are $(0, b)$ and $(0, -b)$.

In general, we have the following results.

Equations of an Ellipse Centered at the Origin

The equation of an ellipse centered at the origin with x-intercepts at $V_1(a, 0)$ and $V_2(-a, 0)$ and with y-intercepts of $(0, b)$ and $(0, -b)$ is

$$\frac{x^2}{a^2} + \frac{y^2}{b^2} = 1 \quad (a > b > 0) \qquad \text{See Figure 10-16(a).}$$

The equation of an ellipse centered at the origin with y-intercepts at $V_1(0, a)$ and $V_2(0, -a)$ and x-intercepts at $(b, 0)$ and $(-b, 0)$ is

$$\frac{x^2}{b^2} + \frac{y^2}{a^2} = 1 \quad (a > b > 0) \qquad \text{See Figure 10-16(b).}$$

In Figure 10-16, the points V_1 and V_2 are the **vertices** of the ellipse, the midpoint of V_1V_2 is the **center** of the ellipse, and the distance between the center of the ellipse and either vertex is a. The segment V_1V_2 is called the **major axis**, and the segment joining either $(0, b)$ and $(0, -b)$ or $(b, 0)$ and $(-b, 0)$ is called the **minor axis**.

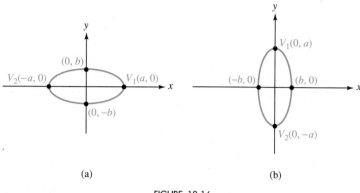

FIGURE 10-16

The equations for ellipses centered at (h, k) are as follows.

Equations of an Ellipse Centered at (h, k)

The equation of an ellipse centered at (h, k) and major axis parallel to the x-axis is

$$\frac{(x - h)^2}{a^2} + \frac{(y - k)^2}{b^2} = 1 \qquad (a > b > 0)$$

The equation of an ellipse centered at (h, k) and major axis parallel to the y-axis is

$$\frac{(x - h)^2}{b^2} + \frac{(y - k)^2}{a^2} = 1 \qquad (a > b > 0)$$

EXAMPLE 1   Graph the ellipse $\dfrac{(x - 2)^2}{16} + \dfrac{(y + 3)^2}{25} = 1$.

Solution   We first write the equation in the form

$$\frac{(x - 2)^2}{4^2} + \frac{[y - (-3)]^2}{5^2} = 1 \qquad (5 > 4)$$

This is the equation of an ellipse centered at $(h, k) = (2, -3)$ with $b = 4$ and $a = 5$. We first plot the center, as shown in Figure 10-17. Since a is the distance from the center to the vertex, we can locate the vertices by counting 5 units above and 5 units below the center. The vertices are at points $(2, 2)$ and $(2, -8)$.

Since $b = 4$, we can locate two more points on the ellipse by counting 4 units to the left and 4 units to the right of the center. The points $(-2, -3)$ and $(6, -3)$ are also on the graph.

Using these four points as guides, we can draw the ellipse.

$$\frac{(x-2)^2}{16} + \frac{(y+3)^2}{25} = 1$$

x	y	(x, y)
2	2	$(2, 2)$
2	−8	$(2, -8)$
6	−3	$(6, -3)$
−2	−3	$(-2, -3)$

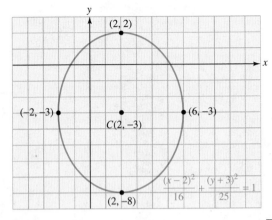

FIGURE 10-17

■ GRAPHING DEVICES

To use a graphing device to graph the equation $\dfrac{(x+2)^2}{4} + \dfrac{(y-1)^2}{25} = 1$, we first multiply both sides by 100 to eliminate the fractions and then solve for y.

$$25(x + 2)^2 + 4(y - 1)^2 = 100 \qquad \text{Multiply both sides by 100.}$$

$$4(y - 1)^2 = 100 - 25(x + 2)^2 \qquad \text{Subtract } 25(x + 2)^2 \text{ from both sides.}$$

$$(y - 1)^2 = \frac{100 - 25(x + 2)^2}{4} \qquad \text{Divide both sides by 4.}$$

$$y - 1 = \pm \frac{\sqrt{100 - 25(x + 2)^2}}{2} \qquad \text{Take the square root of both sides.}$$

$$y = 1 \pm \frac{\sqrt{100 - 25(x + 2)^2}}{2} \qquad \text{Add 1 to both sides.}$$

If we use window settings $[-6, 6]$ for x and $[-6, 6]$ for y and graph the functions

$$y = 1 + \frac{\sqrt{100 - 25(x + 2)^2}}{2} \qquad \text{and} \qquad y = 1 - \frac{\sqrt{100 - 25(x + 2)^2}}{2}$$

we get the ellipse shown in Figure 10-18.

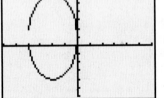

FIGURE 10-18

■ PROBLEM SOLVING

EXAMPLE 2

Landscape design   A landscape architect is designing an elliptical pool that will fit in the center of a 20-by-30-foot rectangular garden, leaving at least 5 feet of space on all sides. Find the equation of the ellipse.

Solution   We place the rectangular garden in a coordinate system, as in Figure 10-19. To maintain 5 feet of clearance at the ends of the ellipse, the vertices must be the points $V_1(10, 0)$ and $V_2(-10, 0)$. Similarly, the y-intercepts are the points $(0, 5)$ and $(0, -5)$.

The equation of the ellipse has the form

$$\frac{x^2}{a^2} + \frac{y^2}{b^2} = 1$$

with $a = 10$ and $b = 5$. The equation of the boundary of the pool is

$$\frac{x^2}{100} + \frac{y^2}{25} = 1$$

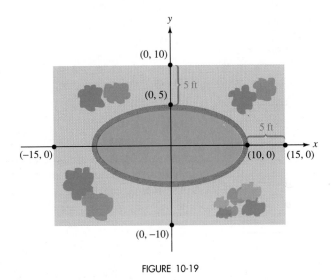

FIGURE 10-19   ■

■ THE HYPERBOLA

The Hyperbola
A **hyperbola** is the set of all points P in the plane for which the difference of the distances of each point from two fixed points is a constant. See Figure 10-20, in which $d_1 - d_2$ is a constant.

Each of the two points is called a **focus**. Midway between the foci is the **center** of the hyperbola.

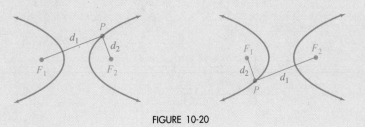

FIGURE 10-20

The graph of the equation

$$\frac{x^2}{25} - \frac{y^2}{9} = 1$$

is a hyperbola. To graph this hyperbola, we make a table of ordered pairs that satisfy the equation, plot each pair, and join the points with a smooth curve as in Figure 10-21.

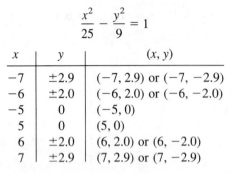

$$\frac{x^2}{25} - \frac{y^2}{9} = 1$$

x	y	(x, y)
-7	± 2.9	$(-7, 2.9)$ or $(-7, -2.9)$
-6	± 2.0	$(-6, 2.0)$ or $(-6, -2.0)$
-5	0	$(-5, 0)$
5	0	$(5, 0)$
6	± 2.0	$(6, 2.0)$ or $(6, -2.0)$
7	± 2.9	$(7, 2.9)$ or $(7, -2.9)$

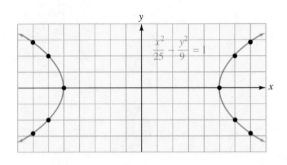

FIGURE 10-21

This graph is centered at the origin and intersects the x-axis at $\left(\sqrt{25}, 0\right)$ and $\left(-\sqrt{25}, 0\right)$. After simplifying, these points are $(5, 0)$ and $(-5, 0)$. We also note that the graph does not intersect the y-axis.

It is possible to draw a hyperbola without plotting many points. For example, if we want to graph the hyperbola with an equation of

$$\frac{x^2}{a^2} - \frac{y^2}{b^2} = 1$$

we first look at the x- and y-intercepts. To find the x-intercepts, we let $y = 0$ and solve for x:

$$\frac{x^2}{a^2} - \frac{0^2}{b^2} = 1$$
$$x^2 = a^2$$
$$x = \pm a$$

Thus, the hyperbola crosses the x-axis at the points $V_1(a, 0)$ and $V_2(-a, 0)$, called the **vertices** of the hyperbola. (See Figure 10-22.)

To attempt to find the y-intercepts, we let $x = 0$ and solve for y:

$$\frac{0^2}{a^2} - \frac{y^2}{b^2} = 1$$
$$y^2 = -b^2$$
$$y = \pm \sqrt{-b^2}$$

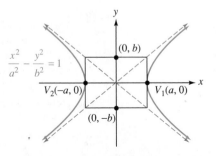

FIGURE 10-22

Since $b^2 > 0$, $-b^2 < 0$ and $\sqrt{-b^2}$ is an imaginary number. This means that the hyperbola does not cross the y-axis.

If we construct a rectangle, called the **fundamental rectangle**, whose sides pass horizontally through $\pm b$ on the y-axis and vertically through $\pm a$ on the x-axis, the extended diagonals of the rectangle will be asymptotes of the hyperbola.

Equation of a Hyperbola Centered at the Origin

Any equation that can be written in the form

$$\frac{x^2}{a^2} - \frac{y^2}{b^2} = 1$$

has a graph that is a hyperbola centered at the origin, as in Figure 10-23. The x-intercepts are the vertices $V_1(a, 0)$ and $V_2(-a, 0)$. There are no y-intercepts.

The asymptotes of the hyperbola are the extended diagonals of the rectangle in the figure.

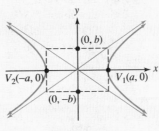

FIGURE 10-23

The branches of the hyperbola in the previous discussion open to the left and to the right. It is possible for hyperbolas to have different orientations with respect to the x- and y-axes. For example, the branches of a hyperbola can open upward and downward. In that case, the following equation applies.

Equation of a Hyperbola Centered at the Origin

Any equation that can be written in the form

$$\frac{y^2}{a^2} - \frac{x^2}{b^2} = 1$$

has a graph that is a hyperbola centered at the origin, as in Figure 10-24. The y-intercepts are the vertices $V_1(0, a)$ and $V_2(0, -a)$. There are no x-intercepts.

The asymptotes of the hyperbola are the extended diagonals of the rectangle shown in the figure.

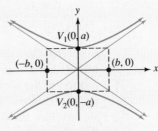

FIGURE 10-24

EXAMPLE 3 Graph the equation $9y^2 - 4x^2 = 36$.

Solution To write the equation in standard form, we divide both sides by 36 to obtain

$$\frac{9y^2}{36} - \frac{4x^2}{36} = 1$$

$$\frac{y^2}{4} - \frac{x^2}{9} = 1 \qquad \text{Simplify each fraction.}$$

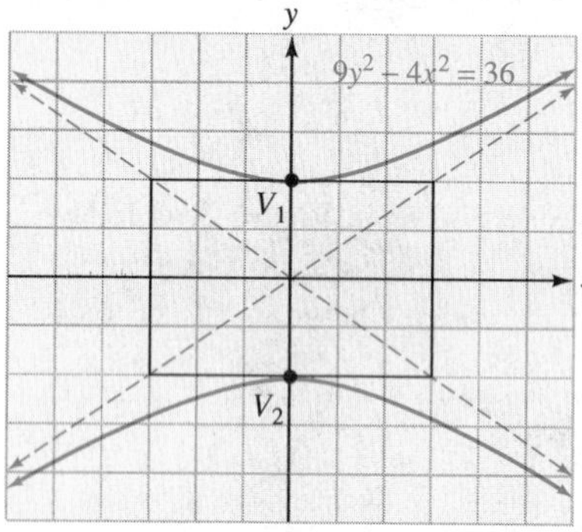

We then find the y-intercepts by letting $x = 0$ and solving for y:

$$\frac{y^2}{4} - \frac{0^2}{9} = 1$$

$$y^2 = 4$$

Thus, $y = \pm 2$, and the vertices of the hyperbola are $V_1(0, 2)$ and $V_2(0, -2)$. (See Figure 10-25.)

Since $\pm\sqrt{9} = \pm 3$, we use the points $(3, 0)$ and $(-3, 0)$ on the x-axis to help draw the fundamental rectangle. We then draw its extended diagonals and sketch the hyperbola. ∎

FIGURE 10-25

If a hyperbola is centered at a point with coordinates (h, k), the following equations apply.

> **Equations of Hyperbolas Centered at (h, k)**
>
> Any equation that can be written in the form
>
> $$\frac{(x - h)^2}{a^2} - \frac{(y - k)^2}{b^2} = 1$$
>
> is a hyperbola with center at (h, k) that opens left and right.
> Any equation of the form
>
> $$\frac{(y - k)^2}{a^2} - \frac{(x - h)^2}{b^2} = 1$$
>
> is a hyperbola with center at (h, k) that opens up and down.

EXAMPLE 4 Graph the hyperbola $\dfrac{(x - 3)^2}{16} - \dfrac{(y + 1)^2}{4} = 1$.

Solution We write the equation in the form

$$\frac{(x - 3)^2}{4^2} - \frac{[y - (-1)]^2}{2^2} = 1$$

to see that the hyperbola is centered at the point $(h, k) = (3, -1)$. Because this hyperbola has its center at $(3, -1)$, its vertices are located at $a = 4$ units to the right and left of the center, at $(7, -1)$ and $(-1, -1)$. Since $b = 2$, we can count 2 units above and below the center to locate points $(3, 1)$ and $(3, -3)$. With these points, we can draw the fundamental rectangle along with its extended diagonals. We can then sketch the hyperbola as shown in Figure 10-26.

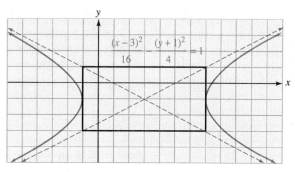

FIGURE 10-26

There is a special type of hyperbola (also centered at the origin) that does not intersect either the *x*- or the *y*-axis. These hyperbolas have equations of the form $xy = k$, where $k \neq 0$.

EXAMPLE 5 Graph the equation $xy = -8$.

Solution We make a table of ordered pairs, plot each pair, and join the points with a smooth curve to obtain the hyperbola in Figure 10-27.

$$xy = -8$$

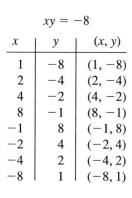

x	y	(x, y)
1	-8	$(1, -8)$
2	-4	$(2, -4)$
4	-2	$(4, -2)$
8	-1	$(8, -1)$
-1	8	$(-1, 8)$
-2	4	$(-2, 4)$
-4	2	$(-4, 2)$
-8	1	$(-8, 1)$

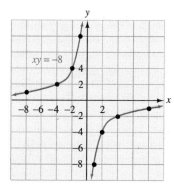

FIGURE 10-27

The result in Example 5 illustrates the following fact.

Equations of Hyperbolas of the Form xy = k
Any equation of the form $xy = k$, where $k \neq 0$, has a graph that is a **hyperbola,** which does not intersect either the *x*- or the *y*-axis.

■ PROBLEM SOLVING

EXAMPLE 6 **Atomic structure** In an experiment that led to the discovery of the atomic structure of matter, Lord Rutherford (1871–1937) shot high-energy alpha particles toward a thin sheet of gold. Because many were reflected, Rutherford showed the existence of the nucleus of a gold atom. The alpha particle in Figure 10-28 is repelled by the nucleus at the origin; it travels along the hyperbolic path given by $4x^2 - y^2 = 16$. How close does the particle come to the nucleus?

Solution To find the distance from the nucleus at the origin, we must find the coordinates of the vertex *V*. To do so, we write the equation of the particle's path in standard form:

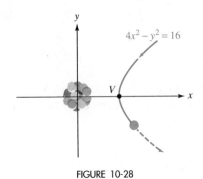

FIGURE 10-28

$$4x^2 - y^2 = 16$$

$$\frac{4x^2}{16} - \frac{y^2}{16} = \frac{16}{16} \qquad \text{Divide both sides by 16.}$$

$$\frac{x^2}{4} - \frac{y^2}{16} = 1 \qquad \text{Simplify.}$$

$$\frac{x^2}{2^2} - \frac{y^2}{4^2} = 1 \qquad \text{Write 4 as } 2^2 \text{ and 16 as } 4^2.$$

This equation is in the form $\dfrac{x^2}{a^2} - \dfrac{y^2}{b^2} = 1$, with $a = 2$. Thus, the vertex of the path is $(2, 0)$. The particle is never closer than 2 units from the nucleus. ■

Orals *Find the x- and y-intercepts of each ellipse.*

1. $\dfrac{x^2}{9} + \dfrac{y^2}{16} = 1$ **2.** $\dfrac{x^2}{25} + \dfrac{y^2}{36} = 1$

Find the center of each ellipse.

3. $\dfrac{(x - 2)^2}{9} + \dfrac{y^2}{16} = 1$ **4.** $\dfrac{x^2}{25} + \dfrac{(y + 1)^2}{36} = 1$

Find the x- or y-intercepts of each hyperbola.

5. $\dfrac{x^2}{9} - \dfrac{y^2}{16} = 1$ **6.** $\dfrac{y^2}{25} - \dfrac{x^2}{36} = 1$

EXERCISE 10.2

In Exercises 1–10, graph each equation.

1. $\dfrac{x^2}{4} + \dfrac{y^2}{9} = 1$ **2.** $x^2 + \dfrac{y^2}{9} = 1$ **3.** $x^2 + 9y^2 = 9$ **4.** $25x^2 + 9y^2 = 225$

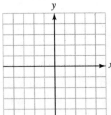

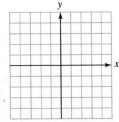

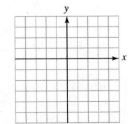

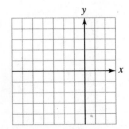

5. $16x^2 + 4y^2 = 64$

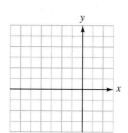

6. $4x^2 + 9y^2 = 36$

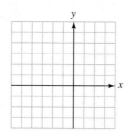

7. $\dfrac{(x-2)^2}{9} + \dfrac{(y-1)^2}{4} = 1$

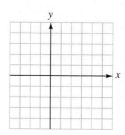

8. $\dfrac{(x-1)^2}{9} + \dfrac{(y-3)^2}{4} = 1$

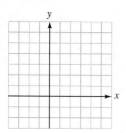

9. $(x+1)^2 + 4(y+2)^2 = 4$

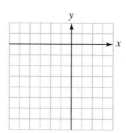

10. $25(x+1)^2 + 9y^2 = 225$

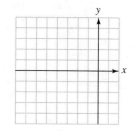

   In Exercises 11–14, use a graphing device to graph each equation.

11. $\dfrac{x^2}{9} + \dfrac{y^2}{4} = 1$

12. $x^2 + 16y^2 = 16$

13. $\dfrac{x^2}{4} + \dfrac{(y-1)^2}{9} = 1$

14. $\dfrac{(x+1)^2}{9} + \dfrac{(y-2)^2}{4} = 1$

In Exercises 15–26, graph each hyperbola.

15. $\dfrac{x^2}{9} - \dfrac{y^2}{4} = 1$

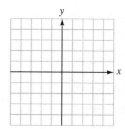

16. $\dfrac{x^2}{4} - \dfrac{y^2}{4} = 1$

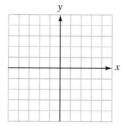

17. $\dfrac{y^2}{4} - \dfrac{x^2}{9} = 1$

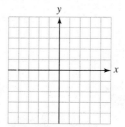

18. $\dfrac{y^2}{4} - \dfrac{x^2}{64} = 1$

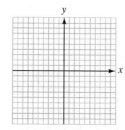

19. $25x^2 - y^2 = 25$

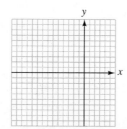

20. $9x^2 - 4y^2 = 36$

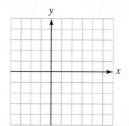

21. $\dfrac{(x-2)^2}{9} - \dfrac{y^2}{16} = 1$

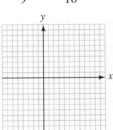

22. $\dfrac{(x+2)^2}{16} - \dfrac{(y-3)^2}{25} = 1$

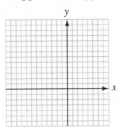

23. $4(x+3)^2 - (y-1)^2 = 4$

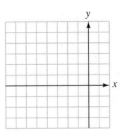

24. $(x+5)^2 - 16y^2 = 16$

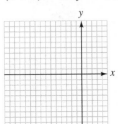

25. $xy = 8$

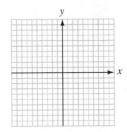

26. $xy = -10$

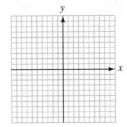

In Exercises 27–30, use a graphing device to graph each equation.

27. $\dfrac{x^2}{9} - \dfrac{y^2}{4} = 1$

28. $16y^2 - x^2 = 16$

29. $\dfrac{(y-1)^2}{9} - \dfrac{x^2}{4} = 1$

30. $\dfrac{(x+1)^2}{9} - \dfrac{(y-2)^2}{4} = 1$

31. Designing an underpass The arch of the underpass in Illustration 1 is a part of an ellipse. Find the equation of the arch.

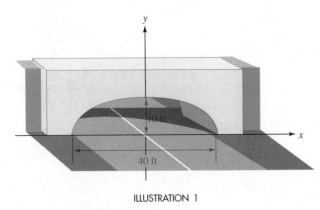

10 ft

40 ft

ILLUSTRATION 1

32. Calculating clearance Find the height of the elliptical arch in Exercise 31 at a point 10 feet from the center of the roadway.

33. Area of an ellipse The area A of the ellipse

$$\frac{x^2}{a^2} + \frac{y^2}{b^2} = 1$$

is given by $A = \pi ab$. Find the area of the ellipse $9x^2 + 16y^2 = 144$.

34. Area of a track The elliptical track in Illustration 2 is bounded by the ellipses

$$4x^2 + 9y^2 = 576 \quad \text{and} \quad 9x^2 + 25y^2 = 900$$

Find the area of the track. (See Exercise 33.)

35. Alpha particles The particle in Illustration 3 approaches the nucleus at the origin along the path $9y^2 - x^2 = 81$. How close does the particle come to the nucleus?

36. LORAN By determining the difference of the distances between the ship in Illustration 4 and two radio transmitters, the LORAN system places the ship on the hyperbola $x^2 - 4y^2 = 576$. If the ship is also 5 miles out to sea, find its coordinates.

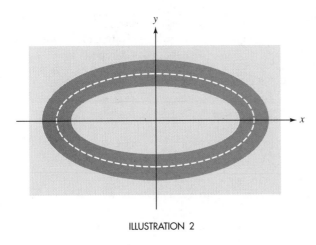

ILLUSTRATION 2

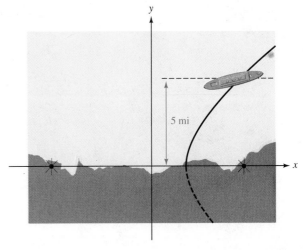

ILLUSTRATION 4

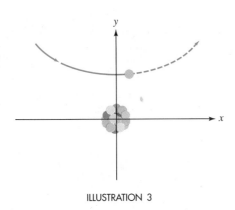

ILLUSTRATION 3

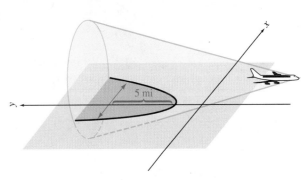

ILLUSTRATION 5

37. Sonic boom The position of the sonic boom caused by the faster-than-sound aircraft in Illustration 5 is the hyperbola $y^2 - x^2 = 25$ in the coordinate system shown. How wide is the hyperbola 5 miles from its vertex?

38. Electrostatic repulsion Two similarly charged particles are shot together for an almost head-on collision, as in Illustration 6. They repel each other and travel the two branches of the hyperbola given by $x^2 - 4y^2 = 4$. How close do they get?

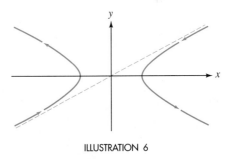

ILLUSTRATION 6

Writing Exercises *Write a paragraph using your own words.*

1. Explain how to find the x- and the y-intercepts of the graph of the ellipse

$$\frac{x^2}{a^2} + \frac{y^2}{b^2} = 1$$

2. Explain why the graph of the hyperbola

$$\frac{x^2}{a^2} - \frac{y^2}{b^2} = 1$$

has no y-intercept.

Something to Think About

1. What happens to the graph of

$$\frac{x^2}{a^2} + \frac{y^2}{b^2} = 1$$

if $a = b$?

2. The hyperbolas $x^2 - y^2 = 1$ and $y^2 - x^2 = 1$ are called **conjugate** hyperbolas. Graph both on the same axes. What do they have in common?

Review Exercises   *Find each product.*

1. $3x^{-2}y^2(4x^2 + 3y^{-2})$

2. $(2a^{-2} - b^{-2})(2a^{-2} + b^{-2})$

Write each expression without using negative exponents.

3. $\dfrac{x^{-2} + y^{-2}}{x^{-2} - y^{-2}}$

4. $\dfrac{2x^{-3} - 2y^{-3}}{4x^{-3} + 4y^{-3}}$

10.3 Solving Simultaneous Second-Degree Equations

■ SOLUTION BY GRAPHING ■ GRAPHING DEVICES ■ SOLUTION BY ELIMINATION

We now discuss ways to solve systems of two equations in two variables where at least one of the equations is of second degree.

■ SOLUTION BY GRAPHING

EXAMPLE 1   Solve the system $\begin{cases} x^2 + y^2 = 25 \\ 2x + y = 10 \end{cases}$ by graphing.

Solution   The graph of the equation $x^2 + y^2 = 25$ is a circle with center at the origin and radius of 5. The graph of the equation $2x + y = 10$ is a straight line. Depending on whether the line is a secant (intersecting the circle at two points) or a tangent (intersecting the circle at one point) or does not intersect the circle at all, there are two, one, or no solutions to the system, respectively.

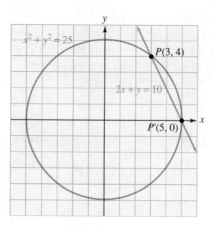

FIGURE 10-29

After graphing the circle and the line, as shown in Figure 10-29, we see that there are two intersection points, $P(3, 4)$ and $P'(5, 0)$. The solutions to the system of equations are

$$\begin{cases} x = 3 \\ y = 4 \end{cases} \quad \text{and} \quad \begin{cases} x = 5 \\ y = 0 \end{cases}$$

Verify that these are exact solutions.

■ GRAPHING DEVICES

To solve Example 1 with a graphing device, we graph the circle and the line on one set of coordinate axes, as shown in Figure 10-30(a). We then use the trace feature to find the coordinates of the intersection points of the graphs, as in Figure 10-30(b) and Figure 10-30(c). Note that no value of y is printed in Figure 10-30(c). This is because most graphing devices cannot connect two points that are nearly vertical.

We can zoom in for better results.

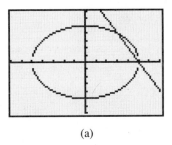

(a)

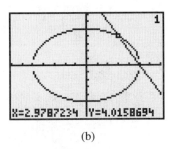

X=2.9787234 Y=4.0158694

(b)

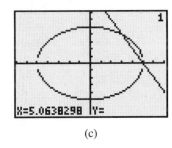

X=5.0638298 Y=

(c)

FIGURE 10-30

■ SOLUTION BY ELIMINATION

Algebraic methods can also be used to solve systems of equations.

EXAMPLE 2 Solve the system $\begin{cases} x^2 + y^2 = 25 \\ 2x + y = 10 \end{cases}$.

Solution This system has one second-degree equation and one first-degree equation. We can solve this type of system by substitution. Solving the linear equation for y gives

$$2x + y = 10$$

1. $\qquad y = -2x + 10$

We can substitute $-2x + 10$ for y in the second-degree equation and solve the resulting quadratic equation for x:

$$x^2 + y^2 = 25$$
$$x^2 + (-2x + 10)^2 = 25$$
$$x^2 + 4x^2 - 40x + 100 = 25 \qquad (-2x + 10)(-2x + 10) = 4x^2 - 40x + 100.$$

$$5x^2 - 40x + 75 = 0 \qquad \text{Combine like terms and subtract } 25 \text{ from both sides.}$$

$$x^2 - 8x + 15 = 0 \qquad \text{Divide both sides by 5.}$$

$$(x - 5)(x - 3) = 0 \qquad \text{Factor } x^2 - 8x + 15.$$

$$x - 5 = 0 \quad \text{or} \quad x - 3 = 0 \qquad \text{Set each factor equal to 0.}$$

$$x = 5 \quad | \quad x = 3$$

If we substitute 5 for x in Equation 1, we get $y = 0$. If we substitute 3 for x in Equation 1, we get $y = 4$. The two solutions are

$$\begin{cases} x = 5 \\ y = 0 \end{cases} \quad \text{or} \quad \begin{cases} x = 3 \\ y = 4 \end{cases} \qquad \blacksquare$$

EXAMPLE 3   Solve the system $\begin{cases} 4x^2 + 9y^2 = 5 \\ y = x^2 \end{cases}$.

Solution   We can solve this system by substitution.

$$4x^2 + 9y^2 = 5$$

$$4y + 9y^2 = 5 \qquad \text{Substitute } y \text{ for } x^2.$$

$$9y^2 + 4y - 5 = 0 \qquad \text{Subtract 5 from both sides.}$$

$$(9y - 5)(y + 1) = 0 \qquad \text{Factor } 9y^2 + 4y - 5.$$

$$9y - 5 = 0 \quad \text{or} \quad y + 1 = 0 \qquad \text{Set each factor equal to 0.}$$

$$y = \frac{5}{9} \quad \bigg| \quad y = -1$$

Since $y = x^2$, the values of x are found by solving the equations

$$x^2 = \frac{5}{9} \quad \text{and} \quad x^2 = -1$$

Because the equation $x^2 = -1$ has no real solutions, this possibility is discarded. The solutions of the equation $x^2 = \frac{5}{9}$ are

$$x = \frac{\sqrt{5}}{3} \quad \text{or} \quad x = -\frac{\sqrt{5}}{3}$$

The solutions of the system are

$$\left(\frac{\sqrt{5}}{3}, \frac{5}{9} \right) \quad \text{and} \quad \left(-\frac{\sqrt{5}}{3}, \frac{5}{9} \right) \qquad \blacksquare$$

EXAMPLE 4   Solve the system $\begin{cases} 3x^2 + 2y^2 = 36 \\ 4x^2 - y^2 = 4 \end{cases}$.

Solution   In this system, both equations are in the form $ax^2 + by^2 = c$. We can solve systems like this by addition.

We can copy the first equation and multiply the second equation by 2 to obtain the equivalent system of equations

$$\begin{cases} 3x^2 + 2y^2 = 36 \\ 8x^2 - 2y^2 = 8 \end{cases}$$

We add the equations to eliminate y and solve the resulting equation for x:

$$11x^2 = 44$$
$$x^2 = 4$$
$$x = 2 \quad \text{or} \quad x = -2$$

To find y, we substitute 2 for x and then -2 for x in the first equation and proceed as follows:

For $x = 2$	**For $x = -2$**
$3x^2 + 2y^2 = 36$	$3x^2 + 2y^2 = 36$
$3(2)^2 + 2y^2 = 36$	$3(-2)^2 + 2y^2 = 36$
$12 + 2y^2 = 36$	$12 + 2y^2 = 36$
$2y^2 = 24$	$2y^2 = 24$
$y^2 = 12$	$y^2 = 12$
$y = +\sqrt{12} \quad \text{or} \quad y = -\sqrt{12}$	$y = +\sqrt{12} \quad \text{or} \quad y = -\sqrt{12}$
$y = 2\sqrt{3} \quad | \quad y = -2\sqrt{3}$	$y = 2\sqrt{3} \quad | \quad y = -2\sqrt{3}$

The four solutions of this system are

$$\left(2, 2\sqrt{3}\right), \quad \left(2, -2\sqrt{3}\right), \quad \left(-2, 2\sqrt{3}\right), \quad \text{and} \quad \left(-2, -2\sqrt{3}\right) \quad \blacksquare$$

Orals *Give the possible number of solutions of a system when the graphs of the equations are*

1. a line and a parabola

2. a line and a hyperbola

3. a circle and a parabola

4. a circle and a hyperbola

EXERCISE 10.3

In Exercises 1–8, solve each system of equations by graphing.

1. $\begin{cases} 8x^2 + 32y^2 = 256 \\ x = 2y \end{cases}$

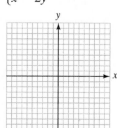

2. $\begin{cases} x^2 + y^2 = 2 \\ x + y = 2 \end{cases}$

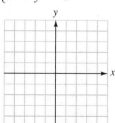

3. $\begin{cases} x^2 + y^2 = 10 \\ y = 3x^2 \end{cases}$

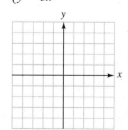

4. $\begin{cases} x^2 + y^2 = 5 \\ x + y = 3 \end{cases}$

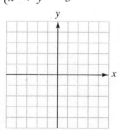

5. $\begin{cases} x^2 + y^2 = 25 \\ 12x^2 + 64y^2 = 768 \end{cases}$

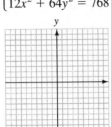

6. $\begin{cases} x^2 + y^2 = 13 \\ y = x^2 - 1 \end{cases}$

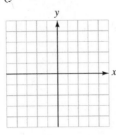

7. $\begin{cases} x^2 - 13 = -y^2 \\ y = 2x - 4 \end{cases}$

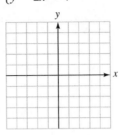

8. $\begin{cases} x^2 + y^2 = 20 \\ y = x^2 \end{cases}$

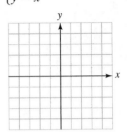

In Exercises 9–10, use a graphing device to solve each system.

9. $\begin{cases} x^2 - 6x - y = -5 \\ x^2 - 6x + y = -5 \end{cases}$

10. $\begin{cases} x^2 - y^2 = -5 \\ 3x^2 + 2y^2 = 30 \end{cases}$

In Exercises 11–36, solve each system of equations algebraically for real values of x and y.

11. $\begin{cases} 25x^2 + 9y^2 = 225 \\ 5x + 3y = 15 \end{cases}$

12. $\begin{cases} x^2 + y^2 = 20 \\ y = x^2 \end{cases}$

13. $\begin{cases} x^2 + y^2 = 2 \\ x + y = 2 \end{cases}$

14. $\begin{cases} x^2 + y^2 = 36 \\ 49x^2 + 36y^2 = 1764 \end{cases}$

15. $\begin{cases} x^2 + y^2 = 5 \\ x + y = 3 \end{cases}$

16. $\begin{cases} x^2 - x - y = 2 \\ 4x - 3y = 0 \end{cases}$

17. $\begin{cases} x^2 + y^2 = 13 \\ y = x^2 - 1 \end{cases}$

18. $\begin{cases} x^2 + y^2 = 25 \\ 2x^2 - 3y^2 = 5 \end{cases}$

19. $\begin{cases} x^2 + y^2 = 30 \\ y = x^2 \end{cases}$

20. $\begin{cases} 9x^2 - 7y^2 = 81 \\ x^2 + y^2 = 9 \end{cases}$

21. $\begin{cases} x^2 + y^2 = 13 \\ x^2 - y^2 = 5 \end{cases}$

22. $\begin{cases} 2x^2 + y^2 = 6 \\ x^2 - y^2 = 3 \end{cases}$

23. $\begin{cases} x^2 + y^2 = 20 \\ x^2 - y^2 = -12 \end{cases}$

24. $\begin{cases} xy = -\dfrac{9}{2} \\ 3x + 2y = 6 \end{cases}$

25. $\begin{cases} y^2 = 40 - x^2 \\ y = x^2 - 10 \end{cases}$

26. $\begin{cases} x^2 - 6x - y = -5 \\ x^2 - 6x + y = -5 \end{cases}$

27. $\begin{cases} y = x^2 - 4 \\ x^2 - y^2 = -16 \end{cases}$

28. $\begin{cases} 6x^2 + 8y^2 = 182 \\ 8x^2 - 3y^2 = 24 \end{cases}$

29. $\begin{cases} x^2 - y^2 = -5 \\ 3x^2 + 2y^2 = 30 \end{cases}$

30. $\begin{cases} \dfrac{1}{x} + \dfrac{1}{y} = 5 \\ \dfrac{1}{x} - \dfrac{1}{y} = -3 \end{cases}$

31. $\begin{cases} \dfrac{1}{x} + \dfrac{2}{y} = 1 \\ \dfrac{2}{x} - \dfrac{1}{y} = \dfrac{1}{3} \end{cases}$

32. $\begin{cases} \dfrac{1}{x} + \dfrac{3}{y} = 4 \\ \dfrac{2}{x} - \dfrac{1}{y} = 7 \end{cases}$

33. $\begin{cases} 3y^2 = xy \\ 2x^2 + xy - 84 = 0 \end{cases}$

34. $\begin{cases} x^2 + y^2 = 10 \\ 2x^2 - 3y^2 = 5 \end{cases}$

35. $\begin{cases} xy = \dfrac{1}{6} \\ y + x = 5xy \end{cases}$

36. $\begin{cases} xy = \dfrac{1}{12} \\ y + x = 7xy \end{cases}$

37. Geometry problem   The area of a rectangle is 63 square centimeters, and its perimeter is 32 centimeters. Find the dimensions of the rectangle.

38. Integer problem   The product of two integers is 32, and their sum is 12. Find the integers.

39. Number problem   The sum of the squares of two numbers is 221, and the sum of the numbers is 212 less. Find the numbers.

40. Investing money   Grant receives $225 annual income from one investment. Jeff invested $500 more than Grant, but at an annual rate of 1% less. Jeff's annual income is $240. What is the amount and rate of Grant's investment?

41. Investing money   Carol receives $67.50 annual income from one investment. John invested $150 more than Carol at an annual rate of $1\frac{1}{2}$% more. John's annual income is $94.50. What is the amount and rate of Carol's investment? (*Hint:* There are two answers.)

42. Artillery   The shell fired from the base of the hill in Illustration 1 follows the parabolic path $y = -\frac{1}{6}x^2 + 2x$, with distances measured in miles. The hill has a slope of $\frac{1}{3}$. How far from the gun is the point of impact? (*Hint:* Find the coordinates of the point and then the distance.)

43. Driving rates   Jim drove 306 miles. Jim's brother made the same trip at a speed of 17 miles per hour slower than Jim did and required an extra $1\frac{1}{2}$ hours. What was Jim's rate and time?

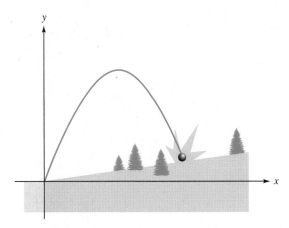

ILLUSTRATION 1

44. Fencing a pasture   A rectangular pasture is to be fenced in along a riverbank, as shown in Illustration 2. If 260 feet of fencing is to enclose an area of 8000 square feet, find the dimensions of the pasture.

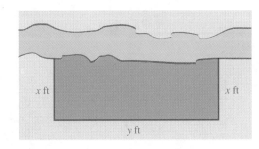

ILLUSTRATION 2

Writing Exercises   *Write a paragraph using your own words.*

1. Describe the benefits of the graphical method for solving a system of equations.

2. Describe the drawbacks of the graphical method.

Something to Think About

1. The graphs of the two equations of a system are parabolas. How many solutions might the system have?

2. The graphs of the two equations of a system are hyperbolas. How many solutions might the system have?

Review Exercises   *Simplify each radical expression. Assume that all variables represent positive numbers.*

1. $\sqrt{200x^2} - 3\sqrt{98x^2}$

2. $a\sqrt{112a} - 5\sqrt{175a^3}$

3. $\dfrac{3t\sqrt{2t} - 2\sqrt{2t^3}}{\sqrt{18t} - \sqrt{2t}}$

4. $\sqrt[3]{\dfrac{x}{4}} + \sqrt[3]{\dfrac{x}{32}} - \sqrt[3]{\dfrac{x}{500}}$

10.4 Piecewise-Defined Functions and the Greatest Integer Function

■ PIECEWISE-DEFINED FUNCTIONS ■ INCREASING AND DECREASING FUNCTIONS ■ THE GREATEST INTEGER FUNCTION

In this section, we will discuss functions that are defined by using different equations for different parts of their domains. Such functions are called **piecewise-defined functions**.

■ PIECEWISE-DEFINED FUNCTIONS

A simple piecewise-defined function is the absolute value function, $f(x) = |x|$, which can be written in the form

$$f(x) = \begin{cases} x & \text{when } x \geq 0 \\ -x & \text{when } x < 0 \end{cases}$$

When x is in the interval $[0, \infty)$, we use the function $f(x) = x$ to evaluate $|x|$. However, when x is in the interval $(-\infty, 0)$, we use the function $f(x) = -x$ to evaluate $|x|$. The graph of the absolute value function is shown in Figure 10-31.

For $x \geq 0$

x	$f(x)$	$(x, f(x))$
0	0	$(0, 0)$
1	1	$(1, 1)$
2	2	$(2, 2)$
3	3	$(3, 3)$

For $x < 0$

x	$f(x)$	$(x, f(x))$
-4	4	$(-4, 4)$
-3	3	$(-3, 3)$
-2	2	$(-2, 2)$
-1	1	$(-1, 1)$

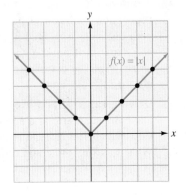

FIGURE 10-31

■ INCREASING AND DECREASING FUNCTIONS

If the values of $f(x)$ increase as x increases on an interval, we say that the function is *increasing on the interval*. (See Figure 10-32(a).) If the values of $f(x)$ decrease

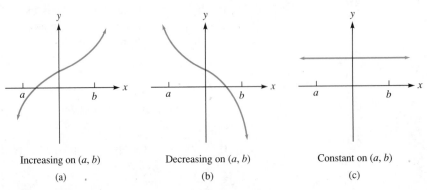

Increasing on (a, b)

(a)

Decreasing on (a, b)

(b)

Constant on (a, b)

(c)

FIGURE 10-32

as x increases on an interval, we say that the function is *decreasing on the interval*. (See Figure 10-32(b).) If the values of $f(x)$ remain constant as x increases on an interval, we say that the function is *constant on the interval*. (See Figure 10-32(c).)

The absolute value function, pictured in Figure 10-31, is decreasing on the interval $(-\infty, 0)$ and is increasing on the interval $(0, \infty)$.

EXAMPLE 1   Graph the piecewise-defined function given by

$$f(x) = \begin{cases} x^2 & \text{when } x \le 0 \\ x & \text{when } 0 < x < 2 \\ -1 & \text{when } x \ge 2 \end{cases}$$

and tell where the function is increasing, decreasing, or constant.

Solution   For each number x, we decide which of these three equations will be used to find the corresponding value of y:

- For numbers $x \le 0$, $f(x)$ is determined by $f(x) = x^2$, and the graph is the left half of a parabola. (See Figure 10-33.) Since the values of $f(x)$ decrease on this graph as x increases, the function is decreasing on the interval $(-\infty, 0)$.
- For numbers x between 0 and 2, $f(x)$ is determined by $f(x) = x$, and the graph is part of a line. Since the values of $f(x)$ increase on this graph as x increases, the function is increasing on the interval $(0, 2)$.
- For numbers $x \ge 2$, $f(x)$ is the constant -1, and the graph is part of a horizontal line. Since the values of $f(x)$ remain constant on this line, the function is constant on the interval $(2, \infty)$.

The use of solid and open circles in the graph indicates that $f(x) = -1$ when $x = 2$.

The domain of this function is the interval $(-\infty, \infty)$, because every number x gives a single value y. The range is the interval $[0, \infty) \cup \{-1\}$.

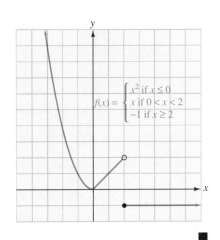

| If $x \le 0$ | | | If $0 < x < 2$ | | | If $x \ge 2$ | | |
| $f(x) = x^2$ | | | $f(x) = x$ | | | $f(x) = -1$ | | |
x	$f(x)$	$(x, f(x))$	x	$f(x)$	$(x, f(x))$	x	$f(x)$	$(x, f(x))$
0	0	$(0, 0)$	$\frac{1}{2}$	$\frac{1}{2}$	$\left(\frac{1}{2}, \frac{1}{2}\right)$	2	-1	$(2, -1)$
-1	1	$(-1, 1)$	1	1	$(1, 1)$	3	-1	$(3, -1)$
-2	4	$(-2, 4)$	$\frac{3}{2}$	$\frac{3}{2}$	$\left(\frac{3}{2}, \frac{3}{2}\right)$	5	-1	$(5, -1)$
-3	9	$(-3, 9)$						

FIGURE 10-33

■ THE GREATEST INTEGER FUNCTION

The **greatest integer function** is important in computer applications. This function is determined by the equation

$$y = f(x) = [\![x]\!] \qquad \text{Read as "} y \text{ equals the greatest integer in } x \text{."}$$

where the value of y that corresponds to x is the greatest integer that is less than or equal to x. For example,

$$[\![4.7]\!] = 4, \qquad \left[\!\left[2\frac{1}{2}\right]\!\right] = 2, \qquad [\![\pi]\!] = 3, \qquad [\![-3.7]\!] = -4, \qquad [\![-5.7]\!] = -6$$

EXAMPLE 2   Graph $f(x) = [\![x]\!]$.

Solution   We list several intervals and the corresponding values of the greatest integer function:

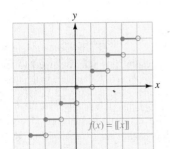

$[0, 1)$ $f(x) = [\![x]\!] = 0$ For numbers from 0 to 1, not including 1, the greatest integer in the interval is 0.

$[1, 2)$ $f(x) = [\![x]\!] = 1$ For numbers from 1 to 2, not including 2, the greatest integer in the interval is 1.

$[2, 3)$ $f(x) = [\![x]\!] = 2$ For numbers from 2 to 3, not including 3, the greatest integer in the interval is 2.

In each interval, the values of y are constant, but they jump by 1 at integer values of x. The graph is shown in Figure 10-34. From the graph, we see that the domain is $(-\infty, \infty)$, and the range is the set of integers $\{\ldots, -3, -2, -1, 0, 1, 2, 3, \ldots\}$. ∎

FIGURE 10-34

Since the greatest integer function is made up of a series of horizontal line segments, it is an example of a group of functions called **step functions**.

EXAMPLE 3   **Ordering stationery**   To print stationery, a printer charges $10 for setup charges, plus $20 for each box. The printer counts any portion of a box as a full box. Graph this step function.

Solution   If we order stationery and cancel before it is printed, the cost will be $10. Thus, the ordered pair $(0, 10)$ will be on the graph.

If we purchase 1 box, the cost will be $10 for setup plus $20 for printing, for a total cost of $30. Thus, the ordered pair $(1, 30)$ will be on the graph.

The cost of $1\frac{1}{2}$ boxes will be the same as the cost of 2 boxes, or $50. Thus, the ordered pairs $(1.5, 50)$ and $(2, 50)$ will be on the graph.

The complete graph is shown in Figure 10-35.

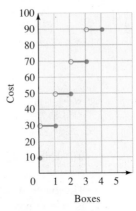

FIGURE 10-35

Orals *Tell whether each function is increasing, decreasing, or constant on the interval* $(-2, 3)$.

1.

2.

3.

4.

EXERCISE 10.4

In Exercises 1–4, give the intervals on which each function is increasing, decreasing, or constant.

1.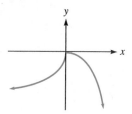

2.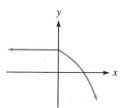

3.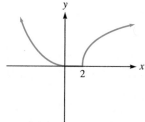

4.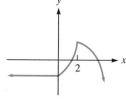

In Exercises 5–8, graph each function and give the intervals on which f is increasing, decreasing, or constant.

5. $f(x) = \begin{cases} -1 & \text{if } x \le 0 \\ x & \text{if } x > 0 \end{cases}$

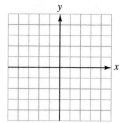

6. $f(x) = \begin{cases} -2 & \text{if } x \le 0 \\ x^2 & \text{if } x > 0 \end{cases}$

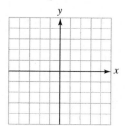

7. $f(x) = \begin{cases} -x & \text{if } x \le 0 \\ x & \text{if } 0 < x < 2 \\ -x & \text{if } x \ge 2 \end{cases}$

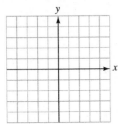

8. $f(x) = \begin{cases} -x & \text{if } x < 0 \\ x^2 & \text{if } 0 \le x \le 1 \\ 1 & \text{if } x > 1 \end{cases}$

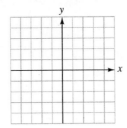

In Exercises 9–12, graph each function.

9. $f(x) = -[\![x]\!]$

10. $f(x) = [\![x]\!] + 2$

11. $f(x) = 2[\![x]\!]$

12. $f(x) = \left[\!\left[\dfrac{1}{2}x\right]\!\right]$

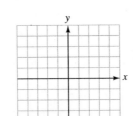

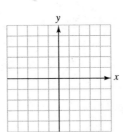

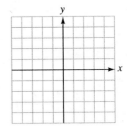

13. Signum function   Computer programmers use a function, denoted by $f(x) = \text{sgn } x$, that is defined in the following way:

$$f(x) = \begin{cases} -1 & \text{if } x < 0 \\ 0 & \text{if } x = 0 \\ 1 & \text{if } x > 0 \end{cases}$$

Graph this function.

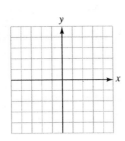

14. Heaviside unit step function   This function, used in calculus, is defined by

$$f(x) = \begin{cases} 1 & \text{if } x > 0 \\ 0 & \text{if } x < 0 \end{cases}$$

Graph this function.

15. Renting a jet ski   A marina charges $20 to rent a jet ski for 1 hour, plus $5 for every extra hour (or portion of an hour). Graph the ordered pairs (h, c), where h represents the number of hours and c represents the cost. Find the cost if the ski is used for 2.5 hours.

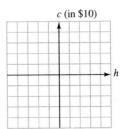

16. Riding in a taxi   A cab company charges $3 for a trip up to 1 mile, and $2 for every extra mile (or portion of a mile). Graph the ordered pairs (m, c), where m represents the number of miles traveled and c represents the cost. Find the cost to ride $10\frac{1}{4}$ miles.

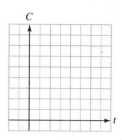

18. Royalties   A publisher has agreed to pay the author of a novel 7% royalties on sales of the first 50,000 copies and 10% on sales thereafter. If the book sells for $10, express the royalty income, I, as a function of s, the number of copies sold, and graph the function. (*Hint:* When sales are into the second 50,000 copies, how much was earned on the first 50,000?)

17. Information access   Computer access to one international data network A costs $10 per day plus $8 per hour or fraction of an hour. Network B charges $15 per day but only $6 per hour or fraction of an hour. For each network, graph the ordered pairs (t, C), where t represents the connect time and C represents the total cost. Find the minimal daily usage at which it would be more economical to use network B.

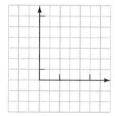

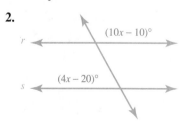

Writing Exercises   *Write a paragraph using your own words.*

1. Tell how to decide whether a function is increasing on the interval (a, b).

2. Describe the greatest integer function.

Something to Think About   **1.** Find a piecewise-defined function that is increasing on the interval $(-\infty, -2)$ and decreasing on the interval $(-2, \infty)$.

2. Find a piecewise-defined function that is constant on the interval $(-\infty, 0)$, increasing on the interval $(0, 5)$, and decreasing on the interval $(5, \infty)$.

Review Exercises   *Find the value of x. Assume that lines r and s are parallel.*

1.

r

$(6x - 10)°$

$(3x + 10)°$

s

2.

r

$(10x - 10)°$

$(4x - 20)°$

s

10.5 The Binomial Theorem

We have discussed how to raise binomials to positive integral powers. For example, we know that

$$(a + b)^2 = a^2 + 2ab + b^2$$

and that

$$(a + b)^3 = (a + b)(a + b)^2$$
$$= (a + b)(a^2 + 2ab + b^2)$$
$$= a^3 + 2a^2b + ab^2 + a^2b + 2ab^2 + b^3$$
$$= a^3 + 3a^2b + 3ab^2 + b^3$$

To show how to raise binomials to positive integral powers without doing the actual multiplications, we consider the following binomial expansions:

$$(a + b)^0 = 1$$
$$(a + b)^1 = a + b$$
$$(a + b)^2 = a^2 + 2ab + b^2$$
$$(a + b)^3 = a^3 + 3a^2b + 3ab^2 + b^3$$
$$(a + b)^4 = a^4 + 4a^3b + 6a^2b^2 + 4ab^3 + b^4$$
$$(a + b)^5 = a^5 + 5a^4b + 10a^3b^2 + 10a^2b^3 + 5ab^4 + b^5$$
$$(a + b)^6 = a^6 + 6a^5b + 15a^4b^2 + 20a^3b^3 + 15a^2b^4 + 6ab^5 + b^6$$

Several patterns appear in these expansions:

1. Each expansion has one more term than the power of the binomial.
2. The degree of each term in each expansion is equal to the exponent of the binomial that is being expanded.
3. The first term in each expansion is a, raised to the power of the binomial.
4. The exponents of a decrease by 1 in each successive term. The exponents of b, beginning with $b^0 = 1$ in the first term, increase by 1 in each successive term. Thus, the variables have the pattern

$$a^n, a^{n-1}b, a^{n-2}b^2, \ldots, ab^{n-1}, b^n$$

■ PASCAL'S TRIANGLE

To see another pattern, we write the coefficients of each binomial expansion in the following triangular array:

```
            1
          1   1
        1   2   1
      1   3   3   1
    1   4   6   4   1
  1   5   10   10   5   1
1   6   15   20   15   6   1
```

In this array, called **Pascal's triangle**, each entry between the 1's is the sum of the closest pair of numbers in the line immediately above it. For example, the first 15 in the bottom row is the sum of the 5 and 10 immediately above it. Pascal's triangle continues with the same pattern forever. The next two lines are

$$1 \quad 7 \quad 21 \quad 35 \quad 35 \quad 21 \quad 7 \quad 1$$
$$1 \quad 8 \quad 28 \quad 56 \quad 70 \quad 56 \quad 28 \quad 8 \quad 1$$

EXAMPLE 1 Expand $(x + y)^5$.

Solution The first term in the expansion is x^5, and the exponents of x decrease by 1 in each successive term. A y first appears in the second term, and the exponents of y increase by 1 in each successive term, concluding when the term y^5 is reached. Thus, the variables in the expansion are

$$x^5, \quad x^4y, \quad x^3y^2, \quad x^2y^3, \quad xy^4, \quad y^5$$

The coefficients of these variables are given in Pascal's triangle in the row whose second entry is 5, the same as the exponent of the binomial being expanded:

$$1 \quad 5 \quad 10 \quad 10 \quad 5 \quad 1$$

Putting these two pieces of information together gives the required expansion:

$$(x + y)^5 = x^5 + 5x^4y + 10x^3y^2 + 10x^2y^3 + 5xy^4 + y^5 \qquad \blacksquare$$

EXAMPLE 2 Expand $(u - v)^4$.

Solution We note that the expression $(u - v)^4$ can be written in the form $[u + (-v)]^4$. The variables in this expansion are

$$u^4, \quad u^3(-v), \quad u^2(-v)^2, \quad u(-v)^3, \quad (-v)^4$$

and the coefficients are given in Pascal's triangle in the row whose second entry is 4:

$$1 \quad 4 \quad 6 \quad 4 \quad 1$$

Hence, the required expansion is

$$(u - v)^4 = u^4 + 4u^3(-v) + 6u^2(-v)^2 + 4u(-v)^3 + (-v)^4$$
$$= u^4 - 4u^3v + 6u^2v^2 - 4uv^3 + v^4 \qquad \blacksquare$$

■ FACTORIAL NOTATION

Although Pascal's triangle gives the coefficients of the terms in a binomial expansion, it is not the best way to expand a binomial. To develop another way to expand a binomial, we introduce **factorial notation**.

> **Factorial Notation**
> If n is a natural number, the symbol $n!$ (read as **n factorial** or as **factorial n**) is defined as
> $$n! = n(n - 1)(n - 2)(n - 3) \cdot \; \cdots \; \cdot (3)(2)(1)$$

(b,d)

EXAMPLE 3 Find **a.** 2!, **b.** 5!, **c.** $-9!$, and **d.** $(n - 2)!$.

Solution **a.** $2! = 2 \cdot 1 = 2$ **b.** $5! = 5 \cdot 4 \cdot 3 \cdot 2 \cdot 1 = 120$

c. $-9! = -9 \cdot 8 \cdot 7 \cdot 6 \cdot 5 \cdot 4 \cdot 3 \cdot 2 \cdot 1 = -362,880$

d. $(n - 2)! = (n - 2)(n - 3)(n - 4) \cdots \cdots 3 \cdot 2 \cdot 1$

WARNING! According to the previous definition, part **d** is meaningful only if $n - 2$ is a natural number. ■

We define zero factorial as follows.

> **0 Factorial**
> $$0! = 1$$

We note that

$$5 \cdot 4! = 5 \cdot 4 \cdot 3 \cdot 2 \cdot 1 = 5!$$
$$7 \cdot 6! = 7 \cdot 6 \cdot 5 \cdot 4 \cdot 3 \cdot 2 \cdot 1 = 7!$$
$$10 \cdot 9! = 10 \cdot 9 \cdot 8 \cdot 7 \cdot 6 \cdot 5 \cdot 4 \cdot 3 \cdot 2 \cdot 1 = 10!$$

These examples suggest the following theorem.

> **Theorem**
> If n is a positive integer, then $n(n - 1)! = n!$.

■ THE BINOMIAL THEOREM

We can now state the binomial theorem.

> **The Binomial Theorem**
> If n is any positive integer, then
> $$(a + b)^n = a^n + \frac{n!}{1!(n - 1)!} a^{n-1}b + \frac{n!}{2!(n - 2)!} a^{n-2}b^2$$
> $$+ \frac{n!}{3!(n - 3)!} a^{n-3}b^3 + \cdots + \frac{n!}{r!(n - r)!} a^{n-r}b^r$$
> $$+ \cdots + b^n$$

In the binomial theorem, the exponents of the variables follow the familiar pattern:

- The sum of the exponents of a and b in each term is n,
- the exponents of a decrease, and
- the exponents of b increase.

Only the method of finding the coefficients is different. Except for the first and last terms, the numerator of each coefficient is $n!$. If the exponent of b in a particular term is r, the denominator of the coefficient of that term is $r!(n - r)!$.

EXAMPLE 4   Use the binomial theorem to expand $(a + b)^3$.

Solution   We can substitute directly into the binomial theorem and simplify:

$$(a + b)^3 = a^3 + \frac{3!}{1!(3 - 1)!} a^2b + \frac{3!}{2!(3 - 2)!} ab^2 + b^3$$

$$= a^3 + \frac{3 \cdot 2 \cdot 1}{1 \cdot 2 \cdot 1} a^2b + \frac{3 \cdot 2 \cdot 1}{2 \cdot 1 \cdot 1} ab^2 + b^3$$

$$= a^3 + 3a^2b + 3ab^2 + b^3 \qquad \blacksquare$$

EXAMPLE 5   Use the binomial theorem to expand $(x - y)^4$.

Solution   We can write $(x - y)^4$ in the form $[x + (-y)]^4$, substitute directly into the binomial theorem, and simplify:

$$(x - y)^4 = [x + (-y)]^4$$

$$= x^4 + \frac{4!}{1!(4 - 1)!} x^3(-y) + \frac{4!}{2!(4 - 2)!} x^2(-y)^2 + \frac{4!}{3!(4 - 3)!} x(-y)^3 + (-y)^4$$

$$= x^4 - \frac{4 \cdot 3!}{1!3!} x^3y + \frac{4 \cdot 3 \cdot 2!}{2!2!} x^2y^2 - \frac{4 \cdot 3!}{3!1!} xy^3 + y^4$$

$$= x^4 - 4x^3y + 6x^2y^2 - 4xy^3 + y^4 \qquad \blacksquare$$

EXAMPLE 6   Use the binomial theorem to expand $(3u - 2v)^4$.

Solution   We write $(3u - 2v)^4$ in the form $[3u + (-2v)]^4$ and let $a = 3u$ and $b = -2v$. Then we can use the binomial theorem to expand $(a + b)^4$.

$$(a + b)^4 = a^4 + \frac{4!}{1!(4 - 1)!} a^3b + \frac{4!}{2!(4 - 2)!} a^2b^2 + \frac{4!}{3!(4 - 3)!} ab^3 + b^4$$

$$= a^4 + 4a^3b + 6a^2b^2 + 4ab^3 + b^4$$

Now we can substitute $3u$ for a and $-2v$ for b and simplify:

$$(3u - 2v)^4 = (3u)^4 + 4(3u)^3(-2v) + 6(3u)^2(-2v)^2 + 4(3u)(-2v)^3 + (-2v)^4$$

$$= 81u^4 - 216u^3v + 216u^2v^2 - 96uv^3 + 16v^4 \qquad \blacksquare$$

■ THE NTH TERM OF A BINOMIAL EXPANSION

To find the fourth term of the expansion of $(a + b)^9$, for example, we could raise the binomial $a + b$ to the 9th power and look at the fourth term. However, this task would be very tedious. By using the binomial theorem, we can construct the fourth term without finding the complete expansion of $(a + b)^9$.

EXAMPLE 7 Find the fourth term in the expansion of $(a + b)^9$.

Solution Since b^1 appears in the second term, b^2 appears in the third term, and so on, the exponent of b in the fourth term is 3. Since the exponent of b added to the exponent of a must equal 9, the exponent of a must be 6. Thus, the variables of the fourth term are

$$a^6b^3 \qquad \text{The sum of the exponents must be 9.}$$

Because of the binomial theorem, the coefficient of the variables must be

$$\frac{n!}{r!(n-r)!} = \frac{9!}{3!(9-3)!}$$

Thus, the complete fourth term is

$$\frac{9!}{3!(9-3)!} a^6b^3 = \frac{9 \cdot 8 \cdot 7 \cdot 6!}{3 \cdot 2 \cdot 1 \cdot 6!} a^6b^3 = 84a^6b^3$$ ∎

EXAMPLE 8 Find the sixth term in the expansion of $(x - y)^7$.

Solution We first find the sixth term of $[x + (-y)]^7$. In the sixth term, the exponent of $(-y)$ is 5. Thus, the variables in the sixth term are

$$x^2(-y)^5 \qquad \text{The sum of the exponents must be 7.}$$

The coefficient of these variables is

$$\frac{n!}{r!(n-r)!} = \frac{7!}{5!(7-5)!}$$

The complete sixth term is

$$\frac{7!}{5!(7-5)!} x^2(-y)^5 = -\frac{7 \cdot 6 \cdot 5!}{5! \cdot 2 \cdot 1} x^2y^5 = -21x^2y^5$$ ∎

EXAMPLE 9 Find the fourth term of the expansion of $(2x - 3y)^6$.

Solution We can let $a = 2x$ and $b = -3y$ and find the fourth term of the expansion of $(a + b)^6$:

$$\frac{6!}{3!(6-3)!} a^3b^3 = \frac{6 \cdot 5 \cdot 4 \cdot 3!}{3! \cdot 3 \cdot 2 \cdot 1} a^3b^3 = 20a^3b^3$$

We can now substitute $2x$ for a and $-3y$ for b and simplify:

$$20a^3b^3 = 20(2x)^3(-3y)^3 = -4320x^3y^3$$

The fourth term in the expansion of $(2x - 3y)^6$ is $-4320x^3y^3$. ∎

Orals *Find each value.*

1. 1! **2.** 4! **3.** 0! **4.** 5!

Expand each binomial.

5. $(m + n)^2$ **6.** $(m - n)^2$

In the expansion of $(x + y)^8$, find the exponent of y in the

7. 3rd term **8.** 4th term **9.** 7th term

In the expansion of $(x + y)^8$, find the exponent of x in the

10. 3rd term **11.** 4th term **12.** 7th term

EXERCISE 10.5

In Exercises 1–16, evaluate each expression.

1. 3! **2.** 7! **3.** $-5!$ **4.** $-6!$

5. $8(7!)$ **6.** $4!(5)$ **7.** $\dfrac{9!}{11!}$ **8.** $\dfrac{13!}{10!}$

9. $\dfrac{49!}{47!}$ **10.** $\dfrac{101!}{100!}$ **11.** $\dfrac{5!}{3!(5-3)!}$ **12.** $\dfrac{6!}{4!(6-4)!}$

13. $\dfrac{7!}{5!(7-5)!}$ **14.** $\dfrac{8!}{6!(8-6)!}$ **15.** $\dfrac{5!(8-5)!}{4!7!}$ **16.** $\dfrac{6!7!}{(8-3)!(7-4)!}$

In Exercises 17–30, use the binomial theorem to expand each expression.

17. $(x + y)^3$ **18.** $(x - y)^3$

19. $(x + y)^4$ **20.** $(x - y)^4$

21. $(2x + y)^3$ **22.** $(x + 2y)^3$

23. $(x - 2y)^3$ **24.** $(2x - y)^3$

25. $(2x + 3y)^3$ **26.** $(3x - 2y)^3$

27. $\left(\dfrac{x}{2} - \dfrac{y}{3}\right)^3$ **28.** $\left(\dfrac{x}{3} + \dfrac{y}{2}\right)^3$

29. $(3 + 2y)^4$ **30.** $(2x + 3)^4$

31. Without referring to the text, write the first ten rows of Pascal's triangle.

32. Find the sum of the numbers in each row of the first ten rows of Pascal's triangle. What is the pattern?

In Exercises 33–58, use the binomial theorem to find the required term of each expansion.

33. $(a + b)^3$; 2nd term **34.** $(a + b)^3$; 3rd term **35.** $(x - y)^4$; 4th term

36. $(x - y)^5$; 2nd term **37.** $(x + y)^6$; 5th term **38.** $(x + y)^7$; 5th term

39. $(x - y)^8$; 3rd term **40.** $(x - y)^9$; 7th term **41.** $(x + 3)^5$; 3rd term

42. $(x - 2)^4$; 2nd term **43.** $(4x + y)^5$; 3rd term **44.** $(x + 4y)^5$; 4th term

45. $(x - 3y)^4$; 2nd term **46.** $(3x - y)^5$; 3rd term **47.** $(2x - 5)^7$; 4th term

48. $(2x + 3)^6$; 6th term

49. $(2x - 3y)^5$; 5th term

50. $(3x - 2y)^4$; 2nd term

51. $\left(\dfrac{x}{2} - \dfrac{y}{3}\right)^4$; 2nd term

52. $\left(\dfrac{x}{3} + \dfrac{y}{2}\right)^5$; 4th term

53. $(a + b)^n$; 4th term

54. $(a + b)^n$; 3rd term

55. $(a - b)^n$; 5th term

56. $(a - b)^n$; 6th term

57. $(a + b)^n$; rth term

58. $(a + b)^n$; $(r + 1)$th term

Writing Exercises

Write a paragraph using your own words.

1. Tell how to construct Pascal's triangle.

2. Tell how to find the variables of the terms in the expansion of $(r + s)^4$.

3. Tell how to find the coefficients of the terms in the expansion of $(x + y)^5$.

4. Explain why the signs alternate in the expansion of $(x - y)^9$.

Something to Think About

1. If we apply the pattern of the coefficients of the binomial theorem to the coefficient of the first term in a binomial expansion, the coefficient would be $\frac{n!}{0!(n - 0)!}$. Show that this expression is 1.

2. If we apply the pattern of the coefficients of the binomial theorem to the coefficient of the last term in a binomial expansion, the coefficient would be $\frac{n!}{n!(n - n)!}$. Show that this expression is 1.

3. Find the constant term in the expansion of $\left(x + \frac{1}{x}\right)^{10}$.

4. Find the coefficient of a^5 in the expansion of $\left(a - \frac{1}{a}\right)^9$.

Review Exercises

Find each value of x.

1. $\log_4 16 = x$

2. $\log_x 49 = 2$

3. $\log_{25} x = \dfrac{1}{2}$

4. $\log_{1/2} \dfrac{1}{8} = x$

Solve each system of equations.

5. $\begin{cases} 3x + 2y = 12 \\ 2x - y = 1 \end{cases}$

6. $\begin{cases} a + b + c = 6 \\ 2a + b + 3c = 11 \\ 3a - b - c = 6 \end{cases}$

Evaluate each determinant.

7. $\begin{vmatrix} 2 & -3 \\ 4 & -2 \end{vmatrix}$

8. $\begin{vmatrix} 1 & 2 & 3 \\ 4 & 5 & 0 \\ -1 & -2 & 1 \end{vmatrix}$

10.6 Arithmetic Sequences

■ ARITHMETIC SEQUENCES ■ ARITHMETIC MEANS ■ THE SUM OF THE FIRST *N* TERMS OF AN
ARITHMETIC SEQUENCE ■ SUMMATION NOTATION ■ ACCENT ON STATISTICS

A **sequence** is a function whose domain is the set of natural numbers. For example, the function $f(n) = 3n + 2$, where n is a natural number, is a sequence. Because a sequence is a function whose domain is the set of natural numbers, it is easy to write its values as a list. If the natural numbers are substituted for n, the function $f(n) = 3n + 2$ generates the list

5, 8, 11, 14, 17, . . .

It is common to call the list, as well as the function, a sequence. Each number in the list is called a **term** of the sequence. Other examples of sequences are

$1^3, 2^3, 3^3, 4^3, \ldots$ — The ordered list of the cubes of the natural numbers.

4, 8, 12, 16, . . . — The ordered list of the positive multiples of 4.

2, 3, 5, 7, 11, . . . — The ordered list of prime numbers.

1, 1, 2, 3, 5, 8, 13, 21, . . . — The Fibonacci sequence.

The **Fibonacci sequence** is named after the 12th-century mathematician Leonardo of Pisa—also known as Fibonacci. Beginning with the 2, each term of the sequence is the sum of the two preceding terms.

■ ARITHMETIC SEQUENCES

One important type of sequence is an **arithmetic sequence**.

> **Arithmetic Sequence**
> An **arithmetic sequence** is a sequence of the form
> $$a, a + d, a + 2d, a + 3d, \ldots, a + (n - 1)d, \ldots$$
> where a is the first term, $a + (n - 1)d$ is the nth term, and d is the common difference.

We note that the second term of an arithmetic sequence has an addend of $1d$, the third term has an addend of $2d$, the fourth term has an addend of $3d$, and the nth term has an addend of $(n - 1)d$. We also note that the difference between any two consecutive terms in an arithmetic sequence is d.

EXAMPLE 1 An arithmetic sequence has a first term of 5 and a common difference of 4.

a. Write the first six terms of the sequence.

b. Write the 25th term of the sequence.

Solution **a.** Because the first term is $a = 5$ and the common difference is $d = 4$, the first six terms are

5, 5 + 4, 5 + 2(4), 5 + 3(4), 5 + 4(4), 5 + 5(4)

or

5, 9, 13, 17, 21, 25

b. The nth term is $a + (n - 1)d$. Because we want the 25th term, we let $n = 25$:

nth term $= a + (n - 1)d$

25th term $= 5 + (25 - 1)4$ Remember that $a = 5$ and $d = 4$.

$= 5 + 24(4)$

$= 5 + 96$

$= 101$

EXAMPLE 2   The first three terms of an arithmetic sequence are 3, 8, and 13. Find **a.** the 67th term and **b.** the 100th term.

Solution   We first find d, the common difference. It is the difference between successive terms:

$$d = 8 - 3 = 13 - 8 = 5$$

a. We substitute 3 for a, 67 for n, and 5 for d in the formula for the nth term and simplify:

$$n\text{th term} = a + (n - 1)d$$
$$67\text{th term} = 3 + (67 - 1)5$$
$$= 3 + 66(5)$$
$$= 333$$

b. We substitute 3 for a, 100 for n, and 5 for d in the formula for the nth term and simplify:

$$n\text{th term} = a + (n - 1)d$$
$$100\text{th term} = 3 + (100 - 1)5$$
$$= 3 + 99(5)$$
$$= 498$$

EXAMPLE 3   The first term of an arithmetic sequence is 12, and the 50th term is 3099. Write the first six terms of the sequence.

Solution   The key is to find the common difference. Because the 50th term of this sequence is 3099, we can let $n = 50$ and solve the following equation for d:

$$50\text{th term} = a + (n - 1)d$$
$$3099 = 12 + (50 - 1)d$$
$$3099 = 12 + 49d \qquad \text{Simplify.}$$
$$3087 = 49d \qquad \text{Subtract 12 from both sides.}$$
$$63 = d \qquad \text{Divide both sides by 49.}$$

The first term of the sequence is 12, and the common difference is 63. Thus, its first six terms are

$$12, 75, 138, 201, 264, 327$$

■ ARITHMETIC MEANS

If numbers are inserted between two numbers a and b to form an arithmetic sequence, the inserted numbers are called **arithmetic means** between a and b.

If a single number is inserted between the numbers a and b, that number is called **the arithmetic mean** between a and b.

EXAMPLE 4   Insert two arithmetic means between 6 and 27.

Solution   In this example, the first term is $a = 6$, and the fourth term (or the last term) is $l = 27$. We must find the common difference such that the terms

$$6, 6 + d, 6 + 2d, 27$$

form an arithmetic sequence. To find d, we can substitute 6 for a and 4 for n in the formula for the nth term:

$$n\text{th term} = a + (n - 1)d$$
$$4\text{th term} = 6 + (4 - 1)d$$
$$27 = 6 + 3d \qquad \text{Simplify.}$$
$$21 = 3d \qquad \text{Subtract 6 from both sides.}$$
$$7 = d \qquad \text{Divide both sides by 3.}$$

The two arithmetic means between 6 and 27 are

$$6 + d = 6 + 7 \quad \text{or} \quad 6 + 2d = 6 + 2(7)$$
$$= 13 \qquad\qquad\qquad = 6 + 14$$
$$\qquad\qquad\qquad\qquad = 20$$

The numbers 6, 13, 20, and 27 are the first four terms of an arithmetic sequence.

■

■ THE SUM OF THE FIRST N TERMS OF AN ARITHMETIC SEQUENCE

There is a formula that gives the sum of the first n terms of an arithmetic sequence. To develop this formula, we let S_n represent the sum of the first n terms of an arithmetic sequence:

$$S_n = \qquad a \qquad + \qquad [a + d] \qquad + \qquad [a + 2d] \qquad + \cdots + [a + (n - 1)d]$$

We write the same sum again, but in reverse order:

$$S_n = [a + (n - 1)d] + [a + (n - 2)d] + [a + (n - 3)d] + \cdots + \qquad a$$

We add these two equations together, term by term, to get

$$2S_n = [2a + (n - 1)d] + [2a + (n - 1)d] + [2a + (n - 1)d] + \cdots + [2a + (n - 1)d]$$

Because there are n equal terms on the right-hand side of the preceding equation, we can write

$$2S_n = n[2a + (n - 1)d]$$
$$2S_n = n[a + a + (n - 1)d]$$
$$2S_n = n[a + l] \qquad\qquad \text{Substitute } l \text{ for } a + (n - 1)d, \text{ because}$$
$$\qquad\qquad\qquad\qquad\qquad a + (n - 1)d \text{ is the last term of the sequence.}$$
$$S_n = \frac{n(a + l)}{2}$$

This reasoning establishes the following fact.

Sum of the First n Terms of an Arithmetic Sequence
The sum of the first n terms of an arithmetic sequence is given by the formula

$$S_n = \frac{n(a + l)}{2} \qquad \text{with } l = a + (n - 1)d$$

where a is the first term, l is the last (or nth) term, and n is the number of terms in the sequence.

EXAMPLE 5   Find the sum of the first 40 terms of the arithmetic sequence 4, 10, 16, ...

Solution   In this example, we let $a = 4$, $n = 40$, $d = 6$, and $l = 4 + (40 - 1)6 = 238$ and substitute these values into the formula for S_n:

$$S_n = \frac{n(a + l)}{2}$$

$$S_{40} = \frac{40(4 + 238)}{2}$$

$$= 20(242)$$

$$= 4840$$

The sum of the first 40 terms is 4840.   ∎

■ SUMMATION NOTATION

There is a shorthand notation for indicating the sum of a finite (ending) number of consecutive terms in a sequence. This notation, called **summation notation**, involves the Greek letter Σ (sigma). The expression

$$\sum_{k=2}^{5} 3k$$ Read as "the summation of $3k$ as k runs from 2 to 5."

designates the sum of all terms obtained if we successively substitute the numbers 2, 3, 4, and 5 for k, called the **index of the summation**. Thus, we have

$$\sum_{k=2}^{5} 3k = 3(2) + 3(3) + 3(4) + 3(5)$$

$$= 6 + 9 + 12 + 15$$

$$= 42$$

EXAMPLE 6   Find each sum: **a.** $\displaystyle\sum_{k=3}^{5} (2k + 1)$,   **b.** $\displaystyle\sum_{k=2}^{5} k^2$,   and   **c.** $\displaystyle\sum_{k=1}^{3} (3k^2 + 3)$.

Solution   **a.** $\displaystyle\sum_{k=3}^{5} (2k + 1) = [2(3) + 1] + [2(4) + 1] + [2(5) + 1]$

$$= 7 + 9 + 11$$

$$= 27$$

b. $\displaystyle\sum_{k=2}^{5} k^2 = 2^2 + 3^2 + 4^2 + 5^2$

$$= 4 + 9 + 16 + 25$$

$$= 54$$

c. $\displaystyle\sum_{k=1}^{3} (3k^2 + 3) = [3(1^2) + 3] + [3(2^2) + 3] + [3(3^2) + 3]$

$$= 6 + 15 + 30$$

$$= 51$$ ∎

■ ■ ■ ■ ■ ■ ■ ■ ■   **Measures of Dispersion**

ACCENT ON   Previously we have discussed three measures of central tendency—the mean, which
STATISTICS   is denoted by the Greek letter μ (mu), the median, and the mode. We now discuss
three measures of dispersion—the range, the variance, and the standard deviation.

Suppose that an inventory clerk records the number of days it takes to sell each of seven cars, with the following results:

1, 2, 5, 5, 6, 8, 15   (in days)

The *range* is the difference between the highest score and the lowest score. Thus, the range is 15 days − 1 day = 14 days.

To define the *variance*, we organize the data as in Table 1.

(1) Original terms	(2) Mean (μ)	(3) Original terms minus the mean	(4) Squares of the differences
1	6	1 − 6 = −5	25
2	6	2 − 6 = −4	16
5	6	5 − 6 = −1	1
5	6	5 − 6 = −1	1
6	6	6 − 6 = 0	0
8	6	8 − 6 = 2	4
15	6	15 − 6 = 9	81
		The sum of the square of the differences	128

TABLE 1

In Table 1, the mean was subtracted from each of the original terms to get the seven differences shown in column (3). Each difference was then squared to get the first seven numbers in column (4). The 128 is the sum of the squares of the differences.

The variance of the distribution (denoted as σ^2) is defined to be the mean of the squares of the differences:

$$\sigma^2 = \frac{\text{the sum of the squares of the differences}}{\text{the number of terms in the distribution}} = \frac{128}{7} \approx 18.285714$$

To write a formula for the variance, we note that the lines of Table 1 have the following form, where the symbol $\Sigma(x - \mu)^2$ represents the sum of the squares of the differences.

(1) Original terms	(2) Mean (μ)	(3) Original terms minus the mean	(4) Squares of the differences
x_1	μ	$x_1 - \mu$	$(x_1 - \mu)^2$
x_2	μ	$x_2 - \mu$	$(x_2 - \mu)^2$
x_3	μ	$x_3 - \mu$	$(x_3 - \mu)^2$
			$\Sigma (x - \mu)^2$

Using this notation, we have the formula for the variance

$$\sigma^2 = \frac{\Sigma(x - \mu)^2}{N}$$

The standard deviation (denoted by σ) is the square root of the variance. Thus,

$$\sigma = \sqrt{\frac{\Sigma(x - \mu)^2}{N}}$$

EXAMPLE 7   A dentist finds that after dental checkups, six of his patients have the following number of cavities, respectively:

3, 2, 5, 1, 3, 4

Find the standard deviation.

Solution   To find $\Sigma(x - \mu)^2$, we fill out the following table

(1) Original terms	(2) Mean (μ)	(3) Original terms minus the mean	(4) Squares of the differences
1	3	$1 - 3 = -2$	4
2	3	$2 - 3 = -1$	1
3	3	$3 - 3 = 0$	0
3	3	$3 - 3 = 0$	0
4	3	$4 - 3 = 1$	1
5	3	$5 - 3 = 2$	4
			$\Sigma(x - \mu)^2 = 10$

We then substitute 10 for $\Sigma(x - \mu)^2$ and 6 for N in the formula

$$\sigma = \sqrt{\frac{\Sigma(x - \mu)^2}{N}} = \sqrt{\frac{10}{6}} = 1.290994449$$

The standard deviation is approximately 1.3.

Orals   *Find the next term in each arithmetic sequence.*

1. 2, 6, 10, . . .   **2.** 10, 7, 4, . . .

Find the common difference in each arithmetic sequence.

3. $-2, 3, 8, . . .$ **4.** $5, -1, -7, . . .$

Find each sum.

5. $\displaystyle\sum_{k=1}^{2} k$ **6.** $\displaystyle\sum_{k=2}^{3} k$

EXERCISE 10.6

In Exercises 1–14, write the first five terms of each arithmetic sequence with the given properties.

1. $a = 3, d = 2$ **2.** $a = -2, d = 3$ **3.** $a = -5, d = -3$ **4.** $a = 8, d = -5$

5. $a = 5$, fifth term is 29   **6.** $a = 4$, sixth term is 39

7. $a = -4$, sixth term is -39 **8.** $a = -5$, fifth term is -37

9. $d = 7$, sixth term is -83

11. $d = -3$, seventh term is 16

13. The 19th term is 131, and the 20th term is 138.

10. $d = 3$, seventh term is 12

12. $d = -5$, seventh term is -12

14. The 16th term is 70, and the 18th term is 78.

15. Find the 30th term of the arithmetic sequence with $a = 7$ and $d = 12$.

17. Find the 37th term of the arithmetic sequence with a 2nd term of -4 and a 3rd term of -9.

19. Find the 1st term of the arithmetic sequence with a common difference of 11 and whose 27th term is 263.

21. Find the common difference of the arithmetic sequence with a 1st term of 40 if its 44th term is 556.

23. Insert three arithmetic means between 2 and 11.

25. Insert four arithmetic means between 10 and 20.

27. Find the arithmetic mean between 10 and 19.

29. Find the arithmetic mean between -4.5 and 7.

16. Find the 55th term of the arithmetic sequence with $a = -5$ and $d = 4$.

18. Find the 40th term of the arithmetic sequence with a 2nd term of 6 and a 4th term of 16.

20. Find the common difference of the arithmetic sequence with a 1st term of -164 if its 36th term is -24.

22. Find the 1st term of the arithmetic sequence with a common difference of -5 and whose 23rd term is -625.

24. Insert four arithmetic means between 5 and 25.

26. Insert three arithmetic means between 20 and 30.

28. Find the arithmetic mean between 5 and 23.

30. Find the arithmetic mean between -6.3 and -5.2.

In Exercises 31–38, find the sum of the first n terms of each arithmetic sequence.

31. $1, 4, 7, \ldots$; $n = 30$

33. $-5, -1, 3, \ldots$; $n = 17$

35. second term is 7, third term is 12; $n = 12$

37. $f(n) = 2n + 1$, nth term is 31; n is a natural number

32. $2, 6, 10, \ldots$; $n = 28$

34. $-7, -1, 5, \ldots$; $n = 15$

36. second term is 5, fourth term is 9; $n = 16$

38. $f(n) = 4n + 3$, nth term is 23; n is a natural number

39. Find the sum of the first 50 natural numbers.

41. Find the sum of the first 50 odd natural numbers.

40. Find the sum of the first 100 natural numbers.

42. Find the sum of the first 50 even natural numbers.

43. **Saving money** Fred puts $60 into a safety deposit box. After each succeeding month, he puts $50 more in the safety deposit box. Write the first six terms of an arithmetic sequence that gives the monthly amounts in his savings, and find his savings after 10 years.

44. **Installment loan** Freda borrowed $10,000, interest-free, from her mother. Freda agreed to pay back the loan in monthly installments of $275. Write the first six terms of an arithmetic sequence that shows the balance due after each month, and find the balance due after 17 months.

45. **Designing a patio** Each row of bricks in the triangular patio in Illustration 1 is to have one more brick than the previous row, ending with the longest row of 150 bricks. How many bricks will be needed?

ILLUSTRATION 1

46. Falling object The equation $s = 16t^2$ represents the distance s in feet that an object will fall in t seconds. After 1 second, the object has fallen 16 feet. After 2 seconds, the object has fallen 64 feet, and so on. Find the distance that the object will fall during the second and third seconds.

47. Falling object Refer to Exercise 46. How far will the object fall during the 12th second?

48. Interior angles The sums of the angles of several polygons are given in Illustration 2. Assuming that the pattern continues, find the sum of the interior angles of an octagon (8 sides) and dodecagon (12 sides).

Figure	Number of sides	Sum of angles
Triangle	3	180°
Quadrilateral	4	360°
Pentagon	5	540°
Hexagon	6	720°

ILLUSTRATION 2

49. Show that the arithmetic mean between a and b is the average of a and b: $\frac{a+b}{2}$.

50. Show that the sum of the two arithmetic means between a and b is $a + b$.

In Exercises 51–56, find each sum.

51. $\displaystyle\sum_{k=1}^{4} 6k$

52. $\displaystyle\sum_{k=2}^{5} 3k$

53. $\displaystyle\sum_{k=3}^{4} (k^2 + 3)$

54. $\displaystyle\sum_{k=2}^{6} (k^2 + 1)$

55. $\displaystyle\sum_{k=4}^{4} (2k + 4)$

56. $\displaystyle\sum_{k=3}^{5} (3k^2 - 7)$

57. Show that $\displaystyle\sum_{k=1}^{5} 5k = 5\sum_{k=1}^{5} k$.

58. Show that $\displaystyle\sum_{k=3}^{6} (k^2 + 3k) = \sum_{k=3}^{6} k^2 + \sum_{k=3}^{6} 3k$.

59. Show that $\displaystyle\sum_{k=1}^{n} 3 = 3n$. (*Hint:* Consider 3 to be $3k^0$.)

60. Show that $\displaystyle\sum_{k=1}^{3} \frac{k^2}{k} \neq \frac{\displaystyle\sum_{k=1}^{3} k^2}{\displaystyle\sum_{k=1}^{3} k}$.

In Exercises 61–62, find the range, variance, and standard deviation of each distribution.

61. The life spans of five dogs: 5, 8, 10, 12, 15.

62. The heights (in inches) of six people: 60, 64, 66, 67, 70, 72.

Writing Exercises *Write a paragraph using your own words.*

1. Define an arithmetic sequence.

2. Develop the formula for finding the sum of first n terms of an arithmetic sequence.

Something to Think About **1.** Write the first 6 terms of the arithmetic sequence given by

$$\sum_{k=1}^{n} \left(\frac{1}{2}k + 1 \right)$$

2. Find the sum of the first 6 terms of the sequence given in Problem 1.

Review Exercises   *Do the operations and simplify if possible.*

1. $3(2x^2 - 4x + 7) + 4(3x^2 + 5x - 6)$ **2.** $(2p + q)(3p^2 + 4pq - 3q^2)$

3. $\dfrac{3a + 4}{a - 2} + \dfrac{3a - 4}{a + 2}$ **4.** $2t - 3\overline{)8t^4 - 12t^3 + 8t^2 - 16t + 6}$

10.7 Geometric Sequences

■ GEOMETRIC SEQUENCES ■ GEOMETRIC MEANS ■ THE SUM OF THE FIRST *N* TERMS OF A
GEOMETRIC SEQUENCE ■ POPULATION GROWTH

Another important type of sequence is called a **geometric sequence**.

■ GEOMETRIC SEQUENCES

Geometric Sequence
A **geometric sequence** is a sequence of the form
$$a, ar, ar^2, ar^3, \ldots, ar^{n-1}, \ldots$$
where a is the first term, ar^{n-1} is the nth term, and r is the common ratio.

We note that the second term of a geometric sequence has a factor of r^1, the third term has a factor of r^2, the fourth term has a factor of r^3, and the nth term has a factor of r^{n-1}. We also note that the quotient obtained when any term is divided by the previous term is r.

EXAMPLE 1   A geometric sequence has a first term of 5 and a common ratio of 3.

a. Write the first five terms of the sequence.

b. Write the ninth term of the sequence.

Solution **a.** Because the first term is $a = 5$ and the common ratio is $r = 3$, the first five terms are
$$5, \quad 5(3), \quad 5(3^2), \quad 5(3^3), \quad 5(3^4)$$
or
$$5, 15, 45, 135, 405$$

b. The nth term is ar^{n-1}, where $a = 5$ and $r = 3$. Because we want the ninth term, we let $n = 9$:

nth term $= ar^{n-1}$
9th term $= 5(3)^{9-1}$
$= 5(3)^8$
$= 5(6561)$
$= 32,805$

■

EXAMPLE 2 The first three terms of a geometric sequence are 16, 4, and 1. Find the seventh term of the sequence.

Solution We substitute 16 for a, $\frac{1}{4}$ for r, and 7 for n in the formula for the nth term and simplify:

$$n\text{th term} = ar^{n-1}$$
$$7\text{th term} = 16\left(\frac{1}{4}\right)^{7-1}$$
$$= 16\left(\frac{1}{4}\right)^{6}$$
$$= 16\left(\frac{1}{4096}\right)$$
$$= \frac{1}{256}$$

∎

■ GEOMETRIC MEANS

If numbers are inserted between two numbers a and b to form a geometric sequence, the inserted numbers are called **geometric means** between a and b.

If a single number is inserted between the numbers a and b, that number is called a **geometric mean** between a and b.

EXAMPLE 3 Insert two geometric means between 7 and 1512.

Solution In this example, the first term is $a = 7$, and the fourth term (or last term) is $l = 1512$. To find the common ratio r such that the terms

$$7, \quad 7r, \quad 7r^2, \quad 1512$$

form a geometric sequence, we substitute 4 for n and 7 for a in the formula for the nth term of a geometric sequence and solve for r.

$$n\text{th term} = ar^{n-1}$$
$$4\text{th term} = 7r^{4-1}$$
$$1512 = 7r^3$$
$$216 = r^3 \qquad \text{Divide both sides by 7.}$$
$$6 = r \qquad \text{Take the cube root of both sides.}$$

The two geometric means between 7 and 1512 are

$$7r = 7(6) = 42$$

and

$$7r^2 = 7(6)^2 = 7(36) = 252$$

The numbers 7, 42, 252, and 1512 are the first four terms of a geometric sequence.

∎

EXAMPLE 4   Find a geometric mean between 2 and 20.

Solution   We want to find the middle term of the three-termed geometric sequence

$$2, \quad 2r, \quad 20$$

with $a = 2$, $l = 20$, and $n = 3$. To find r, we substitute these values into the formula for the nth term of a geometric sequence:

$$n\text{th term} = ar^{n-1}$$
$$3\text{rd term} = 2r^{3-1}$$
$$20 = 2r^2$$
$$10 = r^2 \qquad \text{Divide both sides by 2.}$$
$$\pm\sqrt{10} = r \qquad \text{Take the square root of both sides.}$$

Because r can be either $\sqrt{10}$ or $-\sqrt{10}$, there are two values for a geometric mean. They are

$$2r = 2\sqrt{10} \qquad \text{and} \qquad 2r = -2\sqrt{10}$$

The numbers 2, $2\sqrt{10}$, 20 and 2, $-2\sqrt{10}$, 20 both form geometric sequences. The common ratio of the first sequence is $\sqrt{10}$, and the common ratio of the second sequence is $-\sqrt{10}$. ■

■ THE SUM OF THE FIRST N TERMS OF A GEOMETRIC SEQUENCE

There is a formula that gives the sum of the first n terms of a geometric sequence. To develop this formula, we let S_n represent the sum of the first n terms of a geometric sequence.

1. $S_n = a + ar + ar^2 + ar^3 + \cdots + ar^{n-1}$

We multiply both sides of Equation 1 by r to get

2. $S_n r = \qquad ar + ar^2 + ar^3 + \cdots + ar^{n-1} + ar^n$

We now subtract Equation 2 from Equation 1 and solve for S_n:

$$S_n - S_n r = a - ar^n$$
$$S_n(1 - r) = a - ar^n \qquad \text{Factor out } S_n \text{ from the left side.}$$
$$S_n = \frac{a - ar^n}{1 - r} \qquad \text{Divide both sides by } 1 - r.$$

This reasoning establishes the following theorem.

Sum of the First *n* Terms of a Geometric Sequence
The sum of the first n terms of a geometric sequence is given by the formula

$$S_n = \frac{a - ar^n}{1 - r} \qquad (r \neq 1)$$

where S_n is the sum, a is the first term, r is the common ratio, and n is the number of terms.

EXAMPLE 5   Find the sum of the first six terms of the geometric sequence 250, 50, 10, ...

Solution   In this geometric sequence, $a = 250$, $r = \frac{1}{5}$, and $n = 6$. We substitute these values into the formula for the sum of the first n terms of a geometric sequence and simplify:

$$S_n = \frac{a - ar^n}{1 - r}$$

$$S_6 = \frac{250 - 250\left(\frac{1}{5}\right)^6}{1 - \frac{1}{5}}$$

$$= \frac{250 - 250\left(\frac{1}{15,625}\right)}{\frac{4}{5}}$$

$$= \frac{5}{4}\left(250 - \frac{250}{15,625}\right)$$

$$= \frac{5}{4}\left(\frac{3,906,000}{15,625}\right)$$

$$= 312.48$$

The sum of the first six terms is 312.48.   ■

■ POPULATION GROWTH

EXAMPLE 6   **Growth of a town**   The mayor of Eagle River (population 1500) predicts a growth rate of 4% each year for the next ten years. Find the population of Eagle River ten years from now.

Solution   Let P_0 be the initial population of Eagle River. After 1 year, there will be a different population, P_1. The initial population (P_0) plus the growth (the product of P_0 and the rate of growth, r) will equal this new population, P_1:

$$P_1 = P_0 + P_0 r = P_0(1 + r)$$

The population after 2 years will be P_2, and

$$
\begin{aligned}
P_2 &= P_1 + P_1 r \\
&= P_1(1 + r) && \text{Factor out } P_1. \\
&= P_0(1 + r)(1 + r) && \text{Remember that } P_1 = P_0(1 + r). \\
&= P_0(1 + r)^2
\end{aligned}
$$

The population after 3 years will be P_3, and

$$
\begin{aligned}
P_3 &= P_2 + P_2 r \\
&= P_2(1 + r) && \text{Factor out } P_2. \\
&= P_0(1 + r)^2(1 + r) && \text{Remember that } P_2 = P_0(1 + r)^2. \\
&= P_0(1 + r)^3
\end{aligned}
$$

The yearly population figures

$$P_0, \quad P_1, \quad P_2, \quad P_3, \ldots$$

or

$$P_0, \quad P_0(1 + r), \quad P_0(1 + r)^2, \quad P_0(1 + r)^3, \ldots$$

form a geometric sequence with a first term of P_0 and a common ratio of $1 + r$. The population of Eagle River after 10 years is P_{10}, which is the 11th term of this sequence:

$$n\text{th term} = ar^{n-1}$$

$$
\begin{aligned}
P_{10} = 11\text{th term} &= P_0(1 + r)^{10} \\
&= 1500(1 + 0.04)^{10} \\
&= 1500(1.04)^{10} \\
&\approx 1500(1.480244285) \qquad \text{Use a calculator.} \\
&\approx 2220
\end{aligned}
$$

The estimated population ten years from now is 2220. ∎

EXAMPLE 7

Amount of an annuity An *annuity* is a sequence of equal payments made periodically over a length of time. The sum of the payments and the interest earned during the *term* of the annuity is called the *amount* of the annuity.

After a sales clerk works six months, her employer will begin an annuity for her and will contribute $500 every six months to a fund that pays 8% annual interest. After she has been employed for two years, what will be the amount of the annuity?

Solution Because the payments are to be made semiannually, there will be four payments of $500, each earning a rate of 4% per six-month period. These payments will occur at the end of 6 months, 12 months, 18 months, and 24 months. The first payment, to be made after 6 months, will earn interest for three interest periods. Thus, the amount of the first payment is $500(1.04)^3$. The amounts of each of the four payments after two years are shown in Figure 10-36.

The amount of the annuity is the sum of the amounts of the individual payments, a sum of $2123.23.

Payment (at the end of period)	Amount of payment at the end of 2 years
1	$500(1.04)^3 = \$ 562.43$
2	$500(1.04)^2 = \$ 540.80$
3	$500(1.04)^1 = \$ 520.00$
4	$\$500 = \$ 500.00$
	$A_n = \$2123.23$

FIGURE 10-36 ∎

Orals   *Find the next term in each geometric sequence.*

1. 1, 3, 9, . . .   **2.** $1, \dfrac{1}{3}, \dfrac{1}{9}, \ldots$

Find the common ratio in each geometric sequence.

3. 0.2, 0.5, 1.25, . . .   **4.** $\sqrt{3}, 3, 3\sqrt{3}, \ldots$

Find x in each geometric sequence.

5. −2, x, −18, 54, . . .   **6.** $3, x, \dfrac{1}{3}, \dfrac{1}{9}, \ldots$

EXERCISE 10.7

In Exercises 1–14, write the first five terms of each geometric sequence with the given properties.

1. $a = 3, r = 2$

2. $a = -2, r = 2$

3. $a = -5, r = \dfrac{1}{5}$

4. $a = 8, r = \dfrac{1}{2}$

5. $a = 2, r > 0$, third term is 32

6. $a = 3$, fourth term is 24

7. $a = -3$, fourth term is −192

8. $a = 2, r < 0$, third term is 50

9. $a = -64, r < 0$, fifth term is −4

10. $a = -64, r > 0$, fifth term is −4

11. $a = -64$, sixth term is −2

12. $a = -81$, sixth term is $\dfrac{1}{3}$

13. The second term is 10, and the third term is 50.

14. The third term is −27, and the fourth term is 81.

15. Find the tenth term of the geometric sequence with $a = 7$ and $r = 2$.

16. Find the 12th term of the geometric sequence with $a = 64$ and $r = \frac{1}{2}$.

17. Find the first term of the geometric sequence with a common ratio of −3 and an eighth term of −81.

18. Find the first term of the geometric sequence with a common ratio of 2 and a tenth term of 384.

19. Find the common ratio of the geometric sequence with a first term of −8 and a sixth term of −1944.

20. Find the common ratio of the geometric sequence with a first term of 12 and a sixth term of $\frac{3}{8}$.

21. Insert three positive geometric means between 2 and 162.

22. Insert four geometric means between 3 and 96.

23. Insert four geometric means between −4 and −12,500.

24. Insert three geometric means (two positive and one negative) between −64 and −1024.

25. Find the negative geometric mean between 2 and 128.

26. Find the positive geometric mean between 3 and 243.

27. Find the positive geometric mean between 10 and 20.

28. Find the negative geometric mean between 5 and 15.

29. Find a geometric mean, if possible, between −50 and 10.

30. Find a negative geometric mean, if possible, between −25 and −5.

In Exercises 31–42, find the sum of the first n terms of each geometric sequence.

31. $2, 6, 18, \ldots; n = 6$

32. $2, -6, 18, \ldots; n = 6$

33. $2, -6, 18, \ldots; n = 5$

34. $2, 6, 18, \ldots; n = 5$

35. $3, -6, 12, \ldots; n = 8$

36. $3, 6, 12, \ldots; n = 8$

37. $3, 6, 12, \ldots; n = 7$

38. $3, -6, 12, \ldots; n = 7$

39. The second term is 1, and the third term is $\frac{1}{5}$; $n = 4$.

40. The second term is 1, and the third term is 4; $n = 5$.

41. The third term is -2, and the fourth term is 1; $n = 6$.

42. The third term is -3, and the fourth term is 1; $n = 5$.

In Exercises 43–52, use a calculator to solve each problem.

43. Population growth The population of Union is predicted to increase by 6% each year. What will be the population of Union 5 years from now if its current population is 500?

44. Population decline The population of Forreston is decreasing by 10% each year. If its current population is 98, what will be the population 8 years from now?

45. Declining savings John has $10,000 in a safety deposit box. Each year he spends 12% of what is left in the box. How much will be in the box after 15 years?

46. Savings growth Sally has $5000 in an investment account earning 12% annual interest. How much will be in her account 10 years from now? (Assume that Sally makes no deposits or withdrawals.)

47. House appreciation A house appreciates by 6% each year. If the house is worth $70,000 today, how much will it be worth 12 years from now?

48. Motorboat depreciation A motorboat that cost $5000 when new depreciates at a rate of 9% per year. How much will the boat be worth in 5 years?

49. Inscribed squares Each inscribed square in Illustration 1 joins the midpoints of the next larger square. If the area of the first square, the largest, is 1, find the area of the 12th square.

50. Genealogy The family tree in Illustration 2 spans three generations and lists seven people. How many names would be listed in a family tree that spans ten generations?

51. Annuities Find the amount of an annuity if $1000 is paid semiannually for two years at 6% annual interest. Assume that the first of the four payments is made immediately.

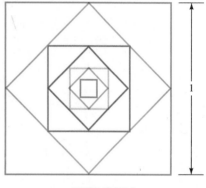

ILLUSTRATION 1

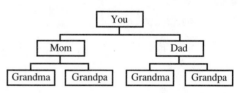

ILLUSTRATION 2

52. Annuities Note that the amounts shown in Figure 10-36 (on page 637) form a geometric sequence. Verify the answer for Example 7 by using the formula for the sum of a geometric sequence.

53. Show that the formula for the sum of the first n terms of a geometric sequence can be written in the form

$$S_n = \frac{a - lr}{1 - r} \qquad \text{where } l = ar^{n-1}$$

54. Show that the formula for the sum of the first n terms of a geometric sequence can be written in the form

$$S_n = \frac{a(1 - r^n)}{1 - r}$$

Writing Exercises   *Write a paragraph using your own words.*

1. Define a geometric sequence.

2. Develop the formula for finding the sum of the first n terms of a geometric sequence.

Something to Think About

1. Show that the arithmetic mean between a and b is $\frac{a+b}{2}$.

2. Show that a geometric mean between a and b is $\sqrt{ab}$.

3. If $a > b$, $a > 0$, and $b > 0$, which is larger: the arithmetic mean between a and b or the geometric mean between a and b?

4. Is there a geometric mean between -5 and 5?

Review Exercises   *Solve each inequality.*

1. $x^2 - 5x - 6 \leq 0$

2. $a^2 - 7a + 12 \geq 0$

3. $\dfrac{x - 4}{x + 3} > 0$

4. $\dfrac{t^2 + t - 20}{t + 2} < 0$

10.8   Infinite Geometric Sequences

■ THE SUM OF AN INFINITE GEOMETRIC SEQUENCE

An **infinite geometric sequence** is a geometric sequence with an infinite number of terms. Two examples of infinite geometric sequences are

$$2,\ 6,\ 18,\ 54,\ 162,\ \ldots \qquad (r = 3)$$

$$\frac{3}{2},\ \frac{3}{4},\ \frac{3}{8},\ \frac{3}{16},\ \frac{3}{32},\ \ldots \qquad \left(r = \frac{1}{2}\right)$$

■ THE SUM OF AN INFINITE GEOMETRIC SEQUENCE

Under certain conditions, we can find the sum of all the terms of an infinite geometric sequence. To define this sum, we consider the geometric sequence

$$a,\ ar,\ ar^2,\ ar^3,\ \ldots,\ ar^{n-1},\ \ldots$$

- The first partial sum, S_1, of the sequence is $S_1 = a$.
- The second partial sum, S_2, of the sequence is $S_2 = a + ar$.
- The third partial sum, S_3, of the sequence is $S_3 = a + ar + ar^2$.
- The nth partial sum, S_n, of the sequence is $S_n = a + ar + ar^2 + \cdots + ar^{n-1}$.

If the nth partial sum, S_n, approaches some number S as n approaches infinity, then S is called the **sum of the infinite geometric sequence**.

To develop a formula for finding the sum of all the terms in an infinite geometric sequence, we consider the formula

$$S_n = \frac{a - ar^n}{1 - r} \qquad (r \neq 1)$$

If $|r| < 1$ and a is constant, then the term ar^n in the above formula approaches 0 as n becomes very large. For example,

$$a\left(\frac{1}{4}\right)^1 = \frac{1}{4}a, \qquad a\left(\frac{1}{4}\right)^2 = \frac{1}{16}a, \qquad a\left(\frac{1}{4}\right)^3 = \frac{1}{64}a$$

and so on. When n is very large, the value of ar^n is negligible, and the term ar^n in the above formula can be ignored. This reasoning justifies the following theorem.

> **Sum of an Infinite Geometric Sequence**
> If a is the first term and r is the common ratio of an infinite geometric sequence, and if $|r| < 1$, then the sum of the terms of the sequence is given by the formula
>
> $$S = \frac{a}{1 - r}$$

EXAMPLE 1   Find the sum of the terms of the infinite geometric sequence 125, 25, 5, ...

Solution   In this geometric sequence, $a = 125$ and $r = \frac{1}{5}$. Because $|r| = \left|\frac{1}{5}\right| = \frac{1}{5} < 1$, we can find the sum of all the terms of the sequence. We do this by substituting 125 for a and $\frac{1}{5}$ for r in the formula $S = \frac{a}{1-r}$ and simplifying:

$$S = \frac{a}{1 - r} = \frac{125}{1 - \dfrac{1}{5}} = \frac{125}{\dfrac{4}{5}} = \frac{5}{4}(125) = \frac{625}{4}$$

The sum of all the terms of the sequence 125, 25, 5, . . . is $\dfrac{625}{4}$. ∎

EXAMPLE 2   Find the sum of the infinite geometric sequence $64, -4, \dfrac{1}{4}, \ldots$.

Solution   In this geometric sequence, $a = 64$ and $r = -\frac{1}{16}$. Because $|r| = \left|-\frac{1}{16}\right| = \frac{1}{16} < 1$, we can find the sum of all the terms of the sequence. We substitute 64 for a and $-\frac{1}{16}$ for r in the formula $S = \frac{a}{1-r}$ and simplify:

$$S = \frac{a}{1 - r} = \frac{64}{1 - \left(-\dfrac{1}{16}\right)} = \frac{64}{\dfrac{17}{16}} = \frac{16}{17}(64) = \frac{1024}{17}$$

The sum of all the terms of the geometric sequence $64, -4, \dfrac{1}{4}, \ldots$ is $\dfrac{1024}{17}$. ∎

EXAMPLE 3   Change $0.\overline{8}$ to a common fraction.

Solution   The decimal $0.\overline{8}$ can be written as the sum of an infinite geometric sequence.

$$0.\overline{8} = 0.888 \ldots = \frac{8}{10} + \frac{8}{100} + \frac{8}{1000} + \cdots$$

where $a = \frac{8}{10}$ and $r = \frac{1}{10}$. Because $|r| = \left|\frac{1}{10}\right| = \frac{1}{10} < 1$, we can find the sum as follows:

$$S = \frac{a}{1-r} = \frac{\dfrac{8}{10}}{1 - \dfrac{1}{10}} = \dfrac{\dfrac{8}{10}}{\dfrac{9}{10}} = \frac{8}{9}$$

Thus, $0.\overline{8} = \frac{8}{9}$. Long division will verify that $\frac{8}{9} = 0.888\ldots$ ■

EXAMPLE 4   Change $0.\overline{25}$ to a common fraction.

Solution   The decimal $0.\overline{25}$ can be written as the sum of an infinite geometric sequence

$$0.\overline{25} = 0.252525\ldots = \frac{25}{100} + \frac{25}{10,000} + \frac{25}{1,000,000} + \cdots$$

where $a = \frac{25}{100}$ and $r = \frac{1}{100}$. Because $|r| = \left|\frac{1}{100}\right| = \frac{1}{100} < 1$, we can find the sum as follows:

$$S = \frac{a}{1-r} = \frac{\dfrac{25}{100}}{1 - \dfrac{1}{100}} = \dfrac{\dfrac{25}{100}}{\dfrac{99}{100}} = \frac{25}{99}$$

Thus, $0.\overline{25} = \frac{25}{99}$. Long division will verify that this is true. ■

Orals   *Find the common ratio in each infinite geometric sequence.*

1. $\dfrac{1}{64}, \dfrac{1}{8}, 1, \ldots$

2. $1, \dfrac{1}{8}, \dfrac{1}{64}, \ldots$

3. $\dfrac{2}{3}, \dfrac{1}{3}, \dfrac{1}{6}, \ldots$

4. $64, 8, 1, \ldots$

Find the sum of the terms in each infinite geometric sequence.

5. $18, 6, 2, \ldots$

6. $12, 3, \dfrac{3}{4}, \ldots$

EXERCISE 10.8

In Exercises 1–12, find the sum of each infinite geometric sequence, if possible.

1. $8, 4, 2, \ldots$

2. $12, 6, 3, \ldots$

3. $54, 18, 6, \ldots$

4. $45, 15, 5, \ldots$

5. $12, -6, 3, \ldots$

6. $8, -4, 2, \ldots$

7. $-45, 15, -5, \ldots$

8. $-54, 18, -6, \ldots$

9. $\dfrac{9}{2}, 6, 8, \ldots$

10. $-112, -28, -7, \ldots$

11. $-\dfrac{27}{2}, -9, -6, \ldots$

12. $\dfrac{18}{25}, \dfrac{6}{5}, 2, \ldots$

In Exercises 13–20, change each decimal to a common fraction. Then check the answer by long division.

13. $0.\overline{1}$ **14.** $0.\overline{2}$ **15.** $-0.\overline{3}$ **16.** $-0.\overline{4}$

17. $0.\overline{12}$ **18.** $0.\overline{21}$ **19.** $0.\overline{75}$ **20.** $0.\overline{57}$

21. Bouncing ball On each bounce, the rubber ball in Illustration 1 rebounds to a height one-half of that from which it fell. Find the total vertical distance the ball travels.

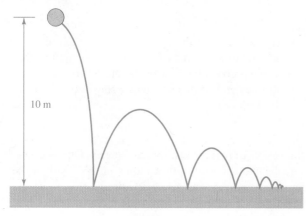

ILLUSTRATION 1

10 m

22. Bouncing ball A golf ball is dropped from a height of 12 feet. On each bounce, it returns to a height two-thirds of that from which it fell. Find the total distance the ball travels.

23. Controlling moths To reduce the population of a destructive moth, biologists release 1000 sterilized male moths into the environment each day. If 80% of these moths alive one day survive until the next, then after a long time the population of sterile males is the sum of the infinite geometric series

$$1000 + 1000(0.8) + 1000(0.8)^2 + 1000(0.8)^3 + \cdots$$

Find the long-term population.

24. Controlling moths If mild weather increases the day-to-day survival rate of the sterile male moths in Exercise 23 to 90%, find the long-term population.

25. Show that $0.\overline{9} = 1$.

26. Show that $1.\overline{9} = 2$.

27. Does $0.999999 = 1$? Explain.

28. If $f(x) = 1 + x + x^2 + x^3 + x^4 + \cdots$, find $f\left(\dfrac{1}{2}\right)$ and $f\left(-\dfrac{1}{2}\right)$.

Writing Exercises   *Write a paragraph using your own words.*

1. Why must the common ratio be less than 1 before an infinite geometric sequence can have a sum?

2. If its common difference is not 0, can an infinite arithmetic sequence have a sum?

Something to Think About   **1.** Find the first 5 terms of the infinite geometric sequence defined by

$$\sum_{k=1}^{\infty}\left(\frac{1}{2}\right)^k$$

2. Find the sum of *all* the terms of the sequence in Problem 1.

Review Exercises   *Tell whether each equation determines y to be a function of x.*

1. $y = 3x^3 - 4$ **2.** $xy = 12$ **3.** $3x = y^2 + 4$ **4.** $x = |y|$

■ ■ ■ ■ ■ ■ ■ ■ ■ ■ **PROBLEMS AND PROJECTS**

1. A pool contractor proposes building a circular pool with a surface area of 36π m^2. However, the homeowner thinks that an elliptical pool would look better. Write the equation and sketch the graph of an elliptical pool that has the same surface area as the proposed circular pool. (*Hint:* Area of an ellipse $= ab\pi$.)

2. On a coordinate system, the distance from the origin to the point (x, y) is $\sqrt{5}$ units, and the distance from point (x, y) to the point $(4, 0)$ is $\sqrt{13}$ units. Find the values of x and y.

3. The perimeter of an equilateral triangle is 240 cm. A second equilateral triangle is inscribed in the first by joining the midpoints of the sides of the first triangle. This process continues until there are ten triangles. Find the perimeter of the tenth triangle.

4. The seating in a theater is arranged so that the number of seats in each row increases by 6 as the rows approach the rear of the theater. If there are 30 rows and 3510 seats, how many seats are in the 16th row?

PROJECT 1

The zillionaire G. I. Luvmoney is known for his love of flowers. On his estate, he recently set aside a circular plot of land with a radius of 100 yards to be made into a flower garden. He has hired your landscape design firm to do the job. If Luvmoney is satisfied, he will hire your firm to do more lucrative jobs. Here is Luvmoney's plan.

The center of the circular plot of land is to be the origin of a rectangular coordinate system. You are to make 100 circles, all centered at the origin, with radii of 1 yard, 2 yards, 3 yards, and so on, up to the outermost circle, which will have a radius of 100 yards. Inside the innermost circle, he wants a fountain with a circular walkway around it. In the ring between the first and second circle, he wants to plant his favorite kind of flower; in the next ring his second favorite, and so on, until you reach the edge of the circular plot. Luvmoney provides you with a list ranking his 99 favorite flowers.

The first thing he wants to know is the area of each ring, so that he will know how many of each plant to order. Then he wants a simple formula that will give the area of any ring just by substituting in the number of the ring.

He also wants a walkway to go through the garden in the form of a hyperbolic path, following the equation $x^2 - \dfrac{y^2}{9} = 1$. Luvmoney wants to know the x- and y-coordinates of the points where the path will intersect the circles, so that those points can be marked with stakes to keep gardeners from planting flowers where the walkway will later be built. He wants a formula (or two) that will enable him to put in the number of a circle and get out the intersection points.

Finally, although cost has no importance for Luvmoney, his accountants will want an estimate of the total cost of all of the flowers.

You go back to your office with Luvmoney's list. You find that because the area of the rings grow from the inside of the garden to the outside, and because of Luvmoney's ranking of flowers, a strange thing happens. The first ring of flowers will cost $360, and the flowers in every ring after that will cost 110% as much as the flowers in the previous ring. That is, the second ring of flowers will cost $360(1.1) = \$396$, the third will cost $435.60, and so on.

Find answers to all of Luvmoney's questions, and show work that will convince him that you are right.

PROJECT 2   Baytown is building an auditorium. The city council has already decided on the layout shown in Illustration 1. Each of the sections A, B, C, D, E is to be 60 feet in length from front to back. The aisle widths cannot be changed due to fire regulations. The one thing left to decide is how many rows of seats to put in each section. Based on the following information regarding each section of the auditorium, help the council decide on a final plan.

Sections A and C each have four seats in the front row, five seats in the second row, six seats in the third row and so on, adding one seat per row as we count from front to back.

Section B has eight seats in the front row and adds one seat per row as we count from front to back.

Sections D and E each have 28 seats in the front row and add two seats per row as we count from front to back.

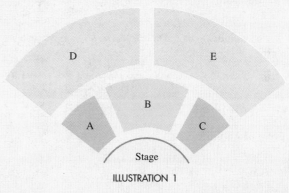

ILLUSTRATION 1

a. One plan calls for a distance of 36 inches, front to back, for each row of seats. Another plan allows for 40 inches (an extra four inches of leg room) for each row. How many seats will the auditorium have under each of these plans?

b. Another plan calls for the higher-priced seats (Sections A, B, and C) to have the extra room afforded by 40-inch rows, but for Sections D and E to have enough rows to make sure that the auditorium holds at least 2700 seats. Determine how many rows Sections D and E would have to contain for this to work. (This answer should be an integer.) How much space to the nearest tenth of an inch) would be allotted for each row in Sections D and E?

■ **Chapter Summary**

Key Words

Key Ideas

(10.1) Equations of a circle:

$$(x - h)^2 + (y - k)^2 = r^2 \quad \text{center } (h, k), \text{ radius } r$$

$$x^2 + y^2 = r^2 \quad \text{center } (0, 0), \text{ radius } r$$

Equations of parabolas:

Parabola opening	Vertex at origin	Vertex at (h, k)
Up	$y = ax^2$	$y = a(x - h)^2 + k$
Down	$y = -ax^2$	$y = -a(x - h)^2 + k$
Right	$x = ay^2$	$x = a(y - k)^2 + h$
Left	$x = -ay^2$	$x = -a(y - k)^2 + h$

(10.2) Equations of an ellipse:

$$\frac{x^2}{a^2} + \frac{y^2}{b^2} = 1 \quad (a > b > 0) \quad \text{center } (0, 0)$$

$$\frac{x^2}{b^2} + \frac{y^2}{a^2} = 1 \quad (a > b > 0) \quad \text{center } (0, 0)$$

$$\frac{(x - h)^2}{a^2} + \frac{(y - k)^2}{b^2} = 1 \quad (a > b > 0) \quad \text{center } (h, k)$$

$$\frac{(x - h)^2}{b^2} + \frac{(y - k)^2}{a^2} = 1 \quad (a > b > 0) \quad \text{center } (h, k)$$

Equations of a hyperbola:

$$\frac{x^2}{a^2} - \frac{y^2}{b^2} = 1 \quad \text{or} \quad \frac{y^2}{a^2} - \frac{x^2}{b^2} = 1 \quad \text{center } (0, 0)$$

$$\frac{(x - h)^2}{a^2} - \frac{(y - k)^2}{b^2} = 1 \quad \text{or}$$

$$\frac{(y - k)^2}{a^2} - \frac{(x - h)^2}{b^2} = 1 \quad \text{center } (h, k)$$

(10.3) Good estimates for solutions to systems of simultaneous second-degree equations can be found by graphing.

Exact solutions to systems of simultaneous second-degree equations can be found with algebraic techniques.

(10.4) A function is increasing on the interval (a, b) if the values of $f(x)$ increase as x increases from a to b.

A function is decreasing on the interval (a, b) if the values of $f(x)$ decrease as x increases from a to b.

A function is constant on the interval (a, b) if the value of $f(x)$ is constant as x increases from a to b.

(10.5) The symbol $n!$ (**n factorial**) is defined as $n! = n(n - 1)(n - 2)(n - 3) \cdots 3 \cdot 2 \cdot 1$, where n is a natural number.

$$0! = 1$$

$n(n - 1)! = n!$, provided that n is a natural number.

The binomial theorem:

$$(a + b)^n = a^n + \frac{n!}{1!(n - 1)!} a^{n-1}b + \frac{n!}{2!(n - 2)!} a^{n-2}b^2$$

$$+ \frac{n!}{3!(n - 3)!} a^{n-3}b^3 + \cdots + b^n$$

(10.6) An **arithmetic sequence** is a sequence of the form $a, a + d, a + 2d, a + 3d, \ldots, a + (n - 1)d$, where a is the first term, $a + (n - 1)d$ is the nth term, and d is the common difference.

If numbers are inserted between two given numbers a and b to form an arithmetic sequence, the inserted numbers are **arithmetic means** between a and b.

The sum of the first n terms of an arithmetic sequence is given by the formula

$$S_n = \frac{n(a + l)}{2} \quad \text{with } l = a + (n - 1)d$$

where a is the first term, l is the last (or nth) term, and n is the number of terms in the sequence.

$$\sum_{k=1}^{n} f(k) = f(1) + f(2) + f(3) + \cdots + f(n)$$

$$\sigma^2 = \frac{\Sigma(x - \mu)^2}{N} \quad \text{and} \quad \sigma = \sqrt{\frac{\Sigma(x - \mu)^2}{N}}$$

(10.7) A **geometric sequence** is a sequence of the form

$$a, ar, ar^2, ar^3, \ldots, ar^{n-1}$$

where a is the first term, ar^{n-1} is the nth term, and r is the common ratio.

If numbers are inserted between two numbers a and b to form a geometric sequence, the inserted numbers are **geometric means** between a and b.

The sum of the first n terms of a geometric sequence is given by the formula

$$S_n = \frac{a - ar^n}{1 - r} \quad (r \neq 1)$$

where S_n is the sum, a is the first term, r is the common ratio, and n is the number of terms in the sequence.

(10.8) If r is the common ratio of an infinite geometric sequence, and if $|r| < 1$, then the sum of the terms of the infinite geometric sequence is given by the formula

$$S = \frac{a}{1 - r}$$

where a is the first term and r is the common ratio.

■ Chapter 10 Review Exercises

In Review Exercises 1–10, graph each equation.

1. $x^2 + y^2 = 16$

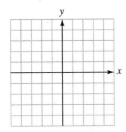

2. $(x - 3)^2 + (y + 2)^2 = 4$

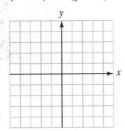

3. $x = -3(y - 2)^2 + 5$

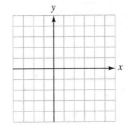

4. $x = 2(y + 1)^2 - 2$

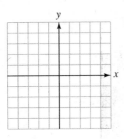

5. $9x^2 + 16y^2 = 144$

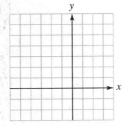

6. $\dfrac{(x - 2)^2}{4} + \dfrac{(y - 1)^2}{9} = 1$

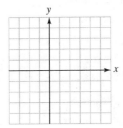

7. $xy = 9$

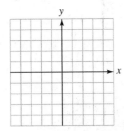

8. $9x^2 - y^2 = -9$

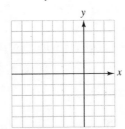

9. $y(x + y) = (y + 2)(y - 2)$

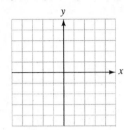

10. $x(x - y) = (x + 1)(x - 1)$

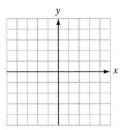

In Review Exercises 11–12, solve each system.

11. $\begin{cases} 3x^2 + y^2 = 52 \\ x^2 - y^2 = 12 \end{cases}$

12. $\begin{cases} \dfrac{x^2}{16} + \dfrac{y^2}{12} = 1 \\ x^2 - \dfrac{y^2}{3} = 1 \end{cases}$

In Review Exercises 13–14, graph each function.

13. $f(x) = \begin{cases} x \text{ if } x \leq 1 \\ -x^2 \text{ if } x > 1 \end{cases}$

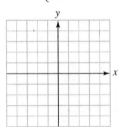

14. $f(x) = 3[\![x]\!]$

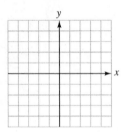

In Review Exercises 15–18, evaluate each expression.

15. $(4!)(3!)$

16. $\dfrac{5!}{3!}$

17. $\dfrac{6!}{2!(6 - 2)!}$

18. $\dfrac{12!}{3!(12 - 3)!}$

In Review Exercises 19–22, use the binomial theorem to find each expansion.

19. $(x + y)^5$

20. $(x - y)^4$

21. $(4x - y)^3$

22. $(x + 4y)^3$

In Review Exercises 23–26, find the required term in each expansion.

23. $(x + y)^4$; 3rd term

24. $(x - y)^5$; 4th term

25. $(3x - 4y)^3$; 2nd term

26. $(4x + 3y)^4$; 3rd term

27. Write the first five terms of the arithmetic sequence whose ninth term is 242 and whose seventh term is 212.

28. Find two arithmetic means between 8 and 25.

29. Find the sum of the first 20 terms of the sequence 11, 18, 25, ...

30. Find the sum of the first ten terms of the sequence 9, $6\frac{1}{2}$, 4, ...

In Review Exercises 31–34, find each sum.

31. $\displaystyle\sum_{k=4}^{6} \frac{1}{2}k$

32. $\displaystyle\sum_{k=2}^{5} 7k^2$

33. $\displaystyle\sum_{k=1}^{4} (3k - 4)$

34. $\displaystyle\sum_{k=10}^{10} 36k$

35. Write the first five terms of the geometric sequence whose fourth term is 3 and whose fifth term is $\frac{3}{2}$.

36. Find two geometric means between -6 and 384.

37. Find the sum of the first eight terms of the sequence $\frac{1}{8}, -\frac{1}{4}, \frac{1}{2}, \ldots$.

38. Find the sum of the first seven terms of the sequence 162, 54, 18, ...

39. Find the sum of the infinite geometric sequence 25, 20, 16, ...

40. Change the decimal $0.\overline{05}$ to a common fraction.

41. Car depreciation   A $5000 car depreciates at the rate of 20% of the previous year's value. How much is the car worth after five years?

42. Stock appreciation   The value of Mia's stock portfolio is expected to appreciate at the rate of 18% per year. How much will the portfolio be worth in ten years if its current value is $25,700?

43. Planting corn   A farmer planted 300 acres in corn this year. He intends to plant an additional 75 acres

in corn each successive year until he has 1200 acres in corn. In how many years will that be?

44. Falling object   If an object is in free fall, the sequence 16, 48, 80, . . . represents the distance in feet that the object falls during the first second, during the second second, during the third second, and so on. How far will the object fall during the first ten seconds?

■ Chapter 10 Test

1. Find the center and the radius of the circle $(x - 2)^2 + (y + 3)^2 = 4$.

In Problems 2–3, graph each equation.

2. $9x^2 + 4y^2 = 36$

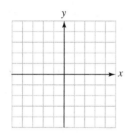

3. $\dfrac{(x - 2)^2}{9} - y^2 = 1$

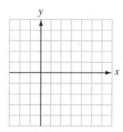

In Problems 4–5, solve each system.

4. $\begin{cases} 2x - y = -2 \\ x^2 + y^2 = 16 + 4y \end{cases}$

5. $\begin{cases} x^2 + y^2 = 25 \\ 4x^2 - 9y = 0 \end{cases}$

6. Graph $f(x) = \begin{cases} -x^2, \text{ when } x < 0 \\ -x, \text{ when } x \geq 0 \end{cases}$.

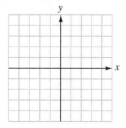

7. Evaluate $\dfrac{7!}{4!2!}$.

8. Find the second term in the expansion of $(x - y)^5$.

9. Find the third term in the expansion of $(x + 2y)^4$.

10. Find the tenth term of an arithmetic sequence whose first three terms are 3, 10, and 17.

11. Find the sum of the first 12 terms of the sequence $-2, 3, 8, \ldots$.

12. Find two arithmetic means between 2 and 98.

13. Evaluate $\displaystyle\sum_{k=1}^{3} (2k - 3)$.

14. Find the seventh term of the geometric sequence whose first three terms are $-\frac{1}{9}$, $-\frac{1}{3}$, and -1.

15. Find the sum of the first six terms of the sequence $\frac{1}{27}, \frac{1}{9}, \frac{1}{3}, \ldots$.

16. Find two geometric means between 3 and 648.

17. Find the sum of all of the terms of the infinite geometric sequence $9, 3, 1, \ldots$.

■ Cumulative Review Exercises (Chapters 1–10)

In Exercises 1–2, do the operations.

1. $(4x - 3y)(3x + y)$

2. $(a^n + 1)(a^n - 3)$

In Exercises 3–4, simplify each fraction.

3. $\dfrac{5a - 10}{a^2 - 4a + 4}$

4. $\dfrac{a^4 - 5a^2 + 4}{a^2 + 3a + 2}$

In Exercises 5–6, do the operations and simplify the result, if possible.

5. $\dfrac{a^2 - a - 6}{a^2 - 4} \div \dfrac{a^2 - 9}{a^2 + a - 6}$

6. $\dfrac{2}{a - 2} + \dfrac{3}{a + 2} - \dfrac{a - 1}{a^2 - 4}$

In Exercises 7–8, tell whether the graphs of the linear equations are parallel or perpendicular.

7. $3x - 4y = 12, \; y = \dfrac{3}{4}x - 5$

8. $y = 3x + 4, \; x = -3y + 4$

In Exercises 9–10, write the equation of each line with the following properties.

9. $m = -2$, passing through $(0, 5)$

10. Passing through $P(8, -5)$ and $Q(-5, 4)$

In Exercises 11–12, graph each inequality.

11. $2x - 3y < 6$

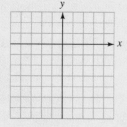

12. $y \geq x^2 - 4$

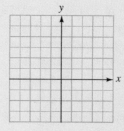

In Exercises 13–14, simplify each expression.

13. $\sqrt{98} + \sqrt{8} - \sqrt{32}$

14. $12\sqrt[3]{648x^4} + 3\sqrt[3]{81x^4}$

In Exercises 15–18, solve each equation.

15. $\sqrt{3a + 1} = a - 1$

16. $\sqrt{x + 3} - \sqrt{3} = \sqrt{x}$

17. $6a^2 + 5a - 6 = 0$

18. $3x^2 + 8x - 1 = 0$

19. If $f(x) = x^2 - 2$ and $g(x) = 2x + 1$, find $(f \circ g)(x)$.

20. Find the inverse function of $y = 2x^3 - 1$.

21. Graph $y = \left(\dfrac{1}{2}\right)^x$.

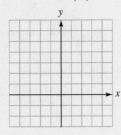

22. Write $y = \log_2 x$ in exponential form.

In Exercises 23–26, find x.

23. $\log_x 25 = 2$

24. $\log_5 125 = x$

25. $\log_3 x = -3$

26. $\log_5 x = 0$

27. Find the inverse of $y = f(x) = \log_2 x$.

28. If $\log_{10} 10^x = y$, find y.

In Exercises 29–32, $\log 7 = 0.8451$ and $\log 14 = 1.1461$. Evaluate each expression without using a calculator.

29. $\log 98$

30. $\log 2$

31. $\log 49$

32. $\log \dfrac{7}{5}$ (*Hint:* $\log 10 = 1$.)

In Exercises 33–34, solve each equation.

33. $2^{x+2} = 3^x$

34. $2 \log 5 + \log x - \log 4 = 2$

In Exercises 35–36, use a calculator.

35. Boat depreciation How much will a $9000 boat be worth after 9 years if it depreciates 12% per year?

36. Find $\log_6 8$ to five decimal places.

In Exercises 37–38, graph each equation.

37. $x^2 + (y + 1)^2 = 9$

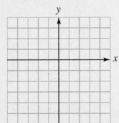

38. $x^2 - 9(y + 1)^2 = 9$

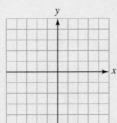

39. $\dfrac{6!7!}{5!}$

40. Use the binomial theorem to expand $(3a - b)^4$.

41. Find the 7th term in the expansion of $(2x - y)^8$.

42. Find the 20th term of an arithmetic sequence with a first term of -11 and a common difference of 6.

43. Find the sum of the first 20 terms of an arithmetic sequence with a first term of 6 and a common difference of 3.

44. Insert two arithmetic means between -3 and 30.

45. Evaluate $\displaystyle\sum_{k=1}^{3} 3k^2$.

46. Evaluate $\displaystyle\sum_{k=3}^{5} (2k + 1)$.

47. Find the 7th term of a geometric sequence with a first term of $\frac{1}{27}$ and a common ratio of 3.

48. Find the sum of the first 10 terms of the sequence $\frac{1}{64}, \frac{1}{32}, \frac{1}{16}, \ldots$.

49. Insert two geometric means between -3 and 192.

50. Find the sum of all the terms of the sequence $9, 3, 1, \ldots$.

SYMMETRIES OF GRAPHS

There are several ways that a graph can exhibit symmetry about the coordinate axes and the origin. It is often easier to draw graphs of equations if we first find the x- and y-intercepts and find any of the following symmetries of the graph:

1. **y-axis symmetry**: If the point $(-x, y)$ lies on a graph whenever the point (x, y) does, as in Figure I-1(a), we say that the graph is **symmetric about the y-axis**.

2. **Symmetry about the origin**: If the point $(-x, -y)$ lies on the graph whenever the point (x, y) does, as in Figure I-1(b), we say that the graph is **symmetric about the origin**.

3. **x-axis symmetry**: If the point $(x, -y)$ lies on the graph whenever the point (x, y) does, as in Figure I-1(c), we say that the graph is **symmetric about the x-axis**.

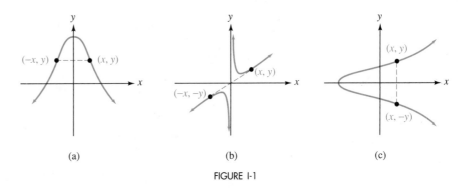

FIGURE I-1

Tests for Symmetry for Graphs in x and y

- To test a graph for y-axis symmetry, replace x with $-x$. If the new equation is equivalent to the original equation, the graph is symmetric about the y-axis. Symmetry about the y-axis will occur whenever x appears with only even exponents.

- To test a graph for symmetry about the origin, replace x with $-x$ and y with $-y$. If the resulting equation is equivalent to the original equation, the graph is symmetric about the origin.

- To test a graph for x-axis symmetry, replace y with $-y$. If the resulting equation is equivalent to the original equation, the graph is symmetric about the x-axis. Except for $f(x) = 0$, the graph of no function is symmetric to the x-axis.

EXAMPLE 1   Find the intercepts and the symmetries of the graph of $y = f(x) = x^3 - 9x$. Then graph the function.

Solution   **x-intercepts**: To find the x-intercepts, we let $y = 0$ and solve for x:

$$y = x^3 - 9x$$
$$0 = x^3 - 9x \qquad \text{Substitute 0 for } y.$$
$$0 = x(x^2 - 9) \qquad \text{Factor out } x.$$
$$0 = x(x + 3)(x - 3) \qquad \text{Factor } x^2 - 9.$$

$$x = 0 \quad \text{or} \quad x + 3 = 0 \quad \text{or} \quad x - 3 = 0 \qquad \text{Set each factor equal to 0.}$$
$$\qquad\qquad\qquad x = -3 \qquad\quad x = 3$$

Since the x-coordinates of the x-intercepts are 0, -3, and 3, the graph intersects the x-axis at $(0, 0)$, $(-3, 0)$, and $(3, 0)$.

y-intercepts: To find the y-intercepts, we let $x = 0$ and solve for y.

$$y = x^3 - 9x$$
$$y = 0^3 - 9(0) \qquad \text{Substitute 0 for } x.$$
$$y = 0$$

Since the y-coordinate of the y-intercept is 0, the graph intersects the y-axis at $(0, 0)$.

Symmetry: We test for symmetry about the y-axis by replacing x with $-x$, simplifying, and comparing the result to the original equation.

1. $y = x^3 - 9x$ $\qquad\qquad\qquad$ The original equation.
$\quad\ y = (-x)^3 - 9(-x)$ $\qquad$ Replace x with $-x$.
2. $y = -x^3 + 9x$ $\qquad\qquad$ Simplify.

Because Equation 2 is not equivalent to Equation 1, the graph is not symmetric about the y-axis.

We test for symmetry about the origin by replacing x and y with $-x$ and $-y$, respectively, and comparing the result to the original equation.

1. $y = x^3 - 9x$ $\qquad\qquad\qquad$ The original equation.
$\quad -y = (-x)^3 - 9(-x)$ $\qquad$ Replace x with $-x$, and y with $-y$.
$\quad -y = -x^3 + 9x$ $\qquad\qquad$ Simplify.
3. $y = x^3 - 9x$ $\qquad\qquad\qquad$ Multiply both sides by -1.

Because Equation 3 is equivalent to Equation 1, the graph is symmetric about the origin.

We test for symmetry about the x-axis by replacing y with $-y$ and comparing the result to the original equation.

1. $y = x^3 - 9x$ $\qquad\qquad\qquad$ The original equation.
$\quad -y = x^3 - 9x$ $\qquad\qquad\quad$ Replace y with $-y$.
4. $y = -x^3 + 9x$ $\qquad\qquad$ Multiply both sides by -1.

Because Equation 4 is not equivalent to Equation 1, the graph is not symmetric about the x-axis.

To graph the equation, we plot the x-intercepts of $(-3, 0)$, $(0, 0)$, and $(3, 0)$ and the y-intercept of $(0, 0)$. We also plot other points for positive values of x and use the symmetry about the origin to draw the rest of the graph, as in Figure I-2(a). (Note that the scale on the x-axis is different from the scale on the y-axis.)

If we graph the equation with a graphing device, with window settings of $[-10, 10]$ for x and $[-10, 10]$ for y, we will obtain the graph shown in Figure I-2(b).

From the graph, we can see that the domain is the interval $(-\infty, \infty)$, and the range is the interval $(-\infty, \infty)$.

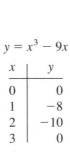

$y = x^3 - 9x$

x	y
0	0
1	-8
2	-10
3	0

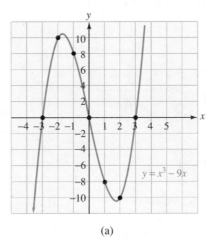

(a)

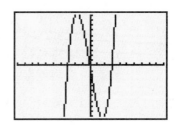

(b)

FIGURE I-2

EXAMPLE 2   Graph the function $y = f(x) = |x| - 2$.

Solution   **x-intercepts**: To find the x-intercepts, we let $y = 0$ and solve for x:

$$y = |x| - 2$$
$$0 = |x| - 2$$
$$2 = |x|$$
$$x = -2 \quad \text{or} \quad x = 2$$

Since -2 and 2 are solutions, the points $(-2, 0)$ and $(2, 0)$ are the x-intercepts, and the graph passes through $(-2, 0)$ and $(2, 0)$.

y-intercepts: To find the y-intercepts, we let $x = 0$ and solve for y:

$$y = |x| - 2$$
$$y = |0| - 2$$

Since $y = -2$, $(0, -2)$ is the y-intercept, and the graph passes through the point $(0, -2)$.

Symmetry: To test for y-axis symmetry, we replace x with $-x$.

5. $y = |x| - 2$ The original equation.

      $y = |-x| - 2$ Replace x with $-x$.

6. $y = |x| - 2$ $|-x| = |x|$.

Since Equation 6 is equivalent to Equation 5, the graph is symmetric about the y-axis. The graph has no other symmetries.

We plot the x- and y-intercepts and several other points (x, y), and use the y-axis symmetry to obtain the graph shown in Figure I-3(a).

If we graph the equation with a graphing device, with window settings of $[-10, 10]$ for x and $[-10, 10]$ for y, we will obtain the graph shown in Figure I-3(b).

From the graph, we see that the domain is the interval $(-\infty, \infty)$, and the range is the interval $[-2, \infty)$.

$y = |x| - 2$

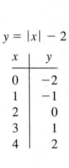

x	y
0	-2
1	-1
2	0
3	1
4	2

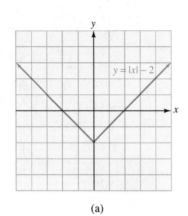

(a)

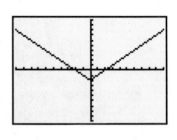

(b)

FIGURE I-3   ■

EXERCISE I.1

In Exercises 1–12, find the symmetries of the graph of each relation. **Do not draw the graph.**

1. $y = x^2 - 1$ **2.** $y = x^3$ **3.** $y = x^5$ **4.** $y = x^4$

5. $y = -x^2 + 2$ **6.** $y = x^3 + 1$ **7.** $y = x^2 - x$ **8.** $y^2 = x + 7$

9. $y = -|x + 2|$ **10.** $y = |x| - 3$ **11.** $|y| = x$ **12.** $y = 2\sqrt{x}$

In Exercises 13–24, graph each function and give its domain and range. Check each graph with a graphing calculator.

13. $y = x^4 - 4$ **14.** $y = \dfrac{1}{2}x^4 - 1$ **15.** $y = -x^3$ **16.** $y = x^3 + 2$

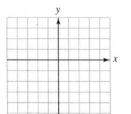   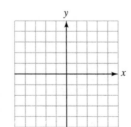   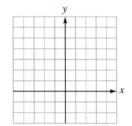

17. $y = x^4 + x^2$

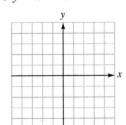

18. $y = 3 - x^4$

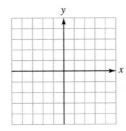

19. $y = x^3 - x$

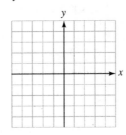

20. $y = x^3 + x$

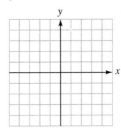

21. $y = \dfrac{1}{2}|x| - 1$

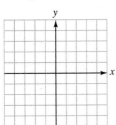

22. $y = -|x| + 1$

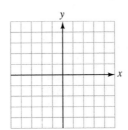

23. $y = -|x + 2|$

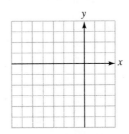

24. $y = |x - 2|$

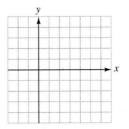

Table A   Powers and Roots

n	n^2	$\sqrt{n}$	n^3	$\sqrt[3]{n}$	n	n^2	$\sqrt{n}$	n^3	$\sqrt[3]{n}$
1	1	1.000	1	1.000	51	2,601	7.141	132,651	3.708
2	4	1.414	8	1.260	52	2,704	7.211	140,608	3.733
3	9	1.732	27	1.442	53	2,809	7.280	148,877	3.756
4	16	2.000	64	1.587	54	2,916	7.348	157,464	3.780
5	25	2.236	125	1.710	55	3,025	7.416	166,375	3.803
6	36	2.449	216	1.817	56	3,136	7.483	175,616	3.826
7	49	2.646	343	1.913	57	3,249	7.550	185,193	3.849
8	64	2.828	512	2.000	58	3,364	7.616	195,112	3.871
9	81	3.000	729	2.080	59	3,481	7.681	205,379	3.893
10	100	3.162	1,000	2.154	60	3,600	7.746	216,000	3.915
11	121	3.317	1,331	2.224	61	3,721	7.810	226,981	3.936
12	144	3.464	1,728	2.289	62	3,844	7.874	238,328	3.958
13	169	3.606	2,197	2.351	63	3,969	7.937	250,047	3.979
14	196	3.742	2,744	2.410	64	4,096	8.000	262,144	4.000
15	225	3.873	3,375	2.466	65	4,225	8.062	274,625	4.021
16	256	4.000	4,096	2.520	66	4,356	8.124	287,496	4.041
17	289	4.123	4,913	2.571	67	4,489	8.185	300,763	4.062
18	324	4.243	5,832	2.621	68	4,624	8.246	314,432	4.082
19	361	4.359	6,859	2.668	69	4,761	8.307	328,509	4.102
20	400	4.472	8,000	2.714	70	4,900	8.367	343,000	4.121
21	441	4.583	9,261	2.759	71	5,041	8.426	357,911	4.141
22	484	4.690	10,648	2.802	72	5,184	8.485	373,248	4.160
23	529	4.796	12,167	2.844	73	5,329	8.544	389,017	4.179
24	576	4.899	13,824	2.884	74	5,476	8.602	405,224	4.198
25	625	5.000	15,625	2.924	75	5,625	8.660	421,875	4.217
26	676	5.099	17,576	2.962	76	5,776	8.718	438,976	4.236
27	729	5.196	19,683	3.000	77	5,929	8.775	456,533	4.254
28	784	5.292	21,952	3.037	78	6,084	8.832	474,552	4.273
29	841	5.385	24,389	3.072	79	6,241	8.888	493,039	4.291
30	900	5.477	27,000	3.107	80	6,400	8.944	512,000	4.309
31	961	5.568	29,791	3.141	81	6,561	9.000	531,441	4.327
32	1,024	5.657	32,768	3.175	82	6,724	9.055	551,368	4.344
33	1,089	5.745	35,937	3.208	83	6,889	9.110	571,787	4.362
34	1,156	5.831	39,304	3.240	84	7,056	9.165	592,704	4.380
35	1,225	5.916	42,875	3.271	85	7,225	9.220	614,125	4.397
36	1,296	6.000	46,656	3.302	86	7,396	9.274	636,056	4.414
37	1,369	6.083	50,653	3.332	87	7,569	9.327	658,503	4.431
38	1,444	6.164	54,872	3.362	88	7,744	9.381	681,472	4.448
39	1,521	6.245	59,319	3.391	89	7,921	9.434	704,969	4.465
40	1,600	6.325	64,000	3.420	90	8,100	9.487	729,000	4.481
41	1,681	6.403	68,921	3.448	91	8,281	9.539	753,571	4.498
42	1,764	6.481	74,088	3.476	92	8,464	9.592	778,688	4.514
43	1,849	6.557	79,507	3.503	93	8,649	9.644	804,357	4.531
44	1,936	6.633	85,184	3.530	94	8,836	9.695	830,584	4.547
45	2,025	6.708	91,125	3.557	95	9,025	9.747	857,375	4.563
46	2,116	6.782	97,336	3.583	96	9,216	9.798	884,736	4.579
47	2,209	6.856	103,823	3.609	97	9,409	9.849	912,673	4.595
48	2,304	6.928	110,592	3.634	98	9,604	9.899	941,192	4.610
49	2,401	7.000	117,649	3.659	99	9,801	9.950	970,299	4.626
50	2,500	7.071	125,000	3.684	100	10,000	10.000	1,000,000	4.642

Table B   (continued)

N	0	1	2	3	4	5	6	7	8	9
5.5	7404	7412	7419	7427	7435	7443	7451	7459	7466	7474
5.6	7482	7490	7497	7505	7513	7520	7528	7536	7543	7551
5.7	7559	7566	7574	7582	7589	7597	7604	7612	7619	7627
5.8	7634	7642	7649	7657	7664	7672	7679	7686	7694	7701
5.9	7709	7716	7723	7731	7738	7745	7752	7760	7767	7774
6.0	7782	7789	7796	7803	7810	7818	7825	7832	7839	7846
6.1	7853	7860	7868	7875	7882	7889	7896	7903	7910	7917
6.2	7924	7931	7938	7945	7952	7959	7966	7973	7980	7987
6.3	7993	8000	8007	8014	8021	8028	8035	8041	8048	8055
6.4	8062	8069	8075	8082	8089	8096	8102	8109	8116	8122
6.5	8129	8136	8142	8149	8156	8162	8169	8176	8182	8189
6.6	8195	8202	8209	8215	8222	8228	8235	8241	8248	8254
6.7	8261	8267	8274	8280	8287	8293	8299	8306	8312	8319
6.8	8325	8331	8338	8344	8351	8357	8363	8370	8376	8382
6.9	8388	8395	8401	8407	8414	8420	8426	8432	8439	8445
7.0	8451	8457	8463	8470	8476	8482	8488	8494	8500	8506
7.1	8513	8519	8525	8531	8537	8543	8549	8555	8561	8567
7.2	8573	8579	8585	8591	8597	8603	8609	8615	8621	8627
7.3	8633	8639	8645	8651	8657	8663	8669	8675	8681	8686
7.4	8692	8698	8704	8710	8716	8722	8727	8733	8739	8745
7.5	8751	8756	8762	8768	8774	8779	8785	8791	8797	8802
7.6	8808	8814	8820	8825	8831	8837	8842	8848	8854	8859
7.7	8865	8871	8876	8882	8887	8893	8899	8904	8910	8915
7.8	8921	8927	8932	8938	8943	8949	8954	8960	8965	8971
7.9	8976	8982	8987	8993	8998	9004	9009	9015	9020	9025
8.0	9031	9036	9042	9047	9053	9058	9063	9069	9074	9079
8.1	9085	9090	9096	9101	9106	9112	9117	9122	9128	9133
8.2	9138	9143	9149	9154	9159	9165	9170	9175	9180	9186
8.3	9191	9196	9201	9206	9212	9217	9222	9227	9232	9238
8.4	9243	9248	9253	9258	9263	9269	9274	9279	9284	9289
8.5	9294	9299	9304	9309	9315	9320	9325	9330	9335	9340
8.6	9345	9350	9355	9360	9365	9370	9375	9380	9385	9390
8.7	9395	9400	9405	9410	9415	9420	9425	9430	9435	9440
8.8	9445	9450	9455	9460	9465	9469	9474	9479	9484	9489
8.9	9494	9499	9504	9509	9513	9518	9523	9528	9533	9538
9.0	9542	9547	9552	9557	9562	9566	9571	9576	9581	9586
9.1	9590	9595	9600	9605	9609	9614	9619	9624	9628	9633
9.2	9638	9643	9647	9652	9657	9661	9666	9671	9675	9680
9.3	9685	9689	9694	9699	9703	9708	9713	9717	9722	9727
9.4	9731	9736	9741	9745	9750	9754	9759	9763	9768	9773
9.5	9777	9782	9786	9791	9795	9800	9805	9809	9814	9818
9.6	9823	9827	9832	9836	9841	9845	9850	9854	9859	9863
9.7	9868	9872	9877	9881	9886	9890	9894	9899	9903	9908
9.8	9912	9917	9921	9926	9930	9934	9939	9943	9948	9952
9.9	9956	9961	9965	9969	9974	9978	9983	9987	9991	9996

Table B   Base-10 Logarithms

N	0	1	2	3	4	5	6	7	8	9
1.0	0000	0043	0086	0128	0170	0212	0253	0294	0334	0374
1.1	0414	0453	0492	0531	0569	0607	0645	0682	0719	0755
1.2	0792	0828	0864	0899	0934	0969	1004	1038	1072	1106
1.3	1139	1173	1206	1239	1271	1303	1335	1367	1399	1430
1.4	1461	1492	1523	1553	1584	1614	1644	1673	1703	1732
1.5	1761	1790	1818	1847	1875	1903	1931	1959	1987	2014
1.6	2041	2068	2095	2122	2148	2175	2201	2227	2253	2279
1.7	2304	2330	2355	2380	2405	2430	2455	2480	2504	2529
1.8	2553	2577	2601	2625	2648	2672	2695	2718	2742	2765
1.9	2788	2810	2833	2856	2878	2900	2923	2945	2967	2989
2.0	3010	3032	3054	3075	3096	3118	3139	3160	3181	3201
2.1	3222	3243	3263	3284	3304	3324	3345	3365	3385	3404
2.2	3424	3444	3464	3483	3502	3522	3541	3560	3579	3598
2.3	3617	3636	3655	3674	3692	3711	3729	3747	3766	3784
2.4	3802	3820	3838	3856	3874	3892	3909	3927	3945	3962
2.5	3979	3997	4014	4031	4048	4065	4082	4099	4116	4133
2.6	4150	4166	4183	4200	4216	4232	4249	4265	4281	4298
2.7	4314	4330	4346	4362	4378	4393	4409	4425	4440	4456
2.8	4472	4487	4502	4518	4533	4548	4564	4579	4594	4609
2.9	4624	4639	4654	4669	4683	4698	4713	4728	4742	4757
3.0	4771	4786	4800	4814	4829	4843	4857	4871	4886	4900
3.1	4914	4928	4942	4955	4969	4983	4997	5011	5024	5038
3.2	5051	5065	5079	5092	5105	5119	5132	5145	5159	5172
3.3	5185	5198	5211	5224	5237	5250	5263	5276	5289	5302
3.4	5315	5328	5340	5353	5366	5378	5391	5403	5416	5428
3.5	5441	5453	5465	5478	5490	5502	5514	5527	5539	5551
3.6	5563	5575	5587	5599	5611	5623	5635	5647	5658	5670
3.7	5682	5694	5705	5717	5729	5740	5752	5763	5775	5786
3.8	5798	5809	5821	5832	5843	5855	5866	5877	5888	5899
3.9	5911	5922	5933	5944	5955	5966	5977	5988	5999	6010
4.0	6021	6031	6042	6053	6064	6075	6085	6096	6107	6117
4.1	6128	6138	6149	6160	6170	6180	6191	6201	6212	6222
4.2	6232	6243	6253	6263	6274	6284	6294	6304	6314	6325
4.3	6335	6345	6355	6365	6375	6385	6395	6405	6415	6425
4.4	6435	6444	6454	6464	6474	6484	6493	6503	6513	6522
4.5	6532	6542	6551	6561	6571	6580	6590	6599	6609	6618
4.6	6628	6637	6646	6656	6665	6675	6684	6693	6702	6712
4.7	6721	6730	6739	6749	6758	6767	6776	6785	6794	6803
4.8	6812	6821	6830	6839	6848	6857	6866	6875	6884	6893
4.9	6902	6911	6920	6928	6937	6946	6955	6964	6972	6981
5.0	6990	6998	7007	7016	7024	7033	7042	7050	7059	7067
5.1	7076	7084	7093	7101	7110	7118	7126	7135	7143	7152
5.2	7160	7168	7177	7185	7193	7202	7210	7218	7226	7235
5.3	7243	7251	7259	7267	7275	7284	7292	7300	7308	7316
5.4	7324	7332	7340	7348	7356	7364	7372	7380	7388	7396

Table C   Base-e Logarithms

N	0	1	2	3	4	5	6	7	8	9
1.0	.0000	.0100	.0198	.0296	.0392	.0488	.0583	.0677	.0770	.0862
1.1	.0953	.1044	.1133	.1222	.1310	.1398	.1484	.1570	.1655	.1740
1.2	.1823	.1906	.1989	.2070	.2151	.2231	.2311	.2390	.2469	.2546
1.3	.2624	.2700	.2776	.2852	.2927	.3001	.3075	.3148	.3221	.3293
1.4	.3365	.3436	.3507	.3577	.3646	.3716	.3784	.3853	.3920	.3988
1.5	.4055	.4121	.4187	.4253	.4318	.4383	.4447	.4511	.4574	.4637
1.6	.4700	.4762	.4824	.4886	.4947	.5008	.5068	.5128	.5188	.5247
1.7	.5306	.5365	.5423	.5481	.5539	.5596	.5653	.5710	.5766	.5822
1.8	.5878	.5933	.5988	.6043	.6098	.6152	.6206	.6259	.6313	.6366
1.9	.6419	.6471	.6523	.6575	.6627	.6678	.6729	.6780	.6831	.6881
2.0	.6931	.6981	.7031	.7080	.7129	.7178	.7227	.7275	.7324	.7372
2.1	.7419	.7467	.7514	.7561	.7608	.7655	.7701	.7747	.7793	.7839
2.2	.7885	.7930	.7975	.8020	.8065	.8109	.8154	.8198	.8242	.8286
2.3	.8329	.8372	.8416	.8459	.8502	.8544	.8587	.8629	.8671	.8713
2.4	.8755	.8796	.8838	.8879	.8920	.8961	.9002	.9042	.9083	.9123
2.5	.9163	.9203	.9243	.9282	.9322	.9361	.9400	.9439	.9478	.9517
2.6	.9555	.9594	.9632	.9670	.9708	.9746	.9783	.9821	.9858	.9895
2.7	.9933	.9969	1.0006	.0043	.0080	.0116	.0152	.0188	.0225	.0260
2.8	1.0296	.0332	.0367	.0403	.0438	.0473	.0508	.0543	.0578	.0613
2.9	.0647	.0682	.0716	.0750	.0784	.0818	.0852	.0886	.0919	.0953
3.0	1.0986	.1019	.1053	.1086	.1119	.1151	.1184	.1217	.1249	.1282
3.1	.1314	.1346	.1378	.1410	.1442	.1474	.1506	.1537	.1569	.1600
3.2	.1632	.1663	.1694	.1725	.1756	.1787	.1817	.1848	.1878	.1909
3.3	.1939	.1969	.2000	.2030	.2060	.2090	.2119	.2149	.2179	.2208
3.4	.2238	.2267	.2296	.2326	.2355	.2384	.2413	.2442	.2470	.2499
3.5	1.2528	.2556	.2585	.2613	.2641	.2669	.2698	.2726	.2754	.2782
3.6	.2809	.2837	.2865	.2892	.2920	.2947	.2975	.3002	.3029	.3056
3.7	.3083	.3110	.3137	.3164	.3191	.3218	.3244	.3271	.3297	.3324
3.8	.3350	.3376	.3403	.3429	.3455	.3481	.3507	.3533	.3558	.3584
3.9	.3610	.3635	.3661	.3686	.3712	.3737	.3762	.3788	.3813	.3838
4.0	1.3863	.3888	.3913	.3938	.3962	.3987	.4012	.4036	.4061	.4085
4.1	.4110	.4134	.4159	.4183	.4207	.4231	.4255	.4279	.4303	.4327
4.2	.4351	.4375	.4398	.4422	.4446	.4469	.4493	.4516	.4540	.4563
4.3	.4586	.4609	.4633	.4656	.4679	.4702	.4725	.4748	.4770	.4793
4.4	.4816	.4839	.4861	.4884	.4907	.4929	.4951	.4974	.4996	.5019
4.5	1.5041	.5063	.5085	.5107	.5129	.5151	.5173	.5195	.5217	.5239
4.6	.5261	.5282	.5304	.5326	.5347	.5369	.5390	.5412	.5433	.5454
4.7	.5476	.5497	.5518	.5539	.5560	.5581	.5602	.5623	.5644	.5665
4.8	.5686	.5707	.5728	.5748	.5769	.5790	.5810	.5831	.5851	.5872
4.9	.5892	.5913	.5933	.5953	.5974	.5994	.6014	.6034	.6054	.6074
5.0	1.6094	.6114	.6134	.6154	.6174	.6194	.6214	.6233	.6253	.6273
5.1	.6292	.6312	.6332	.6351	.6371	.6390	.6409	.6429	.6448	.6467
5.2	.6487	.6506	.6525	.6544	.6563	.6582	.6601	.6620	.6639	.6658
5.3	.6677	.6696	.6715	.6734	.6752	.6771	.6790	.6808	.6827	.6845
5.4	.6864	.6882	.6901	.6919	.6938	.6956	.6974	.6993	.7011	.7029

Table C   (continued)

N	0	1	2	3	4	5	6	7	8	9
5.5	1.7047	.7066	.7084	.7102	.7120	.7138	.7156	.7174	.7192	.7210
5.6	.7228	.7246	.7263	.7281	.7299	.7317	.7334	.7352	.7370	.7387
5.7	.7405	.7422	.7440	.7457	.7475	.7492	.7509	.7527	.7544	.7561
5.8	.7579	.7596	.7613	.7630	.7647	.7664	.7681	.7699	.7716	.7733
5.9	.7750	.7766	.7783	.7800	.7817	.7834	.7851	.7867	.7884	.7901
6.0	1.7918	.7934	.7951	.7967	.7984	.8001	.8017	.8034	.8050	.8066
6.1	.8083	.8099	.8116	.8132	.8148	.8165	.8181	.8197	.8213	.8229
6.2	.8245	.8262	.8278	.8294	.8310	.8326	.8342	.8358	.8374	.8390
6.3	.8405	.8421	.8437	.8453	.8469	.8485	.8500	.8516	.8532	.8547
6.4	.8563	.8579	.8594	.8610	.8625	.8641	.8656	.8672	.8687	.8703
6.5	1.8718	.8733	.8749	.8764	.8779	.8795	.8810	.8825	.8840	.8856
6.6	.8871	.8886	.8901	.8916	.8931	.8946	.8961	.8976	.8991	.9006
6.7	.9021	.9036	.9051	.9066	.9081	.9095	.9110	.9125	.9140	.9155
6.8	.9169	.9184	.9199	.9213	.9228	.9242	.9257	.9272	.9286	.9301
6.9	.9315	.9330	.9344	.9359	.9373	.9387	.9402	.9416	.9430	.9445
7.0	1.9459	.9473	.9488	.9502	.9516	.9530	.9544	.9559	.9573	.9587
7.1	.9601	.9615	.9629	.9643	.9657	.9671	.9685	.9699	.9713	.9727
7.2	.9741	.9755	.9769	.9782	.9796	.9810	.9824	.9838	.9851	.9865
7.3	.9879	.9892	.9906	.9920	.9933	.9947	.9961	.9974	.9988	2.0001
7.4	2.0015	.0028	.0042	.0055	.0069	.0082	.0096	.0109	.0122	.0136
7.5	2.0149	.0162	.0176	.0189	.0202	.0215	.0229	.0242	.0255	.0268
7.6	.0281	.0295	.0308	.0321	.0334	.0347	.0360	.0373	.0386	.0399
7.7	.0412	.0425	.0438	.0451	.0464	.0477	.0490	.0503	.0516	.0528
7.8	.0541	.0554	.0567	.0580	.0592	.0605	.0618	.0631	.0643	.0656
7.9	.0669	.0681	.0694	.0707	.0719	.0732	.0744	.0757	.0769	.0782
8.0	2.0794	.0807	.0819	.0832	.0844	.0857	.0869	.0882	.0894	.0906
8.1	.0919	.0931	.0943	.0956	.0968	.0980	.0992	.1005	.1017	.1029
8.2	.1041	.1054	.1066	.1078	.1090	.1102	.1114	.1126	.1138	.1150
8.3	.1163	.1175	.1187	.1199	.1211	.1223	.1235	.1247	.1258	.1270
8.4	.1282	.1294	.1306	.1318	.1330	.1342	.1353	.1365	.1377	.1389
8.5	2.1401	.1412	.1424	.1436	.1448	.1459	.1471	.1483	.1494	.1506
8.6	.1518	.1529	.1541	.1552	.1564	.1576	.1587	.1599	.1610	.1622
8.7	.1633	.1645	.1656	.1668	.1679	.1691	.1702	.1713	.1725	.1736
8.8	.1748	.1759	.1770	.1782	.1793	.1804	.1815	.1827	.1838	.1849
8.9	.1861	.1872	.1883	.1894	.1905	.1917	.1928	.1939	.1950	.1961
9.0	2.1972	.1983	.1994	.2006	.2017	.2028	.2039	.2050	.2061	.2072
9.1	.2083	.2094	.2105	.2116	.2127	.2138	.2148	.2159	.2170	.2181
9.2	.2192	.2203	.2214	.2225	.2235	.2246	.2257	.2268	.2279	.2289
9.3	.2300	.2311	.2322	.2332	.2343	.2354	.2364	.2375	.2386	.2396
9.4	.2407	.2418	.2428	.2439	.2450	.2460	.2471	.2481	.2492	.2502
9.5	2.2513	.2523	.2534	.2544	.2555	.2565	.2576	.2586	.2597	.2607
9.6	.2618	.2628	.2638	.2649	.2659	.2670	.2680	.2690	.2701	.2711
9.7	.2721	.2732	.2742	.2752	.2762	.2773	.2783	.2793	.2803	.2814
9.8	.2824	.2834	.2844	.2854	.2865	.2875	.2885	.2895	.2905	.2915
9.9	.2925	.2935	.2946	.2956	.2966	.2976	.2986	.2996	.3006	.3016

Use the properties of logarithms and $\ln 10 \approx 2.3026$ to find logarithms of numbers less than 1 or greater than 10.

Orals (page 11)

1. {1, 2, 3, 4, 5}   **2.** {4}   **3.** no   **4.** yes   **5.** 2, 3, 5, 7   **7.** 6   **8.** 10

Exercise 1.1 (page 11)

1. ⊆   **3.** ∈   **5.** ⊆   **7.** ⊆   **9.** {1, 2, 3, 4, 5, 6, 8, 10}   **11.** ∅   **13.** ∅   **15.** {1, 3, 5}   **17.** 2, 3, 5, 7   **19.** no elements
21. 1, 2, 9   **23.** −3, 0, 1, 2, 9   **25.** $\sqrt{3}$ **27.** 2   **29.** 2   **31.** 9   **33.** (number line graph: filled dots at 2, 4, 5, 7)

35. (number line graph 10–20)   **37.** 0.875, terminating   **39.** $-0.7\overline{3}$, repeating   **41.** <   **43.** >   **45.** <
47. >   **49.** 12 < 19   **51.** −5 ≥ −6   **53.** −3 ≤ 5   **55.** 0 > −10   **57.** (graph)   **59.** (graph)

61. (graph 0 to 5)   **63.** (graph −2)   **65.** (graph −6 to 9)   **67.** (graph −2 to 4)   **69.** (graph 2 to 4)
71. (graph −4 to 8)   **73.** (graph 2 to 4)   **75.** (graph −4 to 6)   **77.** 20   **79.** −6   **81.** 7   **83.** 20   **85.** 3 or −3
87. $x \ge 0$ **89.** 99

Review Exercises (page 13)

1. $\frac{3}{4}$ **2.** $\frac{3}{4}$ **3.** $\frac{4}{5}$ **4.** $\frac{7}{9}$ **5.** $\frac{3}{20}$ **6.** $\frac{4}{9}$ **7.** $\frac{14}{9}$ **8.** 1   **9.** 1   **10.** 2   **11.** $\frac{22}{15}$ **12.** $\frac{17}{45}$

Orals (page 23)

1. −2   **2.** −7   **3.** −28   **4.** 28   **5.** −3   **6.** −3

Exercise 1.2 (page 23)

1. −8   **3.** −5   **5.** −7   **7.** 0   **9.** −12   **11.** 21   **13.** −2   **15.** 4   **17.** $\frac{1}{6}$ **19.** $\frac{11}{10}$ **21.** $-\frac{1}{6}$ **23.** $-\frac{6}{7}$ **25.** −2   **27.** $\frac{24}{25}$ **29.** $62
31. +4°   **33.** 12°   **35.** 6900 gal   **37.** 23   **39.** 0   **41.** 2   **43.** −13   **45.** 1   **47.** 4   **49.** −1   **51.** −4   **53.** −12   **55.** −20
57. +1325 m   **59.** $421.88   **61.** 8   **63.** 9   **65.** 12   **67.** $1211   **69.** 80   **71.** not really   **75.** −8   **77.** −9   **79.** $-\frac{5}{4}$ **81.** $-\frac{1}{8}$
83. 19,900   **85.** 30 cm   **87.** 100.4 ft   **89.** symmetric property   **91.** reflexive property   **93.** transitive property
95. clos. prop. of mult.   **97.** comm. prop. of add.   **99.** dist. prop.   **101.** additive identity prop.   **103.** mult. inverse prop.
105. assoc. prop. of add.   **107.** comm. prop. of mult.   **109.** assoc. prop. of add.   **111.** dist. prop.

Review Exercises (page 27)

1. (graph 4)   **2.** (graph −5)   **3.** (graph 2 to 10)   **4.** (graph −4 to 4)   **5.** $45.53   **6.** $13,295.10

Orals (page 35)

1. 16   **2.** 27   **3.** x^5 **4.** y^7 **5.** 1   **6.** x^6 **7.** a^6b^3 **8.** $\dfrac{b^2}{a^4}$ **9.** $\frac{1}{25}$ **10.** x^2 **11.** x^3 **12.** $\dfrac{1}{x^3}$

Exercise 1.3 (page 36)

1. base is 5, exponent is 3   **3.** base is x, exponent is 5   **5.** base is b, exponent is 6   **7.** base is $-mn^2$, exponent is 3   **9.** 9
11. -9 **13.** 9   **15.** $-32x^5$ **17.** $\frac{1}{25}$ **19.** $-\frac{1}{25}$ **21.** $\frac{1}{25}$ **23.** 1   **25.** -1 **27.** 1   **29.** $-128x^7$ **31.** $64x^6$ **33.** x^5 **35.** k^7
37. x^{10} **39.** p^{10} **41.** a^4b^5 **43.** x^5y^4 **45.** x^{28} **47.** $\frac{1}{b^{72}}$ **49.** $x^{12}y^8$ **51.** $\frac{s^3}{r^9}$ **53.** a^{20} **55.** $-\frac{1}{d^3}$ **57.** $27x^9y^{12}$ **59.** $\frac{1}{729}m^6n^{12}$
61. x^2y^{10} **63.** $\frac{1}{b^{14}}$ **65.** $\frac{a^{15}}{b^{10}}$ **67.** $\frac{a^6}{b^4}$ **69.** a^5 **71.** c^7 **73.** m **75.** a^4 **77.** $3m$ **79.** $\frac{64b^{12}}{27a^9}$ **81.** 1   **83.** $\frac{-b^3}{8a^{21}}$ **85.** $\frac{27}{8a^{18}}$
87. $\frac{1}{9x^3}$ **89.** $\frac{-3y^2}{x^2}$ **91.** a^{n-1} **93.** b^{3n-9} **95.** $\frac{1}{a^{n+1}}$ **97.** a^{2-n} **99.** 133.6336   **101.** 0.000008315   **103.** -244.140625
105. -9962506.263 **115.** 108   **117.** $-\frac{1}{216}$ **119.** $\frac{1}{324}$ **121.** $\frac{27}{8}$ **123.** 15 m²   **125.** 113 cm²   **127.** 45 cm²   **129.** 300 cm²
131. 343 m³   **133.** 360 ft³   **135.** 168 ft³   **137.** 2714 m³

Review Exercises (page 38)

1. 7   **2.** 3   **3.** 1   **4.** $-\frac{3}{5}$

Orals (page 42)

1. 3.52×10^2 **2.** 5.13×10^3 **3.** 2×10^{-3} **4.** 2.5×10^{-4} **5.** 350   **6.** 4300   **7.** 0.27   **8.** 0.085

Exercise 1.4 (page 42)

1. 3.9×10^3 **3.** 7.8×10^{-3} **5.** -4.5×10^4 **7.** -2.1×10^{-4} **9.** 1.76×10^7 **11.** 9.6×10^{-6} **13.** 3.23×10^7
15. 6.2×10^{-4} **17.** 5.27×10^3 **19.** 3.17×10^{-4} **21.** 270   **23.** 0.00323   **25.** 796,000   **27.** 0.00037   **29.** 5.23
31. 23,650,000   **33.** 2×10^{11} **35.** 1.44×10^7 **37.** 0.04   **39.** 6000   **41.** 0.6   **43.** 1.1916×10^8 cm/hr   **45.** 1.67248×10^{-18} g
47. 1.49×10^{10} in.   **49.** 5.9×10^4 mi   **51.** 3×10^8 **53.** almost 23 yr   **55.** 2.5×10^{13} mi   **57.** 1.2874×10^{13}
59. 5.671×10^{10} **61.** 3.6×10^{25}

Review Exercises (page 44)

1. 0.75   **2.** 0.8   **3.** $1.\overline{4}$ **4.** $1.\overline{27}$ **5.** 89   **6.** -5

Orals (page 52)

1. $9x$ **2.** $2s^2$ **3.** no   **4.** yes   **5.** no   **6.** no   **7.** 3   **8.** 12   **9.** 5   **10.** 3

Exercise 1.5 (page 52)

1. yes   **3.** yes   **5.** 2   **7.** 25   **9.** 3   **11.** 28   **13.** $\frac{2}{3}$ **15.** 6   **17.** -8 **19.** $\frac{8}{3}$ **21.** yes, $8x$ **23.** no   **25.** yes, $-2x^2$ **27.** no
29. 3   **31.** 13   **33.** 9   **35.** -4 **37.** -6 **39.** -11 **41.** 13   **43.** -8 **45.** -2 **47.** 24   **49.** 6   **51.** 4   **53.** 3   **55.** 0   **57.** 6
59. identity   **61.** -6 **63.** impossible   **65.** identity   **67.** $\frac{1}{3}$ **69.** $-\frac{691}{1980}$ **71.** $w = \frac{4}{7}$ **73.** $B = \frac{3V}{h}$ **75.** $t = \frac{I}{pr}$ **77.** $w = \frac{p-2l}{2}$
79. $B = \frac{2A}{h} - b$ **81.** $x = \frac{y-b}{m}$ **83.** $n = \frac{l-a+d}{d}$ **85.** $l = \frac{a-S+Sr}{r}$ **87.** $l = \frac{2S-na}{n}$ **89.** $m = Fd^2/(GM)$ **91.** 0°, 21.1°, 100°
93. $n = (C - 6.50)/0.07$; 621, 1000, 1692.9 kwh   **95.** $R = E/I$; $R = 8$ ohms   **97.** $n = 360°/(180° - a)$; 8 sides

Review Exercises (page 54)

1. -64 **2.** -27 **3.** 1   **4.** x^{20} **5.** x^8 **6.** $\frac{y^6}{x^3}$ **7.** $\frac{1}{8x^3}$ **8.** $\frac{y^{20}}{x^8}$

Orals (page 61)

1. 100   **2.** 200   **3.** $270   **4.** $54x   **5.** 24 m²   **6.** $l(l - 5)$ m²

Exercise 1.6 (page 61)

1. 32 ft   **3.** 7 ft, 15 ft   **5.** $355   **7.** 20%   **9.** 233%   **11.** 300 shares of BB, 200 shares of SS   **13.** 35 $15 calculators,
50 $67 calculators   **15.** 13   **17.** 60 mi, 120 mi   **19.** 12 m by 24 m   **21.** 156 ft by 312 ft   **23.** 10 ft   **25.** 72.5°   **27.** 56°
29. 50   **31.** 10   **33.** 60°   **35.** 12 in.   **37.** 5 ft   **39.** 90 lb   **41.** 4 ft   **43.** $-40°$

Review Exercises (page 65)

1. $\dfrac{256x^{20}}{81}$ **2.** $\dfrac{r^{40}}{s^{30}}$ **3.** a^{m+1} **4.** b^{3n-9}

Orals (page 69)

1. \$90   **2.** $0.05x$ **3.** $\$(30,000 - x)$ **4.** $\$20x$ **5.** 0.08 gal

Exercise 1.7 (page 69)

1. \$2000 at 8%, \$10,000 at 9%   **3.** 10.8%   **5.** \$8250   **7.** 25   **9.** $3\frac{1}{2}$ hr   **11.** $1\frac{1}{2}$ hr   **13.** $1\frac{1}{2}$ hr   **15.** 4 hr at each rate
17. 20 lb of 95¢ candy; 10 lb of \$1.10 candy   **19.** 10 oz   **21.** 2 gal   **23.** 15   **25.** 30   **27.** \$50   **29.** 6 in.

Review Exercises (page 71)

1. 1   **2.** 2   **3.** 8   **4.** 16

Problems and Projects (page 72)

1. \$31.85   **2.** 108 g/mole   **3.** \$210, \$420   **4.** 1400 lb   **5.** 3.6 ft   **6.** $10\frac{2}{3}$ oz pecans, $5\frac{1}{3}$ oz cashews, 40 bags

Chapter 1 Review Exercises (page 75)

1. true   **2.** false; $B \not\subset C$ **3.** false; $B \subseteq A$ **4.** true   **5.** true   **6.** false; $\emptyset \subset C$ **7.** true   **8.** false; $\{4, 8\} \subseteq A$ **9.** $\{1, 3, 4, 5, 6\}$
10. $\{5\}$ **11.** $\{1, 3, 5\}$ **12.** $\{5\}$ **13.** 0, 1, 2, 4   **14.** 1, 2, 4   **15.** $-4, 0, 1, 2, 4, -\frac{2}{3}$ **16.** $-4, 0, 1, 2, 4$ **17.** π
18. $-4, 0, 1, 2, 4, -\frac{2}{3}, \pi$ **19.** $-4, -\frac{2}{3}$ **20.** 1, 2, 4, π **21.** 2   **22.** 4   **23.** $-4, 0, 2, 4$ **24.** 1
25. 20 21 22 23 24 25 26 27 28 29 30   **26.** 5 6 7 8 9 10 11 12 13   **27.** -4
28. -2 6   **29.** -2 3   **30.** 2 6   **31.** 2   **32.** -1
33. -2 6   **34.** 0 4   **35.** 0   **36.** 1   **37.** 8   **38.** -8 **39.** -2 **40.** 2   **41.** 3   **42.** -8 **43.** -5
44. 16   **45.** -12 **46.** -24 **47.** 12   **48.** 27   **49.** -4 **50.** -2 **51.** 4   **52.** 5   **53.** -1 **54.** 4   **55.** 15.4   **56.** 15   **57.** 15
58. yes   **59.** $-\frac{19}{6}$ **60.** $\frac{2}{15}$ **61.** $-\frac{4}{3}$ **62.** $\frac{11}{5}$ **63.** dist. prop.   **64.** symm. prop.   **65.** comm. prop. of add.   **66.** asso. prop. of add.
67. add. identity prop.   **68.** add. inverse prop.   **69.** reflexive prop.   **70.** comm. prop. of $\times$ **71.** asso. prop. of $\times$ **72.** mult.
identity prop.   **73.** mult. inverse prop.   **74.** trans. prop.   **75.** 729   **76.** -64 **77.** -64 **78.** $-\frac{1}{625}$ **79.** $-6x^6$ **80.** $-3x^8$

81. $\frac{1}{x}$ **82.** x^2 **83.** $27x^6$ **84.** $256x^{16}$ **85.** $-32x^{10}$ **86.** $243x^{15}$ **87.** $\dfrac{1}{x^{10}}$ **88.** x^{20} **89.** $\dfrac{x^6}{9}$ **90.** $\dfrac{16}{x^{16}}$ **91.** x^2 **92.** x^5 **93.** $\dfrac{1}{a^5}$

94. $\dfrac{1}{a^3}$ **95.** $\dfrac{1}{y^7}$ **96.** y^9 **97.** $\dfrac{1}{x}$ **98.** x^3 **99.** $9x^4y^6$ **100.** $\dfrac{1}{81a^{12}b^8}$ **101.** $\dfrac{64y^9}{27x^6}$ **102.** $\dfrac{64y^3}{125}$ **103.** 1.93×10^{10}
104. 2.73×10^{-8} **105.** 72,000,000   **106.** 0.0000000083   **107.** 5   **108.** -9 **109.** 8   **110.** 7   **111.** 19   **112.** 8   **113.** 12

114. 5   **115.** $r^3 = \dfrac{3V}{4\pi}$ **116.** $h = \dfrac{3V}{\pi r^2}$ **117.** $x = \dfrac{6v}{ab} - y$ **118.** $r = \dfrac{V}{\pi h^2} + \dfrac{h}{3}$ **119.** 5 ft from one end   **120.** 45 m²   **121.** $2\frac{2}{3}$ ft
122. \$18,000 at 10%, \$7000 at 9%   **123.** 10 liters   **124.** $\frac{1}{3}$ hr

Chapter 1 Test (page 78)

1. $\in$ **2.** $\subseteq$ **3.** $\subseteq$ **4.** $\subseteq$ **5.** $\{1, 2, 3, 4, 5, 6\}$ **6.** $\emptyset$ **7.** 1, 2, 5   **8.** $\sqrt{7}$ **9.** -4 -3 -2 -1 0 1 2 3 4 5 6
10. 1 2 3 4 5 6 7 8 9 10 11 12   **11.** -2 4   **12.** -1 3   **13.** -8 **14.** 5
15. 2   **16.** 20   **17.** -4 **18.** 1   **19.** 0.3   **20.** 0.5   **21.** -6 **22.** -10 **23.** 6   **24.** 1   **25.** comm. prop. of add.   **26.** dist. prop.
27. x^8 **28.** x^6y^9 **29.** $\dfrac{1}{m^8}$ **30.** $\dfrac{m^4}{n^{10}}$ **31.** 4.7×10^6 **32.** 2.3×10^{-7} **33.** 653,000   **34.** 0.0245   **35.** -12 **36.** 6

37. $i = \dfrac{f(P - L)}{s}$ **38.** 36 cm²   **39.** \$4000   **40.** 80 liters

Exercise 2.1   (page 88)

1. $6.90   **3.** $4140   **5.** $2542.50   **7.** nonconstruction laborers   **9.** 4.8%   **11.** 1990   **13.** 1970   **15.** 46%   **17.** 80,000
19. 62%   **21.** 62%   **23.** 1   **25.** 1   **27.** 1 is running, 2 is stopped   **29.** $\frac{5}{12}$ **31.** 2, 4, 3, 1

Review Exercises   (page 91)

1. 22   **2.** 16   **3.** $\frac{5}{2}$ **4.** $\frac{5}{6}$ **5.** 11, 13, 17, 19, 23, 29   **6.** 4, 6, 8, 9, 10, 12, 14, 15, 16, 18

Orals   (page 100)

1. $(3, 0), (0, 3)$ **2.** $(2, 0), (0, 6)$ **3.** $(8, 0), (0, 2)$ **4.** $(4, 0), (0, -3)$ **5.** vertical   **6.** horizontal   **7.** $(4, 6)$ **8.** $(0, -1)$

Exercise 2.2   (page 101)

1–8.

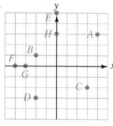

9. $(2, 4)$ **11.** $(-2, -1)$ **13.** $(4, 0)$ **15.** $(0, 0)$ **17.** 5, 4, 2   **19.** $-5, -3, 3$

21.

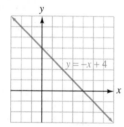

23.

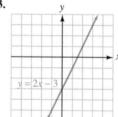

25.

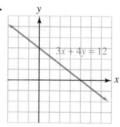

27.

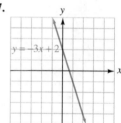

29.

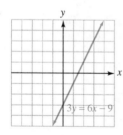

31.

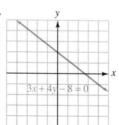

33.

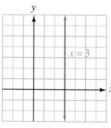

35.

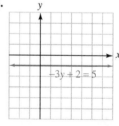

37. $48   **39.** $3000   **41.** $162,500   **43.** 200   **45.** 100 rpm   **47.** $(3, 4)$ **49.** $(9, 12)$ **51.** $\left(\frac{7}{2}, 6\right)$ **53.** $\left(\frac{1}{2}, -2\right)$ **55.** $(-4, 0)$
57. $(4, 1)$ **59.** 1.22   **61.** 4.67

Review Exercises   (page 104)

1. 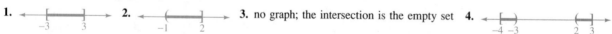   **2.**   **3.** no graph; the intersection is the empty set   **4.**

Orals   (page 110)

1. 3   **2.** 2   **3.** yes   **4.** 5   **5.** yes

Exercise 2.3   (page 110)

1. 3   **3.** -1 **5.** $-\frac{1}{3}$ **7.** 0   **9.** undefined   **11.** -1 **13.** $-\frac{3}{2}$ **15.** $\frac{3}{4}$ **17.** $\frac{1}{2}$ **19.** 0   **21.** 3.5 students per yr   **23.** $-$642.86 per
year   **25.** negative   **27.** positive   **29.** undefined   **31.** perpendicular   **33.** neither   **35.** parallel   **37.** parallel   **39.** perpendicular
41. neither   **43.** not the same line   **45.** not the same line   **47.** same line   **49.** $y = 0, m = 0$ **57.** 4

Review Exercises (page 113)

1. x^9y^6 **2.** x^6 **3.** $\dfrac{x^{12}}{y^8}$ **4.** $x^{24}y^{12}$ **5.** 1   **6.** $x^{16}y^2$

Orals (page 121)

1. $y - 3 = 2(x - 2)$ **2.** $y - 8 = 2(x + 3)$ **3.** $y = -3x + 5$ **4.** $y = -3x - 7$ **5.** parallel   **6.** perpendicular

Exercise 2.4 (page 121)

1. $5x - y = -7$ **3.** $3x + y = 6$ **5.** $2x - 3y = -11$ **7.** $y = x$ **9.** $y = \frac{7}{3}x - 3$ **11.** $y = -\frac{9}{5}x + \frac{2}{5}$ **13.** $y = 3x + 17$
15. $y = -7x + 54$ **17.** $y = -4$ **19.** $y = -\frac{1}{2}x + 11$
21. $1, (0, -1)$ **23.** $\frac{2}{3}, (0, 2)$ **25.** $-\frac{2}{3}, (0, 6)$

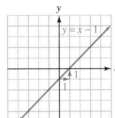
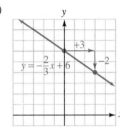

27. $\frac{3}{2}, (0, -4)$ **29.** $-\frac{1}{3}, \left(0, -\frac{5}{6}\right)$ **31.** $\frac{7}{2}, (0, 2)$ **39.** parallel   **41.** perpendicular   **43.** parallel   **45.** perpendicular
47. perpendicular   **49.** perpendicular   **51.** $y = 4x$ **53.** $y = 4x - 3$ **55.** $y = \frac{4}{5}x - \frac{26}{5}$ **57.** $y = -\frac{1}{4}x$ **59.** $y = -\frac{1}{4}x + \frac{11}{2}$
61. $y = -\frac{5}{4}x + 3$ **63.** perpendicular   **65.** parallel   **67.** $x = -2$ **69.** $x = 5$ **71.** $y = -\frac{A}{B}x + \frac{C}{B}$ **73.** $y = -3200x + 24,300$
75. $y = 47{,}500x + 475{,}000$ **77.** $y = -\frac{710}{3}x + 1900$ **79.** \$90   **81.** \$890   **83.** \$37,200   **85.** \$230   **87.** about 838

Review Exercises (page 125)

1. 6   **2.** -3 **3.** -1 **4.** 16   **5.** 20 oz   **6.** 60 lb

Orals (page 132)

1. yes   **2.** no   **3.** no   **4.** 1   **5.** 3   **6.** -3

Exercise 2.5 (page 132)

1. {Maris, Ruth, Foxx, Greenberg, Wilson}, {56, 58, 60, 61}, yes   **3.** {green, red, yellow, blue}, {1, 5, 10, 15}, no   **5.** yes
7. yes   **9.** yes   **11.** no   **13.** 9, -3 **15.** 3, -5 **17.** 22, 2   **19.** 3, 11   **21.** 4, 9   **23.** 7, 26   **25.** 9, 16   **27.** 6, 15   **29.** 4, 4
31. 2, 2   **33.** $\frac{1}{5}$, 1   **35.** $-2, \frac{2}{5}$ **37.** $2w, 2w + 2$ **39.** $3w - 5, 3w - 2$ **41.** 12   **43.** $2b - 2a$ **45.** $2b$ **47.** 1
49. $D = \{-2, 4, 6\}, R = \{3, 5, 7\}$ **51.** $D = (-\infty, 4) \cup (4, \infty), R = (-\infty, 0) \cup (0, \infty)$ **53.** $D = (-\infty, 1], R = (-\infty, \infty)$, not a
function   **55.** $D = (-\infty, \infty), R = (-\infty, \infty)$, a function
57.   **59.**

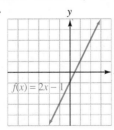

$D = (-\infty, \infty), R = (-\infty, \infty)$

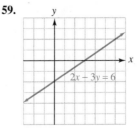

$D = (-\infty, \infty), R = (-\infty, \infty)$

61. no   **63.** yes   **65.** 624 ft   **67.** 77° F   **69.** 192

Review Exercises (page 135)

1. -2 **2.** -3 **3.** 6   **4.** no solution

Exercise 2.6 (page 138)

1.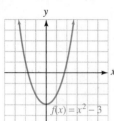

3.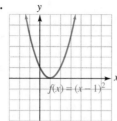

5.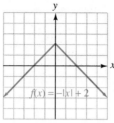

7.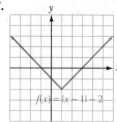

17.

19.

21.

23.

25.

27.

29.

31.

33. -2 **35.** 4   **37.** -3 **39.** $(-2, -4)$ **41.** $(3, 9)$ **43.** $(-2, -9)$ **45.** $(-2, 6)$

Review Exercises (page 140)

1. 41, 43, 47   **2.** $(a + b) + c = a + (b + c)$ **3.** $a \cdot b = b \cdot a$ **4.** 0   **5.** 1   **6.** $\frac{3}{5}$

Problems and Projects (page 140)

1. $V = -115t + 795$, \$335, $-$\$115   **2.** \$50,750, 2250 units   **3.** 8.33 units   **4.** 1755 ft, 2195 ft, 2278 ft, 23.6 sec

Chapter 2 Review Exercises (page 143)

1. midsize   **2.** 2%   **3.** 31%   **4.** Companies require employees to rent them.   **5.** investments/savings   **6.** 42%   **7.** 27%   **8.** no

9.

Hours	Frequency
0	2
1	0
2	2
3	12
4	8
5	5
6	1

10.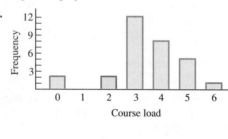

11.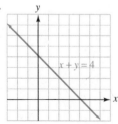

12.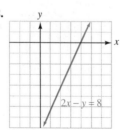

13.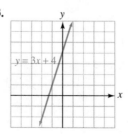

14.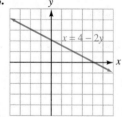

15. 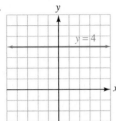   **16.** 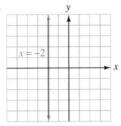   **17.** 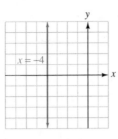   **18.**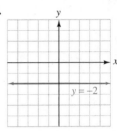

19. 1   **20.** $\frac{14}{9}$ **21.** 5   **22.** $\frac{5}{11}$ **23.** $\frac{2}{3}$ **24.** -2 **25.** undefined   **26.** 0   **27.** perpendicular   **28.** parallel   **29.** neither
30. perpendicular   **31.** $3x - y = -29$ **32.** $13x + 8y = 6$ **33.** $3x - 2y = 1$ **34.** $2x + 3y = -21$ **35.** yes   **36.** yes   **37.** no
38. no   **39.** -7 **40.** 60   **41.** 0   **42.** 17   **43.** $D = (-\infty, \infty)$, $R = (-\infty, \infty)$ **44.** $D = (-\infty, \infty)$, $R = (-\infty, \infty)$ **45.** $D = (-\infty, \infty)$,
$R = [1, \infty)$ **46.** $D = (-\infty, 2) \cup (2, \infty)$, $R = (-\infty, 0) \cup (0, \infty)$ **47.** $D = (-\infty, 3) \cup (3, \infty)$, $R = (-\infty, 1) \cup (1, \infty)$ **48.** $D = [0, \infty)$,
$R = (-\infty, \infty)$ **49.** a function   **50.** not a function   **51.** not a function   **52.** a function   **53.** yes   **54.** yes   **55.** yes   **56.** no
57. 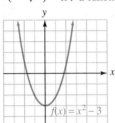   **58.** 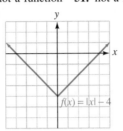   **59.** 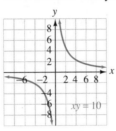   **60.**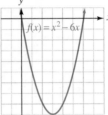

Chapter 2 Test (page 146)

1. small-company stock   **2.** short-term government bills   **3.** 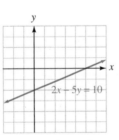   **4.** x-intercept $(3, 0)$, y-intercept $\left(0, -\frac{3}{5}\right)$

5. $\frac{1}{2}$ **6.** $\frac{2}{3}$ **7.** undefined   **8.** 0   **9.** $y = \frac{2}{3}x - \frac{23}{3}$ **10.** $8x - y = -22$ **11.** $m = -\frac{1}{3}$, $b = -\frac{3}{2}$ **12.** neither   **13.** perpendicular
14. $y = \frac{3}{2}x$ **15.** $y = \frac{3}{2}x + \frac{21}{2}$ **16.** no   **17.** $D = (-\infty, \infty)$, $R = [0, \infty)$ **18.** $D = (-\infty, \infty)$, $R = (-\infty, \infty)$ **19.** 10   **20.** -2
21. $3a + 1$ **22.** $x^2 - 2$ **23.** yes   **24.** no   **25.** 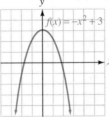   **26.**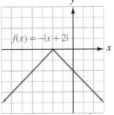

Cumulative Review Exercises (Chapters 1–2) (page 147)

1. 1, 2, 6, 7   **2.** 0, 1, 2, 6, 7   **3.** $-2, 0, 1, 2, \frac{13}{12}, 6, 7$ **4.** $\sqrt{5}, \pi$ **5.** -2 **6.** $-2, 0, 1, 2, \frac{13}{12}, 6, 7, \sqrt{5}, \pi$ **7.** 2, 7   **8.** 6
9. $-2, 0, 2, 6$ **10.** 1, 7   **11.** ⟵───(───┤───── $\begin{smallmatrix}-2 & & 5\end{smallmatrix}$ **12.** ⟵───├───)───⟶ $\begin{smallmatrix}-3 & & 0\end{smallmatrix}$ **13.** -2 **14.** -2 **15.** 22   **16.** -2 **17.** -4
18. -3 **19.** 4   **20.** -5 **21.** transitive prop.   **22.** dist. prop.   **23.** commut. prop. of add.   **24.** assoc. prop. of mult.   **25.** $x^8 y^{12}$
26. c^2 **27.** $-\dfrac{b^3}{a^2}$ **28.** 1   **29.** 4.97×10^{-6} **30.** 932,000,000   **31.** 8   **32.** -27 **33.** -1 **34.** 6   **35.** $\frac{7}{8}$ **36.** $\frac{5}{11}$

37. $a = \dfrac{2S}{n} - l$ **38.** $h = \dfrac{2A}{b_1 + b_2}$ **39.** 28, 30, 32 **40.** 14 cm by 42 cm **41.** yes **42.** $-\dfrac{7}{5}$ **43.** $y = -\dfrac{7}{5}x + \dfrac{11}{5}$

44. $y = -3x - 3$ **45.** 5 **46.** -1 **47.** $2t - 1$ **48.** $3r^2 + 2$

49. yes, $D = (-\infty, \infty)$, $R = (-\infty, 1]$

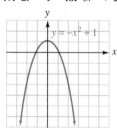

50. yes, $D = (-\infty, \infty)$, $R = [0, \infty)$

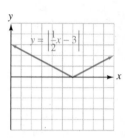

Orals (page 156)

1. no solution **2.** infinitely many **3.** one solution **4.** no solution

Exercise 3.1 (page 156)

1. yes **3.** no

5.

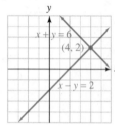

7.

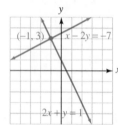

9.

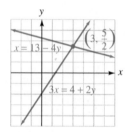

11.

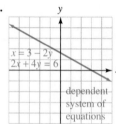

13.

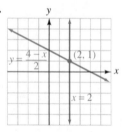

15.

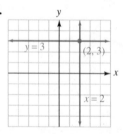

17.

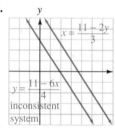

19.

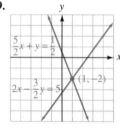

21.

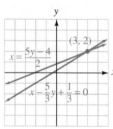

23.

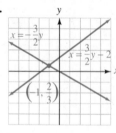

25. $(-0.37, -2.69)$ **27.** $(-7.64, 7.04)$

Review Exercises (page 158)

1. 9.3×10^7 **2.** 2.36×10^{-8} **3.** 3.45×10^4 **4.** 7.52×10^{-3}

Orals (page 168)

1. 2 **2.** 4 **3.** 4 **4.** 3

Exercise 3.2 (page 168)

1. $(2, 2)$ **3.** $(5, 3)$ **5.** $(-2, 4)$ **7.** no solution   **9.** $\left(5, \frac{3}{2}\right)$ **11.** $\left(-2, \frac{3}{2}\right)$ **13.** $(5, 2)$ **15.** $(-4, -2)$ **17.** $(1, 2)$ **19.** $\left(\frac{1}{2}, \frac{2}{3}\right)$
21. dependent equations   **23.** no solution   **25.** $(4, 8)$ **27.** $(20, -12)$ **29.** $\left(\frac{2}{3}, \frac{3}{2}\right)$ **31.** $\left(\frac{1}{2}, -3\right)$ **33.** $(2, 3)$ **35.** $\left(-\frac{1}{3}, 1\right)$ **37.** \$57
39. 625 and 750 ohms   **41.** 16 m by 20 m   **43.** \$3000 at 10%, \$5000 at 12%   **45.** 40 oz of 8% solution, 60 oz of 15% solution
47. 55 mph   **49.** 85 racing bikes, 120 mountain bikes   **51.** 200 plates   **53.** 21   **55.** 750   **57.** 6500 gal per month
59. A (a smaller loss)   **61.** 590 units per month   **63.** A (a smaller loss)   **65.** A **67.** $35°, 145°$ **69.** $x = 22.5, y = 67.5$

71. $f^2 = \dfrac{1}{4\pi^2 LC}$

Review Exercises (page 172)

1. a^{22} **2.** $\dfrac{d^9}{a^3 b^3 c^3}$ **3.** $\dfrac{1}{81 x^{32} y^4}$ **4.** $\dfrac{4}{7}$

Orals (page 178)

1. yes   **2.** no

Exercise 3.3 (page 178)

1. yes   **3.** $(1, 1, 2)$ **5.** $(0, 2, 2)$ **7.** $(3, 2, 1)$ **9.** inconsistent system   **11.** dependent equations   **13.** $(2, 6, 9)$ **15.** $-2, 4, 16$
17. $A = 40°, B = 60°, C = 80°$ **19.** $1, 2, 3$ **21.** 30 expensive, 50 middle-priced, 100 inexpensive   **23.** 250 \$5 tickets,
375 \$3 tickets, 125 \$2 tickets   **25.** 3 totem poles, 2 bears, 4 deer   **27.** $y = x^2 - 4x$ **29.** $x^2 + y^2 - 2x - 2y - 2 = 0$

Review Exercises (page 181)

1. $\frac{9}{5}$ **2.** $9x - 5y = 2$ **3.** 1 **4.** 9 **5.** $2s^2 + 1$ **6.** $8t^2 + 1$

Orals (page 186)

1. $\begin{bmatrix} 3 & 2 \\ 4 & -3 \end{bmatrix}$ **2.** $\left[\begin{array}{cc|c} 3 & 2 & 8 \\ 4 & -3 & 6 \end{array}\right]$ **3.** yes   **4.** no

Exercise 3.4 (page 186)

1. 0 **3.** 8 **5.** $(1, 1)$ **7.** $(2, -3)$ **9.** $(0, -3)$ **11.** $(1, 2, 3)$ **13.** $(-1, -1, 2)$ **15.** $(2, 1, 0)$ **17.** $(1, 2)$ **19.** $(2, 0)$
21. no solution   **23.** $(1, 2)$ **25.** $(-6 - z, 2 - z, z)$ **27.** $(2 - z, 1 - z, z)$ **29.** $22°, 68°$ **31.** $40°, 65°, 75°$ **33.** $y = 2x^2 - x + 1$

Review Exercises (page 188)

1. 9.3×10^7 **2.** 4.5×10^{-4} **3.** 6.3×10^4 **4.** 3.3×10^2

Orals (page 194)

1. 1 **2.** -2 **3.** 0 **4.** $\begin{vmatrix} 1 & 2 \\ 2 & -1 \end{vmatrix}$ **5.** $\begin{vmatrix} 5 & 2 \\ 4 & -1 \end{vmatrix}$ **6.** $\begin{vmatrix} 1 & 5 \\ 2 & 4 \end{vmatrix}$

Exercise 3.5 (page 194)

1. 8 **3.** -2 **5.** $x^2 - y^2$ **7.** 0 **9.** -13 **11.** 26 **13.** 0 **15.** $10a$ **17.** 0 **19.** $(4, 2)$ **21.** $(-1, 3)$ **23.** $\left(-\frac{1}{2}, \frac{1}{3}\right)$ **25.** $(2, -1)$
27. no solution   **29.** $\left(5, \frac{14}{5}\right)$ **31.** $(1, 1, 2)$ **33.** $(3, 2, 1)$ **35.** no solution   **37.** $(3, -2, 1)$ **39.** dependent equations
41. $(-2, 3, 1)$ **43.** no solution   **45.** 2 **47.** 2 **49.** \$5000 in HiTech, \$8000 in SaveTel, \$7000 in HiGas   **51.** -23 **53.** 26

Review Exercises (page 197)

1. -3 **2.** -14 **3.** 0 **4.** 2

Problems and Projects   (page 197)

1. yes; break-even point is (12.06, 494)   **2.** 25 m by $16\frac{2}{3}$ m   **3.** about 4.26 pt of thinner and 8.54 pt of paint
4. $y = -0.0013x^2 + 7.7236x + 2105$

Chapter 3 Review Exercises   (page 199)

1. 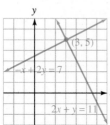   **2.** 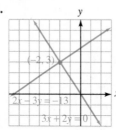   **3.** 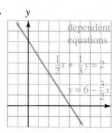   **4.**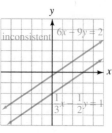

5. $(-1, 3)$ **6.** $(-3, -1)$ **7.** $(3, 4)$ **8.** $(-4, 2)$ **9.** $(-3, 1)$ **10.** $(1, -1)$ **11.** $(9, -4)$ **12.** $\left(4, \frac{1}{2}\right)$ **13.** $(1, 2, 3)$
14. inconsistent system   **15.** $(2, 1)$ **16.** $(1, 3, 2)$ **17.** $(1, 2)$ **18.** $(3z, 1 - 2z, z)$ **19.** 18   **20.** 38   **21.** -3 **22.** 28   **23.** $(2, 1)$
24. $(-1, 3)$ **25.** $(1, -2, 3)$ **26.** $(-3, 2, 2)$

Chapter 3 Test   (page 200)

1. 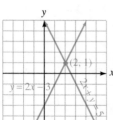   **2.** $(7, 0)$ **3.** $(2, -3)$ **4.** $(-6, 4)$ **5.** dependent   **6.** consistent   **7.** 6   **8.** -8

9. $\begin{bmatrix} 1 & 1 & 1 & \vdots & 4 \\ 1 & 1 & -1 & \vdots & 6 \\ 2 & -3 & 1 & \vdots & -1 \end{bmatrix}$ **10.** $\begin{bmatrix} 1 & 1 & 1 \\ 1 & 1 & -1 \\ 2 & -3 & 1 \end{bmatrix}$ **11.** $(2, 2)$ **12.** $(-1, 3)$ **13.** 22   **14.** -17 **15.** 4   **16.** 13

17. $\begin{vmatrix} -6 & -1 \\ -6 & 1 \end{vmatrix}$ **18.** $\begin{vmatrix} 1 & -1 \\ 3 & 1 \end{vmatrix}$ **19.** -3 **20.** 3   **21.** 3   **22.** -1

Orals   (page 210)

1. $x < 2$ **2.** $x \geq 3$ **3.** $x < -4$ **4.** $x \geq -8$ **5.** $-1 < x < 4$ **6.** $9 \leq x \leq 12$

Exercise 4.1   (page 210)

1. $(-\infty, 1)$ **3.** $[-2, \infty)$ **5.** $(2, \infty)$ **7.** $(-\infty, 20]$ **9.** $(-\infty, 10)$ **11.** $[-36, \infty)$ **13.** $(-\infty, 45/7)$

15. $(-2, 5)$ **17.** $(8, 11)$ **19.** $[-4, 6)$ **21.** no solution   **23.** $[-2, 4]$

25. $(2, 3)$ **27.** $[1, 9/4]$ **29.** $(-\infty, -15)$ **31.** $(-\infty, 2) \cup (7, \infty)$ **33.** $(-\infty, 1)$

35. no solution   **37.** no   **39.** 5 hr   **41.** more than \$5000   **43.** 18   **45.** 88 or higher   **47.** 13   **49.** anything over \$900
51. $x < 1$ **53.** $x \geq -4$ **55.** 139

Review Exercises   (page 213)

1. $\dfrac{1}{t^{12}}$ **2.** $\dfrac{a^{12}}{b^{24}}$ **3.** 471 or more   **4.** \$10,000

Orals   (page 220)

1. 5   **2.** -5 **3.** -6 **4.** -4 **5.** 8 or -8 **6.** no solution   **7.** $-8 < x < 8$ **8.** $x < -8$ or $x > 8$ **9.** $x \le -4$ or $x \ge 4$
10. $-7 \le x \le 7$

Exercise 4.2   (page 221)

1. 8   **3.** -2 **5.** -30 **7.** $4 - \pi$ **9.**   **11.**   **13.** 4, -4 **15.** 9, -3

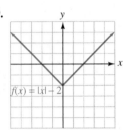

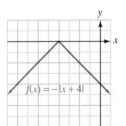

17. 4, -1 **19.** $\frac{14}{3}$, -6 **21.** no solution   **23.** 8, -4 **25.** 2, $-\frac{1}{2}$ **27.** -8 **29.** -4, -28 **31.** 0, -6 **33.** $\frac{20}{3}$ **35.** -2, $-\frac{4}{5}$
37. 3, -1 **39.** 0, -2 **41.** 0   **43.** $\frac{4}{3}$ **45.** no solution   **47.** $(-4, 4)$ **49.** $[-21, 3]$ **51.** no solution

53. $[-3/2, 2]$ **55.** $(-2, 5)$ **57.** $(-\infty, -1) \cup (1, \infty)$ **59.** $(-\infty, -12) \cup (36, \infty)$ **61.** $(-\infty, -16/3) \cup (4, \infty)$

63. $(-\infty, \infty)$ **65.** $(-\infty, -2] \cup [10/3, \infty)$ **67.** $(-\infty, -2) \cup (5, \infty)$ **69.** $(-\infty, 3/8) \cup (3/8, \infty)$ **71.** $[-10, 14]$

73. $(-5/3, 1)$ **75.** $(-\infty, -4] \cup [-1, \infty)$ **77.** no solution   **79.** $(-\infty, -24) \cup (-18, \infty)$ **81.** $(-\infty, 25) \cup (25, \infty)$

83. $[-7, -7]$ **85.** $[5, 5]$ **87.** $|x| < 4$ **89.** $|x + 3| > 6$ **91.** $70° \le t \le 86°$ **93.** $|c - 0.6°| \le 0.5°$

Review Exercises   (page 224)

1. $\frac{3}{4}$ **2.** 6   **3.** 6   **4.** 1   **5.** $t = \dfrac{A - p}{pr}$ **6.** $l = \dfrac{P - 2w}{2}$

Orals   (page 228)

1. yes   **2.** no   **3.** no   **4.** yes   **5.** no   **6.** no   **7.** yes   **8.** yes

Exercise 4.3   (page 228)

1.   **3.**   **5.**   **7.**

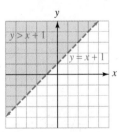

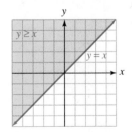

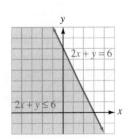

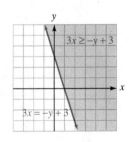

9.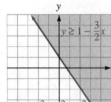

11.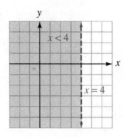

13.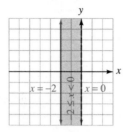

15.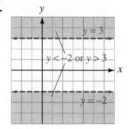

17. $3x + 2y > 6$ **19.** $x \le 3$ **21.** $y \le x$ **23.** $-2 \le x \le 3$ **25.** $y > -1$ or $y \le -3$
27. $(1, 1), (2, 1), (2, 2)$ y

29. $(2, 2), (3, 3), (5, 1)$

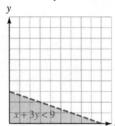

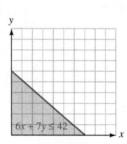

31. $(40, 20), (60, 40), (80, 20)$

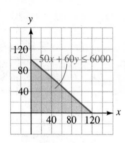

33.

35.

Review Exercises (page 231)

1. $(3, 1)$ **2.** $(-1, 2)$ **3.** $(2, -3)$ **4.** $(-1, -2)$

Orals (page 234)

1. yes   **2.** yes

Exercise 4.4 (page 234)

1.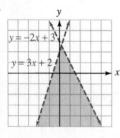

3.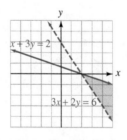

5.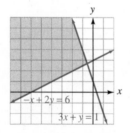

7.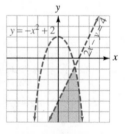

9.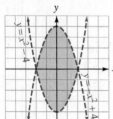

11.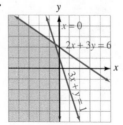

13.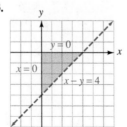

15.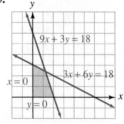

17. 1 $10 CD and 2 $15 CDs,
    4 $10 CDs and 1 $15 CD

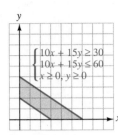

19. 2 desk chairs and 4 side chairs,
    1 desk chair and 5 side chairs

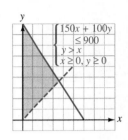

Review Exercises (page 236)

1. $r = \dfrac{A - p}{pt}$ **2.** $F = \dfrac{9}{5}C + 32$ **3.** $x = z\sigma + \mu$ **4.** $w = \dfrac{P - 2l}{2}$ **5.** $d = \dfrac{l - a}{n - 1}$ **6.** $\mu = x - \sigma z$

Orals (page 243)

1. 25   **2.** 9   **3.** $(0, 0), (0, 3), (3, 0)$ **4.** $(0, 0), (0, 2), (4, 0)$

Exercise 4.5 {page 243}

1. $P = 12$ at $(0, 4)$ **3.** $P = \frac{13}{6}$ at $\left(\frac{5}{3}, \frac{4}{3}\right)$ **5.** $P = \frac{18}{7}$ at $\left(\frac{3}{7}, \frac{12}{7}\right)$ **7.** $P = 3$ at $(1, 0)$ **9.** $P = 0$ at $(0, 0)$ **11.** $P = 0$ at $(0, 0)$
13. $P = -12$ at $(-2, 0)$ **15.** $P = -2$ at $(1, 2)$ and $(-1, 0)$ **17.** 3 tables, 12 chairs, $1260   **19.** 30 IBMs, 30 Macs, $2700
21. in batches of 20 oz from A and 40 oz from B; minimum fat per batch is 14 oz   **23.** 10 66-MHz chips and 100 50-MHz chips
25. $150,000 in stocks, $50,000 in bonds, $17,000   **27.** 2 buses, 2 trucks, $1100

Review Exercises (page 245)

1. $\frac{3}{7}$ **2.** $3x - 7y = -34$ **3.** $y = \frac{3}{7}x + \frac{34}{7}$ **4.** $y = \frac{3}{7}x$

Problems and Projects (page 246)

1. Plan B is best for 0–12 visits, then Plan A is best.   **2.** 83.3   **3.** You must sell over 30,000 pills.   **4.** 50 ft above ground, 175 ft
below ground; $660

Chapter 4 Review Exercises (page 248)

1. $(-\infty, 3]$ **2.** $(2, \infty)$ **3.** $(-\infty, -24]$ **4.** $(-\infty, -51/11)$ **5.** $(-1/3, 2)$ **6.** $(2, \infty)$ **7.** 7   **8.** 8

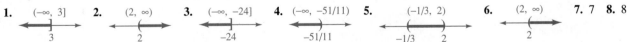

9. -7 **10.** -12 **11.**

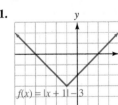

12.

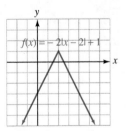

13. $3, -\frac{11}{3}$ **14.** $\frac{26}{3}, -\frac{10}{3}$ **15.** $14, -10$

16. $\frac{1}{5}, -5$ **17.** $-1, 1$ **18.** $\frac{13}{12}$ **19.** $(-5, -2)$

20. $(-\infty, 4/3] \cup [4, \infty)$

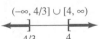

21. $(-\infty, \infty)$

22. no solutions

23. $[-3, 19/3]$

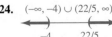

24. $(-\infty, -4) \cup (22/5, \infty)$

25.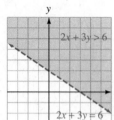

26.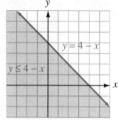

27.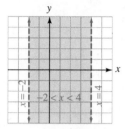

28.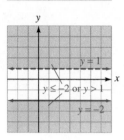

29.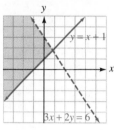

30.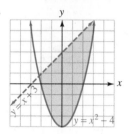

31. max. of 6 at $(3, 0)$ **32.** 1000 bags of X, 1400 bags of Y

Chapter 4 Test   (page 250)

1. $(-\infty, -5]$ **2.** $(-2, 16)$ **3.** 3   **4.** $4\pi - 4$ **5.** 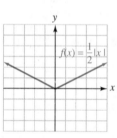   **6.**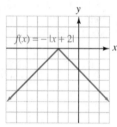

7. $4, -7$ **8.** $-5, \frac{23}{3}$ **9.** $4, -4$ **10.** 0   **11.** $[-7, 1]$ **12.** $(-\infty, -9) \cup (13, \infty)$ **13.** $(-\infty, 1) \cup (3, \infty)$

14. $[1, 3]$

15. 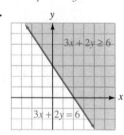   **16.** 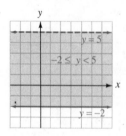   **17.** 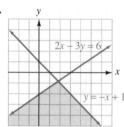   **18.**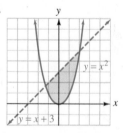

19. max is 2 at $(1, 1)$

Cumulative Review Exercises (Chapters 1–4)   (page 251)

1. 53 59   **2.** 5   **3.** $[-5, 7]$ **4.** $(-2, 5)$ **5.** 10   **6.** -6 **7.** x^{10} **8.** x^{14} **9.** x^4

10. $a^{2-n}b^{n-2}$ **11.** 3.26×10^7 **12.** 1.2×10^{-5} **13.** $\frac{26}{3}$ **14.** 3   **15.** 6   **16.** impossible   **17.** perpendicular   **18.** parallel

19. $y = \frac{1}{3}x + \frac{11}{3}$ **20.** $h = \dfrac{2A}{b_1 + b_2}$ **21.** 10   **22.** 14   **23.** (2, 1)

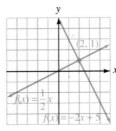

24. (1, 1)   **25.** (2, −2)

26. (3, 1)   **27.** (−1, −1, 3)   **28.** (0, −1, 1)   **29.** −1   **30.** 16   **31.** (−1, −1)   **32.** (1, 2, −1)   **33.** $x \le 11$ **34.** $-3 < x < 3$
35. $3, -\frac{3}{2}$ **36.** $-5, -\frac{3}{5}$ **37.** $-\frac{2}{3} \le x \le 2$ **38.** $x < -4$ or $x > 1$
39.

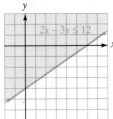

40.

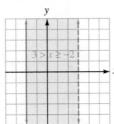

41.

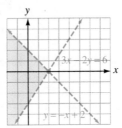

42.

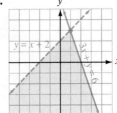

43. 24

Orals   (page 260)

1. 3   **2.** 4   **3.** 3   **4.** 3   **5.** 1   **6.** 5   **7.** −1   **8.** −3

Exercise 5.1   (page 260)

1. monomial   **3.** trinomial   **5.** binomial   **7.** monomial   **9.** 2   **11.** 8   **13.** 10   **15.** 0   **17.** $-2x^4 - 5x^2 + 3x + 7$
19. $7a^3x^5 - ax^3 - 5a^3x^2 + a^2x$ **21.** $7y + 4y^2 - 5y^3 - 2y^5$ **23.** $2x^4y - 5x^3y^3 - 2y^4 + 5x^3y^6 + x^5y^7$ **25.** 2   **27.** 8   **29.** 0 ft
31. 64 ft   **33.** 63 ft   **35.** 198 ft   **37.** 10 in.^{2} **39.** 18 in.^{2} **41.** 13   **43.** 35   **45.** −90   **47.** −48   **49.** −34.225
51. 0.17283171   **53.** −0.12268448
55.

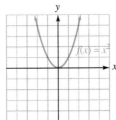

57.

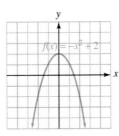

59.

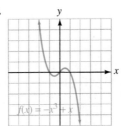

61.

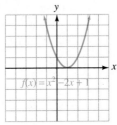

63.

65.

Review Exercises   (page 262)

1. a^5 **2.** b^2 **3.** $3y^{23}$ **4.** $2x^7$ **5.** 1.14×10^8 **6.** 1×10^{-7}

Orals   (page 265)

1. $9x^2$ **2.** $-2y^2$ **3.** $3x^2 + 2$ **4.** $-x^2 + 4x$ **5.** $3x^2 - 4x - 4$ **6.** $x^2 + 2x - 2$

Exercise 5.2 (page 265)

1. like terms, $10x$ **3.** unlike terms **5.** like terms, $-5r^2t^3$ **7.** unlike terms **9.** $12x$ **11.** $2x^3y^2z$ **13.** $-7x^2y^3 + 3xy^4$
15. $10x^4y^2$ **17.** $x^2 - 5x + 6$ **19.** $-5a^2 + 4a + 4$ **21.** $-y^3 + 4y^2 + 6$ **23.** $4x^2 - 11$ **25.** $3x^3 - x + 13$ **27.** $4y^2 - 9y + 3$
29. $6x^3 - 6x^2 + 14x - 17$ **31.** $-9y^4 + 3y^3 - 15y^2 + 20y + 12$ **33.** $x^2 - 8x + 22$ **35.** $-3y^3 + 18y^2 - 28y + 35$ **37.** $5x - 4$
39. $-11t + 29$ **41.** $8x^3 - x^2$ **43.** $-4m - 3n$ **45.** $-16x^3 - 27x^2 - 12x$ **47.** $-8z^2 - 40z + 54$ **49.** $14a^2 + 16a - 24$

Review Exercises (page 267)

1. $(-\infty, 4]$ **2.** $(9, \infty)$ **3.** $(-1, 9)$ **4.** $(-\infty, -4] \cup [3, \infty)$

Orals (page 272)

1. $-6a^3b^3$ **2.** $-8x^2y^3$ **3.** $6a^3 - 3a^2$ **4.** $-16mn^2 + 4n^3$ **5.** $2x^2 + 3x + 1$ **6.** $6y^2 - y - 2$

Exercise 5.3 (page 272)

1. $-6a^3b$ **3.** $-15a^2b^2c^3$ **5.** $-120a^9b^3$ **7.** $12x^5y^9$ **9.** $25x^8y^8$ **11.** $-405x^5y^6$ **13.** $100x^{10}y^6$ **15.** $3x + 6$ **17.** $-a^2 + ab$
19. $3x^3 + 9x^2$ **21.** $-6x^3 + 6x^2 - 4x$ **23.** $10a^6b^4 - 25a^2b^6$ **25.** $7r^3st + 7rs^3t - 7rst^3$ **27.** $-12x^4y^3 + 16x^3y^4 - 4x^2y^5$
29. $-12m^4n^2 - 12m^3n^3$ **31.** $x^2 + 5x + 6$ **33.** $z^2 - 9z + 14$ **35.** $2a^2 - 3a - 2$ **37.** $6t^2 + 5t - 6$ **39.** $6y^2 - 5yz + z^2$
41. $2x^2 + xy - 6y^2$ **43.** $9x^2 - 6xy - 3y^2$ **45.** $8a^2 + 14ab - 15b^2$ **47.** $x^2 + 4x + 4$ **49.** $a^2 - 8a + 16$ **51.** $4a^2 + 4ab + b^2$
53. $4x^2 - 4xy + y^2$ **55.** $x^2 - 4$ **57.** $a^2 - b^2$ **59.** $4x^2 - 9y^2$ **61.** $x^3 - y^3$ **63.** $6y^3 + 11y^2 + 9y + 2$ **65.** $8a^3 - b^3$
67. $2a^2 + ab - b^2 - 3bc - 2c^2$ **69.** $x^2 + 4xy + 6xz + 4y^2 + 12yz + 9z^2$ **71.** $4x^3 - 8x^2 - 9x + 6$ **73.** $a^3 - 3a^2b - ab^2 + 3b^3$
75. $x^2 - 4y^2 + 4y - 1$ **77.** $2x^5 + x$ **79.** $\dfrac{3x}{y^4z} - \dfrac{x^6}{y^{10}z^2}$ **81.** $\dfrac{1}{x^2} - y^2$ **83.** $2x^3y^3 - \dfrac{2}{x^3y^3} + 3$ **85.** $x^{3n} - x^{2n}$ **87.** $x^{2n} - 1$
89. $x^{2n} - \dfrac{x^n}{y^n} - x^ny^n + 1$ **91.** $x^{4n} - y^{4n}$ **93.** $x^{2n} + x^n + x^ny^n + y^n$ **95.** $3x^2 + 12x$ **97.** $-p^2 + 4pq$ **99.** $m^2 - mn + 2n^2$
101. $3x^2 + 3x - 11$ **103.** $5x^2 - 36x + 7$ **105.** $21x^2 - 6xy + 29y^2$ **107.** $24y^2 - 4yz + 21z^2$
109. $9.2127x^2 - 7.7956x - 36.0315$ **111.** $299.29y^2 - 150.51y + 18.9225$

Review Exercises (page 275)

1. 10 **2.** 24 **3.** -15 **4.** 1 **5.** $\$51{,}025$ **6.** 5.87×10^{12}

Orals (page 281)

1. $x(3x - 1)$ **2.** $7t^2(t + 2)$ **3.** $-3a(a + 2)$ **4.** $-4x(x - 3)$ **5.** $(a + b)(3 + x)$ **6.** $(m - n)(a - b)$

Exercise 5.4 (page 281)

1. $2 \cdot 3$ **3.** $3^3 \cdot 5$ **5.** 2^7 **7.** $5^2 \cdot 13$ **9.** 12 **11.** 2 **13.** $4a^2$ **15.** $6xy^2z^2$ **17.** 4 **19.** 1 **21.** $2(x + 4)$ **23.** $2x(x - 3)$
25. prime **27.** $5x^2y(3 - 2y)$ **29.** $9x^2y^2(7x + 9y^2)$ **31.** prime **33.** $3z(9z^2 + 4z + 1)$ **35.** $6s(4s^2 - 2st + t^2)$
37. $9x^7y^3(5x^3 - 7y^4 + 9x^3y^7)$ **39.** prime **41.** $8a^3b^3(3b^2 + 4a^2 - 8a^2b^2c^5)$ **43.** $-3(a + 2)$ **45.** $-x(3x + 1)$ **47.** $-3x(2x + y)$
49. $-6ab(3a + 2b)$ **51.** $-7u^2v^3z^2(9uv^3z^7 - 4v^4 + 3uz^2)$ **53.** $x^2(x^n + x^{n+1})$ **55.** $y^n(2y^2 - 3y^3)$ **57.** $2t^{2n}(t^{2n} + 2t^n - 4)$
59. $x^{-2}(x^6 - 5x^8)$ **61.** $t^{-3}(t^8 + 4t^{-3})$ **63.** $4y^{-2n}(2y^{4n} + 3y^{2n} + 4)$ **65.** $r^{-3n}s^2(r^{12n}s^2 - 3r^{6n}s - 2r^{3n})$ **67.** $(x + y)(4 + t)$
69. $(a - b)(r - s)$ **71.** $(m + n + p)(3 + x)$ **73.** $(x + y)(x + y + z)$ **75.** $(u + v)(u + v - 1)$ **77.** $-(x + y)(a - b)$
79. $x[x(2x + 5) - 2] + 8$ **81.** $(x + y)(a + b)$ **83.** $(x + 2)(x + y)$ **85.** $(3 - c)(c + d)$ **87.** $(a + b)(a - 4)$ **89.** $(a + b)(x - 1)$
91. $(x + y)(x + y + z)$ **93.** $x(m + n)(p + q)$ **95.** $y(x + y)(x + y + 2z)$ **97.** $n(2n - p + 2m)(n^2p - 1)$ **99.** yes **101.** no
103. yes **105.** yes **107.** $r_1 = \dfrac{rr_2}{r_2 - r}$ **109.** $f = \dfrac{d_1d_2}{d_2 + d_1}$ **111.** $a^2 = \dfrac{b^2x^2}{b^2 - y^2}$ **113.** $r = \dfrac{S - a}{S - l}$ **115.** $a = \dfrac{Hb}{2b - H}$
117. $y = \dfrac{4x + 1}{3x - 2}$

Review Exercises (page 283)

1. $a^2 - 16$ **2.** $4b^2 - 9$ **3.** $16r^4 - 9s^2$ **4.** $25a^2 - 4b^6$ **5.** $m^3 + 64$ **6.** $p^3 - q^3$

Orals (page 288)

1. $(x + 1)(x - 1)$ **2.** $(a^2 + 4)(a + 2)(a - 2)$ **3.** $(x + 1)(x^2 - x + 1)$ **4.** $(a - 2)(a^2 + 2a + 4)$ **5.** $2(x + 2)(x - 2)$ **6.** prime

Exercise 5.5 (page 288)

1. $(x + 2)(x - 2)$ **3.** $(3y + 8)(3y - 8)$ **5.** prime   **7.** $(25a + 13b^2)(25a - 13b^2)$ **9.** $(9a^2 + 7b)(9a^2 - 7b)$
11. $(6x^2y + 7z^2)(6x^2y - 7z^2)$ **13.** $(x + y + z)(x + y - z)$ **15.** $(a - b + c)(a - b - c)$ **17.** $(x^2 + y^2)(x + y)(x - y)$
19. $(16x^2y^2 + z^4)(4xy + z^2)(4xy - z^2)$ **21.** $2(x + 12)(x - 12)$ **23.** $2x(x + 4)(x - 4)$ **25.** $5x(x + 5)(x - 5)$
27. $t^2(rs + x^2y)(rs - x^2y)$ **29.** $(r + s)(r^2 - rs + s^2)$ **31.** $(x - 2y)(x^2 + 2xy + 4y^2)$ **33.** $(4a - 5b^2)(16a^2 + 20ab^2 + 25b^4)$
35. $(5xy^2 + 6z^3)(25x^2y^4 - 30xy^2z^3 + 36z^6)$ **37.** $(x^2 + y^2)(x^4 - x^2y^2 + y^4)$ **39.** $5(x + 5)(x^2 - 5x + 25)$
41. $4x^2(x - 4)(x^2 + 4x + 16)$ **43.** $2u^2(4v - t)(16v^2 + 4vt + t^2)$ **45.** $(a + b)(x + 3)(x^2 - 3x + 9)$
47. $6(ab - z)(a^2b^2 + abz + z^2)$ **49.** $(x^m + y^{2n})(x^m - y^{2n})$ **51.** $(10a^{2m} + 9b^n)(10a^{2m} - 9b^n)$ **53.** $(x^n - 2)(x^{2n} + 2x^n + 4)$
55. $(a^m + b^n)(a^{2m} - a^mb^n + b^{2n})$ **57.** $2(x^{2m} + 2y^m)(x^{4m} - 2x^{2m}y^m + 4y^{2m})$ **59.** $(a + b)(a - b + 1)$ **61.** $(a - b)(a + b + 2)$
63. $(2x + y)(1 + 2x - y)$

Review Exercises (page 290)

1. $x^2 + 2x + 1$ **2.** $2m^2 - 7m + 6$ **3.** $4m^2 + 4mn + n^2$ **4.** $9m^2 - 12mn + 4n^2$ **5.** $a^2 + 7a + 12$ **6.** $6b^2 - 11b - 10$
7. $8r^2 - 10rs + 3s^2$ **8.** $15a^2 + 14ab - 8b^2$

Orals (page 298)

1. $(x + 2)(x + 1)$ **2.** $(x + 4)(x + 1)$ **3.** $(x - 3)(x - 2)$ **4.** $(x + 1)(x - 4)$ **5.** $(2x + 1)(x + 1)$ **6.** $(3x + 1)(x + 1)$

Exercise 5.6 (page 298)

1. $(x + 1)^2$ **3.** $(a - 9)^2$ **5.** $(2y + 1)^2$ **7.** $(3b - 2)^2$ **9.** $(3z + 4)^2$ **11.** factorable   **13.** not factorable   **15.** $(x + 3)(x + 2)$
17. $(x - 2)(x - 5)$ **19.** prime   **21.** $(y - 7)(y + 3)$ **23.** $3(x + 7)(x - 3)$ **25.** $b^2(a - 11)(a - 2)$ **27.** $x^2(b - 7)(b - 5)$
29. $-(a - 8)(a + 4)$ **31.** $-3(x - 3)(x - 2)$ **33.** $-4(x - 5)(x + 4)$ **35.** $(3y + 2)(2y + 1)$ **37.** $(4a - 3)(2a + 3)$
39. $(3x - 4)(2x + 1)$ **41.** prime   **43.** $(4x - 3)(2x - 1)$ **45.** $(3z - 4)(2z + 5)$ **47.** $(a + b)(a - 4b)$ **49.** $(2y - 3t)(y + 2t)$
51. $x(3x - 1)(x - 3)$ **53.** $-(3a + 2b)(a - b)$ **55.** $(3t + 2)(3t - 1)$ **57.** $(3x - 2)^2$ **59.** $-(2x - 3)^2$ **61.** $(5x - 1)(3x - 2)$
63. $5(a - 3b)^2$ **65.** $z(8x^2 + 6xy + 9y^2)$ **67.** $(15x - 1)(x + 5)$ **69.** $x^2(7x - 8)(3x + 2)$ **71.** $(3xy - 4z)(2xy - 3z)$
73. $(x^2 + 5)(x^2 + 3)$ **75.** $(y^2 - 10)(y^2 - 3)$ **77.** $(a + 3)(a - 3)(a + 2)(a - 2)$ **79.** $(z^2 + 3)(z + 2)(z - 2)$
81. $(x^3 + 3)(x + 1)(x^2 - x + 1)$ **83.** $(x^n + 1)^2$ **85.** $(2a^{3n} + 1)(a^{3n} - 2)$ **87.** $(x^{2n} + y^{2n})^2$ **89.** $(3x^n - 1)(2x^n + 3)$
91. $(x + 2)^2$ **93.** $(a + b + 4)(a + b - 6)$ **95.** $(3x + 3y + 4)(2x + 2y - 5)$ **97.** $(x + 2 + y)(x + 2 - y)$
99. $(x + 1 + 3z)(x + 1 - 3z)$ **101.** $(c + 2a - b)(c - 2a + b)$ **103.** $(a + 4 + b)(a + 4 - b)$ **105.** $(2x + y + z)(2x + y - z)$
107. $(a - 16)(a - 1)$ **109.** $(2u + 3)(u + 1)$ **111.** $(5r + 2s)(4r - 3s)$ **113.** $(5u + v)(4u + 3v)$

Review Exercises (page 301)

1. 31   **2.** 10   **3.** 12   **4.** 6   **5.** -3 **6.** -3

Orals (page 303)

1. $(x + y)(x - y)$ **2.** $2x^3(1 - 2x)$ **3.** $(x + 2)^2$ **4.** $(x - 3)(x - 2)$ **5.** $(x - 2)(x^2 + 2x + 4)$ **6.** $(x + 2)(x^2 - 2x + 4)$

Exercise 5.7 (page 303)

1. $(x + 4)^2$ **3.** $(2xy - 3)(4x^2y^2 + 6xy + 9)$ **5.** $(x - t)(y + s)$ **7.** $(5x + 4y)(5x - 4y)$ **9.** $(6x + 5)(2x + 7)$
11. $2(3x - 4)(x - 1)$ **13.** $(8x - 1)(7x - 1)$ **15.** $y^2(2x + 1)(2x + 1)$ **17.** $(x + a^2y)(x^2 - a^2xy + a^4y^2)$
19. $2(x - 3)(x^2 + 3x + 9)$ **21.** $(a + b)(f + e)$ **23.** $(2x + 2y + 3)(x + y - 1)$ **25.** $(25x^2 + 16y^2)(5x + 4y)(5x - 4y)$
27. $36(x^2 + 1)(x + 1)(x - 1)$ **29.** $2(x^2 + y^2)(x^4 - x^2y^2 + y^4)$ **31.** $(a + 3)(a - 3)(a + 2)(a - 2)$ **33.** $(x + 3 + y)(x + 3 - y)$
35. $(2x + 1 + 2y)(2x + 1 - 2y)$ **37.** $(z + 4 + 4y)(z + 4 - 4y)$ **39.** $(x + 1)(x^2 - x + 1)(x + 1)(x - 1)$
41. $(x + 3)(x - 3)(x + 2)(x^2 - 2x + 4)$ **43.** $2z(x + y)(x - y)^2(x^2 + xy + y^2)$ **45.** $(x^m - 3)(x^m + 2)$
47. $(a^n - b^n)(a^{2n} + a^nb^n + b^{2n})$ **49.** $\left(\frac{1}{x} + 1\right)^2$ **51.** $\left(\frac{3}{x} + 2\right)\left(\frac{2}{x} - 3\right)$ **53.** $(a + 2)(a^3 + a^2 + 1)$

Review Exercises (page 304)

1. $7a^2 + a - 7$ **2.** $-7b^2 - b - 7$ **3.** $4y^2 - 11y + 3$ **4.** $15x^2 + 3x$ **5.** $m^2 + 2m - 8$ **6.** $6p^2 - pq - 12q^2$

Orals (page 309)

1. 2, 3   **2.** -4, 2   **3.** 2, 3, -1 **4.** -3, -2, 5, 6

Exercise 5.8 (page 309)

1. $0, -2$ **3.** $4, -4$ **5.** $0, -1$ **7.** $0, 5$ **9.** $-3, -5$ **11.** $1, 6$ **13.** $3, 4$ **15.** $-2, -4$ **17.** $-\frac{1}{3}, -3$ **19.** $\frac{1}{2}, 2$ **21.** $1, -\frac{1}{2}$
23. $2, -\frac{1}{3}$ **25.** $3, 3$ **27.** $\frac{1}{4}, -\frac{3}{2}$ **29.** $2, -\frac{5}{6}$ **31.** $\frac{1}{3}, -1$ **33.** $2, \frac{1}{2}$ **35.** $\frac{1}{5}, -\frac{5}{3}$ **37.** $0, 0, -1$ **39.** $0, 7, -7$ **41.** $0, 7, -3$
43. $3, -3, 2, -2$ **45.** $0, -2, -3$ **47.** $0, \frac{5}{6}, -7$ **49.** $1, -1, -3$ **51.** $3, -3, -\frac{3}{2}$ **53.** $2, -2, -\frac{1}{3}$ **55.** $16, 18$ or $-18, -16$
57. $6, 7$ **59.** 40 m **61.** 15 ft by 25 ft **63.** 5 ft by 12 ft **65.** 20 ft by 40 ft **67.** 10 sec **69.** 11 sec and 19 sec **71.** 50 m
73. 3 ft **75.** no solution **77.** 1.00

Review Exercises (page 312)

1. $2, 3, 5, 7$ **2.** $8, 9, 10, 12, 14, 15, 16$ **3.** $40,081.00$ cm^3 **4.** 2340.30 m^3

Problems and Projects (page 313)

2. $3x^2 + 9x + 7$ **3.** $180,000$ ft^2 **4.** 3

Chapter 5 Review Exercises (page 316)

1. 6 **2.** 9 **3.** $-t^2 - 4t + 6$ **4.** $-z^2 + 4z + 6$ **5.** 5 **6.** 8 **7.** $5x^2 + 3x$ **8.** $x^3 + x^2 + x + 9$ **9.** $4x^2 - 9x + 19$
10. $-7x^3 - 30x^2 - 4x + 3$ **11.** $-16a^3b^3c$ **12.** $-6x^2y^2z^4$ **13.** $2x^4y^3 - 8x^2y^7$ **14.** $a^4b + 2a^3b^2 + a^2b^3$ **15.** $16x^2 + 14x - 15$
16. $6x^3 - 12x^2 + 4x - 8$ **17.** $15x^4 - 22x^3 + 73x^2 - 50x + 50$ **18.** $3x^4 - 2x^3 + 3x^2 + 4x - 4$ **19.** $4(x + 2)$ **20.** $3x(x - 2)$
21. $5xy^2(xy - 2)$ **22.** $7a^3b(ab + 7)$ **23.** $-4x^2y^3z^2(2z^2 + 3x^2)$ **24.** $3a^2b^4c^2(4a^4 + 5c^4)$ **25.** $9x^2y^3z^2(3xz + 9x^2y^2 - 10z^5)$
26. $-12a^2b^3c^2(3a^3b - 5a^5b^2c + 2c^5)$ **27.** $x^n(x^n + 1)$ **28.** $y^{2n}(1 - y^{2n})$ **29.** $x^{-2}(x^{-2} - 1)$ **30.** $a^{-3}(a^9 + a^3)$
31. $5x^2(x + y)^3(1 - 3x^2 - 3xy)$ **32.** $-7a^2b^2(a - b)^3(7a^2 - 7ab - 9b^2)$ **33.** $(x + 2)(y + 4)$ **34.** $(a + b)(c + 3)$
35. $(x^2 + 4)(x^2 + y)$ **36.** $(a^3 + c)(a^2 + b^2)$ **37.** $(z + 4)(z - 4)$ **38.** $(y + 11)(y - 11)$ **39.** $(xy^2 + 8z^3)(xy^2 - 8z^3)$ **40.** prime
41. $(x + z + t)(x + z - t)$ **42.** $(c + a + b)(c - a - b)$ **43.** $2(x^2 + 7)(x^2 - 7)$ **44.** $3x^2(x^2 + 10)(x^2 - 10)$
45. $(x + 7)(x^2 - 7x + 49)$ **46.** $(a - 5)(a^2 + 5a + 25)$ **47.** $8(y - 4)(y^2 + 4y + 16)$ **48.** $4y(x + 3z)(x^2 - 3xz + 9z^2)$
49. $(y + 20)(y + 1)$ **50.** $(z - 5)(z - 6)$ **51.** $-(x + 7)(x - 4)$ **52.** $(y - 8)(y + 3)$ **53.** $(4a - 1)(a - 1)$ **54.** prime **55.** prime
56. $-(5x + 2)(3x - 4)$ **57.** $y(y + 2)(y - 1)$ **58.** $2a^2(a + 3)(a - 1)$ **59.** $-3(x + 2)(x + 1)$ **60.** $4(2x + 3)(x - 2)$
61. $3(5x + y)(x - 4y)$ **62.** $5(6x + y)(x + 2y)$ **63.** $(8x + 3y)(3x - 4y)$ **64.** $(2x + 3y)(7x - 4y)$ **65.** $x(x - 1)(x + 6)$
66. $3y(x + 3)(x - 7)$ **67.** $(z - 2)(z + x + 2)$ **68.** $(x + 1 + p)(x + 1 - p)$ **69.** $(x + 2 + 2p^2)(x + 2 - 2p^2)$
70. $(y + 2)(y + 1 + x)$ **71.** $(x^m + 3)(x^m - 1)$ **72.** $\left(\frac{1}{x} - 2\right)\left(\frac{1}{x} + 1\right)$ **73.** $h = \dfrac{S - 2wl}{2w + 2l}$ **74.** $l = \dfrac{S - 2wh}{2w + 2h}$ **75.** $0, \frac{3}{4}$ **76.** $6, -6$
77. $\frac{1}{2}, -\frac{5}{6}$ **78.** $\frac{2}{7}, 5$ **79.** $0, -\frac{2}{3}, \frac{4}{5}$ **80.** $-\frac{2}{3}, 7, 0$ **81.** 7 cm **82.** 17 m by 20 m

Chapter 5 Test (page 318)

1. 5 **2.** 13 **3.** -9 **4.** -6 **5.** $5y^2 + y - 1$ **6.** $-4u^2 + 2u - 14$ **7.** $10a^2 + 22$ **8.** $-x^2 + 15x - 2$ **9.** $-6x^4yz^4$
10. $-15a^3b^4 + 10a^3b^5$ **11.** $z^2 - 16$ **12.** $12x^2 + x - 6$ **13.** $3xy(y + 2x)$ **14.** $3abc(4a^2b - abc + 2c^2)$ **15.** $y^n(x^2y^2 + 1)$
16. $b^n(a^n - ab^{-2n})$ **17.** $(u - v)(r + s)$ **18.** $(a - y)(x + y)$ **19.** $(x + 7)(x - 7)$ **20.** $2(x + 4)(x - 4)$
21. $4(y^2 + 4)(y + 2)(y - 2)$ **22.** $(b + 5)(b^2 - 5b + 25)$ **23.** $(b - 3)(b^2 + 3b + 9)$ **24.** $3(u - 2)(u^2 + 2u + 4)$
25. $(a - 6)(a + 1)$ **26.** $(3b + 2)(2b - 1)$ **27.** $3(u + 2)(2u - 1)$ **28.** $5(4r + 1)(r - 1)$ **29.** $(x^n + 1)^2$
30. $(x + 3 + y)(x + 3 - y)$ **31.** $r = \dfrac{r_1 r_2}{r_2 + r_1}$ **32.** $6, -1$ **33.** 25 **34.** 11 ft by 22 ft

Orals (page 329)

1. -1 **2.** 1 **3.** $(-\infty, 2) \cup (2, \infty)$ **4.** $(-\infty, -3) \cup (-3, 3) \cup (3, \infty)$ **5.** $\frac{5}{6}$ **6.** $\frac{x}{y}$ **7.** 2 **8.** -1

Exercise 6.1 (page 329)

1. 20 hr **3.** 12 hr **5.** $\$5555.56$ **7.** $\$50,000$ **9.** $c = f(x) = 1.25x + 700$ **11.** $\$1325$ **13.** $\$1.95$ **15.** $c = f(n) = 0.09n + 7.50$

17. $\$77.25$ **19.** $9.75¢$ **21.** almost 8 days **23.** about 2.55 hr **25.** $(-\infty, 2) \cup (2, \infty)$ **27.** $(-\infty, -2) \cup (-2, 2) \cup (2, \infty)$ **29.** $\dfrac{2}{3}$

31. $-\dfrac{28}{9}$ **33.** $\dfrac{12}{13}$ **35.** $-\dfrac{122}{37}$ **37.** $4x^2$ **39.** $-\dfrac{4y}{3x}$ **41.** x^2 **43.** $-\dfrac{x}{2}$ **45.** $\dfrac{3y}{7(y - z)}$ **47.** 1 **49.** $\dfrac{1}{x - y}$ **51.** $\dfrac{5}{x - 2}$

53. $\dfrac{-3(x + 2)}{x + 1}$ **55.** 3 **57.** $x + 2$ **59.** $\dfrac{x + 1}{x + 3}$ **61.** $\dfrac{m - 2n}{n - 2m}$ **63.** $\dfrac{x + 4}{2(2x - 3)}$ **65.** $\dfrac{3(x - y)}{x + 2}$ **67.** $\dfrac{2x + 1}{2 - x}$ **69.** $\dfrac{a^2 - 3a + 9}{4(a - 3)}$

71. lowest terms **73.** $m + n$ **75.** $-\dfrac{m + n}{2m + n}$ **77.** $\dfrac{x - y}{x + y}$ **79.** $\dfrac{2a - 3b}{a - 2b}$ **81.** $\dfrac{1}{x^2 + xy + y^2 - 1}$ **83.** $x^2 - y^2$ **85.** $\dfrac{(x + 1)^2}{(x - 1)^3}$

87. $\dfrac{3a + b}{y + b}$ **89.** $\dfrac{c + 1}{2 - a}$ **91.** $\frac{1}{6}$ **93.** 0 **95.** $\frac{1}{2}$ **97.** $\frac{1}{13}$ **99.** $\frac{1}{6,000,000}$ **101.** $\frac{1}{8}$

Review Exercises (page 332)

1. $3x(x - 3)$ **2.** $(3t + 2)(2t - 3)$ **3.** $(3x^2 + 4y)(9x^4 - 12x^2y + 16y^2)$ **4.** $(x + a)(x + 2)$

Orals (page 337)

1. $\frac{9}{8}$ **2.** $\frac{1}{2}$ **3.** $\frac{x - 2}{x + 2}$ **4.** $\frac{9}{16}$ **5.** 5 **6.** $\frac{y}{2}$

Exercise 6.2 (page 337)

1. $\dfrac{10}{7}$ **3.** $-\dfrac{5}{6}$ **5.** $\dfrac{xy^2d}{c^3}$ **7.** $-\dfrac{x^{10}}{y^2}$ **9.** $x + 1$ **11.** 1 **13.** $\dfrac{x - 4}{x + 5}$ **15.** $\dfrac{(a + 7)^2(a - 5)}{12x^2}$ **17.** $\dfrac{t - 1}{t + 1}$ **19.** $\dfrac{n + 2}{n + 1}$ **21.** $\dfrac{1}{x + 1}$

23. $(x + 1)^2$ **25.** $x - 5$ **27.** $\dfrac{x + y}{x - y}$ **29.** $-\dfrac{x + 3}{x + 2}$ **31.** $\dfrac{a + b}{(x - 3)(c + d)}$ **33.** $-\dfrac{x + 1}{x + 3}$ **35.** $-\dfrac{2x + y}{3x + y}$ **37.** $-\dfrac{x^7}{18y^4}$

39. $x^2(x + 3)$ **41.** $\dfrac{3x}{2}$ **43.** $\dfrac{t + 1}{t}$ **45.** $\dfrac{x - 2}{x - 1}$ **47.** $\dfrac{x + 2}{x - 2}$ **49.** $\dfrac{x - 1}{3x + 2}$ **51.** $\dfrac{x - 7}{x + 7}$ **53.** 1

Review Exercises (page 339)

1. $-6a^5 + 2a^4$ **2.** $4t^2 - 4t + 1$ **3.** $m^{2n} - 4$ **4.** $\dfrac{3}{b^{2n}} - \dfrac{2c}{b^n} - c^2$

Orals (page 345)

1. x **2.** $\frac{a}{2}$ **3.** 1 **4.** 1 **5.** $\dfrac{7x}{6}$ **6.** $\dfrac{5y - 3x}{xy}$

Exercise 6.3 (page 345)

1. $\dfrac{5}{2}$ **3.** $-\dfrac{1}{3}$ **5.** $\dfrac{11}{4y}$ **7.** $\dfrac{3 - a}{a + b}$ **9.** 2 **11.** 3 **13.** 3 **15.** $\dfrac{6x}{(x - 3)(x - 2)}$ **17.** 72 **19.** $x(x + 3)(x - 3)$

21. $(x + 3)^2(x^2 - 3x + 9)$ **23.** $(2x + 3)^2(x + 1)^2$ **25.** $\dfrac{5}{6}$ **27.** $-\dfrac{16}{75}$ **29.** $\dfrac{9a}{10}$ **31.** $\dfrac{21a - 8b}{14}$ **33.** $\dfrac{17}{12x}$ **35.** $\dfrac{9a^2 - 4b^2}{6ab}$

37. $\dfrac{10a + 4b}{21}$ **39.** $\dfrac{8x - 2}{(x + 2)(x - 4)}$ **41.** $\dfrac{7x + 29}{(x + 5)(x + 7)}$ **43.** $\dfrac{x^2 + 1}{x}$ **45.** 2 **47.** $\dfrac{9a^2 - 11a - 10}{(3a + 2)(3a - 2)}$ **49.** $\dfrac{2x^2 + x}{(x + 3)(x + 2)(x - 2)}$

51. $\dfrac{-x^2 + 11x + 8}{(3x + 2)(x + 1)(x - 3)}$ **53.** $\dfrac{-4x^2 + 14x + 54}{x(x + 3)(x - 3)}$ **55.** $\dfrac{x^3 + x^2 - 1}{x(x + 1)(x - 1)}$ **57.** $\dfrac{2x^2 + 5x + 4}{x + 1}$ **59.** $\dfrac{x^2 - 5x - 5}{x - 5}$

61. $\dfrac{-x^3 - x^2 + 5x}{x - 1}$ **63.** $\dfrac{-y^2 + 48y + 161}{(y + 4)(y + 3)}$ **65.** $\dfrac{2}{x + 1}$ **67.** $\dfrac{3x + 1}{x(x + 3)}$ **69.** $\dfrac{2x^3 + x^2 - 43x - 35}{(x + 5)(x - 5)(2x + 1)}$ **71.** $\dfrac{3x^2 - 2x - 17}{(x - 3)(x - 2)}$

73. $\dfrac{2a}{a - 1}$ **75.** $\dfrac{x^2 - 6x - 1}{2(x + 1)(x - 1)}$ **77.** $\dfrac{2b}{a + b}$ **79.** $\dfrac{7mn^2 - 7n^3 - 6m^2 + 3mn + n}{(m - n)^2}$ **81.** $\dfrac{2}{m - 1}$ **83.** 0 **85.** 1

Review Exercises (page 348)

1. **2.** **3.** $w = \dfrac{P - 2l}{2}$ **4.** $a = S - Sr + lr$

Orals (page 353)

1. $\frac{3}{5}$ **2.** $\frac{a}{d}$ **3.** 2 **4.** $\dfrac{x + y}{x - y}$ **5.** $\dfrac{x - y}{x}$ **6.** $\dfrac{b + a}{a}$

Exercise 6.4   (page 353)

1. $\dfrac{2}{3}$ **3.** -1 **5.** $\dfrac{10}{3}$ **7.** $-\dfrac{1}{7}$ **9.** $\dfrac{2y}{3z}$ **11.** $125b$ **13.** $-\dfrac{1}{y}$ **15.** $\dfrac{y-x}{x^2y^2}$ **17.** $\dfrac{b+a}{b}$ **19.** $\dfrac{y+x}{y-x}$ **21.** $y-x$ **23.** $\dfrac{-1}{a+b}$

25. x^2+x-6 **27.** $\dfrac{5x^2y^2}{xy+1}$ **29.** -1 **31.** $\dfrac{x+2}{x-3}$ **33.** $\dfrac{a-1}{a+1}$ **35.** $\dfrac{y+x}{x^2y}$ **37.** $\dfrac{xy^2}{y-x}$ **39.** $\dfrac{y+x}{y-x}$ **41.** xy **43.** $\dfrac{x^2(xy^2-1)}{y^2(x^2y-1)}$

45. $\dfrac{(b+a)(b-a)}{b(b-a-ab)}$ **47.** $\dfrac{x-1}{x}$ **49.** $\dfrac{5b}{5b+4}$ **51.** $\dfrac{3a^2+2a}{2a+1}$ **53.** $\dfrac{(-x^2+2x-2)(3x+2)}{(2-x)(-3x^2-2x+9)}$ **55.** $\dfrac{2(x^2-4x-1)}{-x^2+4x+8}$

57. $\dfrac{-2}{x^2-3x-7}$ **59.** $\dfrac{k_1k_2}{k_2+k_1}$

Review Exercises   (page 355)

1. 8   **2.** $\frac{4}{3}, -\frac{3}{2}$ **3.** $2, -2, 3, -3$ **4.** $5, -4$

Orals   (page 362)

1. 2   **2.** 3   **3.** 1   **4.** 3   **5.** 1   **6.** 2

Exercise 6.5   (page 362)

1. 12   **3.** 40   **5.** $\frac{1}{2}$ **7.** no solution   **9.** $\frac{17}{25}$ **11.** 0   **13.** 1   **15.** 2   **17.** 2   **19.** $\frac{1}{3}$ **21.** 0   **23.** $2, -5$ **25.** $-4, 3$ **27.** $6, \frac{17}{3}$

29. $1, -11$ **31.** $f=\dfrac{pq}{q+p}$ **33.** $r=\dfrac{S-a}{S-l}$ **35.** $R=\dfrac{r_1r_2r_3}{r_1r_3+r_1r_2+r_2r_3}$ **37.** $4\frac{8}{13}$ in.   **39.** $1\frac{7}{8}$ days   **41.** $5\frac{5}{6}$ min   **43.** $1\frac{1}{5}$ days

45. $2\frac{4}{13}$ weeks   **47.** 3 mph   **49.** 60 mph and 40 mph   **51.** 3 mph   **53.** 60 mph   **55.** 2   **57.** 7 motors   **59.** 12 days
61. about 3,500,000 in. lb/rad

Review Exercises   (page 365)

1. $\dfrac{n^6}{m^4}$ **2.** $\dfrac{1}{1+a}$ **3.** 0   **4.** $\dfrac{8}{x}-\dfrac{16}{x^2}+6x-12$

Orals   (page 372)

1. 1   **2.** 9   **3.** $\frac{14}{5}$ **4.** $a=kb$ **5.** $a=\frac{k}{b}$ **6.** $a=kbc$ **7.** $a=\frac{kb}{c}$

Exercise 6.6   (page 372)

1. 3   **3.** 5   **5.** -3 **7.** 5   **9.** $4, -1$ **11.** $2, -2$ **13.** 39   **15.** $-\frac{5}{2}, -1$ **17.** \$62.50   **19.** $7\frac{1}{2}$ gal   **21.** 32 ft   **23.** 80 ft
25. 0.18 g   **27.** 42 ft   **29.** $46\frac{7}{8}$ ft   **31.** 6750 ft   **33.** $A=kp^2$ **35.** $v=k/r^3$ **37.** $B=kmn$ **39.** $P=ka^2/j^3$ **41.** $F=km_1m_2/d^2$
43. L varies jointly with m and n.   **45.** E varies jointly with a and the square of b.   **47.** X varies directly with x^2 and inversely
with y^2.   **49.** R varies directly with L and inversely with d^2.   **51.** 36π in.^{2} **53.** 432 mi   **55.** 25 days   **57.** 12 in.^{3} **59.** 85.3
61. 12   **63.** 26,437.5 gal   **65.** 3 ohms   **67.** 0.275 in.   **69.** 546 Kelvin

Review Exercises   (page 376)

1. x^{10} **2.** a^{30} **3.** -1 **4.** 8   **5.** 3.5×10^4 **6.** 3.5×10^{-4} **7.** 0.0025   **8.** 25,000

Orals   (page 381)

1. $3xy$ **2.** $2b+4a$ **3.** $x+1$ **4.** $x+2$

Exercise 6.7   (page 381)

1. $\dfrac{y}{2x^3}$ **3.** $\dfrac{3b^4}{4a^4}$ **5.** $\dfrac{-5}{7xy^7t^2}$ **7.** $\dfrac{13a^nb^{2n}c^{3n-1}}{3}$ **9.** $\dfrac{2x}{3}-\dfrac{x^2}{6}$ **11.** $\dfrac{xy^2}{2}+\dfrac{x^2y}{8}$ **13.** $\dfrac{x^4y^4}{2}-\dfrac{x^3y^9}{4}+\dfrac{3}{4xy^2}$ **15.** $\dfrac{b^2}{4a}-\dfrac{a^3}{2b^4}+\dfrac{3}{4ab}$
17. $1-3x^ny^n+6x^{2n}y^{2n}$ **19.** $x+2$ **21.** $x+7$ **23.** $3x-5+\frac{3}{2x+3}$ **25.** $3x^2+x+2+\frac{8}{x-1}$ **27.** $2x^2+5x+3+\frac{4}{3x-2}$
29. $3x^2+4x+3$ **31.** $a+1$ **33.** $2y+2$ **35.** $6x-12$ **37.** $3x^2-x+2$ **39.** $4x^3-3x^2+3x+1$ **41.** $a^2+a+1+\frac{2}{a-1}$
43. $5a^2-3a-4$ **45.** $6y-12$ **47.** $16x^4-8x^3y+4x^2y^2-2xy^3+y^4$ **49.** x^4+x^2+4 **51.** x^2+x+1 **53.** x^2+x+2
55. $9.8x+16.4+\frac{-36.5}{x-2}$

Review Exercises   (page 382)

1. $8x^2 + 2x + 4$ **2.** $-2a^2 - 25a + 34$ **3.** $-2y^3 - 3y^2 + 6y - 6$ **4.** $2y^3 + 11y^2 + 7y + 10$

Orals   (page 387)

1. 9   **2.** -3 **3.** yes   **4.** no

Exercise 6.8   (page 387)

1. $x + 2$ **3.** $x - 3$ **5.** $x + 2$ **7.** $x - 7 + \frac{28}{x+2}$ **9.** $3x^2 - x + 2$ **11.** $2x^2 + 4x + 3$ **13.** $6x^2 - x + 1 + \frac{3}{x+1}$
15. $7.2x - 0.66 + \frac{0.368}{x-0.2}$ **17.** $2.7x - 3.59 + \frac{0.903}{x+1.7}$ **19.** $9x^2 - 513x + 29{,}241 + \frac{-1{,}666{,}762}{x+57}$ **21.** -1 **23.** -37 **25.** 23   **27.** -1
29. 2   **31.** -1 **33.** 18   **35.** 174   **37.** -8 **39.** 59   **41.** 44   **43.** $\frac{29}{32}$ **45.** yes   **47.** no   **49.** 64

Review Exercises   (page 389)

1. 4   **2.** 7   **3.** $12a^2 + 4a - 1$ **4.** $3t^2 - 2t - 1$ **5.** $8x^2 + 2x + 4$ **6.** $-2y^3 - 3y^2 + 6y - 6$

Problems and Projects   (page 389)

1. 4 m by 16 m   **2.** 8.4   **3.** 12, 5 talents   **4.** $r = 15$

Chapter 6 Review Exercises   (page 392)

1. $\dfrac{31x}{72y}$ **2.** $\dfrac{53m}{147n^2}$ **3.** $\dfrac{x-7}{x+7}$ **4.** $\dfrac{1}{x-6}$ **5.** $\dfrac{1}{2x+4}$ **6.** -1 **7.** -2 **8.** $\dfrac{-a-b}{c+d}$ **9.** $\frac{1}{18}$ **10.** $\frac{1}{8}$ **11.** 1   **12.** 1   **13.** $\dfrac{5y-3}{x-y}$
14. $\dfrac{6x-7}{x^2+2}$ **15.** $\dfrac{5x+13}{(x+2)(x+3)}$ **16.** $\dfrac{4x^2+9x+12}{(x-4)(x+3)}$ **17.** $\dfrac{3x(x-1)}{(x-3)(x+1)}$ **18.** 1   **19.** $\dfrac{5x^2+11x}{(x+1)(x+2)}$ **20.** $\dfrac{2(3x+1)}{x-3}$
21. $\dfrac{5x^2+23x+4}{(x+1)(x-1)(x-1)}$ **22.** $\dfrac{-x^4-4x^3+3x^2+18x+16}{(x-2)(x+2)^2}$ **23.** $\dfrac{2x^3+13x^2+31x+31}{(x+3)(x+2)(x-3)}$ **24.** $\dfrac{x^2+26x+3}{(x+3)(x-3)^2}$ **25.** $\dfrac{3y-2x}{x^2y^2}$
26. $\dfrac{y+2x}{2y-x}$ **27.** $\dfrac{2x+1}{x+1}$ **28.** $\dfrac{3x+2}{3x-2}$ **29.** $\dfrac{1}{x}$ **30.** $\dfrac{x-2}{x+3}$ **31.** $\dfrac{x^2y^2}{(x-y)^2(y^2-x^2)}$ **32.** $\dfrac{1}{x}$ **33.** $\dfrac{1}{x^2}$ **34.** $\dfrac{y-x}{y+x}$ **35.** 5
36. $-1, -2$ **37.** no solution   **38.** $-1, -12$ **39.** $y^2 = \dfrac{x^2b^2-a^2b^2}{a^2}$ **40.** $b = \dfrac{Ha}{2a-H}$ **41.** 50 mph   **42.** 200 mph   **43.** $14\frac{2}{5}$ hr
44. $18\frac{2}{3}$ days   **45.** 5   **46.** $-4, -12$ **47.** 72   **48.** 6   **49.** 2   **50.** 16   **51.** $-\dfrac{x^3}{2y^3}$ **52.** $-3x^2y + \dfrac{3x}{2} + y$ **53.** $x + 5y$
54. $x^2 + 2x - 1 + \dfrac{6}{2x+3}$ **55.** yes   **56.** no

Chapter 6 Test   (page 394)

1. $\dfrac{-2}{3xy}$ **2.** $\dfrac{2}{x-2}$ **3.** -3 **4.** $\dfrac{2x+1}{4}$ **5.** $\dfrac{1}{8}$ **6.** $\dfrac{2}{13}$ **7.** $\dfrac{xz}{y^4}$ **8.** $\dfrac{x+1}{2}$ **9.** 1   **10.** $\dfrac{(x+y)^2}{2}$ **11.** $\dfrac{2}{x+1}$ **12.** -1 **13.** 3
14. 2   **15.** $\dfrac{2s+r^2}{rs}$ **16.** $\dfrac{2x+3}{(x+1)(x+2)}$ **17.** $\dfrac{u^2}{2vw}$ **18.** $\dfrac{2x+y}{xy-2}$ **19.** $\dfrac{5}{2}$ **20.** 5, 3   **21.** $a^2 = \dfrac{x^2b^2}{b^2-y^2}$ **22.** $r_2 = \dfrac{rr_1}{r_1-r}$
23. 10 days   **24.** \$5000 at 6% and \$3000 at 10%   **25.** 6, -1 **26.** $\frac{44}{3}$ **27.** $\dfrac{-6x}{y} + \dfrac{4x^2}{y^2} - \dfrac{3}{y^3}$ **28.** $3x^2 + 4x + 2$ **29.** -7
30. 47

Cumulative Review Exercises (Chapters 1–6)   (page 395)

1. a^8b^4 **2.** $\dfrac{b^4}{a^4}$ **3.** $\dfrac{81b^{16}}{16a^8}$ **4.** $x^{21}y^3$ **5.** 42,500   **6.** 0.000712   **7.** 1   **8.** -2 **9.** $\frac{5}{6}$ **10.** $-\frac{3}{4}$ **11.** 3   **12.** $-\frac{1}{3}$ **13.** 0   **14.** 8
15. $-\frac{16}{25}$ **16.** $t^2 - 4t + 3$ **17.** $y = \dfrac{kxz}{r}$ **18.** no   **19.** ←—[———→ 5/2   **20.** ←—[———[——→ 3 11/3   **21.** trinomial   **22.** 7   **23.** 18

24.

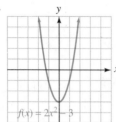

$f(x) = 2x^2 - 3$

25. $4x^2 - 4x + 14$ **26.** $-3x^2 - 3$ **27.** $6x^2 - 7x - 20$ **28.** $2x^{2n} + 3x^n - 2$ **29.** $3rs^3(r - 2s)$

30. $(x - y)(5 - a)$ **31.** $(x + y)(u + v)$ **32.** $(9x^2 + 4y^2)(3x + 2y)(3x - 2y)$ **33.** $(2x - 3y^2)(4x^2 + 6xy^2 + 9y^4)$
34. $(2x + 3)(3x - 2)$ **35.** $(3x - 5)^2$ **36.** $(5x + 3)(3x - 2)$ **37.** $(3a + 2b)(9a^2 - 6ab + 4b^2)$ **38.** $(2x + 5)(3x - 7)$
39. $(x + 5 + y^2)(x + 5 - y^2)$ **40.** $(y + x - 2)(y - x + 2)$ **41.** $0, 2, -2$ **42.** $-\frac{1}{3}, -\frac{7}{2}$ **43.** $\dfrac{2x - 3}{3x - 1}$ **44.** $\dfrac{x - 2}{x + 3}$ **45.** $\dfrac{4}{x - y}$

46. $\dfrac{a^2 + ab^2}{a^2b - b^2}$ **47.** 0   **48.** -17 **49.** $x + 4$ **50.** $-x^2 + x + 5 + \dfrac{8}{x - 1}$

Orals   (page 408)

1. 3   **2.** -4 **3.** -2 **4.** 2   **5.** $8|x|$ **6.** $-3x$ **7.** not real   **8.** $(x + 1)^2$

Exercise 7.1   (page 408)

1. $3x^2$ **3.** $a^2 + b^3$ **5.** 11   **7.** -8 **9.** $\frac{1}{3}$ **11.** $-\frac{5}{7}$ **13.** not real   **15.** 0.4   **17.** 4   **19.** not real   **21.** 3.4641   **23.** 26.0624
25. $4|x|$ **27.** $|t + 5|$ **29.** $5|b|$ **31.** $|a + 3|$ **33.** 0   **35.** 4   **37.** 4.1231   **39.** 2.5539
41.

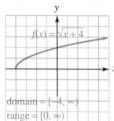

$f(x) = \sqrt{x + 4}$

domain $= [-4, \infty)$
range $= [0, \infty)$

43.

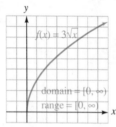

$f(x) = 3\sqrt{x}$

domain $= [0, \infty)$
range $= [0, \infty)$

45. 3 units   **47.** 4 sec   **49.** about 7.4 amperes   **51.** 1
53. -5 **55.** $-\frac{2}{3}$ **57.** 0.4   **59.** $2a$ **61.** $-10pq$ **63.** $-\frac{1}{2}m^2n$
65. $0.2z^3$ **67.** 3   **69.** -3 **71.** -2 **73.** $\frac{2}{5}$ **75.** $\frac{1}{2}$
77. not real   **79.** $2|x|$ **81.** $2a$ **83.** $\frac{1}{2}|x|$ **85.** $|x^3|$
87. $-x$ **89.** $-3a^2$ **91.** $x + 2$ **93.** $0.1x^2|y|$ **95.** 1.67
97. 11.8673

Review Exercises   (page 410)

1. $\dfrac{x + 3}{x - 4}$ **2.** $\dfrac{-(a^2 + ab + b^2)}{a + b}$ **3.** 1   **4.** $\dfrac{(x - 4)(x + 1)}{(x - 3)^2}$ **5.** $\dfrac{3(m^2 + 2m - 1)}{(m + 1)(m - 1)}$ **6.** $\dfrac{x^2 + 21x - 1}{(3x - 1)(2x + 1)}$

Orals   (page 416)

1. 2   **2.** 3   **3.** 3   **4.** 1   **5.** 8   **6.** 4   **7.** $\dfrac{1}{2}$ **8.** 2   **9.** $2x$ **10.** $2x^2$

Exercise 7.2   (page 417)

1. $\sqrt[3]{7}$ **3.** $\sqrt[4]{3x}$ **5.** $\sqrt[4]{\frac{1}{2}x^3y}$ **7.** $\sqrt{x^2 + y^2}$ **9.** $11^{1/2}$ **11.** $(3a)^{1/4}$ **13.** $\left(\frac{1}{7}abc\right)^{1/6}$ **15.** $(a^2 - b^2)^{1/3}$ **17.** 2   **19.** 2   **21.** 2
23. 2   **25.** $\frac{1}{2}$ **27.** $\frac{1}{2}$ **29.** -2 **31.** -3 **33.** not real   **35.** 0   **37.** $5|y|$ **39.** $2|x|$ **41.** $3x$ **43.** not real   **45.** 216   **47.** 27
49. 1728   **51.** $\dfrac{1}{4}$ **53.** $125x^6$ **55.** $\dfrac{4x^2}{9}$ **57.** $\dfrac{1}{2}$ **59.** $\dfrac{1}{8}$ **61.** $\dfrac{1}{64x^3}$ **63.** $\dfrac{1}{9y^2}$ **65.** $\dfrac{1}{4p^2}$ **67.** 8   **69.** $\dfrac{16}{81}$ **71.** $-\dfrac{3}{2x}$ **73.** $5^{5/7}$
75. $4^{3/5}$ **77.** $9^{1/5}$ **79.** $7^{1/2}$ **81.** $\dfrac{1}{36}$ **83.** $2^{2/3}$ **85.** a **87.** $a^{2/9}$ **89.** $a^{3/4}b^{1/2}$ **91.** $\dfrac{n^{2/5}}{m^{3/5}}$ **93.** $\dfrac{2x}{3}$ **95.** $\dfrac{1}{3}x$ **97.** $y + y^2$
99. $x^2 - x + x^{3/5}$ **101.** $x - 4$ **103.** $x^{4/3} - x^2$ **105.** $x^{4/3} + 2x^{2/3}y^{2/3} + y^{4/3}$ **107.** $a^3 - 2a^{3/2}b^{3/2} + b^3$ **109.** $\sqrt{p}$ **111.** $\sqrt{5b}$

Review Exercises   (page 418)

1. $x < 3$ **2.** $t \geq 3$ **3.** $r > 28$ **4.** $0 < x \leq 6$ **5.** $1\frac{2}{3}$ pints   **6.** 10

Orals (page 424)

1. 7   **2.** 4   **3.** 3   **4.** $3\sqrt{2}$ **5.** $2\sqrt[3]{2}$ **6.** $\dfrac{\sqrt[3]{3x^2}}{4b^2}$ **7.** $7\sqrt{3}$ **8.** $3\sqrt{7}$ **9.** $5\sqrt[3]{9}$ **10.** $8\sqrt[5]{4}$

Exercise 7.3 (page 424)

1. 6   **3.** t **5.** $5x$ **7.** 10   **9.** $7x$ **11.** $6b$ **13.** 2   **15.** $3a$ **17.** $2\sqrt{5}$ **19.** $-10\sqrt{2}$ **21.** $2\sqrt[3]{10}$ **23.** $-3\sqrt[3]{3}$ **25.** $2\sqrt[4]{2}$
27. $2\sqrt[5]{3}$ **29.** $\dfrac{\sqrt{7}}{3}$ **31.** $\dfrac{\sqrt[3]{7}}{4}$ **33.** $\dfrac{\sqrt[4]{3}}{10}$ **35.** $\dfrac{\sqrt[5]{3}}{2}$ **37.** $5x\sqrt{2}$ **39.** $4\sqrt{2b}$ **41.** $-4a\sqrt{7a}$ **43.** $5ab\sqrt{7b}$ **45.** $-10\sqrt{3xy}$
47. $-3x^2\sqrt[3]{2}$ **49.** $2x^4y\sqrt[3]{2}$ **51.** $2x^3y\sqrt[4]{2}$ **53.** $\dfrac{z}{4x}$ **55.** $\dfrac{\sqrt[4]{5x}}{2z}$ **57.** $10\sqrt{2x}$ **59.** $\sqrt[5]{7a^2}$ **61.** $4\sqrt{3}$ **63.** $-\sqrt{2}$ **65.** $2\sqrt{2}$
67. $9\sqrt{6}$ **69.** $3\sqrt[3]{3}$ **71.** $-\sqrt[3]{4}$ **73.** -10 **75.** $-17\sqrt[4]{2}$ **77.** $16\sqrt[4]{2}$ **79.** $-4\sqrt{2}$ **81.** $3\sqrt{2}+\sqrt{3}$ **83.** $-11\sqrt[3]{2}$ **85.** $y\sqrt{z}$
87. $13y\sqrt{x}$ **89.** $12\sqrt[3]{a}$ **91.** $-7y^2\sqrt{y}$ **93.** $4x\sqrt[5]{xy^2}$ **95.** $2x+2$

Review Exercises (page 425)

1. $\dfrac{-15x^5}{y}$ **2.** $-8b^2+\dfrac{4a^4}{b}-6a^5$ **3.** $9t^2+12t+4$ **4.** $25r^2-5rs-6s^2$ **5.** $3p+4+\dfrac{-5}{2p-5}$ **6.** $2m^2-mn+n^2$

Orals (page 431)

1. 3   **2.** 2   **3.** $3\sqrt{3}$ **4.** $a^2|b|$ **5.** $6+3\sqrt{2}$ **6.** 1   **7.** $\dfrac{\sqrt{2}}{2}$ **8.** $\dfrac{\sqrt{3}+1}{2}$

Exercise 7.4 (page 431)

1. 4   **3.** $5\sqrt{2}$ **5.** $6\sqrt{2}$ **7.** 5   **9.** 18   **11.** $2\sqrt[3]{3}$ **13.** ab^2 **15.** $5a\sqrt{b}$ **17.** $r\sqrt[3]{10s}$ **19.** $2a^2b^2\sqrt[3]{2}$ **21.** $x^2(x+3)$
23. $3x(y+z)\sqrt[3]{4}$ **25.** $12\sqrt{5}-15$ **27.** $12\sqrt{6}+6\sqrt{14}$ **29.** $-8x\sqrt{10}+6\sqrt{15x}$ **31.** $-1-2\sqrt{2}$ **33.** $8x-14\sqrt{x}-15$
35. $5z+2\sqrt{15z}+3$ **37.** $3x-2y$ **39.** $6a+5\sqrt{3ab}-3b$ **41.** $18r-12\sqrt{2r}+4$ **43.** $-6x-12\sqrt{x}-6$ **45.** $f/4$ **47.** $\dfrac{\sqrt{7}}{7}$
49. $\dfrac{\sqrt{6}}{3}$ **51.** $\dfrac{\sqrt{10}}{4}$ **53.** 2   **55.** $\dfrac{\sqrt[3]{4}}{2}$ **57.** $\sqrt[3]{3}$ **59.** $\dfrac{\sqrt[3]{6}}{3}$ **61.** $2\sqrt{2x}$ **63.** $\dfrac{\sqrt{5y}}{y}$ **65.** $\dfrac{\sqrt[3]{2ab^2}}{b}$ **67.** $\dfrac{\sqrt[4]{4}}{2}$ **69.** $\dfrac{\sqrt[5]{2}}{2}$
71. $\sqrt{2}+1$ **73.** $\dfrac{3\sqrt{2}-\sqrt{10}}{4}$ **75.** $2+\sqrt{3}$ **77.** $\dfrac{9-2\sqrt{14}}{5}$ **79.** $\dfrac{2(\sqrt{x}-1)}{x-1}$ **81.** $\dfrac{x(\sqrt{x}+4)}{x-16}$ **83.** $\sqrt{2z}+1$
85. $\dfrac{x-2\sqrt{xy}+y}{x-y}$ **87.** $\dfrac{1}{\sqrt{3}-1}$ **89.** $\dfrac{x-9}{x(\sqrt{x}-3)}$ **91.** $\dfrac{x-y}{\sqrt{x}(\sqrt{x}-\sqrt{y})}$

Review Exercises (page 433)

1. 1   **2.** 4   **3.** $\frac{1}{3}$ **4.** $\frac{1}{3}$

Orals (page 438)

1. 7   **2.** 3   **3.** 0   **4.** 9   **5.** 17   **6.** 31

Exercise 7.5 (page 438)

1. 2   **3.** 4   **5.** 0   **7.** 4   **9.** 8   **11.** $\frac{5}{2},\frac{1}{2}$ **13.** 1   **15.** 16   **17.** 14, $\cancel{6}$ **19.** 4, 3   **21.** 2, $\cancel{7}$ **23.** 9, -25 **25.** 2, -1 **27.** -1, $\cancel{1}$
29. $\cancel{1}$, no solutions   **31.** 0, $\cancel{4}$ **33.** $-\cancel{3}$, no solutions   **35.** 0   **37.** 1, 9   **39.** 4, $\cancel{0}$ **41.** 2, $\cancel{142}$ **43.** 2   **45.** $\cancel{6}$, no solutions
47. 0, $-\frac{12}{\cancel{11}}$ **49.** 1   **51.** 4, $-\cancel{9}$ **53.** 2, $\frac{24}{\cancel{2}}$ **55.** 2010 ft   **57.** about 29 mph   **59.** $5

Review Exercises (page 440)

1. 2   **2.** 41   **3.** 6   **4.** $\frac{3}{4}$

Orals (page 445)

1. 5   **2.** 10   **3.** 13   **4.** 5   **5.** 10   **6.** 13   **7.** 4   **8.** 3   **9.** 5

Exercise 7.6 (page 446)

1. 10 ft **3.** 80 m **5.** 13 ft **7.** about 127 ft **9.** about 135 ft **11.** $h = 2.83$, $x = 2.00$ **13.** $x = 8.66$, $h = 10.00$
15. $x = 4.69$, $y = 8.11$ **17.** $x = 12.11$, $y = 12.11$ **19.** $7\sqrt{2}$ cm **21.** 5 **23.** 5 **25.** 13 **27.** 10 **29.** $2\sqrt{26}$ **33.** $x = 7$
35. $(7, 0)$ and $(3, 0)$ **37.** not quite **39.** yes **41.** 173 yd **43.** 0.05 ft **45.** 8 cm^3

Review Exercises (page 449)

1. $12x^2 - 14x - 10$ **2.** $6y^2 - y - 15$ **3.** $15t^2 + 2ts - 8s^2$ **4.** $8r^3 + 6r^2 - 25r + 12$

Problems and Projects (page 449)

1. about 15.5 in., 29 **2.** about 27.8 ft^2 **3.** about \$60,131 **4.** \$429, \$88.79 **5.** 8 **6.** $\sqrt{2}$ m by $\sqrt{2}$ m, about \$771.33

Chapter 7 Review Exercises (page 453)

1. 7 **2.** -11 **3.** -6 **4.** 15 **5.** -3 **6.** -6 **7.** 5 **8.** -2 **9.** $5|x|$ **10.** $|x + 2|$ **11.** $3a^2b$ **12.** $4x^2|y|$
13. 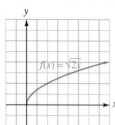 **14.** 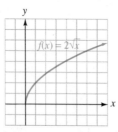 **15.** 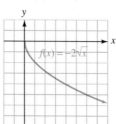 **16.**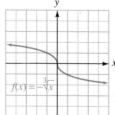

17. 12 **18.** about 5.7 **19.** 5 **20.** -6 **21.** 27 **22.** 64 **23.** -2 **24.** -4 **25.** $\dfrac{1}{4}$ **26.** $\dfrac{1}{2}$ **27.** $-16,807$ **28.** $\dfrac{1}{3125}$ **29.** 8

30. $\dfrac{27}{8}$ **31.** $3xy^{1/3}$ **32.** $3xy^{1/2}$ **33.** $125x^{9/2}y^6$ **34.** $\dfrac{1}{4u^{4/3}v^2}$ **35.** $5^{3/4}$ **36.** $a^{5/7}$ **37.** $u - 1$ **38.** $v + v^2$

39. $x + 2x^{1/2}y^{1/2} + y$ **40.** $a^{4/3} - b^{4/3}$ **41.** $\sqrt[5]{5}$ **42.** $\sqrt{x}$ **43.** $\sqrt[3]{3ab^2}$ **44.** $\dfrac{\sqrt{5ab}}{5b}$ **45.** $4\sqrt{15}$ **46.** $3\sqrt[3]{2}$ **47.** $2\sqrt[4]{2}$

48. $2\sqrt[5]{3}$ **49.** $2|x|\sqrt{2y}$ **50.** $3x^2|y|\sqrt{2}$ **51.** $2xy\sqrt[3]{2x^2y}$ **52.** $3x^2y\sqrt[3]{2x}$ **53.** $4|x|$ **54.** $2x$ **55.** $\dfrac{\sqrt[3]{2a^2b}}{3x}$ **56.** $\dfrac{\sqrt{17xy}}{8a^2}$

57. $3\sqrt{2}$ **58.** $\sqrt{5}$ **59.** 0 **60.** $8\sqrt[4]{2}$ **61.** $29x\sqrt{2}$ **62.** $32a\sqrt{3a}$ **63.** $13\sqrt[3]{2}$ **64.** $-4x\sqrt[4]{2x}$ **65.** $6\sqrt{10}$ **66.** 72 **67.** $3x$
68. 3 **69.** $-2x$ **70.** $-20x^3y^3\sqrt{xy}$ **71.** $4 - 3\sqrt{2}$ **72.** $2 + 3\sqrt{2}$ **73.** $\sqrt{10} - \sqrt{5}$ **74.** $3 + \sqrt{6}$ **75.** 1 **76.** $5 + 2\sqrt{6}$

77. $x - y$ **78.** $6u - 12 + \sqrt{u}$ **79.** $\dfrac{\sqrt{3}}{3}$ **80.** $\dfrac{\sqrt{15}}{5}$ **81.** $\dfrac{\sqrt{xy}}{y}$ **82.** $\dfrac{\sqrt[3]{u^2}}{u^2v^2}$ **83.** $2(\sqrt{2} + 1)$ **84.** $\dfrac{\sqrt{6} + \sqrt{2}}{2}$ **85.** $2(\sqrt{x} - 4)$

86. $\dfrac{a + 2\sqrt{a} + 1}{a - 1}$ **87.** $\dfrac{3}{5\sqrt{3}}$ **88.** $\dfrac{1}{\sqrt[3]{3}}$ **89.** $\dfrac{9 - x}{2(3 + \sqrt{x})}$ **90.** $\dfrac{a - b}{a + \sqrt{ab}}$ **91.** 22 **92.** 16, 9 **93.** 3, 9 **94.** $\dfrac{9}{16}$ **95.** 2
96. 0, -2 **97.** 7.07 in. **98.** 8.66 cm **99.** 13 **100.** 10 **101.** 3 mi **102.** 8.2 ft **103.** 88 yd **104.** 16,000 yd, or about 9 mi

Chapter 7 Test (page 456)

1. $4\sqrt{3}$ **2.** $5xy^2\sqrt{10xy}$ **3.** $2x^5y\sqrt[3]{3}$ **4.** $\frac{1}{4a}$ **5.** $|x|$ **6.** $2|y|\sqrt{2x}$ **7.** $3x$ **8.** $3x^2y^4\sqrt{2y}$ **9.** 10 **10.** about 2.5 **11.** 2
12. 9 **13.** $\frac{1}{216}$ **14.** $\frac{9}{4}$ **15.** $2^{4/3}$ **16.** $8xy$ **17.** $-\sqrt{3}$ **18.** $14\sqrt[3]{5}$ **19.** $2y^2\sqrt{3y}$ **20.** $6z\sqrt[4]{3z}$ **21.** $-6x\sqrt{y} - 2xy^2$
22. $3 - 7\sqrt{6}$ **23.** $\dfrac{\sqrt{5}}{5}$ **24.** $2\sqrt[3]{3}$ **25.** $\sqrt{2}(\sqrt{5} - 3)$ **26.** $\sqrt{3t} + 1$ **27.** $\dfrac{3}{\sqrt{21}}$ **28.** $\dfrac{a - b}{a - 2\sqrt{ab} + b}$ **29.** 10
30. $\cancel{4}$, no solutions **31.** 9.24 cm **32.** 8.67 cm **33.** 10 **34.** 25 **35.** 28 in. **36.** 1.25 m

Orals (page 467)

1. ± 7 **2.** $\pm\sqrt{10}$ **3.** 4 **4.** 9 **5.** $\frac{9}{4}$ **6.** $\frac{25}{4}$ **7.** 3, -4, 7 **8.** -2, 1, -5

Exercise 8.1   (page 467)

1. $0, -2$ **3.** $5, -5$ **5.** $-2, -4$ **7.** $6, 1$ **9.** $2, \frac{1}{2}$ **11.** $\frac{2}{3}, -\frac{5}{2}$ **13.** $6, -6$ **15.** $\sqrt{5}, -\sqrt{5}$ **17.** $\pm\frac{4\sqrt{3}}{3}$ **19.** $0, -2$

21. $4, 10$ **23.** $-5 \pm \sqrt{3}$ **25.** $2, -4$ **27.** $2, 4$ **29.** $-1, -4$ **31.** $1, -\frac{1}{2}$ **33.** $-\frac{1}{3}, -\frac{3}{2}$ **35.** $\frac{3}{4}, -\frac{3}{2}$ **37.** $\frac{-7 \pm \sqrt{29}}{10}$

39. $-1, -2$ **41.** $-6, -6$ **43.** $\frac{-5 \pm \sqrt{5}}{10}$ **45.** $-\frac{3}{2}, -\frac{1}{2}$ **47.** $\frac{1}{4}, -\frac{3}{4}$ **49.** $\frac{-5 \pm \sqrt{17}}{2}$ **51.** $8.98, -3.98$ **53.** $16, 18$

55. $6, 7$ **57.** 8 ft by 12 ft   **59.** 4 units   **61.** $\frac{4}{3}$ cm   **63.** 30 mph   **65.** \$4.80 or \$5.20   **67.** 4000   **69.** 2.26 in.
71. $x^2 - 8x + 15 = 0$ **73.** $x^3 - x^2 - 14x + 24 = 0$ **75.** about 6.13×10^{-3} M

Review Exercises   (page 469)

1. 1   **2.** 5   **3.** $t \leq 4$ **4.** $y \leq \frac{9}{10}$ **5.** $B = \frac{-Ax + C}{y}$ **6.** $L = \frac{Rd^2}{k}$

Orals   (page 478)

1. down   **2.** up   **3.** up   **4.** down   **5.** $(3, -1)$ **6.** $(-2, 2)$

Exercise 8.2   (page 478)

1. 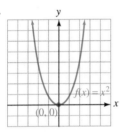   **3.** 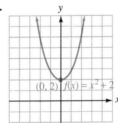   **5.**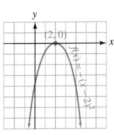

7. 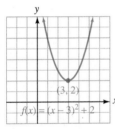   **9.** 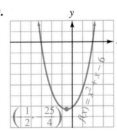   **11.**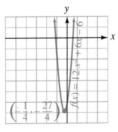

13. $(1, 2), x = 1$ **15.** $(-3, -4), x = -3$ **17.** $(0, 0), x = 0$ **19.** $(1, -2), x = 1$ **21.** $(2, 21), x = 2$ **23.** $\left(\frac{5}{12}, \frac{143}{24}\right), x = \frac{5}{12}$
25. $(5, 2)$ **27.** $(0.25, 0.88)$ **29.** $(0.50, 7.25)$ **31.** $\{2, -3\}$ **33.** $\{-1.85, 3.25\}$ **35.** 36 ft, 3 sec   **37.** 50 ft by 50 ft, 2500 ft^2
39. 5000   **41.** 3276, \$14,742   **43.** \$35   **45.** 0.25 and 0.75

Review Exercises   (page 480)
1. 10   **2.** 5   **3.** $3\frac{3}{5}$ hr   **4.** \$15,000

Orals   (page 488)

1. $-i$ **2.** -1 **3.** 1   **4.** i **5.** $7i$ **6.** $8i$ **7.** $10i$ **8.** $9i$ **9.** 5   **10.** 13

Exercise 8.3   (page 488)

1. $\pm 3i$ **3.** $\pm\frac{4\sqrt{3}}{3}i$ **5.** $-1 \pm i$ **7.** $-\frac{1}{4} \pm \frac{\sqrt{7}}{4}i$ **9.** $\frac{2}{3} \pm \frac{\sqrt{2}}{3}i$ **11.** $\frac{1}{3} \pm \frac{2\sqrt{2}}{3}i$ **13.** i **15.** $-i$ **17.** 1   **19.** i **21.** yes
23. no   **25.** no   **27.** $8 - 2i$ **29.** $3 - 5i$ **31.** $15 + 7i$ **33.** $6 - 8i$ **35.** $2 + 9i$ **37.** $-15 + 2\sqrt{3}i$ **39.** $3 + 6i$ **41.** $-25 - 25i$

43. $7 + i$ **45.** $14 - 8i$ **47.** $8 + \sqrt{2}i$ **49.** $-20 - 30i$ **51.** $3 + 4i$ **53.** $-5 + 12i$ **55.** $7 + 17i$ **57.** $5 + 5i$ **59.** $16 + 2i$
61. $0 - i$ **63.** $0 + \frac{4}{5}i$ **65.** $\frac{1}{8} + 0i$ **67.** $0 + \frac{3}{5}i$ **69.** $2 + i$ **71.** $\frac{1}{2} + \frac{5}{2}i$ **73.** $-\frac{42}{25} - \frac{6}{25}i$ **75.** $\frac{1}{4} + \frac{3}{4}i$ **77.** $\frac{5}{13} - \frac{12}{13}i$ **79.** $\frac{11}{10} + \frac{3}{10}i$
81. $\dfrac{1}{4} - \dfrac{\sqrt{15}}{4}i$ **83.** $-\frac{5}{169} + \frac{12}{169}i$ **85.** $\frac{3}{5} + \frac{4}{5}i$ **87.** $-\frac{6}{13} - \frac{9}{13}i$ **89.** 10 **91.** 13 **93.** $\sqrt{74}$ **95.** 1

Review Exercises   (page 489)

1. -1 **2.** $\dfrac{4x^2 + 4x + 16}{(x + 2)(x - 2)}$ **3.** 20 mph   **4.** about 11.4 mph faster

Orals   (page 494)

1. -3 **2.** -7 **3.** real and irrational   **4.** complex conjugates   **5.** no   **6.** yes

Exercise 8.4   (page 494)

1. rational, equal   **3.** complex conjugates   **5.** irrational, unequal   **7.** rational, unequal   **9.** $6, -6$ **11.** $12, -12$ **13.** 5
15. $12, -3$ **17.** yes   **19.** $k < -\frac{4}{3}$ **21.** $1, -1, 4, -4$ **23.** $1, -1, \sqrt{2}, -\sqrt{2}$ **25.** $1, -1, \sqrt{5}, -\sqrt{5}$ **27.** $1, -1, 2, -2$
29. 1 **31.** no solution   **33.** $-8, -27$ **35.** $-1, 27$ **37.** $-1, -4$ **39.** $4, -5$ **41.** $0, 2$ **43.** $-1, -\dfrac{27}{13}$ **45.** $1, 1, -1, -1$
47. $1 \pm i$ **49.** $x = \pm\sqrt{r^2 - y^2}$ **51.** $d = \pm\sqrt{\dfrac{k}{l}} = \pm\dfrac{\sqrt{kl}}{l}$ **53.** $y = \dfrac{-3x \pm \sqrt{9x^2 - 28x}}{2x}$ **55.** $\mu^2 = \dfrac{\Sigma x^2}{N} - \sigma^2$ **57.** $\dfrac{2}{3}, -\dfrac{1}{4}$
59. $\dfrac{-5 \pm \sqrt{17}}{4}$ **61.** $\dfrac{1 \pm i\sqrt{11}}{3}$ **63.** $-1 \pm 2i$

Review Exercises   (page 496)

1. -2 **2.** 6 **3.** $\frac{9}{5}$ **4.** $9x - 5y = 2$

Orals   (page 500)

1. $x = 2$ **2.** $x > 2$ **3.** $x < 2$ **4.** $x = -3$ **5.** $x > -3$ **6.** $x < -3$ **7.** $1 < 2x$ **8.** $1 > 2x$

Exercise 8.5   (page 500)

1. $(1, 4)$ **3.** $(-\infty, 3) \cup (5, \infty)$ **5.** $[-4, 3]$ **7.** $(-\infty, -5] \cup [3, \infty)$ **9.** no solution

11. $(-\infty, -3] \cup [3, \infty)$ **13.** $(-5, 5)$ **15.** $(-\infty, 0) \cup (1/2, \infty)$ **17.** $(0, 2]$ **19.** $(-\infty, -5/3) \cup (0, \infty)$

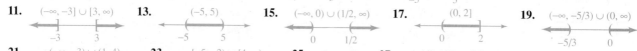

21. $(-\infty, -3) \cup (1, 4)$ **23.** $[-5, -2) \cup [4, \infty)$ **25.** $(-\infty, -4)$ **27.** $(-1/2, 1/3) \cup (1/2, \infty)$

29. $(0, 2) \cup (8, \infty)$ **31.** $(-\infty, -2) \cup (2, 18]$ **33.** $[-34/5, -4) \cup (3, \infty)$ **35.** $(-4, -2] \cup (-1, 2]$

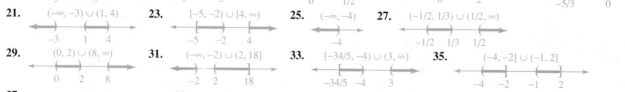

37. $(-\infty, -16) \cup (-4, -1) \cup (4, \infty)$ **39.** $(-\infty, -2) \cup (-2, \infty)$ **41.** $(-1, 3)$ **43.** $(-\infty, -3) \cup (2, \infty)$

45.   **47.**   **49.**

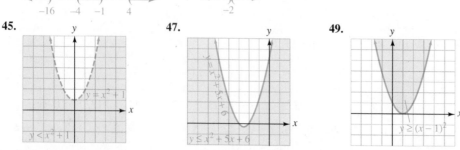

51.

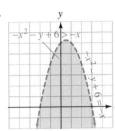

53.

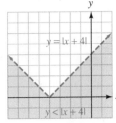

55.

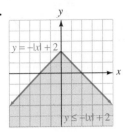

Review Exercises   (page 502)

1. $x = ky$ **2.** $y = \dfrac{k}{t}$ **3.** $t = kxy$ **4.** $d = \dfrac{kt}{u^2}$ **5.** 3   **6.** 2

Orals   (page 508)

1. $5x$ **2.** x **3.** $8x^2$ **4.** $\frac{3}{2}$ **5.** 2   **6.** $12x^2$ **7.** $2x + 1$ **8.** $2x + 2$

Exercise 8.6   (page 508)

1. $7x$, $(-\infty, \infty)$ **3.** $12x^2$, $(-\infty, \infty)$ **5.** x, $(-\infty, \infty)$ **7.** $\frac{4}{3}$, $(-\infty, 0) \cup (0, \infty)$ **9.** $3x - 2$, $(-\infty, \infty)$ **11.** $2x^2 - 5x - 3$, $(-\infty, \infty)$
13. $-x - 4$, $(-\infty, \infty)$ **15.** $\frac{x-3}{2x+1}$, $\left(-\infty, -\frac{1}{2}\right) \cup \left(-\frac{1}{2}, \infty\right)$ **17.** $-2x^2 + 3x - 3$, $(-\infty, \infty)$ **19.** $(3x - 2)/(2x^2 + 1)$, $(-\infty, \infty)$
21. 3, $(-\infty, \infty)$ **23.** $(x^2 - 4)/(x^2 - 1)$, $(-\infty, -1) \cup (-1, 1) \cup (1, \infty)$ **25.** 7   **27.** 24   **29.** -1 **31.** $-\frac{1}{2}$ **33.** $2x^2 - 1$
35. $16x^2 + 8x$ **37.** 58   **39.** 110   **41.** 2   **43.** $9x^2 - 9x + 2$ **49.** $2x + h$ **51.** $3x^2 + 3xh + h^2$ **53.** $C(t) = \frac{5}{9}(2668 - 200t)$

Review Exercises   (page 509)

1. $-\dfrac{3x + 7}{x + 2}$ **2.** $-\dfrac{2x^2(x + 7)}{x + 1}$ **3.** $\dfrac{x - 4}{3x^2 - x - 12}$ **4.** $\dfrac{x - 2}{2}$

Orals   (page 516)

1. $\{(2, 1), (3, 2), (10, 5)\}$ **2.** $\{(1, 1), (8, 2), (64, 4)\}$ **3.** $f^{-1}(x) = 2x$ **4.** $f^{-1}(x) = \frac{1}{2}x$ **5.** no   **6.** yes

Exercise 8.7   (page 517)

1. $\{(2, 3), (1, 2), (0, 1)\}$, yes   **3.** $\{(2, 1), (3, 2), (3, 1), (5, 1)\}$, no   **5.** $\{(1, 1), (4, 2), (9, 3), (16, 4)\}$, yes   **7.** $y = f^{-1}(x) = \frac{x-1}{3}$, yes
9. $y = f^{-1}(x) = 5x - 4$, yes   **11.** $y = f^{-1}(x) = 5x + 4$, yes   **13.** $y = f^{-1}(x) = \frac{5x + 20}{4}$, yes
15.

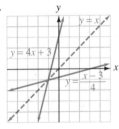

17.

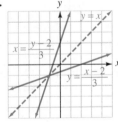

19.

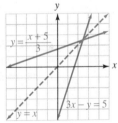

21.

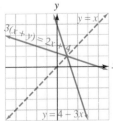

23.

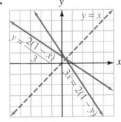

25. $y = \pm\sqrt{x - 4}$; no   **27.** $y = x^2 - 4$; yes   **29.** $y = \sqrt[3]{x}$; yes   **31.** $y = x^2$; yes   **33.** $y = \sqrt{x}$; yes   **35.** $y = f^{-1}(x) = \sqrt[3]{\dfrac{x + 3}{2}}$

37.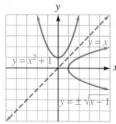

39.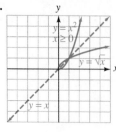

41.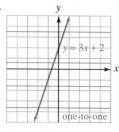

43.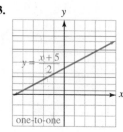

45.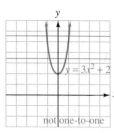

47.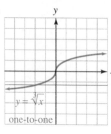

49.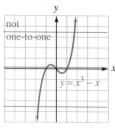

Review Exercises (page 519)

1. $3 - 8i$ **2.** $6 + 2i$ **3.** $18 - i$ **4.** $-\dfrac{2}{5} + \dfrac{9}{5}i$ **5.** 10 **6.** $\dfrac{\sqrt{2}}{2}$

Problems and Projects (page 519)

1. 42, \$1687.25   **2.** $R(x) = -0.01x^2 + 54x$, \$27   **3.** $P(t) = 21t - 140$ **4.** $P(t) = 351t + 619.5$

Chapter 8 Review Exercises (page 522)

1. $\frac{2}{3}, -\frac{3}{4}$ **2.** $-\frac{1}{3}, -\frac{5}{2}$ **3.** $\frac{2}{3}, -\frac{4}{5}$ **4.** $4, -8$ **5.** $-4, -2$ **6.** $\frac{7}{2}, 1$ **7.** $9, -1$ **8.** $0, 10$ **9.** $\frac{1}{2}, -7$ **10.** $7, -\frac{1}{3}$ **11.** 4 cm by 6 cm
12. 2 ft by 3 ft   **13.** 7 sec   **14.** 196 ft
15.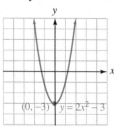

16.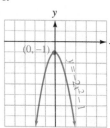

17.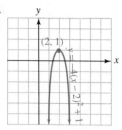

18.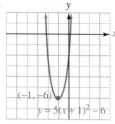

19. $12 - 8i$ **20.** $2 - 68i$ **21.** $-96 + 3i$ **22.** $-2 - 2\sqrt{2}i$ **23.** $22 + 29i$ **24.** $-16 + 7i$ **25.** $-12 + 28\sqrt{3}i$
26. $118 + 10\sqrt{2}i$ **27.** $0 - \frac{3}{4}i$ **28.** $0 - \frac{2}{5}i$ **29.** $\frac{12}{5} - \frac{6}{5}i$ **30.** $\frac{21}{10} + \frac{7}{10}i$ **31.** $\frac{15}{17} + \frac{8}{17}i$ **32.** $\frac{4}{5} - \frac{3}{5}i$ **33.** $\frac{15}{29} - \frac{6}{29}i$ **34.** $\frac{1}{3} + \frac{1}{3}i$
35. $15 + 0i$ **36.** $26 + 0i$ **37.** irrational, unequal   **38.** complex conjugates   **39.** 12, 152   **40.** $k \geq -\frac{7}{3}$ **41.** 1, 144   **42.** 8, -27
43. 1   **44.** 1, $-\frac{8}{5}$ **45.** $\frac{14}{3}$ **46.** 1   **47.** $(-\infty, -7) \cup (5, \infty)$ **48.** $(-9, 2)$ **49.** $(-\infty, 0) \cup [3/5, \infty)$

47.

48. $(-9, 2)$

49. $(-\infty, 0) \cup [3/5, \infty)$

50. $(-7/2, 1) \cup (4, \infty)$

55.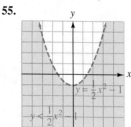

56.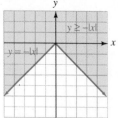

57. $(f + g)(x) = 3x + 1$ **58.** $(f - g)(x) = x - 1$ **59.** $(f \cdot g)(x) = 2x^2 + 2x$ **60.** $\left(f/g\right)(x) = \frac{2x}{x+1}$ **61.** 6   **62.** -1
63. $2(x + 1)$ **64.** $2x + 1$ **65.** $y = f^{-1}(x) = \frac{x+3}{6}$ **66.** $y = f^{-1}(x) = \frac{x-5}{4}$ **67.** $y = f^{-1}(x) = \sqrt{\frac{x+1}{2}}$ **68.** $x = |y|$

69. yes
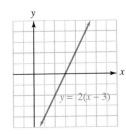

70. no
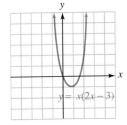

71. no
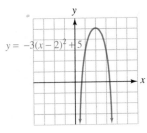

72. no
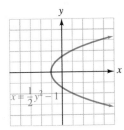

Chapter 8 Test   (page 525)

1. 3, −6   **2.** $-\frac{3}{2}, -\frac{5}{3}$ **3.** 144   **4.** 625   **5.** $-2 \pm \sqrt{3}$ **6.** $\dfrac{5 \pm \sqrt{37}}{2}$ **7.** $-1 + 11i$ **8.** $4 - 7i$ **9.** $8 + 6i$ **10.** $-10 - 11i$

11. $0 - \dfrac{\sqrt{2}}{2}i$ **12.** $\frac{1}{2} + \frac{1}{2}i$ **13.** nonreal   **14.** 2   **15.** 10 in.   **16.** $1, \frac{1}{4}$

17.
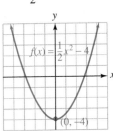

18.
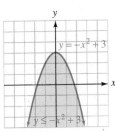

19. $(-\infty, -2) \cup (4, \infty)$

20. $(-3, 2]$

21. $(g + f)(x) = 5x - 1$ **22.** $(f - g)(x) = 3x + 1$ **23.** $(g \cdot f)(x) = 4x^2 - 4x$ **24.** $(g/f)(x) = \frac{x-1}{4x}$ **25.** 3   **26.** −4   **27.** −8
28. −9   **29.** $4(x - 1)$ **30.** $4x - 1$ **31.** $y = \frac{12 - 2x}{3}$ **32.** $y = -\sqrt{\frac{x-4}{3}}$

Cumulative Review Exercises (Chapters 1–8)   (page 526)

1. $D = (-\infty, \infty)$; $R = [-3, \infty)$ **2.** $D = (-\infty, \infty)$; $R = (-\infty, 0]$ **3.** $y = 3x + 2$ **4.** $y = -\frac{2}{3}x - 2$ **5.** $-4a^2 + 12a - 7$
6. $6x^2 - 5x - 6$ **7.** $(x^2 + 4y^2)(x + 2y)(x - 2y)$ **8.** $(3x + 2)(5x - 4)$ **9.** $6, -1$ **10.** $0, \frac{2}{3}, -\frac{1}{2}$ **11.** $\dfrac{x + 1}{x - 1}$ **12.** $\dfrac{3x^2 + 7x}{(x + 4)(x - 1)}$
13. $4t\sqrt{3t}$ **14.** $-3x$ **15.** $4x$ **16.** 16   **17.** y^2 **18.** $x^{17/12}$ **19.**
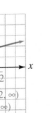
20.
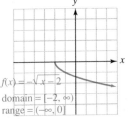

21. $x^{4/3} - x^{2/3}$ **22.** $\frac{1}{x} + 2 + x$ **23.** $7\sqrt{2}$ **24.** $-12\sqrt[4]{2} + 10\sqrt[4]{3}$ **25.** $-18\sqrt{6}$ **26.** $\dfrac{5\sqrt[3]{x^2}}{x}$ **27.** $\dfrac{x + 3\sqrt{x} + 2}{x - 1}$ **28.** $\sqrt{xy}$

29. 2, 7   **30.** $\frac{1}{4}$ **31.** $3\sqrt{2}$ in.   **32.** $2\sqrt{3}$ in.   **33.** 10   **34.** 9   **35.** $1, -\dfrac{3}{2}$ **36.** $\dfrac{-2 \pm \sqrt{7}}{3}$

37.
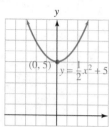

38.
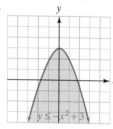

39. $7 + 2i$ **40.** $-5 - 7i$ **41.** 13   **42.** $12 - 6i$ **43.** $-12 - 10i$

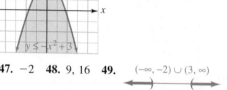

44. $\frac{3}{2} + \frac{1}{2}i$ **45.** $\sqrt{13}$ **46.** $\sqrt{61}$ **47.** -2 **48.** 9, 16   **49.** $(-\infty, -2) \cup (3, \infty)$ **50.** $[-2, 3]$ **51.** 5   **52.** 27

53. $12x^2 - 12x + 5$ **54.** $6x^2 + 3$ **55.** $f^{-1}(x) = \frac{x-2}{3}$ **56.** $f^{-1}(x) = \sqrt[3]{x - 4}$

Orals (page 539)

1. 4   **2.** 25   **3.** 18   **4.** 3   **5.** $\frac{1}{4}$ **6.** $\frac{1}{25}$ **7.** $\frac{2}{9}$ **8.** $\frac{1}{27}$

Exercise 9.1 (page 539)

1. 2.6651   **3.** 36.5548

5.
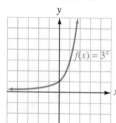

7.
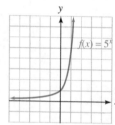

9.
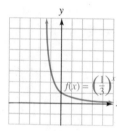

11.
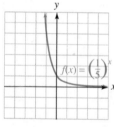

17.
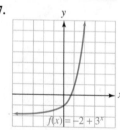

19.
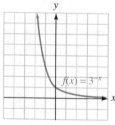

21. 5   **23.** no value of b **25.** $\frac{1}{2}$ **27.** no value of b **29.** \$22,080.40   **31.** \$32.03   **33.** \$2,273,996.13   **35.** 3645   **37.** 10.30 billion   **39.** 1.679046×10^8 **41.** $\frac{32}{243}A_0$ **43.** \$3094.15   **45.** \$1115.33

Review Exercises (page 542)

1. 40   **2.** 60°   **3.** 120°   **4.** 60°

Orals (page 547)

1. 3   **2.** 2   **3.** 5   **4.** 2   **5.** 2   **6.** 2   **7.** $\frac{1}{4}$ **8.** $\frac{1}{2}$ **9.** 2

Exercise 9.2 (page 548)

1. $3^4 = 81$ **3.** $\left(\frac{1}{2}\right)^3 = \frac{1}{8}$ **5.** $4^{-3} = \frac{1}{64}$ **7.** $\left(\frac{1}{2}\right)^3 = \frac{1}{8}$ **9.** $\log_8 64 = 2$ **11.** $\log_4 \frac{1}{16} = -2$ **13.** $\log_{1/2} 32 = -5$ **15.** $\log_x z = y$
17. 3   **19.** 3   **21.** 3   **23.** $\frac{1}{2}$ **25.** -3 **27.** 64   **29.** 7   **31.** 5   **33.** $\frac{1}{25}$ **35.** $\frac{1}{6}$ **37.** $-\frac{3}{2}$ **39.** $\frac{2}{3}$ **41.** 5   **43.** $\frac{3}{2}$ **45.** 4   **47.** 8
49. 5   **51.** 4   **53.** 3   **55.** 100   **57.** 0.5119   **59.** -2.3307 **61.** -0.0726 **63.** 3.6348   **65.** -0.2752 **67.** impossible
69. 25.25   **71.** 69.41   **73.** 0.00   **75.** 120.07

77. 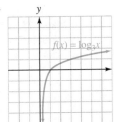   **79.** 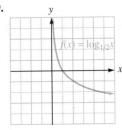   **81.** 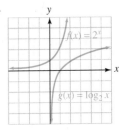   **83.**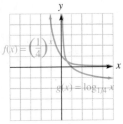

85. no value of b **87.** 3   **89.** no value of b **91.** $2\sqrt{2}$

Review Exercises (page 550)

1. 10   **2.** $\frac{3}{5}$ does not check   **3.** 0; $-\frac{5}{9}$ does not check   **4.** 4

Orals (page 555)

1. 2   **2.** 5   **3.** 7^7 **4.** $\frac{1}{4}$ **5.** 2   **6.** 2   **7.** -2 **8.** 3

Exercise 9.3 (page 555)

1. 0   **3.** 7   **5.** 10   **7.** 1   **15.** $\log_b x + \log_b y + \log_b z$ **17.** $\log_b 2 + \log_b x - \log_b y$ **19.** $3 \log_b x + 2 \log_b y$
21. $\frac{1}{2}(\log_b x + \log_b y)$ **23.** $\log_b x + \frac{1}{2} \log_b z$ **25.** $\frac{1}{3} \log_b x - \frac{1}{4} \log_b y - \frac{1}{4} \log_b z$ **27.** $\log_b \frac{x+1}{x}$ **29.** $\log_b x^2 y^{1/2}$

31. $\log_b \frac{z^{1/2}}{x^3 y^2}$ **33.** $\log_b \dfrac{\frac{x}{z}+x}{\frac{y}{z}+y} = \log_b \frac{x}{y}$ **35.** false   **37.** false   **39.** true   **41.** false   **43.** true   **45.** true   **47.** false   **49.** false

51. true   **53.** true   **55.** 1.4472   **57.** 0.3521   **59.** 1.1972   **61.** 2.4014   **63.** 2.0493   **65.** 0.4682   **67.** 1.7712   **69.** -1.0000
71. 1.8928   **73.** 2.3219

Review Exercises (page 557)

1. $-\frac{7}{6}$ **2.** $\sqrt{85}$ **3.** $\left(1, -\frac{1}{2}\right)$ **4.** $y = -\frac{7}{6}x + \frac{2}{3}$

Orals (page 560)

1. $pH = -\log [H^+]$ **2.** db gain $= 20 \log \frac{E_O}{E_I}$ **3.** $R = \log \frac{A}{P}$ **4.** $t = \frac{\ln 2}{r}$ **5.** $L = k \ln I$

Exercise 9.4 (page 561)

1. 4.77   **3.** from 2.5119×10^{-8} to 1.585×10^{-7} **5.** 0.71 V   **7.** 49.5 db   **9.** 4.4   **11.** 2500 μm   **13.** 5.8 yr   **15.** 9.2 yr
17. It will increase by $k \ln 2$.   **19.** The intensity must be cubed.   **21.** 3 yr old   **23.** 10.8 yr

Review Exercises (page 562)

1. $x^2 + 3x + 1$ **2.** $-x^2 + 3x - 5$ **3.** $3x^3 - 2x^2 + 9x - 6$ **4.** $\dfrac{x^2 + 3}{3x - 2}$ **5.** 67   **6.** $3x^2 + 7$

Orals (page 569)

1. $x = \frac{\log 5}{\log 3}$ **2.** $x = \frac{\log 3}{\log 5}$ **3.** $x = -\frac{\log 7}{\log 2}$ **4.** $x = -\frac{\log 1}{\log 6} = 0$ **5.** $x = 2$ **6.** $x = \frac{1}{2}$ **7.** $x = 10$ **8.** $x = 10$

Exercise 9.5   (page 569)

1. 1.1610   **3.** 1.2702   **5.** 1.7095   **7.** 0   **9.** ± 1.0878   **11.** 0, 1.0566   **13.** 3, −1   **15.** −2, −2   **17.** 0   **19.** 0.2789   **21.** 1.8
23. 8.8, 0.2   **25.** 2   **27.** 3   **29.** −7   **31.** 4   **33.** 10, −10   **35.** 50   **37.** 20   **39.** 10   **41.** 10   **43.** 1, 100   **45.** no solution
47. 6   **49.** 9   **51.** 4   **53.** 1, 7   **55.** 20   **57.** 8   **59.** 5.1 yr   **61.** 42.7 days   **63.** about 4200 yr   **65.** 5.6 yr   **67.** 5.4 yr
69. because ln 2 ≈ 0.7   **71.** 25.3 yr   **73.** 2.828 times larger   **75.** 13.3   **77.** $\frac{\ln 0.75}{3}$

Review Exercises   (page 572)

1. 0, 5   **2.** $\frac{5}{2}, -\frac{5}{2}$ **3.** $\frac{2}{3}, -4$ **4.** $\left(-3 \pm \sqrt{5}\right)/4$

Problems and Projects   (page 572)

1. $17,094.43   **2.** 20.9 months   **3.** 577.62 yr   **4.** 2.8 days

Chapter 9 Review Exercises   (page 575)

1. 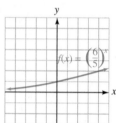   **2.** 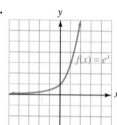   **3.** 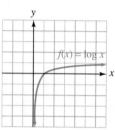   **4.**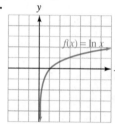

5. 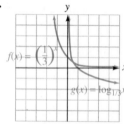   **6.** 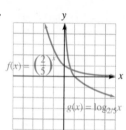   **7.** 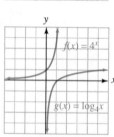   **8.**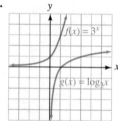

9. 2   **10.** $-\frac{1}{2}$ **11.** 0   **12.** −2   **13.** $\frac{1}{2}$ **14.** $\frac{1}{3}$ **15.** 4   **16.** 0   **17.** 7   **18.** 3   **19.** 4   **20.** 9   **21.** 8   **22.** $\frac{1}{9}$ **23.** 3   **24.** 2
25. 1   **26.** $\frac{1}{2}$ **27.** $\frac{1}{6}$ **28.** 2   **29.** −2   **30.** 0   **31.** 27   **32.** $\frac{1}{5}$ **33.** 32   **34.** 9   **35.** 27   **36.** −1   **37.** $\frac{1}{8}$ **38.** 2   **39.** 4   **40.** 2
41. 10   **42.** $\frac{1}{25}$ **43.** 5   **44.** 3   **45.** $2 \log_b x + 3 \log_b y - 4 \log_b z$ **46.** $\frac{1}{2} \log_b x - \frac{1}{2} \log_b y - \log_b z$ **47.** $\log_b \dfrac{x^3 z^7}{y^5}$
48. $\log_b \dfrac{y^3 \sqrt{x}}{z^7}$ **49.** 3.36   **50.** 1.56   **51.** 2.64   **52.** −6.72   **53.** $\dfrac{\log 7}{\log 3}$ **54.** 2   **55.** $\dfrac{\log 3}{\log 3 - \log 2}$ **56.** −3, −1   **57.** 25, 4
58. 4   **59.** −2, 4   **60.** 4, 3   **61.** 6   **62.** 31   **63.** $\dfrac{\ln 9}{\ln 2}$ **64.** no solution   **65.** $\dfrac{e}{e - 1}$ **66.** 1   **67.** about 3300 yr
68. about 7.94×10^{-4} gram-ions/liter   **69.** about 34.2 yr

Chapter 9 Test   (page 577)

1. 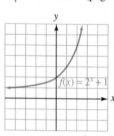   **2.** 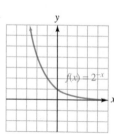   **3.** $\frac{3}{64}$ g   **4.** $1060.90   **5.**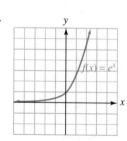

6. $4451.08   **7.** 2   **8.** 3   **9.** $\frac{1}{27}$ **10.** 10   **11.** 2   **12.** $\frac{27}{8}$

13.
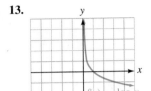

14.
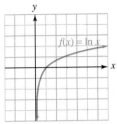

15. $2 \log a + \log b + 3 \log c$ **16.** $\frac{1}{2}(\ln a - 2 \ln b - \ln c)$ **17.** $\log \dfrac{b\sqrt{a+2}}{c^3}$ **18.** $\log \dfrac{\sqrt[3]{a}}{c\sqrt[3]{b^2}}$ **19.** 1.3801   **20.** 0.4259

21. $\dfrac{\log 3}{\log 7}$ **22.** $\dfrac{\log e}{\log \pi}$ **23.** true   **24.** false   **25.** false   **26.** false   **27.** 6.4   **28.** 46   **29.** $\dfrac{\log 3}{\log 5}$ **30.** $\dfrac{\log 3}{(\log 3) - 2}$ **31.** 1   **32.** 10

Orals   (page 588)

1. $(0, 0)$, 12   **2.** $(0, 0)$, 11   **3.** $(2, 0)$, 4   **4.** $(0, -1)$, 3   **5.** down   **6.** up   **7.** left   **8.** right

Exercise 10.1   (page 588)

1.
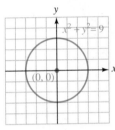

3.
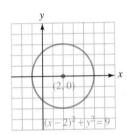

5.
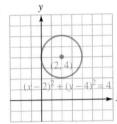

7.
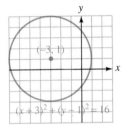

9.
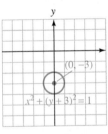

11.

13.

15. $x^2 + y^2 = 1$ **17.** $(x - 6)^2 + (y - 8)^2 = 25$
19. $(x + 2)^2 + (y - 6)^2 = 144$ **21.** $x^2 + y^2 = 8$

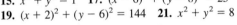

23.
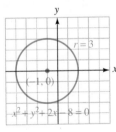

25.
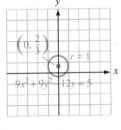

27.
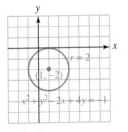

29.
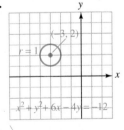

31.
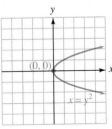

33.
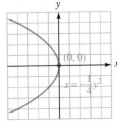

35.
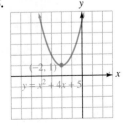

37.
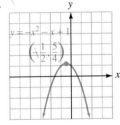

39.
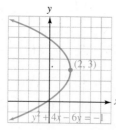

41.
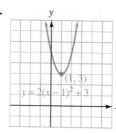

43.
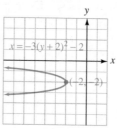

45.

47.

49. $(x-7)^2 + y^2 = 9$ **51.** no   **53.** 30 ft away   **55.** 2 AU

Review Exercises   (page 592)

1. $5, -\frac{7}{3}$ **2.** $\frac{64}{3}, -\frac{56}{3}$ **3.** $3, -\frac{1}{4}$ **4.** $\frac{4}{5}, \frac{8}{3}$

Orals   (page 602)

1. $(\pm 3, 0), (0, \pm 4)$ **2.** $(\pm 5, 0), (0, \pm 6)$ **3.** $(2, 0)$ **4.** $(0, -1)$ **5.** $(\pm 3, 0)$ **6.** $(0, \pm 5)$

Exercise 10.2   (page 602)

1.
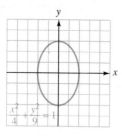

3.
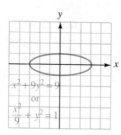

5.
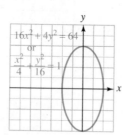

7.
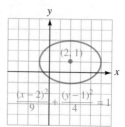

9.
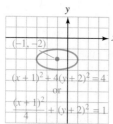

11.

13.

15.
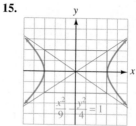

17.
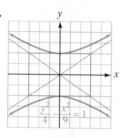

19.
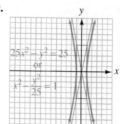

21.
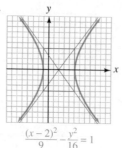

$$\frac{(x-2)^2}{9} - \frac{y^2}{16} = 1$$

23.
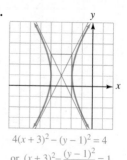

$$4(x+3)^2 - (y-1)^2 = 4$$
$$\text{or } (x+3)^2 - \frac{(y-1)^2}{4} = 1$$

25.

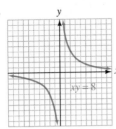

$xy = 8$

27.

29.

31. $\dfrac{x^2}{400} + \dfrac{y^2}{100} = 1$

33. 12π sq. units

35. 3 units **37.** $10\sqrt{3}$ mi

Review Exercises (page 606)

1. $12y^2 + \dfrac{9}{x^2}$ **2.** $\dfrac{4}{a^4} - \dfrac{1}{b^4}$ **3.** $\dfrac{y^2 + x^2}{y^2 - x^2}$ **4.** $\dfrac{y^3 - x^3}{2(y^3 + x^3)}$

Orals (page 609)

1. 0, 1, 2 **2.** 0, 1, 2 **3.** 0, 1, 2, 3, 4 **4.** 0, 1, 2, 3, 4

Exercise 10.3 (page 609)

1.

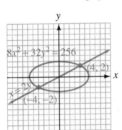

3.

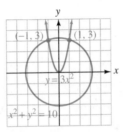

5.

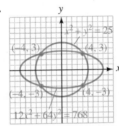

7.

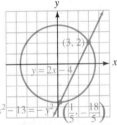

9. $(1, 0), (5, 0)$

11. $(3, 0), (0, 5)$ **13.** $(1, 1)$ **15.** $(1, 2), (2, 1)$ **17.** $(-2, 3), (2, 3)$ **19.** $\left(\sqrt{5}, 5\right), \left(-\sqrt{5}, 5\right)$

21. $(3, 2), (3, -2), (-3, 2), (-3, -2)$ **23.** $(2, 4), (2, -4), (-2, 4), (-2, -4)$ **25.** $\left(-\sqrt{15}, 5\right), \left(\sqrt{15}, 5\right), (-2, -6), (2, -6)$
27. $(0, -4), (-3, 5), (3, 5)$ **29.** $(-2, 3), (2, 3), (-2, -3), (2, -3)$ **31.** $(3, 3)$ **33.** $(6, 2), (-6, -2), \left(\sqrt{42}, 0\right), \left(-\sqrt{42}, 0\right)$
35. $\left(\frac{1}{2}, \frac{1}{3}\right), \left(\frac{1}{3}, \frac{1}{2}\right)$ **37.** 7 cm by 9 cm **39.** 14 and -5 **41.** either \$750 at 9% or \$900 at 7.5% **43.** 68 mph, 4.5 hr

Review Exercises (page 611)

1. $-11x\sqrt{2}$ **2.** $-21a\sqrt{7a}$ **3.** $\dfrac{t}{2}$ **4.** $\dfrac{13\sqrt[3]{2x}}{20}$

Orals (page 615)

1. increasing **2.** decreasing **3.** constant **4.** increasing

Exercise 10.4 (page 615)

1. increasing on $(-\infty, 0)$, decreasing on $(0, \infty)$ **3.** decreasing on $(-\infty, 0)$, constant on $(0, 2)$, increasing on $(2, \infty)$
5. constant on $(-\infty, 0)$, increasing on $(0, \infty)$ **7.** decreasing on $(-\infty, 0)$, increasing on $(0, 2)$, decreasing on $(2, \infty)$

9.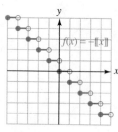

11.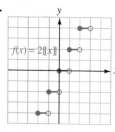

13.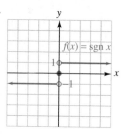

15.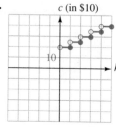

17. After 2 hours, network B is cheaper.

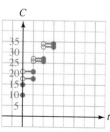

$30

Review Exercises (page 617)

1. 20 **2.** 15

Orals (page 623)

1. 1 **2.** 24 **3.** 1 **4.** 120 **5.** $m^2 + 2mn + n^2$ **6.** $m^2 - 2mn + n^2$ **7.** 2 **8.** 3 **9.** 6 **10.** 6 **11.** 5 **12.** 2

Exercise 10.5 (page 623)

1. 6 **3.** -120 **5.** 40,320 **7.** $\frac{1}{110}$ **9.** 2352 **11.** 10 **13.** 21 **15.** $\frac{1}{168}$ **17.** $x^3 + 3x^2y + 3xy^2 + y^3$
19. $x^4 + 4x^3y + 6x^2y^2 + 4xy^3 + y^4$ **21.** $8x^3 + 12x^2y + 6xy^2 + y^3$ **23.** $x^3 - 6x^2y + 12xy^2 - 8y^3$
25. $8x^3 + 36x^2y + 54xy^2 + 27y^3$ **27.** $\frac{x^3}{8} - \frac{x^2y}{4} + \frac{xy^2}{6} - \frac{y^3}{27}$ **29.** $81 + 216y + 216y^2 + 96y^3 + 16y^4$ **33.** $3a^2b$ **35.** $-4xy^3$
37. $15x^2y^4$ **39.** $28x^6y^2$ **41.** $90x^3$ **43.** $640x^3y^2$ **45.** $-12x^3y$ **47.** $-70,000x^4$ **49.** $810xy^4$ **51.** $-\frac{1}{6}x^3y$ **53.** $\frac{n!}{3!(n-3)!}a^{n-3}b^3$
55. $\frac{n!}{4!(n-4)!}a^{n-4}b^4$ **57.** $\frac{n!}{(r-1)!(n-r+1)!}a^{n-r+1}b^{r-1}$

Review Exercises (page 624)

1. 2 **2.** 7 **3.** 5 **4.** 3 **5.** $(2, 3)$ **6.** $(3, 2, 1)$ **7.** 8 **8.** -12

Orals (page 630)

1. 14 **2.** 1 **3.** 5 **4.** -6 **5.** 3 **6.** 5

Exercise 10.6 (page 630)

1. 3, 5, 7, 9, 11 **3.** $-5, -8, -11, -14, -17$ **5.** 5, 11, 17, 23, 29 **7.** $-4, -11, -18, -25, -32$ **9.** $-118, -111, -104, -97,$
-90 **11.** 34, 31, 28, 25, 22 **13.** 5, 12, 19, 26, 33 **15.** 355 **17.** -179 **19.** -23 **21.** 12 **23.** $\frac{17}{4}, \frac{13}{2}, \frac{35}{4}$ **25.** 12, 14, 16, 18
27. $\frac{29}{2}$ **29.** $\frac{5}{4}$ **31.** 1335 **33.** 459 **35.** 354 **37.** 255 **39.** 1275 **41.** 2500 **43.** 60, 110, 160, 210, 260, 310; $6060
45. 11,325 **47.** 368 ft **51.** 60 **53.** 31 **55.** 12 **61.** $R = 10$, $\sigma^2 = 11.6$, $\sigma = 3.4$

Review Exercises (page 633)

1. $18x^2 + 8x - 3$ **2.** $6p^3 + 11p^2q - 2pq^2 - 3q^3$ **3.** $\dfrac{6a^2 + 16}{(a + 2)(a - 2)}$ **4.** $4t^3 + 4t - 2$

Orals (page 638)

1. 27 **2.** $\frac{1}{27}$ **3.** 2.5 **4.** $\sqrt{3}$ **5.** 6 **6.** 1

Exercise 10.7   (page 638)

1. 3, 6, 12, 24, 48   **3.** $-5, -1, -\frac{1}{5}, -\frac{1}{25}, -\frac{1}{125}$ **5.** 2, 8, 32, 128, 512   **7.** $-3, -12, -48, -192, -768$ **9.** $-64, 32, -16, 8, -4$
11. $-64, -32, -16, -8, -4$ **13.** 2, 10, 50, 250, 1250   **15.** 3584   **17.** $\frac{1}{27}$ **19.** 3   **21.** 6, 18, 54   **23.** $-20, -100, -500,$
-2500 **25.** -16 **27.** $10\sqrt{2}$ **29.** No geometric mean exists.   **31.** 728   **33.** 122   **35.** -255 **37.** 381   **39.** $\frac{156}{25}$ **41.** $-\frac{21}{4}$
43. about 669   **45.** \$1469.74   **47.** \$140,853.75   **49.** $\left(\frac{1}{2}\right)^{11} \approx 0.0005$ **51.** \$4309.14

Review Exercises   (page 640)

1. $[-1, 6]$ **2.** $(-\infty, 3] \cup [4, \infty)$ **3.** $(-\infty, -3) \cup (4, \infty)$ **4.** $(-\infty, -5) \cup (-2, 4)$

Orals   (page 642)

1. 8   **2.** $\frac{1}{8}$ **3.** $\frac{1}{2}$ **4.** $\frac{1}{8}$ **5.** 27   **6.** 16

Exercise 10.8   (page 642)

1. 16   **3.** 81   **5.** 8   **7.** $-\frac{135}{4}$ **9.** no sum   **11.** $-\frac{81}{2}$ **13.** $\frac{1}{9}$ **15.** $-\frac{1}{3}$ **17.** $\frac{4}{33}$ **19.** $\frac{25}{33}$ **21.** 30 m   **23.** 5000   **27.** no;
$0.999999 = \frac{999,999}{1,000,000} < 1$

Review Exercises   (page 643)

1. yes   **2.** yes   **3.** no   **4.** no

Problems and Projects   (page 644)

1. Answers will vary. One answer is $\dfrac{x^2}{16} + \dfrac{y^2}{81} = 1$.   **2.** $x = 1, y = 2$ or $x = 1, y = -2$ **3.** $\frac{15}{32}$ cm   **4.** 120

Chapter 10 Review Exercises   (page 647)

1.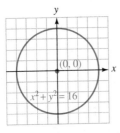

2.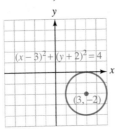

3.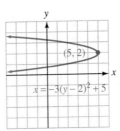

4.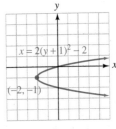

5.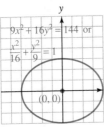

6.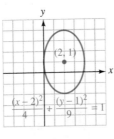

7.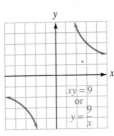

8.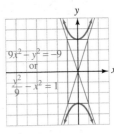

11. $(4, 2), (4, -2), (-4, 2), (-4, -2)$
12. $(2, 3), (2, -3), (-2, 3), (-2, -3)$

9.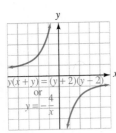

10.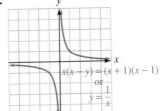

13.

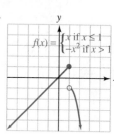

14.

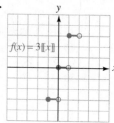

15. 144   **16.** 20   **17.** 15   **18.** 220   **19.** $x^5 + 5x^4y + 10x^3y^2 + 10x^2y^3 + 5xy^4 + y^5$ **20.** $x^4 - 4x^3y + 6x^2y^2 - 4xy^3 + y^4$
21. $64x^3 - 48x^2y + 12xy^2 - y^3$ **22.** $x^3 + 12x^2y + 48xy^2 + 64y^3$ **23.** $6x^2y^2$ **24.** $-10x^2y^3$ **25.** $-108x^2y$ **26.** $864x^2y^2$
27. 122, 137, 152, 167, 182   **28.** $\frac{41}{3}, \frac{58}{3}$ **29.** 1550   **30.** $-\frac{45}{2}$ **31.** $\frac{15}{2}$ **32.** 378   **33.** 14   **34.** 360   **35.** 24, 12, 6, 3, $\frac{3}{2}$
36. 24, -96 **37.** $-\frac{85}{8}$ **38.** $\frac{2186}{9}$ **39.** 125   **40.** $\frac{5}{99}$ **41.** \$1638.40   **42.** \$134,509.57   **43.** 12 yr   **44.** 1600 ft

Chapter 10 Test   (page 649)

1. $(2, -3); 2$ **2.**

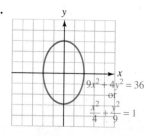

3.

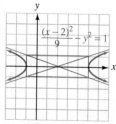

4. $(2, 6), (-2, -2)$ **6.**

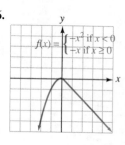

5. $(3, 4), (-3, 4)$

7. 105   **8.** $-5x^4y$ **9.** $24x^2y^2$ **10.** 66   **11.** 306   **12.** 34, 66
13. 3   **14.** -81 **15.** $\frac{364}{27}$ **16.** 18, 108   **17.** $\frac{27}{2}$

Cumulative Review Exercises (Chapters 1–10)   (page 650)

1. $12x^2 - 5xy - 3y^2$ **2.** $a^{2n} - 2a^n - 3$ **3.** $\frac{-5}{a-2}$ **4.** $a^2 - 3a + 2$ **5.** 1   **6.** $\dfrac{4a-1}{(a+2)(a-2)}$ **7.** parallel   **8.** perpendicular
9. $y = -2x + 5$ **10.** $y = -\frac{9}{13}x + \frac{7}{13}$ **11.**

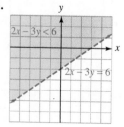

12.

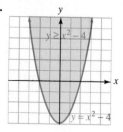

13. $5\sqrt{2}$ **14.** $81x\sqrt[3]{3x}$ **15.** $0, 5$ **16.** 0   **17.** $\frac{2}{3}, -\frac{3}{2}$ **18.** $\dfrac{-4 \pm \sqrt{19}}{3}$ **19.** $4x^2 + 4x - 1$ **20.** $f^{-1}(x) = \sqrt[3]{\dfrac{x+1}{2}}$

21.

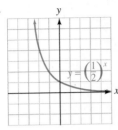

$y = \left(\frac{1}{2}\right)^x$

22. $2^y = x$ **23.** 5 **24.** 3 **25.** $\frac{1}{27}$ **26.** 1 **27.** $y = 2^x$ **28.** x **29.** 1.9912 **30.** 0.3010
31. 1.6902 **32.** 0.1461 **33.** $\frac{2 \log 2}{\log 3 - \log 2}$ **34.** 16 **35.** \$2848.31 **36.** 1.16056

37.

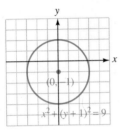

$(0, -1)$

$x^2 + (y + 1)^2 = 9$

38.

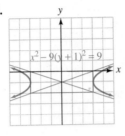

$x^2 - 9(y + 1)^2 = 9$

39. 30,240 **40.** $81a^4 - 108a^3b + 54a^2b^2 - 12ab^3 + b^4$
41. $112x^2y^6$ **42.** 103 **43.** 690 **44.** 8 and 19 **45.** 42
46. 27 **47.** 27 **48.** $\frac{1023}{64}$ **49.** 12, -48 **50.** $\frac{27}{2}$

Exercise I.1   (page A-4)

1. y-axis **3.** origin **5.** y-axis **7.** none **9.** none **11.** x-axis

13.

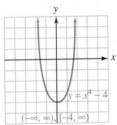

$y = x^4 - 4$

$(-\infty, \infty), [-4, \infty)$

15.

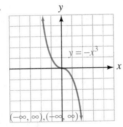

$y = -x^3$

$(-\infty, \infty), (-\infty, \infty)$

17.

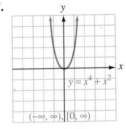

$y = x^4 + x^2$

$(-\infty, \infty), [0, \infty)$

19.

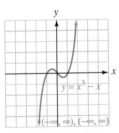

$y = x^3 - x$

$(-\infty, \infty), (-\infty, \infty)$

21.

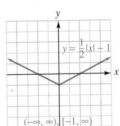

$y = \frac{1}{2}|x| - 1$

$(-\infty, \infty), [-1, \infty)$

23.

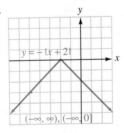

$y = -|x + 2|$

$(-\infty, \infty), (-\infty, 0]$

PROBLEMS AND PROJECTS

Chapter 1: Managing Hook n' Slice, 73
Chapter 1: Parchdale City Planning Council, 72
Chapter 2: Representing Buy-from-Us Corporation, 141
Chapter 2: Working for Boondocks County, 140
Chapter 3: Goodstuff Produce Company, 198
Chapter 3: HeckuvaDeal Soda Pop, Inc., 197
Chapter 4: Adding on to the barn, 246
Chapter 4: Points on a parabola, 247
Chapter 5: Bait shop on the Snarly River, 313
Chapter 5: Fluid through a pipe, 314
Chapter 6: Hybrid corn, 390
Chapter 6: Visiting your relatives, 389
Chapter 7: The bicycle race, 451
Chapter 7: Working at Widget Industries, 451
Chapter 8: Amazing Glendo, 520
Chapter 8: Park Bounded by Parabolic Boulevard, 521
Chapter 9: Dread Pirates of Hancock Isle, 573
Chapter 9: Fearless Freda, the Skydiving Daredevil, 572
Chapter 10: An auditorium in Baytown, 645
Chapter 10: The garden of G. I. Luvmoney, 644

MATHEMATICS IN THE WORKPLACE

Actuary, 580
Chemist, 459
Computer programmer, 255
Computer systems analyst, 2
Economist, 81

Electrical/electronic engineer, 151
Mechanical engineer, 321
Medical laboratory worker, 530
Photographer, 399
Statistician, 203

APPLICATIONS

Boldface page numbers refer to examples; lightface numbers refer to exercises.

Business and Industry

Alloys, 509
Analyzing ads, 25
Break-even analysis, **166**
Carpentry, 77
Chain saw sculpting, 180
Charges for computer repair, 124
Computing selling price, 70
Costs of a trucking company, 375
Cutting a beam, 61
Cutting a board, 61
Demand equation, 103
Depreciating a word processor, 124
Depreciation, 541, 562
Discount buying, 365
Earning money, **227**
Finding the annual rate of depreciation, 124
Finding an appreciation equation, 124
Finding a depreciation equation, 124
Finding the percent of markdown, **57**
Finding printer charges, 125
Finding salvage value, 124
Finding the value of antiques, 124
Hourly wages, 102
Inventories, 244
Landscaping, **233**
Making bicycles, 170
Making clothing, 180
Making crafts, 244
Making furniture, 71

Making sporting goods, 230
Making statues, 179
Managing a beauty shop, 170
Managing a computer store, 170
Manufacturing footballs, 179
Manufacturing hammers, **176**
Merchandising, 169
Meshing gears, 591
Milling brass plates, 170
Mixing candies, 70, 169
Mixing coffee, 125
Mixing nuts, **67**, 180
Mixing solutions, 70
Planning a work schedule, 212
Price increase, 365
Printing books, 170
Production problem, 244, 245
Production-schedule problem, 242
Rate of decrease, 111
Rate of growth, 111
Reading proof, 363
Retail sales, **163**
Royalties, 617
Running a record company, 170
Running a small business, 170
Salvage value, 541
Satellite antenna, 592
Scheduling equipment, 212
Selling clothes, 25
Selling shirts, 373
Selling stuffed animals, 62
Selling tape recorders, 134
Selling tires, 134
Setting bus fares, 468
Supply equation, 103
Tolerance of a sheet of steel, 223
Value of a lathe, **118**

Finance and Investing

Amount of an annuity, **637**
Annuities, 639
Assets of a pension fund, 62
Choosing salary plans, 171
Comparing saving plans, 540
Compound interest, 540, 571
Computing profit, 468
Computing salaries, 62

Continuous compound interest, 571
Declining savings, 639
Depreciation, **98**
Doubling money, 54, 561
Doubling time, **560**
Earning interest, 54, 102
Earning money, 24
Figuring taxes, 230
Financial planning, 245
Finding interest rates, 468
Frequency of compounding, 540
Inheriting money, 69
Installment loan, 631
Investing, 395
Investing in stocks, 230
Investing money, **65**, 69, 212, 611
Investment income, 169
Investment problem, 78
Making investments, 196
Managing a checkbook, 25
Marketing, 440
Maximizing income, **239**
Maximizing revenue, 479, 480
Metal fabrication, 469
Portfolio analysis, **57**
Rule of seventy, 571
Saving money, 631
Savings growth, 639
Simple interest, **280**
Stock appreciation, 649
Stock averages, 25
Supplemental income, 69
Time for money to grow, 562
Time for money to grow tenfold, 562
Tripling money, 561
Value of an IRA, 62

Geometry

Angles of an equilateral triangle, 64
Angles of a polygon, 54
Angles of a quadrilateral, 64
Angles in a triangle, **59**
Area of a circle, 374
Area of an ellipse, 604

7.1 RADICAL EXPRESSIONS

If n is an even natural number, then $\sqrt[n]{a^n} = |a|$.

If n is an odd natural number, then $\sqrt[n]{a^n} = a$.

7.2 RATIONAL EXPONENTS

If m and n are positive integers, $x > 0$, and $\frac{m}{n}$ is in simplest form, then

$$x^{m/n} = \sqrt[n]{x^m} = \left(\sqrt[n]{x}\right)^m$$

$$x^{-m/n} = \frac{1}{x^{m/n}} \quad \text{and} \quad \frac{1}{x^{-m/n}} = x^{m/n} \ (x \neq 0)$$

7.3 SIMPLIFYING AND COMBINING RADICAL EXPRESSIONS

$$\sqrt[n]{ab} = \sqrt[n]{a}\sqrt[n]{b} \quad \text{and} \quad \sqrt[n]{\frac{a}{b}} = \frac{\sqrt[n]{a}}{\sqrt[n]{b}} \quad (b \neq 0)$$

7.5 RADICAL EQUATIONS

The power rule: If $x = y$, then $x^n = y^n$.

7.6 APPLICATIONS OF RADICALS

Pythagorean theorem: If a and b are the lengths of the legs of a right triangle and c is the length of the hypotenuse, then $a^2 + b^2 = c^2$.

In an isosceles right triangle, the length of the hypotenuse is the length of either leg times $\sqrt{2}$.

The shorter leg of a 30°–60°–90° right triangle is half as long as the hypotenuse.

The length of the longer leg of a 30°–60°–90° right triangle is the length of the shorter leg times $\sqrt{3}$.

Distance formula: $d = \sqrt{(x_2 - x_1)^2 + (y_2 - y_1)^2}$

8.1 COMPLETING THE SQUARE AND THE QUADRATIC FORMULA

Square root property: $x^2 = c \ (c > 0)$ has two real solutions:

$$x = \sqrt{c} \quad \text{and} \quad x = -\sqrt{c}$$

Quadratic formula: $x = \dfrac{-b \pm \sqrt{b^2 - 4ac}}{2a} \quad (a \neq 0)$

8.2 GRAPHS OF QUADRATIC FUNCTIONS

If f is a function and k is a positive number, then

- The graph of $y = f(x) + k$ is identical to the graph of $y = f(x)$, except that it is shifted k units upward.
- The graph of $y = f(x) - k$ is identical to the graph of $y = f(x)$, except that it is shifted k units downward.
- The graph of $y = f(x - h)$ is identical to the graph of $y = f(x)$, except that it is shifted h units to the right.
- The graph of $y = f(x + h)$ is identical to the graph of $y = f(x)$, except that it is shifted h units to the left.

If $a \neq 0$, the graph of the equation $y = a(x - h)^2 + k$ is a parabola with vertex at (h, k). It opens upward when $a > 0$ and downward when $a < 0$.

8.3 COMPLEX NUMBERS

If a, b, c, and d are real numbers and $i^2 = -1$,

$$a + bi = c + di \text{ if and only if } a = c \text{ and } b = d$$

$$(a + bi) + (c + di) = (a + c) + (b + d)i$$

$$(a + bi)(c + di) = (ac - bd) + (ad + bc)i$$

$$|a + bi| = \sqrt{a^2 + b^2}$$

8.4 THE DISCRIMINANT AND EQUATIONS THAT CAN BE WRITTEN IN QUADRATIC FORM

If	the solutions of $ax^2 + bx + c = 0$ are
$b^2 - 4ac > 0$	unequal real numbers.
$b^2 - 4ac = 0$	equal real numbers.
$b^2 - 4ac < 0$	complex conjugates.

If r_1 and r_2 are solutions of $ax^2 + bx + c = 0$, then

$$r_1 + r_2 = -\frac{b}{a} \quad \text{and} \quad r_1 r_2 = \frac{c}{a}$$

8.6 ALGEBRA AND COMPOSITION OF FUNCTIONS

$$(f + g)(x) = f(x) + g(x) \qquad (f - g)(x) = f(x) - g(x)$$

$$(f \cdot g)(x) = f(x)g(x) \qquad (f/g)(x) = \frac{f(x)}{g(x)} \ (g(x) \neq 0)$$

$$(f \circ g)(x) = f(g(x))$$

9.1 EXPONENTIAL FUNCTIONS

$e = 2.718281828.\ .\ .$

Formula for compound interest:

$$A = P_0\left(1 + \frac{r}{k}\right)^{kt} \quad \text{or} \quad FV = PV(1 + i)^n$$

Formula for continuous compound interest: $A = Pe^{rt}$